Knowledge Discovery in Big Data from Astronomy and Earth Observation

Knowledge Discovery in Big Data from Astronomy and Earth Observation

AstroGeoInformatics

Edited by

Petr Škoda
Stellar Department
Astronomical Institute of the Czech Academy of Sciences
Ondřejov, Czech Republic

Fathalrahman Adam
Earth Observation Center
German Remote Sensing Data Center
DLR German Aerospace Center
Wessling, Germany

Elsevier
3251 Riverport Lane
St. Louis, Missouri 63043

Knowledge Discovery in Big Data from Astronomy and Earth Observation ISBN: 978-0-12-819154-5

Publisher: Candice Janco
Acquisitions Editor: Marisa LaFleur
Editorial Project Manager: Andrea Dulberger
Production Project Manager: Sreejith Viswanathan
Designer: Alan Studholme

Contents

LIST OF CONTRIBUTORS, *vii*

A WORD FROM THE BIG-SKY-EARTH CHAIR, *xi*

PREFACE, *xiii*

ACKNOWLEDGMENTS, *xvii*

PART I
DATA

1 **Methodologies for Knowledge Discovery Processes in Context of AstroGeoInformatics**, *1*
Peter Butka, Peter Bednár, Juliana Ivančáková

2 **Historical Background of Big Data in Astro and Geo Context**, *21*
Christian Muller

PART II
INFORMATION

3 **AstroGeoInformatics: From Data Acquisition to Further Application**, *31*
Bianca Schoen-Phelan

4 **Synergy in Astronomy and Geosciences**, *39*
Mikhail Minin, Angelo Pio Rossi

5 **Surveys, Catalogues, Databases, and Archives of Astronomical Data**, *57*
Irina Vavilova, Ludmila Pakuliak, Iurii Babyk, Andrii Elyiv, Daria Dobrycheva, Olga Melnyk

6 **Surveys, Catalogues, Databases/Archives, and State-of-the-Art Methods for Geoscience Data Processing**, *103*
Lachezar Filchev, Lyubka Pashova, Vasil Kolev, Stuart Frye

7 **High-Performance Techniques for Big Data Processing**, *137*
Philipp Neumann, Julian Kunkel

8 **Query Processing and Access Methods for Big Astro and Geo Databases**, *159*
Karine Zeitouni, Mariem Brahem, Laurent Yeh, Atanas Hristov

9 **Real-Time Stream Processing in Astronomy**, *173*
Veljko Vujčić, Darko Jevremović

PART III
KNOWLEDGE

10 **Time Series**, *183*
Ashish Mahabal

11 **Advanced Time Series Analysis of Generally Irregularly Spaced Signals: Beyond the Oversimplified Methods**, *191*
Ivan L. Andronov

12 **Learning in Big Data: Introduction to Machine Learning**, *225*
Khadija El Bouchefry, Rafael S. de Souza

13 **Deep Learning – an Opportunity and a Challenge for Geo- and Astrophysics**, *251*
Christian Reimers, Christian Requena-Mesa

14 **Astro- and Geoinformatics – Visually Guided Classification of Time Series Data**, *267*
Roman Kern, Tarek Al-Ubaidi, Vedran Sabol, Sarah Krebs, Maxim Khodachenko, Manuel Scherf

15 **When Evolutionary Computing Meets Astro- and Geoinformatics**, *283*
Zaineb Chelly Dagdia, Miroslav Mirchev

PART IV
WISDOM

16 **Multiwavelength Extragalactic Surveys: Examples of Data Mining**, *307*
Irina Vavilova, Daria Dobrycheva, Maksym Vasylenko, Andrii Elyiv, Olga Melnyk

17 **Applications of Big Data in Astronomy and Geosciences: Algorithms for Photographic Images Processing and Error Elimination**, *325*
Ludmila Pakuliak, Vitaly Andruk

18 **Big Astronomical Datasets and Discovery of New Celestial Bodies in the Solar System in Automated Mode by the CoLiTec Software**, *331*
Sergii Khlamov, Vadym Savanevych

19 **Big Data for the Magnetic Field Variations in Solar-Terrestrial Physics and Their Wavelet Analysis**, *347*
Bozhidar Srebrov, Ognyan Kounchev, Georgi Simeonov

20 **International Database of Neutron Monitor Measurements: Development and Applications**, *371*
D. Sapundjiev, T. Verhulst, S. Stankov

21 **Monitoring the Earth Ionosphere by Listening to GPS Satellites**, *385*
Liubov Yankiv-Vitkovska, Stepan Savchuk

22 **Exploitation of Big Real-Time GNSS Databases for Weather Prediction**, *405*
Nataliya Kablak, Stepan Savchuk

23 **Application of Databases Collected in Ionospheric Observations by VLF/LF Radio Signals**, *419*
Aleksandra Nina

24 **Influence on Life Applications of a Federated Astro-Geo Database**, *435*
Christian Muller

INDEX, *445*

List of Contributors

Tarek Al-Ubaidi, MSc
DCCS – IT Business Solutions, Graz, Austria

Ivan L. Andronov, DSc, Prof
Department of Mathematics, Physics and Astronomy, Odessa National Maritime University, Odessa, Ukraine

Vitaly Andruk
Main Astronomical Observatory of the National Academy of Sciences of Ukraine, Kyiv, Ukraine

Iurii Babyk, Dr
Main Astronomical Observatory of the National Academy of Sciences of Ukraine, Kyiv, Ukraine

Peter Bednár, PhD
Department of Cybernetics and Artificial Intelligence, Technical University of Košice, Košice, Slovakia

Mariem Brahem, PhD
DAVID Lab., University of Versailles Saint-Quentin-en-Yvelines, Université Paris-Saclay, Versailles, France

Peter Butka, PhD
Department of Cybernetics and Artificial Intelligence, Technical University of Košice, Košice, Slovakia

Zaineb Chelly Dagdia, Dr
Université de Lorraine, CNRS, Inria, LORIA, F-54000 Nancy, France
LARODEC, Institut Supérieur de Gestion de Tunis, Tunis, Tunisia

Rafael S. de Souza, PhD
Department of Physics & Astronomy, University of North Carolina at Chapel Hill, Chapel Hill, NC, United States

Daria Dobrycheva, Dr
Main Astronomical Observatory of the National Academy of Sciences of Ukraine, Kyiv, Ukraine

Khadija El Bouchefry, PhD
South African Radio Astronomy Observatory, Rosebank, JHB, South Africa

Andrii Elyiv, Dr
Main Astronomical Observatory of the National Academy of Sciences of Ukraine, Kyiv, Ukraine

Lachezar Filchev, Assoc Prof, PhD
Space Research and Technology Institute, Bulgarian Academy of Sciences, Sofia, Bulgaria

Stuart Frye, MSc
National Aeronautics and Space Administration, Washington, DC, United States

Atanas Hristov, PhD
University of Information Science and Technology "St. Paul the Apostle", Ohrid, North Macedonia

Juliana Ivančáková, MSc
Department of Cybernetics and Artificial Intelligence, Technical University of Košice, Košice, Slovakia

Darko Jevremović, Dr
Astronomical Observatory Belgrade, Belgrade, Serbia

Nataliya Kablak, DSc, Prof
Uzhhorod National University, Uzhhorod, Ukraine

Roman Kern, PhD
Institute of Interactive Systems and Data Science, Technical University of Graz, Graz, Austria

Sergii Khlamov, PhD
Institute of Astronomy, V. N. Karazin Kharkiv National University, Kharkiv, Ukraine
Main Astronomical Observatory of the NAS of Ukraine, Kyiv, Ukraine

Maxim Khodachenko, PhD
Space Research Institute, Austrian Academy of Sciences, Graz, Austria
Skobeltsyn Institute of Nuclear Physics, Moscow State University, Moscow, Russia

Vasil Kolev, MSc
Institute of Information and Communication Technologies, Bulgarian Academy of Sciences, Sofia, Bulgaria

Ognyan Kounchev, Prof, Dr
Institute of Mathematics and Informatics, Bulgarian Academy of Sciences, Sofia, Bulgaria

Sarah Krebs, MSc
Know-Center, Graz, Austria

Julian Kunkel, Dr
University of Reading, Reading, United Kingdom

Ashish Mahabal, PhD
California Institute of Technology, Pasadena, CA, United States

Olga Melnyk, Dr
Main Astronomical Observatory of the National Academy of Sciences of Ukraine, Kyiv, Ukraine

Mikhail Minin, MSc
Jacobs University Bremen, Bremen, Germany

Miroslav Mirchev, Dr
Faculty of Computer Science and Engineering, Ss. Cyril and Methodius University in Skopje, Skopje, North Macedonia

Christian Muller, Dr
Royal Belgian Institute for Space Aeronomy, Belgian Users Support and Operation Centre, Brussels, Belgium

Philipp Neumann, Prof, Dr
Helmut-Schmidt-Universität Hamburg, Hamburg, Germany

Aleksandra Nina, PhD
Institute of Physics Belgrade, University of Belgrade, Belgrade, Serbia

Ludmila Pakuliak, Dr
Main Astronomical Observatory of the National Academy of Sciences of Ukraine, Kyiv, Ukraine

Lyubka Pashova, Assoc Prof, PhD
National Institute of Geophysics, Geodesy and Geography, Bulgarian Academy of Sciences, Sofia, Bulgaria

Christian Reimers, MSc
Computer Vision Group, Faculty of Mathematics and Computer Science, Friedrich Schiller University Jena, Jena, Germany
Climate Informatics Group, Institute of Data Science, German Aerospace Center, Jena, Germany

Christian Requena-Mesa, MSc
Computer Vision Group, Faculty of Mathematics and Computer Science, Friedrich Schiller University Jena, Jena, Germany
Climate Informatics Group, Institute of Data Science, German Aerospace Center, Jena, Germany
Department Biogeochemical Integration, Max Planck Institute for Biogeochemistry, Jena, Germany

Angelo Pio Rossi, Dr, PhD
Jacobs University Bremen, Bremen, Germany

Vedran Sabol, PhD
Know-Center, Graz, Austria

D. Sapundjiev, PhD
Royal Meteorological Institute (RMI), Brussels, Belgium

Vadym Savanevych, DSc
Main Astronomical Observatory of the NAS of Ukraine, Kyiv, Ukraine

Stepan Savchuk, DSc, Prof
Lviv Polytechnic National University, Lviv, Ukraine

Manuel Scherf, MSc
Space Research Institute, Austrian Academy of Sciences, Graz, Austria

Bianca Schoen-Phelan, PhD
Technical University Dublin, School of Computer Science, Dublin, Ireland

Georgi Simeonov, Assistant
Institute of Mathematics and Informatics, Bulgarian Academy of Sciences, Sofia, Bulgaria

Bozhidar Srebrov, Assoc Prof, Dr
Institute of Mathematics and Informatics, Bulgarian Academy of Sciences, Sofia, Bulgaria

S. Stankov, PhD
Royal Meteorological Institute (RMI), Brussels, Belgium

Maksym Vasylenko, MSc
Main Astronomical Observatory of the National Academy of Sciences of Ukraine, Kyiv, Ukraine

Irina Vavilova, Dr
Main Astronomical Observatory of the National Academy of Sciences of Ukraine, Kyiv, Ukraine

T. Verhulst, PhD
Royal Meteorological Institute (RMI), Brussels, Belgium

Veljko Vujčić
Astronomical Observatory Belgrade, Belgrade, Serbia

Liubov Yankiv-Vitkovska, Assoc Prof, Dr
Lviv Polytechnic National University, Lviv, Ukraine

Laurent Yeh, Assoc Prof, PhD
DAVID Lab., University of Versailles Saint-Quentin-en-Yvelines, Université Paris-Saclay, Versailles, France

Karine Zeitouni, Prof, PhD
DAVID Lab., University of Versailles Saint-Quentin-en-Yvelines, Université Paris-Saclay, Versailles, France

A Word from the BIG-SKY-EARTH Chair

In the summer of 2013, a small group of astronomers and Earth observation experts gathered around an intriguing idea – why not assemble a network of experts from those two disciplines that have so much in common, but essentially do not communicate much as professionals? An opportunity opened for this idea to be realized as the COST funding scheme opened a call for transdisciplinary networks. A proposal was put together and the Big Data Era in Sky and Earth Observation (BIG-SKY-EARTH) project was born.[1] Funded by COST, the project officially started in January 2015 and ended in January 2019. In the end, the BIG-SKY-EARTH network included 28 COST countries and hundreds of researchers from all around the world. The project organized four training schools, two workshops and two conferences, exchange of numerous scientists between research groups, many working group meetings, and formal and informal gatherings. The lively exchange of ideas and a mixture of people of various expertise fueled a plethora of collaborations between participants. It has been a mixture of experts from academia and the private sector, with special attention given to supporting young researchers still in the early stages of their careers.

Toward the end of the COST project, a suggestion emerged to put together a book that would be an interesting reading for astronomers curious about remote sensing and vice versa. Essentially, the book would be about AstroGeoInformatics – an amalgam of computer science, astronomy, and Earth observation. The suggestion was enthusiastically supported by lots of participants and now, finally, you have it in front of you – the first book written for such transdisciplinary readers. The book was a challenge for authors to write, since the readers will surely come with a large range of background knowledge, and texts had to be balanced between the disciplines. But everyone should find at least some parts of the book intriguing and interesting, hopefully as much as we found our BIG-SKY-EARTH network inspiring and motivating.

Dejan Vinković
Chair of the BIG-SKY-EARTH network

This book is based upon work from COST Action TD1403 - BIG-SKY-EARTH, supported by the European Cooperation in Science and Technology (COST).

COST[2] is a funding agency for research and innovation networks. Our actions help connect research initiatives across Europe and enable scientists to grow their ideas by sharing them with their peers. This boosts their research, career, and innovation.

Funded by the Horizon 2020 Framework Programme of the European Union

[1] https://bigskyearth.eu.

[2] www.cost.eu.

Preface

WHAT'S IN THIS BOOK?

This book has several parts reflecting various stages of Big Data processing and machine learning following the Data–Information–Knowledge–Wisdom (DIKW) pyramid explained below. There are several main parts reflecting the particular stages of the pyramid.

Part I is an introductory section about the origin of Big Data and its history. It shows that Big Data is not an entirely new concept, as humans have tended for a long time to collect ever-growing amounts of data, and summarizes the general principles and recent developments of knowledge discovery.

Part II discusses the different stages of data acquisition, preprocessing, and interpretation and data repositories. We want to point out what basic steps have to be made to reach end-to-end systems allowing an automated extraction of results based on instrument data and auxiliary databases.

Part III addresses the primary goal of this book – the extraction of new wisdom about the Universe (including our Earth) from Big Data. It presents various approaches of data analysis, such as genetic programming and machine learning techniques, including the commercially overhyped deep learning.

Part IV addresses the specific readership that we want to reach. We think of a wide range of experts ranging from students of different disciplines to practitioners as well as theory-oriented researchers. Here, our aim is to acquaint these readers with the wide variety of tasks that can be solved by modern approaches of Big Data processing and give them some examples of an interdisciplinary approach using smart untraditional methods and data sources. The advanced combination of apparently nonrelevant data may even help reveal unexpected correlations and relationships having an impact on our everyday life. So, for instance, the real-time monitoring of GPS signals helps in predicting the weather, and listening to very long-wave transmitters can be crucial in predictions of natural disasters such as hurricanes or earthquakes.

MOTIVATION AND SCOPE

Fundamental science at the beginning of our civilization was naturally treated as a multidisciplinary task. Most important ancient Greek scientists and philosophers focused on mathematics, physics, astronomy, geometry, biology, mineralogy, meteorology, and medicine, as well as on social and political matters. The discovery of nature per se was primarily driven by curiosity and not by the intention of exploiting it. Particularly famous are many philosophical discussions in the Pythagorean school or the Platonic Academy. The principles of "Pansophism" as an educational goal were proclaimed by Comenius and the didactic principles he had introduced are probably the top achievement in teaching that seem to continue almost unchanged until these days. Our goal is to maintain some of his principles in this book, as well. That is why we put emphasis on the understandability and on the joy of gaining new knowledge together with using (some) practical examples. We also accentuate (in agreement with Comenius) the role of color pictures that try to associate the read text with something familiar to the reader.

The problem with current science is mainly in breaking with all the principles given above. The contemporary scientific communities are very narrow-focused, having their own terminology which presents a high entry barrier for the intruders into their "sacred land." An enthusiastic researcher searching similarities between his research and a completely different field of science is quickly redirected into the "proper" corridors by a number of ways (e.g., by funding agencies, referees of scientific journals, or leaders of their home institution).

The result is that interdisciplinarity, although strongly proclaimed by science policymakers, is very difficult to practice in reality. Most researchers hardly get sufficient funds to visit essential conferences and workshops in their own narrow field, and here they usually meet the same colleagues they already know quite well, and everybody knows what topics (and even which objects) will be presented. The big symposia are more promising to promote interdisciplinarity at least within the boundaries of one branch of science (e.g., astronomy, space physics, and geophysics). However, the multitude of parallel sessions and the shortage of available time enforces the attendees again to visit only sessions about their own and closely related fields. So it seems that many scientists remain locked in their small, highly specialized communities. Fortunately, on the other hand, there are also quite large meetings not based on limited research subjects but particular methodologies and technology.

So, for instance, in astronomy, there is a traditional Astronomical Data Analysis Software and Systems (ADASS), conference where the common uniting subject is modern computer technology in fields such as satellite and telescope control systems, advanced mathematical algorithms, high-performance computing and databases, software development methodology, or the analysis of Big Data. The same holds for Earth observation, where the International Geoscience and Remote Sensing Symposium (IGARSS) conferences are targeted to participants from numerous geoscientific and remote sensing disciplines.

Another example of a broad community from all fields of astronomy are interoperability meetings of the International Virtual Observatory Alliance, where the main goal is to define data formats, models, and protocols allowing the global interoperability of all astronomical databases and archives. Their "terrestrial" counterpart are lots of standards and conventions published by the Open Geospatial Consortium (OGC), national and international space agencies (e.g., NASA, ESA, and DLR), and even industrial consortia that are highly specialized in image processing, spectroscopy, precision farming, database technologies, computer networks, etc.

Nowadays, we can see that more and more quantitative physical and chemical parameters can be retrieved from advanced instruments and their processing chains; a typical criterion is the attainable signal-to-noise level of an instrument. Here we can see considerable improvements when we compare traditional instruments with innovative devices. However, we must also consider the specific requirements of astro- and geoscience applications such as the long-term data availability and traceability of the results (data curation, reproducibility), proven scientific data quality with quantitative error levels, and appropriate tools to validate them. The importance of archives of historical records, namely, photographic plates, is rising in accordance with the implementation of FAIR principles of data management (striving to make data findable, accessible, interoperable, and reusable). Digitization of legacy sources brings about a considerable amount of Big Data volumes but also new, unexpected opportunities of their analysis. A typical astronomical example are old astronomical plates with spectra of bright hot stars (secured at the beginning of the 20th century) to measure ozone concentrations in the Earth's stratosphere at that time.

In the future, we expect that more complex scientific problems will lead to common "astro-/geo-" approaches to be applied in synergy. There are already several cases where astronomical instrumentation was used for analyzing terrestrial phenomena. For example, the Earth's aurora X-ray radiation was observed in 2015 by ESA Integral mission, ordinarily busy with observing black holes, during calibration of the diffuse cosmic X-ray background. Another example is the world's largest radio telescope array LOFAR, which, when monitoring the three-dimensional propagation of a lightning 18 km above the Netherlands with microsecond time resolution, discovered interesting needle-shaped plasma structures, probably responsible for multiple strikes of the same lightning within seconds.

The astronomical observations also contribute to meteorology and vice versa. In astronomical spectroscopy, almost every spectrum is contaminated by atmospheric water vapor absorption lines, which provides a direct way to measure the line-of-sight water content in the atmosphere. This water contamination must be removed from the recorded spectra by complicated methods, although in many cases, it is used to correct the wavelength calibration of the spectrograph as a reference source with negligible radial velocity in comparison to stellar ones.

Current astronomy is making many new discoveries thanks to the possibility of aggregating the observations in the whole electromagnetic spectrum (with recent extensions to astroparticle physics – cosmic rays, neutrinos, and also gravitational wave astronomy) and also by cross-matching of gigantic multi-petabyte scaled sky surveys. The new astronomical instruments LSST, SKA, and EUCLID foresee to produce tens of petabytes of raw data every year, and their future archives are expected to be operating at the edge of contemporary IT technology, with the embedded state-of-the-art machine learning functionality.

As for geosciences, we have to consider additional application tasks with specific constraints, such as real-time traffic monitoring and rapid disaster analysis (e.g., of flooding, lava flows, storm warnings, and volcanic plume dynamics) or the handling of large data volumes that we encounter in weather predictions or crop yield calculations. In particular, some applications call for highly reliable and carefully calibrated input data, for instance, for climate research and risk assessments, while other forms put more emphasis on keeping up with the data volume or access to cloud computing.

The geosciences benefit a lot from satellite observations and remote sensing, where the continuous flow of new hyperspectral data, motion and displacement measurements (based on speed and phase measurements derived from synthetic-aperture radars or optical instruments) have to be combined with accurate "ground truth" reference data obtained by geophysics, hydrology, and terrain investigations. This presents challenging opportunities for agriculture, forestry, water resource planning, and ore mining as well as new approaches to infrastructure development and urban planning. The importance of aggregated Earth observation databases and so-called Virtual Earths is continuously increasing in the case of natural disasters for rescue and humanitarian aid purposes.

The current Fourth Paradigm of science, characterized by Big Data, is data-driven research and exploratory analysis exploiting innovative machine learning principles to obtain new knowledge, e.g., about a vast tsunami threatening a coastal area, thus offering a qualitatively new way of scientific methodologies, e.g., by simulation runs and comparisons with existing databases. The Big Data phenomenon is common for every scientific discipline and presents a serious obstacle to conducting efficient research by established methods. Therefore, a new multidisciplinary approach is needed, which integrates the particular discipline knowledge with advanced statistics, high-performance computing and data analytics, new computer software design principles, machine learning and other fields belonging rather to computer science, artificial intelligence, and data science. This change also requires a new kind of a career path which is not yet fully established despite first experimental courses at renowned universities, where the graduates are "data scientists."

The same term is used in business for the "sexiest job of the 21st century," and is understood rather as a commercial data analysis task – using statistics and simple machine learning for the analysis of data warehouses. The real scientific data scientist must, however, have a considerably broader knowledge of his scientific branch, in addition to excellent knowledge of statistics and machine learning.

The first successful integration happened in bioinformatics, which recently became an officially accepted branch of science. In astronomy, the Astroinformatics originated around the year 2010 and is still being treated by the majority of astronomers like a kind of sorcery, despite the public outreach effort of the International Astrostatistics Association, the working groups in astrostatistics and astroinformatics of both the International Astronomical Union and the American Astronomical Society, and also the recently established International AstroInformatics Association. In contrast, Geoinformatics for Earth observation suffers from the misunderstanding of being confounded with geomatics and geographical information systems; however, the importance of machine learning and Big Data processing is already well understood and it is applied routinely.

Despite the lack of a proper taxonomy for these newly established fields, there are attempts to transfer some methods that were successfully applied in one scientific field into a completely different one. Within this context, a key methodology for transferring digitized knowledge is transfer learning, allowing us to train a machine learning system in one domain and then apply the fully trained system within a completely different application area of data analytics. However, this approach calls for an appropriate design of the required databases. Currently, the first results could already be demonstrated for image content classification in remote sensing applications.

There are already world-renowned institutions where such research is being conducted. For instance, the Center for Data-Driven Discovery at Caltech and some groups in NASA successfully applied methods developed for astronomical analysis in medicine (e.g., prediction of autism in EEG records or the identification of cancer metastases in histologic samples.).

Another field where the methodological similarities with astronomy are even more striking is the complete range of geosciences and remote sensing disciplines. Image analysis using data recorded by CCD detectors, multispectral analysis, hyperdata cubes, time series measurements, data streams, various coordinate systems, deep learning, classification techniques, and federalization of resources – all these fields are applied in similar ways both in astronomy and geosciences. So it seems to be very useful for one community to learn from the other one, and all of them should acquire a lot of practical skills from computer science and data science. We are convinced that both communities ("astro" and "geo") will benefit from understanding the more comprehensive interdisciplinary view. An interesting topic could be advanced algorithms to identify (and remove) the varying atmospheric background of satellite images.

The idea of this book originated during very productive meetings of the transdisciplinary European COST Action TD1403 called BIG-SKY-EARTH. It would not be possible without the great enthusiasm of many people who devoted a considerable amount of time to the preparation of the state-of-the-art reviews of typical topics that, according to their feelings, will be important for future data scientists.

The main goal of the book, which is a kind of experiment trying to show the potential synergy between astronomy and geosciences, is to give some first-time overview of the technologies and methodologies together with references to the practical usage of many important fields related to knowledge discovery in astro- and geo-Big Data. Its purpose is to give a very general background and some ideas with numerous references that could be helpful in the design of fully operational data analysis systems – and of an experimental data science book.

The book tries to follow the pyramid called *Data* (from data acquisition to distributed databases), *Information* (data processing), *Knowledge* (machine learning), and *Wisdom* (applied knowledge for applications) (DIKW). We intend to present a global picture seen from history, data gathering, data processing and knowledge extraction to inferences of new wisdom. This has to be understood in conjunction with the pros and cons of every processing step and its concatenations (including user interfaces).

In the last part of the book, we address a fundamental question, namely, what kind of new knowledge we could get if the data from both astronomical and geo-research will be properly processed in close synergy, and what impact this will have on human health, environmental problems, economic prosperity, etc. In the following chapters, we present somewhat preliminary topics of a newly emerging scientific field which we try to define as AstroGeoInformatics.

As we are aware that there are limits in understanding the terminologies of other fields, we asked all authors to avoid complex mathematics as well as deep details and the common jargon of each discipline, and try to treat their contribution more on an educational or outreach level (also by using more illustrations) while still maintaining a rigorous structure with proper references for every important statement. Thus, we can avoid detailed discussions about very specific problems such as the avoidance of overfitting during image classification; however, we have to be aware of typical decision making problems and imminent limitations: Shall we believe in earthquake prediction based on deep learning and evacuate a full region?

We hope that a wide scope of readers will find this book interesting, and that it will serve them as a starter for an interdisciplinary way of individual thinking. This should be an important characteristic of this book. The future of humankind is dependent on a close collaboration between many scientific disciplines in synergy. Big Data is one example of global problems that must be overcome by changes of paradigms of how the research was done in each discipline so far.

Data scientists will be indispensable leaders of these changes. We hope that our book will help educate new graduates in this emerging field of science.

Petr Škoda
Fathalrahman Adam
Editors

With great help from Gottfried Schwarz, DLR.

Acknowledgments

We want to thank all the people who participated in the preparation of this book. The book is not just another collection of papers like typical conference proceedings, but the result of long-term planning, inspiring discussions with experts in many fields of natural sciences, asking personally more than fifty people we knew from various conferences, making open calls in different discussion groups and communities, and exchanging a large number of e-mails with the world's leaders in related fields. So in addition to the authors who finally wrote some chapters, we also thank those who established contacts with potential authors or contributed by providing links to interesting articles, as well as those who have enthusiastically promised writing the chapter, but later were not able to do this due to more urgent matters.

We are also grateful to the European Union for its funding of the COST Action TD1403 BIG-SKY-EARTH, which succeeded in attracting such a nice group of experts that benefited from the interdisciplinary nature of this action. This COST action has also shown the need of personal contacts in preparing exciting research ideas. The fruitful COST meetings helped amalgamate the initially heterogeneous group of researchers from different fields and countries into a real task force capable of accepting each other's visions, methodologies, and technologies. This book is the result of such a new interdisciplinary collaboration that tries to present the benefits of synergetic empowerment in natural sciences.

We are also indebted to Elsevier's representatives, who were in direct contact with other authors and us, editors, during the whole process of book preparation.

It was Marisa LaFleur who followed the book project from its start and managed to arrange things in a comfortable but determined way. Lots of people, including a part of the referees during the initial review phase of the book's table of contents, did not believe in our vision of mixing astronomy, geosciences, and computer science in all chapters. They would have preferred the more classic approach – to split the book into two parts – one for astronomers and another second for geo-experts. Despite their skepticism, the Elsevier people trusted the strength of our vision and helped us finish the work successfully.

So we thank Ashwathi Aravindakshan, our copyrights coordinator, who was keeping an eye on the proper copyright status of each figure, as well as the tables and examples which were already published elsewhere, and Subramaniam Jaganathan, the contract coordinator, who helped us arrange the initial paperwork for signing the contract.

Finally, we would like to express our deep gratitude to Andrea Dulberger, the editorial project manager, as well as Sreejith Viswanathan, the project manager, for leading the project to a successful end.

CHAPTER 1

Methodologies for Knowledge Discovery Processes in Context of AstroGeoInformatics

PETER BUTKA, PHD • PETER BEDNÁR, PHD • JULIANA IVANČÁKOVÁ, MSC

1.1 INTRODUCTION

Whenever someone wants to apply data mining techniques to specific problem or data, it is useful to see anything done in a broader and more organized way. Therefore, successful data science projects usually follow some methodology which can provide data scientist with basic guidelines on how to challenge the problem and how to work with data, algorithms, or models. This methodology is then a structured way to describe the knowledge discovery process. Without a flexible structure of steps data science projects can be unsuccessful, or at least it will be hard to achieve a result that can be easily applied and shared. A better understanding of at least an overview of the process is quite beneficial both to the data scientist and to anyone who needs to discuss results or steps of the process (such as data engineers, customers, or managers). Moreover, in some domains, including those working with data from astronomy and geophysics, steps used in preprocessing and analysis of data are crucial to understanding provided products.

From the 1990s, research in this area started to define its terms more precisely, with the definition of knowledge discovery (or knowledge discovery in databases [KDD]) (Fayyad et al., 1996) as a synonym for knowledge discovery process (KDP). It included data mining as one of the steps in the knowledge acquisition effort. KDD (or KDP) and data mining are even today often seen as equal terms, but data mining is a subpart (step) of the whole process dedicated to the application of algorithms able to extract patterns from data. Moreover, KDD also becomes the first description of KDP as a formalized methodology. During the next years, new efforts bring more attempts which lead to other methodologies and their applications. We will describe more details on selected cases later.

For a better understanding of KDPs, we can shortly describe how basic terms about data, information, or knowledge are defined. We have to say that there are many attempts to explain them more precisely. One example is the DIKW pyramid (Rowley, 2007). This model represents and characterizes information-based levels (according to the area of information engineering) in the chain of grading informativeness known as Data–Information–Knowledge–Wisdom (see Fig. 1.1). Similar models often apply such chains, even if some parts are removed or combined. For example, very often such a model is simplified to Data–Information–Knowledge or even Data–Knowledge, but semantics is usually the same or similar as in the case of the DIKW pyramid. Moreover, there are many models which describe not only its objects but also processes for their transitions, e.g., Bloom's taxonomy (Anderson and Krathwohl, 2001), decision process models (Bouyssou et al., 2010), or knowledge management – SECI models (Nonaka et al., 2000). The description of methodology usually defines what we understand under data, information, and knowledge level.

While methodologies started from a more general view, logically more and more attempts were transformed into a more structured way. Also, many of them became more tool-specific. When we try to look at the evolution of the KDP, the main further steps after the creation of more general methodologies are two basic concepts. First, in order to have more precise and formalized processes, many of them were transformed into standardized process-based definitions with the automation of their steps. Such effort is logically achieved more easily by the application in specific domains (such as industry, medicine, science), with clear standards for exchanging documents and often with the support of

Knowledge Discovery in Big Data from Astronomy and Earth Observation. https://doi.org/10.1016/B978-0-12-819154-5.00010-2

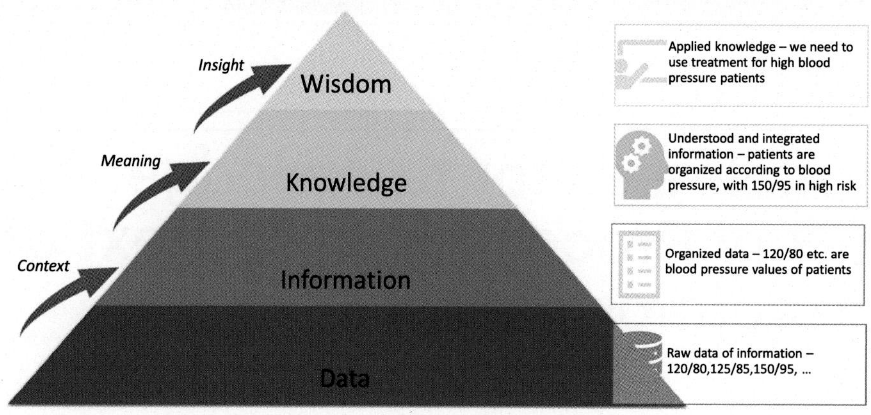

FIG. 1.1 DIKW pyramid – understanding the difference between Data, Information, Knowledge, and Wisdom.

specific tools used for the automation of processes. Second, when we have several standardized processes in different domains, it is often not easy to apply methods from one area directly in another one. One of the solutions is to support better cross-domain understanding of steps using some shared terminology. This solution leads to the creation of formalized semantic models like ontologies that are helpful in better understanding of terminology between domains. Moreover, another step towards a new view of methodologies and sharing of information about them was proposed based on the ontologies of KDPs, like OntoDM (Panov et al., 2013).

Therefore, if we summarize, generalized methodologies are basic concepts related to KDPs. More specific versions of them provide standards and automation in specific domains, and on the other hand, cross-domain models share domain-specific knowledge between different domains. This basic overview also describes the structure of sections for this chapter. In the next section, we provide some details on data–information–knowledge definitions and KDPs. In the following section, we describe existing more general methodologies. In Section 1.4 we provide a look at methodologies in a more precise way, through standardization and automation efforts, as well as attempts to share knowledge in cross-domain view. In the following section, the astro/geo context is discussed, mainly focusing on their specifics and shared aspects, and the possible transfer of knowledge.

1.2 KNOWLEDGE DISCOVERY PROCESSES

Currently, we can store and access large amounts of data. One of the main problems is to transform raw data into some useful artifacts. Hence, the real benefit is in our ability to extract such useful artifacts, which can be in the form of reports, policies, decisions, or recommended actions. Before we provide more details on processes that transform raw data into these artifacts, we can start with the basic notion of data, information, or knowledge.

As we already mentioned in the previous section, there are different definitions with a different scope, i.e., from the DIKW pyramid with a more granular view, to simpler definitions when there are only two levels of data–knowledge relations. For our purposes we would stay with simpler versions of DIKW, where we define Data–Information–Knowledge relations in this way, adapted from broader Beckman definitions (Beckman, 1997):

- Data – facts, numbers, pictures, recorded sound, or another raw source usually describing real-world objects and their relations;
- Information – data with added interpretation and meaning, i.e., formatted, filtered, and summarized data;
- Knowledge – information with actions and applications, i.e., ideas, rules, and procedures, which lead to decisions and actions.

While there are also extended versions of such relations, this basic view is quite sufficient with all methodologies for KDPs. It is because raw data gathering (Data part), their processing and manipulation (Information part), and creation of models that are suitable for support of decisions and further actions (Knowledge part) are all necessary aspects of standard data analytical tasks. Hence, transformations in this Data–Information–Knowledge chain represent a very general understanding of the KDP or a simple version of methodology. We have the input dataset (raw sources – Data part), which is transferred using several steps (often including data manipulation to get more interpreted and meaningful data – Information part) to knowledge (models containing rules or patterns – Knowledge part).

For example, data from a customer survey are in the raw form of Yes/No answers, values on an ordinal scale, or numbers. If we put these data about customers in the context of questions, combine them in infographics, and analyze their relations with each other, we transform raw data into information. In practice, we mine some rules on how these customers and their subgroups usually react in specific cases discussed in the survey. We can try to understand their behavior (what they prefer, buy), predict their future reactions (if they will be interested in a new product) in similar cases, and provide actionable knowledge in the form of a recommendation to the responsible actor (apply these rules to get higher income).

The presented view of Data–Information–Knowledge relations is also comparable to the view of business analytics. In this case, we have three options in analytics according to our expectations (Evans, 2015):

- Descriptive analytics – uses data aggregation and descriptive data mining techniques to see what happened in the system (business), so the question "What has happened?" is answered. The main idea is to use descriptive analytics if we want to understand at an aggregate level what is going on, summarize such information, and describe different aspects of the system in that way (to understand present and historical data). The methods here lead us to exploration analysis, visualizations, periodic or ad hoc reporting, trend analysis, data warehousing, and creation of dashboards.
- Predictive analytics – basically tasks from this part examine the future of the system. They answer the question "What could happen according to historical data?" We can see this as a predictor of states according to all historical information. It is an estimation of the normal development of the characteristics of our system. This part of analytical tasks is closest to the traditional view of KDPs. The methods here are the same as in the case of any KDP methodology, statistical analysis, and data mining methods.
- Prescriptive analytics – here are all attempts when we select some model about the system and try to optimize its possible outcomes. It means that we analyze what we have to do if we want to get the best efficiency for some output model values. The name came from the word prescribe, so it is prescription or advice for actions to be done. The set of methods applied here is large, including methods from data mining, machine learning (whenever output models are also applicable as actions), operation research, optimization, computational modeling, or expert (knowledge-based) systems.

A nice feature of business analytics is that every option can be applied separately, or we can combine them in the chain as a step-by-step process. In this case, we can see descriptive analytics mainly responsible for transformation between Data and Information. With the addition of predictive analytics, we can enhance the process of transformation to get Knowledge of our system. Our extracted knowledge is then applicable and actionable simply as is, or we can extend it and make it part of the decision making process using methods from the area of prescriptive analytics. Hence, we can see Data–Information–Knowledge in a narrow view as part of predictive analytics in, let us say, traditional understanding (with KDPs as KDD), or we can see it in broader scope with all analytics involved in transformation.

Now we can show differences in this example. Imagine that a company has several hotels with casinos, and they want to analyze customers and optimize their profit. Within descriptive analytics they use data warehousing techniques to make reports about hotel packing in time, activities in the casino and its incomes, and infographics of profit according to different aspects. These methods will help them to understand what is happening in their casinos and hotels. Within predictive analytics, they can create a predictive model that forecasts hotel and casino packing in the future, or they can use data about customers and segment them into groups according to their behavior in casinos. The result is a better understanding of what will happen in the future, what will be occupancy of the hotel in different months, and what is the expected behavior of customers when they come to the casino. Moreover, within prescriptive analytics, they can identify which decision-based input setup (and how) to optimize their profit. It means that according to the prediction of hotel occupancy they can change prices accordingly, set up the allocation of rooms, or provide benefits to some segments of customers. For example, if someone is playing a lot, we can provide him/her with some benefits to support his/her return like a better apartment for a lower price or free food.

As we already mentioned, people often exchange the KDP with data mining, which is only one step. Moreover, for knowledge discovery, some other names were also used in literature, like knowledge extraction, information harvesting, information discovery, data pattern processing, or even data archeology. The mostly used synonym for KDPs is then obviously KDD, which is logical due to the beginnings of KDP with the processing of structured data stored in standard databases. The basic properties are even nowadays the same as or similar to KDD basics from the 1990s. Therefore we can summarize them accordingly (Fayyad et al., 1996):

- The main objective of KDP is to seek new knowledge in the selected application domain.
- Data are a set of facts. The pattern is the expression in some suitable language (part of the outcome model, e.g., rule written in some rule-based language) about a subset of facts.
- KDP is a nontrivial process of identifying valid, novel, potentially useful, and ultimately understandable patterns in data. A process is simply a multistep approach of transformations from data to patterns. The pattern (knowledge) mentioned before is:
  - valid – pattern should be true on new data with some certainty,
  - novel – we did not know about this pattern before,
  - useful – pattern should lead to actions (the pattern is actionable),
  - comprehensible – the process should produce patterns that lead to a better understanding of the underlying data for human (or machine).

KDP is easily generalized also to sources of data which are not in databases or not in structured form, which induces methodology aspects of a similar type also to the area of text mining, Big Data analysis, or data streams processing. Knowledge discovery involves the entire process, including storage and access of data, application of efficient and scalable data processing algorithms to analyze large datasets, interpretation and visualization of outcome results, and support of the human–machine or human–computer interaction, as well as support for learning and analyzing the domain. The KDP model, which is then called methodology, consists of a set of processing steps followed by the data analyst or scientist to run a knowledge discovery project. The KDP methodology usually describes procedures for each step of such a project. The model helps organizations (represented by the data analyst) to understand the process and create a project roadmap. The main advantage is reduced costs for any ad hoc analysis, time savings, better understanding, and acceptation of the advice coming from the results of the analysis. While there are still data analysts who apply ad hoc steps to their projects, most of them apply some common framework with the help of (commercial or open source) software tools for particular steps or one unified analytical platform of tools.

Before we move to a description of selected methodologies in the next section, we summarize the motivation for the use of standardized KDP models (methodologies) (Kurgan and Musilek, 2006):

- The output product (knowledge) must be useful for the user, and ad hoc solutions more often failed in yielding valid, novel, useful, and understandable results.
- Understanding of the process itself is important. Humans often lack a perception of large amounts of untapped and potentially valuable data. A process model that is well structured and logical will help to avoid these issues.
- An often underestimated factor is providing support for management problems (this also includes cases of a larger project in the science area, which needs efficient management). Whenever KDP projects involve large teams, requiring careful planning and scheduling, a management specialist in such projects is often unfamiliar with terms from the data mining area – KDP methodology can then be helpful in managing the whole project.
- Standardization of KDP provides a unified view of current process description and allows an appropriate selection and usage of technology to solve current problems in practice, mostly on an industrial level.

1.3 METHODOLOGIES FOR KNOWLEDGE DISCOVERY PROCESSES

In this section, we provide more details on selected methodologies. From the 1990s, several of them were developed, starting basically from academic research, but they very quickly moved on to an industry level. As we already mentioned, the first more structured way was proposed as KDD in Fayyad et al. (1996). Their approach was later modified and improved by both the research and the industry community. The processes always share a multistep sequential way in processing input data, where each step starts after accessing the result of the successful completion of the previous step as its input. Also, it is common that activities within steps cover understanding of the task, data, preprocessing or preparation of data, analysis, evaluation, understanding of results, and their application. All methodologies also emphasize their iterative nature by introducing feedback loops throughout the process. Moreover, they are often processed with a strong influence of human data scientists and therefore acknowledge its interactivity. The main differences between the methodologies are in the number and scope of steps, the characteristics of their inputs and outputs, and the usage of various formats.

Several studies compared existing methodologies, their advantages and disadvantages, the scope of their application, the relation to software tools and standards, and any other aspects. Probably the most extensive comparisons of methodologies can be found in Kurgan and Musilek (2006) and Mariscal et al. (2010). Other papers also bring ideas and advice, including their applicability in different domains; see, for example, Cios et al. (2007), Ponce (2009), Rogalewicz and Sika (2016).

Before we describe details of some selected methodologies, we provide some information on two aspects, i.e., the evolution of methodologies and practical usage of them by data analysts.

According to the history of methodologies, in Mariscal et al. (2010) one can find quite a thorough description of such evolution. As we already mentioned, the first attempts were fulfilled by Fayyad's KDD process between the years 1993–1996, which we will also describe in the next subsection. This approach inspired several other methodologies, which came in the years after the KDD process, like SEMMA (SAS Institute Inc., 2017), Human-Centered (Brachman and Anand, 1996), or approaches described in Cabena et al. (1998) and Anand and Buchner (1998). On the other hand, also some other ideas evolved into methodologies including the 5As or Six Sigma. Of course, some issues were identified during those years and an answer to them was in the development of CRISP-DM standard methodology, which we will also describe in one of the following subsections. CRISP-DM became the leading methodology and quite a reasonable solution for a start in any data mining project, including new projects with Big Data and data streams processing. Any new methodology or some standardized description of processes usually follows a similar approach to one defined by CRISP-DM (some of them are available in the review papers mentioned before).

The influential role of CRISP-DM is evident by the polls evaluated on KDnuggets,[1] a well-known and widely accepted community-based web site related to knowledge discovery and data mining. Gregory Piatetsky-Shapiro, one of the authors of the KDD process methodology, showed in his article[2] that according to the result of polls from years 2007 and 2014, more than 42% of data analysts (most of all votes) are using CRISP-DM methodology in their analytics, data mining, or data science projects, and the usage of the methodology seems to be stable.

1.3.1 First Attempt to Generalize Steps – Research-Based Methodology

Within the starting field of knowledge discovery in the 1990s, researchers defined the multistep process, which guides users of data mining tools in their knowledge discovery effort. The main idea was to provide a sequence of steps that would help to go through the KDP in an arbitrary domain. As mentioned before, in Fayyad et al. (1996) the authors developed a model known as KDD process.

In general, KDD provides a nine-step process, mainly considered as a research-based methodology. It involves both the evaluation and interpretation of the patterns (possibly knowledge) and the selection of preprocessing, sampling, and projections of the data before the data mining step. While some of these nine steps focus on decisions or analysis, other steps are data transitions within the data–information–knowledge chain. As mentioned before, KDD is a "nontrivial process of identifying valid, novel, potentially useful, and ultimately understandable patterns in data" (Fayyad et al., 1996). The KDD process description also provides an outline of its steps, which is available in Fig. 1.2.

The model of the KDD process consists of the following steps (input of each step is output from the previous one), in an iterative (analysts apply feedback loops if necessary) and interactive way:

1. Developing and understanding the application domain, learning relevant prior knowledge, identifying of the goals of the end-user (input: problem to be solved/our goal, output: understanding of the problem/domain/goal).
2. Creation of a target dataset – selection (querying) of the dataset, identification of subset variables (data attributes), and the creation of data samples for the KDP (output: target data/dataset).
3. Data cleaning and preprocessing – dealing with outliers and noise removal, handling the missing data, collecting data on time sequences, and identifying known changes to data (output: preprocessed data).
4. Data reduction and projection – finding useful features that represent the data (according to goal), including dimension reductions and transformations (output: transformed data).
5. Selection of data mining task – the decision on which methods to apply for classification, cluster-

[1] https://www.kdnuggets.com/.

[2] https://www.kdnuggets.com/2014/10/crisp-dm-top-methodology-analytics-data-mining-data-science-projects.html.

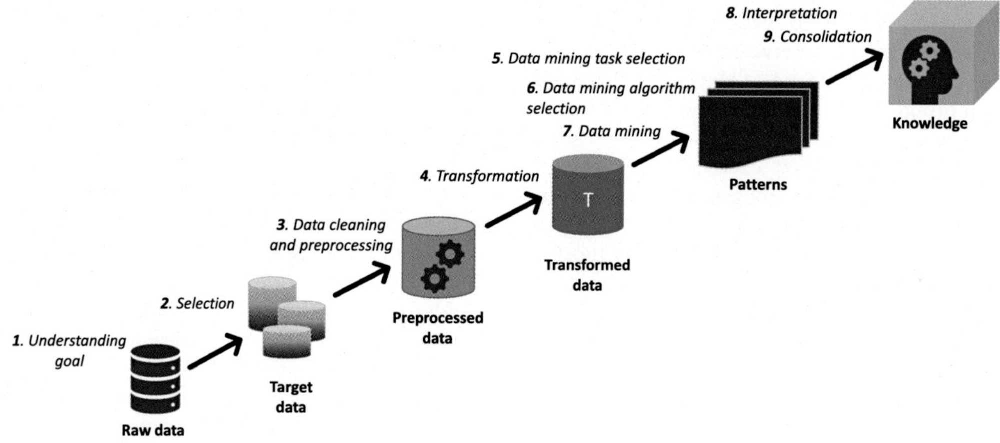

FIG. 1.2 The KDD process.

ing, regression, or another task (output: selected method[s]).

6. Selection of data mining algorithm(s) – select method for pattern search, deciding on appropriate models and their parameters, and matching methods with the goal of the process (output: selected algorithms).
7. Data mining – searching for patterns of interest in specific form like classification rules, decision trees, regression models, trends, clusters, and associations (output: patterns).
8. Interpretation of mined patterns – understanding and visualizations of patterns based on the extracted models (output: interpreted patterns).
9. Consolidation of discovered knowledge – use of discovered patterns into a system analyzed by the KDD process, documenting and reporting knowledge to end-users, and checking and resolving conflicts if needed (output: knowledge, actions/decisions based on the results).

The authors of this model declared its iterative fashion, but they gave no specific details. The KDD process is a simple methodology and quite a natural model for the discussion of KDPs. There are two significant drawbacks of this model. First, lower levels are too abstract and not explicit and formalized. This lack of detail was changed in later methodologies using more formalized step descriptions (in some cases using standards, automation of processes, or specific tools or platforms). The second drawback is its lack of business aspects description, which is logical due to the research-based idea at the start of its development.

1.3.2 Industry-Based Standard – the Success of CRISP-DM

Shortly after the KDD process definition, the industry produced methodologies more suitable for their needs. One of them is CRISP-DM (CRoss-Industry Standard Process for Data Mining) (Chapman et al., 2000), which became the standard for many years and is still widely used in both the industry and the research area. CRISP-DM was originally developed by a project consortium under the ESPRIT EU funding initiative in 1997. The project involved several large companies, which cooperated in its design: SPSS, Teradata, Daimler AG, NCR Corporation, and OHRA. Thanks to the different knowledge of companies, the consortium was able to cover all aspects, like IT technologies, case studies, data sources, and business understanding.

CRISP-DM is an open standard and is available for anyone to follow. Some of the software tools (like SPSS Modeler/SPSS Clementine) have CRISP-DM directly incorporated. As we already mentioned, CRISP-DM is the most widely used KDP methodology. While it still has some drawbacks, it became a part of the most successful story in the data mining industry. The central fact behind this success is that CRISP-DM is industry-based and neutral according to tools and application. One of the drawbacks of this model is that it does not perform project management activities. One major factor behind the success of CRISP-DM is that it is an industry tool, and it is application-neutral (Mariscal et al., 2010).

The CRISP-DM model (see Fig. 1.3) consists of the following six steps, which are then described in more details and can be iteratively applied, including feedback in some places (where necessary):

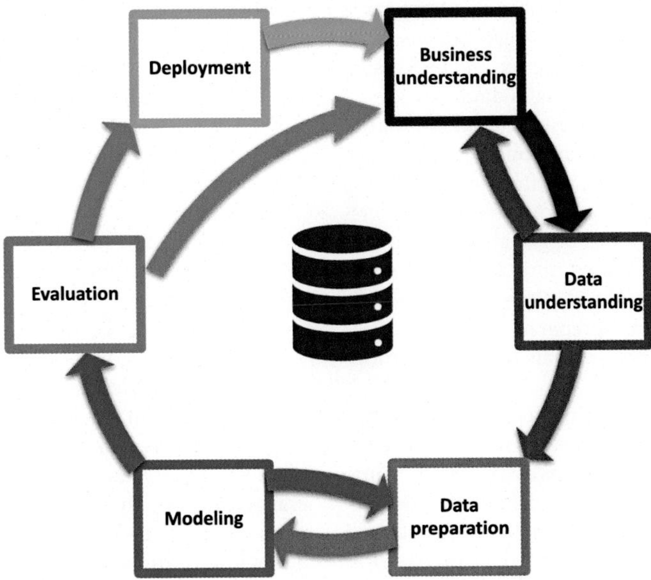

FIG. 1.3 Methodology CRISP-DM.

1. Business understanding – focuses on the understanding of objectives and requirements from a business perspective, and also converts them into the technical definition and prepares the first version of the project plan to achieve the objectives. Therefore, substeps here are:
   a. determination of business objectives – here it is important to define what we expect as business goals (costs, profits, better support of customers, and higher quality of the data product),
   b. assessment of the situation – understanding the actual situation within the objectives, defining the criteria of success for business goals,
   c. determination of technical (data mining) goals – business goals should be transformed into technical goals, i.e., what data mining models we need to achieve business goals, what the technical details of these models are, how we will measure it,
   d. generation of a project plan – the analyst creates the first version of the plan, where details on next steps are available. The analysts should address different issues, from business aspects (how to discuss and transform data mining results, deployment issues from a management point of view) to technical aspects (how to achieve data, data formats, security, anonymization of data, software tools, technical deployment).
2. Data understanding – initial collection of data, understanding the data quality issues, exploration analysis, detection of interesting data subsets. If understanding shows a need to reconsider business understanding substeps, we can move back to the previous step. Hence, the substeps of data understanding are:
   a. collection of initial data – the creation of the first versions of the dataset or its parts,
   b. description of data – understanding the meaning of attributes in data, summary of the initial dataset(s), extraction of basic characteristics,
   c. exploration of data – visualizations, descriptions of relations between attributes, correlations, simple statistical analysis on attributes, exploration of the dataset,
   d. verification of data quality – analysis of missing values, anomalies, or other issues in data.
3. Data preparation – after finishing the first steps, the most important step is the preparation of data for data mining (modeling), i.e., the preparation of the final dataset for modeling using data manipulation methods which can be applied. We can divide them into:
   a. selection of data – a selection of tables, records, and attributes, according to goal needs and reduction of dimensionality,

   b. integration of data – identification of the same entities within more tables, aggregations from more tables, redundancy checks, and processing of and detection of conflicts in data,
   c. cleansing of data – processing of missing values (remove records or imputation of values), processing of anomalies, removing inconsistencies,
   d. construction (transformation) of data – the creation of new attributes, aggregations of values, transformation of values, normalizations of values, and discretization of attributes,
   e. formatting of data – preparation of data as input to the algorithm/software tool for the modeling step.
4. Modeling – various modeling techniques are applied, and usually more types of algorithms are used, with different setup parameters (often with some metaapproach for optimization of parameters). Because methods have different formats of inputs and other needs, the previous step of data preparation could be repeated in a small feedback loop. In general, this step consists of:
   a. selection of modeling technique(s) – choose the method(s) for modeling and examining their assumptions,
   b. generation of test design – plan for training, testing, and evaluating the models,
   c. creation of models – running the selected methods,
   d. assessment of generated models – analysis of models and their qualities, revision of parameters, and rebuild.
5. Evaluation – with some high-quality models (according to the data analysis goal), such models are evaluated from a business perspective. The analyst reviews the process of model construction (to find insufficiently covered business issues) and also decides on the next usage of data mining results. Therefore, we have:
   a. evaluation of the results – assessment of results and identification of approved models,
   b. process review – summarize the process, identify activities which need another iteration,
   c. determination of the next step – a list of further actions is provided, including their advantages and disadvantages,
   d. decision – describe the decision as to how to proceed.
6. Deployment – discovered knowledge is organized and presented in the form of reports or some complex deployment is done. Also, this can be a step that finishes one of the cycles if we have an iterative application of KDP (lifecycle applications). This step consists of:
   a. plan deployment – the deployment strategy is provided, including the necessary steps and how to perform them,
   b. plan monitoring and maintenance – strategy for the monitoring and maintenance of deployment,
   c. generation of the final report – preparation of the final report and final presentation (if expected),
   d. review of the process substeps – summary of experience from the project, unexpected problems, misleading approaches, interesting solutions, and externalization of best practices.

CRISP-DM is relatively easy to understand and has good vocabulary and documentation. Thanks to its generalized nature, this methodology is a very successful and extensively used model. In practice, many advanced analytic platforms are based on this methodology, even if they do not call it the same way.

In order to help in understanding the process, we can provide a simple example. One of the possible applications of the CRISP-DM methodology is to provide tools in support of clinical diagnosis in medicine. For example, our goal is to improve breast cancer diagnostics using data about patients. In terms of CRISP-DM methodology we can describe the KDP in the following way:

1. Business understanding – from a business perspective, our business objective goal is to improve the effectiveness of breast cancer diagnostics. Here we can provide some expectation in numbers related to diagnostics effectiveness and costs of additional medical tests, in order to set up business goals – for example, if our diagnosis using some basic setup will be more effective, it reduces the costs by 20%. Then data mining goals are defined. In terms of data mining, it is a classification task with the binary target attribute, which will be tested using a confusion matrix, and according to business goals we want to achieve at least 95% accuracy of the classifier to fulfill the business goal. According to the project plan, we know that data are available in CSV format, and data and models are processed in R using RStudio, with the Rshiny web application (on available server infrastructure) providing the interface for doctors in their diagnostic process.
2. Data understanding – in this example, let us say we have data collected from the Wisconsin Diagnosis Breast Cancer (WDBC) database. We need to understand the data themselves, and what are their at-

tributes and what is their meaning. In this case, we have 569 records with 32 attributes, which mostly describe original images with/without breast cancer. The first attribute is ID and the second attribute is target class (binary – the result of diagnosis). The other 30 real-valued attributes describe different aspects of cells in the image (shape, texture, radius). We also find no missing values, and we do not need any procedure to clean or transform data. We also explore data, visualize them, and describe relations between attributes and correlations, in order to have enough information for the next steps.

3. Data preparation – any integration, cleaning, and transformation issues are solved here. In our example, there are no missing values other issues in WDBC. There is only one data table, we will select all records, and we will not remove/add an attribute. The data format is CSV, suitable for input in RStudio for the modeling step. We can also select subsets of data according to expected modeling and evaluation, in this case, let us say a simple hold-out method with different ratios for the size of training and test samples (80:20, 70:30, 60:40).
4. Modeling – data mining models are created. In our case, we want classification models (algorithms), i.e., C4.5, Random Forests, neural networks, k-NN, SVM, and naive Bayes. We create models for different hold-out selections and parameters of algorithms to achieve the best models. Then we evaluate models according to test subsets and select the best of them for further deployment, i.e., the SVM-based model with more than 97% accuracy with 70:30 hold-out.
5. Evaluation – the best models are analyzed from a business point of view, i.e., whether we can achieve the business goal using such a model and its sufficiency for application in the deployment phase. We decide on how to proceed with the best model, and what the advantages and disadvantages are. For example, in this case, the application of the selected model can support doctors and remove one intrusive and expensive test out of diagnostics, in some of the new cases.
6. Deployment – a web-based application (based on Rshiny) is created and deployed on the server, which contains an extracted model (SVM classifier) and a user interface for the doctor in order to input results of image characteristics from new patients (records) and provide him/her with a diagnosis of such new samples.

1.3.3 Proprietary Methodologies – Usage of Specific Tools

While the research or open standard methodologies are more general and tool-free, some of the leaders in the area of data analysis also provide to their customers proprietary solutions, usually based on the usage of their software tools.

One of such examples is the SEMMA methodology from the SAS Institute, which provided a process description on how to follow its data mining tools. SEMMA is a list of steps that guide users in the implementation of a data mining project. While SEMMA provides still quite a general overview of KDP, authors claim that it is a most logical organization of their tools to cover core data mining tasks (known as SAS Enterprise Miner). The main difference of SEMMA with the traditional KDD overview is that the first steps of application domain understanding (or business understanding in CRISP-DM) are skipped. SEMMA also does not include the knowledge application step, so the business aspect is out of scope for this methodology (Azevedo and Santos, 2008). Both these steps are in the knowledge discovery community considered as crucial for the success of projects. Moreover, applying this methodology outside SAS software tools is not easy. The phases of SEMMA and related tasks are the following:

1. Sample – the first step is data sampling – a selection of the dataset and data partitioning for modeling; the dataset should be large enough to contain representative information and content, but still small enough to be processed efficiently.
2. Explore – understanding the data, performing exploration analysis, examining relations between the variables, and checking anomalies, all using simple statistics and mostly visualizations.
3. Modify – methods to select, create, and transform variables (attributes) in preparation for data modeling.
4. Model – the application of data mining techniques on the prepared variables, the creation of models with (possibly) the desired outcome.
5. Assess – the evaluation of the modeling results, and analysis of reliability and usefulness of the created models.

IBM Analytics Services have designed a new methodology for data mining/predictive analytics named Analytics Solutions Unified Method for Data Mining/Predictive Analytics (also known as ASUM-DM),[3] which is a refined and extended CRISP-DM. While strong points

[3] https://developer.ibm.com/predictiveanalytics/2015/10/16/have-you-seen-asum-dm/.

of CRISP-DM are on the analytical part, due to its open standard nature CRISP-DM does not cover the infrastructure or operations side of implementing data mining projects, i.e., it has only few project management activities, and has no templates or guidelines for such tasks.

The primary goal of ASUM-DM creation was to solve the disadvantages mentioned above. It means that this methodology retained CRISP-DM and augmented some of the substeps with missing activities, tasks, guidelines, and templates. Therefore, ASUM-DM is an extension or refinement of CRISP-DM, mainly in the more detailed formalization of steps and application of (IBM-based) analytics tools. ASUM-DM is available in two versions – an internal IBM version and an external version. The internal version is a full-scale version with attached assets, and the external version is a scaled-down version without attached assets. Some of these ASUM-DM assets or a modified version are available through a service engagement with IBM Analytics Services. Like SEMMA, it is a proprietary-based methodology, but more detailed and with a broad scope of covered steps within the analytical project.

At the end of this section, we also mention that KDPs can be easily extended using agile methods, initially developed for software development. The main application of agile-based aspects is logically in larger teams in the industrial area. Many approaches are adapted explicitly for some company and are therefore proprietary. Generally, KDP is iterative, and the inclusion of more agile aspects is quite natural (Nascimento and de Oliveira, 2012). The AgileKDD method fulfills the OpenUP lifecycle, which implements Agile Manifesto. The project consists of sprints with fixed deadlines (usually a few weeks). Each sprint must deliver incremental value. Another example of an agile process description is also ASUM-DM from IBM, which combines project management and agility principles.

1.3.4 Methodologies in Big Data Context

Traditional methodologies are usually applied also in Big Data projects. The problem here is that none of the traditional standards support the description of the execution environment or workflow lifecycle aspects. In the case of Big Data projects, it is an important issue due to the complex cluster of distributed services implemented using the various technologies (distributed databases, frameworks for distributed processing, message queues, data provenance tools, coordination, and synchronization tools). An interesting paper discussing these aspects is Ponsard et al. (2017). One of the mentioned methodologies related to Big Data in this paper is Architecture-centric Agile Big data Analytics (AABA) (Chen et al., 2016), which addresses technical and organizational challenges of Big Data with the application of agile delivery. It integrates Big Data system Design (BDD) and Architecture-centric Agile Analytics (AAA) with the architecture-supported DevOps model for effective value discovery and continuous delivery of value. The authors validated the method based on case studies from different domains and summarized several recommendations for Big Data analytics:

- Data analysts should be involved already in the business analysis phase.
- There should be continuous architecture support.
- Agile steps are important and helpful due to fast technology and requirements changes in this area.
- Whenever possible, it is better to follow the reference architecture to make development and evolution of data processing much easier.
- Feedback loops need to be open and should include both technical and business aspects.

As we already mentioned, processing of data and their lifecycle is quite an important aspect in this area. Moreover, the setup of processing architecture and technology stack is probably of the same importance in the Big Data context. One approach for solving such issues is related to the Big Data Integrator (BDI)Platform (Ermilov et al., 2017), developed within the Big Data Europe H2020 flagship project, which provides distribution of Big Data components as one platform with easy installation and setup. While there are several other similar distributions, authors of this platform also provided to potential users a methodology for developing Big Data stack applications and several use cases from different domains. One of their inspirations was to use the CRISP-DM structure and terminology and apply them to a Big Data context, like in Grady (2016), where the author extends CRISP-DM to process scientific Big Data. In the scope of the BDI Platform, authors proposed a BDI Stack Lifecycle methodology, which supports the creation, deployment, and maintenance of the complex Big Data applications. The BDI Stack Lifecycle consists of the following steps (they developed documentation and tools for each of the steps):

1. Development – templates for technological frameworks, most common programming languages, different IDEs applied, distribution formalized for the needs of users (data processing task).
2. Packaging – dockerization and publishing of the developed or existing components, including best practices that can help the user to decide.

3. Composition – assembly of a BDI stack, integration of several components to address the defined data processing task.
4. Enhancement – an extension of BDI stack with enhancement tools (daemons, logging) that provides monitoring.
5. Deployment – instantiation of a BDI stack on physical or virtual servers.
6. Monitoring – observing the status of a running BDI stack, repetition of BDI components, and architecture development when need.

1.4 METHODOLOGIES IN ACTION

In practice, when it is necessary to apply the methodology, specific views and needs are expected for users (data analyst). The general data analysis methodologies are not very formalized, i.e., their direct application for machine-readable sharing or automation of data analysis processes is not easy. We must look at ways how analysts brought methodologies in action within a more precise context. This section will look at such aspects, especially on the automation of KDP and understanding of their steps through shared ontologies.

1.4.1 Standardization and Automation of Processes – Process Models

The primary goal of process modeling is to represent the process in such a way that it can be analyzed, improved, or automatized. In the scope of data analysis, the data analytical work is organized itself as the process consisting of the various steps, such as process and data understanding, data preprocessing, and modeling. Process modeling is also crucial for the data provenance, where it is necessary to capture how the data were transformed using the sequence of operations represented as the data flow process model. Additionally, the analyzed domain can be process-oriented, as is the case for example in the process industries, i.e., process models can be an essential part of the domain knowledge shared by the domain experts and data scientist. Depending on the complexity of the model, the process modeling is typically performed by the process analysts, who provide expertise in the modeling discipline in cooperation together with the domain experts. In the case of data analysis, data scientists typically perform the process analysis. Models based on some formalism, like in the form of the sequence diagrams, can be designed directly by the domain experts. For the alternative, the process model can be derived directly from the observed process events using the process mining tools.

The primary standards designed for the process modeling are flowcharting techniques which represent the process using a graph diagram. Nodes of the diagram correspond to the performed process activities, and edges represent control flow. This flowchart represents the execution ordering of the activities or data flow, i.e., how the data objects pass from one operation to another one. Examples of the standards based on the graphical notation include the Business Process Model and Notation (BPMN[4]) or the Unified Modeling Language (UML[5]) notation. BPMN models consist of simple diagrams constructed from a limited set of graphical elements with the flow objects (graph nodes) and connecting objects (graph edges). The flow objects represent activities and gateways which determine forking and merging of connection paths, depending on the conditions expressed. We can group flow objects using the swim lanes representing, for example, the organization units or different roles in the process. A part of the process model for data processing can be additional annotations representing the data objects generated or received by the activities. Activities can be atomic tasks, or they can consist of further decomposed subprocesses. BPMN is a graphical modeling notation, but version 2.0 also specifies the basic execution semantics, and the workflow engines can directly execute BPMN diagram modeling in order to automatize the processes. Additionally, BPMN models can be directly mapped to workflow execution languages, such as Web Services Business Process Execution Language (WS-BPEL[6]). The main disadvantage of the BPMN is the lack of direct support for knowledge creation processes and support for decision rules, and some ambiguity in the sharing of BPMN models.

In comparison to BPMN, UML is a general purpose modeling language which provides many types of diagrams from two categories: types representing the structural information and types representing the general types of behavior, including types representing different aspects of interactions. The behavior types can be directly used for the process modeling using the activity diagrams or in some cases sequence diagrams. A UML activity diagram generally describes step-by-step operational activities of the components in a modeled system using a similar flowcharting technique like BPMN diagrams. The activity diagram consists of nodes representing the activities or decision gateway with support for choice, iteration, and concurrency. For data flow modeling, diagrams can be additionally annotated with the

[4]http://www.bpmn.org.
[5]http://www.uml.org.
[6]http://docs.oasis-open.org/wsbpel/2.0/wsbpel-v2.0.html.

```
▼<PMML xmlns="http://www.dmg.org/PMML-4_0" version="4.0">
  ▼<Header copyright="DMG">
      <Application name="KNIME" version="2.6.2"/>
    </Header>
  ▼<DataDictionary numberOfFields="7">
    ▼<DataField name="latitude" optype="continuous" dataType="double">
        <Interval closure="closedClosed" leftMargin="-8.28" rightMargin="8.97"/>
      </DataField>
    ▼<DataField name="longitude" optype="continuous" dataType="double">
        <Interval closure="closedClosed" leftMargin="-179.97" rightMargin="179.8"/>
      </DataField>
    ▼<DataField name="zon_winds" optype="continuous" dataType="double">
        <Interval closure="closedClosed" leftMargin="-8.9" rightMargin="7.0"/>
      </DataField>
    ▼<DataField name="mer_winds" optype="continuous" dataType="double">
        <Interval closure="closedClosed" leftMargin="-6.4" rightMargin="7.1"/>
      </DataField>
    ▼<DataField name="humidity" optype="continuous" dataType="double">
        <Interval closure="closedClosed" leftMargin="27.58" rightMargin="99.4"/>
      </DataField>
    ▼<DataField name="airtemp" optype="continuous" dataType="double">
        <Interval closure="closedClosed" leftMargin="22.72" rightMargin="30.04"/>
      </DataField>
    ▼<DataField name="s_s_temp" optype="continuous" dataType="double">
        <Interval closure="closedClosed" leftMargin="22.44" rightMargin="30.34"/>
      </DataField>
    </DataDictionary>
  ▼<RegressionModel functionName="regression" algorithmName="LinearRegression" modelName="KNIME Linear Regression" targetFieldName="airtemp">
    ▼<MiningSchema>
        <MiningField name="latitude" invalidValueTreatment="asIs"/>
        <MiningField name="longitude" invalidValueTreatment="asIs"/>
        <MiningField name="zon_winds" invalidValueTreatment="asIs"/>
        <MiningField name="mer_winds" invalidValueTreatment="asIs"/>
        <MiningField name="humidity" invalidValueTreatment="asIs"/>
        <MiningField name="s_s_temp" invalidValueTreatment="asIs"/>
        <MiningField name="airtemp" invalidValueTreatment="asIs" usageType="predicted"/>
      </MiningSchema>
    ▼<RegressionTable intercept="6.008706171265235">
        <NumericPredictor name="latitude" coefficient="3.363167396766842E-4"/>
        <NumericPredictor name="longitude" coefficient="1.238009786077277E-4"/>
        <NumericPredictor name="zon_winds" coefficient="-0.07364295448649694"/>
        <NumericPredictor name="mer_winds" coefficient="-0.04315230485415502"/>
        <NumericPredictor name="humidity" coefficient="-0.011583900555823673"/>
        <NumericPredictor name="s_s_temp" coefficient="0.7840777698224044"/>
      </RegressionTable>
    </RegressionModel>
</PMML>
```

FIG. 1.4 Example of a PMML file (from DMG PMML examples).

references to the structural entities. The main diagram for structural modeling is the class diagram, which allows modeling classes and instances of entities together with the data types of their properties and interdependencies. The main advantage of the UML is its general applicability to the different aspects which can be modeled, including the structural aspects not included in the BPMN.

The flowcharting techniques were also directly incorporated into the tools for data analytics. Tools such as IBM SPSS Modeler or RapidMiner provide a visual interface which allows users to leverage statistical and data mining algorithms without programming. The data processing and modeling process is represented as the graph chart with the nodes representing data sources, transformation operations, machine learning algorithms, or build models applied to the data. The data flowchart (or data stream) is stored using the proprietary format, but most of the modeling tools also support standards for exchanging of the predictive models such as Predictive Model Markup Language (PMML[7]) or Portable Format for Analytics (PFA[8]). The core of the PMML standard is a structural description of the input and output data and parameters of the models, but the format also allows to specify a sequence of data processing transformation which allow for the mapping of user data into a more desirable form to be used by the mining model. This sequence together with the data dictionary (structural specification of the input and output data) can be used to represent data flow and data provenance. An example of a structure of one PMML file representing a regression task is shown in Fig. 1.4.

In comparison to PMML, the PFA standard is a more generic functional execution model which provides control structures, such as conditionals and loops (like a typical programming language). In PFA, data processing operations are represented as the function

[7] http://dmg.org/pmml/v4-3/GeneralStructure.html.
[8] http://dmg.org/pfa.

with inputs and outputs. The standard provides the vocabulary of common operations for the basic data types and the language to specify user-defined functions. The data analysis process is then the composition of the functions. Although this approach is very flexible, it lacks comprehensiveness of the graphical models.

1.4.2 Understanding Each Other - Semantic Models

During the data analysis process, domain experts have to share the domain knowledge with the data scientists in order to understand business or research problem, identify goals of the data analysis tasks, identify relevant data, and understand relations between them. Data analysts also exchange knowledge in the opposite direction, i.e., they communicate with the domain experts during the interpretation and the validation of the data analysis results. In order to capture the exchanged knowledge, data analysts use various knowledge representations for the externalization process.

Currently, the most elaborated knowledge representation techniques are ontologies known from the Semantic web area. Semantic web technologies cover the whole stack for the representation of both knowledge about structure representing classes and instances of the entities and their relationship or procedural knowledge in the form of the inference or production rules. The structural formalisms are based on the application of logic and were standardized as the ontology languages such as Ontology Web Language (OWL[9]), developed by the World Wide Web Consortium (W3C). Its predecessor is an RDF scheme developed as a standard language for representing ontology. The highest priority during the design phase of OWL was to achieve better extensibility, modifiability, and interoperability. OWL is now striving to achieve a good compromise between scalability and expressive power.

Semantic models based on the ontologies can be generally divided depending on the scope and level of specificity of the knowledge into the following three levels:

- Upper-level ontologies – Upper ontologies describe the most common entities, contain only general specifications, and are used as a basis for specializations. Typical entries in top ontology are, e.g., "entity," "object," and "situation," which include more specific concepts. Boundaries expressed by the top levels of ontologies consist of general world knowledge that is not acquired by language.
- Middle-level ontologies – Mid-level ontologies serve as a bridge between the general entities defined in the upper-level ontology and specific entities defined in the domain ontologies. Upper-level and mid-level ontologies are designed to be able to provide a mechanism for mapping subjects across domains. Mid-level ontologies usually provide a more specific expression of abstract entities found in upper-level ontologies.
- Domain ontologies – Domain ontologies specify entities relevant to the domain and represent the most specific knowledge from the perspective of one domain.

The upper-level ontologies are especially important for the integration and sharing of the knowledge across multiple domains and provide a framework through which different systems can use a common base. The entities in upper-level ontologies are basic and universal and are usually limited to meta, generic, abstract, and philosophical concepts. The following list describes some commonly used generic upper ontologies:

- Suggested Upper Merged Ontology (SUMO) – SUMO (Niles and Pease, 2001) was created by merging several public ontologies into one coherent structure. Ontology is used for search and applied research, linguistics, and logic. The SUMO core contains approximately 1000 classes and 4000 axioms. It consists of SUMO core ontology, mid-level ontology, and a set of domain ontologies such as communication, economics, finance, and physical elements.
- CYC ontology – CYC (Lenat, 1995) provides a common ontology, an upper-level language for defining and creating arguments using ontology. CYC ontology is used in the field of natural language processing, word comprehensibility, answers to questions, and others.
- Descriptive Ontology for Linguistic and Cognitive Engineering (DOLCE) – A very significant top ontology is DOLCE (Gangemi et al., 2002), which focuses on capturing the ontological categories needed to represent the natural language and human reason. Established upper-level categories are considered as cognitive artifacts that depend on human perception, cultural impulses, and social conventions. Categories include abstract quality, abstract area, physical object, the quantity of matter, physical quality, physical area, and process. DOLCE ontology applications include searching for multilingual information, web systems and services, and e-learning.
- Basic Formal Ontology (BFO) – The BFO focuses on the role of providing a true upper-level ontology that can be used to support domain ontologies developed, for example, for scientific research such as biomedicine (Smith et al., 2007). The BFO rec-

[9]https://www.w3.org/OWL.

ognizes the basic distinction between the following two types of entities: the essential entities that persist over time while preserving their identity and the procedural entities that represent the entities that are becoming and developing over time. The characteristic feature of process entities is that they are expanded both in space and in time (Grenon and Smith, 2004).

- General Formal Ontology (GFO) – GFO (Herre et al., 2006) is a basic ontology integrating objects and processes. GFO has a three-layer ontological architecture consisting of an abstract top level, an abstract core level, and a basic level. This ontology involves objects as well as processes integrated into one coherent framework. GFO ontology is designed to support interoperability based on ontological mapping and reduction principles. GFO is designed for applications, especially in the medical, biological, and biomedical fields, but also in the field of economics and sociology.
- Yet Another More Advanced Top-level Ontology (YAMATO) – The YAMATO ontology (Mizoguchi, 2010) was developed mainly to address the deficiencies of other upper-level ontologies, such as DOLCE, BFO, GFO, SUMO, and CYC. It concerns the solutions of qualities and quantities dealing with the representation and content of things and the differentiation of the processes of the process. The current version of YAMATO has been widely used in several projects, such as the development of medical ontology.

Regarding the mid-level ontologies, in recent years, there is an increased need for formalized representations of the data analytics processes and formal representation of outcomes of research in general. Several formalisms for describing scientific investigations and outcomes of research are available, but most of them are specific for the particular domain (e.g., biomedicine). Examples of such formalisms include Ontology of Biomedical Investigations (OBI), or ontology of experiments (EXPO). These ontologies specify useful concepts, which describe general processes producing output data given some input data and formalize outputs and results of the data analytics investigations.

The goal of OBI ontology (Schober et al., 2007) is to provide a standard for the representation of biological and biomedical examinations. OBI is entirely in line with existing formalizations in biomedical areas. Ontology promotes consistent annotation of biomedical research regardless of the specific field of study. OBI defines the investigation as a multipart process, including the design of a general design study, the implementation of the proposed study, and documentation of the results achieved. The OBI ontology uses rigid logic and semantics because it uses higher levels of ontology relationships to define higher levels and a set of relationships. OBI defines processes and contexts (materials, tasks, tools, functions, properties) relevant to biomedical areas.

In comparison to OBI, EXPO (Soldatova and King, 2006) is a more generic ontology and is not specific to the biological domain. The EXPO ontology includes general knowledge of scientific experimental design, methodology, and representation of results. The investigator, method, outcome, and conclusion are the main results with which EXPO defines the types of two main investigations, i.e., computational investigations and physical investigations. Ontology uses a subset of SUMO as the highest classes and minimizes the set of relationships to ensure compliance with existing formalisms. The EXPO ontology is a valuable resource for a description of the experiments from various research areas. The authors used EXPO ontology to describe high energy and phylogenetic experiments.

The EXPO ontology was further extended to the LABORS ontology, which defines research units such as investigation, study, testing, and repetition. These are needed to describe the complex multilayer examinations performed by the robot in a fully automatic way. LABORS is used to create experimental robot survey descriptions that result in the formalization of more than 10,000 research units in a tree structure that is 10 levels deep. Formalization describes how the robot has contributed to the discovery of new science-related knowledge through the process (Soldatova and King, 2006).

Semantic technologies were also applied directly to formalize knowledge about the data analytics processes and KDD. The initial goal of this effort was to build an intelligent data mining assistant that combines planning and metalearning for automatic design of data mining workflows. The assistant relies on the formalized ontologies of data mining operators which specify constraints, required inputs, and provided outputs for various operations in data preprocessing, modeling, and validation phases. Examples of the ontologies for data mining/data analytics are the Data Mining OPtimization Ontology (DMOP) (Hilario et al., 2011) and the Ontology of Data Mining (OntoDM) (Panov et al., 2013). The main concepts describe the following entities:

- Datasets, consisting of data records of the specified type, which can be primitive (nominal, Boolean, numeric) or structured (set, sequence, tree, graph).

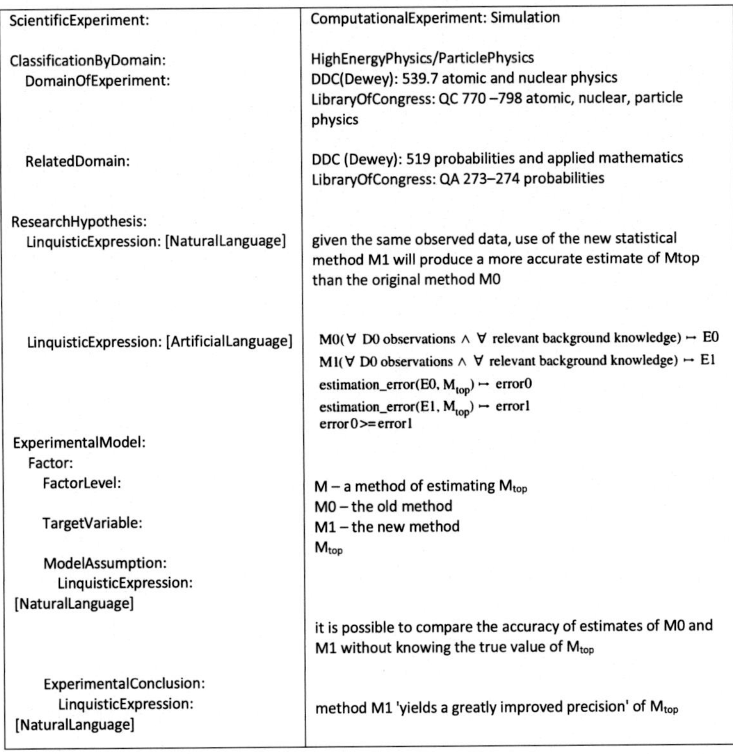

ScientificExperiment:	ComputationalExperiment: Simulation
ClassificationByDomain: DomainOfExperiment:	HighEnergyPhysics/ParticlePhysics DDC(Dewey): 539.7 atomic and nuclear physics LibraryOfCongress: QC 770 –798 atomic, nuclear, particle physics
RelatedDomain:	DDC (Dewey): 519 probabilities and applied mathematics LibraryOfCongress: QA 273–274 probabilities
ResearchHypothesis: LinquisticExpression: [NaturalLanguage]	given the same observed data, use of the new statistical method M1 will produce a more accurate estimate of Mtop than the original method M0
LinquisticExpression: [ArtificialLanguage]	M0(∀ D0 observations ∧ ∀ relevant background knowledge) → E0 M1(∀ D0 observations ∧ ∀ relevant background knowledge) → E1 estimation_error(E0, M_{top}) → error0 estimation_error(E1, M_{top}) → error1 error0>=error1
ExperimentalModel: Factor: FactorLevel:	M – a method of estimating M_{top} M0 – the old method M1 – the new method
TargetVariable:	M_{top}
ModelAssumption: LinquisticExpression: [NaturalLanguage]	it is possible to compare the accuracy of estimates of M0 and M1 without knowing the true value of M_{top}
ExperimentalConclusion: LinquisticExpression: [NaturalLanguage]	method M1 'yields a greatly improved precision' of M_{top}

FIG. 1.5 Example of an EXPO ontology instance from the domain of high energy physics.

- Data mining tasks, which include predictive modeling, pattern discovery, clustering, and probability distribution estimation.
- Generalization, the output of a data mining algorithm, which can be: predictive modeling, pattern discovery, clustering, and probability distribution estimation.
- Data mining algorithms, which solve a data mining task, produce generalizations from a dataset, and include components of algorithms such as distance functions, kernel functions, or refinement operators.

1.4.2.1 Example – EXPO

Scientific experiments and their results are usually described in papers published in the scientific journals. In these papers, the main aspects needed for the precise interpretation and reproducibility of the experiments are presented in the natural language free text ambiguously or implicitly. Therefore, it is difficult to search for the relevant experiments automatically, interpret their results, or capture their reproducibility. The EXPO ontology enables one to describe experiments in a more explicit and unambiguous structured way. Besides the discovery and traceability/reproducibility of the scientific experiments, EXPO also allows the reasoning (logical inference) about the consistency and validity of the conclusions stated in the articles or automatic generation of the new hypothesis for the further research.

The following example from the use of EXPO ontology, with the main properties of the structured record and their values, is illustrated in Fig. 1.5. The figure describes the fragment of the EXPO structured record (ontology instance) created by the annotation of the scientific paper from the domain of high energy physics describing the new estimate of the mass of the top quark (M_{top}) authored by the "D0 Collaboration" (approximately 350 scientists). The experiment was unusual as no new observational data were generated. Instead, it presented the results of applying a new statistical analysis method to existing data. No explicit hypothesis was put forward in the paper. However, the structured record includes the formalized description of the paper's implicit experimental hypothesis, i.e., given the same observed data, the use of the new statistical method will produce a more accurate estimate of M_{top} than the original method.

The record consists of three main parts. The first part is the experiment classification according to the type (ComputationalExperiment: Simulation) and domain. The classification terms are specified as the entries from the controlled vocabularies of existing classification schemes or external ontologies (e.g., Dewey Decimal Classification [DDC]). This part of the EXPO record allows efficient retrieval of the experiments relevant for the specified domain or problem. The next property describes the research hypothesis of the experiment in the natural language and (optionally) in the artificial (logic) language which allows structural matching of the hypotheses during the retrieval/comparison of experiments and automatic validation or generation of new potential hypotheses for further research. The third part describes the procedure or applied method of the experiment, i.e., in this case, the applied method was a statistical factored model. EXPO records explicitly describe its inputs, target variable, and assumptions, which are necessary for the traceability and reproducibility of the experiment process. It also describes the conclusion of the experiment in the natural language, but it is possible to use also an artificial language for structural matching.

1.4.2.2 Example – OntoDM

EXPO ontology provides concepts for the description of the scientific experiments on the upper level and describes hypotheses, assumptions, and results. It also defines elements for descriptions of the experimental methods for processing of the measured/simulated data, statistical testing of defined hypotheses, or prediction of the expected results. The OntoDM ontology extends the description of the experimental method for the application of data mining methods.

When applying the data mining method on the data, it is necessary to describe input datasets, data mining tasks (i.e., if we are dealing with predictive or descriptive data mining), and the operation applied on data during the preprocessing and methods used for the modeling (i.e., applied algorithms and their parameters settings). The description of data mining algorithms in OntoDM covers three different aspects. The first aspect is the data mining algorithm specification (e.g., the C4.5 algorithm for decision tree induction), which describes declarative elements of an algorithm, e.g., it specifies that analysts can use the algorithm for solving a predictive modeling data mining task. The second aspect is the implementation of an algorithm in some tool or software library (e.g., WekaJ48 algorithm implementation). The third aspect is the process aspect, which describes how to apply a particular data mining algorithm to a dataset with specified algorithm parameter settings. All these aspects have to be covered in the description to achieve traceability and reproducibility of the data mining process. Fig. 1.6 presents the process aspect of a data mining algorithm in more detail.

Each process has defined input and output entities which are linked to the process via has-specified-input and has-specified-output relations. An input to an application of a data mining algorithm is a dataset and parameter values, and as output, we get a generalization, i.e., a data mining model such as a decision tree. A dataset has as parts data items that are characterized with a data type which can be primitive (i.e., nominal values, numeric values, Boolean values) or combined into rows/ tuples. Besides the description of the building of data mining models, the OntoDM also supports the description of applying the model to the new dataset (e.g., prediction using a decision tree). It allows us to describe very complex experimental processes built by the composition of multiple models and steps mixing the data mining method with other scientific methods, such as simulations.

1.4.3 Knowledge Discovery Processes in Astro/Geo Context

In research and practice of both domains related to sky and Earth observations, data analysis usually follows similar steps as previously defined methodologies, even if their application is in a more ad hoc way and terminology differs between particular cases. Usage of a standard methodology (like CRISP-DM) is quite rare. Such ad hoc implementation could bring problems when analysts do not recognize some of the issues usually addressed by methodology. On the other hand, in recent years it looks like more experienced data scientists are extending project teams (especially for large projects) and their processes evolve closer to the standard application of KDP.

While there are of course standard applications of data mining techniques, from data to extracted knowledge as a scientific result or engineering output for businesses, both domains share one specific type of projects. In both astronomy and remote sensing, one of the specified outputs of some data processing process can be a data product. This concept is similar to the data warehousing area (Ponniah, 2010), where the company provides data from its infrastructure through data mart – client-based structured access to a subpart of the available data storage, often with rich query and search interface.

Creation and maintenance of data product can be a complex process and represents KDP itself, even if it

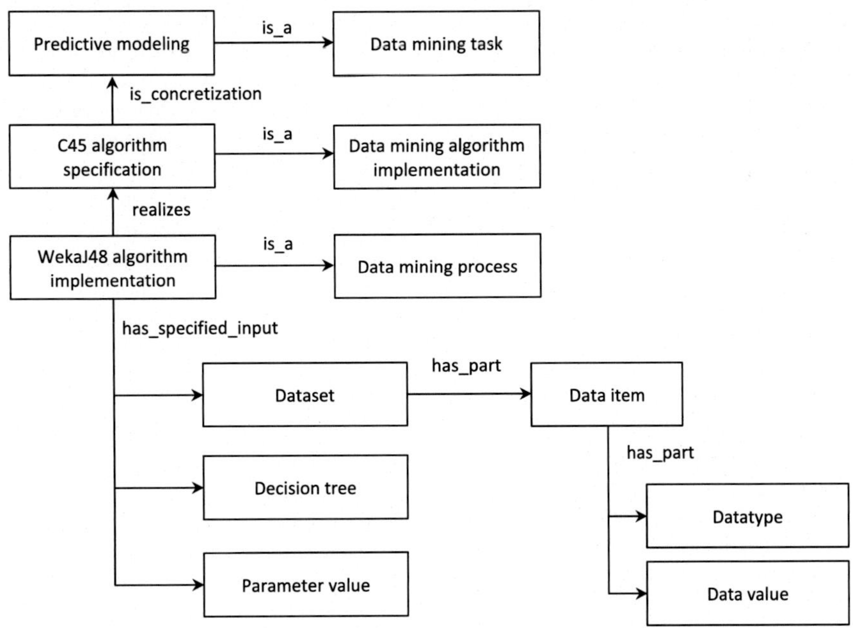

FIG. 1.6 Example of OntoDM instantiation.

needs additional analysis to provide new knowledge. Moreover, classic KDP is usually extended with lifecycle management aspects (see Section 1.3.4 for an example). In both astronomy and remote sensing, new projects (new large telescopes and satellites with high-resolution data) bring data-intensive research to a new level, and even data product projects include almost all steps of KDP. For example, raw data of images are often preprocessed using calibrations, denoising, and transformations, and often some basic patterns are recognized (in the modeling step using some machine learning) to provide better and more reliable data products for analysts, who will use the product in data mining for new knowledge.

Therefore we have three different KDP instantiations in both domains, i.e., first, full KDP project from raw data to extracted knowledge (scientific result or business output); second, data product projects, which reduce traditional KDP at the end of the process, i.e., the final result is the reliable data source for other analysts; third, slightly reduced KDP at the start, where some level of data product is used, i.e., some of the preprocessing steps are not needed, but usually querying the data is still an important step. It seems that the first case, which splits into the second and third case, will be even more rare with more data coming in the next years in both areas. In the projects combining data and methods for AstroGeoInformatics, it will be even more important to apply methods for querying data products, set up particular processes, and share their understanding well. This means that both process modeling and ontology-related aspects might be helpful in the combined effort of data scientists from both areas. We will address some of them in the following subsections.

1.4.3.1 Process Modeling Aspects

From the process point of view, astronomy and remote sensing are going in a very similar way. The most crucial aspect in process modeling is to support workflow-based approaches to Big Data produced by the instruments and their lifecycle management. In the 2020s, it will be the most critical data analytical process for both areas and it will affect any AstroGeoInformatics data mining project.

Regarding the process modeling aspects, there are three main lines in both domains:

- Some types of projects are supported by more individual software packages, suitable for querying data products, visualization, and modeling, thus providing simple (often ad hoc or user-centric) process-based analysis of particular data. This approach brings two types of tools:
  - general purpose packages – usually some toolkit(s) with a connection to different data

products from different projects or data product sites; a good example is an idea of the virtual observatory, extensively applied in astronomy (see the CDS set of tools[10]), but also in the remote sensing area (see TELEIOS[11]); many of the packages support not only virtual observatory querying functions, but also analysis and visualization;
  - project-specific packages – especially with large projects, much effort is made to provide users not only with the data product but also with necessary tools which help data scientists with the analysis of the project data, e.g., Sentinel Toolboxes[12] in remote sensing.
- Instead of the previous case, which provides a more classical KDP approach (without some standards of process modeling), an effort in data processing with grid/cloud services and high-performance computing has been made. The most prominent examples here were related to the execution of workflows, known under the term "scientific workflow systems" (e.g., DiscoveryNet, Apache Airavata, and Apache Taverna). Here analysts were able to create (often through a graphical designer), maintain, and share their workflows for processing of data. Again we have here both:
  - general purpose systems – tools that are not connected to some specific domain(s); here we can also put many of the classical business workflow systems;
  - domain- or project-specific systems – again, some scientific workflow systems more specific to some domain(s) or project.
- As an evolution of scientific workflow systems, the third line can be seen in more advanced platforms, which instead of only workflows also support analysts with more features related to data science and KDP. It means that more traditional scientific workflow systems evolved into e-Science platforms, which assist scientists to see the evolution of data, their processes, intermediate results, and also final results and potential publication of results. We can see this evolution as the advancement of both previous cases, or simply as their enhanced combination.

While previous aspects are shared, there are also some differences. Remote sensing provides more domains of applications and also has a large number of business-oriented applications, e.g., from science-related topics to applied research in agriculture, urban development and land topography, weather forecast, and security monitoring. In astronomy, the main focus is on scientific results and the domain is quite narrow. This leads to a more diverse view of tools and standards within remote sensing than in astronomy; also business-oriented applications force more business process modeling tools in remote sensing. On the other hand, thanks to advancements in Big Data processing architectures on the industry level, both areas will soon share similar software and distributed systems as an industry. This effort will help in the development of any AstroGeoInformatics projects.

1.4.3.2 Ontology-Related Aspects

The usage of the ontologies for KDP in astronomy or remote sensing is sporadic, especially if we think about the complete process and the description of its steps as provided in ontologies like OntoDM. It is different from some other domains, like bioinformatics, where ontologies are also used to understand KDP steps in the analysis itself. In both areas of astronomy and remote sensing, ontologies are usually used in the following cases:

- Light-weight taxonomies for classification of objects – the most frequent application of ontologies for a better understanding of objects in particular domains. In astronomy, it means taxonomy about astronomical objects, data of their observation and their properties. It is quite straightforward and includes efforts like ontology for virtual observatory objects, or space object ontology, or others. In remote sensing, it is more diverse, with domain-specific semantic models related to the different application domains, e.g., an ontology for geoscience, agriculture, climate monitoring, and many others. A more specific domain usually provides some semantic standard on how to share data, models, methods, and application data for its purposes. The light-weight taxonomies bring specific data formats (e.g., GeoJSON[13] in remote sensing for geographic features), as well as controlled vocabulary for annotations in order to achieve better understanding, composition, and execution of queries on real data.
- Semantic models for processing architectures – some ontology-based standards can be used in the processing of workflows, especially for the composition of services. It is not specific for these two areas, but it generally follows the application of dynamic workflow models in Big Data and data stream processing.

[10] Centre de Données astronomiques de Strasbourg – http://cds.u-strasbg.fr/.

[11] http://www.earthobservatory.eu/.

[12] https://sentinel.esa.int/web/sentinel/toolboxes.

[13] http://geojson.org/.

Especially in the remote sensing area, ontologies now became interesting due to the effort to better understand hyperspectral imaging and the combination of different (and often dynamic) data products in applications. It is also crucial due to large initiatives towards the open data; in geodata one of the most prominent initiatives is the Open Geospatial Consortium,[14] which is responsible for the creation, maintenance, and provision of standards, including some semantic models for annotations. This effort for standardization and sharing of knowledge between application domains can be beneficial for any attempt to create an AstroGeoInformatics project. Moreover, any attempt to reuse approaches which tries to share more KDP-like information (e.g., ontologies like EXPO, LABORS, OntoDM) can be beneficial for analysts and researchers from both domains in their work on such projects (because even terminology in KDP steps may differ).

In conclusion, we can say that KDP methodologies are a concept which can help especially in the preparation of cross-domain models and projects. Such ambition is also the case of AstroGeoInformatics projects proposed in this book. In the next chapters, most of the methodology steps will be addressed both from an astronomy and a remote sensing point of view; many of them will bring shared ideas, methods, and tools, but also describe some differences and specifics. Moreover, some of the examples of AstroGeoInformatics ideas and projects will be shown at the end of the book on selected use cases. In the future, such synergy can be supported by the methodologies for KDPs with the ontologies used for better understanding of particular steps, data, methods, and models.

REFERENCES

Anand, S.S., Buchner, A.G., 1998. Decision Support Using Data Mining. Financial Times Management, London, UK.

Anderson, L.W., Krathwohl, D.R. (Eds.), 2001. A Taxonomy for Learning, Teaching, and Assessing. A Revision of Bloom's Taxonomy of Educational Objectives, 2 edn. Allyn & Bacon, New York.

Azevedo, A., Santos, M.F., 2008. KDD, semma and CRISP-DM: a parallel overview. In: Abraham, A. (Ed.), IADIS European Conf. Data Mining. IADIS, pp. 182–185.

Beckman, T., 1997, International Association of Science and Technology for Development. A Methodology for Knowledge Management. IASTED.

Bouyssou, D., Dubois, D., Prade, H., Pirlot, M., 2010. Decision Making Process: Concepts and Methods. ISTE, Wiley.

Brachman, R.J., Anand, T., 1996. The process of knowledge discovery in databases: a human-centered approach. In: Fayyad, U., Piatetsky-Shapiro, G., Smyth, P., Uthurusamy, R. (Eds.), Advances in Knowledge Discovery and Data Mining. AAAI Press, Menlo Park, CA, USA, pp. 37–58.

Cabena, P., Hadjinian, P., Stadler, R., Verhees, J., Zanasi, A., 1998. Discovering Data Mining: From Concept to Implementation. Prentice-Hall, Inc., Upper Saddle River, NJ, USA.

Chapman, P., Clinton, J., Kerber, R., Khabaza, T., Reinartz, T., Shearer, C., Wirth, R., 2000. Crisp-dm 1.0 step-by-step data mining guide. Technical report. The CRISP-DM Consortium.

Chen, H.-M., Kazman, R., Haziyev, S., 2016. Agile big data analytics development: an architecture-centric approach. In: Proceedings of the 2016 49th Hawaii International Conference on System Sciences (HICSS). HICSS '16. IEEE Computer Society, Washington, DC, USA, pp. 5378–5387.

Cios, K.J., Swiniarski, R.W., Pedrycz, W., Kurgan, L.A., 2007. The Knowledge Discovery Process. Springer US, Boston, MA, pp. 9–24.

Ermilov, I., Ngonga Ngomo, A.-C., Versteden, A., Jabeen, H., Sejdiu, G., Argyriou, G., Selmi, L., Jakobitsch, J., Lehmann, J., 2017. Managing lifecycle of big data applications. In: Rozewski, P., Lange, C. (Eds.), Knowledge Engineering and Semantic Web. Springer International Publishing, Cham, pp. 263–276.

Evans, J.R., 2015. Business Analytics. Pearson.

Fayyad, U., Piatetsky-Shapiro, G., Smyth, P., 1996. From data mining to knowledge discovery in databases. AI Magazine 17, 37–54.

Gangemi, A., Guarino, N., Masolo, C., Oltramari, A., Schneider, L., 2002. Sweetening ontologies with DOLCE. In: Knowledge Engineering and Knowledge Management: Ontologies and the Semantic Web. Springer Berlin Heidelberg, Berlin, Heidelberg, pp. 166–181.

Grady, N.W., 2016. Kdd meets big data. In: 2016 IEEE International Conference on Big Data. Big Data, pp. 1603–1608.

Grenon, P., Smith, B., 2004. Snap and span: towards dynamic spatial ontology. Spatial Cognition and Computation 4 (1), 69–103.

Herre, H., et al., 2006. General Formal Ontology (GFO): A Foundational Ontology Integrating Objects and Processes. Part I: Basic Principles. OntoMed, Leipzig.

Hilario, M., Nguyen, P., Do, H., Woznica, A., Kalousis, A., 2011. Ontology-Based Meta-Mining of Knowledge Discovery Workflows. Springer Berlin Heidelberg, Berlin, Heidelberg, pp. 273–315.

Kurgan, L.A., Musilek, P., 2006. A survey of knowledge discovery and data mining process models. Knowledge Engineering Review 21 (1), 1–24.

Lenat, D.B., 1995. CYC: a large-scale investment in knowledge infrastructure. Communications of the ACM 38 (11), 33–38.

Mariscal, G., Marban, O., Fernandez, C., 2010. A survey of data mining and knowledge discovery process models and methodologies. Knowledge Engineering Review 25 (2), 137–166.

Mizoguchi, R., 2010. YAMATO: Yet Another More Advanced Top-level Ontology.

[14] http://www.opengeospatial.org/.

Nascimento, G.S., de Oliveira, A.A., 2012. An agile knowledge discovery in databases software process. In: Data and Knowledge Engineering. Springer Berlin Heidelberg, pp. 56–64.

Niles, I., Pease, A., 2001. Towards a standard upper ontology. In: Proceedings of the International Conference on Formal Ontology in Information Systems - Volume 2001. FOIS '01. ACM, New York, NY, USA, pp. 2–9.

Nonaka, I., Toyama, R., Konno, N., 2000. SECI, Ba and leadership: a unified model of dynamic knowledge creation. Long Range Planning 33, 5–34.

Panov, P., Soldatova, L.N., Dzeroski, S., 2013. OntoDM-KDD: ontology for representing the knowledge discovery process. In: Discovery Science - 16th International Conference. Proceedings. DS 2013, Singapore, October 6–9, 2013, pp. 126–140.

Ponce, Julio (Ed.), 2009. Data Mining and Knowledge Discovery in Real Life Applications. IntechOpen.

Ponniah, P., 2010. Data Warehousing Fundamentals for IT Professionals. Wiley, New Jersey, USA.

Ponsard, C., Touzani, M., Majchrowski, A., 2017. Combining process guidance and industrial feedback for successfully deploying big data projects. Open Journal of Big Data 3 (1), 26–41.

Rogalewicz, M., Sika, R., 2016. Methodologies of knowledge discovery from data and data mining methods in mechanical engineering. Management and Production Engineering Review 7 (4), 97–108.

Rowley, J., 2007. The wisdom hierarchy: representations of the DIKW hierarchy. Journal of Information Science 33 (2), 163–180.

SAS Institute Inc., 2017. SAS Enterprise Miner 14.3: Reference Help. SAS Institute Inc., Cary, NC, USA.

Schober, D., Kusnierczyk, W., Lewis, S., Lomax, J., 2007. Towards naming conventions for use in controlled vocabulary and ontology engineering. In: Proceedings of BioOntologies SIG ISMB. Oxford University Press, pp. 2–9.

Smith, B., Ashburner, M., Rosse, C., Bard, J., Bug, W., Ceusters, W., Goldberg, L.J., Eilbeck, K., Ireland, A., Mungall, C.J., Leontis, N., Rocca-Serra, P., Ruttenberg, A., Sansone, S.-A., Scheuermann, R.H., Shah, N., Whetzel, P.L., Lewis, S., 2007. The OBO Foundry: coordinated evolution of ontologies to support biomedical data integration. Nature Biotechnology 25 (11), 1251–1255.

Soldatova, L.N., King, R.D., 2006. An ontology of scientific experiments. Journal of the Royal Society Interface 3 (11), 795–803.

CHAPTER 2

Historical Background of Big Data in Astro and Geo Context

CHRISTIAN MULLER, DR

2.1 HISTORY OF BIG DATA AND ASTRONOMY

2.1.1 Big Data Before Printing and the Computer Age

Astronomy began in prehistoric times by the observation of the apparent motions of the sun, moon, stars, and planets in the Earth's sky. Big Data was not easy to define before the computer age. According to Zhang and Zhao (2015), Big Data is defined by ten characteristics, each beginning with a V: volume, variety, velocity, veracity, validity, value, variability, venue, vocabulary, and vagueness. This list comes from previous studies of Big Data in other domains, such as marketing and quality control. They consider that the four terms volume, variety, velocity, and value apply to astronomy.

Velocity did not exist before the computer age, as the acquisition of observations and their recording and publishing were entirely manual until the printing and the electric telegraph.

However, volume was already present in the early descriptions of the night sky: "as the stars of heaven" is a biblical equivalent of infinite, meaning "which cannot be counted."

Variety corresponds to the different observed objects; the ancients had only optical observations, but the conditions of these observations differed between observer sites. Their eye sight was also probably better trained. An ethnological study (Griaule and Dieterlen, 1950) revealed that the cosmogony of the Dogon people in present-day Mali indicated two companions of Sirius, four satellites of Jupiter, a ring around Jupiter, and knowledge of Venus phases. This kind of oral tradition will probably be difficult to verify in other ethnic groups, as more and more literacy is spreading over the entire world and oral transmission is disappearing.

Value corresponds to the progress in general knowledge associated with the observations and to their practical application in navigation or the calendar. For example, the heliacal rise of Sirius had an extreme importance for the Egyptian calendar related to its coincidence with the flooding of the Nile (Nickiforov and Petrova, 2012). In this respect, the corpus of Egyptian observations led to the Egyptian calendar, which was adopted by Julius Caesar when he became the ruler of Egypt and which, with a minor revision in the late 16th century, became our current calendar.

Big Data cannot exist without data preservation. The first steps were the compilation of star catalogues, which began in Hellenistic times when the infrastructure of the Alexandria museum and library were available (Thurmond, 2003). Star catalogues not only give the name of stars but also their position. Hipparcos combined previous Babylonian observations, Greek geometrical knowledge, and his own observations in Rhodes, and he was the first to correct his data for precession, but as his manuscripts are lost, he is essentially known by his distant successor, Claudius Ptolemy from Alexandria, whose catalogue, the Almagest, has been preserved (Grasshoff, 1990). The number of stars of the original manuscript is not absolutely clear; Thurmond indicates a number of 1028. See Fig. 2.1.

The precision of the observations was sometime unequal as different observation sites had been used and refraction was not corrected for. The Almagest became the main reference until the end of the Middle Ages, when several versions made their way to the Arab world and Arab astronomers both added new observations and adapted the book to their own epochs. Finally, the Almagest came back to the Western world by the Latin translation from an Arabic version of Gerard of Cremona in 1175. None of the Arabic versions increased the number of stars; some, due to the observation latitude, even mention less stars than Ptolemy. The first new catalogue to appear was endeavored by Ulugh Beg in Samarkand in the 15th century using only original observations from an especially designed large observatory, correcting the errors made by Ptolemy in converting the Hipparcos observations. This time, only 996 stars were observed. This catalogue was fully translated in Europe only in 1665, but it was known in the Arab, Persian, and Ottoman worlds.

Knowledge Discovery in Big Data from Astronomy and Earth Observation. https://doi.org/10.1016/B978-0-12-819154-5.00011-4

Octaua 86

Longitudo et Latitudo ac Magnitudo stellarum fixarum

Forme et Stelle — natur Jeuze. Imago Trigesimaquinta	Lõgitudo s	g	m	Pars	Latitudo g	m	mag	
Septentrionalis que est in capite sublimati siue audacis	1	27	0	M	18	50	nebulosa	
Lucida q̄ ē sr hũeꝝ dextꝛ: ꝛ ipa tẽdit ad rapinã qꝛ appꝛopinq̃t ad ter(rã in humero orionis	2	2	0	M	17	0	1	.e.l.
Que est super humerum sinistrum	1	20	20	M	17	30	2	.e.m.
Sequens que est sub istis duabus	1	25	0	M	18	0	4	.e.l.
Que est super cubitum dextrum	2	4	20	M	14	30	4	
Que est super brachium dextrum	2	6	20	M	11	50	6	
Sequens duplex meridionalis quadrilateri qđ est in palma dextra	2	6	30	M	10	40	4	
Antecedens lateris meridionalis	2	6	0	M	9	45	4	
Sequens lateris septentrionalis	2	7	20	M	8	15	6	
Antecedens lateris septentrionalis	2	6	40	M	8	15	6	
Antecedens duarum que sunt in figura pineali	2	1	40	M	3	45	5	
Sequens earum	2	4	20	M	3	15	5	
Sequens quattuor que sunt quasi super lineam rectam sup dorsum	1	27	30	M	19	40	4	
Antecedens hanc	1	26	20	M	20	0	6	
Antecedens etiam hanc	1	25	20	M	20	20	6	
Reliqua ꝛ est antecedens quattuor	1	24	10	M	20	40	5	
Longior nouem que sunt in dorso manus sinistre in septentrionem	1	20	30	M	8	0	4	
Secunda post istam in septentrione	1	19	20	M	8	10	4	
Tertia post eam in septentrione	1	18	0	M	10	15	4	
Quarta post eam in septentrione	1	16	20	M	12	50	4	
Quinta post eam in septentrione	1	15	10	M	14	15	4	
Sexta post eam in septentrione	1	14	30	M	15	50	3	
Septima post eam in septentrione	1	14	50	M	17	10	3	
Octaua post eam etiam in septentrione	1	15	20	M	20	20	3	
Reliqua ex nouem vltima a meridie	1	16	20	M	21	30	3	
Antecedens trium que sunt super cingulum	1	25	20	M	24	10	2	
Media earum	1	27	20	M	24	50	2	
Sequens trium	1	28	10	M	25	40	2	
Que est apud capulum ensis	1	23	50	M	25	50	3	
Septentrionalis trium coniunctarum cum capite ensis	1	26	50	M	28	40	4	.e.l.
Media earum	1	26	40	M	29	40	3	
Meridionalis trium	1	27	0	M	29	50	3	
Sequens duarum que sunt sub extremitate ensis	1	27	40	M	30	40	4	
Antecedens earum	1	26	10	M	30	50	4	
Lucida que est in pede sinistro: ꝛ est cõmunis ei ꝛ aque	1	19	10	M	31	30	1	
Que est sup decliuiorẽ ea ad septentrionẽ: ꝛ est sup calcaneum	1	21	0	M	30	15	4	.e.m.
Que est super calcaneum sinistrum exterius	1	23	20	M	31	10	4	
Que est super genu dextrum septentrionale	1	0	10	M	33	30	3	

Illarum trigintaocto stellarum in magnitudine pꝛima sunt due.in secũda quattuoꝛ.in tertia octo. in quarta quindecem.in quinta tres.in sexta quinqꝫ.ꝛ nebulosa vna.

Stellatio Fluuij. Imago Trigesimasexta	s	g	m	Pars	g	m	mag
Que ē post illã q̄ ē i pede sublimati siue audacis sup p̄ncipiũ fluminis	1	18	20	M	31	50	4
Que ē decliuioꝛ hac ad sept|.ꝛ ē i toꝛtuositate apđ ꝯphẽdẽtẽ crus sub/(limati siue audacis	1	18	50	M	28	15	4
Sequẽs duaꝝ ꝯtinuaꝝ que sunt post hanc	1	18	0	M	29	50	4
Antecedens earum	1	14	40	M	28	15	4
Sequens duarum continuarum etiam	1	13	10	M	29	15	4
Antecedens earum	1	20	10	M	25	20	4
Sequens trium que sunt post istam	1	6	20	M	26	0	4
Media earum	1	5	30	M	27	0	4
	1	[illegible]	50	M	[illegible]	50	4

FIG. 2.1 Almagest zodiac in the first complete printing by Petri Liechtenstein (1515), United States Library of Congress. Printing had two advantages: the multiplication of copies and thus a better dissemination, and protection against copyist's errors or "improvements."

2.1.2 The Printing and Technological Renaissance Revolution

The 16th century was marked by three important evolutions: the generalization of open sea navigation using astronomical positioning techniques, the appearance of accurate mechanical clocks, and the development of astrology. All these necessitated better star catalogues and planetary ephemerides. At the same time, the printing technology allowed the diffusion of the astronomical writings and was followed by a real explosion of the number of publications (Houzeau and Lancaster, 1887). Printing secured two important elements of Big Data: preservation of controlled copies and availability to a larger number of users.

Astronavigation was already used in the 15th century by the Portuguese, Arab, and Chinese navigators, but proved to be very risky during the first intercontinental explorations. It is in this context that the Ottoman sultan Murad III ordered the construction of a large observatory in Constantinople superior to the Ulugh Beg observatory and equipped with mechanical clocks. The chief astronomer, Taqi ad-Din, wanted to correct the previous catalogues and ephemerides to promote improvement in cartography (Ayduz, 2004). He improved and designed new instruments much superior to previous versions. Unfortunately, the observatory was destroyed in 1570 due to a religious decree condemning astrology.

Almost simultaneously, Tycho Brahe equipped a huge observatory in Denmark with instruments and not only used up to 100 assistants, but also spent for 30 years about 1% of the total budget of Denmark (Couper et al., 2007), Tycho Brahe was the first to take refraction into account and to analyze observational errors. His huge accomplishments were transferred to Prague where he became the astronomer of emperor Rodolph II and was assisted by Johannes Kepler, who succeeded him. Kepler demonstrated the existence of the heliocentric system and determined the parameters of the planetary elliptical orbits using Tycho's data. The quantity of data measured and reduced by Tycho Brahe and their accuracy were an order of magnitude greater than what existed before, representing maybe the first instance of Big Data improving science.

Astrology was the main application of this scientific project and the tables produced by Kepler. The Rudolphine Tables are still used by present-day astrologers, who usually do not have the means to adapt the epoch. Astrology at the time was the equivalent of present-day business intelligence and was commonly used for any kind of forecasts. Galileo taught medical students the art of establishing the horoscopes of their patients. Galileo was in this respect accused in a first inquisition trial of fatalism, which is the catholic sin of believing that the future can be certainly known to human intelligence (Westman, 2011). At the same time, Lloyd's of London were determining marine insurance fees by the expected technique of inspecting the ships and crew records, but the last judgment was left to astrologers (Thomas, 2003). Astrology cannot be considered a precursor of Big Data and their role in business intelligence, as large-scale statistical treatments of economic data were first given by Adam Smith (1778) at the end of the 18th century. Astrologers would rely on a feeling based on their knowledge which they could not quantify for everything outside their analysis of the sky. Astrology became suspected of being linked to superstition during the English Reformation, but luckily, astronomy became a respected science in Great Britain. For example, the founder of the London stock exchange, Thomas Gresham, established Gresham College in the late 16th century for the education of the young bankers and traders with the following professorships: astronomy, divinity, geometry, law, music, physics, and rhetoric. "The astronomy reader is to read in his solemn lectures, first the principles of the sphere, and the theory of the planets, and the use of the astrolabe and the staff, and other common instruments for the capacity of mariners." This program did not make any mention of astrology and its use as a predictive tool in commodity trading.

Robert Hooke, who was professor at Gresham college, insisted on the use of telescopic observations in order to increase the number of stars and their positional accuracy, but this important progress was only initiated by John Flamsteed, the first Astronomer Royal who exceeded the precision of Tycho Brahe's observations and published a catalogue of 2866 stars in 1712 (Thurmond, 2003). At that time, a marine chronometer accurate by one minute in six hours existed and an able seaman was for the first time able to determine an accurate position by using the sextant without any other information. Better marine chronometers were progressively developed (Landes, 1983), but due to their high price, their generalization had to wait until the 19th century. Flamsteed got a commission to build the Greenwich observatory in close connection with the British Admiralty; the accurate chronometers designed by John Harrison for this observatory were essential to the exploration of the Southern hemisphere oceans by Captain Cook and his followers.

Later, in the 18th and 19th century saw the astronomical observations being extended to the Southern hemisphere. At the end of the 19th century, the photographic technique allowed to win again an order of magnitude in the number of stars. At the beginning of the 20th century, about 500,000 stars had been identified and several catalogues were under development. The last catalogue before the space age was the Smithsonian Astrophysical Observatory catalogue in 1965, with 258,997 stars listed with 15 description elements for each. The SAO catalogue used electronic data treatment since the middle of the 1950s and is the first to fully meet the definition of Big Data given in the first paragraph.

FIG. 2.2 Frontispiece of the Rudolphine Tables: Tabulae Rudolphinae, quibus Astronomicae scientiae, temporum longinquitate collapsae Restauratio continetur by Johannes Kepler (1571–1630) (Jonas Saur, Ulm, 1627).

It is now succeeded by the new efforts based on space age techniques and the massive use of large databases which constitute the basis of the BigSkyEarth COST action. See Fig. 2.2.

2.2 BIG DATA AND METEOROLOGY: A LONG HISTORY

2.2.1 Early Meteorology

The study and comparison of large amounts of observations constituted the early base of meteorology. The repetition of phenomena proved very early to be less regular than astronomical events, and even extreme events were the unpredictable action of the gods. The Babylonians and Egyptians compiled a lot of observations without relating them. A big step forward was the classifications and typologies assembled by Aristotle and the structuring of these early sources. Aristotle was also the successor of the Greek natural philosophers who attempted to relate the observations to their causes so that they could explain them and even attempt forecasts. Aristotle was the first to describe the hydrologic cycle. His knowledge of prevailing winds as a function of season proved to be essential to the conquest of Greece by the Macedonian army, the Greek islands being unable to send troops to their allies in the continental cities in time due to contrary winds. The meteorology of Aristotle covered a wider context than now because it included everything in the terrestrial sphere up to the orbit of the moon and thus would have included geology and what is now called space weather (Frisinger, 1972).

Unfortunately, Aristotle's efforts were not continued for long. His successor Theophrastus compiled signs which in combination could lead to a weather forecast. These progresses did not prevent most of the population to attribute weather to divine intervention and when Christianity and Islam took over, the pagan gods were replaced by demons. No systematic records of weather were kept, and present climate historians have to resort to agricultural records or indications in chronicles. During the Renaissance, the revival of Hippocratic medicine led physicians to consider the relation between the environment and health and record meteorological data again; similarly the logbooks of the ships at sea became more systematic, leading in the 18th century to the first large set of meteorological data which began to follow a standardization process, as exemplified by the Societas Meteorologica Palatina (Meteorological Society of Mannheim) (Kington, 1974) which started in 1780, and established a network of 39 weather observation stations; 14 in Germany, and the rest in other countries, including four in the United States, all equipped with comparable and calibrated instruments: barometers, thermometers, hygrometers, and some with a wind vane. During the 19th century, more meteorological observatories were established in Europe, North America, and in the British Empire. The progress of telegraphic communications led to consider the establishment of a synoptic database of identical meteorological parameters measured at different observatories.

2.2.2 Birth of International Synoptic Meteorology

The breakthrough occurred with Leverrier in 1854. Leverrier was a French astronomer who reached celebrity by predicting the position of Neptune from perturbations of the Uranus orbit. Gelle at the Berlin observatory was then able to observe the planet at the predicted position. Following a disastrous storm in the Black Sea during the Anglo-French siege of Sebastopol, the French government commissioned Leverrier to determine if with an extensive network of stations, the storm could have been predicted. After analysis, he determined that the storm had originated in the Atlantic several days before the disaster and that a synoptic network would have allowed to follow it and to make a raw forecast of its arrival in the Black Sea (Lequeux, 2013). Unfortunately, Leverrier could never assemble the legions of laborers necessary to study the long-term physical causes of weather and climate. His efforts were however the first steps to the creation of an international synoptic network in parallel to the geomagnetic network already developed by Gauss, Sabine, and Quetelet (Kamide and Chian, 2007). The extension of the geophysical network to meteorology was rapid due to the establishment of meteorological services in most observatories and the development of the electric telegraph. These founding fathers made an unpreceded effort to internationalize the effort, and most notably, the Dutch meteorologist Buys-Ballot, founder of the Royal Dutch Meteorological Institute, published the empirically discovered relation between cyclones, anticyclones, and wind direction, introducing fluid physics to meteorology and the basis of future forecasting models (WMO, 1973).

These early networks hardly fit the definition of Big Data: the telegraphic systems of the different countries were not standard, the archiving of the data was not uniform, and a lot of parameters were station- or operator-dependant. The exchange of processed data as hourly averages was not evident. However, around 1865, the generalization of the Morse telegraphic protocols together with the application of the newly discovered Maxwell equations improved the reliability of the telegraph, and regular exchange of data between stations became the norm. International meteorological conferences regularly met, beginning in 1853 at the initiative of the United States Naval Observatory, the first one presided by the director of the Brussels Observatory, Adolphe Quetelet. Even though fewer than 15 countries were represented, no explicit resolutions came from this first meeting because any recommendation would have led to modifications of the practice of the signatories; the wording was very general, e.g., "that some uniformity in the logs for recording marine meteorological observations is desirable." Anyway, a process was started, which led in 1873 to the foundation of the International Meteorological Organization at a Vienna international conference led by Buys-Ballot (WMO, 1973). This new organization proved to be strong enough to standardize practices in the entire world. It established a permanent scientific committee who began by adopting common definitions of the meteorological parameters. See Fig. 2.3.

2.2.3 Next Step: Extension of Data Collection to the Entire Globe

The distribution of stations of this first network was heavily biased to Western Europe and the Eastern United States. It was clear at the beginning that a real network should extend to the entire world, including the Southern hemisphere. As a permanent extension was beyond the means of the early International Meteorological Organization, periodic campaigns for the study of polar regions were proposed by several countries, combining exploration and maritime observations. The first one, in 1883, was concentrated on the Arctic ocean and a few sub-Antarctic stations. The observations took place between 1881 and 1884 and demonstrated the feasibility of a network extension.

The success of the first campaign led to the second International Polar Year in 1932–1933. This campaign was initiated and led by the International Meteorological Organization and extended to geomagnetism and ionospheric studies; more countries participated, and the program included simultaneous observations at low latitudes. This campaign should have included a network of Antarctic stations, but the financial crisis of the time limited the funding means of the participating countries. The collection and use of Big Data was already envisaged by the establishment of World Data Centers centralizing data by themes.

The Second World War extended to the entire Northern hemisphere and parts of the Southern Pacific. Meteorological forecasts were essential, and the allies decided on a very wide synoptic network. This effort was led by the UK Met Office, which exfiltrated qualified meteorologists from Norway and several other occupied European countries. The Germans took a more theoretical approach, demanding less stations. The Anglo-American meteorological forecasts, with a better time resolution, were essential in planning successful amphibious operations at the end of the war, as well as air force support. After the war, the extension of this network to the Southern hemisphere led to the 1947 US Navy Highjump operation, combining the exploration of Antarctica and the establishment of stations. This expedition led to numerous accidents, which confirmed that military claim and occupation of Antarctica were beyond the means of any nation. Most of these accidents were related to errors in the positioning of ships and aircrafts related to the proximity of the South Pole and weather conditions. The staff of this huge expedition included the ionospheric scientist Lloyd Berkner,

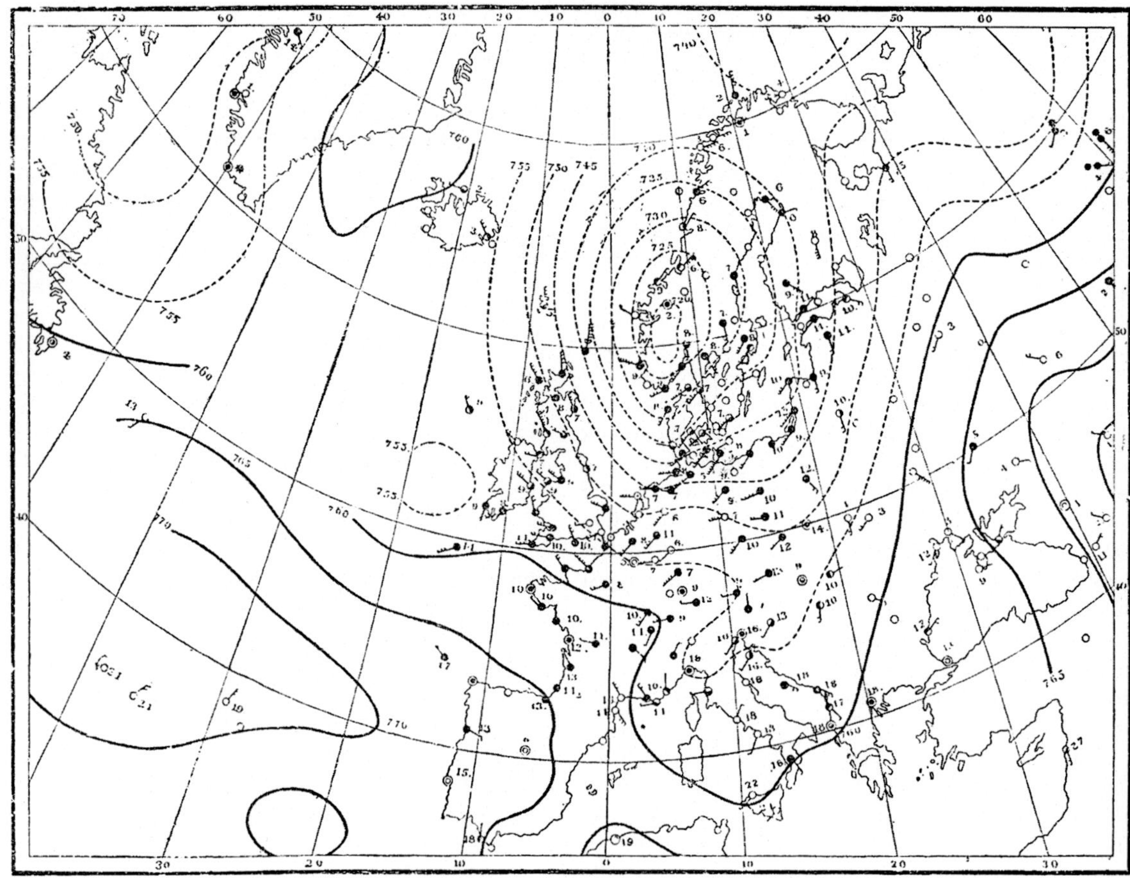

FIG. 2.3 Early synoptic map of Swedish origin (https://en.wikipedia.org/wiki/Timeline_of_meteorology#/media/File:Synoptic_chart_1874.png). Sea level pressure is indicated, as well as an indication of surface winds, demonstrating the success of the International Meteorological Organization at its foundation in 1873. Until the early 1970s, isobar lines were drafted by hand to fit the results of the stations; it was only in the last quarter of the 20th century that they were automatically plotted integrating data from other origins as airplanes and satellites.

who after designing the radio communications of the 1929 Byrd Antarctica expedition multiplied the executive roles in scientific unions while continuing research. He later played an important role in coordinating electronic operations for the US Navy in the Second World War. His positions as a rear admiral, a presidential adviser, and the president of the International Council of Scientific Unions (ICSU) helped him to initiate in 1950 the International Geophysical Year (IGY) project and to take the first steps of the Antarctic treaty. The purpose of IGY was to extend the observations to the entire globe with the cooperation of the Soviet Union and all other scientifically active countries (National Academy Press, 1992). See Fig. 2.5.

The Second World War had seen an increase in the number of weather ships, as these supported also transatlantic air traffic. This network was officialized, and these stations are shown in Fig. 2.4. Unfortunately, their high cost led to their progressive retirement after IGY when their function was taken over by instrumented merchant ships and commercial airliners. Also, beginning in 1960, experimental satellites were devoted to meteorological observations until evolving into the present network of civilian operational meteorological satellites operated by both EUMETSAT and NOAA.

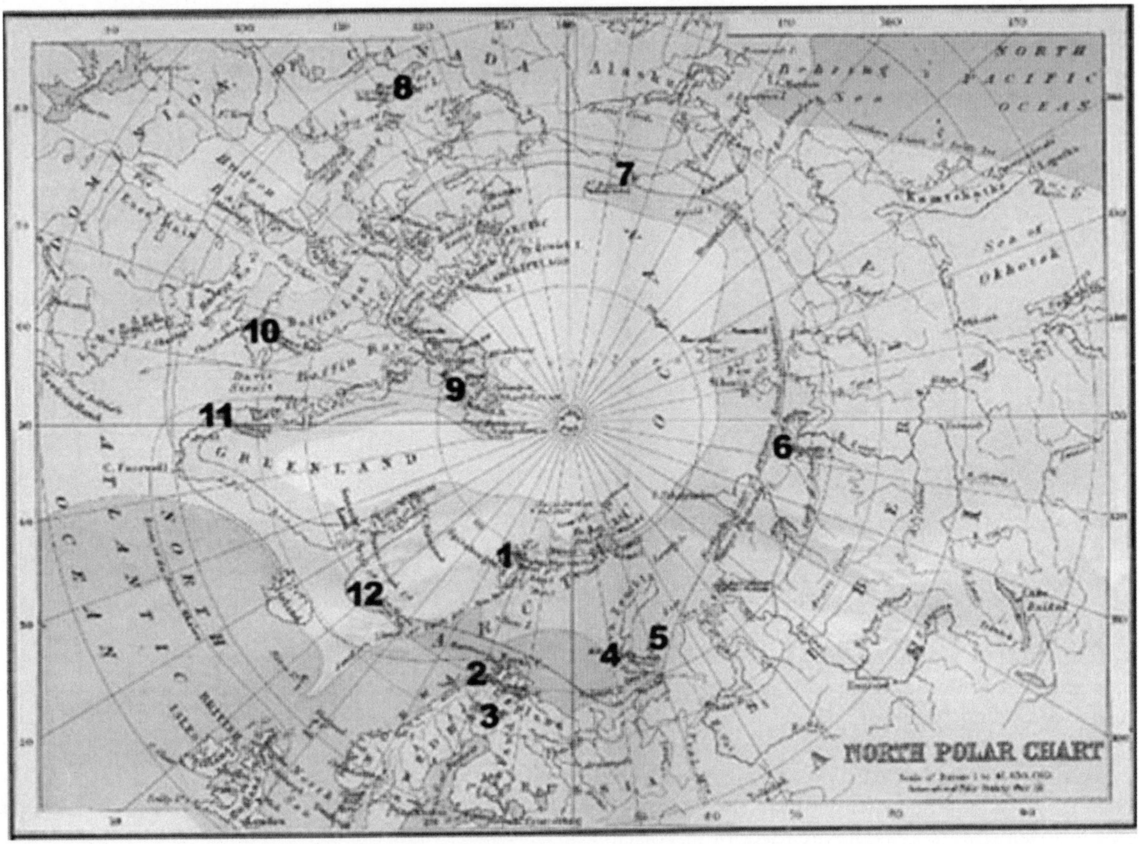

FIG. 2.4 The 12 Arctic stations of the 1883 International Polar Year, NOAA, https://www.pmel.noaa.gov/arctic-zone/ipy-1/index.htm.

FIG. 2.5 Photograph of one of the first preparatory meetings of the IGY at the US Naval Air Weapons Station at China Lake (California) in 1950. The scientists present around Lloyd Berkner and Sydney Chapman on this image represent three quarters of the world authorities on ionosphere and upper atmosphere at the time. A similar group today would include much more than 10,000 participants (Pr. Nicolet private archive).

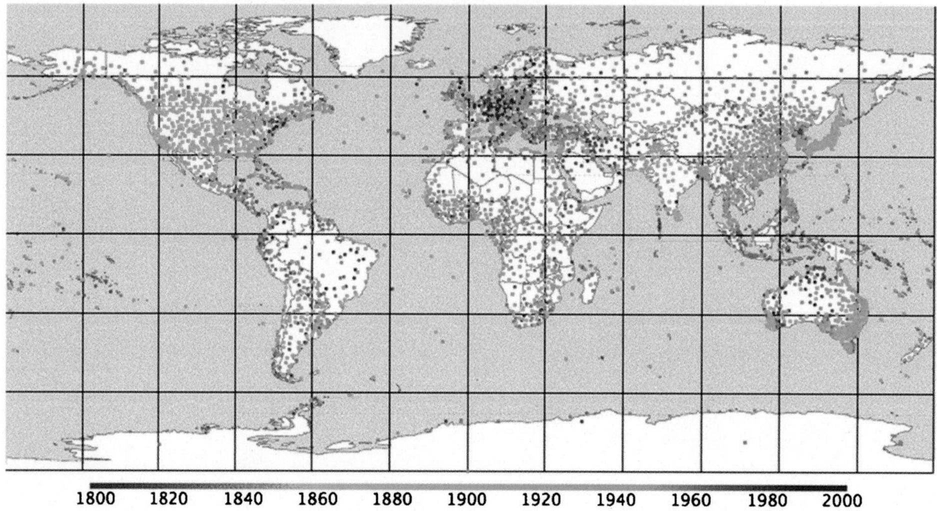

FIG. 2.6 Extension of the network of WMO stations from a European network in the middle of the 19th century to the current network. The stations are color-coded to indicate the first year in which they provided 12 months of data (Hashemi, 2009).

EUMETSAT is a consortium of meteorological organizations regrouping most of Western and Central Europe, including Turkey. It operates both its own network of geostationary METEOSAT satellites and METOP in polar orbit. Since the 2010s, it collaborates with COPERNICUS Sentinel satellites managed by ESA for the European Union. The data are used for forecasts by the European Centre for Medium Range Weather Forecasts (ECMWF) to produce forecast maps for the entire world. In 2019 these have a 20 km resolution, and should reach the 5 km resolution during the 2020s. See Fig. 2.6.

The total amount of data coming from all these sources is difficult to estimate as the definition of data covers all aspects of the raw and processed data. Currently, COPERNICUS, which is not yet in complete operation, generates about 10 petabytes per year; ECMWF claimed in 2017 to have archived more than 130 petabytes of meteorological data, beginning essentially in the 1980s, when EUMETSAT and NOAA data flows started their exponential increase.[1]

Big Data have clearly become a part of the observational database. More and more, Big Data enter the world of forecasts by techniques as assimilation, where the model is tuned to minimize the gaps between observations and the forecast and the ensemble techniques in which a large number of instances of one or several models are run in parallel and in which the final analysis uses statistical techniques (WMO, 2012).

[1] http://copernicus.eu/news/what-can-you-do-130-petabytes-data.

REFERENCES

Ayduz, S., 2004. Science and Related Institutions in the Ottoman Empire During the Classical Period. Foundation for Science, Technology and Civilisation, London.

Couper, H., Henbest, N., Clarke, A.C., 2007. The History of Astronomy. Firefly Books, Richmond Hill, Ontario.

Frisinger, H.H., 1972. Aristotle and his meteorology. Bulletin of the American Meteorological Society 53, 634–638. https://doi.org/10.1175/1520-0477(1972)053<0634:AAH>2.0.CO;2.

Grasshoff, G., 1990. The History of Ptolemy's Star Catalogue. Springer Verlag.

Griaule, M., Dieterlen, G., 1950. Un Systéme soudanais de Sirius. Journal de la Société des Africanistes 20 (2), 273–294.

Hashemi, K., 2009. http://homeclimateanalysis.blogspot.be/2009/12/station-distribution.html. climate blog. Brandeis University.

Houzeau, J.C., Lancaster, A., 1887. Bibliographie Générale de L'Astronomie, Hayez, Bruxelles.

Kamide, Y., Chian, A., 2007. Handbook of the Solar-Terrestrial Environment. Springer Science & Business Media.

Kington, J.A., 1974. The Societas Meteorologica Palatina: an eighteenth-century meteorological society. Weather 29, 416–426. https://doi.org/10.1002/j.1477-8696.1974.tb04330.x.

Landes, David S., 1983. Revolution in Time. Belknap Press of Harvard University Press, Cambridge, Massachusetts. ISBN 0-674-76800-0.

Lequeux, J., 2013. Le Verrier and meteorology. In: Le Verrier—Magnificent and Detestable Astronomer. In: Astrophysics and Space Science Library, vol. 397. Springer, New York, NY.

National Academy Press, 1992. Biographical Memoirs, V.61. ISBN 978-0-309-04746-3.

Nickiforov, M.G., Petrova, A.A., 2012. Heliacal rising of Sirius and flooding of the Nile. Bulgarian Astronomical Journal 18 (3), 53.

Smith, A., 1778. An Inquiry Into the Nature and Causes of the Wealth of Nations. W. Strahan and T. Cadell, London.

Thomas, K., 2003. Religion and the Decline of Magic: Studies in Popular Beliefs in Sixteenth and Seventeenth-Century England. Penguin=History.

Thurmond, R., 2003. A history of star catalogues. http://rickthurmond.com/HistoryOfStarCatalogs.pdf.

Westman, R.S., 2011. The Copernican Question, Prognostication, Skepticism and Celestial Order. University of California Press, Berkeley.

WMO, 1973. One hundred years of international co-operation in meteorology (1873-1973): a historical review. https://library.wmo.int/opac/doc_num.php?explnum_id=4121. World Meteorological Organization.

WMO, 2012. Guidelines on Ensemble Prediction Systems and Forecasting. http://www.wmo.int/pages/prog/www/Documents/1091_en.pdf. Publication 1091. World Meteorological Organization.

Zhang, Y., Zhao, Y., 2015. Astronomy in the big data era. Data Science Journal 14 (11), 1–9. https://doi.org/10.5334/dsj-2015-011.

CHAPTER 3

AstroGeoInformatics: From Data Acquisition to Further Application

BIANCA SCHOEN-PHELAN, PHD

3.1 INTRODUCTION

This chapter investigates different methods of data acquisition that are common to the geoinformatics as well as the astroinformatics area. Many of these techniques are shared and cross-influence each other. This chapter aims to highlight similarities and challenges between these two areas and hopes to stimulate further investigation into shared techniques and facilitate cross-pollination. Additionally, this chapter presents various case studies on current approaches using the recently operational Galileo satellite navigation system, and presents recent or ongoing projects that range from data acquisition to actual data-based applications. A particular focus of this chapter is the issue of our so-called "data tsunami" age. Large quantities of data are commonly generated in both astro- and geoinformatics and similar techniques are employed for storage and meaningful analysis using machine learning approaches.

3.2 BACKGROUND

While the last century in the astronomy and geosciences focused heavily on improving data collection instruments, the challenge now is how we can store, analyze, and interpret these data (Siemiginowska et al., 2019). The term "data tsunami" is commonly used in order to describe the challenges associated with the masses of data, which are often in the multitera or petabyte scale. In the 2000s and 2010s the geoinformatics community has invested considerable efforts into constructing a Big Earth representation, i.e., a combination of Digital Earth data and big datasets (Guo, 2017). A current ERC project run by TU Berlin in Germany called "BigEarth" (2013–2023) aims to develop a "scalable and accurate Earth Observation (EO) image search and retrieval system [...] from Big EO archives" (Bereta et al., 2018). Similarly, the astroinformatics community has made considerable efforts to provide a virtual observatory infrastructure for access and discovery of astronomical phenomena and insights (Djorgovski et al., 2011). Generally, the two domains suffer from a lack of comprehensive tools for knowledge discovery. Machine learning techniques are used in both in order to assist the evaluation of large amounts of heterogeneous data. These tasks typically consist of the following key areas (Biehl et al., 2018):

- Big Data mining,
- image processing,
- filtering the streams of data,
- outlier and novelty detection,
- classification and clustering,
- statistical inference, data modeling, and simulation.

Time-series analyses are equally prominent in astro- and geoinformatics applications. In astroinformatics a one-pass sky survey used to be a common survey strategy. However, new synoptic sky surveys provide a broad view of the whole sky image every few days, resulting in a large body of data that requires prioritization in order to decide on follow-up surveys. Real-time evaluations of these time series are critical within the astroinformatics area given the short timeframe of events, for example the detection of gravitational waves (Biehl et al., 2018). At the same time, time series analyses have become increasingly relevant within the geoinformatics area as well. Specifically within the area of natural hazard analyses and management, real-time aspects of analysis capabilities are crucial (Munoz et al., 2018; Goswami et al., 2018; Chanard et al., 2018). The Web Time Series Service (WTSS) provides JSON access and connects to a multidimensional array database for building time series, which would have traditionally been a very time-costly process (Sánchez et al., 2017).

Knowledge Discovery in Big Data from Astronomy and Earth Observation. https://doi.org/10.1016/B978-0-12-819154-5.00013-8

TABLE 3.1
Overview of the wavelength spectrum.

Visible light	400–700 nm
Infrared	700 nm–1 mm
Mid infrared	1100–3000 nm
Thermal	3000 nm–1 mm
Microwave	1 mm–300 cm

3.3 REMOTE SENSING

Remote sensing relies on capturing and interpreting the electromagnetic radiation from the Earth's surface, or in the case of astroinformatics also the capturing and interpreting of other objects in space. This electromagnetic radiation is segregated into different radiation wavelengths on a spectrum from 400 nm to 300 cm. For example, visible light uses a wavelength of 400–700 nm, while infrared (IR) uses 700 nm–1 mm. The Earth itself emits energy mostly at 10 μm, which relates to the IR spectrum. A further breakdown is available in Table 3.1 (Elert, 1998).

Typical sources of remote sensing data are of the following type (Zhu et al., 2018):

- multispectral,
- optical,
- IR measurements,
- satellite imagery,
- synthetic-aperture radar (SAR), or
- light detection and ranging (LiDAR).

These data find application in many widespread areas, ranging from photographic imagery to seismic tomography and multibeam sonars. Satellite-based geodesy typically relies on radar altimetry and SAR interferometry.

Each year we see new remote sensing instruments developed (Sandwell, 2007).

The quality of remote sensing data is considered to be based on the following attributes:

- spatial resolution,
- sensor footprint,
- sampling rate,
- dwell rate,
- spectral resolution, and
- signal-to-noise (SIN) ratio.

These attributes are further explained as follows. The *spatial resolution* determines the smallest possible feature that can be detected, while the *instantaneous field of view* (IFOV) is a measure of the spatial resolution of a sensing system and is an angle measure. The IFOV together with the distance to a target determine the actual *spatial resolution*. The *sensor footprint* on the other hand determines the amount of distortion observed in the data that are being collected. For example, a small sensor footprint typically results in objects appearing more distorted and in an oblong shape. The *sampling rate* determines how much overlap there is between samples. For example, if the sampling rate is low an area might be represented by the previous sample. The *dwell rate* typically lies somewhere on a spectrum between 0.17 ms and 0.1 ms and has an effect on how much time the instrument has to empty old data. If the *dwell rate* is very low, for example 0.1 ms, the resulting data are very noisy as the sensing device did not have enough time to empty old data. The *spectral resolution* is an indicator of the device's ability to detect very fine wavelength intervals. In the case of satellite technology used for sensing, this factor is influenced by the atmosphere, as some energy can be absorbed by it.

Remote sensing instruments can broadly be categorized into *passive* sensing and *active* sensing instruments. Each are further explained as follows.

3.3.1 Passive Sensing

Passive sensing techniques focus on collecting data from the thermal emissions that are reflected by the Earth as the sun's rays hit. This type of remote sensing mostly collects data from the visible light spectrum; see Table 3.1 for a breakdown of wavelengths. Additionally, passive sensing includes techniques that collect the information from thermal emissions from the Earth directly, without influence from any other stellar objects. This pertains to thermal IR.

3.3.2 Active Sensing

Active sensing on the other hand relies on an instrument to actively emit a beam, such as from a laser, in order to reflect off the Earth's surface. Information is gathered on the return of the beam. Technologies are laser detection (visible light spectrum), IR which also uses technology, and microwave technology where a radar instrument is used instead.

Remote sensing techniques can further be broken down by location of the sensing device. A broad categorization used within this chapter is whether the sensing device is stationed on Earth (which includes UAV) or the sensing device orbits around the Earth, i.e., if a satellite is used.

For satellite technology we typically see two different options (Rao, 2010):

- low Earth orbit (LEO): typically 500 miles or 879 km, and

- geosynchronous Earth orbit (GEO): typically 22,300 miles or 36,000 km.

LEO satellites orbit Earth from pole to pole and collect data as the Earth orbits beneath. Data from LEO satellites are typically of low temporal resolution; for example, any location on the Earth is typically passed over only a maximum of twice per day. As they tend to pass nearly over the poles they provide very high-quality images at the poles. A typical resolution of a polar LEO is around 1 km. The IR spectrum is often the chosen sensed spectrum. However, LEOs can be set up with a higher resolution of about 250 m. These often find application in the monitoring of fires and use multiple wavelength channels.

Geosynchronous satellites orbit Earth by hovering over a single point as the Earth spins. They reside much higher in orbit compared to LEO satellites, with an average of 36,000 km. A typical resolution is 1 km, and they provide a very high temporal resolution, which is useful for monitoring hurricane developments at up to 1-minute intervals. Examples are the Global Positioning System (GPS) and the Molniya orbit used by the Russian communication satellites.

Landsat and Sentinel are two programs that currently deliver the majority of Earth observation (EO) imagery data. Landsat is the longest running program of this type and is a joint operation by NASA and the USGS. Landsat are LEO systems that have been providing EO content since 1972. Since 1999 the TERRA satellite is another example of NASA's EO systems. On board are several sensors for different EO applications, namely, land composition (ASTER), reflected energy (CERES), vegetation, snow, and ice (MODIS), aerosols (MISR), and carbon monoxide (MOPITT). Another program that continuously provides EO data is Sentinel 2, which is a European Commission (EC) and European Space Agency (ESA) program with a specific mission to monitor changes in the Earth's surface. Sentinel 2 is characterized by multispectral sensor equipment, a wide swath width (290 km) and a high revisit time (approximately 2–3 days at mid-latitude).

The remote sensing device may also be positioned on Earth, either on the Earth surface as a stationary unit or a sensing unit mounted onto a vehicle, or by sensing the Earth's surface from an aircraft or an unmanned aerial vehicle (UAV). Recently, we have seen a significant increase in the use of UAV technology for sensing with a myriad of application areas and setups. This chapter will focus on this promising technology going forward.

Within the geoinformatics community it should further be noted that the remotely sensed data are rarely used in their "raw" form, but rather analyzed and visualized based a so-called digital elevation model (DEM). A DEM is a visualization of the relief of the Earth's surface which is constructed by interpolating between known elevation points. A rectangular grid shape is the most commonly used type. DEMs can be constructed based on any type of remote sensing data, for example from aerial photography, satellite images, airborne laser scanning (ALS), or LiDAR. High-resolution DEMs support a variety of studies on natural phenomena, particularly natural hazards, for example landslides (Pirasteh and Li, 2017).

3.4 BIG DATA IN ASTRO- AND GEOINFORMATICS

This book focuses predominantly on the cross-over issues between astro- and geoinformatics. Both domains are driven by data and accrue large datasets during the data collection phase. To illustrate the scale of the problem, NASA's Earth Observing System Data and Information System (EOSDIS) generates 10 terabytes of imagery every day. Several reduction pipelines require around 5000 operations per pixel and multiple processing steps to save and resample these data (Kokoulin et al., 2017). Approximately since 2007 we have observed a significant surge of devices in the spatial domain that are capable of producing time-tagged spatial data. This coincides with a move towards largely web-based infrastructures and associated web-based applications (Körting et al., 2016). The new devices together with more traditional sources such as satellite imagery form a body of data of varying quality but vast quantity. Different storage solutions with a myriad of licensing options accompany this data challenge on a global scale. In terms of visualizations for analysis of such data, representations of map-based analyses using the 2D Euclidean space are still the prevalent choice of visualization technique. The popularity of these traditional representations are considered stifling to new and more innovative approaches in this domain (Körting et al., 2016).

Many organizations rely on a scalable distributed data structure (SDDS) in order to store, analyze, and query the data. For example, LH*RS is a general purpose, high-availability SDDS that can be optimized for use with multidimensional and imagery data (Kokoulin et al., 2017). In terms of data storage it is not so much the input of data itself that presents the challenge, but rather reliable and performant querying and analysis of such data presents a roadblock, as appropriate storage structures and strategies typically depend on the manner in which the data will be used (Tiede et al., 2017). For example, in order to generate an effective

TABLE 3.2
Comparison overview of different global satellite positioning systems.

	GPS	Glonass	Galileo
Owned/provided by	USA/US Air Force	Russian government	European Union
Start date	1973 (civilian use since 1980)	1976	2011
Date fully operational	1995	2001 (fully global since 2007)	2016
Number of satellites	24 in 6 circular orbits	24 in 3 orbital planes	24 (30 planned target by 2020) in 3 orbital planes
Altitude	20,200 km	19,100 km	23,200 km
Inclination	55°	64,8°	56°
Orbital period	11 hrs 58 min	11 hrs 15 min	14 hrs 7 min
Repeat cycle	1 day	8 days	10 days

data cube structure for the storage and querying of large EO data sets it is essential to know in advance the types of queries that will be asked (Strobl et al., 2017).

On top of the newly acquired sensing data we can also avail of several decades of satellite imagery. Working on a large set of data and including prior information and expert information into machine learning algorithms poses a strong challenge for the foreseeable future of astro- and geoinformatics, as black box machine learning solutions are not able to "preserve interpretability" or provide novel insight (Biehl et al., 2018). Global datasets are often either too large or unavailable, creating a new phenomenon of this decade, which is the inability to reproduce scientific work (Baker, 2016; Fanelli, 2018). Web service front ends that provide access to data and analysis environments while hiding internal structures promise to alleviate access and reproducibility bottlenecks (Sánchez et al., 2017). The National Institute for Space Research (INPE) in Brazil provides the e-sensing platform, which sorts satellite imagery into multidimensional space-time arrays and relies on distributed storage (SciDB, Hadoop) and can be accessed from different language APIs. The CEOS data cube platform provides storing, accessing, and management capabilities of metadata of remotely sensed data and is built on top of the Australian Geoscience Data Cube and provides access via Python (Sánchez et al., 2017).

Next to the technical challenges of data storage, analysis, and querying, we experience a semantic gap between asking questions of our data and formulating a query on the data. Semantic gaps of technical terms are not uncommon in cross-over areas where diverse sets of stakeholders mesh. It has been argued that this gap is due to the fact that the recent developments have been mainly technology-driven, as opposed to user-driven (Sudmanns et al., 2018).

3.5 FROM DATA ACQUISITION TO APPLICATIONS

Remote sensing technology is closely knit together with GPS and geographic information system (GIS) in order to produce accurate geo-referencing and harvesting of spatial information.

While GIS has traditionally been a niche area interesting only to expert users of GIS systems, this has significantly changed since 2004 when TomTom launched the first personal navigation system (PNS). The company offered a standalone device that assisted navigation for business purposes, for example truck drivers, as well as personal use for everyday road participants. The need for personal navigation assistance resulted in significant developments in the PNS domain. After a strong period of market saturation of these devices, companies such as Google Maps and Apple Maps started distribution as part of smartphones, which are now the de facto standard for mobile phones. Navigational assistance has since become an expected part of everyday life via the help of a smartphone.

The common denominator in those devices has been the source of information, which is typically a combination of GPS and Global Navigation Satellite System (Glonass). In the following these technologies will be compared with a specific focus on the new player on the global positioning market, Galileo.

Table 3.2 provides a general overview of three main global satellite positioning systems. The table is intended as an overview only and there are certainly more distinguishing characteristics between the systems, such as frequency bands. Additionally, when discussing global satellite positioning systems, the corresponding ground error correction systems should be considered as well, such as European Geostationary Navigation Overlay (EGNOS) in Europe or Wide Area Augmentation System (WAAS) in the US. However, for the purpose of this chapter these will not be discussed further. It is also noteworthy that Table 3.2 does not present a comprehensive list of all global satellite positioning systems available to date. Other systems that are currently in operation include the Chinese BeiDou (BDS) and the Indian NavIC, which has become operational recently.

Each system provides an interesting approach to the actual positioning strategy within the Earth's orbit. For example, GPS suffers from so-called orbital perturbations as a direct consequence of its orbital period strategy, which means that satellites have to be regularly maneuvered in order to maintain their nominal orbit positions. Glonass avoids this effect by using a strategy whereby each satellite repeats the position of the previous satellite. Therefore, the position of each satellite in the system remains unchanged, but the resonance effect that GPS experiences could be avoided. The Galileo system has been built with a view to avoid resonance and with a target of one maneuver maximum per satellite lifetime each (Geng et al., 2018).

GPS is owned by the US Air Force. Although the system has been open to civilian use since 1980, the US government maintains full control and may decrease or revoke the service, as has happened with India in 1999 over disagreement in the Kargil conflict. This may explain the drive towards independence from GPS for global satellite positioning. Most PNS devices sold on the European market these days rely on a combination of GPS and Glonass, but are preparing to include Galileo signals. Google Maps can work with a variety of satellite systems, whichever is supported by the host mobile device, and typically uses two simultaneously. Many devices already contain Galileo-ready chips as the 17 major producers of chips, such as Qualmcom and STM, cover approximately 95% of the global satellite navigation supply market.

3.6 GALILEO APPLICATIONS

The GSA, which is the European Global Navigation Satellite System (GNSS) agency, with its headquarters in Prague, encourages the uptake of Galileo services through a variety of outreach activities on several levels of the academic sector as well as industry participation. For example, the GSA run an annual innovation contest called ActInSpace, which aims to bring together students, entrepreneurs, developers, and creative minds in general in order to solve a certain set of challenges. In addition, the GSA oversees Galileo- and EGNOS-related projects that have been funded under the European Union's funding program for research and innovation, called Horizon 2020. These projects represent the current cutting edge of development and research in several categories, ranging from agriculture to search and rescue and surveying and mapping projects.

Although initially only developed for location-based information, GNSS have fast found application for remote sensing purposes (Zavorotny et al., 2014). Some of these applications use the GNSS signals directly for remote sensing, for example of the ionosphere, and more recently UAVs make use of GNSS for autonomous flight decisions while also performing remote sensing tasks. Both of these application areas will be discussed in the following.

The idea of using the actual GNSS signals for EO originates from as early as the 1980s (Yunck et al., 1988). Among common uses of GNSS for remote sensing are atmospheric and ionospheric measurements, reflectometry, and remote sensing of the oceans (Jin et al., 2014). An interesting application area of remote sensing from both ground-based GNSS and airborne GNSS is earthquake prediction. In this case, the signals read the ionospheric total electron count, which can in turn be used to detect any disturbances and perturbations within the ionosphere of the Earth, which is an indicator for earthquake activity. The ionosphere is a layer of the Earth's atmosphere, which lies 75–1000 km above the Earth. The ionosphere gets its name from its positively charged environment.

Dense GNSS observations are capable of detecting patterns within the ionosphere. It has been shown that acoustic-gravity waves are generated closely to the epicenter on the 2008 Wenchuan earthquake of Mw = 8.0 of the seismic moment magnitude scale and the 2011 Tohoku earthquake of Mw = 9.0 (Jin et al., 2015). This is an interesting observation considering that for example tsunamis generate both acoustic and gravity waves in the ionosphere that generate perturbations in the plasma density and velocity. It has been argued before that these effects are most likely observable only in cases of earthquakes showing a significant dip-slip component. This is due to the large vertical displacement needed to produce the effect. However, it could be shown that in cases of very large strike-slip earthquakes, i.e., earthquakes with a low vertical component but that are high on the Mw scale, also show significant perturbations in the ionosphere (Astafyeva et al., 2014).

Ionospheric monitoring is useful and it makes sense to want a continuous near-real-time monitoring of it. However, the ionosphere is a dispersive medium and produces error that needs to be filtered out. One approach that uses error parameters and is able to process signals from Galileo, BDS, or GPS, respectively, is using only carrier-phase observations combined with Precise Point Positioning and produce an ionospheric delay series. These are then compared with the global ionospheric map for analysis and for filtering out delays that enable constant ionospheric monitoring (Liu et al., 2018). The need for real-time monitoring or even real-time analysis of phenomena is a requirement that the astro- and geoinformatics share, as further discussed in later sections.

In addition to using the GNSS signals directly for sensing, they are more traditionally used for location-based services (LBSs). With the rise in popularity of UAV technology, interesting new projects and products have emerged. One such example is the EU Horizon 2020-funded and GSA-supervised project mapKite (Molina et al., 2015). Here, an autonomous, unpiloted UAV is equipped with a LiDAR remote sensing apparatus that acquires data in tandem with a terrestrial unit that also captures LiDAR information. Together, they are capable of producing highly exact corridor information. The positing information is provided via Galileo.

Corridor mapping is not the only novel application area empowered through using UAVs as remote sensing vehicles. Many other application areas either have already embraced this relatively cost-effective means of remote sensing or are in the process of evaluating its usefulness. For example, global glacier monitoring has traditionally been an area dominated by satellite remote sensing techniques. The high cost of acquisition of information is one of the main motivators for switching to UAV technology (Bhardwaj et al., 2016). UAVs can also be equipped with a variety of sensors, are relatively easy to deploy without the need for specialist equipment or training, and can be deployed flexibly. However, the strongest advantage over traditional sensing mechanisms is the increased resolution of sensed data. Traditional glaciology DEMs often suffer from low resolution and the lack of ground control points in mountainous areas. Consequently, any conclusions drawn from these data, for example estimating geodetic mass balance, are often questionable. Sensing via UAVs is likely the future for glaciology sensing activities.

Low cost of UAVs is a factor cited repeatedly for areas that UAV technology is now being introduced to. Since the early 2000s the term "precision farming" is a new trend that envelopes all means to technologically improve the effectiveness of farming. The economic viability of farming is often closely tied to benefits such as reduced environmental impact and increased availability of data (Auernhammer, 2001) and UAVs are uniquely positioned to deliver on both. An interesting use case has been presented in Indonesia where the main operational attributes required focus on easy operability of local workers, low cost, and submeter accuracy of sensing outputs (Rokhmana, 2015). In this case a UAV has been equipped with several point and shoot cameras and after collection the data are processed using open source GIS software. They achieve a resolution of >10 cm and the data are used for efficiently assessing the condition of vegetation and topography.

3.7 GALILEO AND SMART CITIES

Smart city has been a trending term since the mid-2010s, and refers to an ecosystem of city-based sensors and data analytics methodologies that aim to manage urban resources more efficiently and generally improve citizen safety and experience (McLaren and Agyeman, 2015). Galileo's increased positioning accuracy has just begun to be harnessed within this context and promising early adopter applications are considered in the following discourse.

Improving the quality of a city's infrastructure to real-time monitoring is not a new concept but has increasingly become more achievable by technologies such as the European geostation navigation system EGNOS and the Galileo satellite system. Real-time monitoring is beneficial for a multitude of application areas, ranging from the monitoring of first-time responders in disaster areas to the tracking of critical transport assets and relief for good distributions. A nonterrestrial solution that offers an independent network with a satellite-based structure proves to be advantageous to disaster response and real-time management in terms of availability and robustness in tough environments where local networks likely become unavailable during crisis situations. These systems can be made available to a larger user group, i.e., not just to the first responders but to victims as well (Casciati et al., 2019).

The emergence of UAVs has seen a significant surge in the areas of surveying and sensing, and even domains such as precision. Within the context of the smart city paradigm this technology has received much attention and several implementations embrace Galileo as its satellite system for navigation due to its superior accuracy for location determination compared to other systems, such as GPS. UAVs within a city environment typically rely on low altitude, which results in a host of challenges that are unique to this particular application area, such as dense urban structures and a large number of other moving objects that typically do not follow a clear pattern of movement. Due to these and other obstacles in an urban environment it has been proposed that a group of UAVs would be more successful at achieving goals compared to the use of a single UAV (Sul'aj et al., 2018). A group of UAVs are expected to be more reactive to a dynamically changing environment and collaborate on task achievement. It has been recognized that safety of the citizens when using drone technology should be the main focus within application development. The Secure, Heterogeneous, Autonomous, and Rotational Knowledge for Swarms (SHARK) protocol focuses specifically on providing increased safety and security for UAV swarms operating within a smart city context (Cooley et al., 2018). This protocol considers initial agent distributions, population sizes, and ideal distances successfully and each drone needs to only consider two rules when focusing on a specific target, which decreases the computational load on individual members of the swarm. Without a designated control point unit, deployed UAVs are able to effectively perform their tasks autonomously, which is in line with how other components of the smart city work, such as crowd control (Kumar et al., 2018), route lighting control (Juntunen et al., 2018), and traffic control systems (Rego et al., 2018).

Finding parking in a busy city has become one of the most frustrating tasks for today's citizens. The effect is a vicious cycle of heuristic probing into free parking spaces, which in turn increases congestion within the city. Smart on-street parking solutions are therefore a crowd pleasing research endeavor. Proposed systems are typically based on the use of multiple stationary sensors and rely on a combination of map matching techniques, real-time kinematic (RTK) GPS, or Differential GPS (DGPS), which can bring the accuracy below 1 m. These systems are inherently expensive to implement, and the innate accuracy of Galileo is expected to provide relief to these applications (Roman et al., 2018).

Autonomous self-driving vehicles are a fringe topic within the smart cities paradigm that has gained increasing attention with their emergence around 2013, not in the least since a self-driving car caused a road fatality in 2018 in Arizona in the US and several fatalities involving Tesla occurred between 2016 and 2019. Autonomous self-driving vehicles are expected to increase in relevance during the 2020s and associated challenges concerning accuracy and safety will become increasingly relevant. Research into topics such as safety, precision localization, and cybersecurity is currently carried out in a number of projects in L'Aquila, Italy on the topic of Cooperative and Intelligent Transportation Systems (C-ITS) (Chiocchio et al., 2018). Chicoccio et al. focus on cooperative and intelligent traffic systems where the Galileo system is combined with 5G. Overwhelmingly, the lack of safety and ease of hacking into autonomous self-driving vehicles has inspired the use of methodologies from other domains, such as blockchain, for autonomous vehicles (Ayvaz and Cetin, 2019).

3.8 CONCLUSION

This chapter provided an overview of current data acquisition and application implementation use cases within the astro- and geoinformatics cross-over area. The chapter examined the basics of common technologies and provided an overview of current trends where the latest technologies are concerned in the age of the data tsunami.

REFERENCES

Astafyeva, E., Rolland, L.M., Sladen, A., 2014. Strike-slip earthquakes can also be detected in the ionosphere. Earth and Planetary Science Letters 405, 180–193.

Auernhammer, H., 2001. Precision farming — the environmental challenge. Computers and Electronics in Agriculture 30 (1–3), 31–43.

Ayvaz, S., Cetin, S.C., 2019. Witness of things. International Journal of Intelligent Unmanned Systems 7 (2), 72–87.

Baker, M., 2016. Is there a reproducibility crisis? A Nature survey lifts the lid on how researchers view the 'crisis rocking science and what they think will help. Nature 533 (7604), 452–455.

Bereta, K., et al., 2018. From big data to big information and big knowledge: the case of Earth observation data. In: Proceedings of the 27th ACM International Conference on Information and Knowledge Management, pp. 2293–2294.

Bhardwaj, A., et al., 2016. UAVs as remote sensing platform in glaciology: present applications and future prospects. Remote Sensing of Environment 175, 196–204.

Biehl, M., et al., 2018. Machine learning and data analysis in astroinformatics. In: European Symposium on Artificial Neural Networks. ESANN.

Casciati, F., et al., 2019. Framing a satellite based asset tracking (SPARTACUS) within smart city technology. Journal of Smart Cities 2 (2), 40–48.

Chanard, K., et al., 2018. Toward a global horizontal and vertical elastic load deformation model derived from GRACE and GNSS station position time series. Journal of Geophysical Research. Solid Earth 123 (4), 3225–3237.

Chiocchio, S., et al., 2018. A comprehensive framework for next generation of cooperative ITSs. In: 2018 IEEE 4th International Forum on Research and Technology for Society and Industry. RTSI. IEEE, pp. 1–6.

Cooley, R., Wolf, S., Borowczak, M., 2018. Secure and decentralized swarm behavior with autonomous agents for smart cities. In: 2018 IEEE International Smart Cities Conference. ISC2. IEEE, pp. 1–8.

Djorgovski, S.G., et al., 2011. Towards an automated classification of transient events in synoptic sky surveys. ArXiv preprint arXiv:1110.4655.

Elert, G., 1998. The Electromagnetic Spectrum, the Physics Hypertextbook. Hypertextbook.

Fanelli, D., 2018. Opinion: is science really facing a reproducibility crisis, and do we need it to? Proceedings of the National Academy of Sciences 115 (11), 2628–2631.

Geng, T., et al., 2018. Comparison of ultra-rapid orbit prediction strategies for GPS, GLONASS, Galileo and BeiDou. Sensors 18 (2), 477.

Goswami, S., et al., 2018. A review on application of data mining techniques to combat natural disasters. Ain Shams Engineering Journal 9 (3), 365–378.

Guo, H., 2017. Big Earth data: a new frontier in Earth and information sciences. Big Earth Data 1 (1–2), 4–20.

Jin, S., Cardellach, E., Xie, F., 2014. GNSS remote sensing. In: Remote Sensing and Digital Image Processing. Springer Netherlands, Dordrecht.

Jin, S., Occhipinti, G., Jin, R., 2015. GNSS ionospheric seismology: recent observation evidences and characteristics. Earth-Science Reviews 147, 54–64.

Juntunen, E., et al., 2018. Smart and dynamic route lighting control based on movement tracking. Building and Environment 142, 472–483.

Kokoulin, A., Yuzhakov, A., Kiryanov, D., 2017. Sparse multidimensional data processing in geoinformatics. In: International Multidisciplinary Scientific GeoConference: SGEM: Surveying Geology & Mining Ecology Management 17, pp. 1053–1060.

Körting, T.S., et al., 2016. Trends in GeoInformatics. Revista Brasileira de Cartografia 68 (6).

Kumar, S., et al., 2018. An intelligent decision computing paradigm for crowd monitoring in the smart city. Journal of Parallel and Distributed Computing 118, 344–358.

Liu, Z., et al., 2018. Near real-time PPP-based monitoring of the ionosphere using dual-frequency GPS/BDS/Galileo data. Advances in Space Research 61 (6), 1435–1443.

McLaren, D., Agyeman, J., 2015. Sharing Cities: A Case for Truly Smart and Sustainable Cities. MIT Press.

Molina, P., et al., 2015. A method for simultaneous aerial and terrestrial geodata acquisition for corridor mapping. The International Archives of the Photogrammetry, Remote Sensing and Spatial Information Sciences 40 (1), 227.

Munoz, S.E., et al., 2018. Climatic control of Mississippi River flood hazard amplified by river engineering. Nature 556 (7699), 95.

Pirasteh, S., Li, J., 2017. Landslides investigations from geoinformatics perspective: quality, challenges, and recommendations. Geomatics, Natural Hazards and Risk 8 (2), 448–465.

Rao, G.S., 2010. Global Navigation Satellite Systems. Tata McGraw-Hill, New Delhi.

Rego, A., et al., 2018. Software defined network-based control system for an efficient traffic management for emergency situations in smart cities. Future Generations Computer Systems 88, 243–253.

Rokhmana, C.A., 2015. The potential of UAV-based remote sensing for supporting precision agriculture in Indonesia. Procedia Environmental Sciences 24, 245–253.

Roman, C., et al., 2018. Detecting on-street parking spaces in smart cities: performance evaluation of fixed and mobile sensing systems. IEEE Transactions on Intelligent Transportation Systems 19 (7), 2234–2245.

Sánchez, A., et al., 2017. Reproducible geospatial data science: exploratory data analysis using collaborative analysis environments. In: GEOINFO, pp. 7–16.

Sandwell, D., 2007. Overview of Remote Sensing.

Siemiginowska, A., et al., 2019. Astro2020 science white paper: the next decade of astroinformatics and astrostatistics. ArXiv preprint arXiv:1903.06796.

Strobl, P., et al., 2017. The six faces of the datacube. In: Proceedings of the Conference on Big Data From Space. BiDS'17, pp. 28–30.

Sudmanns, M., et al., 2018. Abstract data types for spatiotemporal remote sensing analysis (short paper). In: 10th International Conference on Geographic Information Science. GIScience 2018.

Sul'aj, P., et al., 2018. UAV management system for the smart city. In: 2018 World Symposium on Digital Intelligence for Systems and Machines. DISA. IEEE, pp. 119–124.

Tiede, D., et al., 2017. Architecture and prototypical implementation of a semantic querying system for big Earth observation image bases. European Journal of Remote Sensing 50 (1), 452–463.

Yunck, T.P., Lindal, G.F., Liu, C.-H., 1988. The role of GPS in precise Earth observation. In: IEEE PLANS '88, Position Location and Navigation Symposium, Record. 'Navigation Into the 21st Century'. IEEE, pp. 251–258.

Zavorotny, V.U., et al., 2014. Tutorial on remote sensing using GNSS bistatic radar of opportunity. IEEE Geoscience and Remote Sensing Magazine 2 (4), 8–45.

Zhu, L., et al., 2018. A Review: Remote Sensing Sensors. IntecOpen.

CHAPTER 4

Synergy in Astronomy and Geosciences

MIKHAIL MININ, MSC • ANGELO PIO ROSSI, DR, PHD

4.1 INTRODUCTION

4.1.1 Basic Data Operations

Fundamentally, all data need to be collected, stored, discovered and accessed, processed, analyzed, and interpreted (Horsburgh et al., 2009). Large volumes of data in astronomy and planetary sciences are collected via remote sensing, and treatment of such data in both industries is similar. One of the key attributes that distinguishes astronomical and geospatial data from many other fields of study is the spatial information associated with each data element. The dataset where point measurements are spaced on a regular rectangular grid are called Raster (Peuquet, 1984), and typically include associated metadata, which allows projecting this grid as an image on a map (Kim, 1999). The datasets where elements have a more complex or nonregularly spaced geometry store it as a sequence of points represented as vector tuples, and therefore they are called vector datasets (Peuquet, 1984). The data are typically stored on a server in a digital format, either as separate files in a hierarchical directory tree or in relational databases. Structured Query Language (SQL) powerfully handles data in relational databases and yet they are limited to a rigid schema and scalability (Gudivada et al., 2016). This limitation is overcome in NoSQL by sacrificing consistency for availability, yet allowing scalable distributed databases (Gudivada et al., 2016). Since remotely sensed data contain geometric information, the SQL databases used in astronomy and geosciences need to be spatially enabled. A popular choice for a spatial relational database is PostgreSQL (Chen and Xie, 2008). A common task in astronomy and geosciences is querying spatiographic information in spherical coordinates. In PostgreSQL this can be achieved using extensions such pgSphere (Chilingarian et al., 2004) or PostGIS (Chen and Xie, 2008). NoSQL databases can also be geospatially enabled, as in the case of the BigTable database powering Google Earth (Chang et al., 2008). A NoSQL implementation on the basis of MongoDB has been successfully demonstrated in the virtual observatory (VO) context by the Chilean VO (Antognini et al., 2015).

The data servers need to provide interfaces to allow data discovery and access, locally or over the internet. PostgreSQL provides its own protocol for front-end and back-end communication over the TCP/IP; however, this is usually used only internally to communicate with a managing tool that serves geospatial data (PostgreSQL, 2010). The geographers use applications such as GeoServer and MapServer, which support SQL with geospatial queries (Ballatore et al., 2011). Astronomers follow their own standards developed by the International Virtual Observatory Alliance (IVOA) (Ohishi, 2007) and supported through data infrastructures such as the German Astronomical Virtual Observatory (GAVO) Data Center Helper Suite (DaCHS) (Demleitner, 2014) and the Canadian Advanced Network For Astronomical Research (CANFAR) (Gaudet et al., 2009). The vector data in VO are supplied by the IVOA Table Access Protocol (TAP). This protocol uses Astronomical Data Query Language (ADQL), derived from SQL and adapted to astronomical needs (Yasuda et al., 2004). Online access to raster maps in the geospatial industry is provided by the Open Geospatial Consortium (OGC) Web Map Service (WMS) and to individual raster coverages via the OGC Web Coverage (Processing) Service (WCS/WCPS). These standards allow for the discovery of service capabilities, retrieving metadata and data subsets from datasets (Fleuren and Müller, 2008). The WCPS is an OGC standard derived from WCS which also specifies server side processing commands for computing aggregate, statistics, and arithmetic operations. Support for WCPS is provided by RasDaMan (Baumann and Chulkov, 2007). In astronomy, basemaps are accessible as Hierarchical Progressive Surveys (HiPS) (Allen et al., 2016). HiPS is based on Hierarchical Equal Area isoLatitude Pixelization (HEALPix) sky tessellation (Fernique et al., 2015). HiPS stores tiles of various resolutions in a hierarchically organized directory tree and can be published as is to an HTTP server, without the need for a relational database to manage user access, which results in a more robust performance compared with WMS, but more limited functionality. This format can be used to serve

Knowledge Discovery in Big Data from Astronomy and Earth Observation. https://doi.org/10.1016/B978-0-12-819154-5.00014-X

global planetary maps, or even panorama photographs. Its benefits are uniform resolution and higher robustness than WMS (Allen et al., 2016). Purely geospatial tasks such as terrain rendering are also possible with HEALPix (Westerteiger et al., 2012); furthermore, it can be extended to oblate spheroids (Gibb, 2016).

The full coverages are accessible through links in TAP tables; however, no subsetting or aggregation is possible. Hierarchical data archives, such as the USA Planetary Data System (PDS) Geosciences Node (Guinness et al., 1996), can be accessed by browsing the directory tree, while querying and searching is enabled through search tools such as the Orbital Data Explorer (ODE), which also provides a REST API for programmatic access (Bennett et al., 2014).

Before the data from different sources can be analyzed, they need to be aligned or projected into a single frame of reference. Digital terrain models and orthoimagery of planetary surfaces can be produced from satellite imagery with Integrated Software for Imagers and Spectrometers (ISIS) (Gaddis et al., 1997). The commonly used tools for reprojection, translation, mosaicing, and data format conversion of geospatial images are provided by the Geospatial Data Abstraction Library (GDAL) (Warmerdam, 2008). In VO, image registration and mosaicing is possible in software such as Montage (Jacob et al., 2009) and Aladin Sky Atlas, which is also capable of generation of HiPS (Fernique et al., 2009). More advanced tasks involving image segmentation, interpolation, and registration can be carried out with specialized image processing toolkits such as the Insight Segmentation and Registration Toolkit (ITK) (Schroeder et al., 2003) and OpenCV (Bradski and Kaehler, 2008), which can also be embedded into Jupyter Notebook (Ragan-Kelley et al., 2014) via a Python wrapper (Mordvintsev and Abid, 2014) to create complex ad hoc workflow solutions to specific problems. For example, super-resolution images can be produced from multiple low-resolution observations, a task common to both geology (Tao and Muller, 2016) and astronomy (Bauer, 2011).

To convert data into meaningful information they must be analyzed. Data analysis is identification of patterns and measurement of statistical relationships between the variables (Abebe et al., 2001). This process consists of a combination of qualitative and/or quantitative methods. Qualitative methods involve nonstatistical, in-depth inspection of select cases to gain insight into the data and identify occurrence of phenomena, whereas quantitative methods aim to apply a rigid statistical approach to determine prevalence of a select phenomenon and establish reliable cause–effect relationships by conducting systematic measurements over large sample sizes and hypothesis testing (Libarkin and Kurdziel, 2002). Therefore, identification of spectral signatures present in an image can be considered a qualitative method, whereas segmentation of an image based on the presence of signatures in each pixel (partitioning of the image based on clustering of pixels in feature space) would be considered a quantitative method.

Since qualitative analysis involves in-depth examination of a few select data points and requires considerable flexibility in the workflow, it is best done manually, as humans are better capable to distinguish signal from noise and identify candidates for potential patterns (Thacker et al., 2008). In spatial sciences this often involves mapping of the data, and exploring spatial relationships in mapping applications such as QGIS in geosciences and Aladin Sky Atlas in astronomy, although both of them are capable of visualizing data of either kind, as will be demonstrated in the case studies of this chapter.

On the other hand, since quantitative analysis generally implies systematic processing of large datasets it is best done programmatically to minimize human bias. The tools of trade for programmatic quantitative data analysis include programming languages R, Python, and MATLAB®. A great variety of analytical libraries are accessible with Python, such as pandas, scipy, numpy (McKinney, 2012), and astropy (Robitaille et al., 2013). These libraries provide wrappers to very efficient subroutines written in other languages, allowing for fast data analysis, while retaining ease of use and convenience of interpretive scripting language (Cai et al., 2005). It also allows combining into a single workflow subroutines from different libraries.

Despite the obvious difficulty of automating qualitative analysis, image segmentation by machine learning alone is nevertheless still possible when classification is done on the basis of unsupervised learning techniques, such as clustering (K-means), latent variable unmixing (principal component analysis), or neural networks (Friedman et al., 2001). Unlike supervised learning, unsupervised learning lacks a direct measure of success, and hence it requires a human review to validate the results (Friedman et al., 2001). Development of qualitative analytical methods based on deep neural networks is an area of ongoing research; see, e.g., Hong et al. (2015).

Following the image segmentation, geometric features may be extracted based on relationships between the pixels (Nixon and Aguado, 2012), thus producing vector datasets from raster. These derived data are then interpreted and published as academic articles. The de-

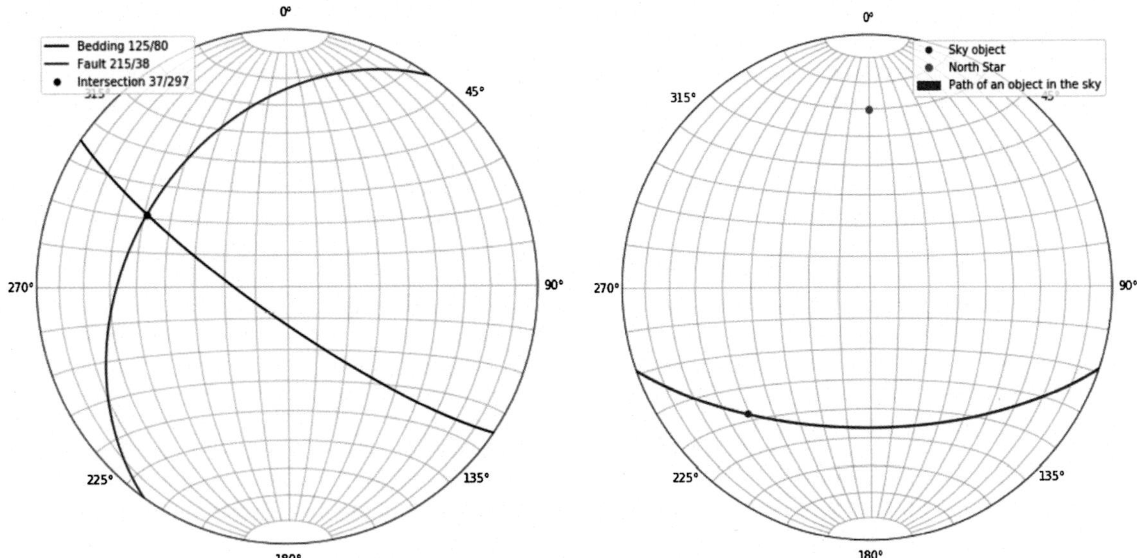

FIG. 4.1 Stereographic projections in geology and astronomy plotted using the mplstereonet Python library (Kington, 2015). Left: Using Stereonet in geology to calculate intersection of two planes. Right: Using Stereonet to plot an apparent path of an astronomical object in the sky.

rived datasets themselves can be published using portals such as Pangaea (Schindler et al., 2012) or Zenodo (Kratz and Strasser, 2014) and exposed to VO with GAVO DaCHS (Demleitner, 2014).

4.1.2 Coordinate Transformations

A key feature distinguishing data in geosciences and astronomy from other spatially enabled datasets like those used in computer design and medical anatomy is that the data are observed on a sphere or a spheroid and need to be mapped in order to be displayed on a two-dimensional surface of a screen or a paper. This is especially the case for large mosaics and surveys, in which space curvature is too significant to be considered locally flat. Geosciences have a rich tradition of using cartographic projections to display geographical data, such as Mercator, Equirectangular, Aitoff, and stereographic projection (Stereonet) (Snyder, 1997). Many analogous projections have been mathematically defined for astronomy with corresponding FITS format representations (Calabretta and Greisen, 2002). Generic conversion functions for coordinate transformations between different coordinate reference systems (CRSs) in geographical context are provided by the PROJ.4 software library (PROJ contributors, 2018). In astronomy, predefined sky coordinate frames and transformations are available from Astropy (Robitaille et al., 2013). A CRS transformation defined as a PROJ.4 string can be applied to a geospatial raster using the gdal_wrap tool in GDAL (Warmerdam, 2008). A similar task can be achieved for FITS files in astronomy using Montage toolkit (Jacob et al., 2009) or Aladin Sky Atlas (Fernique et al., 2009).

In addition to mapping remote sensed data in spherical geometry, structural geologists often rely on visualization of lines and planes in Stereonet in order to quickly perform tasks in spherical trigonometry such as computation of rake angles, plane intersections, and attitude of fold axis (Ragan, 2009). The basic methods involved in Stereonet analysis involve drawing of small and great circles, rotations around the center and around the poles by moving points along the small circles (Ragan, 2009). Since astronomical coordinate system conversions and transformations involve a lot of spherical trigonometry (Roy and Clarke, 2003), and while they can be performed algebraically (Greisen, 1993), some of astronomical calculations and visualizations can also be done using Stereonets (Fig. 4.1) or the corresponding spherical trigonometry methods (Casey, 1889).

4.1.3 Distance Measurements

In both astronomy and geosciences it is often desirable to measure distances to remote objects. Land surveying often relies on optical three-dimensional measuring systems that employ triangulation of a remote point knowing the orientation of projection rays when the object is observed from two different perspectives (Stevens et al., 2012). The similar method is used in astronomy to calculate distances to the stars relative to the orbit of Earth from their parallax, which is the apparent displacement on the celestial sphere when observed at different times of the year (Seeds and Backman, 2012). When pictures of an planetary surface are taken by a satellite or a drone, parallax may present considerable difficulties when mosaicing, especially when the mosaic is stitched from aerial photos taken at low altitude (Qi and Cooperstock, 2007). However, if the positions and orientations of the camera are known, the parallax effect can be used to triangulate distances to the image pixels, thus producing a colorized point cloud (Remondino et al., 2013) in a technique called stereogrammetry (Rosenfeld, 1969). Stereogrammetry produces point clouds by aligning the images, taking a small subset from one image (a moving window), and then finding a location of a similar subset on a second image, while recording the displacement of the moving window (Remondino et al., 2013). This point cloud can then be used to interpolate a surface (Remondino, 2003) from which digital models and orthoimages can be produced (Baltsavias and Käser, 1998). The NASA Ames Stereo Pipeline (ASP) is a stereogrammetry toolkit which can produce DTMs and orthoimages from satellite stereo imagery (Moratto et al., 2010). By interfacing with the USGS Integrated Software for Imagers and Spectrometers (ISIS) (Gaddis et al., 1997), NASA ASP can determine camera geometry of satellite planetary imaging experiments from corresponding SPICE kernels and thus produce accurate cartographic models from a single stereopair (Re et al., 2012). In situations where the orthoimagery is collected by drone photography, orthorectification and generation of digital terrain models can be done with software such as Pix4D or with open source toolkits such as MicMac, which can calculate relative camera orientations by cross-correlating multiple photographs (Niederheiser et al., 2016). Unfortunately, the same software can not yet be applied in astronomy to calculate three-dimensional shapes of distant nebulae due to the fact that atmospheric perturbations limit reliability of this technique (Reinsch et al., 2009; Keller, 1956); however, stereogrammetry of distant astronomical objects is still possible with space-based telescopes such as the Hubble (ESA, 2015).

Another method of measuring distances is to send out a signal and measure how long it takes for it to return after reflecting from the target. This method is employed widely in geosciences in technologies such as radar, LiDAR, and seismic survey. Seismic survey uses low-frequency sound waves to obtain density profiles of the subsurface by observing reflections due to density changes (Hedström and Kollert, 1949). Ground penetrating radar operates on a similar principle to a seismic survey; however, it uses low-frequency (radio) electromagnetic waves instead of sound for subsurface sounding, and thus allows conducting subsurface measurements in remote sensing (Davis and Annan, 1989). It is also important in areas where optical observations is impossible, e.g., due to cloud cover on Venus. LiDAR uses a narrow high-energy laser beam (usually in a visible or IR spectrum) instead of radio waves, which allows it to measure distances to a very high accuracy, yet since visible light cannot penetrate deep into the subsurface it is limited only to conducting very high-precision surface mapping (Lefsky et al., 2002). In astronomy, LiDAR is used for photometric atmospheric corrections (Gimmestad et al., 2007). LiDARs are also used for measuring distances to the lunar retroreflectors (Bender et al., 1973). Radars have larger swath and longer range than LiDARs (Green et al., 2017) and are used in astronomy for locating near-Earth objects and for ground-based observations of planetary surfaces in the solar system (Black, 2002).

Increasing imaging aperture allows for recording of observations of objects on the plane of focus at a greater spatial resolution (Strutt, 1880). Aperture can be increased synthetically by combining several images, a technique which is common in radar remote sensing – synthetic-aperture radar (SAR) (Cutrona, 1970). The phase shift in combined SAR images can be measured in a technique known as interferometry, which in geosciences is able to measure planetary ground surface deformation (e.g., subsidence and uplift) (Strozzi et al., 2001). In astronomy interferometry is employed to produce synthetic-aperture images from radio telescopes, in which case software is used to reconstruct an image from interference patterns obtained by a combination of images from simultaneous observations of the same object from telescopes far apart (Baldwin and Haniff, 2002). Doppler shift measured in SAR can be used to determine ocean currents (Rufenach et al., 1983) while in astronomy it is used to calculate speed of movement of stellar objects towards or away from Earth, which can be obtained from a shift in spectral lines (Zwicky, 2009).

4.1.4 One-Dimensional Series

Time series in astronomy are used to detect periodic signals in noisy data (Scargle, 1982), for instance transiting planet photometry (Brown et al., 2001), measurement of stellar rotation (Basri et al., 2011), and study of pulsars (Kramer et al., 2006) and galactic centers (Aharonian et al., 2009). The geoscience community similarly uses time series in areas of climatology and meteorology, for example to look at variations in temperature and precipitation where, being linked to astronomical phenomena, signals are also periodic (Huybers and Curry, 2006; Landscheidt, 2000). Since climatic conditions have significant impact on geological processes affecting mineral deposit formation, enrichment, and erosion, the analysis of time series in geology and environmental science is a question of great economic importance (Cronin et al., 1983). The analytical tools for time series in astronomy and meteorology/climatology are similar and deal with identifying periodicities, wavelength unmixing (spectral analysis), detection of critical points (extrema, inflection), high-order methods for nonlinear waves (e.g., gravity waves, ocean waves), wavelet analysis, and computation of fractal dimensions in spatiotemporal context (Rao et al., 1998).

Another type of one-dimensional sequences are spectral series. Light spectra are the recordings of light intensity at different frequencies or wavelengths. They are used in astronomy and geosciences to derive compositional information from spectral features such as absorption bands (van der Meer, 2004) and spectral slope (Osterloo et al., 2008). Astronomy uses the emission spectra to determine the elemental composition of the stars from absorption bands or Fraunhofer lines (Gray, 2005). The transmission spectra in the visible and IR spectra are used in astronomy to tell the molecular composition of gases and dispersion-dominated reflectance in IR for composition of solid phase cosmic dust in interstellar media (Kawaguchi, 2006). Reflectance spectra are useful for determining molecular composition, especially for hydrogen containing molecules, such as phyllosilicates and hydrated minerals, due to characteristic absorption bands formed by resonance of chemical bonds (Clark and Roush, 1984).

Geology typically uses reflectance spectra, with remote sensing relying primarily on the light from the sun reflected off a surface. Since in geosciences the differences between orientations of light source, reflecting surface, and the observer are not insignificant, reflectance spectra obtained from remote sensing of planetary surfaces has associated geometric data such as incidence, phase, and emission angles which affect the shape of the spectra and may provide additional useful information (Lohr and Kortüm, 2012). The extent to which composition can be determined by spectroscopy is limited by the fact that only spectra which have distinct and unique signatures can be identified and unmixed. A compound whose spectra are flat and featureless in the observed range will not be detected, and their proportion in the sample cannot be measured. Additionally, several compounds can sometimes be mixed in different ways to produce the same reflectance signature; hence mixtures of three or more compounds sometimes cannot be unambiguously unmixed (Keshava and Mustard, 2002). Furthermore, the presence of absorption in a transmission medium may hide features on the target of observation (Ryerson et al., 1999). As a result of these difficulties, reflective spectroscopy, while able to identify the presence of the compounds, can provide only a very rough and ambiguous estimate of samples' relative composition. The industry standard for spectroscopic work in geosciences is the IDL/ENVI commercial software (Canty, 2014); however, some free software alternatives such as GDL (Coulais et al., 2011), Tetracorder, ISIS, and Multispec (Boardman et al., 2006) are capable of performing some of the same tasks, although functionality and support for them is often limited. The astronomical community, on other hand, has a wide variety of free and well-developed and maintained tools, such as VOSpec (Osuna et al., 2005) and CASSIS (Vastel et al., 2015). The work is ongoing to enable VO spectral tools to analyze reflectance spectra in geospatial context with a massive SSHADE database of reflectance spectra becoming exposed to the VO community through the efforts of VESPA (Erard et al., 2018).

Both spectra and time series are one-dimensional continuously differentiable data types and can be displayed by the same plotting tools (TIGER-NET, 2018) and analyzed using similar algorithms. Common data operations on one-dimensional series include computing derivatives such as slope and curvature and local extrema such as peaks and valleys, as well as filtering via wavelet and Fourier transforms (Mukherjee et al., 2016). When one-dimensional data are collected on a grid such an arrangement is called a data cube, which could even have more than three dimensions (Baumann et al., 1998), for instance when a spectral cube over the same location is recorded over time.

4.2 STATE OF THE ART: VESPA INITIATIVE OF BRINGING TOGETHER IVOA, IPDA (PDS), AND OGC

4.2.1 Standards and Software

In data science, tools communicate with each other through interfaces. When different teams in the same industry want their products to be able to talk to each other they must first reach a consensus to define a common language so that data encoded by one tool can be interpreted by any other adherent to the same communication protocol (Bochmann and Sunshine, 1980). Consequently, the desire to enable interoperability necessitates the development of standards (Ledrick and Spring, 1990). One example of such a standard is SQL, which allows different tools to access different databases in a systematic and consistent way (Groff et al., 2008). A fundamental standard provider in precise sciences is the International Bureau of Weights and Measures (BIPM), responsible for the units of the metric system (Page and Vigoureux, 1975).

Some of the most important standard providers for the network communications include the International Organization for Standardization (ISO), which provides data model standards for digital data and metadata archives (Hughes et al., 2018); the Institute of Electrical and Electronics Engineers (IEEE) Standards Association, which develops industrial standards for electronics (IEEE, 2018); and the World Wide Web Consortium (W3C), which develops web standards (W3C, 2018).

Geography and geology rely on the geospatial standards for plotting geographical data, and engineering standards for performing Computer-assisted design (CAD) tasks on civil and industrial objects. The geospatial standards are developed by the Open Geospatial Consortium (OGC) (Percivall, 2010), while the CAD standards are largely industry-driven (Howard and Björk, 2008). Nevertheless, due to a rising interest in three-dimensional geography, the OGC has taken up a task through its 3DIM Working Group to develop Open Standards in CAD (Lapierre and Cote, 2007). Mature geospatial standards of OGC eventually become ISO standards (Di, 2004). The standards for geological spectral libraries and spectral cubes are also industry-driven, and hence dominated by ASD and IDL/ENVI (Chisholm and Hueni, 2015).

For astronomy the technical standards are maintained by the IVOA (Ohishi, 2007), while the fundamental standards and naming of the celestial bodies are maintained by the International Astronomical Union (Blaauw, 2012). The standards for controlling astronomical hardware and equipment through the Windows Common Object Model environment are maintained through the Astronomy Common Object Model (ASCOM) (Denny, 2002).

A very important standard for the scientific literature is the digital object identifier (DOI) syntax, defined by the ISO, which is used by the digital catalogs to track scientific publications and datasets. These DOIs are generated for new digital peer-reviewed publications but can also be produced for any digital objects such as images, datasets, or GitHub repositories through services like Zenodo (Sicilia et al., 2017).

A modern historical phenomenon of artificial intelligence being able to make correct yet unconscious decisions through the use of algorithms independent of the field of knowledge (Harari, 2016) motivates the convergence of standards between different industries. The unification of standards promises to deliver a new paradigm of science, where the same algorithms can be applied to data regardless of its source, and data of different origin can be associated based on its context and analyzed together on a very large scale (Hey et al., 2009). One such example is the convergence of CAD and GIS (Zlatanova and Prosperi, 2005), another is the convergence of astronomy and geosciences. The main similarity between astronomy and geography is that both plot data on a spheroid, and while geography usually plots data on an oblate spheroid with positive nonzero major and minor radii (Iliffe, 2000), the celestial sphere used in astronomy is an abstract sphere without a radius (Kaler, 2002). Furthermore, astronomy looks up, while geography looks down, which results in reverse coordinate orientation with right ascension increasing West rather than East as is common for latitude (Erard et al., 2014). Despite these differences, approaches to data discovery and the data provider solutions are similar. For example, OGC and W3C standards for servicing spatial data on the web, such as Web Map/Feature/Coverage (Processing) Services (Percivall, 2010), could be employed to display celestial images in GeoTIFF format, just like the astronomical Table Access Protocol (TAP) can be used to serve geospatial data (Erard et al., 2014).

A typical raster format used in geosciences is GeoTIFF, and one in astronomy is FITS. Conversion between the two formats is an ongoing effort (Marmo et al., 2016). Typical vector formats used in geosciences are Shapefiles (ESRI, 1998) and GeoJSON (Butler et al., 2016), as well as a number of XML-based formats such as well-known text (Lott, 2015), Geography Markup Language, and Scalable Vector Graphics (Zaslavsky, 2003). For spectral work, geoscientists typically use ENVI standards, in which a spectral library stored in binary format (.sli) has an associated ASCII header (.hdr) (Corporation, 2009). The astronomical community uses a flexible format based on XML called VOTable as a standard for storing vector and tabular data of any kind, including the spectra. This format is very flexible and extendable, and also allows storing metadata together with the data in one file (Ochsenbein et al., 2004). Conversion of other data tables to VOTable is fairly straightforward; however, lossless conversion of VOTable to a more restrictive format may sometimes be impossible, since formats like GeoJSON do not allow for storing of metadata (Butler et al., 2016). Wider adoption of the VOTable format in spatially intensive industries would foster greater interoperability between engineering, geology, geography, and astronomy.

Data conversion to and from the VOTable format can be done with Astropy. Astropy is an open source and community-developed Python package providing a core framework for astronomy functionality, cosmological transformations, and conventions, such as support of FITS, VOTables, conversions of units and physical quantities, celestial coordinates and time transformations, and support of the World Coordinate System (Robitaille et al., 2013). It provides a Python interface to access VO through vo.samp and the affiliated PyVO Python package (Graham et al., 2014).

Jupyter Notebook is an interactive browser-based development environment that can provide an umbrella environment for the development and demonstrations of unified reproducible workflows mixing-and-matching tools and subroutines from astronomical and geographic industries (Ragan-Kelley et al., 2014). Jupyter Notebook, by virtue of supporting Python can integrate with existing libraries providing VO interfaces such as astropy.vo.samp and pyvo.dal.tap for purposes of interoperability and data discovery. Furthermore, the maps can be plotted in Jupyter Notebook using IPyLeaflet.

4.2.2 VESPA – Virtual Observatory for Planetary Science

The Planetary Data System (PDS) (Guinness et al., 1996) and Planetary Science Archive (PSA) (Heather et al., 2013) are distributed planetary data archives of NASA and ESA, respectively. The International Planetary Data Alliance (IPDA) has defined a Planetary Data Access Protocol (PDAP) for a standardized access to these archives. The JRA-4 activity of the EuroPlanet Research Infrastructure (RI) was tasked with development of access tools to interrogate IPDA data centers, but found the existing PDAP protocol insufficient for certain cases, and instead suggested the use of the astronomical TAP as a more flexible solution for querying planetary data (Gangloff et al., 2011). Consequently, the Virtual European Solar and Planetary Access (VESPA) data access environment has been developed as a community-driven effort to apply VO standards and tools to solar system and planetary science (Erard et al., 2018). This environment consists of a distributed collaborative infrastructure created through standardization of data models and interfaces necessary for accessing planetary data while inheriting from IVOA standards (Hare et al., 2018). VESPA provides a uniform search interface that enables interdisciplinary research by combining data from different sources, for instance to study interaction of the Sun and the Earth (Hughes et al., 2018). It is an ongoing effort of VESPA to provide mirrored access to data exposed on the PDS Geosciences Node Orbital Data Explorer (ODE) (Bennett et al., 2014).

VESPA infrastructure interacts with the data providers through a generic data access protocol called EPN-TAP, which is an implementation of the TAP protocol containing a set of columns standardized for the use with planetary data, allowing planetary geographic data to be served using interfaces of the VO (Erard et al., 2014). The geometry of vector features served through EPN-TAP is stored in the s_region field of TAP as Space Time Coordinate String (STC-S). Unfortunately, the current implementation of STC-S is only designed for polygon features, and while point features can be served via other fields, the polylines can not yet be transmitted and queried via EPN-TAP (Dowler et al., 2010). The VO tool TOPCAT provides means of connecting to TAP as well as some visualization and statistical analysis capabilities (Taylor, 2005).

Through VESPA, several VO tools were adapted for planetary applications. Firstly, the (GAVO) Data Center Helper Suite (DaCHS), which is the main VO data publishing suite (Demleitner, 2014), has received support for EPN-TAP protocol, and several tables exposing planetary geoscientific data have been published with it, exposing them to the VO (Erard et al., 2014, 2018). Consequently, the Aladin Sky Atlas received support for planetary CRS, while several planetary basemaps have been converted to HiPS and published to the VO. The list of the available planetary HiPS can be found at USGS (2018). The primary raster data format of the VO, the Flexible Image Transport System (FITS) has given support for the geographic CRS through the definition of new geo-specific keywords (Marmo et al., 2016). Additionally, GDAL has being adapted to include support for the FITS format which allows to display geographic FITS files natively in QGIS (EPN-VESPA, 2018).

While VO consists of an open framework of applications developed by many different teams, all these applications are able to communicate with each other directly through the use of the Simple Application Messaging Protocol (SAMP) (Erard et al., 2018). This protocol enables interoperability through XML Remote Procedure Call (RPC) messages (Taylor et al., 2012). SAMP works by having one hub routing mes-

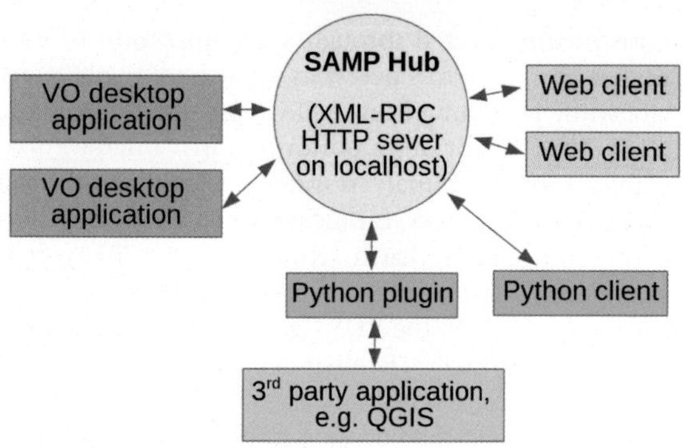

FIG. 4.2 Different VO applications can exchange commands and data via XML-RPC calls routed through SAMP Hub.

sages between many applications connected to it (see Fig. 4.2). Whereas VO tools have a built-in support for SAMP, many existing third-party tools such as QGLS can have SAMP interface implemented through plugin extensions. Extending web-based applications is straightforward via the SAMPJS library (Fitzpatrick et al., 2013), while Python plugins can be built using astropy.vo.samp (Robitaille et al., 2013). Several third-party tools have already received a SAMP interface through the efforts of VESPA, such as the geography application QGIS (Erard et al., 2018), the MATISSE application for visualization of three-dimensional surfaces (Longobardo et al., 2017), and the 3DView application for plotting data along the spacecraft trajectory (Génot et al., 2018).

Work on supporting geospatial and solid-state analysis in VO spectroscopy tools such as VOSpec (Osuna et al., 2005), CASSIS (Vastel et al., 2015), and SPLAT-VO (Škoda et al., 2014) is ongoing. Of these, CASSIS has already received support for reflected spectra, although it is still in early stages of development (Erard et al., 2018). At the same time a massive Solid Spectroscopy Hosting Architecture of Databases and Expertise (SSHADE) database infrastructure has been brought online (Schmitt et al., 2015) with work on implementing the EPN-TAP interface currently ongoing (Erard et al., 2018).

In parallel and independently from VO, GoogleTM has adapted Google Earth to work with celestial data, releasing Google Sky in 2007 (Kanipe, 2009). At the time of writing, Google Sky has partial support for KML and KMZ files – the same formats as used in Google Earth (Connolly et al., 2008). Compared with VO, Google Sky has the same shortcomings as Google Earth when compared with GIS – there is only limited functionality meant mostly for public outreach rather than professional and scientific study, while the current trend in development of Google products is increased deprecation of useful features (such as ability to connect to a custom database) in favor of enhancing a select few – for example the ability to use Google Sky database in the Android Google Earth app has been deprecated in version 8.

Several samples of surface data have been made accessible to VO as part of the VESPA activity. M3 and CRISM data from the PlanetServer (Marco Figuera et al., 2018), the USGS Astrogeology WMS layers (Hare et al., 2014) in cylindrical projection, and the Mars Crater Catalog by (Robbins and Hynek, 2012) have been exposed on Minin (2018). Furthermore, potential for integration with external data repositories such as Zenodo has been demonstrated by publication of the Pangaea-x dataset on EPN1 (Unnithan et al., 2017). The subgranule access for WMS/WCPS was enabled via an auxiliary page containing a minimal and hence very light-weight web viewer providing capability for panning and zooming on the coverage, displaying spectra on HTML5 canvas, and sending it to other VO applications through SAMP. The methods used in EPN1 have been implemented by other institutions such as the Freie Universität Berlin in the publication of HRSC WMS layers (Walter et al., 2018).

4.3 CASE STUDIES: INTEROPERABILITY OF VIRTUAL OBSERVATORY AND GEOGRAPHICAL INFORMATION SYSTEMS

4.3.1 Geographical Data and Virtual Observatory

PlanetServer is an online open source spectral analysis and visualization tool, based on a RasDaMan back-end, implementing WCPS access to planetary surface data. The data are obtained from PDS Archives, which after preprocessing and projection is ingested into RasDaMan (Marco Figuera et al., 2018). The back-end is supported through several Tomcat (Brittain and Darwin, 2007) servers running on different ports running the apps for semantic coordinate resolver (SECORE) (Misev et al., 2012), PETASCOPE access point (Aiordăchioaie and Baumann, 2010), and geoservers providing access to elevations and basemaps accessible as a Web Map Service (Marco Figuera et al., 2018).

The front-end of the PlanetServer is the NASA WebWorldWind client based on JavaScript, which combines data from different servers in the browser interactive environment to plot the requested data on a virtual globe.

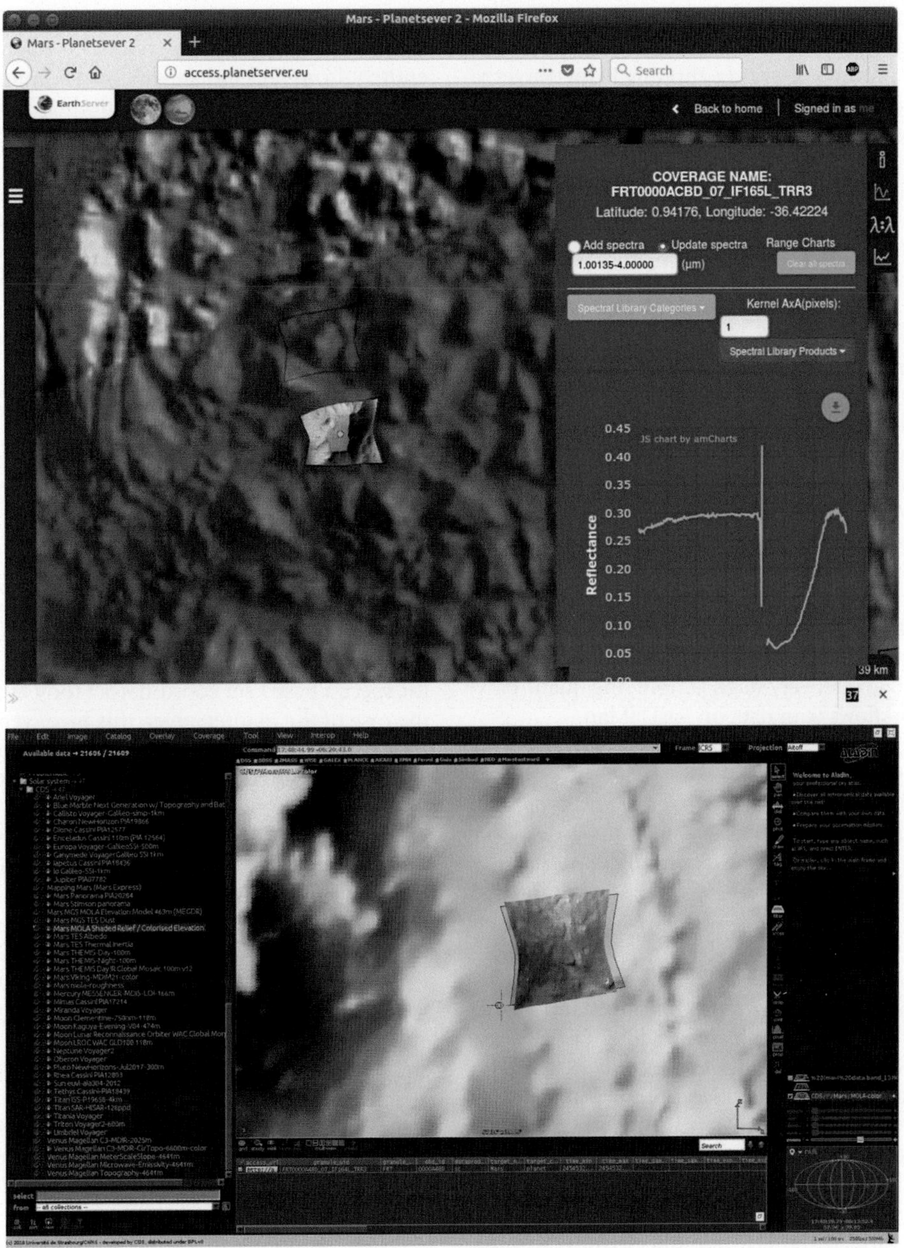

FIG. 4.3 Displaying geospatial datasets in (top) PlanetServer and (bottom) Aladin.

The WebWorldWind platform provides means of plotting WMS over the elevations together with the footprints of the WCPS coverages. The coverages can be accessed and spectra plotted from within the PlanetServer web interface. The accessed coverages can be plotted together with reference spectra obtained from the splib6a (Clark et al., 2007) spectral library (Marco Figuera et al., 2018).

Despite the convenience of having all data in one place, the project ran into scalability limitations due to sluggishness of the WebWorldWind client when displaying large numbers of polygons. With over 20,000 polygons at the time of writing, running PlanetServer requires a high-speed internet connection and a desktop computer with a strong graphics card. Furthermore, the interface is designed with qualitative rather than quantitative research in mind.

Through the efforts of VESPA's Surfaces Task, the problems of accessing large volumes of geospatial data have efficiently been addressed by implementing solutions from VO. The datasets available on PlanetServer have been exposed to VO as tables accessible through EPN-TAP protocol via Minin (2018). This enables integration with tools like Aladin Sky Atlas (Fig. 4.3), while opening possibilities to development of visualization applications based on AladinLite. AladinLite-based front-end may resolve performance problems encountered with WebWorldWind at an expense of functionality – for instance, displaying topography in AladinLite is currently not possible. Furthermore, through the support for complex ADQL queries EPN-TAP allows selection not only by geometry but also by metadata (Dowler et al., 2010). This enables a great degree of automation during the data discovery stage, simplifying quantitative research workflows with data on RasDaMan.

4.3.2 Astronomical Data and Geographical Information Systems

Similarly to how geographical data can be displayed using astronomical software, astronomical data can be displayed using geographical software. For instance, astronomical data cubes can be displayed in QGIS using a modified GDAL driver (EPN-VESPA, 2018) (Fig. 4.4). This driver implements the CFITSIO library, which is a subroutine interface written in C for reading and writing FITS files (Pence, 1999). Implementation of CFITSIO in GDAL allows the FITS header to be converted to TIFF tags using gdal_translate, while also enabling support for FITS in QGIS, since GDAL is a fundamental component used by QGIS to manage raster formats (Ose, 2018).

An alternative way to import astronomical rasters into GIS is to load the FITS file into Python with astropy.io and use the FITS header to create a GDAL virtual raster containing basic information, such as bit type and bit depth, a geographic transformation matrix, and the pointer to the data array. This approach has been implemented in the fits2vrt Python script (Marmo et al., 2018). While this script is targeted for conversion of FITS files in geographical context, addition to this script support for celestial coordinate frames would enable conversion of astronomical images into TIFFs. Although processing of astronomical data with GIS is possible, support for astronomical reference frames is missing entirely from GIS industry. A workaround could be to change the right ascension and declination to longitude and latitude and use a made-up value for the radius of the sphere.

For vector datasets EPN-TAP makes no distinction between the Ra/Dec and Lat/Lon in STC-S used by ADQL aside from using a CRS-specific keyword and the footprint is supplied as a sequence of coordinate pairs regardless of the CRS (Dowler et al., 2010); such datasets can be imported with an appropriate plugin (Minin and Marmo, 2018).

4.4 PERSPECTIVES AND POSSIBILITIES

Geosciences can benefit from astronomical data technology in many ways. For example, the current specification of the WCPS requires a coverage ID in order to query for a spectrum, so the protocol does not allow to query for all available spectra at a particular geographic location (Baumann, 2010). On other hand, the astronomical SSAP allows querying celestial spheres for spectra based on coordinates (Tody et al., 2012). In the case of the PlanterServer, this inconvenience can be resolved by implementing a Simple Spectral Access Protocol (SSAP) interface to RasDaMan through an app on a Tomcat or a Python server. Such an app could query VO through EPN-TAP for a list of coverages present at a geographical location, query RasDaMan for a spectrum at this point on each of those coverages, and then return the list of spectra as a VOTable. Wider adaption of EPN-TAP will also benefit the geosciences community with the flexibility of distributed data storage in VESPA – a unified framework of independent data providers adherent to the same standard. Furthermore, if WCS and WCPS servers could output files as FITS it would enable seamless integration of geospatial multiband rasters into VO through Aladin. This can be achieved by implementing a Python server app to translate TIFF to FITS using the GDAL development branch (EPN-VESPA, 2018).

Wider adoption of HiPS in geosciences will allow serving global basemaps much faster than with WMS, while also reducing geodesic artifacts in surface derivative data (Westerteiger et al., 2012). HealPix on a spheroid has been demonstrated before (Gibb, 2016). Since HiPS have a rigid hierarchical structure, the individual tiles (typically already in png format) have predefined positioning and projection (Fernique et al., 2015); knowing the number of the HiPS tile it is possible to determine the position of its centroid and then

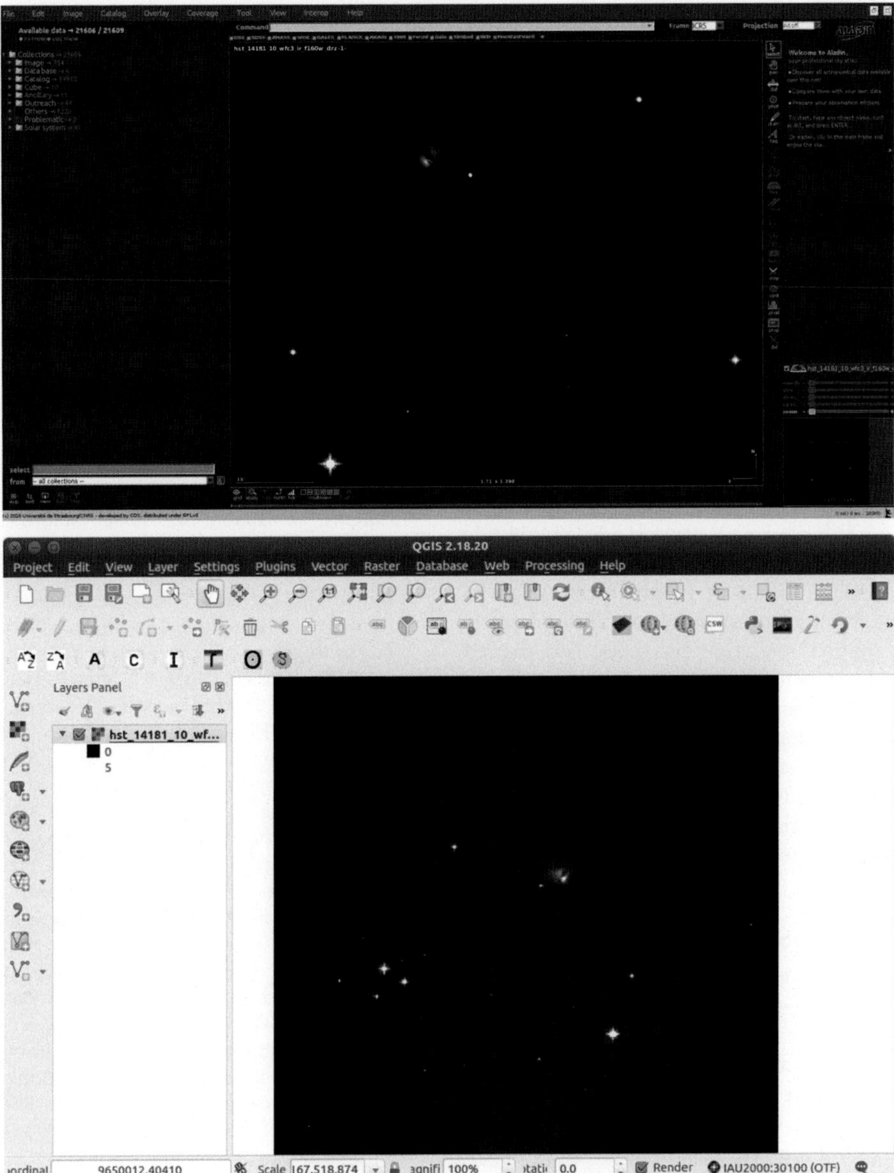

FIG. 4.4 Displaying astronomical images in (top) Aladin Sky Atlas and (bottom) QGIS.

convert to another projection using gdalwarp with an appropriate proj4 string (Gibb, 2016). On the other hand, since all tiles are precut images, arbitrary subsetting similar to WMS is impossible.

Making astronomy data accessible to the GIS would allow using GRASS to process astronomical imagery. Pattern recognition subroutines used in geosciences can be used to employ machine vision to sky mapping tasks. The astronomical community can likewise benefit from adoption of geoscientific protocols such as the Web Coverage Processing Service to effectively serve large-scale temporal/spectral sky surveys. This would allow on-the-fly condensing, i.e., basic statistical operations such as mean, maximum, and minimum, and pixel counts (Baumann, 2010) of a raster cube subset of a sky survey, which in turn will allow for a greater capacity for automation of workflows. Unfortunately the implementation of RasDaMan for sky surveys will have problems with a Semantic Resolver for Coordinate Reference Systems (SECORE) (Misev et al., 2012).

The principal task for integrating astronomical data into GIS would involve the development of appropriate proj4 strings for defining projections for astronomical coordinate systems as no coordinate system appropriate for celestial sphere exists in geosciences. The geographical sphere always has associated major and minor radius, whereas sky can be considered a perfect sphere with no radius. Although in simple cases world files describing a matrix transformation for projecting a small image on a map could be produced from astronomical FITS keywords, the comparison between longitude and right ascension, while close, is not entirely perfect. Right ascension is increasing Westward while longitude is increasing Eastward (Erard et al., 2014), so plotting astronomical images in GIS or geographical images in VO will make them appear inverted until appropriate correction is applied. Also, right ascension is measured in hours, unlike longitude, which is in degrees, and it is also not independent from the dimension of time due to precession (Roy and Clarke, 2003). Additionally, relativistic effects such as aberration of light can be significant in astronomy, have no analog in Geoscience industry, and are distinct from the parallax effect while influencing parallax measurements (Phipps, 1989).

4.5 CONCLUSIONS

Collection, storage, sharing, and processing of data in astronomy and geosciences is similar in many aspects. Both industries produce, share, and analyze large volumes of spatially enabled data, sometimes with an added time dimension and spectral information. Both industries can benefit from adoption of each other's methods and techniques. For instance, astronomical experience in distributed data storage can help improve planetary surface data discovery, and adaption of OGC standards for use in astronomy can help astronomers to better process data on the server (e.g., WMS, WCPS standards) or locally with geoprocessing software. The two industries are converging, yet the main hurdle to enabling interoperability is difference in standards of file formats, data access protocols, and the treatment of coordinate systems on the sky as compared to the planetary spheroid. Work on standards and software interoperability is currently being done by the VESPA initiative. At present, tools and services developed by the surfaces task of VESPA allow displaying geographical data in VO and astronomical data in QGIS. Enabling SAMP in QGIS allows sending data to it directly from the VO using XML-RPC calls. The future development would likely involve greater use of Python, with development of various workflows through collaborative efforts in Jupyter Notebooks.

REFERENCES

Abebe, A., Daniels, J., McKean, J., Kapenga, J., 2001. Statistics and Data Analysis. Western Michigan University, Michigan.

Aharonian, F., Akhperjanian, A.G., Anton, G., Barres de Almeida, U., Bazer-Bachi, A.R., Becherini, Y., Behera, B., Bernlöhr, K., Boisson, C., Bochow, A., Borrel, V., Braun, I., Brion, E., Brucker, J., Brun, P., Bühler, R., Bulik, T., Büsching, I., Boutelier, T., Chadwick, P.M., Charbonnier, A., Chaves, R.C.G., Cheesebrough, A., Chounet, L.-M., Clapson, A.C., Coignet, G., Dalton, M., Daniel, M.K., Davids, I.D., Degrange, B., Deil, C., Dickinson, H.J., Djannati-Ataï, A., Domainko, W., O'C.Drury, L., Dubois, F., Dubus, G., Dyks, J., Dyrda, M., Egberts, K., Emmanoulopoulos, D., Espigat, P., Farnier, C., Feinstein, F., Fiasson, A., Förster, A., Fontaine, G., Füßling, M., Gabici, S., Gallant, Y.A., Gérard, L., Giebels, B., Glicenstein, J.F., Glück, B., Goret, P., Hauser, D., Hauser, M., Heinz, S., Heinzelmann, G., Henri, G., Hermann, G., Hinton, J.A., Hoffmann, A., Hofmann, W., Holleran, M., Hoppe, S., Horns, D., Jacholkowska, A., de Jager, O.C., Jung, I., Katarzýnski, K., Katz, U., Kaufmann, S., Kendziorra, E., Kerschhaggl, M., Khangulyan, D., Khélifi, B., Keogh, D., Komin, Nu., Kosack, K., Lamanna, G., Lenain, J.-P., Lohse, T., Marandon, V., Martin, J.M., Martineau-Huynh, O., Marcowith, A., Maurin, D., McComb, T.J.L., Medina, M.C., Moderski, R., Moulin, E., Naumann-Godo, M., de Naurois, M., Nedbal, D., Nekrassov, D., Niemiec, J., Nolan, S.J., Ohm, S., Olive, J.-F., de Oña Wilhelmi, E., Orford, K.J., Ostrowski, M., Panter, M., Paz Arribas, M., Pedaletti, G., Pelletier, G., Petrucci, P.-O., Pita, S., Pühlhofer, G., Punch, M., Quirrenbach, A., Raubenheimer, B.C., Raue, M., Rayner, S.M., Renaud, M., Rieger, F., Ripken, J., Rob, L., Rolland, L., Rosier-Lees, S., Rowell, G., Rudak, B., Rulten, C.B., Ruppel, J., Sahakian, V., Santangelo, A., Schlickeiser, R., Schöck, F.M., Schröder, R., Schwanke, U., Schwarzburg, S., Schwemmer, S., Shalchi, A., Skilton, J.L., Sol, H., Spangler, D., Stawarz, L., Steenkamp, R., Stegmann, C., Superina, G., Szostek, A., Tam, P.H., Tavernet, J.-P., Terrier, R., Tibolla, O., van Eldik, C., Vasileiadis, G., Venter, C., Venter, L., Vialle, J.P., Vincent, P., Vivier, M., Völk, H.J., Volpe, F., Wagner, S.J., Ward, M., Zdziarski, A.A., Zech, A., 2009. Spectrum and variability of the galactic center VHE source HESS j1745-290. Astronomy & Astrophysics 503 (3), 817–825.

Aiordăchioaie, A., Baumann, P., 2010. Petascope: an open-source implementation of the OGC WCS geo service standards suite. In: Gertz, M., Ludäscher, B. (Eds.), Scientific and Statistical Database Management. Springer Berlin Heidelberg, Berlin, Heidelberg, pp. 160–168.

Allen, M.G., Fernique, P., Boch, T., Durand, D., Oberto, A., Merin, B., Stoehr, F., Genova, F., Pineau, F., Salgado, J., 2016. An hierarchical approach to big data. arXiv preprint arXiv:1611.01312.

Antognini, J., Araya, M., Solar, M., Valenzuela, C., Lira, F., 2015. Evaluating a NoSQL alternative for Chilean virtual observatory services. In: Taylor, A.R., Rosolowsky, E. (Eds.), Astronomical Data Analysis Software and Systems XXIV (ADASS XXIV). In: Astronomical Society of the Pacific Conference Series, vol. 495, p. 409.

Baldwin, J.E., Haniff, C.A., 2002. The application of interferometry to optical astronomical imaging. Philosophical Transactions of the Royal Society of London A: Mathematical, Physical and Engineering Sciences 360 (1794), 969–986.

Ballatore, A., Tahir, A., McArdle, G., Bertolotto, M., 2011. A comparison of open source geospatial technologies for web mapping. International Journal of Web Engineering and Technology 6 (4), 354–374.

Baltsavias, E.P., Käser, C., 1998. DTM and orthoimage generation – a thorough analysis and comparison of four digital photogrammetric systems. In: ISPRS Commission IV Symposium "GIS-Between Visions and Applications". Swiss Federal Institute of Technology, Institute of Geodesy and Photogrammetry.

Basri, G., Walkowicz, L.M., Batalha, N., Gilliland, R.L., Jenkins, J., Borucki, W.J., Koch, D., Caldwell, D., Dupree, A.K., Latham, D.W., Marcy, G.W., Meibom, S., Brown, T., 2011. Photometric variability in Kepler target stars. II. An overview of amplitude, periodicity, and rotation in first quarter data. The Astronomical Journal 141 (1), 20.

Bauer, T., 2011. Super-resolution imaging: the use case of optical astronomy. In: Proceedings of the IADIS International Conference Computer Graphics, Visualization, Computer Vision and Image Processing.

Baumann, P., 2010. The OGC web coverage processing service (WCPS) standard. Geoinformatica 14 (4), 447–479.

Baumann, P., Chulkov, G., 2007. Web coverage processing service (WCPS) implementation specification.

Baumann, P., Dehmel, A., Furtado, P., Ritsch, R., Widmann, N., 1998. The multidimensional database system RasDaMan. In: Proceedings of the 1998 ACM SIGMOD International Conference on Management of Data. SIGMOD'98. ACM, New York, NY, USA, pp. 575–577.

Bender, P.L., Currie, D.G., Poultney, S.K., Alley, C.O., Dicke, R.H., Wilkinson, D.T., Eckhardt, D.H., Faller, J.E., Kaula, W.M., Mulholland, J.D., Plotkin, H.H., Silverberg, E.C., Williams, J.G., 1973. The lunar laser ranging experiment. Science 182 (4109), 229–238.

Bennett, K., Wang, J., Scholes, D., 2014. Accessing PDS data in pipeline processing and websites through PDS geosciences orbital data explorer's web-based API (REST) interface. In: Lunar and Planetary Science Conference, Vol. 45, p. 1026.

Blaauw, A., 2012. History of the IAU: The Birth and First Half-Century of the International Astronomical Union. Springer, Netherlands.

Black, G.J., 2002. Planetary radar astronomy. In: Stanimirovic, S., Altschuler, D., Goldsmith, P., Salter, C. (Eds.), Single-Dish Radio Astronomy: Techniques and Applications. In: Astronomical Society of the Pacific Conference Series, vol. 278, pp. 271–290.

Boardman, J., Biehl, L., Clark, R., Kruse, F., Mazer, A., Torson, J., Staenz, K., 2006. Development and implementation of software systems for imaging spectroscopy. In: 2006 IEEE International Symposium on Geoscience and Remote Sensing, pp. 1969–1973.

Bochmann, G., Sunshine, C., 1980. Formal methods in communication protocol design. IEEE Transactions on Communications 28 (4), 624–631.

Bradski, G., Kaehler, A., 2008. Learning OpenCV: Computer Vision With the OpenCV Library. O'Reilly Media, Inc..

Brittain, J., Darwin, I., 2007. Tomcat: The Definitive Guide: The Definitive Guide. O'Reilly Media.

Brown, T.M., Charbonneau, D., Gilliland, R.L., Noyes, R.W., Burrows, A., 2001. Hubble space telescope time-series photometry of the transiting planet of HD 209458. The Astrophysical Journal 552 (2), 699.

Butler, H., Daly, M., Doyle, A., Gillies, S., Hagen, S., Schaub, T., 2016. The GeoJSON format, RFC7946. Technical report. Internet Engineering Task Force.

Cai, X., Langtangen, H.P., Moe, H., 2005. On the performance of the Python programming language for serial and parallel scientific computations. Scientific Programming 13 (1), 31–56.

Calabretta, M.R., Greisen, E.W., 2002. Representations of celestial coordinates in fits. Astronomy & Astrophysics 395 (3), 1077–1122.

Canty, M., 2014. Image Analysis, Classification and Change Detection in Remote Sensing: With Algorithms for ENVI/IDL and Python, 3rd edition. CRC Press.

Casey, J., 1889. A Treatise on Spherical Trigonometry and Applications to Geodesy and Astronomy. Hodges, Figgis & Co., Dublin.

Chang, F., Dean, J., Ghemawat, S., Hsieh, W.C., Wallach, D.A., Burrows, M., Chandra, T., Fikes, A., Gruber, R.E., 2008. Bigtable: a distributed storage system for structured data. ACM Transactions on Computer Systems 26 (2), 4:1–4:26.

Chen, R., Xie, J., 2008. Open source databases and their spatial extensions. In: Hall, G.B., Leahy, M.G. (Eds.), Open Source Approaches in Spatial Data Handling. Springer Berlin Heidelberg, Berlin, Heidelberg, pp. 105–129.

Chilingarian, I., Bartunov, O., Richter, J., Sigaev, T., 2004. PostgreSQL: the suitable dbms solution for astronomy and astrophysics. In: Astronomical Data Analysis Software and Systems (ADASS) XIII, vol. 314, p. 225.

Chisholm, L., Hueni, A., 2015. The spectroscopy dataset lifecycle: best practice for exchange and dissemination. In: Held, A., Phinn, S., Soto-Berelov, M., Jones, S. (Eds.), AusCover Good Practice Guidelines: a Technical Handbook Supporting Calibration and Validation Activities of Remotely Sensed Data Products, pp. 234–248. Terrestrial Ecosystem Research Network TERN, St Lucia (Australia), Version 1.1, Kapitel 14.

Clark, R., Swayze, G., Wise, R., Livo, E., Hoefen, T., Kokaly, R., Sutley, S., 2007. USGS digital spectral library splib06a. http://speclab.cr.usgs.gov/spectral.lib06.

Clark, R.N., Roush, T.L., 1984. Reflectance spectroscopy: quantitative analysis techniques for remote sensing applications. Journal of Geophysical Research: Solid Earth 89 (B7), 6329–6340.

Connolly, A., Scranton, R., Ornduff, T., 2008. Google sky: a digital view of the night sky. In: Gibbs, M.G., Barnes, J., Manning, J.G., Partridge, B. (Eds.), Preparing for the 2009 International Year of Astronomy: a Hands-On Symposium. In: Astronomical Society of the Pacific Conference Series, vol. 400, p. 96.

Corporation, I., 2009. ENVI user's guide. ITT Visual Information Solutions.

Coulais, A., Schellens, M., Gales, J., Arabas, S., Boquien, M., Chanial, P., Messmer, P., Fillmore, D., Poplawski, O., Maret, S., et al., 2011. Status of GDL-GNU data language. arXiv preprint arXiv:1101.0679.

Cronin, T. M, Cannon, W.F., Poore, R.Z. (Eds.), 1983. Paleoclimate and Mineral Deposits. Geological Survey Circular, vol. 822. United States Geological Survey.

Cutrona, L., 1970. Syntetic aperture radar. In: Skolnik, M. (Ed.), Radar Handbook. McGraw–Hill, New York. Chapter 21.

Davis, J.L., Annan, A.P., 1989. Ground-penetrating radar for high-resolution mapping of soil and rock stratigraphy. Geophysical Prospecting 37 (5), 531–551.

Demleitner, M., 2014. The DaCHS multi-protocol VO server. In: Astronomical Data Analysis Software and Systems XXIII, vol. 485, p. 309.

Denny, R.B., 2002. Software interoperation and compatibility: ASCOM update. Society for Astronomical Sciences Annual Symposium 21, 39.

Di, L., 2004. The development of geospatially-enabled grid technology for Earth science applications. In: Proceedings of NASA Earth Science Technology Conference.

Dowler, P., Rixon, G., Tody, D., Andrews, K., Good, J., Hanisch, R., Lemson, G., McGlynn, T., Noddle, K., Ochsenbein, F., et al., 2010. Table access protocol version 1.0. IVOA Recommendation, 27.

EPN-VESPA, 2018. https://github.com/epn-vespa/gdal/releases/tag/v2.3.0.

Erard, S., Cecconi, B., Sidaner, P.L., Berthier, J., Henry, F., Molinaro, M., Giardino, M., Bourrel, N., André, N., Gangloff, M., Jacquey, C., Topf, F., 2014. The EPN-TAP protocol for the planetary science virtual observatory. Astronomy and Computing 7 (8), 52–61. Special Issue on the Virtual Observatory: I.

Erard, S., Cecconi, B., Sidaner, P.L., Rossi, A., Capria, M., Schmitt, B., Génot, V., André, N., Vandaele, A., Scherf, M., Hueso, R., Määttänen, A., Thuillot, W., Carry, B., Achilleos, N., Marmo, C., Santolik, O., Benson, K., Fernique, P., Beigbeder, L., Millour, E., Rousseau, B., Andrieu, F., Chauvin, C., Minin, M., Ivanoski, S., Longobardo, A., Bollard, P., Albert, D., Gangloff, M., Jourdane, N., Bouchemit, M., Glorian, J.-M., Trompet, L., Al-Ubaidi, T., Juaristi, J., Desmars, J., Guio, P., Delaa, O., Lagain, A., Soucek, J., Pisa, D., 2018. VESPA: a community-driven virtual observatory in planetary science. Planetary and Space Science 150, 65–85. Enabling Open and Interoperable Access to Planetary Science and Heliophysics Databases and Tools.

ESA, 2015. Stereo image of the veil nebula. http://sci.esa.int/hubble/56524-stereo-image-of-the-veil-nebula/. (Accessed 27 August 2018).

ESRI, 1998. Shapefile technical description, jul. 1998. https://www.esri.com/library/whitepapers/pdfs/shapefile.pdf.

Fernique, P., Allen, M.G., Boch, T., Oberto, A., Pineau, F.-X., Durand, D., Bot, C., Cambrésy, L., Derriere, S., Genova, F., Bonnarel, F., 2015. Hierarchical progressive surveys. Multi-resolution HEALPix data structures for astronomical images, catalogues, and 3-dimensional data cubes. Astronomy & Astrophysics 578, A114.

Fernique, P., Boch, T., Bonnarel, F., 2009. Aladin sky atlas release 6. In: Astronomical Data Analysis Software and Systems XVIII, vol. 411, p. 559.

Fitzpatrick, M., Laurino, O., Paioro, L., Taylor, M.B., 2013. Application interoperability with SAMP. In: Friedel, D.N. (Ed.), Astronomical Data Analysis Software and Systems XXII. In: Astronomical Society of the Pacific Conference Series, vol. 475, p. 395.

Fleuren, T., Müller, P., 2008. BPEL workflows combining standard OGC web services and grid-enabled OGC web services. In: 2008 34th Euromicro Conference Software Engineering and Advanced Applications, pp. 337–344.

Friedman, J., Hastie, T., Tibshirani, R., 2001. The Elements of Statistical Learning. Springer Series in Statistics, vol. 1. Springer, New York, NY, USA.

Gaddis, L., Anderson, J., Becker, K., Becker, T., Cook, D., Edwards, K., Eliason, E., Hare, T., Kieffer, H., Lee, E., et al., 1997. An overview of the integrated software for imaging spectrometers (ISIS). In: Lunar and Planetary Science Conference, volume 28, p. 387.

Gangloff, M., Erard, S., Cecconi, B., Le Sidaner, P., Jacquey, C., Berthier, J., Bourrel, N., André, N., Pallier, E., Aboudarham, J., Capria, M.T., Khodachenko, M., Manaud, N., Schmidt, W., Schmitt, B., Topf, F., Trautan, F., Sarkissian, A., 2011. An assessment of the IPDA/PDAP protocol to access planetary data. In: EPSC-DPS Joint Meeting 2011, p. 1148.

Gaudet, S., Dowler, P., Goliath, S., Hill, N., Kavelaars, J., Peddle, M., Pritchet, C., Schade, D., 2009. The Canadian advanced network for astronomical research. In: Astronomical Data Analysis Software and Systems XVIII, vol. 411, p. 185.

Génot, V., Beigbeder, L., Popescu, D., Dufourg, N., Gangloff, M., Bouchemit, M., Caussarieu, S., Toniutti, J.-P., Durand, J., Modolo, R., André, N., Cecconi, B., Jacquey, C., Pitout, F., Rouillard, A., Pinto, R., Erard, S., Jourdane, N., Leclercq, L., Hess, S., Khodachenko, M., Al-Ubaidi, T., Scherf, M., Budnik, E., 2018. Science data visualization in planetary and heliospheric contexts with 3DView. Planetary and Space Science 150, 111–130. Enabling Open and Interoperable Access to Planetary Science and Heliophysics Databases and Tools.

Gibb, R., 2016. The Rhealpix Discrete Global Grid System. IOP Conference Series: Earth and Environmental Science, vol. 34. IOP Publishing.

Gimmestad, G.G., Roberts, D.W., Stewart, J.M., Wood, J.W., 2007. LIDAR system for monitoring turbulence profiles. Proceedings - SPIE 6551, 6551.

Graham, M. Plante, R. Tody, D. Fitzpatrick, M., 2014. PyVO: Python access to the Virtual Observatory. Astrophysics Source Code Library.

Gray, D., 2005. The Observation and Analysis of Stellar Photospheres. Cambridge University Press.

Green, K., Congalton, R., Tukman, M., 2017. Imagery and GIS: Best Practices for Extracting Information from Imagery. Esri Press.

Greisen, E.W., 1993. Non-linear coordinate systems in AIPS, reissue of November 1983 version. Technical report, AIPS Memo.

Groff, J., Weinberg, P., Oppel, A., 2008. SQL The Complete Reference. The Complete Reference. 3rd edition. McGraw–Hill Education.

Gudivada, V.N., Rao, D., Raghavan, V.V., 2016. Renaissance in database management: navigating the landscape of candidate systems. Computer 49 (4), 31–42.

Guinness, E.A., Arvidson, R.E., Slavney, S., 1996. The planetary data system geosciences node. Planetary and Space Science 44 (1), 13–22. Planetary data system.

Harari, Y., 2016. Homo Deus: A Brief History of Tomorrow. Harper Collins.

Hare, T.M., Keszthelyi, L., Gaddis, L., Kirk, R.L., 2014. Online planetary data and services at USGS astrogeology. In: Lunar and Planetary Science Conference, Vol. 45, p. 2487.

Hare, T.M., Rossi, A.P., Frigeri, A., Marmo, C., 2018. Interoperability in planetary research for geospatial data analysis. Planetary and Space Science 150, 36–42. Enabling Open and Interoperable Access to Planetary Science and Heliophysics Databases and Tools.

Heather, D.J., Barthelemy, M., Manaud, N., Martinez, S., Szumlas, M., Vazquez, J.L., Osuna, P., 2013. ESA's planetary science archive: status, activities and plans. European Planetary Science Congress 8. EPSC2013-626.

Hedström, H., Kollert, R., 1949. Seismic sounding of shallow depths. Tellus 1 (4), 24–36.

Hey, A., Tansley, S., Tolle, K., 2009. The Fourth Paradigm: Data-Intensive Scientific Discovery. Microsoft Research.

Hong, S., Noh, H., Han, B., 2015. Decoupled deep neural network for semi-supervised semantic segmentation. In: Cortes, C., Lawrence, N.D., Lee, D.D., Sugiyama, M., Garnett, R. (Eds.), Advances in Neural Information Processing Systems, vol. 28. Curran Associates, Inc., pp. 1495–1503.

Horsburgh, J.S., Tarboton, D.G., Piasecki, M., Maidment, D.R., Zaslavsky, I., Valentine, D., Whitenack, T., 2009. An integrated system for publishing environmental observations data. Environmental Modelling & Software 24 (8), 879–888.

Howard, R., Björk, B.-C., 2008. Building information modelling – experts' views on standardisation and industry deployment. Advanced Engineering Informatics 22 (2), 271–280. Network Methods in Engineering.

Hughes, J.S., Crichton, D.J., Raugh, A.C., Cecconi, B., Guinness, E.A., Isbell, C.E., Mafi, J.N., Gordon, M.K., Hardman, S.H., Joyner, R.S., 2018. Enabling interoperability in planetary sciences and heliophysics: the case for an information model. Planetary and Space Science 150, 43–49. Enabling Open and Interoperable Access to Planetary Science and Heliophysics Databases and Tools.

Huybers, P., Curry, W., 2006. Links between annual, Milankovitch and continuum temperature variability. Nature 441, 329–332.

IEEE, 2018. https://standards.ieee.org/. (Accessed 30 August 2018).

Iliffe, J., 2000. Datums and Map Projections: For Remote Sensing, GIS, and Surveying. CRC Press.

Jacob, J.C., Katz, D.S., Berriman, G.B., Good, J.C., Laity, A., Deelman, E., Kesselman, C., Singh, G., Su, M.-H., Prince, T., Williams, R., 2009. Montage: a grid portal and software toolkit for science-grade astronomical image mosaicking. International Journal of Computational Science and Engineering 4 (2), 73–87.

Kaler, J., 2002. The Ever-Changing Sky: A Guide to the Celestial Sphere. Cambridge University Press.

Kanipe, J., 2009. The universe in your computer. Communications of the ACM 52 (1), 12–14.

Kawaguchi, K., 2006. Astronomical vibrational spectroscopy. In: Handbook of Vibrational Spectroscopy. American Cancer Society.

Keller, G., 1956. Astronomical seeing and scintillation. Smithsonian Contributions to Astrophysics 1, 9.

Keshava, N., Mustard, J.F., 2002. Spectral unmixing. IEEE Signal Processing Magazine 19 (1), 44–57.

Kim, T.J., 1999. Metadata for geo-spatial data sharing: a comparative analysis. The Annals of Regional Science 33 (2), 171–181.

Kington, J., 2015. Mplstereonet v0.5.

Kramer, M., Lyne, A.G., O'Brien, J.T., Jordan, C.A., Lorimer, D.R., 2006. A periodically active pulsar giving insight into magnetospheric physics. Science 312 (5773), 549–551.

Kratz, J., Strasser, C., 2014. Data publication consensus and controversies [version 3; referees: 3 approved]. F1000Research 3 (94).

Landscheidt, T., 2000. Solar forcing of El Niño and La Niña. In: Wilson, A. (Ed.), The Solar Cycle and Terrestrial Climate, Solar and Space Weather. In: ESA Special Publication, vol. 463, p. 135.

Lapierre, A., Cote, P., 2007. Using Open Web Services for urban data management: a testbed resulting from an OGC initiative for offering standard CAD/GIS/BIM services. In: Rumor, M., Coors, V., Fendel, E., Zlatanova, S. (Eds.), Urban and Regional Data Management: UDMS 2007 Annual. CRC Press.

Ledrick, D.P., Spring, M.B., 1990. International standardized profiles. Computer Standards & Interfaces 11 (2), 95–103.

Lefsky, M.A., Cohen, W.B., Parker, G.G., Harding, D.J., 2002. Lidar remote sensing for ecosystem studies: Lidar, an emerging remote sensing technology that directly measures the three-dimensional distribution of plant canopies, can accurately estimate vegetation structural attributes and should be of particular interest to forest, landscape, and global ecologists. BioScience 52 (1), 19–30.

Libarkin, J.C., Kurdziel, J.P., 2002. Research methodologies in science education: the qualitative-quantitative debate. Journal of Geoscience Education 50 (1), 78–86.

Lohr, J., Kortüm, G., 2012. Reflectance Spectroscopy: Principles, Methods, Applications. Springer, Berlin Heidelberg.

Longobardo, A., Zinzi, A., Capria, M., Erard, S., Giardino, M., Ivanovski, S., Fonte, S., Palomba, E., Persio, G.D., Antonelli, L., 2017. Production and 3d visualization of high-level data of minor bodies: the MATISSE tool in the framework of VESPA-Europlanet 2020 activity. Advances in Space Research.

Lott, R., 2015. Geographic information – Well-known text representation of coordinate reference systems.

Marco Figuera, R., Pham Huu, B., Rossi, A.P., Minin, M., Flahaut, J., Halder, A., 2018. Online characterization of planetary surfaces: PlanetServer, an open-source analysis and visualization tool. Planetary and Space Science 150, 141–156.

Marmo, C., Hare, T., Minin, M., 2018. WCS fits conversion to virtual GDAL header. https://github.com/epn-vespa/fits2vrt.

Marmo, C., Hare, T.M., Erard, S., Cecconi, B., Costard, F., Schmidt, F., Rossi, A.P., 2016. FITS format for planetary surfaces: bridging the gap between FITS world coordinate systems and geographical information systems. In: Lunar and Planetary Science Conference, vol. 47, p. 1870.

McKinney, W., 2012. Python for Data Analysis: Data Wrangling With Pandas, NumPy, and IPython. O'Reilly Media, Inc.

Minin, M., 2018. Epn1 VESPA data centre. http://epn1.epn-vespa.jacobs-university.de/.

Minin, M., Marmo, C., 2018. Epn-vespa/vo_qgis_plugin: QGIS plugins providing interface to VESPA.

Misev, D., Rusu, M., Baumann, P., 2012. A semantic resolver for coordinate reference systems. In: Di Martino, S., Peron, A., Tezuka, T. (Eds.), Web and Wireless Geographical Information Systems. Springer Berlin Heidelberg, Berlin, Heidelberg, pp. 47–56.

Moratto, Z.M., Broxton, M.J., Beyer, R.A., Lundy, M., Husmann, K., 2010. Ames stereo pipeline, NASA's open source automated stereogrammetry software. In: Lunar and Planetary Science Conference. In: Lunar and Planetary Science Conference, vol. 41, p. 2364.

Mordvintsev, A., Abid, K., 2014. OpenCV-Python Tutorials Documentation.

Mukherjee, B., Srivardhan, V., Roy, P., 2016. Identification of formation interfaces by using wavelet and Fourier transforms. Journal of Applied Geophysics 128, 140–149.

Niederheiser, R., Mokroš, M., Lange, J., Petschko, H., Prasicek, G., Oude Elberink, S., 2016. Deriving 3d point clouds from terrestrial photographs – comparison of different sensors and software. In: ISPRS – International Archives of the Photogrammetry, Remote Sensing and Spatial Information Sciences, pp. 685–692.

Nixon, M., Aguado, A.S., 2012. Feature Extraction & Image Processing for Computer Vision, 3rd edition. Academic Press, Inc., Orlando, FL, USA.

Ochsenbein, F., Williams, R., Davenhall, C., Durand, D., Fernique, P., Hanisch, R., Giaretta, D., McGlynn, T., Szalay, A., Wicenec, A., 2004. VOTable: tabular data for the virtual observatory. In: Quinn, P.J., Górski, K.M. (Eds.), Toward an International Virtual Observatory. Springer Berlin Heidelberg, Berlin, Heidelberg, pp. 118–123.

Ohishi, M., 2007. International virtual observatory alliance. Highlights of Astronomy 14, 528–529.

Ose, K., 2018. Introduction to GDAL Tools in QGIS. Wiley-Blackwell, pp. 19–65. Chapter 2.

Osterloo, M.M., Hamilton, V.E., Bandfield, J.L., Glotch, T.D., Baldridge, A.M., Christensen, P.R., Tornabene, L.L., Anderson, F.S., 2008. Chloride-bearing materials in the southern highlands of Mars. Science 319 (5870), 1651–1654.

Osuna, P., Barbarisi, I., Salgado, J., Arviset, C., 2005. VOSpec: a tool for handling virtual observatory compliant spectra. In: Shopbell, P., Britton, M., Ebert, R. (Eds.), Astronomical Data Analysis Software and Systems XIV. In: Astronomical Society of the Pacific Conference Series, vol. 347, p. 198.

Page, C., Vigoureux, P., 1975. The International Bureau of Weights and Measures, 1875–1975: Translation of the BIPM Centennial Volume. NBS special publication U.S. Dept. of Commerce, National Bureau of Standards. For sale by the Supt. of Docs., U.S. Govt. Print. Off.

Pence, W., 1999. CFITSIO, v2.0: a new full-featured data interface. In: Mehringer, D.M., Plante, R.L., Roberts, D.A. (Eds.), Astronomical Data Analysis Software and Systems VIII. In: Astronomical Society of the Pacific Conference Series, vol. 172, p. 487.

Percivall, G., 2010. The application of open standards to enhance the interoperability of geoscience information. International Journal of Digital Earth 3 (sup1), 14–30.

Peuquet, D.J., 1984. A conceptual framework and comparison of spatial data models. Cartographica: The International Journal for Geographic Information and Geovisualization 21 (4), 66–113.

Phipps, T.E., 1989. Relativity and aberration. American Journal of Physics 57 (6), 549–551.

PostgreSQL, 2010. PostgreSQL 8.1.23 Documentation. The PostgreSQL Global Development Group.

PROJ contributors, 2018. PROJ Coordinate Transformation Software Library. Open Source Geospatial Foundation.

Qi, Z., Cooperstock, J.R., 2007. Overcoming parallax and sampling density issues in image mosaicing of non-planar scenes. In: Proceedings of the British Machine Vision Conference. BMVA Press, pp. 36.1–36.10.

Ragan, D.M., 2009. Structural Geology: An Introduction to Geometrical Techniques. Cambridge University Press.

Ragan-Kelley, M., Perez, F., Granger, B., Kluyver, T., Ivanov, P., Frederic, J., Bussonnier, M., 2014. The Jupyter/IPython architecture: a unified view of computational research, from interactive exploration to communication and publication. AGU Fall Meeting Abstracts. pages H44D-07.

Rao, T.S., Priestly, M.B., Lessi, O., 1998. Applications of time series analysis in astronomy and meteorology. Technometrics 40 (2), 163.

Re, C., Roncella, R., Forlani, G., Cremonese, G., Naletto, G., 2012. Evaluation of area-based image matching applied to DTM generation with hirise images. ISPRS Annals of Photogrammetry, Remote Sensing and Spatial Information Sciences, 209–214.

Reinsch, K., Delfs, M., Junker, E., Völker, P., Roth, G.D., Treichel, C., Böhnhardt, H., Schmeidler, F., Leinert, C., 2009. Earth and solar system. In: Roth, G.D. (Ed.), Handbook of Practical Astronomy. Springer Berlin Heidelberg, Berlin, Heidelberg, pp. 309–571.

Remondino, F., 2003. From point cloud to surface: the modeling and visualization problem. In: International Archives of Photogrammetry, Remote Sensing and Spatial Information Sciences, p. 34.

Remondino, F., Spera, M.G., Nocerino, E., Menna, F., Nex, F., Gonizzi-Barsanti, S., 2013. Dense image matching: comparisons and analyses. In: 2013 Digital Heritage International Congress (DigitalHeritage), Vol. 1, pp. 47–54.

Robbins, S.J., Hynek, B.M., 2012. A new global database of mars impact craters ≥ 1 km: 1. Database creation, properties, and parameters. Journal of Geophysical Research: Planets 117 (E5), E05004.

Robitaille, T.P., Tollerud, E.J., Greenfield, P., Droettboom, M., Bray, E., Aldcroft, T., Davis, M., Ginsburg, A., Price-Whelan, A.M., Kerzendorf, W.E., et al., 2013. Astropy: a community python package for astronomy. Astronomy & Astrophysics 558, A33.

Rosenfeld, A., 1969. Picture processing by computer. ACM Computing Surveys 1 (3), 147–176.

Roy, A.E., Clarke, D., 2003. Astronomy: Principles and Practice, (PBK). CRC Press.

Rufenach, C.L., Shuchman, R.A., Lyzenga, D.R., 1983. Interpretation of synthetic aperture radar measurements of ocean currents. Journal of Geophysical Research: Oceans 88 (C3), 1867–1876.

Ryerson, R., Rencz, A., 1999. Manual of Remote Sensing, Remote Sensing for the Earth Sciences, third edition. Wiley.

Scargle, J.D., 1982. Studies in astronomical time series analysis. II – statistical aspects of spectral analysis of unevenly spaced data. The Astrophysical Journal 263, 835–853.

Schindler, U., Diepenbroek, M., Grobe, H., 2012. Pangaea® – research data enters scholarly communication. In: EGU General Assembly Conference Abstracts, vol. 14, p. 13378.

Schmitt, B., Albert, D., Bollard, P., Bonal, L., Gorbacheva, M., Mercier, L., SSHADE Consortium Partners, 2015. SSHADE in H2020: development of a European database infrastructure in solid spectroscopy. European Planetary Science Congress 10. EPSC2015-628.

Schroeder, W., Ng, L., Cates, J., et al., 2003. The ITK Software Guide. The Insight Consortium.

Seeds, M., Backman, D., 2012. The Solar System. Cengage Learning.

Sicilia, M.-A., García-Barriocanal, E., Sánchez-Alonso, S., 2017. Community curation in open dataset repositories: insights from zenodo. Procedia Computer Science 106, 54–60. 13th International Conference on Current Research Information Systems, CRIS2016, Communicating and Measuring Research Responsibly: Profiling, Metrics, Impact, Interoperability.

Škoda, P., Draper, P., Neves, M., Andrešič, D., Jenness, T., 2014. Spectroscopic analysis in the virtual observatory environment with SPLAT-VO. Astronomy and Computing 7 (8), 108–120. Special Issue on the Virtual Observatory: I.

Snyder, J.P., 1997. Flattening the Earth: Two Thousand Years of Map Projections. University of Chicago Press.

Stevens, J., Smith, J.M., Bianchetti, R.A., 2012. Land surveying and conventional techniques for measuring positions on the Earth's surface. In: MacEachren, A.M., Peuquet, D.J. (Eds.), Mapping Our Changing World. Department of Geography, The Pennsylvania State University. Chapter 5.5.

Strozzi, T., Wegmüller, U., Tosl, L., Bitelll, G., Spreckeis, V., 2001. Land subsidence monitoring with differential SAR interferometry. Photogrammetric Engineering and Remote Sensing 67 (11), 1261–1270.

Strutt, R., 1880. Xvii. On the resolving-power of telescopes. The London, Edinburgh, and Dublin Philosophical Magazine and Journal of Science 10 (60), 116–119.

Tao, Y., Muller, J.-P., 2016. A novel method for surface exploration: super-resolution restoration of mars repeat-pass orbital imagery. Planetary and Space Science 121, 103–114.

Taylor, M.B., 2005. TOPCAT & STIL: Starlink Table/VOTable Processing Software. In: Shopbell, P., Britton, M., Ebert, R. (Eds.), Astronomical Data Analysis Software and Systems XIV. In: Astronomical Society of the Pacific Conference Series, vol. 347, p. 29.

Taylor, M.B., Boch, T., Fay, J., Fitzpatrick, M., Paioro, L., 2012. SAMP: application messaging for desktop and web applications. In: Ballester, P., Egret, D., Lorente, N.P.F. (Eds.), Astronomical Data Analysis Software and Systems XXI. In: Astronomical Society of the Pacific Conference Series, vol. 461, p. 279.

Thacker, N.A., Clark, A.F., Barron, J.L., Beveridge, J.R., Courtney, P., Crum, W.R., Ramesh, V., Clark, C., 2008. Performance characterization in computer vision: a guide to best practices. Computer Vision and Image Understanding 109 (3), 305–334.

TIGER-NET, 2018. QGIS temporal/spectral profile tool.

Tody, D., Dolensky, M., McDowell, J., Bonnarel, F., Budavari, T., Busko, I., Micol, A., Osuna, P., Salgado, J., Škoda, P., et al., 2012. Simple Spectral Access Protocol Version 1.1. IVOA.

Unnithan, V., Pio, R.A., Jaehrig, T., 2017. Drone-based photogrammetric survey raw data from ESA Pangaea-x 2017 planetary analogue campaign – data collected on 2017-11-19. Data were collected in the framework of the ESA PANGAEA-X testing campaign held in November 2017: we acknowledge ESA for organizing the campaign and providing scientific and logistic assistance on site. The authors would like also to thank the Geopark of Lanzarote, the touristic center of Cueva de Los Verdes. the Cabildo of Lanzarote, the National Park of Timanfaya. and the IGEO-CSIC-UCM for providing the necessary permits.

USGS, 2018. HiPS planet maps. http://aladin.u-strasbg.fr/java/nph-aladin.pl?frame=aladinHpxList#hipsplanet.

van der Meer, F., 2004. Analysis of spectral absorption features in hyperspectral imagery. International Journal of Applied Earth Observation and Geoinformation 5 (1), 55–68.

Vastel, C., Bottinelli, S., Caux, E., Glorian, J.-M., Boiziot, M., 2015. CASSIS: a tool to visualize and analyse instrumental and synthetic spectra. In: Martins, F., Boissier, S., Buat, V., Cambrésy, L., Petit, P. (Eds.), SF2A-2015: Proceedings of the Annual Meeting of the French Society of Astronomy and Astrophysics, pp. 313–316.

W3C, 2018. https://www.w3.org. (Accessed 30 August 2018).

Walter, S.H.G., Muller, J.-P., Sidiropoulos, P., Tao, Y., Gwinner, K., Putri, A.R.D., Kim, J.-R., Steikert, R., Gasselt, S., Michael, G.G., Watson, G., Schreiner, B.P., 2018. The web-based interactive Mars analysis and research system for HRSC and the iMars project. Earth and Space Science 5 (7), 308–323.

Warmerdam, F., 2008. The geospatial data abstraction library. In: Hall, G.B., Leahy, M.G. (Eds.), Open Source Approaches in Spatial Data Handling. Springer Berlin Heidelberg, Berlin, Heidelberg, pp. 87–104.

Westerteiger, R., Gerndt, A., Hamann, B., 2012. Spherical Terrain Rendering using the hierarchical HEALPix grid. In: Garth, C., Middel, A., Hagen, H. (Eds.), Visualization of Large and Unstructured Data Sets: Applications in Geospatial Planning, Modeling and Engineering – Proceedings of IRTG 1131 Workshop 2011. In: OpenAccess Series in Informatics (OASIcs), vol. 27. Schloss Dagstuhl–Leibniz-Zentrum für Informatik, Germany, pp. 13–23.

Yasuda, N., Mizumoto, Y., Ohishi, M., O'Mullane, W., Budavári, T., Haridas, V., Li, N., Malik, T., Szalay, A., Hill, M., et al., 2004. Astronomical data query language: simple query protocol for the virtual observatory. In: Astronomical Data Analysis Software and Systems (ADASS) XIII, vol. 314, p. 293.

Zaslavsky, I., 2003. Chapter 11 – online cartography with xml. In: Peterson, M. (Ed.), Maps and the Internet, International Cartographic Association. Elsevier Science, Oxford, pp. 171–196.

Zlatanova, S., Prosperi, D., 2005. Large-Scale 3D Data Integration: Challenges and Opportunities. CRC Press.

Zwicky, F., 2009. Republication of: the redshift of extragalactic nebulae. General Relativity and Gravitation 41 (1), 207–224.

CHAPTER 5

Surveys, Catalogues, Databases, and Archives of Astronomical Data

IRINA VAVILOVA, DR • LUDMILA PAKULIAK, DR • IURII BABYK, DR • ANDRII ELYIV, DR • DARIA DOBRYCHEVA, DR • OLGA MELNYK, DR

5.1 INTRODUCTION

The night sky has always served as a source of wonder and mystery to people. However, it has only been in the past few decades that we have truly begun to observe the universe in all its glory over the entire electromagnetic spectrum due to advances in space science and technology. The orbital telescopes and receiving devices sensitive to the infrared, ultraviolet, X-ray, and gamma ray wavelengths above the Earth's atmosphere were developed since the space exploration era.

Our universe contains objects that produce a vast range of radiation with wavelengths either too short or too long for our eyes to see, ranging from gamma rays to radio waves (top panel of Fig. 5.1). Some celestial bodies emit mostly infrared radiation, others mostly visible light, and still others mostly ultraviolet radiation or X-rays. What determines the type of electromagnetic radiation emitted by astronomical objects? The simple answer is the *temperature*. The hotter a solid or gas, the more rapid the motion of the molecules or atoms, and temperature is just a measure of the average energy of those particles. The nature of astronomical sources can be obtained only through detecting different kinds of radiation coming from them. But the Earth's atmosphere is quite opaque; as a result, we are able to observe only a tiny proportion of all the radiation present in space. The bottom panel of Fig. 5.1 illustrates the transparency of the Earth's atmosphere for electromagnetic wavelengths.

Therefore, multiwavelength astronomy provides a huge amount of data of different sources over the whole electromagnetic spectrum. All-sky and large-area surveys and their catalogued databases enriched and continue to enrich our knowledge of the universe. Astronomy has entered the Big Data era, as these data are combined into numerous archives. Modern astronomy is the data-oriented science that allows classifying astroinformatics as a new academic research field, which covers various multidisciplinary applications of e-astronomy. Among them are data modeling, data mining, metadata standards development, data access, digital astronomical databases, image archives and visualization, machine learning, statistics, and other computational methods and software for work with astronomical surveys and catalogues with their tera- to peta-scale astroinformation resources.

"Many scientific disciplines are developing formal subdisciplines that are information-rich and data-based, to such an extent that these are now standalone research and academic programs recognized on their own merits. In astronomy, peta-scale sky surveys will soon challenge our traditional research approaches and will radically transform how we train the next generation of astronomers, whose experiences with data are now increasingly more virtual (through online databases) than physical (through trips to mountaintop observatories). "We describe astroinformatics as a rigorous approach to these challenges" (cited by "Astroinformatics: A 21st Century Approach to Astronomy," Borne et al., 2009; see, also, Cavuoti et al., 2012; Vavilova et al., 2012; Goodman, 2012; Borne, 2013).

However, modern astrophysics requires studying an object across the whole electromagnetic spectrum, because different physical processes can be studied at different wavelengths (different substructures of a celestial body radiate at different wavelengths). A greatest zest in identifying such structures and processes relates to high-energy astrophysical objects (especially gamma sources) and radio sources, wherein gamma ray, X-ray, and radio sources need to be identified exactly with their optical counterparts. Namely, the optical electromagnetic band was our first window into the universe due to its transparency of the Earth's atmosphere.

Knowledge Discovery in Big Data from Astronomy and Earth Observation. https://doi.org/10.1016/B978-0-12-819154-5.00015-1

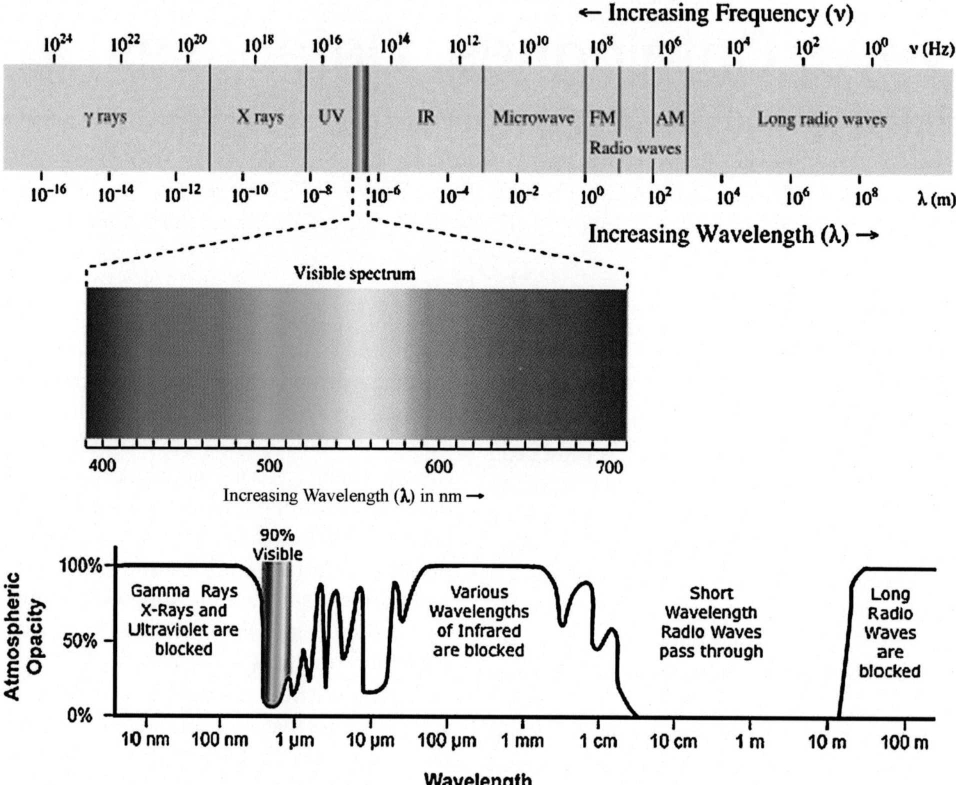

FIG. 5.1 (Top) Electromagnetic spectrum. Image credit: https://en.wikipedia.org/wiki/ (public domain). Original author: Philip Ronan. (Bottom) The transparency of the Earth's atmosphere for electromagnetic waves. Image credit: https://www.quora.com/ (public domain).

5.2 FROM THE FIRST STAR PHOTOGRAPHIC CATALOGUES TO THE MODERN DIGITAL SKY SURVEYS. OPTICAL AND NEAR-INFRARED ASTRONOMY

5.2.1 First Important Visual Surveys and Catalogues

Since the last half of the 19th century, astronomy has become a Big Data science. The first projects which could be referred to as "Big Data" were comprised of dozens to hundreds of thousands of celestial objects. Each object was obtained and treated manually and, mostly, was considered independently of others. The main challenges faced by astronomers of that time were:

- protracted observations;
- a large amount of protracted manual work;
- different approaches of different groups of observers to the processing of observational data;
- as a consequence, the heterogeneity of resulting data and the need to bring them to one reference system;
- the problem of choosing the reference system in the absence of standards and the need for their prior creation.

We can only be astonished by and admire the thoroughness of the work of astronomers, which allowed them to obtain accurate and reliable data for decades and centuries.

The first "big datasets" of astronomical objects were obtained as result of visual observations. The beginning was laid by F. Argelander with his *Bonner Durchmusterung* (BD, 1859–1862) survey of stars down to about visual magnitude 9^m–10^m. The BD firstly contained 324,188 stars from the North celestial pole to 2° South of the celestial equator. The catalogue was periodically updated, supplemented, and reissued again in 1989 being comprised of 325,037 stars. The star list included all the stars visible in the 3-inch Bonn refractor. During the observation an observer kept the telescope

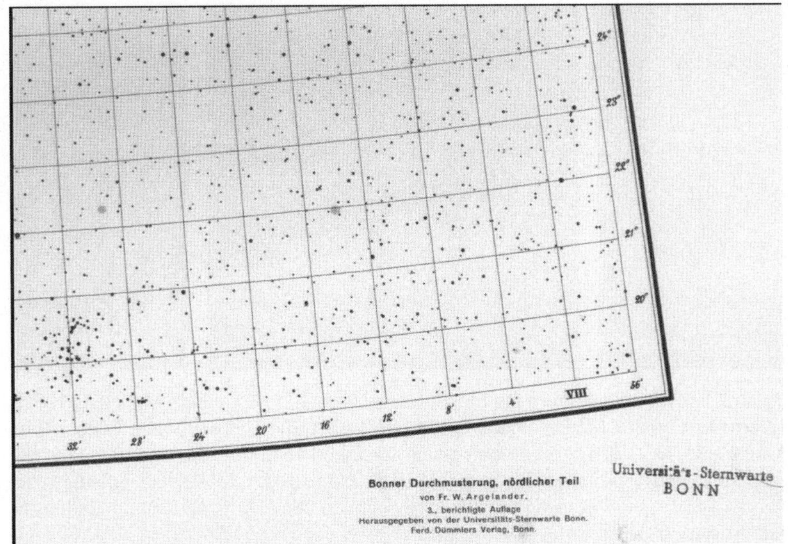

FIG. 5.2 BD star atlas. Picture from https://astro.uni-bonn.de/geffert/ge/arg/arg.htm (reproduced with permission of the Argelander-Institut für Astronomie).

fixed and registered every star appearing in the field of view in each 1° zone on declination. The star charts made on its basis and published after the catalogue in 1863 were the most complete and accurate compared with others created at that time (Fig. 5.2). In 1886 the BD was enhanced with *Sudliche Bonner Durchmusterung* (SBD) down to −23°. The SBD added about 138,000 stars observed with a 6-inch Bonn refractor using the same technique as the BD.

The BD does not cover the entire sky, but only the part that is visible from Germany. Therefore, it was supplemented with two surveys made at observatories located in the Southern hemisphere. The first one was the *Cordoba Durchmusterung* (CD, 1892–1932) covering the Southern stars in declination zones from −22° to −89°. The CD list of objects contains around 614,000 records for stars brighter than visual magnitude 10^m. The survey was made using visual observations just like the BD has done. The observational period of CD stars is ten times longer than that of the BD. The conditions of observations changed so much that it led to the heterogeneity of CD data in comparison with its Northern counterpart.

The history of cataloguing the extragalactic objects began with the first catalogue by Charles Messier, a French astronomer of the 18th century. However, Messier himself did not care about the nebulae. The comets were his main scientific interest, and precisely in order not to confuse the comet with the nebulae, he made his famous catalogue in 1784, 2nd edition, *Messier Catalogue* (M) which contained 102 dissimilar sky objects observed in the Northern hemisphere (Messier, 1781). William Herschel was the first to make a systematic visual survey of the weak nebulae and to identify patterns in their distribution. He discovered more than 2500 nebulae and stellar clusters, and about 80% of these objects were later, in the 20th century, identified as other galaxies. William Herschel published three catalogues in 1786–1802. In the 19th century, the work of William Herschel was continued by his only son John Herschel, who supplemented these discoveries by 1754 new celestial objects and published the *General Catalogue of Nebulae and Clusters* (GC) (Herschel, 1864). This catalogue was later edited by John Dreyer, who included discoveries of other 19th-century astronomers and published the *New General Catalogue* (NGC) (Dreyer, 1888). Later he supplemented it with the Index Catalogues (Dreyer, 1895, 1910). Together, these nonspecialized catalogues contained the data of 13,226 celestial objects (galaxies, star clusters, planetary nebulae, etc.). Fig. 5.3 demonstrates present-day images of several well-known objects from the Messier catalogue and NGC, which are available for observation also using binoculars and amateur telescopes.

Almost at the same time, another attempt of a Southern sky inventory was undertaken using a quite different technique. For the first time in astronomy, photography was applied to obtain photographic images of the

FIG. 5.3 The well-known objects from the Messier catalogue and NGC. (Left) M1, NGC 1952, the Crab Nebula, a supernova remnant. Mosaic image taken with NASA/ESA Hubble Space Telescope (public domain). (Middle) M31, NGC 224, Andromeda galaxy, distance 157,000 light years. Mosaic image taken in ultraviolet band with Galaxy Evolution Explorer (public domain). (Right) M76, NGC 650+651, the Little Dumbbell Planetary Nebula in Perseus constellation, distance 2550 light years. Image taken by Robert J. Vanderbei (public domain).

star fields. In 1896–1900, the *Cape Photographic Durchmusterung* (CPD) appeared. The CPD contained about 454,000 Southern stars down to 11 photographic magnitude South of the declination $-19°$.

5.2.2 Photographic Observations. Stellar and Extragalactic Surveys

The new method of observations allowed both to get "the permanent record of the sky at the epoch of observation",[1] a kind of sky events' preservation, and the possibility for improving the processing technique using state-of-the-art facilities to produce new knowledge from old data. One of the generally used nowadays catalogues – the *TYCHO*, which is based on the star mapper of the European Space Agency Hipparcos (Hoeg et al., 1997) – contains proper motions of about one million stars down to 14 magnitude, improved in half due to the unprecedented photographic project of the 19th and 20th centuries – the stellar atlas *Carte du Ciel*. The Carte du Ciel project was initiated practically at the same time as CPD, but it was more ambitious and involved 20 observatories from around the world to expose and measure 22,000 photographic plates obtained with normal astrographs and covered the entire sky with a twofold overlap (Fig. 5.4).

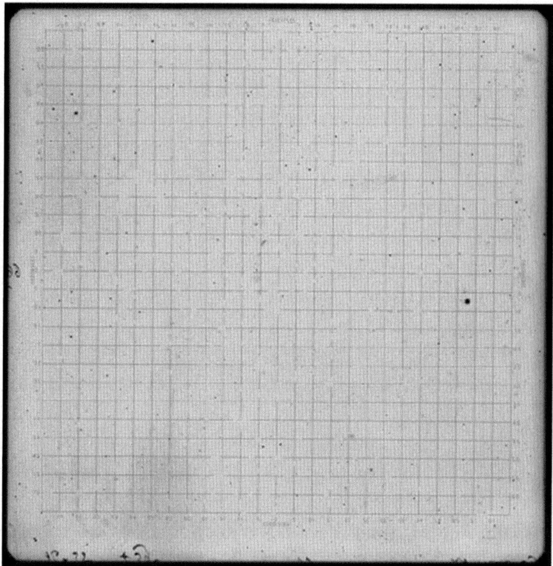

FIG. 5.4 Original picture from the Carte du Ciel taken at Uccle. Each star is represented by three points placed on the edges of a small equilateral triangle of 18/100 mm side length (1 point per exposure, the plate being slightly moved between each exposure). Image credit: https://www.astro.oma.be/wp-content/uploads/2014/09/cartedciel.gif.

All the measurements were also made manually using special procedures developed and tested with the aim to estimate possible errors and to provide the homogeneity of the resulting data. The survey, dealing with around 5 million stars (9 million images in the twofold overlap), occurred to be so laborious extending for several decades that it has never been fully completed. The results of measurements in the form of rectangular coordinates of stars and estimations of

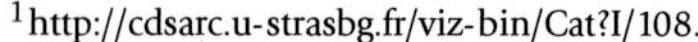

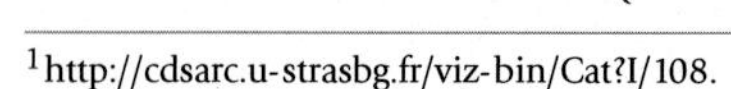

[1] http://cdsarc.u-strasbg.fr/viz-bin/Cat?I/108.

their magnitudes were published in 254 volumes in 1902–1964 as the original *Astrographic Catalog* (AC). Only half of the participating observatories published the Carte du Ciel charts. "The catalogue was difficult to use because of the inconvenience and inaccuracy of converting the rectangular coordinates to Right Ascension and Declination and inhomogeneities in limiting magnitude, accuracy, and presentation" (Jones, 2000). So, one of the main goals of the project – homogeneity of the resulting data – was not attained.

In the last decade of the 20th century, interest in old data of the AC emerged again after being triggered by the Hipparcos satellite results due to the lack of data for determination of accurate proper motions of the great number of stars. Two teams of astronomers undertook the successful attempts to digitize AC data. Volumes of the AC were key-punched, checked, and rereduced at the US Naval Observatory (Urban et al., 1998) and at the Sternberg Astronomical Institute in Moscow (Kuzmin et al., 1999).

After discovering the extragalactic nature of the weak nebulae, it became clear that for studying the large-scale structure of the universe it is necessary to compile and analyze large samples of extragalactic objects. This work was begun by E. Hubble and then continued by other researchers as Shapley and Ames with their specialized Harvard survey of the external galaxies brighter than the 13th magnitude (Shapley and Ames, 1932), as well as Humason, Lundmark, Holmberg, and others with their catalogues in the 1930s. So, Hubble performed a laborious survey of 40,000 galaxies in 1283 regions of the sky, located in both sky hemispheres. The main results of this work were as follows: the number of galaxies continues to grow up to the limit value of the survey; the quantitative growth is consistent with the uniform distribution of galaxies in space (later, Hubble found that when moving to weaker objects, the number of galaxies increases more slowly than with a homogeneous distribution); there is a strong dependence of the number of galaxies per unit area, depending on the galactic latitude (this is due to the absorption in the Milky Way); the distribution of the number of galaxies per unit area has a normal (Gaussian) distribution in magnitude of log N, and not in N. The results of Hubble differed greatly from the counting of stars: for the stars the boundary of our Galaxy is reached, while observations of galaxies do not give a hint at the boundary of the extragalactic world.

Later several general catalogues of extragalactic objects were created, which allowed the discovery of the large-scale galaxy distribution of the universe using *Lick counts* by Shane and Wirtanen (1954), to estimate its averaged characteristics and to develop the hierarchical clustering scenario of structure formation (Peebles, 1980). Because the number of galaxies with known distances (redshifts) was very small at that time, these conclusions were based on counts of galaxies, where the Shane–Virtanen catalogue has played a crucial role.

The next step in the Big Data astronomy was made in the 1950s–1970s, when the photographic *Palomar Observatory Sky Survey I* (POSS I) and POSS II appeared. POSS I, released in 1954–1958, covers the sky area down to -33° available for observation from Mount Palomar in Southern California. Observations were conducted with a 1.22 m Oschin Schmidt telescope in blue and red colors. The reached limiting photographic magnitude is 21, far fainter than in any other photographic projects. The result was a collection of 935 pairs of blue- and red-sensitive plates, each covering 6.5 deg^2 sky.

The complement to the POSS I was made in the 1970s, when the 1.24 m Schmidt telescope of the United Kingdom (UKST) in Australia and 1 m Schmidt of ESO in Chile conducted a coverage of the Southern sky with penetration to stars fainter than magnitude 22^m and with higher resolution than the Northern POSS I. The more advanced telescope equipment allowed the UKST to extend the survey to the near-infrared electromagnetic spectrum. After upgrading the Palomar Oschin Schmidt telescope in the early 1980s, POSS II was initiated. During the 1980s–1990s, the 894 areas of the Northern hemisphere were covered by plates in three electromagnetic bands – blue and red in optical and in near-infrared ones (Reid et al., 1991).

POSS I and POSS II together with surveys of the Southern sky became the origin of a number of specialized catalogues not only with positional data but intended for astrophysical purposes (see, for example, Fig. 5.5, which illustrates the discovery of dwarf galaxies of low surface brightness (K61) with the POSS survey by V. Karachentseva). We note the *Uppsala General Catalogue of galaxies* (UGC) by Nilson (1973, 1974) supplemented by the *ESO/Uppsala Survey of the ESO (B) Atlas*, ESO/Uppsala (Lauberts, 1982). The UGC is based mainly on the POSS prints and designed to be essentially complete to the limiting diameter 1 arcmin on the blue print. It covers the sky to the North of $\delta = -02^\circ 30'$ and includes all galaxies to the limiting apparent magnitude 14.5 in the CGCG by *Zwicky*.

Among other extragalactic catalogues created at the same time we mention two revised catalogues: the *Revised New General Catalogue of Nonstellar Astronomical Objects* (Rev NGC) (Sulentic et al., 1973) with its list of fundamental data on most bright nonstellar objects of

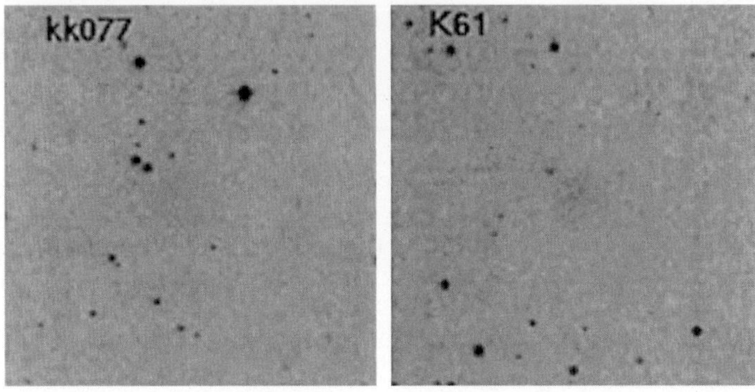

FIG. 5.5 Image of dwarf galaxy of low surface brightness (K61), which was discovered with the POSS I survey by V. Karachentseva in 1968 (size of K61 galaxy is 1.5 mm, size of plate is 36×36 cm).

interest to both professional and amateur astronomers, and the *Complete New General Catalogue and Index Catalogues of Nebulae* (Rev NGC/IC) (Sinnott, 1988), which includes all the objects in the original NGC/IC catalogues compiled by Dreyer.

The atlases of extragalactic objects printed on photographic paper have served as a good accomplishment to numerous catalogues and first databases (later they were digitized) as well as other specialized photometrical and positional catalogues of galaxies, galaxy groups, galaxy clusters and superclusters, quasars, and interstellar and intergalactic media. We mention several of such atlases, catalogues of galaxy systems, and specialized catalogues: *The Hubble Atlas of Galaxies* issued by Sandage (1961); *Atlas and Catalogue of Interacting Galaxies* (VV) conducted in 1962–1974 by Vorontsov-Velyaminov (1959); *Atlas of peculiar galaxies* by Arp (1966); *Atlas of Interacting Galaxies* by Vorontsov-Velyaminov (1977); *Nearby Galaxies Atlas* by Tully and Fisher (1987); *The de Vaucouleurs Atlas of galaxies* by Buta et al. (2007); *A catalog of dwarf galaxies* (DDO) by van den Bergh (1959, 1966); *Catalogue of selected Compact Galaxies and of Post-Eruptive Galaxies* (CSCG) by Zwicky and Zwicky (1971); *The Catalog of Isolated Pairs of Galaxies in Northern Hemisphere* (CIPG) by Karachentsev (1997); *The Catalogue of Isolated Galaxies* (CIG) by Karachentseva (1973); *Compact groups of compact galaxies*, Shakhbasian groups[2]; *Nearby groups of galaxies* by de Vaucouleurs (1975); *Galaxies of high surface brightness* by Arakelian (1975); *Systematic properties of compact groups of galaxies*, Hickson's groups, by Hickson (1982); *A General Catalogue of HI Observations of External Galaxies* by Huchtmeier et al. (1983); *A catalogue of dwarf galaxies south of* $\delta = -17.5°$ (FG) by Feitzinger and Galinski (1985); *A new optical catalog of quasi-stellar objects*, QSO catalogue, by Hewitt and Burbidge (1987) (see, also, the 13th edition of the *Catalogue of quasars and active nuclei* by Véron-Cetty and Véron (2010)); *The Catalogue of low surface brightness dwarf galaxies* (LSB dwarfs) by Karachentseva and Sharina (1988); *Northern dwarf and low surface brightness galaxies*. I. The Arecibo neutral hydrogen survey by Schneider et al. (1990) and II. The Green Bank neutral hydrogen survey by Schneider et al. (1992); *The 1000 brightest HIPASS galaxies: HI properties* (HIPASS) by Koribalski et al. (2004).

In the past, the Big Data left the study of the universe as a collection of individual objects and provided the basis to detect the similarity of the properties of individual groups of objects, introduce their classification, and move from studying a separate object to studying the regularities both in galaxy systems and the universe as an organic whole.

It mostly concerns the clustering of galaxies and clusters of galaxies by visual inspection of the POSS plates. "The first of these catalogues was prepared by Abell (1958) for clusters of galaxies. This catalogue covers the sky north of declination −27°. This cluster catalogue was extended to the southern sky by Abell et al. (1989). Both these catalogues together contained 4074 galaxy clusters. Distances of clusters were estimated on the basis of the brightness of the 10th brightest galaxy. A much larger catalogue was compiled by Zwicky with co-authors in 1961–1968, **Catalogue of Galaxies and Clusters of Galaxies**, CGCG (Zwicky et al., 1961); in this catalogue all galaxies brighter than photographic magnitude 15.7 as well as galaxy clusters north of declination −2.5° are listed. Since the definition of clusters in the Abell catalogue is more exact, this catalogue has

[2]https://heasarc.gsfc.nasa.gov/W3Browse/all/shk.html.

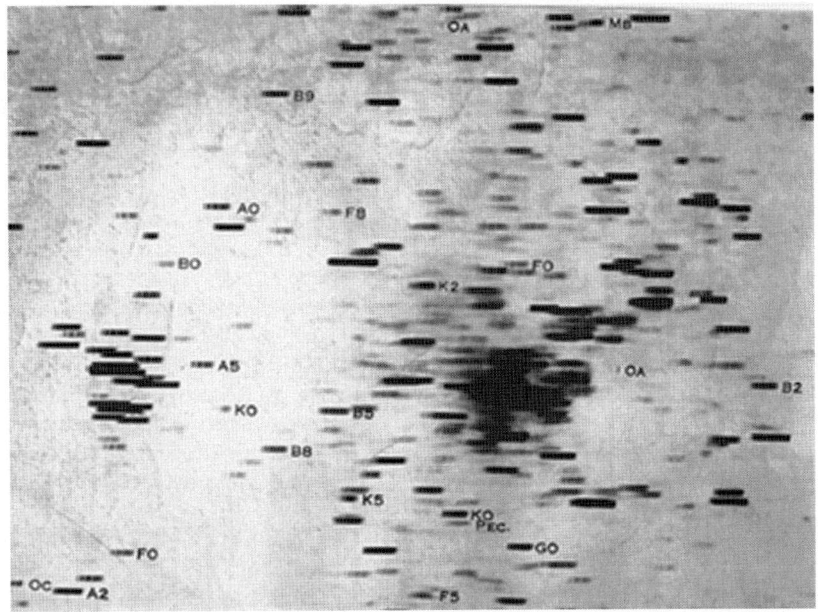

FIG. 5.6 A Bache telescope objective prism photograph used for the Henry Draper Catalogue, showing spectra of stars in Carina, recorded May 1893. The exposure time was 140 minutes (Hearnshaw, 2009). Reproduced with permission of the Licensor through PLSclear.

served for a large number of studies of the structure of the Universe. On the other hand, the Zwicky catalogue of galaxies was the basic source of targets for redshift determinations" (cited from Einasto, 2001).

5.2.3 Spectral Photographic Surveys

In the early 20th century the photographic surveys were carried out not only for catalogues of star positions but for spectral determination also. The first star photographic spectra applicable for scientific purposes were obtained by Henry Draper in the late 19th century with an 11 inch Draper instrument of Harvard College Observatory (HCO) equipped with the objective prism cell which contained two quartz prisms. Later the director of the HCO E. C. Pickering made efforts to equip several instruments with objective prisms and plate holders in order to make photographs of star fields with spectra (see Fig. 5.6). Thanks to him, the extensive survey of stellar spectra over both hemispheres to 9th visual magnitude began (MacConnell, 1995).

All the plates were processed manually. All the objects registered in the frames were classified and occupied their places in the star hierarchy according to specific features of their spectra. That tremendous job resulted in the *Henry Draper Catalogue* (HD) of stellar spectra and the Harvard system of spectral classification of stars depending on their stellar temperatures. The HD was published in 1918–1924. It contains 225,000 stars extending down to 9^m. Altogether more than 390,000 stars were classified at Harvard.

Spectral photographic observations with objective prism remain actual until now but their goals transformed from the comprehensive classification of objects in star fields into the surveys of specific groups of astronomical objects. The number of objects fixed on photographic plates increased to tens of millions. The manual processing of observations is history. The extraction of spectra is performed on digitized images in automatic or semiautomatic modes.

In the second half of the 20th century, the *First Byurakan Survey* (FBS) or Markarian Survey, 1965–1980, covering 17,000 deg^2 (Markarian et al., 1989) and its continuation, the *Second Byurakan Survey* (SBS), 1978–1991, 965 deg^2, covering some 3,000,000 low-dispersion spectra (Markaryan et al., 1983; Stepanian, 2005), are the largest photographic spectral surveys in the Northern sky initially targeted on a search of galaxies with ultraviolet excess. Both have been carried out with the Byurakan Observatory 102/132/213 cm Schmidt telescope using an objective prism. The spectra on the plates are about 1 mm long. Objects with $13^m < B < 16.5^m$ have useful spectra. Each FBS plate

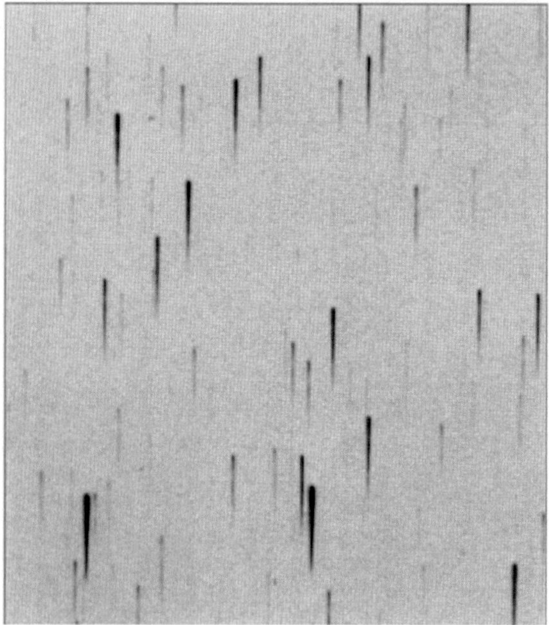

FIG. 5.7 A piece of the digitized FBS image showing its low-dispersion spectra (Mickaelian et al., 2009). The figure is published under the License CC BY 4.0 https://creativecommons.org/licenses/by/4.0/.

contains low-dispersion spectra of some 15,000–20,000 objects, so there are about 20 million objects in the whole survey (Nesci et al., 2009). The plate collection of FBS was digitized using the commercial scanner EPSON 1680 Pro (Fig. 5.7). Yet one survey related to the objective prism spectroscopy is *Case Low-Dispersion Northern Sky Survey* (Pesch et al., 1995); see also the updated versions of *Markarian galaxies* (Mrk), catalogues by Mazzarella and Balzano (1986), Petrosian et al. (2007).

Another two wide-angle Hamburg surveys for the Northern and Southern sky were targeted onto quasar detection and investigation. The *Hamburg Quasar Survey* (HQS) covers 14,000 deg^2 in the Northern sky (1985–1997, Engels et al., 1988; Hagen et al., 1999). The next one is its enhancement to the Southern sky. The *Hamburg-ESO Survey* (HES) is based on digitized objective-prism photographs taken with the ESO Schmidt telescope, covering essentially the entire Southern extragalactic sky (Wisotzki et al., 2000) (1990–1996). Targets of this survey are bright QSOs and Seyfert galaxies. It covers 9000 deg^2 in the Southern sky with the magnitude range $13 < B < 18$. All spectroscopic plates of both collections were digitized using the Hamburg PDS 1010G microdensitometer with further extraction of spectra by application of the specially designed numerical algorithm.

Among the extragalactic catalogues created at the same time we mention the three *Reference Catalogues of Bright Galaxies* (RC1, RC2, RC3) by de Vaucouleurs et al. (1964, 1976, 1991); *Southern Galaxy Catalogue* (SGC) (Corwin et al., 1985); *A Revised Shapley–Ames Catalogue of Bright Galaxies* (Rev SA) (Sandage and Tammann, 1987), and *Catalogue of Principal Galaxies* (PCG) by Paturel (1989). So, a revised version of the Shapley–Ames catalogue contained data on galaxies brighter than 13.5^m, including its redshifts. This catalogue and the compilation of all available data on bright galaxies by de Vaucouleurs catalogues became the sources of distances which made it possible to obtain the first three-dimensional distributions of galaxies in the universe. Much more detailed information on the spatial distribution of galaxies was obtained on the basis of redshifts, measured at the Harvard Center for Astrophysics (CfA catalogue) for all Zwicky galaxies brighter than $m_{ph} = 14.5$. Later this survey was extended to galaxies brighter than $m_{ph} = 15.5$ (the *Second CfA catalogue*) and to galaxies of the Southern sky *(Southern Sky Redshift Survey)*; see *A survey of galaxy redshifts* (z-catalogue) by Huchra et al. (1983). At this time the first catalogues of superclusters of galaxies were prepared: *Superclusters* by Oort (1983), *A supercluster catalogue* by Bahcall (1983), and *All-sky catalogs of superclusters of Abell-Aco clusters* by Zucca et al. (1993).

But the modern era of galaxy redshift catalogues started with the *Las Campanas Redshift Survey* (LCRS). Here, for the first time, multiobject spectrographs were used to measure simultaneously redshifts of 50–120 galaxies (Shectman et al., 1996). The LCRS covers six slices of size $1.5° \times 80°$, the total number of galaxies with redshifts is ~26,000, and the limiting magnitude is 18.8^m. "Generally the construction of a redshift survey involves two phases: first the selected area of the sky is imaged with a wide-field telescope, then galaxies brighter than a defined limit are selected from the resulting images as nonpointlike objects; optionally, colour selection may also be used to assist discrimination between stars and galaxies. Secondly, the selected galaxies are observed by spectroscopy, most commonly at visible wavelengths, to measure the wavelengths of prominent spectral lines; comparing observed and laboratory wavelengths then gives the redshift for each galaxy" (cited by "SkyServer – Algorithms").

These first redshift databases and catalogues of galaxy clusters allowed for the discovery of the three-dimensional cell structure of the universe (filamentary distribution of galaxies and galaxy clusters form super-

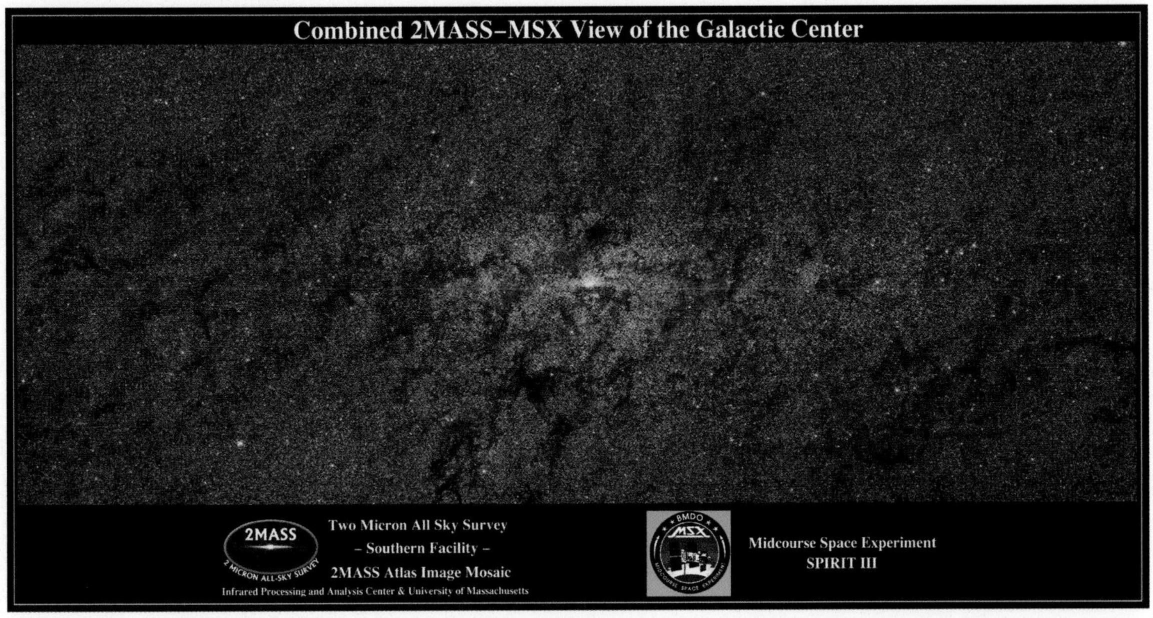

FIG. 5.8 Combined 2MASS-MSX mosaic view of the Galactic Center (IPAC public domain).

clusters, while there are huge voids without galaxies). These results demonstrated that "the pancake scenario of structure formation by Zeldovich et al. (1982) fits observations better than the hierarchical clustering scenario by Peebles (1980). More detailed studies of the structure formation by numerical simulations showed that the original pancake scenario by Zeldovich also has weak points – there is no fine structure in large voids between superclusters observed in the real universe (Zeldovich et al., 1982) and the structure forms too late, thus a new scenario of structure formation was suggested based on the dominating role of the cold dark matter in structure evolution. In a sense the new scenario is a hybrid between the original Peebles and Zeldovich scenarios: structure forms by hierarchical clustering of small structures within large filamentary structures – superclusters" (Einasto, 2001).

5.2.4 CCD Surveys

The surveys discussed above are based on photographic observations and therefore they have the standard defects of photo emulsions (low sensitivity, limited dynamic range, nonlinearity, etc.). Next, we will talk about projects that are truly digital, because the detectors use modern solid-state receivers. Application of CCDs instead of photographic emulsions allows to exclude the digitization of images from the Big Data acquisition process and significantly cut the period between observations and final product creation. Astronomical observations became fully automated and opened the question how to process such a huge astroinformation resource. For example, only 4 years (1997–2001) were needed for the *Two Micron All Sky Survey*, 2MASS, to collect 25.4 terabyte of raw imaging data covering 99.998% of the celestial sphere in the near-infrared electromagnetic band (see, for example, a mosaic image of the Galactic Center in Fig. 5.8).

2MASS is one of the most famous digital surveys, which was the result of joint efforts by the University of Massachusetts and the California Institute of Technology's Infrared Processing and Analysis Center.[3] 2MASS is a purely photometric survey, in which the images of the entire sky were constructed in three infrared filters J (1.24 μm), H (1.66 μm), and Ks (2.16 μm). Observations were conducted on two identical highly automated 1.3 m telescopes (in Arizona, USA, and in Chile). Each telescope was equipped with a three-channel camera, which allowed simultaneously to observe the sky with the help of matrices from 256 to 256 solid-state infrared detectors (the size of one element is $2''$). Bright source extractions have photometric uncertainty of $\leq 0.03^m$ and astrometric accuracy of order 100 mas.

[3] www.ipac.caltech.edu/2mass.

The *2MASS All-Sky Data Release* includes 4.1 million compressed FITS images covering the entire sky, 471 million source extractions in a *Point Source Catalog*. The image of any sky area in J, H, and Ks filters is available on several sites, such as http://irsa.ipac.caltech.edu/. Based on the 2MASS, the *Large Galaxy Atlas* (LGC) of about 600 bright galaxies and the *eXtended Source Catalog* (XSC), containing about 1.6 million objects with angular sizes larger than 10″–15″ were compiled (Skrutskie et al., 2006). The vast majority of such objects are galaxies, others are regions of ionized hydrogen HII, stellar clusters, planetary nebulae, and others. Limit values for extended objects with XSC are 13.5^m (2.9 mJy) and 15.0^m (1.6 mJy) in Ks and J bands, respectively. The *DEep Near Infrared Survey* (DENIS) gives the data on IJKs photometry for 355,220,325 objects over 16,700 deg^2 of the Southern sky to 18.5^m in I, 16.5^m in J, and 14.0^m in Ks (DENIS Consortium, 2005).

Several additional catalogues were developed on the 2MASS basis, for example, the *2MASS-selected Flat Galaxy Catalog*, 2MFGC by Mitronova et al. (2004) and the *Catalog of near-Infrared Isolated Galaxies* (2MIG) by Karachentseva et al. (2010). An effective method for the study of environmental influence on galaxy properties is to compare the properties of galaxies from dense environments with corresponding properties of isolated galaxies, i.e., objects that have not been noticeably affected by the surrounding galaxies. In order to have such a reference galaxy sample homogeneously distributed in the sky, the 2MIG was compiled on the data selected from the 2MASS XSC (Jarrett et al., 2000). The 2MIG catalogue consists of 3227 galaxies brighter than Ks $= 12^m$ and with angular diameters $a_{Ks} \geq 30$ arcsec, so approximately 6% of isolated galaxies are among extended sources of the 2MASS XSC survey. Colors and star formation rates of isolated galaxies from this catalogue in comparison with corresponding properties of galaxies in denser regions were studied by Melnyk et al. (2014, 2015). It was found that the galaxy environment helps to trigger the star formation in the highest mass galaxies. Pulatova et al. (2015) and Melnyk et al. (2015) analyzed the X-ray and infrared active galactic nucleus (AGN) fractions in this catalogue, respectively. Both studies reached the similar conclusion that AGN phenomenon is probably defined by secular galaxy evolution.

The *2-degree Field Galaxy Redshift Survey* (2dFGRS) is a major spectroscopic survey of ~5% of the sky conducted by British and Australian astronomers, who created the 3.9 m telescope of the Australian Astronomical Observatory (Fig. 5.9). The selection of objects is compiled on the basis of the *APM Galaxy Survey*. The multiobjective spectrograph allows simultaneous receiving of spectra of 400 objects in the field of view of 2°. One can get photometric and spectroscopic catalogues of 245,591 redshift objects (its median value is $z = 0.11$) at http://www.mso.anu.edu.au/2dFGRS/. A 2dF subproject, the *2dF QSO Redshift Survey* (2QZ), is a survey of the redshifts of quasars in two regions of the Northern and Southern galactic poles. Spectra of 23,338 quasars are available at http://www.2dfquasar.org/. The 2dFGRS survey and its database are described in detail in Colless et al. (2001) and are available at http://www.2dfgrs.net/Public/Release/Database/index.shtml.

The *6-Degree Field Galaxy Survey* (6dFGS) is a survey of redshifts and peculiar velocities of galaxies selected from the XSC 2MASS Survey with using 1.2 m Schmidt spectrograph of the Australian Astronomical Observatory (150 objects in the field of 6°). Spectra for more than 136,304 galaxies have been obtained and 110,256 among them have been identified for the first time. The catalogue contains 125,000 galaxies and virtually covers the entire Southern sky, and the redshift $z < 0.05$ (the near universe). The purpose of the survey is to study local deviations from the Hubble expansion. For 11,000 galaxies, the distance and peculiar velocities are determined by the Fundamental Plan for early galaxies.[4] All redshifts and spectra are available through the 6dFGS Online Database, hosted at the Royal Observatory, Edinburgh. An online 6dFGS atlas is available through the University of Cape Town as well as at http://www-wfau.roe.ac.uk/6dFGS/.

The 6dF name is a reference to the Six-degree Field instrument, a nifty device that uses optical fibers and robotic positioning technology. It increases the observational power of the Anglo-Australian Observatory's UK Schmidt Telescope to more than 100 times its original capability. Dozens of tiny fingertip-sized glass prisms (left) let the telescope see up to 150 stars or galaxies at the same time – making 6dF the ultimate machine for mapping the nearby universe.

Sloan Digital Sky Survey (SDSS) is the most magnificent astronomical project, the name of which is associated with Alfred Sloan, who made a large financial contribution to the project. Since the late 1980s, hundreds of researchers from the US, Japan, and Europe have been working on it.

The purpose of the first phase of the project, *SDSS-I*, was a photometric and spectral study of about a quarter of the sky (a large area near the Northern galactic pole and several bands in the Southern hemisphere). Observations were made on a specially constructed 2.5 m telescope at the Apache Point Observatory (New Mexico,

[4] http://www.aao.gov.au/local/www/6df/.

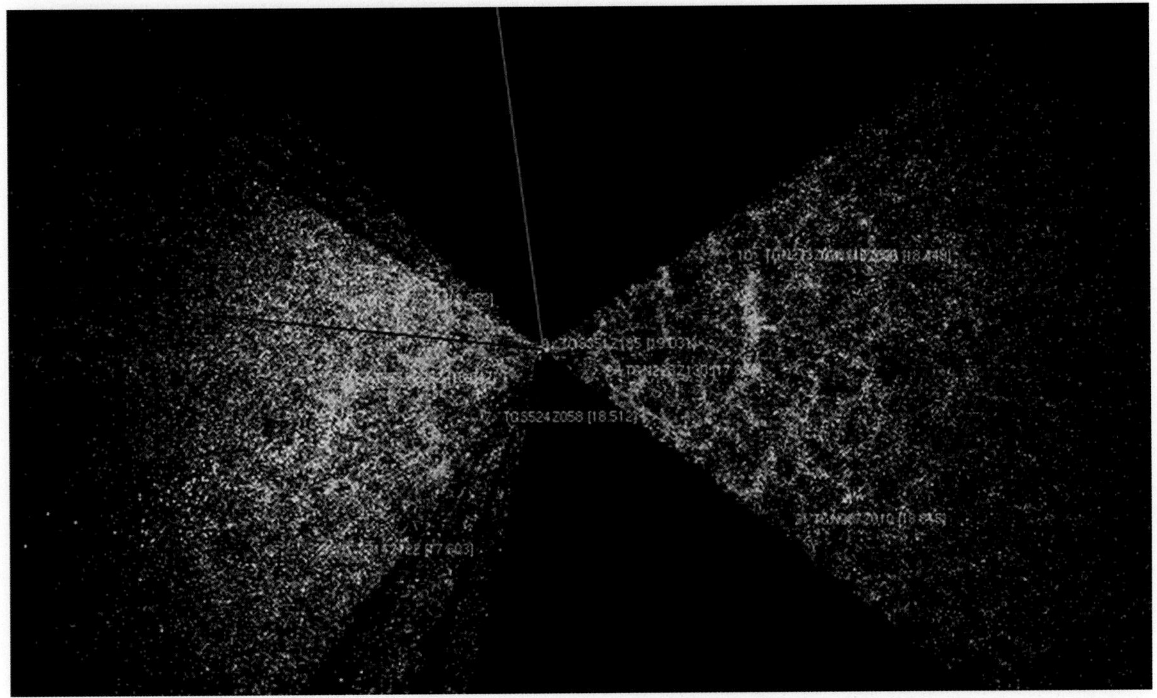

FIG. 5.9 Rendering of the 2dFGRS data (public domain).

USA). Equipment includes a 120 megapixel CCD camera and two multiobjective spectrographs allowing simultaneously receiving a spectrum of 640 objects. Auxiliary works are also carried out with other telescopes. The goal of photometric observations is a database for about 10^8 galaxies and as many stars with exact coordinates (Doi et al., 2010). The survey is carried out in five filters (u, g, r, i, z) from 350 μm to 910 μm, the marginal magnitude $B = 23^m$, and exposure time is 54 s in each filter. The objective of spectral observations in a database of 10^6 galaxies, 10^5 quasars, and 10^5 stars from a photometric survey was reached in 2005.

The second phase of the project, *SDSS II* (2005–2008), was designed to compile three different reviews: the *SDSS Legacy Survey* (or the Sloan Legacy Survey), the *Sloan Extension for Galactic Understanding and Exploration* (SEGUE), as the extension of the SDSS for the galaxy study, and the *Sloan Supernova Survey*. The purpose of these surveys is to solve fundamental questions about the nature of the universe, the origin of galaxies and quasars, and the formation and evolution of our Galaxy.

Sloan Legacy Survey covers 7500 deg^2 of imagery in the Northern galactic region with data on 2 million objects and spectra of more than 800,000 galaxies and 100,000 quasars. The SEGUE project gives spectra of 240,000 stars for the compilation of a three-dimensional map of the Milky Way. Within the framework of the project, a galaxy, a new satellite of our Galaxy, was discovered at a distance of 23 kpc from the Sun with a record (in 2011) ratio of mass to luminosity of 3400 $M_\odot/L_\odot$, which suggests it is almost entirely composed of dark matter. SEGUE results were included in the SDSS Data Release 7. By 2007, under the *Sloan Supernova Survey* project, by multiplying the area of the sky in 300 deg^2 the supernova type Ia was searched to determine the distance to distant objects. In total, more than 300 supernovae were discovered.

From mid-2008 until 2014, the third phase of the project, SDSS III, was passed with a significant upgrade of the SDSS spectrographs to conduct the four surveys. The purpose was to study (1) clustering of galaxies and intergalactic gas in the distant universe (*BOSS*), (2–3) the dynamics and chemical evolution of the Milky Way (*SEGUE-2* and *APOGEE*), and (4) the population of nonsolar giant planets (*MARVELS*). The *Baryon Oscillation Spectroscopic Survey* (BOSS) studied the spatial distribution of so-called luminous red galaxies (LRGs) and quasars. The survey allows us to investigate the nonhomogeneous distribution of the masses caused by acoustic baryon oscillations in the early universe.

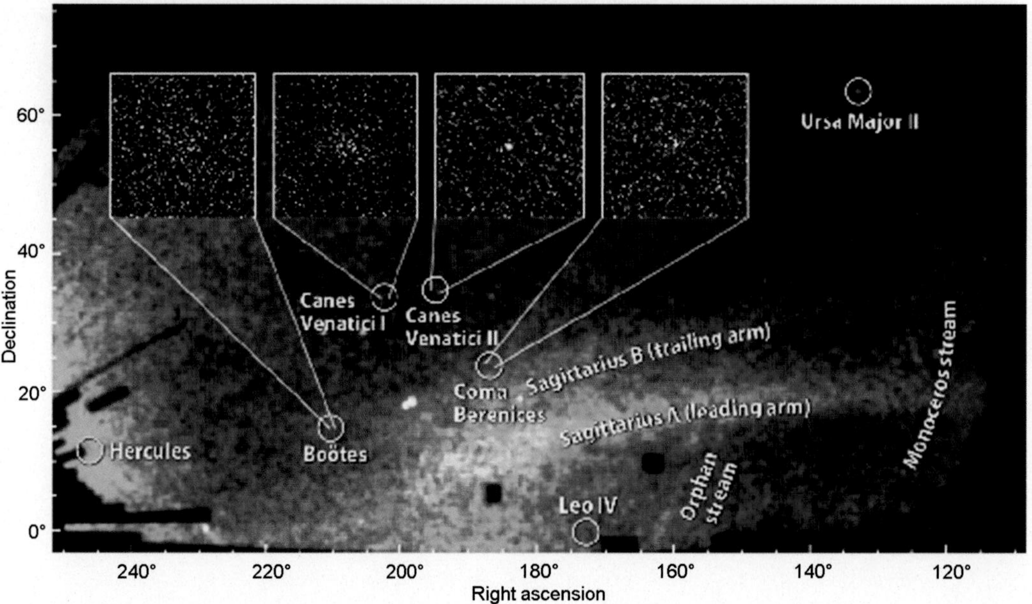

FIG. 5.10 SDSS stellar map of the Northern sky, showing trails and streams of stars torn from disrupted Milky Way satellites. Insets show new dwarf companions discovered by the SDSS (image credit: V. Belokurov, https://www.sdss.org/surveys/segue/).

The *SEGUE-2* project is a continuation of the SEGUE-1 project to obtain spectra about 120,000 stars of our Galaxy at distances from 10 to 60 kpc from the Earth. Under the *Apache Point Observatory Galactic Evolution Experiment* (APOGEE, 2011–2014), high-resolution, low-level infrared spectroscopy is used to observe the inner spaces of our Galaxy by hidden space dust. About 100,000 sources of different populations (bulge, bar, disk, halo) have been investigated, which has allowed to increase the number of sources of the Galaxy by more than a hundred times with high-precision infrared spectra (see, for example, Fig. 5.10). During the *Multi-object APO Radial Velocity Exoplanet Large-area Survey* (MARVELS), radial velocities of 11,000 bright stars were measured to detect planets around them.

In July 2014, the fourth phase of the SDSS project began. In this phase (SDSS–IV, 2014–2020), the accuracy of cosmological measurements in the critical early stage of space history (*eBOSS project*, extended BOSS) and infrared spectroscopic studies of the Galaxy in the Northern and Southern hemispheres (*APOGEE-2*) will be improved. For the first time, using Sloan spectrographs, the spatially separated maps of individual galaxies (*MaNGA*) will be made (Alam et al., 2015). Two smaller reviews are running as an eBOSS routine: the *Time Domain Spectroscopic Survey* (TDSS), which is the first systematic large-scale spectroscopic overview of variable sources, and the *Spectroscopic Identification of eROSITA Sources* (SPIDERS), which provides a unique complete and uniform description of optical spectroscopic data for X-ray sources of space missions ROSAT, XMM, and *extended ROentgen Survey with an Imaging Telescope Array* (eROSITA) on-board the Russian satellite Spectrum-Roentgen-Gamma, which was launched in 2019. All information is available at http://www.sdss.org.

Photographic spectroscopy of stellar spectra is now hopelessly behind the multiwave surveys grounded on modern technological facilities, which allow registering not tens but hundreds of millions of objects in significantly shorter terms with immeasurably higher accuracy. The *Large Sky Area Multi-Object Fibre Spectroscopic Telescope* (LAMOST) project or *Guo Shoujing Telescope* project (China) was initiated to meet the challenges of the accelerating observational data accumulation in spectroscopy. LAMOST is a quasimeridian reflecting Schmidt telescope with an effective aperture of about 4 m and a field of view of 5° in diameter. Recording 4000 celestial object spectra simultaneously with a parallel fully controllable fiber positioning system, the LAMOST provides the highest rate of spectral acquisition of tens of thousands of spectra per night. The

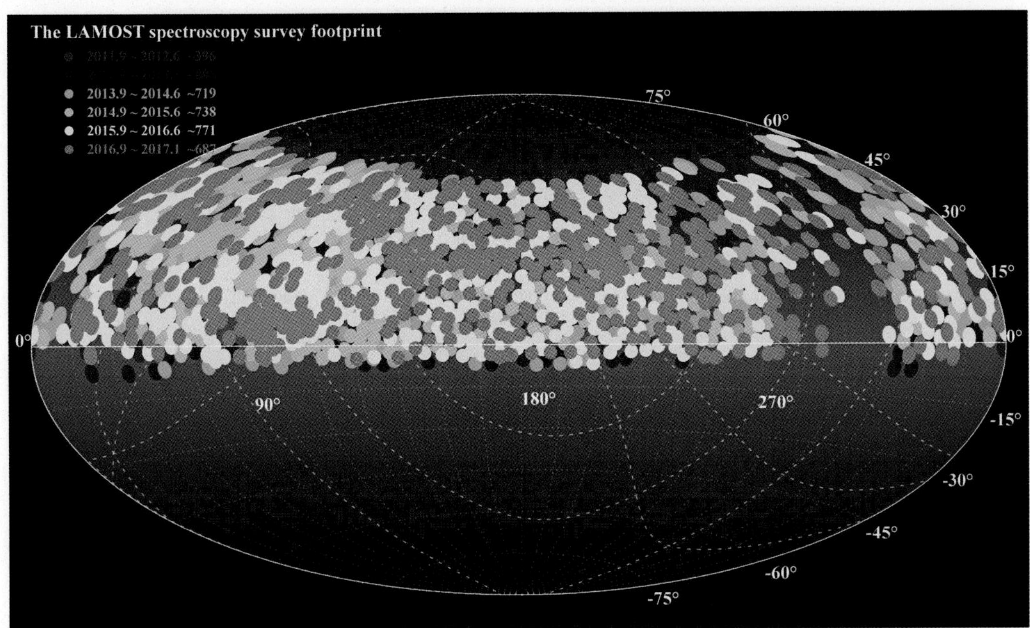

FIG. 5.11 Footprint of the LAMOST DR5 spectroscopy survey. Credit to NAOC.

survey has two major components: the *LAMOST Extra-GAlactic Survey* (LEGAS) and the *LAMOST Experiment for Galactic Understanding and Exploration survey* (LEGUE), survey of the Milky Way stellar structure. The limiting magnitude of the survey is 20.5^m. (See Fig. 5.11.)

The LAMOST optical system consists of a reflecting Schmidt mirror at the Northern end, a spherical primary mirror at the Southern end, and a focal surface in between. The light collected is reflected from reflecting mirror to the primary one, is again reflected by the primary mirror, and forms an image of the observed sky on the focal surface. The light of individual objects is fed into the front ends of optical fibers accurately positioned on the focal surface, and then transferred into the spectrographs fixed in the room underneath, to be dispersed into spectra and simultaneously recorded on the CCD detectors. The LAMOST data volume obtainable per night will be several gigabytes. LAMOST has an automatic software system for its observation and data processing.[5] The LAMOST data include three major types: "(I) Raw Data: all original data as well as original provenance information (for example, the observing log files, calibration files, software versions used, etc.), and the batch reduced two-dimensional spectra; (II) 1D Spectral Data: one-dimensional spectra of observed objects, reduced through standardized reduction pipelines; (III) Catalog Data: objective physical quantities with errors, derived from the spectral data and input catalog. The catalog includes the coordinates, magnitudes, radial velocities, effective temperature, surface gravity, elemental abundances, warning flags and so on".[6] To this time, the LAMOST collected the data on more than 10 million spectra of stars, galaxies, quasars, etc. For example, the LAMOST Data Release 6 version 1 (survey for 2017–2018 years of observations) consists of astroinformation of about 889,947 spectra (Low Resolution Statistics) and 1,344,289 spectra (Medium Resolution Statistics) on 422 observed plates for 781,620 stars, 20,003 galaxies, and 7718 quasars.[7] The LAMOST provides many kinds of research related to the physics of these objects: the white dwarf binary pathways survey (Parsons et al., 2016), the properties of red giant stars (Zhang et al., 2017), the stars with rocky exoplanets (Mulders et al., 2016), the largest galaxy pairs catalogue to date, which is recently built on SDSS and LAMOST data by Shen et al. (2016),[8] and other celestial bodies.

Galaxy and Mass Assembly (GAMA) is a digital spectroscopic project to exploit the latest generation of

[5] http://www.lamost.org/public/science?locale=en.

[6] http://dr.lamost.org/ucenter/doc/lssdp.

[7] For more details, see http://dr6.lamost.org/.

[8] http://cluster.shao.ac.cn/~shen/Gal_pair/.

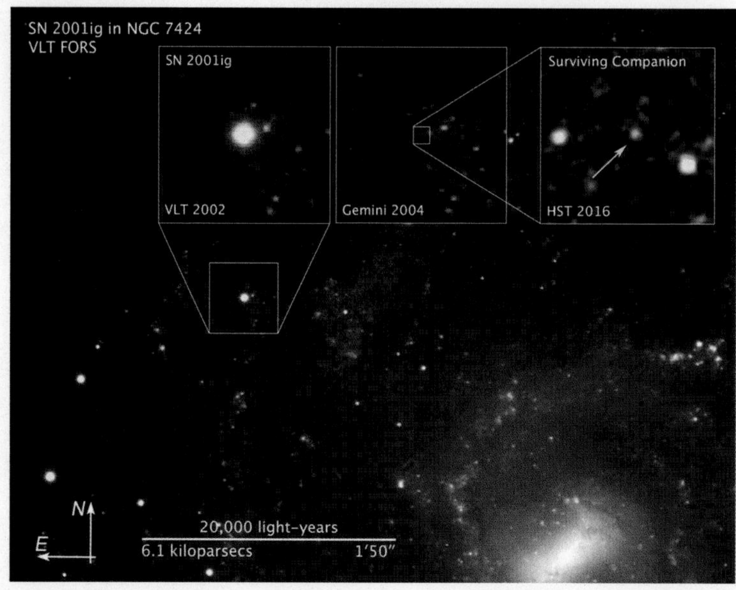

FIG. 5.12 Compass image for NGC 7424 and Supernova SN 2001IG (distance is 40 million light years, in the direction of the Southern constellation Grus, the Crane). In 2002, shortly after SN 2001IG exploded, it was photographed with the ESO Very Large Telescope. In 2004, the supernova was studied with the Gemini South Observatory, when the presence of a surviving binary companion was hinted. In 2016, as the SN 2001IG glow faded, it was pinpointed and photographed with the Hubble Space Telescope. The surviving companion image was possible only due to the Hubble Space Telescope exquisite resolution and ultraviolet sensitivity. The Hubble Space Telescope observations of SN 2001IG provide the best evidence yet that some supernovae originate in double-star systems. Image credits: NASA, European Space Agency, S. Ryder (Australian Astronomical Observatory), and O. Fox (STScI).

ground-based and space-borne survey facilities to study cosmology and galaxy formation and evolution. At the heart of this project lies the GAMA spectroscopic survey of ~300,000 galaxies down to $r < 19.8^m$ over ~286 deg^2, carried out using the AAO megamultiobject spectrograph on the Anglo-Australian Telescope by the GAMA team. This project was conducted 210 nights over 7 years (2008–2014) and the observations are now completed. This survey builds on, and is augmented by, previous spectroscopic surveys such as the SDSS, the 2dFGRS, and the Millennium Galaxy Catalogue (MGC).

Modern space missions likely the GAIA and the Hubble Space Telescope, as well as planned or currently constructed ground-based automated observation systems (LSST, PanSTARRS, and others) built using the latest technologies have increased the number of registered objects to billions, establishing new challenges for astronomers and programmers.

Hubble Space Telescope. Hubble is the first major optical telescope to be placed in space. Hubble has an unobstructed view of the universe. Scientists have used Hubble to observe the most distant stars and galaxies as well as the planets in our Solar System (see Fig. 5.12). All data are combined in the Hubble Legacy Archive (HLA). The HLA is designed to optimize science from the Hubble Space Telescope by providing online, enhanced Hubble products and advanced browsing capabilities. This archive consists of many different catalogues, e.g., Hubble Source Catalog, which is designed to optimize science from the Hubble Telescope by combining the tens of thousands of visit-based source lists in the HLA into a single master catalogue.

Operated in 1989–1993, *Hipparcos* was the first space experiment totally dedicated to astrometric purposes, which are the precise measurements of positions, proper motions, and parallaxes of stars. In 1997 the European Space Agency published a set of resulting catalogues – the *Hipparcos catalogue* and *Tycho catalogue* – including a three-volume *Millennium Star Atlas*. The processing of observations by the main instrument generated the *Hipparcos Catalogue* of 118,218 stars charted in the Atlas with the highest precision. An auxiliary star mapper pinpointed many more stars with lesser but still unprecedented accuracy in the *Tycho Catalogue* of 1,058,332 stars. The *Tycho 2 Catalogue* comprises 2,539,913 stars with 85 mas rms positions, and includes 99% of all stars down to 11^m (see Høg, 2000 and https://www.cosmos.esa.int/web/hipparcos).

The positional accuracy of these resulting catalogues is significantly better than that of ground projects, in particular, by up to a factor 4 in the accuracies of extremely bright stars. Final standard errors for stars brighter than 9^m, at epoch 1991.25, are as follows: positions: 0.77/0.64 milliarcsec (two coordinates); parallaxes: 0.97 milliarcsec; proper motions: 0.88/0.74 milliarcsec/year (two coordinates). Besides, the mission provided the essential improvement in the quality and quantity of photometric data in two spectral bands, B and V. The most outstanding result of the Hipparcos mission is the building of the superaccurate reference frame for the whole sky. Amongst the key achievements are also "refining the cosmic distance scale, characterising the large-scale kinematic motions in the solar neighbourhood, providing precise luminosities for stellar modelling, and confirming Einstein's prediction of the effect of gravity on starlight" (Kovalevsky, 2009).

The successor mission to Hipparcos is the *Gaia*, operated by the European Space Agency and launched in December 2013. "The final catalogue of the mission will contain full astrometric (position, distance, and motion) and photometric (brightness and colour) parameters for over one billion stars as well as extensive additional information including a classification of the sources and lists of variable stars, multiple stellar systems and exoplanet-hosting stars. For more than 150 million stars there will be measurements of the radial-velocity. The final Gaia catalogue will be a census of our Galaxy of such superb precision and detail that it will redefine the fundamental reference frame used for all astronomical coordinate systems".[9]

Current results of the mission are given in two releases issued in 2016 and 2018. *Gaia Data Release 1* contains data covering 14 months of observations. The content includes positions and measured G-magnitudes for 1.1 billion stars using only Gaia data; the five-parameter astrometric solution (positions, parallaxes, and proper motions) for more than 2 million stars in common between the Tycho-2 Catalogue and Gaia; light curves and characteristics for about 3000 variable stars; and positions and G-magnitudes for 2152 ICRF quasars (Gaia Collaboration, 2016).

Gaia Data Release 2 covers 22 months of observations. The content of the release contains positions and G-magnitudes for nearly 1.7 billion stars; parallaxes, proper motions and BP/RP (blue/red photometer) colors for more than 1.3 billion stars with a limiting magnitude of $G = 21$ and a bright limit of $G \sim 3$; median radial velocities for more than 6 million stars with a mean G-magnitude between about 4 and 13 and an effective temperature in the range of about 3550 to 6900 K; astrophysical parameters such as surface temperature (161 million stars brighter than 17th G-magnitude), extinction and reddening for 87 million stars, radius and luminosity (76 million stars); light curves and classification for about 0.5 million variable sources consisting of Cepheids, RR Lyrae, Mira, and Semi-Regular Candidates as well as High-Amplitude Delta Scuti, BY Draconis candidates, SX Phoenicis Candidates and short-timescale phenomena; positions and epoch of observation of 14,099 known Solar System objects – mainly asteroids – based on more than 1.5 million observations (Fig. 5.13); positions and G-magnitudes for more than 0.5 million quasars – which allows the celestial reference frame to be fully defined for the first time using only optical observations of extragalactic sources[10] (Gaia Collaboration, 2018).

[9]http://sci.esa.int/gaia/58277-towards-the-final-gaia-catalogue/.

[10]http://sci.esa.int/gaia/60243-data-release-2/.

5.3 NEW LIFE OF OLD ASTRONOMICAL DATA

5.3.1 Digitization of Photographic Sky Surveys

The tremendous number of plates and objects fixed in the observational material of vast surveys made it possible to get digital data not only in a manual manner but also using automated and semiautomated coordinate measuring machines. To transform the photographic plate into the catalogued list of objects the special scanners devised and developed for digitization of astronomical images were used.

The Precision Measuring Machine (PMM) constructed and operated in USNO Flagstaff Station digitized more than 20,000 photographic plates of large format, among them, more than 7000 plates of both POSS surveys. The result was 13 terabyte of the digital library of images and the final *USNO-B1 catalog*. USNO-B is the next step in the sequence of catalogues that started with UJ1.0 (Monet et al., 1994), USNO-A1.0 (Monet, 1996), and USNO-A2.0 (Monet, 1998). USNO-A2.0 contains 526,280,881 objects and is still useful because photometric data differ from USNO-B1.0 and sometimes are better.

USNO-B is an all-sky catalogue that presents positions, proper motions, magnitudes in various optical

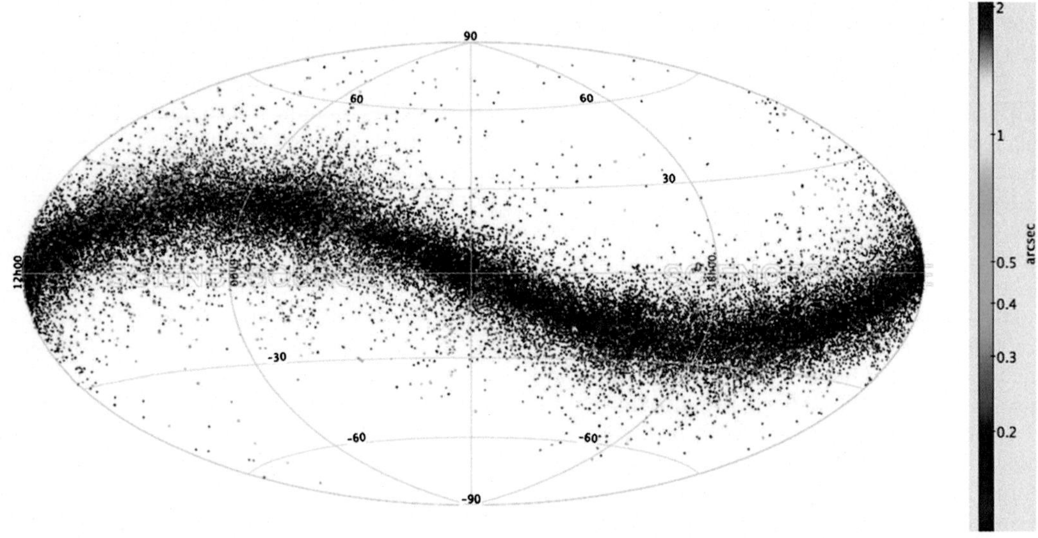

FIG. 5.13 The image shows Gaia's detections of asteroids in eight months' worth of data, compared with the positions on the sky of a sample of 50,000 known asteroids. The color of the data points is an indication of the accuracy of the detections, showing the separation on the sky between the observed position of Gaia's detection and the expected position of each asteroid: blue indicates higher accuracy, whereas green and red indicate lower accuracy. The regions showing lower accuracy of asteroid detections correspond to patches of the sky where the stellar density is very high, thus complicating the identification process (image credit: European Space Agency/Gaia/DPAC/CU4, L. Galluccio, F. Mignard, P. Tanga (Observatoire de la Côte d'Azur)/Science Source).

passbands, and star/galaxy estimators for 1,042,618,261 objects derived from 3,643,201,733 separate observations. The data were obtained from scans of 7435 Schmidt plates taken for the various sky surveys during the 50 years. USNO-B1.0 is believed to provide completeness down to $V = 21^m$, and $0.2''$ astrometric accuracy at J2000, 0.3^m photometric accuracy in up to five colors, and 85% accuracy for distinguishing stars from nonstellar objects (Monet et al., 2003). It gives proper motions for all 1,045,913,669 objects present in POSS I/SERC-J and POSS II/AAO-SES over the whole sky, limited to 22.5^m in B and 20.8^m in R.

USNO-B1.0 is one of the largest among all available catalogues by the number of objects. The catalogue's size is about 80 gigabytes. There is a problem to distribute it entirely. To get a fast access to the catalogue data the proposed solution is to make a service on access *in situ*. For example, one of the USNO web portals proposes next service options for users of USNO-B1: extract the catalogue data, plot up finder charts, view raw images, overplot the catalogue data onto the images, and overplot the user's own markers.[11] There has been the next trend in Big Data administration and operation without needing to transfer the large amounts of data over the network: Big Data + service on access *in situ*.

The scientific importance of POSS II is not limited to USNO catalogues. Another team in the late 1990s initiated the *Digital Palomar Observatory Sky Survey* (DPOSS) project. The survey covers the entire sky north of $\delta = -3°$ in three bands, calibrated to the Gunn *gri* system, reaching to equivalent limiting magnitude of $B_{lim} \sim 22^m$. As a result of the state-of-the-art technique of plate digitization, detailed processing of the scans, and a very extensive CCD calibration program, the data quality exceeds that of the previous photograph-based efforts. The end product of the survey will be the *Palomar-Norris Sky Catalog*, anticipated to contain > 50 million galaxies and > 2 billion stars (Djorgovski et al., 1998).

Photographic surveys played an important role in the development of modern astronomy. For example, copies of the POSS in the form of glass plates, films, or prints of them on photographic paper were distributed among most of the world's astronomical institutions and were actively used for development of all the research fields of astronomy: from studies of the Solar System to the study of galaxies and quasars. For a long

[11] https://www.usno.navy.mil/USNO/astrometry/optical-IR-prod/icas/icas-intro.

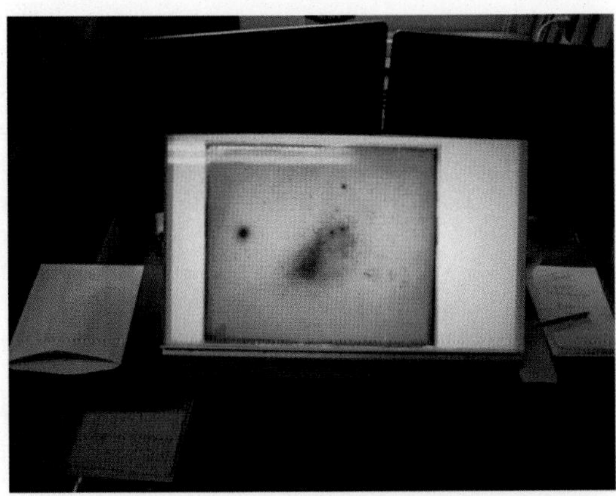

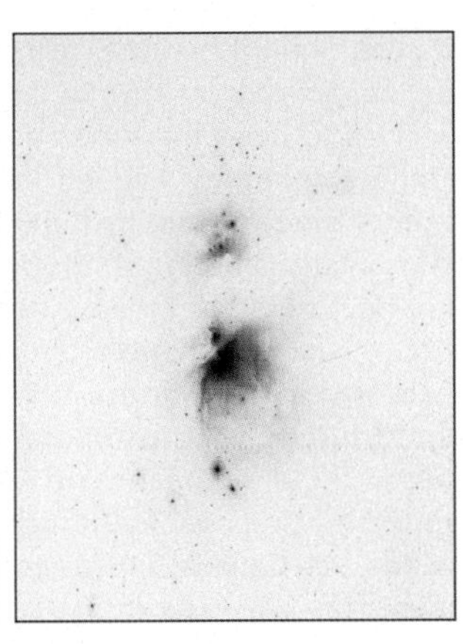

FIG. 5.14 The digitization of astroplate collections provides the capability for systematic study of the celestial bodies on about 100 year timescales. (Left) Image of the 11 × 14 inch Plate of the Magellanic Clouds as the example of the Stellar Glass Plate Collection founded by E. Pickering in Harvard College Observatory in 1880 (image credit: L. Halloran, http://www.liahalloran.com/blog/2016/3/21/9mciuvnc70v6cky2mtv64gennqkibf). (Middle) Images of the Moon dated on 25 May 1912 (exposure 3 sec) taken with the Merts–Repsold Double Astrograph at the observatory of the University of St. Vladimir, Kyiv, Ukraine, as the example of the UkrVO Astroplate Digital Collection (image credit: Joined Digitized Archive of the Ukrainian Virtual Observatory http://ukr-vo.org/). (Right) Image of the Orion Nebula (M42) taken with the Hamburg Observatory's Great Schmidt Mirror on 9 March 1959 (image credit: Digital Plate Archives of Hamburger Sternwarte).

time, photographic plates were the only effective mechanism for recording surveys. They had a fairly long time of storage – more than a hundred years – and contained very high-quality information (see Fig. 5.14). However, they were not directly admissible for processing. For example, one photographic plate from the large Schmidt telescope contains 10^5–10^6 stars and galaxies. Thus, the early work was reduced to a simple visual inspection of the brightest objects and a detailed study of only small areas on the plate. Effective extraction of information from the astroplates became possible only after the appearance of high-speed measuring machines, which allowed the digitized image to be transferred to electronic media and processed on a computer. This is the way the first digital surveys of the sky (such as APM and DSS) appeared in the early 1990s and later this "scanning motion" began to progress in all the observatories, which were holders of astronomical glass collections.

APM Survey. Within the framework of this project the 185 plates obtained on the 1.2 m telescope of the Schmidt Anglo-Australian Observatory (Australia) on the Southern galactic pole were scanned in the English Cambridge on the Automatic Plate Measuring (APM) machine (scan step 0.5). The inspection plates cover approximately 4300 deg^2 in the sky. The coordinates, the apparent magnitude, and also more than a dozen other parameters characterizing the shape and distribution of brightness were determined for each object. APM survey gives the 0.600 rms positions, b and r photometry for 166,466,987 POSS-I objects over 21,000 deg^2 limited to 21.0^m in b and 20.0^m in r (McMahon et al., 2000).

On the basis of the analysis of photometric profiles of objects, their classification and the formed sample of extragalactic objects were performed, which was marked by almost 100% completeness and contained about 2 million galaxies. On the basis of the APM-survey, a number of important studies of large-scale distribution of galaxies were performed and the first objective catalogue of galaxy clusters was compiled. The results of the APM survey and its extensions (in particular to the Northern hemisphere) became the basis of one of the most interesting survey of recent years – 2dF – in view of the redshifts of galaxies.

Digital Sky Survey (DSS) is the first high-quality and publicly available digital image of the entire celestial sphere in the optical range. DSS arose from the design of the Space Telescope Scientific Institute (STScI) to create a catalogue that was used to accurately navigate the Hubble Space Telescope to the desired object and to control during observation. To create such a catalogue, the scans of the "blue" photographic plates of the POSS-I and European South Observatory/Science and Engineering Research Council (ESO/SERC) surveys were started. Soon it became clear that the value of digitizing images goes far beyond the original task, and it was decided to make the survey accessible to the wider scientific community. The full volume of the original version of the DSS-I reached 600 gigabytes (DSS, ESO).[12] Such a large amount of information at that time could not be easily transported. After a special 10× compression of the information, the review was fitted on 102 CDs, which, beginning in 1994, began to be distributed through the Pacific Astronomical Society. Later, free access to DSS-I was opened on STScI.[13] Now access to the survey is also possible through a number of other sites in the US, Canada, Japan, and some European countries.

The *DSS-II digital survey* emerged as a natural development of the DSS-I work on the second Palomar overview of POSS-II. In step 1, the POSS-II for the Northern sky was scanned, as well as SERC and other surveys for the Southern sky. The digitized plates were presented into three colored bands. The total amount of information is about 5 terabyte; access to it is usually done through the same web sites as POSS-I.[14] For example, the GSC 2.3.2 gives 0.300 rms positions and j/V/F/N photometry for 945,592,683 objects presented in DSS-II, limited to 22.5^m in j and 20.8^m in F (Lasker et al., 2008). Since its inception, DSS has become one of the most useful and used tools of modern astronomy. With the use of images from the DSS, a huge amount of work was devoted to the study of individual galaxies, their groups, and their distribution and geometric characteristics.

The *Digitized Palomar Observatory Sky Survey* (DPOSS)[15] is also based on the scanning of POSS-II inspection photographic plates in three color bands, but the calibration and processing procedures developed at the California Institute of Technology are different from those used in DSS. This survey covers the whole Northern sky. DPOSS consists of a database of images (~3 terabyte) and several catalogues. The ultimate goal of DPOSS is to compile the *Palomar Norris Sky Catalog* (PNSC) database for all objects up to 22^m. The catalogue will contain more than one hundred measured parameters for each source. It is expected that the PNSC will include data for more than 50 million extragalactic objects, including 10^5 quasars and more than 10^9 stars.

APS (or MAPS). The Minnesota Automated Plate Scanner (MAPS) is made of glassy blue (O) and red (E) copies of POSS-I for 632 galactic latitude fields $|b| > 20°$. The scanning procedures are similar to those performed by the APM. Pixel size is 0.8. The MAPS database contains coordinates, magnitudes, colors, and other calculated image source parameters for approximately 89,234,404 POSS-I objects (stars and galaxies) over 21,000 deg^2 to a value less than 21^m in a blue filter (Cabanela et al., 2003). Stellar and nonstellar objects were separated using the neural network image classifier described by Odewahn et al. (1992, 1993). The MAPS is located at http://aps.umn.edu.

The *Coordinates, Sizes, Magnets, Orientations, and Shapes* (COSMOS) machine is an automated microdensitometer that worked in Edinburgh, UK, from 1975 to the end of 1993 and was replaced by SuperCOSMOS. It used the COSMOS/UKST database based on the 1-color VJ Southern sky survey on the glass copies of the SERC-J/EJ atlas ($|b| > 10°$) and UK Schmidt's "short" red originals in the low (Southern) galactic latitudes ($|b| > 10°$). The objects of the catalogue contain coordinates and morphological, photometric, and classification information in one color (BJ or R).

SuperCOSMOS is a fast, high-precision machine with a 0.67-pixel resolution for digitizing photographic plates. Part of the project is the systematic scanning of the sky surveys taken on the UK Schmidt telescope, the ESO Schmidt telescope, and the Palomar Observatory, as well as the provision of data to a wide range of users. The result of the scan is an overview of the SuperCOSMOS Sky Survey (SSS), which was completed in 2001 with free access.[16]

Not only large international projects need special equipment for plate digitization. Local plate collections of national observatories all over the world comprise the photographic material that can satisfy the inquiries of time-domain astronomy on timescales as long as a century or longer. The main demands for digitizers include time saving and a high precision of object fixation. The high-speed scanner was designed and constructed to obtain plate scans of the *Harvard College Observatory plate collection*. The total collection consists of ~450,000 direct plates on both hemispheres and covers the time interval between 1885 and 1995. The Digital Access to a Sky Century Harvard (DASCH) project was initiated to digitize all the plates and make their digital images (~400 terabyte) and reduced photometric data available online (Grindlay et al., 2011). The fully automated process from cleaning the plate to unloading it from scanner allowed to scan 400 plates a day. The full DASCH output database of ~450,000 plate images and

[12] http://archive.eso.org/dss/dss.
[13] http://archive.stsci.edu/dss.
[14] http://archive.stsci.edu/cgi-bin/dss_form/, http://archive.eso.org/dss/.
[15] http://www.astro.caltech.edu/~george/dposs/dposs_pop.html.
[16] http://www-wfau.roe.ac.uk/sss.

derived magnitudes for each resolved object will reach ~1 petabyte in total.

The digitizer built on the same principles as the Harvard scanner was used also at the Royal Observatory in Belgium (De Cuyper et al., 2012). It was involved in the EU FP7 European Satellite Partnership for Computing Ephemerides (ESPaCE) project to digitize the photographic glass plates of the natural satellites of Mars, Jupiter, and Saturn taken over 30 years. Both digitizers use the method of block scanning to take a series of frames of the plate which are then stitched together in a mosaic to create an image of the whole plate. They are equipped with an XY-table that has an air bearing and linear motor, with micron accuracy, a double-sided telecentric lens, and a stable light system composed of LED arrays, to ensure the quality of digitization in astronomical plates.

A digitizer with high precision and high measuring speed was developed jointly by Shanghai Astronomical Observatory, China, and Nishimura Co. Ltd, Japan, to digitize 30,000 astronomical plates of glass collections in observatories of China (Yu et al., 2017). The digitizer at the Shanghai Astronomical Observatory uses the method of line scanning in several steps. The final step is the stitching of mosaic strips into the whole image. After two years of development, the machine achieved a precision of better than 1 μm in digitization position, and 10 minutes are needed to digitize a plate with dimensions of 300 mm × 300 mm.

In the late 1990s, the Commission 9 of the IAU Working Group on Sky Surveys initiated the development of the *Wide-Field Plate Database* (WFPDB). The WFPDB was supposed to collect information about the wide-field photographic archives all over the world. The WFPDB currently includes information about 610,000 plates from 144 plate archives. The latest version (7.1, November 2015) of the Catalogue of Wide-Field Plate Archives (CWFPAs, www.wfpdb.org/catalogue.html) contains descriptions practically for all of the existing professional wide-field photographic observations stored in 509 archives with more than 2,500,000 plates from 163 observatories, available in CSV, ASCII, and VOT file format (Tsvetkova et al., 2018). The WFPDB provides access to more than 25% of all wide-field plate archives. Following the requirements of the Centre de Données astronomiques de Strasbourg (CDS) and the International Virtual Observatory Alliance (IVOA), the WFPDB offers digitized plate preview images along with plate raw data. The WFPDB team continues to fill the database with information submitted or retrieved from the astroplates, which would eventually enable the astronomical community to complement their studies with data going more than 130 years back in time.

5.3.2 Scientific Objectives for Old Data Involving

Astronomy research is changing from being hypothesis-driven to being data-driven to being data-intensive. To cope with the various challenges and opportunities offered by the exponential growth of astronomical data volumes, rates, and complexity, the new disciplines of astrostatistics and astroinformatics have emerged. *Astrostatistics* is an interdisciplinary field of astronomy/astrophysics and statistics that applies statistics to the study and analysis of astronomical data. *Astroinformatics* is an interdisciplinary field of astronomy/astrophysics and informatics that uses information/communication technologies to solve the Big Data problems faced in astronomy (see Table 5.1).

The old data involving allows us to save the unique astronomical observational heritage accumulated in observatories from the 1890s, opening the wide online access to the joint database of digitized astronomic negatives and spectra for the national/foreign scientific community. Some examples of how the old photographic astroplates and spectra can be served are given in Fig. 5.15, illustrating that there are no obsolete observational data in astronomy.

Exoplanets. The Kepler mission during its over 9.5 years of service (launched on 7 March 2009) observed 530,506 stars and discovered more than 2300 confirmed exoplanets, providing precision light curves for their host stars. Some of these systems can be complemented with data from old archives. In particular, the *Digital Access to a Sky Century Harvard* (DASCH) project allowed complementing that database with much longer timescale 100 year light curves for 261 planet host stars that have at least 10 good measurements on DASCH plates. DASCH light curves over 100 year timescales not only provide unique constraints for planet host stars but any other interesting objects in the Kepler field (Tang et al., 2013).

It is interesting to note that in 2019 the International Astronomical Union (IAU) organized the IAU100 NameExoWorlds global competition that allows any country in the world to give a popular name to a selected exoplanet and its host star. The IAU is the authority responsible for assigning official designations and names to celestial bodies, and this campaign will contribute "to the fraternity of all people with a significant token of global identity. With this aim the planetary systems for naming (composed of single stars with only one known planet orbiting around them) could be ob-

TABLE 5.1
Data volumes of different sky survey projects (Zhang and Zhao, 2015).

Sky survey project	Data volume
DPOSS (The Palomar Digital Sky Survey)	3 terabyte
2MASS (The Two Micron All-Sky Survey)	10 terabyte
GBT (Green Bank Telescope)	20 petabyte
GALEX (The Galaxy Evolution Explorer)	30 terabyte
SDSS (The Sloan Digital Sky Survey)	40 terabyte
SkyMapper Southern Sky Survey	500 terabyte
PanSTARRS (The Panoramic Survey Telescope and Rapid Response System)	~40 petabyte expected
LSST (The Large Synoptic Survey Telescope)	~200 petabyte expected
SKA (The Square Kilometer Array)	~4.6 exabyte expected

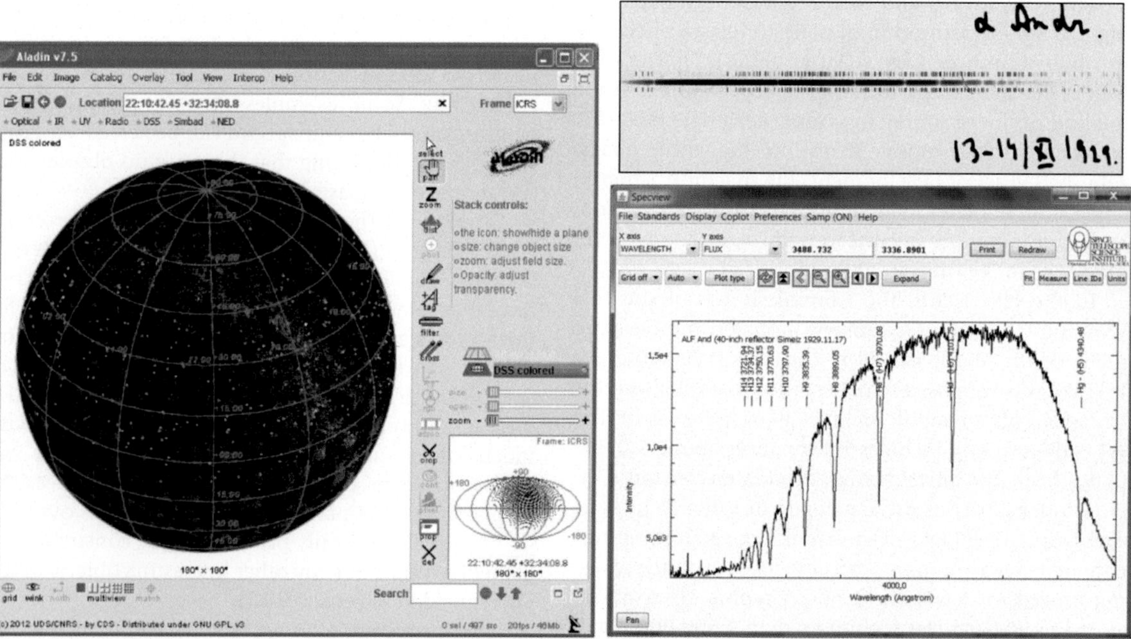

FIG. 5.15 The digitized old photographic astroplates and spectra with Aladin service. (left) Distribution of celestial objects that were observed with a 1 m reflector at the Simeiz Observatory, Crimea, from 1929 to 1941. Image credit: A. A. Shlyapnikov, UkrVO JDA (Shlyapnikov, 2007). (right) The α Andromedae spectrogram obtained with 1 m reflector at the Simeiz Observatory, Crimea, on the night of 13–14 November 1929, and an example of its interpretation by the Specview software. Image credit: Pakuliak et al. (2014). The α And with apparent visual magnitude $+2.06^m$ is actually a binary system, where the brighter of the two stars is the brightest mercury–manganese star known. Yet Ptolemy considered α And to be shared by Pegasus (δ Peg) and only in 1928 by the IAU decision the modern Andromedae constellation boundaries were fixed with its α star and the Pegasus constellation lost its δ star.

served with a small telescope in the latitude of the capital of each country were selected. In many cases, the system has a link with the assigned country, such as the facilities used and the scientists involved in the discovery of the planet. Some years of research are required to scrutinize the planetary system in order to confirm the planet's existence. For this reason, the sample is focused on exoplanets revealed during the first two decades of exoplanet exploration, that is, most of them have been discovered before about 2010," including in the Kepler mission field.[17]

Variable stars. The DASCH project allowed the discovery of three unusual long-term variables with $\sim 1^m$ variations in their light curves on $\sim$10–100 year timescales. They are all spectroscopically identified as K2III giant stars, probably in the thick disk. Their light curves do not match any previously measured for known types of variable stars, or any theoretical model reported for red giants, and instead suggest a new dust formation mechanism or the first direct observation of "short" timescale evolution-driven variability (Tang et al., 2010).

Binaries. Another field of interest of the DASCH project has been an analysis of the 100 year light curves of galactic high-mass X-ray binaries using the Harvard photographic plate collection. For the first time, the four objects of HMXBs that are currently well covered were analyzed over 100 years using optical light curves. They are the supergiant X-ray binary Cyg X-1 (V1357 Cyg) and the Be X-ray binaries 1H 1936+541 (BD+53 2262), RX J1744.7-2713 (HD 161103), and RX J2030.5+4751 (SAO 49725). This study reveals a new kind of variability, poorly studied for HMXBs (Tang et al., 2011).

Cross-identification of objects. New knowledge of the object under study can be obtained by comparing its characteristics observed in different wavelength ranges. The state-of-the-art technology allowed us to cast a glance at the universe at different wavelengths. Many new objects have been discovered at first in nonoptical ranges, and then their optical counterparts were identified; radio galaxies, quasars, pulsars, cosmic background, molecular clouds, the hot intracluster medium, ultraluminous starbursts and AGN (ULIRGs), brown dwarfs, hidden X-ray AGN, etc. (Mickaelian, 2016a; Pulatova et al., 2015; Melnyk et al., 2013). Cross-correlation of different surveys can even produce new science if new previously undiscovered unique objects are found in the process. Hence, the importance of the accurate identification of an object in various surveys follows.

Secure cross-identifications of objects require accurate positions and accurate proper motions. Proper motions not only have their own intrinsic worth in the study of stellar kinematics and galactic dynamics but with a baseline of a century or more, they provide a precise reference coordinate system for decades. For example, the astroplates give us a unique opportunity to identify the optical counterparts of the gamma ray bursts (GRBs) and study all the objects in the vicinity of the registered GRB. Because the results of GRB observations are published in the GCN Circulars in real-time, it is useful to extract lists of the faint stars in the sky areas around the GRBs within a circle with the radius of dozens of arcminutes (Beardmore et al., 2015; Vavilova et al., 2014).

The "Vanishing and Appearing Sources during a Century of Observations" (VASCO) is a cross-disciplinary project (Villarroel et al., 2019) aimed at finding some of the most unusual variable phenomena and other astrophysical anomalies with the help of existing large surveys, machine learning, and citizen science. The project is primarily centered around searches for vanishing objects in the sky and our Milky Way. Unless a star directly collapses into a black hole, there is no known reason why a star would physically entirely vanish, which makes it also interesting for searching for new exotic phenomena or even signs of technologically advanced civilizations (SETI program). The VASCO project aims to find also objects showing extreme variability on extended timescales (decades) by comparing near-century-old sky scans with modern-day astronomical surveys like PanSTARRS. For example, hypervariable AGN (Lawrence et al., 2016) was discovered by comparing two astronomical surveys separated by a 10 year time gap (see also on GRBs statistics[18]).

Comets and asteroids. Old photographic archives can be the source of earliest images of minor bodies of the Solar System, including potentially dangerous objects. The old data allow improving of the orbit on the long timescale.

For example, a catalogue of approximately 90 equatorial coordinates and stellar magnitudes of the Pluto–Charon system (Kazantseva et al., 2015) and a catalogue of 1385 astrometric positions of Saturn's satellites (Yizhakevych et al., 2017) based on the observations during 1961–1990 were extracted from the data of the Ukrainian Virtual Observatory Joint Digital Archive (UkrVO JDA). The observational material of the UkrVO JDA wide-field plates made it possible to obtain the

[17] http://www.nameexoworlds.iau.org/.

[18] https://vasconsite.wordpress.com/2018/06/21/new-questions-and-old-observations-how-the-front-line-of-astronomy-benefits-from-its-history/.

catalogue of positions and stellar magnitudes of 2162 asteroids and 11 comets, which were registered on astroplates at the time of other program observations (Shatokhina et al., 2018). It was found that the moment of observation for 54 asteroids preceded the moment of discovery of the asteroid, and observations of four of them can be considered the earliest of all known observations of these asteroids in the world. The same search was conducted at the Baldone observatory of the University of Latvia: seven images of Pluto observations and 50 positions of asteroids fainter than 14.5^m were identified in the U-band collection of the 1.2 m Schmidt telescope photographic archive obtained in 1967–1991. Among them the 11 positions for seven asteroids are the earliest in the world at the time of their observations.

5.4 MULTIWAVELENGTH GROUND-BASED AND SPACE-BORN SURVEYS, ARCHIVES, AND DATABASES

5.4.1 Gamma Ray Astronomy

The *Energetic Gamma-Ray Experiment Telescope* (EGRET), which operated on-board the Compton-Gamma Ray Observatory from April 1991 to May 2000, detected photons in the 20 MeV to 30 GeV range. The *Revised EGRET catalog* (EGR) lists positions of 188 sources, 14 of which are marked as confused, in contrast to the 271 entries of the *Third EGRET catalogue* (3EG) (Hartman et al., 1999).

The *BeppoSAX X-ray astronomy satellite* was a major program of the Italian Space Agency (ASI) with participation of the Netherlands Agency for Aerospace Programs (NIVR). The main scientific goal of the mission was to perform spectroscopic and timing studies of several classes of X-ray sources in a very broad energy band (0.1–300 keV). The satellite was launched on 30 April 1996 and observations were carried out until 30 April 2002. The *BeppoSAX catalog* is the catalogue of GRBs detected with the Gamma-Ray Burst Monitor (GRBM) aboard the BeppoSAX satellite (Boella et al., 1997). It includes 1082 GRBs with 40–700 keV energy in the range from 1.3×10^{-7} to 4.5×10^{-4} erg/cm^2, and 40–700 keV peak fluxes from 3.7×10^{-8} to 7.0×10^{-5} erg/cm^2/s. Some relevant parameters of each GRB are reported in the catalogue.

The *INTErnational Gamma-Ray Astrophysics Laboratory* (INTEGRAL) is a currently operational space telescope for observing gamma rays. It was launched by the European Space Agency into Earth orbit in 2002, and it is designed to detect some of the most energetic radiation that comes from space. It was the most sensitive gamma ray observatory in space before NASA's Fermi was launched in 2008. It has had some notable successes, for example in detecting a mysterious "iron quasar." It has also had success in investigating GRBs and evidence for black holes. The INTEGRAL Reference Source Catalogue (Bird et al., 2010) contains all sources that have been brighter than 0.1 mCrab above 1 keV. It also includes (flagged separately) a number of high-energy (>10 keV) sources candidate for detection. The last version of the catalogue includes 2394 sources, of which 1213 were detected by IBIS/ISGRI and 189 by JEM-X.

The *Fermi Gamma-ray Space Telescope* (originally called the Gamma-Ray Large Area Space Telescope (GLAST) and renamed for Enrico Fermi, an outstanding American physicist of Italian origin) was launched on 11 June 2008. The main task is to study the high-energy universe: gamma ray sources and diffuse emission, pulsars, active galactic nuclei, the Galactic Center, energetic binary star systems and novae, supernova remnants, solar flares and terrestrial gamma ray flashes, etc. One of the important results obtained with the Fermi Large Area Telescope (LAT) issued in January, 2017, was the detection of high-redshift ($z > 3$) blazars: light from the most distant object, NVSS J151002+570243 ($z = 4.31$), was emitted when the universe was 1.4 billion years old (Fig. 5.16, bottom).

During the Fermi-GLAST operation several catalogues were created, for example, the *2nd Fermi-LAT Gamma-Ray Pulsar Catalog*, which consists of 117 high-confidence ≤ 0.1 GeV gamma ray pulsar detections (half are neutron stars) (Abdo et al., 2013) and the *Fermi LAT Third Source Catalog* (Fermi 3FGL) of 3033 sources above 4 σ significance in the 100 MeV–300 GeV range (Acero et al., 2015). A useful tool is "The Fermi All-sky Variability Analysis" (Fermi FAVA) to systematically study the variability of the gamma ray sky measured by the Fermi GLAST LAT, which lists 215 flaring gamma ray sources, including the 27 possibly extragalactic sources found at galactic latitudes $\pm 10°$ (Ackermann et al., 2013).

5.4.2 X-Ray Astronomy

The first X-ray satellite was launched by NASA in 1970 and was called X-ray Explorer, but later was renamed Uhuru. Uhuru operated in orbit for 2.5 years, and sent a lot of significant scientific information. Uhuru detected 339 X-ray sources, including the famous source in the constellation Cygnus which later became the first contender in the history of astronomy for the role of the black hole (Cyg-1). The first X-ray telescope was launched in November 1978 and was called High Energy Astronomy Observatory-2 (later renamed Einstein). Einstein operated in the 0.2–20.0 keV energy

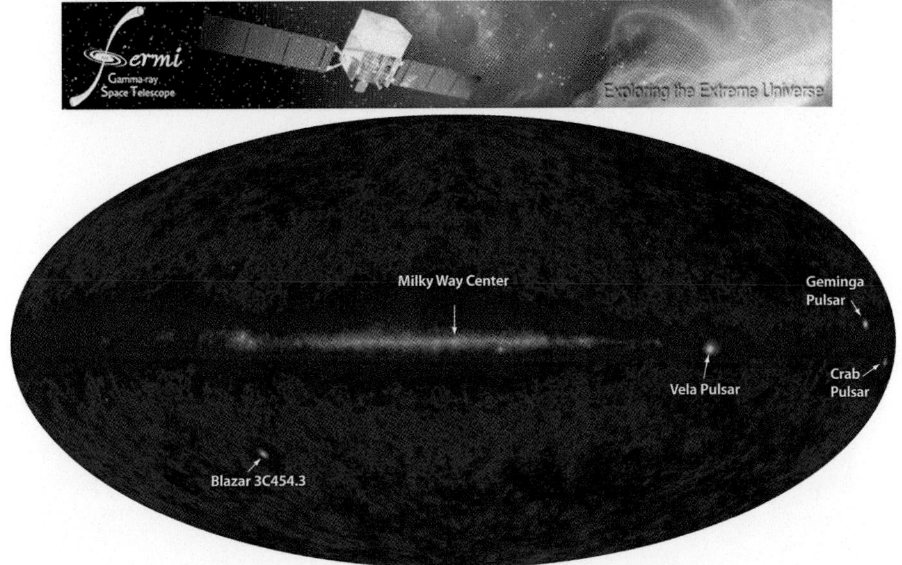

FIG. 5.16 (Top) The Fermi/LAT telescope. (Bottom) All-sky view from GLAST reveals bright gamma-ray emission in the plane of the Milky Way (center), bright pulsars and super-massive black holes, including in the 3C454.3 blazar. NASA/DOE/International LAT Team (public domain).

band with about 5 arcsec resolution. Einstein made high-quality spectroscopic images of supernova remnants, and discovered a lot of very weak extragalactic X-ray sources. During the 1980s, many X-ray satellites and telescopes were launched, e.g., ROSAT, SAX, ASCA, etc. Currently, the third generation of X-ray space telescopes are monitoring the sky.

The *EXOSAT* satellite was the European Space Agency mission launched on 26 May 1983. The mission lifetime was ultimately limited by orbital decay, which was projected to occur within 3 years. On 9 April 1986 a failure in the attitude control system caused the loss of the spacecraft. The natural decay of the orbit caused EXOSAT to reenter on 6 May 1986. The three instruments on-board the EXOSAT observatory (LE, ME, GSPC) were complementary and were designed to give complete coverage over a wide energy band pass of 0.05–50 keV. The *Slew Survey Catalog*, EXMS, in 1–8 keV, gives 1210 sources including 992 identified ones (Warwick et al., 1988).

The *Roentgen Satellite* ROSAT, a Germany/US/UK collaboration, was launched on 1 June 1990 and operated for almost 9 years. The first 6 months of the mission were dedicated to the all-sky survey (using the Position Sensitive Proportional Counter detector) followed by the pointed phase. The survey obtained by ROSAT was the first X-ray and XUV all-sky survey using an imaging telescope with an X-ray sensitivity of about a factor of 1000 better than that of the Uhuru satellite. The total number of new X-ray sources discovered by ROSAT is larger than 150,000, which is more than a factor of 20 larger compared with the number of X-ray sources known before ROSAT (Voges et al., 1999). The source lists were produced automatically by the Standard Analysis Software System (SASS). The problem is that ROSAT sources are difficult to identify due to inaccurate positions errors ($\sim 1'$). The ROSAT consists of a number of catalogues, including the First-1RXP, WGACAT, ROSHRICAT, ROSAT All-Sky Surveys maps (Truemper, 1992), and RASS-BSC-1RXS. For example, ROSAT FSC gives X-ray positions and 0.07–2.4 keV photon counts and two hardness ratios for 105,924 sources (Voges et al., 2000). The image of the catalogue of bright sources is presented in Fig. 5.17.

ASCA (formerly named Astro-D) is Japan's fourth cosmic X-ray astronomy mission. The satellite was successfully launched on 20 February 1993. The first eight months of the ASCA mission were devoted to performance verification. Having established the quality of performance of all ASCA's instruments, the project changed to a general/guest observer for the remainder of the mission. The ASCA satellite made 107 pointing observations on a 5×5 deg^2 region around the center of our Milky Way (the Galactic Center) from 1993 to

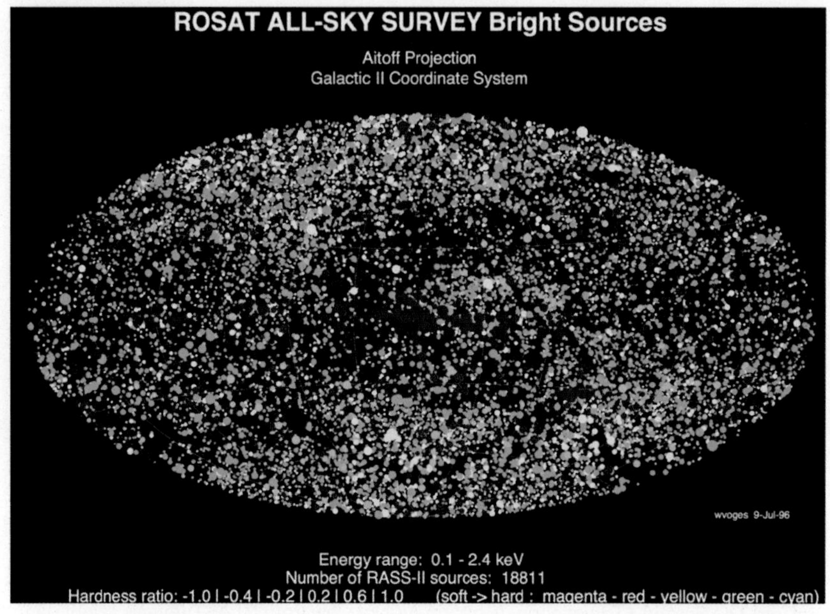

FIG. 5.17 The map of ROSAT bright sources. Image credit: official ROSAT web page.

1999 (Sakano et al., 2001). The database includes data on 1190 sources.

The *X-ray Multi-Mirror Mission*, XMM-Newton, was launched by the European Space Agency in December 1999. XMM-Newton is named in honor of Sir Isaac Newton. To date, it is one of the most popular and successful X-ray astronomy missions. The main aims of the XMM-Newton Observatory are to observe accurately faint point and extended sources, to conduct deep surveys of the sky in X-rays with the possibility to obtain images and spectra of X-ray sources simultaneously. XMM-Newton includes both X-ray and optical/ultraviolet telescopes. The XMM-Newton has observed more than 500,000 sources (Rosen et al., 2016). The *XMM-Newton Science Archive* (XSA) provides simple and flexible access to data from the XMM-Newton mission, including searches on the associated EPIC and OM catalogues. The XSA has been developed for the XMM-Newton project by the European Space Agency Science Data Center (ESDC) with requirements provided by the XMM-Newton Science Operations Center. An image of the *Third XMM-Newton source catalogue* is given in Fig. 5.18.

We underline once again that X-ray surveys constitute an important part of wide-field surveys of extragalactic objects. The weak absorption of keV photons in the intergalactic medium makes observations of distant sources via X-ray possible. The most prominent observational results have been achieved by the XMM-Newton and Chandra space-based observatories. The bulk of all detected objects in X-ray are point-like sources, mainly AGNs. Above 5% are the extended sources: clusters of galaxies and nearby galaxies. AGNs can be detected over a wide range of redshifts, and they are good tracers of the large-scale structure of the universe as well as a tool for cosmological probes. We give some examples below.

The most famous and the largest extragalactic survey performed by the ESO XMM-Newton X-ray telescope is the *XMM-XXL survey* over 50 deg^2. The main goal of 10 years of projects was to discover hundreds of galaxy clusters and tens of thousands of AGNs. XXL survey is composed with two equal-size fields in different parts of the hemisphere. Observations were made in the soft (0.5–2 keV) and hard (2–10 keV) bands. Typical expositions of individual 30 arcmin pointing of the EPIC camera were 10–20 ksec. The sensitivity limits are near 10^{-15} and 3×10^{-15} erg s^{-1} cm^{-2} for the soft and hard bands, respectively (Pierre et al., 2016). A prototype of the XXL survey was a pseudocontiguous 4.2 deg^2 survey centered on RA = 36°, Dec = −4° (Gandhi et al., 2006). These authors found a weak positive correlation signal for ~1130 AGNs at angular scale 6.3 arcsec. Extension of this field was 11.2 deg^2 (*XMM-LSS survey*) (Elyiv et al., 2012). Rich statistics of AGNs in this field allowed finding significant differences in correlation properties of AGNs with different hardness ratios. It was shown

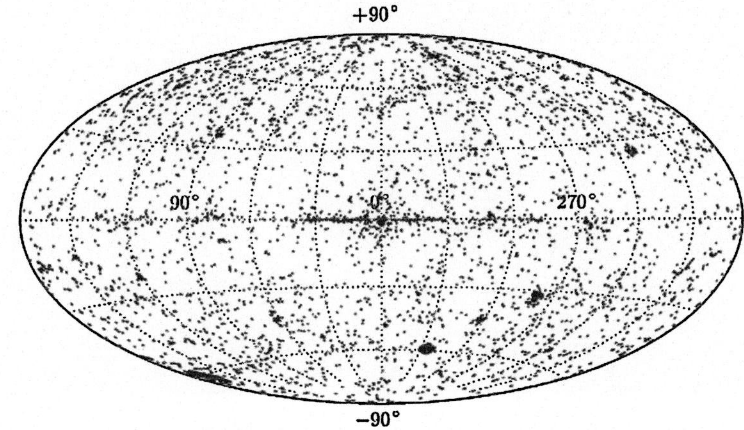

FIG. 5.18 The XMM-Newton serendipitous survey. The third XMM-Newton source catalogue. Image credit: Rosen et al. (2016).

that the sources with a hard spectrum are more clustered than soft-spectrum ones. Therefore, the two main types of AGNs populate different environments. Studies of the overdensity of X-ray AGNs in 33 galaxy clusters with $z < 1.05$ from the XMM-LSS field were presented by Koulouridis et al. (2014). At the same time, Melnyk et al. (2013) have shown that X-ray sources in the XMM-LSS field mainly reside in overdense regions and X-ray-selected galaxies are in higher galaxy environments. As concerns X-ray-selected galaxy clusters, the correlations between their optical and X-ray properties were examined, and peculiar compact structures were found (very distant clusters and fossil groups) in the XMM-LSS field (Adami et al., 2011). The *XXL Galaxy clusters catalogue* was published by Pacaud et al. (2016). It contains 100 of the brightest galaxy clusters in X-ray in redshift range from $z = 0.1$ to 0.5 and gives strong evidence of large-scale structures in the XXL survey. It was concluded that the obtained angular density of clusters is below that predicted by the *WMAP9* model and significantly below the *Planck 2015* cosmology model.

The *Chandra X-ray observatory* was launched on 23 July 1999. Chandra has the highest resolution (0.492 arcsec), a better spatial resolution ($<1''$) than all past and current X-ray telescopes. The observatory covers the 0.3–10.0 keV energy range, corresponding to soft X-rays. An example of a Chandra X-ray image is given in Fig. 5.19. The *Chandra Source Catalog* (CSC) is ultimately intended to be the definitive catalogue of X-ray sources detected by the Chandra X-ray Observatory (Evans et al., 2010). Additionally, a lot of electronic databases are given in the *VisieR* online library (Babyk et al., 2014), which includes huge samples of already analyzed data. For example, *ACCEPT*[19] and *REJECT*[20] are databases of galaxy clusters, groups, and elliptical galaxies.

The *Nuclear Spectroscopic Telescope Array* (NuSTAR) mission has deployed the first orbiting telescopes to focus light in the high-energy X-ray (3–79 keV) range of the electromagnetic spectrum. Our view of the universe in this spectral window has been limited because previous orbiting telescopes have not employed true focusing optics, but rather have used coded apertures that have intrinsically high backgrounds and limited sensitivity. The observatory provides a combination of sensitivity, spatial, and spectral resolution factors of 10 to 100 improved over previous missions that have operated at these X-ray energies. An image from the NuStar catalogue is given in Fig. 5.20.

The main task of the *Swift observatory* is to observe GRBs and their afterflares. The Burst Alert Telescope (BAT) is intended for monitoring of GRBs in the energy range 15–150 keV and has a wide field of view (about 1.4 sr). The XRT uses different operating modes to observe the sources. Depending on the brightness of the source, X-ray observation can be performed using Imaging mode (IM), Photo-diode (PD) mode, Windowed Timing (WT) mode, or Photon Counting (PC) mode. An image of the *Swift survey* is presented in Fig. 5.21.

5.4.3 Ultraviolet Astronomy

The *Extreme Ultraviolet Explorer* (EUVE) was a space telescope for ultraviolet astronomy, launched on 7 June

[19] https://web.pa.msu.edu/astro/MC2/accept/.
[20] https://ancient-sands-40156.herokuapp.com/.

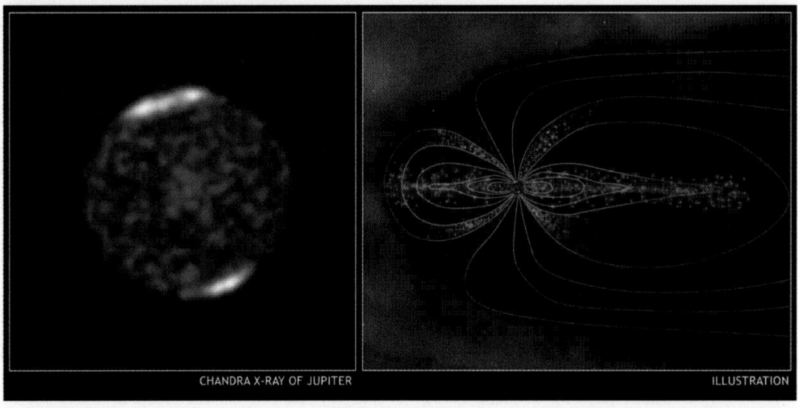

FIG. 5.19 Jupiter shows intense X-ray emission associated with auroras in its polar regions (Chandra image on the left). Extended monitoring by Chandra showed that the auroral X-rays (illustration on the right) are caused by highly charged particles crashing into the atmosphere above Jupiter's poles (public domain, http://chandra.harvard.edu/).

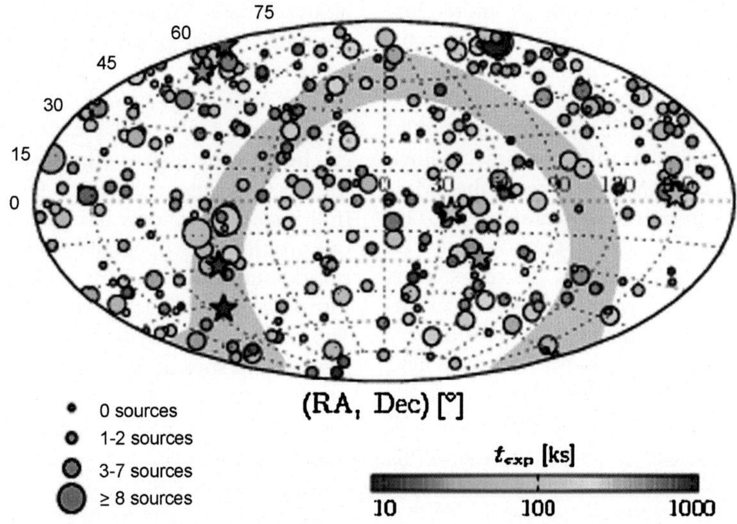

FIG. 5.20 The NuStar survey: 40 month observation catalogue. Image credit: official NuStar web page.

1992. With instruments for ultraviolet radiation between wavelengths of 7 and 76 nm, the EUVE was the first satellite mission especially for the short-wave ultraviolet range (Fig. 5.22). The satellite compiled an all-sky survey of 801 astronomical targets before being decommissioned on 31 January 2001.

The *Far Ultraviolet Spectroscopic Explorer* (FUSE), 1999, was a NASA astrophysics satellite/telescope, whose purpose was to explore the universe using the technique of high-resolution spectroscopy in the far-ultraviolet spectral region.[21] Over 8 years of operations, FUSE acquired over 6000 observations of nearly 3000 separate astronomical targets. All archived data are now public and no longer require user registration. Astronomers used FUSE to observe a tremendous range of object types, from planets and comets in our Solar System to hot and cool stars in our Milky Way and nearby galaxies, and even to distant active galaxies and quasars. However, FUSE's real claim to fame was its abil-

[21] http://fuse.pha.jhu.edu.

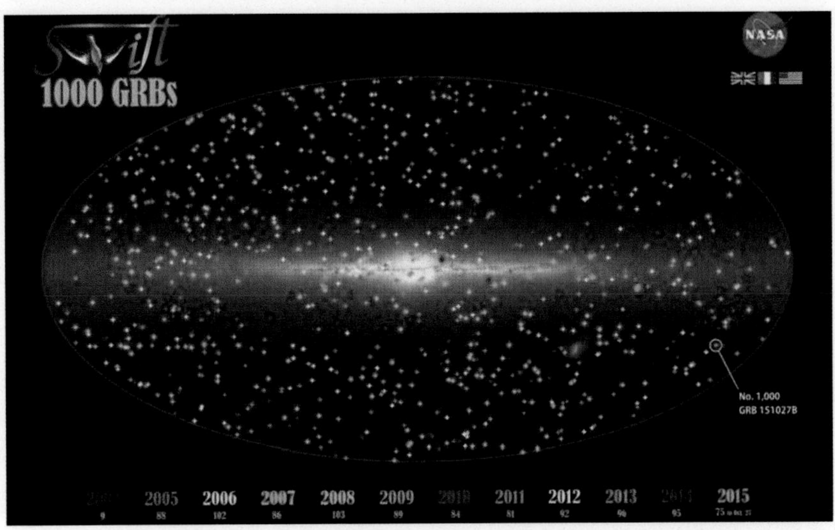

FIG. 5.21 The Swift survey. Image credit: official Swift web page.

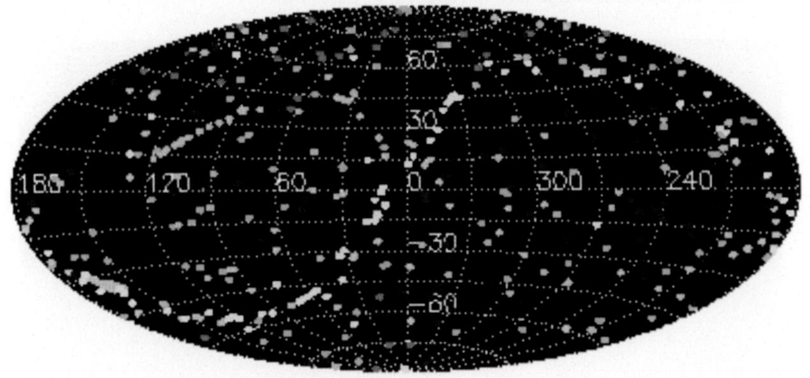

FIG. 5.22 EUVE-collected data of over 350 unique astronomical targets. Image credit: NASA.

ity to sense and diagnose the physical conditions in the tenuous regions of interstellar and intergalactic space, regions that are often considered to be empty.

The *Galaxy Evolution Explorer* (GALEX) is an orbiting space telescope observing galaxies in ultraviolet light across 10 billion years of cosmic history. A Pegasus rocket launched GALEX into orbit on 28 April 2003. Although originally planned as a 29-month mission, the NASA Senior Review Panel in 2006 recommended that the mission lifetime be extended.[22] Surveys *GALEX AIS* and *GALEX MIS* contain 65,266,291 sources and 12,597,912 sources, respectively. There 200 rms accurate positions and fluxes in FUV 1528Å and NUV 2271Å bands are obtained for about 78 million sources over the whole sky to 20.8^m for AIS and 22.7^m for MIS surveys (Bianchi et al., 2011) (see Fig. 5.23).

[22]http://www.galex.caltech.edu/, http://galex.stsci.edu/GR6/.

5.4.4 Mid- and Far-Infrared Astronomy

The *Infrared Astronomical Satellite* (IRAS) was the first-ever space telescope intended for a full-sky survey at infrared wavelengths. IRAS was launched on 25 January 1983. During its mission, which lasted ten months, over 250,000 infrared sources were observed, many of which are still awaiting identification. Among the identified ones are normal stars with dust around them, starburst galaxies, and objects of the Milky Way's core as well as the discovered asteroids and comets (see, for example, Fig. 5.24).

FIG. 5.23 Hot stars burn brightly in this new image from NASA's Galaxy Evolution Explorer, showing the ultraviolet side of a familiar face. Image credit: NASA/JPL-Caltech.

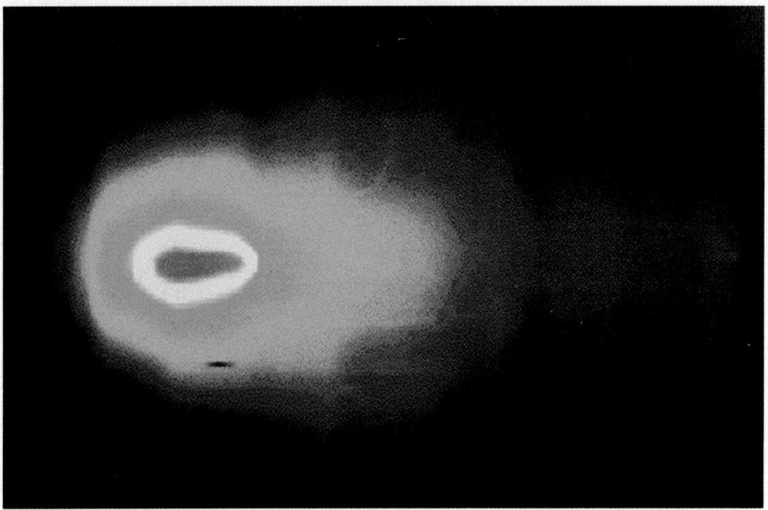

FIG. 5.24 False color image of a long-period comet IRAS-Araki-Alcock (C/1983 H1) discovered by IRAS in 1983.

The *InfraRed Astronomical Satellite Point Source Catalog* (IRAS PSC) and the *Faint Source Catalog* (IRAS FSC) contain positions and fluxes in 12 μm (0.4 Jy sensitivity) and 25 μm (0.5 Jy) bands for 245,889 sources and 173,044 sources (Moshir et al., 1990), respectively (see, also, the IRAS joint catalogue of 345,162 sources by Abrahamyan et al. (2015)). Among other IRAS catalogues we mention IRAS SSSC by Helou and Walker (1985) and the Catalogue of Infrared Observations by Gezari et al. (1987); the *AKARI-IRC Point Source Catalogue* contains data on 0.300 rms accurate positions and 9 μm (50 mJy sensitivity) and 18 μm (120 mJy sensitivity) fluxes for 870,973 sources over the whole sky (Ishihara et al., 2010).

The *Wide-field Infrared Survey Explorer* (WISE) includes data on the 0.500 rms accurate positions and 3.4 μm-band (80 μJy sensitivity), 4.6 μm (110 μJy), 12 μm (1 mJy), 22 μm (6 mJy) photometry for

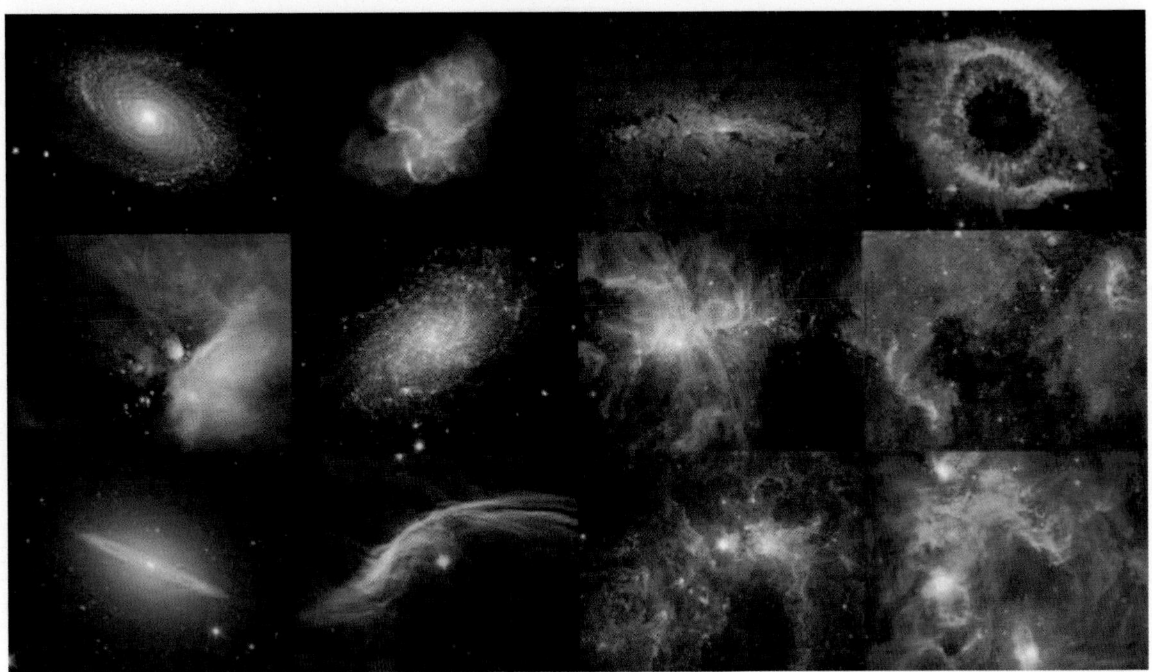

FIG. 5.25 NASA's Spitzer Space Telescope is celebrating 12 years in space with a new digital calendar of 2015 (public domain: NASA/JPL-Caltech, http://www.jpl.nasa.gov/images/spitzer/20150820/Spitzer12thAnniversaryCalendar.pdf).

563,921,584 sources over the whole sky (Cutri et al., 2012).

The *Spitzer Space Telescope* (SST), formerly the Space Infrared Telescope Facility (SIRTF), was launched in 2003 and is planned to be retired in January 2020. This Sun orbiting telescope is the last of four space telescopes in NASA's Great Observatories program (following the Hubble Space Telescope [1990], the Compton Gamma Ray Observatory [1991], and the Chandra X-ray Observatory [1999]). The SST has an Infrared Array Camera (IRAC), Infrared Spectrograph (IRS), and Multiband Imaging Photometer for Spitzer (MIPS) (at this time, only the two shortest-wavelength modules of the IRAC camera are still operable as the Spitzer Warm Mission).

The *Galactic Legacy Infrared Mid-Plane Survey Extraordinaire* (GLIMPSE) is a survey of approximately 444,000 images covering 300° of our Galaxy taken with the SST IRAC at four separate wavelengths. The other *MIPSGAL* is a similar survey covering 278° of the galactic disk at longer wavelengths. The best SST images of the nearest and distant Universe, which were released in a digital calendar to mark the occasion of SST's 12 year celebration and highlight some of the most important discoveries, are given in Fig. 5.25.

The *Herschel Space Observatory* was a space observatory built and operated by the European Space Agency. It was active from 2009 to 2013, and was the largest infrared telescope ever launched, carrying a 3.5 meter (11.5 ft) mirror and instruments sensitive to the far infrared and submillimeter wavebands (55–672 μm). Herschel was the fourth and final cornerstone mission in the Horizon 2000 program, following SOHO/Cluster II, XMM-Newton, and Rosetta. NASA was a partner in the Herschel mission, with US participants contributing to the mission, providing mission enabling instrument technology and sponsoring the NASA Herschel Science Center (NHSC) at the Infrared Processing and Analysis Center and the Herschel Data Search at the *Infrared Science Archive.*

5.4.5 Submillimeter/Millimeter Astronomy

The *Planck space observatory* was operated by the European Space Agency from 2009 to 2013. It has mapped the anisotropies of the cosmic microwave background (CMB) at microwave and infrared frequencies, with high sensitivity and small angular resolution (Planck Collaboration, 2011). The mission substantially improved upon observations made by the NASA Wilkin-

FIG. 5.26 A full-sky map produced by the Wilkinson Microwave Anisotropy Probe (WMAP) showing cosmic background radiation, a very uniform glow of microwaves emitted by the infant universe more than 13 billion years ago. Color differences indicate tiny fluctuations in the intensity of the radiation, a result of tiny variations in the density of matter in the early universe. According to inflation theory, these irregularities were the "seeds" that became the galaxies. WMAP's data support the big bang and inflation models.

son Microwave Anisotropy Probe (WMAP). Planck provided a major source of information relevant to several cosmological and astrophysical issues, such as testing theories of the early universe and the origin of cosmic structure. Since the end of its mission, Planck has defined the most precise measurements of several key cosmological parameters, including the average density of ordinary barionic matter and dark matter in the universe and the age of the universe.

The *Wilkinson Microwave Anisotropy Probe* (WMAP) is named after Dr. David Wilkinson, a member of the science team and pioneer in the study of cosmic background radiation. The science goals of the WMAP broadly dictate that the relative CMB temperature be measured accurately over the full sky with high angular resolution and sensitivity (Gold et al., 2011). The overriding priority in the design was the need to control systematic errors in the final maps. The specific goal of WMAP is to map the relative CMB temperature over the full sky with an angular resolution of at least 0.3°, a sensitivity of 20 μK per 0.3° square pixel, with systematic artifacts limited to 5 μK per pixel (Fig. 5.26). To achieve these goals, WMAP used differential microwave radiometers that measure temperature differences between two points on the sky.

"Data from WMAP showed temperature variations of 0.0002 K caused by intense sound waves echoing through the dense early universe, about 380,000 years after the big bang. This anisotropy hinted at density variations where matter would later coalesce into the stars and galaxies that form today's universe. WMAP determined the age of the universe to be 13.8 billion years. WMAP also measured the composition of the early, dense universe, showing that it started at 63 percent dark matter, 12 percent atoms, 15 percent photons, and 10 percent neutrinos. As the universe expanded, the composition shifted to 23 percent dark matter and 4.6 percent atoms. The contribution of photons and neutrinos became negligible, while dark energy, a poorly understood field that accelerates the expansion of the universe, is now 72 percent of the content. Although neutrinos are now a negligible component of the universe, they form their own cosmic background, which was discovered by WMAP. WMAP also showed that the first stars in the universe formed half a billion years after the big bang. The European Space Agency's Planck satellite, which was launched in 2009, is designed to map the CMB in even greater detail than WMAP" (cited by https://www.britannica.com/topic/Wilkinson-Microwave-Anisotropy-Probe).

ALMA. The National Radio Astronomy Observatory (NRAO) in the United States conducts a project called the *Millimeter Array* (MMA), which includes the construction of 40 antennas with 8 meter diameter, reaching an atmospheric window of 30 to 350 GHz to receive millimeter wavelengths.

5.4.6 Centimeter/Meter/Decameter Radio Astronomy

NRAO VLA Sky Survey (NVSS) is a 1.4 GHz continuum survey covering the entire sky north of −40° declination (Condon and Kaplan, 1998).

Faint Images of the Radio Sky at Twenty centimeters (FIRST) (Helfand et al., 2015) is a project designed to produce the radio equivalent of the Palomar Observatory Sky Survey over 10,000 deg^2 of the North and South Galactic Caps. Using the NRAO Very Large Array (VLA) and an automated mapping pipeline, we produce images with 1.8″ pixels, a typical rms of 0.15 milliJanskies (mJy), and a resolution of 5″. At the 1 mJy source detection threshold, there are ∼90 sources per deg^2, ∼35% of which have a resolved structure on scales from 2–30″; 30% of the sources have counterparts in the Sloan Digital Sky Survey.

A source catalogue including peak and integrated flux densities and sizes derived from fitting a two-dimensional Gaussian to each source is generated from the coadded images. The astrometric reference frame of the maps is accurate to 0.05″, and individual sources have 90% confidence error circles of radius < 0.5″ at the 3 mJy level and 1″ at the survey threshold. Approximately 15% of the sources have optical counterparts at the limit of the POSS I plates (E∼20.0); unambiguous optical identifications (< 5% false rates) are achievable to mv∼24. The survey area has been chosen to coincide with that of the SDSS; at the mv∼23 limit of SDSS, ∼40% of the optical counterparts to FIRST sources were detected.

The *Sydney University Molonglo Sky Survey* (SUMSS) (Mauch et al., 2013). The Molonglo Observatory Synthesis Telescope operating at 843 MHz with a 5 deg^2 field of view is carrying out a radio imaging survey of the sky South of declination −30°. This survey produces images with a resolution of 45″ × 45″ cosec δ and an rms noise level of ∼1 mJy beam^{-1}. The SUMSS is therefore similar in sensitivity and resolution to the Northern NRAO VLA Sky Survey. The survey is progressing at a rate of about 1000 deg^2 per year, yielding individual and statistical data for many thousands of weak radio sources.

The *7 Cambridge Survey* (7C) of radio sources was performed by the Cavendish Astrophysics Group using the Cambridge Low-Frequency Synthesis Telescope at Mullard Radio Astronomy Observatory (Hales et al., 2007).

The *Westerbork Northern Sky Survey* (WENSS) is a low-frequency radio survey that covers the whole sky North of declination +30° at a $\lambda = 92$ cm to a limiting flux density of approximately 18 mJy at the 5σ level (de Bruyn, 1996). WENSS is a collaboration between the Netherlands Foundation for Research in Astronomy (NFRA/ASTRON) and the Leiden Observatory. The version of the WENSS Catalog as implemented at the HEASARC is a union of two separate catalogues obtained from the WENSS web site: the WENSS Polar Catalog (18,186 sources above +72° declination) and the WENSS Main Catalog (211,234 sources in the declination region from +28° to +76°).

The *Ukrainian T-shape Radiotelescope* (UTR-2) is the largest in the world array operated at decametric wavelength (Fig. 5.27, right). It was constructed in the early 1970s at the Radio astronomy observatory near the village of Hrakovo, 15 km West-South-West from Shevchenkove, Kharkiv region, Ukraine (now S. Ya. Braude Radioastronomical Observatory, subdivision of the Institute of Radio Astronomy of the NAS of Ukraine).

At present the UTR-2 is the greatest and the most perfect tool at the decameter band of the wavelengths. Its characteristic features are the great linear sizes (2 × 1 km) and the effective area (150,000 m^2), the high directivity (the beamwidth is 25′ at the frequency 25 MHz), the low level of sidelobes, and the wide frequency range (8–35 MHz), the many-beaming, the electronic control of the beam at the wide sector in both coordinates, the great dynamic range, the interference stability, and the reliability in operation. Combination of the high efficiency of the radio telescope UTR-2 and the new registration facilities allow it to carry out a large volume of investigations in different programs obtaining new results and discoveries. Most of the universe's objects from the nearest vicinity of the Earth, the Solar System, and our Galaxy to the most distant radio galaxies and quasars are accessible for studies by the decameter radio astronomy methods. The telescope is a part of the *Ukrainian Radio Interferometer of NASU* (URAN), decametric VLBI system, which includes another four significantly smaller low-frequency radio telescopes with a base from 40 to 900 km.

During 25 years (1978–2002) the most complete catalogue *Very-low frequency sky survey of discrete sources of the Northern sky* has been obtained with the UTR-2 radio telescope at a number of the lowest frequencies used in contemporary radio astronomy within the range from 10 to 25 MHz (Braude et al., 2007). The analysis of the statistical complete survey of about 4000 extragalactic sources allows for the determination and classification of their radio spectra, to reveal great variations of the low-frequency spectral indices, and to identify the catalogued objects. It is available at http://rian.kharkov.ua/index.php/ru/decameter-catalogs/95-onr/research/151-utr2. One of the remarkable discoveries made with the

FIG. 5.27 The largest ground-based radio low-frequency arrays in the world. (Left) LOFAR, The Netherlands. (Right) UTR-2, Ukraine.

UTR-2 radiotelescope already in 1978 was the detection of carbon radio recombination lines with quantum numbers $n > 600$ at $\nu < 30$ MHz in the direction of Cas A (Konovalenko and Sodin, 1981). This discovery of highly excited carbon at extremely low frequencies alongside with other observation methods of ionized carbon (in the ultraviolet and infrared ranges) opened a new window for studying characteristics of low-density interstellar plasma.

The *Low-Frequency Array* (LOFAR) is the largest radio telescope network, located mainly in the Netherlands, completed in 2012 by "ASTRonomisch Onderzoek in Nederland" (ASTRON). To make radio surveys of the sky with adequate resolution, the antennas are arranged in clusters that are spread out over an area of more than 1000 km in diameter (Fig. 5.27, left). The LOFAR stations in The Netherlands reach baselines of about 100 km. LOFAR currently receives data from 24 core stations (in Exloo), 14 "remote" stations in the Netherlands, and 12 international stations.

It has prepared, or is in the process of carrying out, a number of sky surveys: the *LOFAR Multifrequency Snapshot Sky Survey* (MSSS) was the first wide-area Northern-sky survey to be carried out with LOFAR (Heald et al., 2015). It surveyed the whole Northern sky to a depth of around 10 mJy/beam with a resolution of 2 arcmin. The ongoing *LOFAR HBA 120–168 MHz wide-area survey* is referred to as the *LOFAR Two-metre Sky Survey* (LoTSS) (Shimwell et al., 2017). It aims to survey the whole Northern sky at the full resolution of the Dutch LOFAR (6 arcseconds) and a declination-dependent sensitivity which will typically be around 100 μJy/beam. This is a factor 10 more sensitive than the current most high-resolution sky survey, FIRST, and will detect over 10 million radio sources, mostly star forming galaxies, but with a large proportion of active galactic nuclei. A long-term goal is to image large areas with the resolution of the International LOFAR Telescope, around 0.5 arcsec, which will give a combination of resolution and sensitivity that will be unsurpassed well into the SKA era. Data from the parts of the wide-area survey that have currently been made public are available to download through the data release page https://www.lofar-surveys.org/surveys.html. *WEAVE-LOFAR* is a spectroscopic survey of the LOTSS sky with the WEAVE fiber spectrograph on the William Herschel Telescope. It will provide redshifts for the wide and deep tiers of LOTSS.

Useful additional information related to the main data for the most important all-sky and large-area astronomical surveys providing multiwavelength photometric data is given in Fig. 5.28 (originally presented by Mickaelian, 2016b).

5.5 MULTIWAVELENGTH DATA ARCHIVES

We presented briefly several important catalogues, databases, and archives of astronomical data in their historical retrospective. One can see that each of these astroinformation resources obtained with the help of ground-based and space-born telescopes has its own web portal, allowing open access for users to get the observational archive. Instead of concluding the chapter, we here give information on some very useful archives of joined astronomical databases, which were not mentioned in this chapter. They allow the user to get data

Survey, catalogue	Years	Spectral range	Sky area (deg²)	Sensitivity (mag/mJy)	Number of sources	Density (obj/deg²)
Fermi-GLAST	2008–2014	10 MeV–100 GeV	All-sky		3033	0.07
CGRO	1991–1999	20 keV–30 GeV	All-sky		1300	0.03
INTEGRAL	2002–2014	15 keV–10 MeV	All-sky		1126	0.03
ROSAT BSC	1990–1999	0.07–2.4 keV	All-sky		18,806	0.46
ROSAT FSC	1990–1999	0.07–2.4 keV	All-sky		105,924	2.57
GALEX AIS	2003–2012	1344–2831Å	21,435	20.8 mag	65,266,291	3044.85
APM	2000	opt b, r	20,964	21.0 mag	166,466,987	7940.61
MAPS	2003	opt O, E	20,964	21.0 mag	89,234,404	4256.55
USNO-A2.0	1998	opt B, R	All-sky	21.0 mag	526,280,881	12,757.40
USNO-B1.0	2003	opt B, R, I	All-sky	22.5 mag	1,045,913,669	25,353.64
GSC 2.3.2	2008	opt j, V, F, N	All-sky	22.5 mag	945,592,683	22,921.79
Tycho-2	1989–1993	opt BT, VT	All-sky	16.3 mag	2,539,913	61.57
SDSS DR12	2000–2014	opt u, g, r, i, z	14,555	22.2 mag	932,891,133	64,094.20
DENIS	1996–2001	0.8–2.4 μm	16,700	18.5 mag	355,220,325	21,270.68
2MASS PSC	1997–2001	1.1–2.4 μm	All-sky	17.1 mag	470,992,970	11,417.46
2MASS ESC	1997–2001	1.1–2.4 μm	All-sky	17.1 mag	1,647,599	39.94
WISE	2009–2013	3–22 μm	All-sky	15.6 mag	563,921,584	13,669.83
AKARI IRC	2006–2008	7–26 μm	38,778	50 mJy	870,973	22.46
IRAS PSC	1983	8-120 μm	39,603	400 mJy	245,889	6.21
IRAS FSC	1983	8–120 μm	34,090	400 mJy	173,044	5.08
IRAS SSSC	1983	8–120 μm	39,603	400 mJy	16,740	0.42
AKARI FIS	2006–2008	50–180 μm	40,428	550 mJy	427,071	10.56
Planck	2009–2011	0.35–10 mm	All-sky	183 mJy	33,566	0.81
WMAP	2001–2011	3–14 mm	All-sky	500 mJy	471	0.01
GB6	1986–1987	6 cm	20,320	18 mJy	75,162	3.70
NVSS	1998	21 cm	33,827	2.5 mJy	1,773,484	52.43
FIRST	1999–2015	21 cm	10,000	1 mJy	946,432	94.64
SUMSS	2003–2012	36 cm	8,000	1 mJy	211,050	26.38
WENSS	1998	49/92 cm	9,950	18 mJy	229,420	23.06
7C	2007	198 cm	2,388	40 mJy	43,683	18.29

FIG. 5.28 Main data for the most important all-sky and large-area astronomical surveys providing multiwavelength photometric data. Catalogues are given in the order of increasing wavelengths (Open Astronomy, Mickaelian, 2016b).

on chosen celestial bodies simultaneously in all multiwavelength ranges as well as to have software tools for making simple astronomical calculations *in situ* of such archives.

The *Planetary Data System* (PDS) is a long-term archive of digital data products returned from NASA's planetary missions, and from other kinds of flight and ground-based data acquisitions.[23] It includes such key nodes as atmospheres, geosciences, cartography and imaging sciences, navigational and ancillary information, planetary plasma interactions, ring–moon systems, and small bodies. We mention below several surveys and archives.

Exoplanets.

KEPLER. The NASA's Kepler space telescope was in operation from 7 March 2009 until 30 October 2018 (active operation from 2009 through 2013) and was designed to discover Earth-size exoplanets in or near habitable zones of the Earth's region of our Galaxy. The construction of a space-born photometer and the embedded planet-transit-star method, the only exoplanets whose orbits are seen edge-on from Earth can be detected. Totally, Kepler observed 530,506 stars and discovered 5000 more exoplanet candidates and 2662 confirmed exoplanets. NASA's Kepler space telescope team has released a mission catalogue of planet candidates, which is the most comprehensive and detailed catalogue release of candidate exoplanets (planets outside our Solar System) (see the NASA Exoplanet Archive).[24] The final Kepler catalogue will serve as a basis to determine the prevalence and demographics of planets in our Galaxy, while the discovery of the two distinct planetary populations shows that about half the planets we know either have no surface, or lie beneath a deep, crushing atmosphere – an environment unlikely to host life.

[23] https://pds.nasa.gov/.

[24] https://exoplanetarchive.ipac.caltech.edu/index.html.

TESS. The Transiting Exoplanet Survey Satellite (TESS) is the next step in the search for planets outside of our Solar System, including those that could support life. The TESS is aimed to survey 200,000 of the brightest stars near the Sun to search for transiting exoplanets. TESS was launched on 18 April 2018, aboard a SpaceX Falcon 9 rocket. TESS scientists expect to create a catalogue of thousands of exoplanet candidates and to increase vastly the current number of known exoplanets.

Moon image surveys.

The *Lunar Consortium Data*. Photographic hardcopy scans of 1970 photographs captured by a Soyuz 7K-L1 unmanned spacecraft on the soviet Zond8 mission (Zond8 Lunar Consortium Historical Data Archive[25]) along with data from Apollo, Lunar Orbiter, and Galileo missions is a part of the collection commonly referred to as the Lunar Consortium Data. The complete archive of datasets from the European Space Agency's 3-year *SMART-1* mission to the moon contains the archive of three-dimensional maps of the lunar poles along with detailed spectroscopic measurements of the lunar surface. Researchers can utilize this information, and cross-refer it with the wider Planetary Science Archive, to further investigate the formation and evolution of our nearest neighbor in space.

Moon Zoo was a project to study the lunar surface in unprecedented detail, by asking volunteers to identify craters and boulders in millions of images from NASA's Lunar Reconnaissance Orbiter (LRO). Moon Zoo was launched in May 2010 and the project was wrapped up in June 2015. Registered users identified, classified, and measured feature shapes on the surface of the Moon using a tailored graphical interface. The interface was also linked to a wide range of education and public outreach material, including a forum and blog, with contributions and moderation by experts and invited specialists (Joy et al., 2016). The key scientific objectives of Moon Zoo relate to the statistical population survey of small craters, boulder distributions, and cataloguing of various geomorphology features across the lunar surface such as linear features, bright fresh craters, bench craters, etc., including the crater survey results around the Apollo-17 region.

The selected huge internet resources about our Sun and Solar System planets, planetary systems, near-Earth objects, extrasolar systems, and SETI project are available through https://www.loc.gov/rr/scitech/selected-internet/astronomy-selected.html#sun, http://tdc-www.harvard.edu/astro.data.html, http://www.eso.org/~ppadovan/astronomical_data_web.html.

The *Hubble Legacy Archive* (HLA) is designed to optimize science from the Hubble Space Telescope by providing online, enhanced Hubble products and advanced browsing capabilities.[26] The HLA is a joint project of the Space Telescope Science Institute (STScI), the Space Telescope European Coordinating Facility (ST-ECF), and the Canadian Astronomy Data Centre (CADC). It contains the *Hubble Source Catalogs*, including the last version released on 27 June 2019.[27] This archive is a part of the *Mikulski Archive for Space Telescope* (MUST) NASA archive of space missions data in optical, ultraviolet, and near-infrared electromagnetic ranges.[28]

The *European Space Agency science archive* is worldwide available and can be requested after the proprietary period by the astronomical community.[29] Its content includes the data from the ESO Imaging Survey and other projects as GOODS, zCOSMOS, etc.; all the raw observations performed at the La Silla Paranal Observatory on the NTT, MPG-ESO 2.2m, ESO 3.6m, VLT/VLTI, and APEX telescopes; the raw data from the UKIRT Infrared Deep Sky Survey (UKIDSS) taken with the Wide-Field Infrared Camera (WFCAM); the Hubble Space Telescope Data; such catalogues as USNO A2.0, Guide Star Catalogue v2.2, 2MASS, Tycho-1 and Tycho-2, Hipparcos, and others; Digitized Sky Survey and ESO Schmidt Plates.

The *Space Science Data Center* (SSDC), a facility of the Italian Space Agency, is a multimission science operations, data processing, and data archiving center.[30] The SSDC has significant responsibilities for a number of high-energy astronomy/astroparticle satellites (e.g., EINSTEIN, EXOSAT, ROSAT, Swift, AGILE, Fermi, NuSTAR, and AMS) and supports at different levels other old and future space mission's archives, like Herschel, Gaia and Planck, BeppoSAX and Chang-E, CHEOPS, EUCLID, and others. One can find the numerous catalogues (among them are the TeGeV Catalog, AGILE Catalogs, Fermi Catalogs, Third EGRET Catalog, BeppoSAX catalogues including The BeppoSAX GRBM catalog of GRBs, the catalogue of White dwarfs in the SDSS, the Plotkin Catalog, Planck, and WMAP catalogues), search tools and calculators (SSDC Sky Explorer, Swift Simulator, NuSTAR Simulator, SSDC Angular Distance Calculator, and various conversion tools) at the SSDC web site.

[25] https://astrogeology.usgs.gov/maps/moon-zond8-historical-data-archive.

[26] https://hla.stsci.edu/hlaview.html.

[27] http://archive.stsci.edu/searches.html.

[28] http://archive.stsci.edu/index.html.

[29] http://archive.eso.org/cms.html.

[30] https://www.ssdc.asi.it/overview.html#.

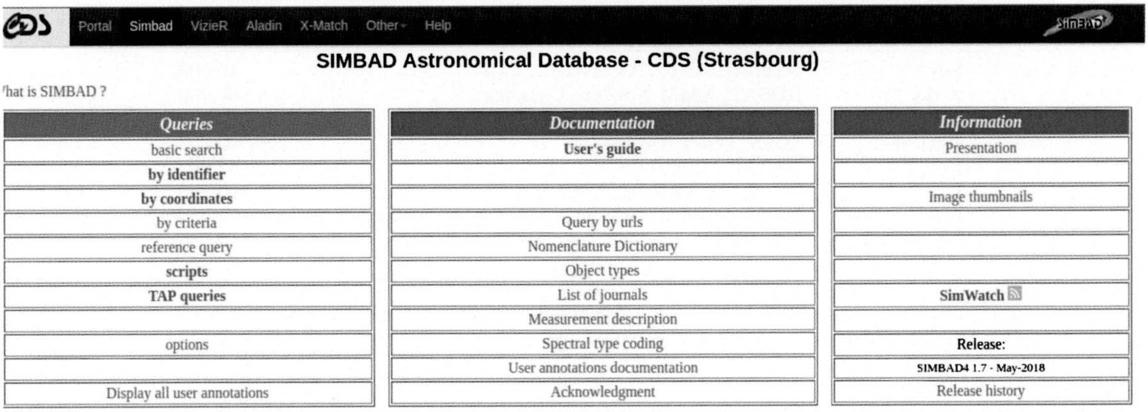

CDS Portal Simbad VizieR Aladin X-Match Other Help SIMBAD

SIMBAD Astronomical Database - CDS (Strasbourg)

'hat is SIMBAD ?

Queries	*Documentation*	*Information*
basic search	**User's guide**	Presentation
by identifier		
by coordinates		Image thumbnails
by criteria	Query by urls	
reference query	Nomenclature Dictionary	
scripts	Object types	
TAP queries	List of journals	**SimWatch**
	Measurement description	
options	Spectral type coding	**Release:**
	User annotations documentation	**SIMBAD4 1.7 - May-2018**
Display all user annotations	Acknowledgment	Release history

FIG. 5.29 The SIMBAD database. Credit: http://simbad.u-strasbg.fr/simbad/.

Astronomer's Bazaar. The *Strasbourg astronomical Data Center* (CDS) maintains the biggest collection of astronomical data catalogues of observations of star, Galaxy, and other galactic and extragalactic objects. The online services currently provided by the CDS include *VizieR Catalogue Service*, which provides access to the most complete library of published astronomical catalogues and data tables available online; *Aladin*, an interactive sky atlas; and *SIMBAD*, a database of astronomical objects (Fig. 5.29).

SIMBAD[31] includes basic information on celestial objects as well as the bibliography. SIMBAD is a meta-compilation database built from what is published in the literature, and from their expertise on cross-identifications. By construction it is highly inhomogeneous, as the data come from any kind of instruments at all wavelengths with any resolution and astrometry, and different names from one publication to another. Nevertheless, it is quite useful for professional astronomers to get full information about catalogued celestial bodies, and this database is supplemented with the *Aladin* service, allowing "the user to visualize digitized astronomical images, superimpose entries from astronomical catalogues or databases, and interactively access related data and information from the SIMBAD database... Compliance with existing or emerging VO standards, interconnection with other visualization or analysis tools, ability to easily compare heterogeneous data are key topics allowing Aladin to be a powerful data exploration and integration tool as well as a science enabler".[32]

Good educational tools are provided through the web site of the *Royal Astronomical Society*, which gives access to 126 astronomical databases and archives[33] as well as the *Data Centre for Astrophysics*, Astronomy Department of the University of Geneva,[34] and to the INTEGRAL, Planck, Gaia, and other data obtained with space-born telescopes. Some comparative understanding about the wavelength coverage of the observed universe and about the multiwavelength archives of databases for stars and extragalactic sources can be found in Fig. 5.30 (originally presented by Mickaelian, 2016b).

During the last decades the newest results on galaxies and galaxy clusters studies have been obtained using X-ray observations. Currently, we are using the third generation of X-ray space observatories, namely, Chandra, XMM-Newton, Swift, and NuSTAR, and we are waiting on the fourth generation, such as eROSITA[35] and Hitomi-2 (XARM, Tashiro et al., 2018), missions that should be launched soon.

In the case of X-ray/high-energy space missions (where only space-born observational data can be collected, see Fig. 5.1) the most important database is the *High Energy Astrophysics Science Archive Research Center* (HEASARC, see Fig. 5.31). This is a multimission archive facility. HEASARC services include providing, via a variety of user interfaces, access to many different types of data and information, including proposal and grant tracking information, astronomical catalogues, observation logs, images, etc.

[31] http://simbad.u-strasbg.fr/simbad/.

[32] https://aladin.u-strasbg.fr/.

[33] https://ras.ac.uk/education-and-careers/for-everyone/.

[34] https://www.isdc.unige.ch/.

[35] https://www.mpe.mpg.de/eROSITA.

Wavelength range	Major surveys/catalogues	Number of catalogued sources
γ-ray	Fermi-GLAST, INTEGRAL, Swift	10,000
X-ray	ROSAT, XMM-Newton, Chandra	1,500,000
UV	GALEX, HST	100,000,000
Optical	SDSS, DSS I, DSS II	1,700,000,000
NIR	2MASS, DENIS	600,000,000
MIR	WISE, AKARI-IRC, SST	600,000,000
FIR	IRAS, AKARI-FIS, SST, Herschel	500,000
sub-mm/mm	Planck, WMAP, SCUBA, Herschel, ALMA	100,000
Radio	NVSS, FIRST, SUMSS	2,000,000

FIG. 5.30 Number of catalogued sources in different wavelength ranges giving a comparative understanding about the wavelength coverage of the observed universe (Open Astronomy, Mickaelian, 2016b).

FIG. 5.31 The HEASARC interface with all facilities. Credit: https://heasarc.gsfc.nasa.gov/.

HEASARC is the primary archive for the NASA (and other space agencies) missions studying electromagnetic radiation from extremely energetic cosmic phenomena ranging from black holes to the big bang. Since its merger with the *Legacy Archive for Microwave Background Data Analysis* (LAMBDA[36]) in 2008, the HEASARC archive contains the data obtained by high-energy astronomy missions observing in the extreme-ultraviolet, X-ray, and gamma ray bands, as well as data from space missions, balloons, and ground-based facilities that have studied the relic CMB radiation in the sub-mm, mm, and cm bands.

Using the HEASARC archive tool and specifying X-ray telescopes we are able to download all the available data of galaxies and galaxy clusters. These data can be used to plot X-ray images, light curves, and spectra required for such studies. Fig. 5.32 illustrates how to combine the data from different high-energy space missions for constructing the spectral energy distribution of a galaxy in a wider X-ray range. We give an example of the isolated galaxy NGC 6300 of Seyfert 2 type (with AGN); the data from XMM-Newton, Integral, and Chandra were used to obtain physical parameters of this galaxy in 0.5–250 keV; an optical image is taken from the SDSS database (Vavilova et al., 2016; Pulatova et al., 2015).

However, X-ray and optical databases are not enough to fully describe a galaxy or galaxy cluster. For example, applying additional radio observations from other databases, we are able "to observe" galaxy clusters much more deeply. Radio observations of galaxy clusters can often help to probe the dynamical state of the intracluster medium and identify any significant contamination of the diffuse thermal X-ray emission by point-like nonthermal X-ray sources such as radio loud galaxies with AGNs. The radio data can be found in VLA, ALMA, NRAO, and many other radio data archives.

Fig. 5.33 illustrates the importance and advantage of multiwavelength astrophysics as compared to single-wavelength observations. The print-screen image for galaxy cluster A2744 taken from NED (Fig. 5.34) demonstrates that this cluster was observed in X-ray

[36]https://lambda.gsfc.nasa.gov/.

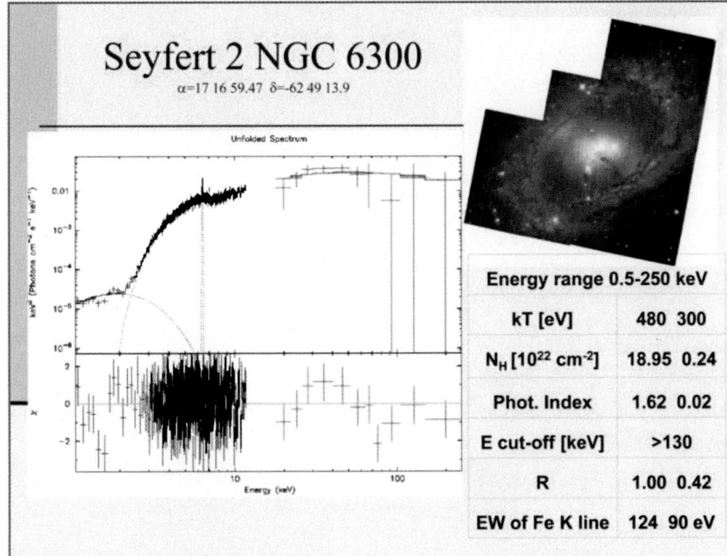

Energy range 0.5-250 keV	
kT [eV]	480 300
N_H [10^{22} cm^{-2}]	18.95 0.24
Phot. Index	1.62 0.02
E cut-off [keV]	>130
R	1.00 0.42
EW of Fe K line	124 90 eV

FIG. 5.32 The isolated AGN galaxy NGC 6300 of Seyfert 2 type as an example (Vavilova et al., 2016) to combine the data in 0.5–250 keV from XMM-Newton, Integral, and Chandra high-energy space missions; an optical image is taken from the SDSS database.

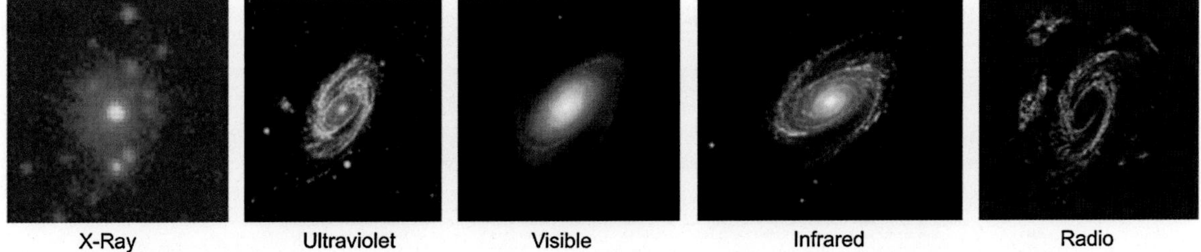

FIG. 5.33 Multiwavelength observations of M81. Image credits: X-ray, ROSAT; UV, GALEX; IR, Spitzer; Radio, NRAO/AUI.

0.1–2.4 keV range by Chandra and ROSAT space observatories and with VLA at 325 MHz in the radio range; it was discovered at the POSS-II F (North) and AAO-SES/SERC ER (South) astroplates. One can find all the available information on this extragalactic object, including the full bibliography, since this object has been studied. One can see also that the absence of the photometric data in ultraviolet, visible, and infrared bands does not allow to construct its spectral energy distribution.

In the case of X-ray galaxy clusters, astronomers are able to use, for examples, ACCEPT and REJECT databases. *ACCEPT* (Cavagnolo et al., 2009) is a research tool for astrophysicists interested in galaxy clusters, and this is a project at Michigan State University (Fig. 5.35). The purpose of this database is to distribute the science-ready data products such as temperature, density, entropy, cooling mass, and other parameters of galaxy clusters as well as the finalized analytic results to the interested researcher. What makes ACCEPT unique is that it is a large, uniformly analyzed database of many galaxy cluster properties. Such a database is useful for many research areas, for example, when making comparisons between simulations and observations.

The *REJECT* database (Hogan et al., 2017; Pulido et al., 2018; Babyk et al., 2018) is a project led by members from the University of Waterloo, Institute of Astronomy, Durham University, Harvard-Smithsonian Center for Astrophysics, International Centre for Radio Astronomy Research, Main Astronomical Observatory of NAS of Ukraine, and Perimeter Institute for Theoretical Physics.

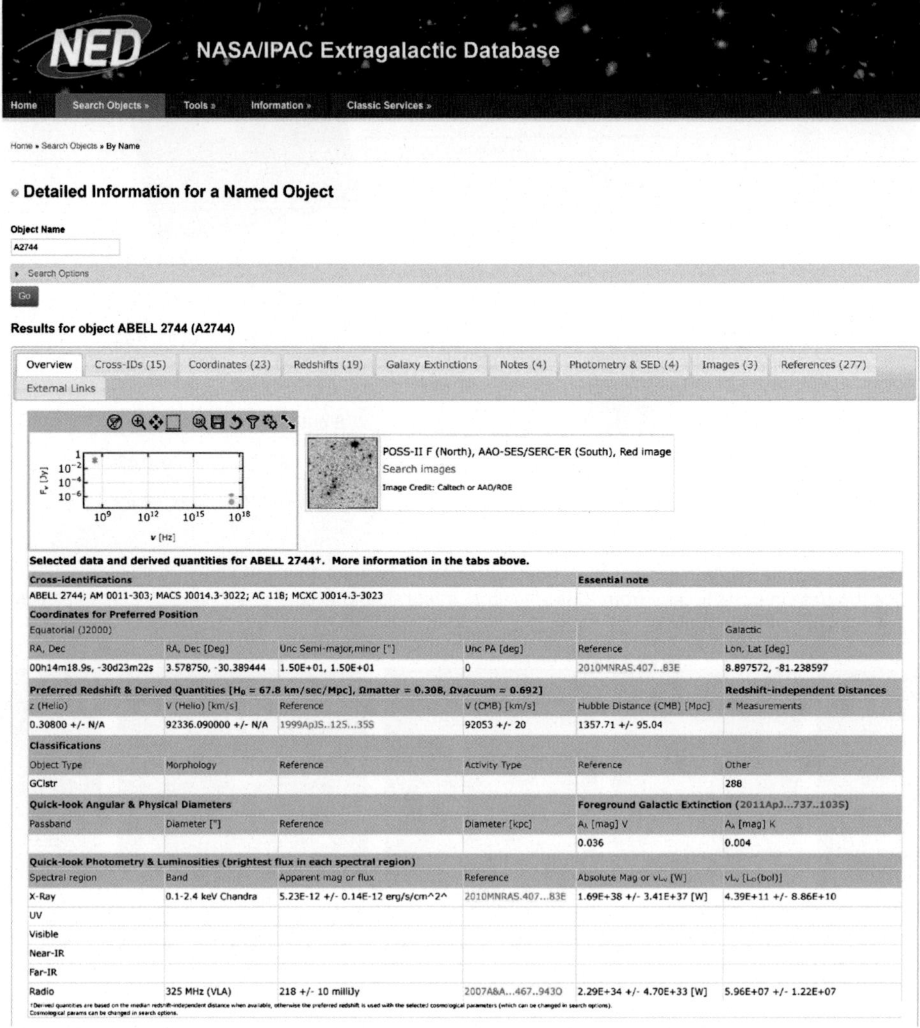

FIG. 5.34 The data on galaxy cluster A2744 from the NED.

The REJECT database was created to study the AGN feedback in galaxy clusters, groups, and galaxies. At the current moment, this database includes thermodynamic products such as temperature, density, entropy, cooling, free-fall time, mass, etc., profiles for 110 clusters and 60 groups/galaxies (Fig. 5.35). These data could be used to explore the energy released by material accreting onto the giant black holes of massive galaxies, which is believed to regulate their growth and surrounding atmospheres. This AGN feedback is most impressive at the centers of large galaxy clusters, where radio jets launched by the AGN carve huge cavities in the surrounding hot gas. However, the details of this feedback process remain poorly understood. The authors of this project have calculated the radial profiles of various thermodynamic and gravitational properties for a large sample of galaxy groups and clusters. Alongside additional ancillary data, these profiles are being used to investigate what governs both the onset of thermally unstable cooling and the response of the AGN.

Other useful databases for extragalactic researches are *NASA/IPAC Extragalactic Database*, NED (Fig. 5.34), and *HyperLEDA* (Fig. 5.36).

FIG. 5.35 The ACCEPT and REJECT databases. Credit: https://web.pa.msu.edu/astro/MC2/.

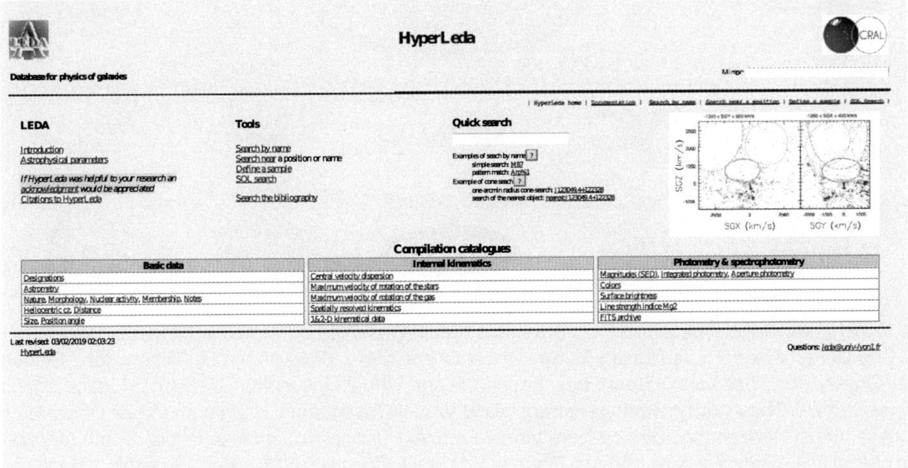

FIG. 5.36 The HYPERLEDA database. Credit: http://leda.univ-lyon1.fr.

HyperLEDA is a database and a collection of tools to study the physics of galaxies and cosmology. It is maintained by a collaboration between Observatoire de Lyon (France) and the Special Astrophysical Observatory of the Russian Academy of Sciences. "The principle behind HyperLEDA is to collect measurements published in literature and modern surveys, and to combine them into a unique homogeneous description of the astronomical objects. This enables the researcher to compare objects located at very different distances. This approach is a continuation of the famous series of Bright Galaxy Catalogues (RC1, RC2, RC3) by de Vaucouleurs and coworkers. The result of the homogenization is the Leda catalogue".[37]

NED[38] includes basic information on the nature of objects, their coordinates, magnitudes, redshift/velocity, size, spectral or morphological type, etc., as well as the reference's lookup. It provides also simple coordinates, velocity, and extinction calculators. Among the new services is the *NED Gravitational Wave Follow-up Service* (GWF), which is intended "to facilitate searches for electromagnetic counterparts to gravitational wave events. Within minutes after the Laser Interferometer Gravitational-Wave Observatory (LIGO)-Virgo collaboration issues an alert using the Gamma-ray Coordinates Network (GCN) operated by the NASA GSFC,[39] this NED service[40] responds by crossmatching in 3D the event's HEALPix map with the galaxies in the local universe and provides an all-sky image of the probability contours, the location of all galaxies in NED within the LIGO 90% probability volume, and the top 20 galaxies sorted by 2MASS absolute Ks-band magnitude."

The concept of the *virtual observatory* (VO) was first described by Szalay and Gray (2001). "The virtual observatory is a collection of interoperating data archives and software tools that utilize the internet to form a scientific research environment in which astronomical research programs can be conducted. The main goal of the Virtual Observatory is to provide transparent and distributed access to data with worldwide availability,

[37] http://leda.univ-lyon1.fr.

[38] http://ned.ipac.caltech.edu/.

[39] https://gcn.gsfc.nasa.gov/.

[40] https://ned.ipac.caltech.edu/gwf/.

FIG. 5.37 IVOA Member Organizations: Argentina Virtual Observatory, Armenian Virtual Observatory, AstroGrid United Kingdom, Australian All-Sky Virtual Observatory, Brazilian Virtual Observatory, Chinese Virtual Observatory, Canadian Virtual Observatory, Chilean Virtual Observatory, European Space Agency, European Virtual Observatory, German Astrophysical Virtual Observatory, Hungarian Virtual Observatory, Japanese Virtual Observatory, Observatoire Virtuel France, Russian Virtual Observatory, South African Astroinformatics Alliance, Spanish Virtual Observatory, Italian Virtual Observatory, Ukrainian Virtual Observatory, US Virtual Observatory Alliance, and Virtual Observatory India.

which helps scientists to discover, access, analyze, and combine nature and lab data from heterogeneous data collections in a user-friendly manner".[41] Zhang and Zhao (2015) consider the VO not like a data warehouse, but rather like an ecosystem of mutually compatible datasets, resources, services, and software tools that use a common set of interoperable technologies and a common set of standards.

The crucial role in promulgating new methods and services into the broader astronomy community belongs to the *International Virtual Observatory Alliance* (IVOA).[42] The IVOA is an organization that formulates the technical standards that make the VO possible. Technical architecture (user layer, VO core layer, resource layer) for processing this resource was proposed by the IVOA Executive Committee. Such approach allows easy access to the metadata due to the interoperability between different astronomical archives and data centers. The IVOA web portal includes a section for astronomers (how to use VO services and tools) and a technical section for deployers (how to build VO services, deploy VO infrastructure at the data center, how IVOA standards are put together, etc.). At the beginning of the era of Big Data the IVOA Executive Committee initiated a creation of the national VOs and development of the infrastructure of the astronomical data center (Fig. 5.37). Such centers give to users handy tools for data selection from large astronomical catalogues for a relatively small region of interest in the sky, data standards and services for image processing (likely to Aladin), and manuals for processing the raw space-born data of various space missions (Chandra, XMM-Newton, Swift, NuSTAR, WISE, Kepler, GAIA, etc.). Besides the VOs, there are other sources for public outreach, for example, the well-known *Astrostatistics and Astroinformatics Portal* (ASAIP),[43] data centers and systems such as NED, HEASARC, SkyHound, NStED, SIMBAD, and Planetary DS, and other aforementioned data archives.

The last but not the least data archive which we should highlight is the *SAO/NASA Astrophysics Data System* (ADS), a Digital Library portal for astronomers and physicists,[44] operated by the Smithsonian Astrophysical

[41] http://en.wikipedia.org/wiki/Virtual_observatory.
[42] http://www.ivoa.net/.
[43] http://asaip.psu.edu.
[44] http://ads.harvard.edu/.

Observatory (SAO) under a NASA grant. The ADS maintains three bibliographic databases (Astronomy and Astrophysics, Physics, and the arXiv e-prints) containing at this time more than 14.7 million publications as well as providing their bibliometric evaluation.

Last remarks. It is important to note that modern astroinformatics tools are not only able to download the observational data that should be further analyzed with specific software, they are able to download already processed data that have been extracted from archived observations. The examples of such data mining will be given in other chapters, and an inquisitive reader can take a fascinating tour of astronomical catalogues from "A" to "Z" compiled on Wikipedia.[45]

ACKNOWLEDGMENTS

The work of authors of this chapter was partially supported in frame of the budgetary program of the NAS of Ukraine "Support for the development of priority fields of scientific research" (CPCEL 6541230).

REFERENCES

Abdo, A.A., et al., 2013. The second Fermi Large Area Telescope catalog of gamma-ray pulsars. The Astrophysical Journal. Supplement Series 208 (1), 17. https://doi.org/10.1088/0067-0049/208/2/17. arXiv:1305.4385 [astro-ph.HE].

Abell, George O., 1958. The distribution of rich clusters of galaxies. The Astrophysical Journal. Supplement Series 3, 211. https://doi.org/10.1086/190036.

Abell, George O., Corwin Jr., Harold G., Olowin, Ronald P., 1989. A catalog of rich clusters of galaxies. The Astrophysical Journal. Supplement Series 70, 1. https://doi.org/10.1086/191333.

Abrahamyan, H.V., Mickaelian, A.M., Knyazyan, A.V., 2015. VizieR online data catalog: IRAS PSC/FSC combined catalogue (Abrahamyan+ 2015). In: VizieR Online Data Catalog, II/338.

Acero, F., et al., 2015. Fermi Large Area Telescope third source catalog. The Astrophysical Journal. Supplement Series 218 (2), 23. https://doi.org/10.1088/0067-0049/218/2/23. arXiv:1501.02003 [astro-ph.HE].

Ackermann, M., et al., 2013. The Fermi All-sky Variability Analysis: a list of flaring gamma-ray sources and the search for transients in our Galaxy. The Astrophysical Journal 771 (1), 57. https://doi.org/10.1088/0004-637X/771/1/57. arXiv:1304.6082 [astro-ph.HE].

Adami, C., et al., 2011. The XMM-LSS survey: optical assessment and properties of different X-ray selected cluster classes. Astronomy & Astrophysics 526, A18. https://doi.org/10.1051/0004-6361/201015182. arXiv:1010.6195 [astro-ph.CO].

[45]https://en.wikipedia.org/wiki/List_of_astronomical_ catalogues.

Alam, Shadab, et al., 2015. The eleventh and twelfth data releases of the Sloan Digital Sky Survey: final data from SDSS-III. The Astrophysical Journal. Supplement Series 219 (1), 12. https://doi.org/10.1088/0067-0049/219/1/12. arXiv:1501.00963 [astro-ph.IM].

Arakelian, M.A., 1975. Galaxies of high surface brightness. Soobshcheniya Byurakanskoj Observatorii Akademiya Nauk Armyanskoj SSR Erevan 47, 3–40.

Arp, Halton, 1966. Atlas of peculiar galaxies. The Astrophysical Journal. Supplement Series 14, 1. https://doi.org/10.1086/190147.

Babyk, Iu.V., Del Popolo, A., Vavilova, I.B., 2014. Chandra X-ray galaxy clusters at $z < 1.4$: constraints on the inner slope of the density profiles. Astronomy Reports 58 (9), 587–610. https://doi.org/10.1134/S1063772914090017. arXiv:1208.2424 [astro-ph.CO].

Babyk, Iu.V., McNamara, B.R., et al., 2018. A universal entropy profile for the hot atmospheres of galaxies and clusters within R_{2500}. The Astrophysical Journal 862 (1), 39. https://doi.org/10.3847/1538-4357/aacce5. arXiv:1802.02589 [astro-ph.CO].

Bahcall, N.A., 1983. Superclusters. Bulletin of the American Astronomical Society 15, 936.

Beardmore, A.P., Page, K.L., Kuulkers, E., 2015. Swift triggers on V404 Cyg. The Astronomer's Telegram, 8455.

Bianchi, L., et al., 2011. GALEX catalogs of UV sources: statistical properties and sample science applications: hot white dwarfs in the Milky Way. Astrophysics and Space Science 335 (1), 161–169. https://doi.org/10.1007/s10509-010-0581-x.

Bird, A.J., et al., 2010. The fourth IBIS/ISGRI soft gamma-ray survey catalog. The Astrophysical Journal. Supplement Series 186 (1), 1–9. https://doi.org/10.1088/0067-0049/186/1/1. arXiv:0910.1704 [astro-ph.HE].

Boella, G., et al., 1997. BeppoSAX, the wide band mission for X-ray astronomy. Astronomy & Astrophysics. Supplement Series 122, 299–307. https://doi.org/10.1051/aas:1997136.

Borne, Kirk, 2013. Virtual observatories, data mining, and astroinformatics. In: Oswalt, Terry D., Bond, Howard E. (Eds.), Planets, Stars and Stellar Systems. Springer Science+Business Media Dordrecht. ISBN 978-94-007-5617-5, p. 403.

Borne, Kirk, et al., 2009. Astroinformatics: a 21st century approach to astronomy. In: astro2010: The Astronomy and Astrophysics Decadal Survey, vol. 2010, p. P6. arXiv:0909.3892 [astro-ph.IM].

Braude, S.Ya., et al., 2007. VizieR online data catalog: the UTR-2 very low-frequency sky survey data (Braude+ 1978-2002). In: VizieR Online Data Catalog, VIII/80.

Buta, Ronald J., Corwin, Harold G., Odewahn, Stephen C., 2007. The de Vaucouleurs Atlas of Galaxies. Cambridge University Press.

Cabanela, Juan E., et al., 2003. The automated plate scanner catalog of the Palomar Observatory Sky Survey. II. The archived database. Publications of the Astronomical Society of the Pacific 115 (809), 837–843. https://doi.org/10.1086/375625.

Cavagnolo, Kenneth W., et al., 2009. Intracluster medium entropy profiles for a Chandra archival sample of galaxy clusters. The Astrophysical Journal. Supplement Series 182 (1), 12–32. https://doi.org/10.1088/0067-0049/182/1/12. arXiv:0902.1802 [astro-ph.CO].

Cavuoti, S., et al., 2012. DAME: a web oriented infrastructure for scientific data mining and exploration. In: Zavidovique, Bertrand, Lo Bosco, Giosue' (Eds.), Science: Image in Action, pp. 241–247.

Colless, Matthew, et al., 2001. The 2dF Galaxy Redshift Survey: spectra and redshifts. Monthly Notices of the Royal Astronomical Society 328 (4), 1039–1063. https://doi.org/10.1046/j.1365-8711.2001.04902.x. arXiv:astro-ph/0106498 [astro-ph].

Condon, J.J., Kaplan, D.L., 1998. Planetary nebulae in the NRAO VLA sky survey. The Astrophysical Journal. Supplement Series 117 (2), 361–385. https://doi.org/10.1086/313128.

Corwin, H.G., de Vaucouleurs, A., de Vaucouleurs, G., 1985. Southern Galaxy Catalogue. Univ. Texas Monogr. Astron., vol. 4, p. 1.

Cutri, R.M., et al., 2012. VizieR online data catalog: WISE all-sky data release (Cutri+ 2012). In: VizieR Online Data Catalog, II/311.

de Bruyn, A.G., 1996. The Westerbork Synthesis Radio Telescope, a second lease on life. In: Raimond, Ernst, Genee, Rene (Eds.), The Westerbork Observatory, Continuing Adventure in Radio Astronomy. In: Astrophysics and Space Science Library, vol. 208, p. 109.

De Cuyper, J., et al., 2012. The digitiser and archive facility at the ROB. In: Ballester, P., Egret, D., Lorente, N.P.F. (Eds.), Astronomical Data Analysis Software and Systems XXI. In: Astronomical Society of the Pacific Conference Series, vol. 461, p. 315.

de Vaucouleurs, G., 1975. Nearby groups of galaxies. In: Sandage, Allan, Sandage, Mary, Kristian, Jerome (Eds.), Galaxies and the Universe. University of Chicago Press, Chicago.

de Vaucouleurs, G., de Vaucouleurs, A., Corwin Jr., H.G., 1976. Second Reference Catalogue of Bright Galaxies. Containing Information on 4364 Galaxies with References to Papers Published Between 1964 and 1975. University of Texas Press, Austin.

de Vaucouleurs, Gerard Henri, de Vaucouleurs, Antoinette, Shapley, Harlow, 1964. Reference Catalogue of Bright Galaxies. University of Texas Monographs in Astronomy. University of Texas Press, Austin.

de Vaucouleurs, Gerard, et al., 1991. Third Reference Catalogue of Bright Galaxies. Springer, New York, NY (USA), 2091P.

DENIS Consortium, 2005. VizieR online data catalog: the DENIS database (DENIS Consortium, 2005). In: VizieR Online Data Catalog, p. 1.

Djorgovski, S.G., et al., 1998. The Palomar Digital Sky Survey (DPOSS). In: Colombi, Stephane, Mellier, Yannick, Raban, Brigitte (Eds.), Wide Field Surveys in Cosmology, vol. 14, p. 89. arXiv:astro-ph/9809187 [astro-ph].

Doi, M., et al., 2010. Photometric response functions of the Sloan Digital Sky Survey imager. The Astronomical Journal 139, 1628–1648. https://doi.org/10.1088/0004-6256/139/4/1628. arXiv:1002.3701 [astro-ph.IM].

Dreyer, J.L.E., 1888. A new general catalogue of nebulæ and clusters of stars, being the catalogue of the late Sir John F. W. Herschel, Bart, revised, corrected, and enlarged. Memoirs of the Royal Astronomical Society 49, 1.

Dreyer, J.L.E., 1895. Index catalogue of nebulæ found in the years 1888 to 1894, with notes and corrections to the new general catalogue. Memoirs of the Royal Astronomical Society 51, 185.

Dreyer, J.L.E., 1910. Second index catalogue of nebulæ and clusters of stars, containing objects found in the years 1895 to 1907; with notes and corrections to the new general catalogue and to the index catalogue for 1888–94. Memoirs of the Royal Astronomical Society 59, 105.

Einasto, Jaan, 2001. Large scale structure. New Astronomy Reviews 45 (4–5), 355–372. https://doi.org/10.1016/S1387-6473(00)00158-5. arXiv:astro-ph/0011332 [astro-ph].

Elyiv, A., et al., 2012. Angular correlation functions of X-ray point-like sources in the full exposure XMM-LSS field. Astronomy & Astrophysics 537, A131. https://doi.org/10.1051/0004-6361/201117983. arXiv:1111.5982 [astro-ph.CO].

Engels, D., et al., 1988. The Hamburg quasar survey. In: Osmer, P., et al. (Eds.), Optical Surveys for Quasars. In: Astronomical Society of the Pacific Conference Series, vol. 2, p. 143.

Evans, Ian N., et al., 2010. The Chandra source catalog. The Astrophysical Journal. Supplement Series 189 (1), 37–82. https://doi.org/10.1088/0067-0049/189/1/37. arXiv:1005.4665 [astro-ph.HE].

Feitzinger, J.V., Galinski, Th., 1985. A catalogue of dwarf galaxies south of $\delta = -17.5°$. Astronomy & Astrophysics. Supplement Series 61, 503–515.

Gaia Collaboration, 2016. VizieR online data catalog: Gaia DR1 (Gaia Collaboration, 2016). In: VizieR Online Data Catalog, I/337.

Gaia Collaboration, 2018. VizieR online data catalog: Gaia DR2 (Gaia Collaboration, 2018). In: VizieR Online Data Catalog, I/345.

Gandhi, P., et al., 2006. The XMM large scale structure survey: properties and two-point angular correlations of point-like sources. Astronomy & Astrophysics 457 (2), 393–404. https://doi.org/10.1051/0004-6361:20065284. arXiv:astro-ph/0607135 [astro-ph].

Gezari, Daniel Y., Schmitz, Marion, Mead, Jaylee M., 1987. Catalog of infrared observations. Tech. Rep.

Gold, B., et al., 2011. Seven-year Wilkinson Microwave Anisotropy Probe (WMAP) observations: galactic foreground emission. The Astrophysical Journal. Supplement Series 192 (2), 15. https://doi.org/10.1088/0067-0049/192/2/15. arXiv:1001.4555 [astro-ph.GA].

Goodman, A.A., 2012. Principles of high-dimensional data visualization in astronomy. Astronomische Nachrichten 333 (5–6), 505. https://doi.org/10.1002/asna.201211705. arXiv:1205.4747 [astro-ph.IM].

Grindlay, J., et al., 2011. Opening the 100-year window for time domain astronomy. Proceedings of the International Astronomical Union 7 (S285), 29–34. https://doi.org/10.1017/S1743921312000166. arXiv:1211.1051 [astro-ph.GA].

Hagen, H.-J., Engels, D., Reimers, D., 1999. The Hamburg Quasar Survey. III. Further new bright quasars. Astronomy & Astrophysics. Supplement Series 134, 483–487.

Hales, S.E.G., et al., 2007. A final non-redundant catalogue for the 7C 151-MHz survey. Monthly Notices of the Royal Astronomical Society 382 (4), 1639–1642. https://doi.org/10.1111/j.1365-2966.2007.12392.x.

Hartman, R.C., et al., 1999. The third EGRET catalog of high-energy gamma-ray sources. The Astrophysical Journal. Supplement Series 123 (1), 79–202. https://doi.org/10.1086/313231.

Heald, G.H., et al., 2015. The LOFAR Multifrequency Snapshot Sky Survey (MSSS). I. Survey description and first results. Astronomy & Astrophysics 582, A123. https://doi.org/10.1051/0004-6361/201425210. arXiv:1509.01257 [astro-ph.IM].

Hearnshaw, John, 2009. Astronomical Spectrographs and Their History. Cambridge University Press. ISBN 9780521882576.

Helfand, David J., White, Richard L., Becker, Robert H., 2015. The last of FIRST: the final catalog and source identifications. The Astrophysical Journal 801 (1), 26. https://doi.org/10.1088/0004-637X/801/1/26. arXiv:1501.01555 [astro-ph.GA].

Helou, George, Walker, D.W., 1985. IRAS Small Scale Structure Catalog. Jet Propulsion Laboratory, Joint IRAS Science Working Group, Pasadena.

Herschel, John Frederick William, 1864. A general catalogue of nebulae and clusters of stars. Philosophical Transactions of the Royal Society of London Series I 154, 1–137.

Hewitt, A., Burbidge, G., 1987. A New Optical Catalog of Quasi-Stellar Objects. University of Chicago Press, Chicago.

Hickson, P., 1982. Systematic properties of compact groups of galaxies. The Astrophysical Journal 255, 382–391. https://doi.org/10.1086/159838.

Hoeg, E., et al., 1997. The TYCHO catalogue. Astronomy & Astrophysics 323, L57–L60.

Høg, E., 2000. Tycho star catalogs: the 2.5 million brightest stars. In: Murdin, Paul (Ed.), Encyclopedia of Astronomy and Astrophysics. Institute of Physics Publishing, Bristol, p. 2862.

Hogan, M.T., et al., 2017. The onset of thermally unstable cooling from the hot atmospheres of giant galaxies in clusters: constraints on feedback models. The Astrophysical Journal 851 (1), 66. https://doi.org/10.3847/1538-4357/aa9af3. arXiv:1704.00011 [astro-ph.GA].

Huchra, J., et al., 1983. X-ray spectra of active galactic nuclei. The Astrophysical Journal. Supplement Series 52, 89. https://doi.org/10.1086/190860.

Huchtmeier, W.K., Richter, O.G., Bohnenstengel, H.D., 1983. A General Catalog of HI Observations of External Galaxies. European Southern Observatory, Garching.

Ishihara, D., et al., 2010. The AKARI/IRC mid-infrared all-sky survey. Astronomy & Astrophysics 514, A1. https://doi.org/10.1051/0004-6361/200913811. arXiv:1003.0270 [astro-ph.IM].

Jarrett, T.H., et al., 2000. 2MASS extended source catalog: overview and algorithms. The Astronomical Journal 119 (5), 2498–2531. https://doi.org/10.1086/301330. arXiv:astro-ph/0004318.

Jones, Derek, 2000. The scientific value of the Carte du Ciel. Astronomy and Geophysics 41 (5), 16. https://doi.org/10.1046/j.1468-4004.2000.41516.x.

Joy, Katherine H., et al., 2016. The Moon: an archive of small body migration in the Solar System. Earth, Moon, and Planets 118 (2–3), 133–158. https://doi.org/10.1007/s11038-016-9495-0.

Karachentsev, I.D., 1997. VizieR online data catalog: isolated pairs of galaxies in northern hemisphere (Karachentsev 1972). In: VizieR Online Data Catalog, VII/77.

Karachentseva, V.E., 1973. The catalogue of isolated galaxies. Astrofizicheskie Issledovaniia Izvestiya Spetsial'noj Astrofizicheskoj Observatorii 8, 3–49.

Karachentseva, V.E., Mitronova, S.N., et al., 2010. Catalog of isolated galaxies selected from the 2MASS survey. Astrophysical Bulletin 65 (1), 1–17. https://doi.org/10.1134/S1990341310010013. arXiv:1005.3191 [astro-ph.CO].

Karachentseva, V.E., Sharina, M.E., 1988. A catalogue of low surface brightness dwarf galaxies. Soobshcheniya Spetsial'noj Astrofizicheskoj Observatorii 57.

Kazantseva, L.V., et al., 2015. Processing results of digitized photographic observations of Pluto from the collections of the Ukrainian Virtual Observatory. Kinematics and Physics of Celestial Bodies 31 (1), 37–54. https://doi.org/10.3103/S0884591315010031.

Konovalenko, A.A., Sodin, L.G., 1981. The 26.13 MHz absorption line in the direction of Cassiopeia A. Nature 294, 135. https://doi.org/10.1038/294135a0.

Koribalski, B.S., et al., 2004. The 1000 brightest HIPASS galaxies: HI properties. The Astronomical Journal 128 (1), 16–46. https://doi.org/10.1086/421744. arXiv:astro-ph/0404436.

Koulouridis, E., et al., 2014. X-ray AGN in the XMM-LSS galaxy clusters: no evidence of AGN suppression. Astronomy & Astrophysics 567, A83. https://doi.org/10.1051/0004-6361/201423601. arXiv:1402.4136 [astro-ph.CO].

Kovalevsky, J., 2009. Michael Perryman: Astronomical applications of Astrometry: Ten years of exploitation of the Hipparcos satellite data. Celestial Mechanics & Dynamical Astronomy 104, 403–405. https://doi.org/10.1007/s10569-009-9221-6.

Kuzmin, A., et al., 1999. Construction of the TYCHO reference catalogue. Astronomy & Astrophysics. Supplement Series 136, 491–508. https://doi.org/10.1051/aas:1999229.

Lasker, Barry M., et al., 2008. The second-generation guide star catalog: description and properties. The Astronomical Journal 136 (2), 735–766. https://doi.org/10.1088/0004-6256/136/2/735. arXiv:0807.2522 [astro-ph].

Lauberts, A., 1982. ESO/Uppsala Survey of the ESO(B) Atlas. European Southern Observatory, Garching.

Lawrence, A., et al., 2016. Slow-blue nuclear hypervariables in PanSTARRS-1. Monthly Notices of the Royal Astronomical Society 463 (1), 296–331. https://doi.org/10.1093/mnras/stw1963. arXiv:1605.07842 [astro-ph.HE].

MacConnell, D.J., 1995. Objective-prism spectroscopy: retrospective and prospective. In: Chapman, J., et al. (Eds.), IAU Colloq. 148: The Future Utilisation of Schmidt Telescopes. In: Astronomical Society of the Pacific Conference Series, vol. 84, p. 323.

Markarian, B.E., et al., 1989. The first Byurakan survey. A catalogue of galaxies with UV-continuum. Soobshcheniya Spetsial'noj Astrofizicheskoj Observatorii 62, 5–117.

Markaryan, B.E., Lipovetskii, V.A., Stepanyan, Dzh A., 1983. Second Byurakan spectral sky survey - part one - quasistellar and Seyfert objects. Astrophysics 19 (1), 14–25. https://doi.org/10.1007/BF01005806.

Mauch, T., et al., 2013. VizieR online data catalog: Sydney University Molonglo Sky Survey (SUMSS V2.1) (Mauch+ 2008). In: VizieR Online Data Catalog, vol. 8081.

Mazzarella, Joseph M., Balzano, Vicki A., 1986. A catalog of Markarian galaxies. The Astrophysical Journal. Supplement Series 62, 751. https://doi.org/10.1086/191155.

McMahon, R.G., Irwin, M.J., Maddox, S.J., 2000. VizieR online data catalog: the APM-North catalogue (McMahon+, 2000). In: VizieR Online Data Catalog, I/267.

Melnyk, O., Karachentseva, V., Karachentsev, I., 2015. Star formation rates in isolated galaxies selected from the Two-Micron All-Sky Survey. Monthly Notices of the Royal Astronomical Society 451 (2), 1482–1495. https://doi.org/10.1093/mnras/stv950. arXiv:1504.07990 [astro-ph.GA].

Melnyk, O., Mitronova, S., Karachentseva, V., 2014. Colours of isolated galaxies selected from the Two-Micron All-Sky Survey. Monthly Notices of the Royal Astronomical Society 438 (1), 548–556. https://doi.org/10.1093/mnras/stt2225. arXiv:1311.4391 [astro-ph.GA].

Melnyk, O., Plionis, M., et al., 2013. Classification and environmental properties of X-ray selected point-like sources in the XMM-LSS field. Astronomy & Astrophysics 557, A81. https://doi.org/10.1051/0004-6361/201220624. arXiv:1307.0527 [astro-ph.CO].

Messier, Charles, 1781. Catalogue des Nébuleuses et des Amas d'Etoiles (Catalog of Nebulae and Star Clusters). Tech. Rep., pp. 227–267.

Mickaelian, A.M., 2016a. Multiwavelength astronomy and big data. Astronomy Reports 60, 857–877. https://doi.org/10.1134/S1063772916090043.

Mickaelian, Areg M., 2016b. Astronomical surveys and big data. Baltic Astronomy 25, 75–88. arXiv:1511.07322 [astro-ph.IM].

Mickaelian, A.M., et al., 2009. The DFBS spectroscopic database and the Armenian Virtual Observatory. Defense Science Journal 8, 152–161.

Mitronova, S.N., et al., 2004. The 2MASS-selected Flat Galaxy Catalog. Bulletin of the Special Astrophysical Observatory 57, 5–163. arXiv:astro-ph/0408257 [astro-ph].

Monet, D., 1996. The general release of USNO-A 1.0. Bulletin of the American Astronomical Society 28, 1282.

Monet, D., 1998. USNO-A2.0. A catalog of astrometric standards. Available at: http://vizier.u-strasbg.fr/viz-bin/VizieR?-source=USNO-A2.0.

Monet, D.G., et al., 2003. The USNO-B catalog. The Astronomical Journal 125, 984–993. https://doi.org/10.1086/345888. arXiv:astro-ph/0210694.

Monet, D., Canzian, B., Henden, A., 1994. Astrometric calibration of UJ1.0. In: American Astronomical Society Meeting Abstracts. Bulletin of the American Astronomical Society 26, 1314.

Moshir, M., et al., 1990. The IRAS faint source catalog, version 2. Bulletin of the American Astronomical Society 22, 1325.

Mulders, Gijs D., et al., 2016. A super-solar metallicity for stars with hot rocky exoplanets. The Astronomical Journal 152 (6), 187. https://doi.org/10.3847/0004-6256/152/6/187. arXiv:1609.05898 [astro-ph.EP].

Nesci, R., Rossi, C., Cirimele, G., 2009. Spectra in the digitized first Byurakan survey. Baltic Astronomy 18, 383–386.

Nilson, Peter, 1973. Uppsala General Catalogue of Galaxies. Almqvist and Wiksell (distr.), 456P.

Nilson, P., 1974. Catalogue of selected non-UGC galaxies. In: Uppsala Astronomical Observatory Reports, vol. 5, 32p.

Odewahn, S.C., Humphreys, R.M., et al., 1993. Star-galaxy separation with a neural network. 2: multiple Schmidt plate fields. Publications of the Astronomical Society of the Pacific 105, 1354–1365. https://doi.org/10.1086/133317.

Odewahn, S.C., Stockwell, E.B., et al., 1992. Automated star/galaxy discrimination with neural networks. The Astronomical Journal 103, 318–331. https://doi.org/10.1086/116063.

Oort, J.H., 1983. Superclusters. Annual Review of Astronomy and Astrophysics 21, 373–428. https://doi.org/10.1146/annurev.aa.21.090183.002105.

Pacaud, F., et al., 2016. The XXL survey. II. The bright cluster sample: catalogue and luminosity function. Astronomy & Astrophysics 592, A2. https://doi.org/10.1051/0004-6361/201526891. arXiv:1512.04264 [astro-ph.CO].

Pakuliak, L., et al., 2014. Spectral photographic archives of observatories of Ukraine: digital versions. In: Astronomical Society of India Conference Series, vol. 11.

Parsons, S.G., et al., 2016. The white dwarf binary pathways survey - I. A sample of FGK stars with white dwarf companions. Monthly Notices of the Royal Astronomical Society 463 (1), 2125–2136. https://doi.org/10.1093/mnras/stw2143. arXiv:1604.01613 [astro-ph.SR].

Paturel, G., 1989. Catalogue of principal galaxies (PGC). Astronomy & Astrophysics. Supplement Series 80 (3), 299.

Peebles, P.J.E., 1980. The Large-Scale Structure of the Universe. Princeton University Press. ISBN 978-0-691-08240-0.

Pesch, P., et al., 1995. The case low-dispersion northern sky survey. XV. A region in Ursa Major and Canes Venatici. The Astrophysical Journal. Supplement Series 98, 41. https://doi.org/10.1086/192154.

Petrosian, Artashes, et al., 2007. Markarian galaxies. I. The optical database and atlas. The Astrophysical Journal. Supplement Series 170 (1), 33–70. https://doi.org/10.1086/511333.

Pierre, M., et al., 2016. The XXL survey. I. Scientific motivations - XMM-Newton observing plan - follow-up observations and simulation programme. Astronomy & Astrophysics 592, A1. https://doi.org/10.1051/0004-6361/201526766. arXiv:1512.04317 [astro-ph.CO].

Planck Collaboration, 2011. Planck early results. I. The Planck mission. Astronomy & Astrophysics 536, A1. https://doi.org/10.1051/0004-6361/201116464. arXiv:1101.2022 [astro-ph.IM].

Pulatova, N.G., et al., 2015. The 2MIG isolated AGNs - I. General and multiwavelength properties of AGNs and host galaxies in the northern sky. Monthly Notices of the Royal Astronomical Society 447 (3), 2209–2223. https://doi.org/10.1093/mnras/stu2556.

Pulido, F.A., et al., 2018. The origin of molecular clouds in central galaxies. The Astrophysical Journal 853 (2), 177. https://doi.org/10.3847/1538-4357/aaa54b. arXiv:1710.04664 [astro-ph.GA].

Reid, I.N., et al., 1991. The second Palomar sky survey. Publications of the Astronomical Society of the Pacific 103, 661. https://doi.org/10.1086/132866.

Rosen, S.R., et al., 2016. The XMM-Newton serendipitous survey. VII. The third XMM-Newton serendipitous source catalogue. Astronomy & Astrophysics 590, A1. https://doi.org/10.1051/0004-6361/201526416. arXiv:1504.07051 [astro-ph.HE].

Sakano, Masaaki, et al., 2001. X-ray source population of the galactic center region obtained with ASCA. In: Inoue, H., Kunieda, H. (Eds.), New Century of X-Ray Astronomy. In: Astronomical Society of the Pacific Conference Series, vol. 251, p. 314. arXiv:astro-ph/0107444 [astro-ph].

Sandage, Allan, 1961. The Hubble Atlas of Galaxies, Later Printing Edition (September 1, 1984). Carnegie Inst. of Washington. ISBN-10: 0872796299, ISBN-13: 978-0872796294.

Sandage, Allan, Tammann, G.A., 1987. A Revised Shapley–Ames Catalog of Bright Galaxies. Carnegie Institution of Washington, Washington, D.C., 157P.

Schneider, Stephen E., Thuan, Trinh X., Magri, Christopher, et al., 1990. Northern dwarf and low surface brightness galaxies. I. The Arecibo neutral hydrogen survey. The Astrophysical Journal. Supplement Series 72, 245. https://doi.org/10.1086/191416.

Schneider, Stephen E., Thuan, Trinh X., Mangum, Jeffrey G., et al., 1992. Northern dwarf and low surface brightness galaxies. II. The Green Bank neutral hydrogen survey. The Astrophysical Journal. Supplement Series 81, 5. https://doi.org/10.1086/191684.

Shane, C.D., Wirtanen, C.A., 1954. The distribution of extragalactic nebulae. The Astronomical Journal 59, 285–304. https://doi.org/10.1086/107014.

Shapley, Harlow, Ames, Adelaide, 1932. A survey of the external galaxies brighter than the thirteenth magnitude. In: Annals of Harvard College Observatory, vol. 88, pp. 41–76.

Shatokhina, S.V., et al., 2018. Catalogue of asteroids from digitized photographic plates of the FON program. Kinematics and Physics of Celestial Bodies 34 (5), 270–276. https://doi.org/10.3103/S0884591318050045.

Shectman, Stephen A., et al., 1996. The Las Campanas Redshift Survey. The Astrophysical Journal 470, 172. https://doi.org/10.1086/177858. arXiv:astro-ph/9604167.

Shen, Shi-Yin, et al., 2016. A sample of galaxy pairs identified from the LAMOST spectral survey and the Sloan Digital Sky Survey. Research in Astronomy and Astrophysics 16 (3), 43. https://doi.org/10.1088/1674-4527/16/3/043. arXiv:1512.02438 [astro-ph.GA].

Shimwell, T.W., et al., 2017. VizieR online data catalog: LOFAR two-metre sky survey (Shimwell+, 2017). In: VizieR Online Data Catalog, J/A+A/598/A104.

Shlyapnikov, A.A., 2007. The "LADAN" project: conception of local data archive of CrAO observations. Izvestiya Ordena Trudovogo Krasnogo Znameni Krymskoj Astrofizicheskoj Observatorii 103, 142–153.

Sinnott, Roger W., 1988. NGC 2000.0: The Complete New General Catalogue and Index Catalogues of Nebulae and Star Clusters by J.L.E. Dreyer. Sky Pub. Corp. ISBN-10: 0933346514, ISBN-13: 978-0933346512.

Skrutskie, M.F., et al., 2006. The Two Micron All Sky Survey (2MASS). The Astronomical Journal 131 (2), 1163–1183. https://doi.org/10.1086/498708.

Stepanian, J.A., 2005. The second Byurakan survey. General catalogue. Revista Mexicana de Astronomia y Astrofisica 41, 155–368.

Sulentic, Jack W., Tifft, William G., Dreyer, John Louis Emil, 1973. The Revised New Catalogue of Nonstellar Astronomical Objects. University of Arizona Press, Tucson.

Szalay, Alexander, Gray, Jim, 2001. The World-Wide Telescope. Science 293 (5537), 2037–2040. https://doi.org/10.1126/science.293.5537.2037.

Tang, Sumin, Grindlay, Jonathan, Los, Edward, Laycock, Silas, 2010. DASCH discovery of large amplitude ~10–100 year variability in K giants. The Astrophysical Journal Letters 710 (1), L77–L81. https://doi.org/10.1088/2041-8205/710/1/L77. arXiv:1001.1395 [astro-ph.SR].

Tang, Sumin, Grindlay, Jonathan, Los, Edward, Servillat, Mathieu, 2011. DASCH on KU Cyg: a ~5 year dust accretion event in ~1900. The Astrophysical Journal 738 (1), 7. https://doi.org/10.1088/0004-637X/738/1/7. arXiv:1106.0003 [astro-ph.SR].

Tang, Sumin, Sasselov, Dimitar, et al., 2013. 100-year DASCH light curves of Kepler planet-candidate host stars. Publications of the Astronomical Society of the Pacific 125 (929), 793. https://doi.org/10.1086/671759. arXiv:1304.7503 [astro-ph.EP].

Tashiro, Makoto, et al., 2018. Concept of the X-ray astronomy recovery mission. In: Proceedings of the SPIE. In: Society of Photo-Optical Instrumentation Engineers (SPIE) Conference Series, vol. 10699, p. 1069922.

Truemper, J., 1992. ROSAT: a new look at the X-ray sky. Quarterly Journal of the Royal Astronomical Society 33, 165.

Tsvetkova, K., et al., 2018. On some Bamberg wide-field plate catalogues recently incorporated into WFPDB. Astronomical and Astrophysical Transactions 30 (4), 467–478.

Tully, R. Brent, Fisher, J. Richard, 1987. Atlas of Nearby Galaxies. Cambridge University Press. ISBN 10: 052130136X, ISBN 13: 9780521301367.

Urban, S.E., et al., 1998. The AC 2000: the astrographic catalogue on the system defined by the HIPPARCOS catalogue. The Astronomical Journal 115 (3), 1212–1223. https://doi.org/10.1086/300264.

van den Bergh, Sidney, 1959. A catalogue of dwarf galaxies. Publications of the David Dunlap Observatory 2 (5), 147–150.

van den Bergh, S., 1966. Luminosity classifications of dwarf galaxies. The Astronomical Journal 71, 922–926. https://doi.org/10.1086/109987.

Vavilova, I., et al., 2014. The scientific use of the UKRVO joint digital archive: GRBs fields, Pluto, and satellites of outer planets. Odessa Astronomical Publications 27, 65.

Vavilova, I.B., Pakulyak, L.K., et al., 2012. Astroinformation resource of the Ukrainian virtual observatory: joint observational data archive, scientific tasks, and software. Kinematics and Physics of Celestial Bodies 28 (2), 85–102. https://doi.org/10.3103/S0884591312020067.

Vavilova, I.B., Vasylenko, A.A., et al., 2016. Multi-wavelength properties and SMBH's masses of the isolated AGNs in the Local Universe. In: Active Galactic Nuclei: What's in a Name?, p. 105.

Véron-Cetty, M.-P., Véron, P., 2010. A catalogue of quasars and active nuclei: 13th edition. Astronomy & Astrophysics 518, A10. https://doi.org/10.1051/0004-6361/201014188.

Villarroel, Beatriz, et al., 2019. The Vanishing & Appearing Sources during a Century of Observations project: I. USNO objects missing in modern sky surveys and follow-up observations of a "missing star". arXiv:1911.05068 [astro-ph.EP].

Voges, W., Aschenbach, B., Boller, Th., et al., 1999. The ROSAT all-sky survey bright source catalogue. Astronomy & Astrophysics 349, 389–405. arXiv:astro-ph/9909315 [astro-ph].

Voges, W., Aschenbach, B., Boller, T., et al., 2000. ROSAT all-sky survey faint source catalogue. In: International Astronomical Union Circular, vol. 7432, p. 3.

Vorontsov-Velyaminov, B.A., 1959. Atlas i Katalog Vzaimodejstvuûsih Galakatik I (Atlas and catalog of interacting galaxies). Sternberg Institute, Moscow State University.

Vorontsov-Velyaminov, B.A., 1977. Atlas of interacting galaxies, part II and the concept of fragmentation of galaxies. Astronomy & Astrophysics. Supplement Series 28, 1–117.

Warwick, R.S., et al., 1988. A survey of the galactic plane with EXOSAT. Monthly Notices of the Royal Astronomical Society 232, 551–564. https://doi.org/10.1093/mnras/232.3.551.

Wisotzki, L., et al., 2000. The Hamburg/ESO survey for bright QSOs. III. A large flux-limited sample of QSOs. Astronomy & Astrophysics 358, 77–87. arXiv:astro-ph/0004162 [astro-ph].

Yizhakevych, O.M., Andruk, V.M., Pakuliak, L.K., 2017. Photographic observations of Saturn's moons at the MAO NAS of Ukraine in 1961–1990. Kinematics and Physics of Celestial Bodies 33 (3), 142–148. https://doi.org/10.3103/S0884591317030035.

Yu, Yong, et al., 2017. Digitizer of astronomical plates at Shanghai Astronomical Observatory and its performance test. Research in Astronomy and Astrophysics 17 (3), 28. https://doi.org/10.1088/1674-4527/17/3/28.

Zeldovich, Ia.B., Einasto, J., Shandarin, S.F., 1982. Giant voids in the Universe. Nature 300 (5891), 407–413. https://doi.org/10.1038/300407a0.

Zhang, X., et al., 2017. The properties of red giant stars along the Sagittarius tidal tails. Astronomy & Astrophysics 597 (A54), A54. https://doi.org/10.1051/0004-6361/201629051.

Zhang, Y., Zhao, Y., 2015. Astronomy in the Big Data era. Data Science Journal 14, 11. https://doi.org/10.5334/dsj-2015-011.

Zucca, Elena, et al., 1993. All-sky catalogs of superclusters of Abell-ACO clusters. The Astrophysical Journal 407, 470. https://doi.org/10.1086/172530.

Zwicky, F., et al., 1961. Catalogue of Galaxies and of Clusters of Galaxies, vol. I. California Institute of Technology, Pasadena.

Zwicky, Fritz, Zwicky, Margrit A., 1971. Catalogue of Selected Compact Galaxies and of Post-Eruptive Galaxies. Zwicky, Guemligen.

CHAPTER 6

Surveys, Catalogues, Databases/Archives, and State-of-the-Art Methods for Geoscience Data Processing

LACHEZAR FILCHEV, ASSOC PROF, PHD • LYUBKA PASHOVA, ASSOC PROF, PHD • VASIL KOLEV, MSC • STUART FRYE, MSC

6.1 GEOSPATIAL SURVEYING

6.1.1 Collecting Geospatial Data Through in Situ, Aerial, and Satellite Surveying

During the 2000s and 2010s, among the many modern applications of big datasets, their number in the field of Earth sciences has grown extremely worldwide thanks to the development of industrial technologies and the space programs of the leading countries in space explorations. The precise and cutting-edge data provided by geospatial technologies empowered modern society to tackle environmental and climate change issues. In the digital age, the rapid growth of processing power and global connectivity of geosensor networks allow to collect, share, and analyze the vast amount of Earth observations (EO) data from geospatial surveying.

Geospatial data are Big Data, the number of which now has grown significantly with the accumulation of satellite images of over 660 satellites performing EOs, as listed by the Union of Concerned Scientists (https://www.ucsusa.org/; https://www.pixalytics.com/). New tools and technologies are now available to deal with Big Geo Data analytics and visualization. Geospatial information is advancing in all the dimensions of Big Data. The high-resolution observations collected by satellites are growing by a few terabytes every day. The worldwide observation information will likely surpass one exabyte according to insights of the Open Geospatial Consortium (OGC) (OGC, 2017).

Advanced Big Data techniques are useful for compressing, clustering, and modeling EO data during analysis, interpretation, and visualization in a variety of applications. The data flow from satellite, airborne, and in situ remote sensing is dramatically increasing and needs to be processing thanks to the "new" Big Data tools, including with a new combination of former classical approaches.

Geospatial information and EO, together with modern data processing and Big Data analytics, offer unprecedented opportunities to modernize national statistical systems and consequently to make a quantum leap in the capacities of countries to efficiently track all facets of sustainable development. The GEO/CEOS study suggests that EO data has a role to play concerning most of the 17 sustainable development goals (SDGs). More specifically, around 40 of the 169 targets (representing about a quarter) and around 30 of the 232 indicators (about an eighth) are supported (CEOS_EOHB, 2018; EO4SDG, 2018).

The benefits of satellite EO are already well understood across many areas of government, industry, and science as a valuable information source in support of many sectors of society. Key benefits of satellite EO data for the SDGs reporting against the indicators are (CEOS_EOHB, 2018):

- satellite EO data making the prospect of a global indicator framework for the SDGs viable;
- the potential to allow more timely statistical outputs, to reduce the frequency of surveys, to reduce respondent burden and other costs, and to provide data at a more disaggregated level for informed decision making;
- improved accuracy in reporting by ensuring that data are more spatially explicit.

Big Data are both a challenge and an opportunity. Big Data require adopting a distributed approach to parallelizing data management, i.e., data storage and processing. According to "ISO/IEC 20546:2019 Information Technology-Big Data-Overview and Vocabulary,"

Knowledge Discovery in Big Data from Astronomy and Earth Observation. https://doi.org/10.1016/B978-0-12-819154-5.00016-3

Big Data are "extensive datasets – primarily in the characteristics of volume, variety, velocity, and/or variability – that require a scalable technology for efficient storage, manipulation, management, and analysis." New calculation methods, algorithms, research infrastructures, and computational resources are highly demanded to handle the big datasets from EO more efficiently for numerous real- and near-time applications (see, e.g., Li et al., 2016).

In March 2012, the United States government proposed the "Big Data" Initiative. It could be the first government project on Big Data that focuses on improving the ability to extract knowledge from large and complex collections of digital data. For remote sensing Big Data, one of the most important US government projects is the Earth Observing System Data and Information System (EOSDIS). It provides end-to-end capabilities for managing the National Aeronautics and Space Administration (NASA) Earth science data from various sources (Peng et al., 2018). In a recent document called "A Decadal Strategy for Earth Observation from Space" published by US National Academies of Sciences, Engineering, and Medicine it is emphasized that the ESAS 2017 steering committee and its study panels carefully considered opportunities to lower the cost of making research quality EO by leveraging advances in technology, international partnerships, and the capabilities emerging in the commercial sector (NASEM, 2018). Attention was also given to the exploitation of "Big Data" for Earth science. Investments in innovative analysis capabilities and methodologies accelerate the ability to convert observations into scientific knowledge. Candidates include (a) data science, including Big Data analytics and other techniques emerging in the commercial world, and (b) a more integrated data analysis system that includes advances in modeling and assimilation of in situ data and data from multiple satellite sensors. NASA and the National Oceanic and Atmospheric Administration (NOAA) are both evaluating the use of Big Data in a manner that facilitates the development of new knowledge and applications (NOAA, Big Data Project, 2018).

For sensor measurements of the Earth to be meaningful, measurement records must be accompanied by geospatial coordinate location data and measurement time/date information. With these metadata, measurement data products can be placed on maps, and changes over time can be quantified for each location. Repeat measurements are made by sensors in the field (in situ), sensors mounted on aerial platforms such as balloons, drones, and airplanes, and sensors flown on satellites in space. The combination of these sensor measurements constitutes surveys of the Earth of various spatial, temporal, spectral, and radiometric resolution.

One of the first satellite surveying platforms was the Landsat series. Landsat-7 (launched in 1999) and Landsat-8 (launched in 2013) are still in operation as of 2019, gathering multispectral information of the Earth's land surface on a 16-day repeat cycle. The sun-synchronous polar orbit of the Landsat satellites is designed such that successive days view different parts of the Earth along the flight path, but bring each point on the Earth into view of the satellite sensors every 16 days.

The CEOS Land Product Validation Team, which comprises several subgroups (http://ceos.org/ourwork/workinggroups/wgcv/), is a team of international calibration and validation scientists who examine the data produced by Landsat and other land remote sensing satellites. The team analyzes also them for measurement accuracy, including geospatial location accuracy, temporal accuracy, and other aspects of the spectral measurements taken by the satellites and processed into products that identify features of the land use and land change characteristics.

In Europe, important initiatives related to Big Data and data mining were developed within the EU R&I Framework Programs, for instance the implementation of the Copernicus core services: Atmosphere (CAMS), Marine (CMEMS), Land (CLMS), Climate (C3S), Emergency (EMS), and Security. The rising importance of the Big Data aspects in the space domain was addressed by establishing a new unit, "Space data for Societal Challenges and Growth," within the DG Internal Market, Industry, Entrepreneurship and SMEs/Space Policy, Copernicus, and Defense, in June 2015. The importance that the European Commission (EC) attributes to Big Data from Space is also highlighted by the several research calls for proposals within the EC's Framework Programs for Research and Technological Development (RTD). More specifically, within Horizon 2020 (the current EU Framework RTD Program), EO activities are considered an essential element to accompany the investments made by the EU in the Copernicus program as well as in GEO. In turn, the European Space Agency (ESA)'s thematic EO exploitation platforms networking has also been focusing on the EU market.

Many other satellites take measurements of the Earth at various frequencies and at various electro-magnetic wavelengths, including microwave, radar, LiDAR, and thermal-infrared, in addition to optical wavelengths in the 400–2400 nm range.

6.1.2 CEOS LPV Group & NASA Aeronet

Satellite agencies are in the process of moving not only their data archive holdings to cloud-based storage but are also transferring their processing and analysis software to the cloud. For example, the NASA Earth Observing Systems (EOS) program has undertaken a project to migrate their 20+-year archive of satellite data products to the Amazon Web Service (AWS). Besides, they are implementing their processing software within cloud-based service offerings under configuration control of the Convex architectural tools. These processing software suites convert the raw data from the satellite sensors to geophysical units, apply calibration and other adjustments to correct the data, and derive higher-level outputs that identify natural and man-made changes on the Earth as maps (Lynnes et al., 2017).

When completed, the archive data products will be searchable and downloadable from the AWS S3 storage. New products that arrive daily as raw data from the satellite downlink stations and higher-level products created and stored using cloud computing services instead of the NASA-supported servers that have been resident at all of the EOS Distributed Active Archive Centers (DAACs) since the inception of the EOS program will be moved there.

Inherent in cloud computing services are software tools that supply data analytics and data mining capabilities for users. Examples of data analytics using the hyperspectral data from NASA's Earth Observing One (EO-1) Hyperion instrument are resident at the U.S. Geological Survey catalog at https://earthexplorer.usgs.gov/. Matsu is a cloud research platform funded by a grant from the US National Science Foundation to explore cloud capabilities for EO data.

Data analytics applied to large datasets such as the satellite hyperspectral Level 1 products provide automated capabilities to identify changes in patterns and anomalies such as hotspots from fires or volcanoes simply by setting threshold values that trigger alerts (or flags) when new data are ingested.

While raw data are delivered from the satellites in binary format, the first-level products are generated as large data files in Hierarchical Data Format (HDF), NetCDF, or GeoTIFF raster formats. The first-level data can be geolocated so they can be viewed on a desktop map, but these products are typically very large, ranging in size from 500 megabytes to several gigabytes. Higher-level map products are beginning to be delivered in vector-based formats (Keyhole Markup Language Zipped [KMZ]; Geo-JSON, Shape) that are much smaller in size and show detection areas as polygons of various colors contained in map overlays.

Users can view these higher-level products directly on the cloud platform (e.g., http://geojson.io/#map=2/20.0/0.0) or download them and open them on their local desktops where they can be combined with other relevant overlays on their local platforms (such as Google Earth or ArcGIS). Standard data products, as well as experimental products, can be generated in the cloud by using cloud-based processing tools or they can be generated on a user desktop and shared via the cloud.

So far, the capabilities that can be provided through cloud-based archive and processing services on data that have already been gathered have been described. However, in addition to providing these services, the cloud also enables linking together multiple satellite capabilities into what is known as a satellite "sensor web," where additional insight into potential upcoming data acquisitions across a fleet of satellites can be obtained.

Typical sensor web services provide internet-based access via common open interfaces for (Fig. 6.1):

- users to query satellite overflights over their area of interest;
- operators to coordinate acquisition strategies to maximize coverage of multiple events;
- rapid notification of delivery status.

The common open interfaces can be just browser-based or can utilize an installed client on the user desktop, such as Whirlwind for search, discovery, and download of Web Map Service holdings, for example.

6.1.3 OGC

Standardization is one of the means to enhance interoperability. In the geospatial data domain, the OGC plays a key role in standardization. The OGC is an international industry consortium of companies, governmental organizations, and universities participating in a consensus process to develop publicly available geoprocessing specifications. Big datasets should be standardized according to recommendations of ISO/TC 211 (de jure formal standards technical committee) and OGC (de facto industry technical specifications).

Concerning service interfaces, it provides specifications for accessing different data: the Web Feature Service (WFS) is tailored to serve vector data, while the Web Coverage Service (WCS) is devoted to multidimensional raster data, point clouds, and meshes; the Web Map Service (WMS) has a special role in that it aims at visualizing vector and raster maps in 2D in the simplest fashion possible.

For spatiotemporal "Big Data," the OGC has defined its unified coverage model (nicknamed GMLCOV)

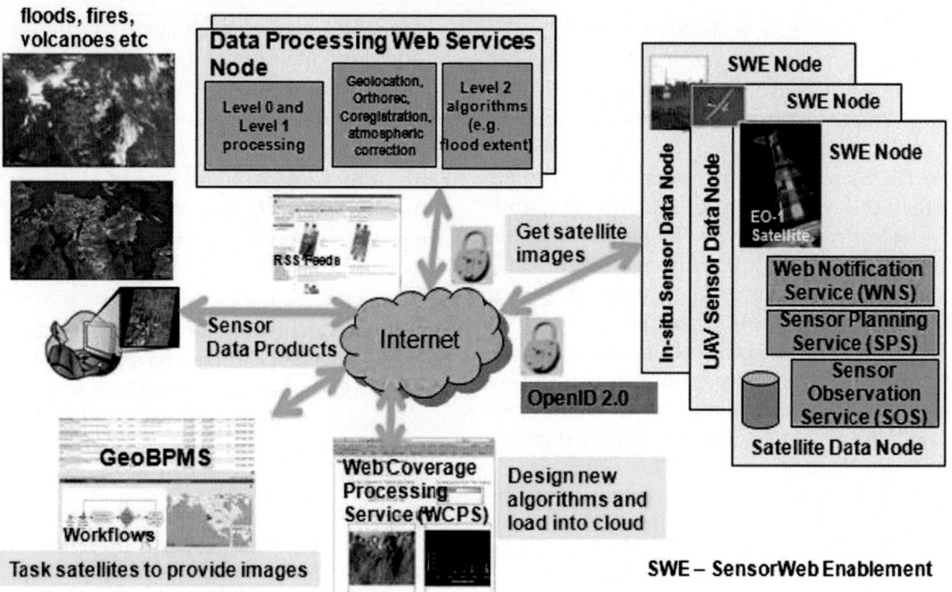

FIG. 6.1 Sensor web diagram (source: Mandl et al., 2013).

which refines the abstract model of ISO 19123 (ISO 2005) to a concrete, interoperable model that can be conformance tested down to the single-pixel level. Coverage is a subtype (i.e., specialization) of a feature, a feature being a geographic object; informally speaking, coverage is a digital representation of some space-time (multidimensional) varying phenomenon (Baumann et al., 2016). Technically, coverage encompasses regular and irregular grids, point clouds, and general meshes. As this notion is not tied to a particular service model, many services can receive or generate coverages, such as OGC WFS, WMS, WCS, WCPS, and WPS. Specifically, the WCS standard provides rich, tailored functionality essential for data access and analysis. For the latter, WCS is closely connected to the Web Coverage Processing Service (WCPS), which defines a query language on coverages, currently on the spatiotemporal raster, i.e., georeferenced arrays addressable by some spatio-temporal coordinate reference system. Generally speaking, this includes an n-D sensor, an image, a simulation output, and statistics data. Over such multidimensional data entities, the WCPS standard offers a leap ahead concerning interoperable processing: it defines a powerful and flexible query language that, on top of data archives, enables coverage data to be used in complex queries. Derived products can thus be built on the fly and combined with other coverages (Baumann et al., 2016).

6.1.4 International Standardization Organization

The International Standardization Organization (ISO) is promoting the use of international standards as a substitute or support to technical regulations. One of the standards and benefit of standardization for geospatial data was established in the 1994 ISO/TC 211 Geographic Information/Geometrics ISO. ISO standardization of OGC specifications defines a simple features access on a web mapping server interface. ISO and OGC jointly developed the imagery and gridded data reference model, Framework and OGC Sensor Markup Language, Geography Markup Language (GML), and ISO Metadata Profile. For geospatial Big Data, standardized methods have been developed to assess, describe, and propagate quality characteristics both quantitatively and qualitatively. For example, ISO 19157 establishes the principles for describing the quality of geographic data. It defines components for describing data quality, specifies components and content structure of a register for data quality measures, describes general procedures for evaluating the quality of geographic data, and establishes principles for reporting data quality (Li et al., 2016).

The Digital Geographic Information Working Group is a NATO cooperative effort resulting in the DIGEST family of standards. It liaises with the international and national standardization institutes, technical committees, and working groups such as CEN/TC 287 and the International Cartographic Association (ICA).

In the USA, the National Spatial Data Infrastructure is an initiative of the US government aimed at the development of policies, standards, and procedures for collection, management, and exchange of geospatial data. The Federal Geographic Data Committee (FGDC), composed of 14 government agencies that produce geospatial information, is the agency responsible for the evolution of the National Spatial Data Infrastructure (NSDI). The Standards Working Group of the FGDC promotes and coordinates FGDC standardization activities, aiming at the coordination of overlapping standards activities, reviews and recommends proposals for FGDC standards, and reviews standards for compliance to policy and procedures (Caprioli et al., 2003).

The European Committee for Standardization (CEN) Technical Committee (CEN/TC 287) was formed to deal with the issue of geospatial standards. One of the main goals of the CEN/TC 287 program in the near past was the development of a quality model (QM) for the geographic information (Caprioli et al., 2003).

The Inter-Agency Committee on Geomatics (IACG) was set up by the Canadian government to enable collaboration between government and industry for the development of an infrastructure to collect, manage, and broadcast geospatial information over the Internet with the relation with the Canadian Geospatial Data Infrastructure as well as several projects, such as Canadian Earth Observation Network (CEONet), GeoExpress, National Atlas of Canada, and Mercator Initiative. The latter is focused chiefly on developing geospatial standards (Caprioli et al., 2003).

DIN is the German standardization strategy – integrate standardization in research and development. It promotes the EU model for adopting ISO and works to establish the EU standardization system in emerging economies and new and future EU member states (Knoop, 2001).

6.2 GEOSPATIAL ARCHIVES, CATALOGS, AND DATABASES

6.2.1 International Archives, Catalogs, and Databases

EO data have been constantly increasing in volume over the last few years, and it is currently reaching petabytes in many satellite archives. It is estimated that up to 95% of the data present in existing archives have never been accessed, so the potential for increasing exploitation is very big. For example, the ESA's Copernicus Missions archive is a ~8 petabyte archive and growing (Cristiano Lopes, ESA, Location Powers Big Linked Geodata, March 2017). The European Centre for Medium-Range Weather Forecasts (ECMWF) currently has 180 petabytes of weather data, with plans to be archiving 1 petabyte/day. The International Data Corporation (IDC) forecasts that by 2025 the so-called global datasphere will have grown to 163 zettabytes (ZB), which is 10 times the 16.1 ZB of data generated in 2016 (Reinsel et al., 2017).

6.2.1.1 UNOOSA

The United Nations Office for Outer Space Affairs (UNOOSA; http://www.unoosa.org) is the United Nations office responsible for promoting international cooperation in the peaceful uses of outer space. UNOOSA serves as the secretariat for the General Assembly's only committee dealing exclusively with international cooperation in the peaceful uses of outer space: UNCOPUOS. It is responsible for implementing the Secretary-General's responsibilities under international space law and maintaining the United Nations Register of Objects Launched into Outer Space. UNOOSA is the current secretariat of the International Committee on Global Navigation Satellite Systems (ICG). It also maintains a 24-hour hotline as the United Nations focal point for satellite imagery requests during disasters and manages the United Nations Platform for Space-based Information for Disaster Management and Emergency Response (UN-SPIDER).

The United Nations Committee of Experts on Global Geospatial Information Management (UN-GGIM) is advancing new approaches to data by implementing a global policy framework that will enable countries to better integrate geospatial and other key information into global development policies and their national plans. Established in 2011, UN-GGIM (http://ggim.un.org/UNGGIM-wg6/) sets directions for the production and use of geospatial information within national and global policy frameworks, and building and strengthening geospatial information capacity of nations, especially of developing countries. Through partnerships within the UN system and with organizations such as the World Bank, the Group on Earth Observations (GEO), the Committee on Earth Observation Satellites (CEOS), and other global actors, our combined ability to contribute towards an interconnected data ecosystem that will allow the member states to properly plan for and implement the SDGs will be realized.

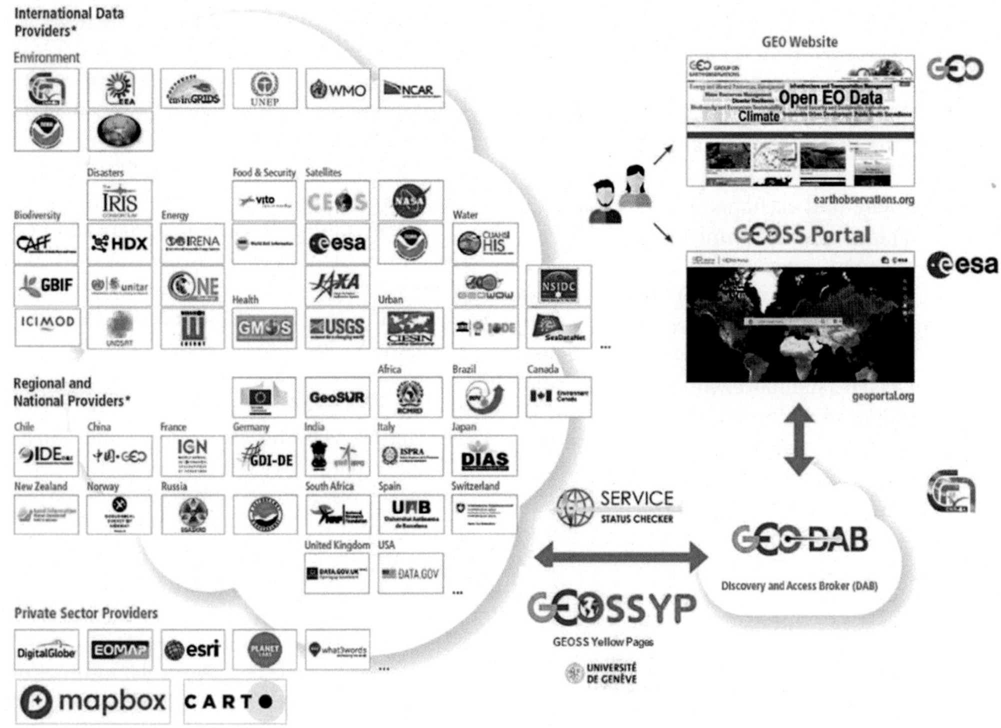

FIG. 6.2 The GEOSS structure (source: Craglia et al., 2017).

6.2.1.2 CEOS

CEOS (http://ceos.org) is the Committee on Earth Observation Satellites, created in 1984 under the aegis of the G7 Economic Summit of Industrialized Nations Working Group on Growth, Technology, and Employment. CEOS was established to provide coordination of the EO provided by satellite missions, recognizing that no single program, agency, or nation can hope to satisfy all of the observational requirements that are necessary for improved understanding of the Earth system. Since its establishment, CEOS has provided a broad framework for international coordination on space-borne EO missions. CEOS membership has reached 32 space agency Members in 2018, comprising almost all of the world's civil agencies responsible for EO satellite programs. GEO and CEOS provide important coordination of the EO community and are joining their efforts in showcasing the value of, and facilitating access to, EO in support of the full realization of the 2030 agenda. The special edition of the CEOS Earth Observation Handbook (2018) presents the main capabilities of satellite EO, its applications, and a systematic overview of present and planned CEOS agency EO satellite missions and their instruments (CEOS_EOHB, 2018).

6.2.1.3 GEO

The Group on Earth Observations (GEO) consists of over 100 member countries plus over 100 contributing organizations that provide access to numerous EO datasets for the benefit of society through their portal at http://www.geoportal.org/. These datasets are derived from ground-based (in situ), aerial, and satellite sensors in addition to other information such as population density and infrastructure data.

The development of GEOSS (Fig. 6.2) has been at the center of achieving the vision of GEO from the very beginning and therefore featured prominently in the initial 10-year implementation plan (2005–2010). GEO is a voluntary initiative, and the building of GEOSS is based on a multitude of individual Earth observing systems from space, air, land, and sea, each continuing to operate independently with their governance structure (Craglia et al., 2017). The task of making these independent systems operate as one in the eyes of the users was entrusted to the GEOSS Architecture and Data Commit-

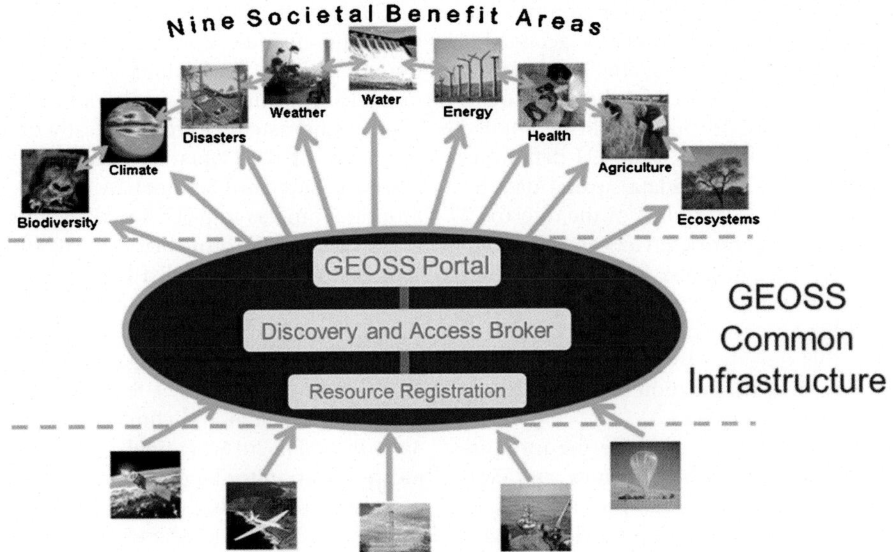

FIG. 6.3 GEOSS components (source: Nativi, 2015).

tee (ADC) first and later to the Infrastructure and Interoperability Board (IIB). They were chaired by USGS and the Joint Research Center (JRC) of the European Commission.

GEO and the GEOSS incorporate EO from diverse sources, including satellite, airborne, and in situ platforms and citizen observatories (Fig. 6.2). There are more than 400 million open data resources at the GEOSS Portal (http://www.geoportal.org) from more than 150 international providers such as national and regional space agencies, international organizations such as the World Meteorological Organization, and commercial satellite data providers.

The initial reference architecture for this System of Systems was that of the NSDI of the US, which had become the model for many similar initiatives across the world (Masser, 2005; Rajabifard et al., 2003). GEOSS applies a brokering service-oriented architecture (B-SOA), implementing a platform whose main components are a GEOSS Portal (managed by ESA), a data discovery and access broker DAB (managed by CNR), a data services status checker (managed by USGS), and a yellow page service (managed by the University of Geneva). The GEOSS platform, as depicted in Fig. 6.3, realized one of the first examples of digital ecosystems (Nativi, 2015; GEO, 2019).

The GEO Plenary and the ESA have organized a "Big Data from Space" conference (Bargellini et al., 2013). The main results of this activity can be summarized in the following list of general requirements:

- EO applications are already facing the Big Data issue, with a need for advanced solutions supporting Big Data handling and Big Data analytics.
- There is a need for flexible solutions enabling ad hoc analytics on Big Data for scientific data exploration on-demand.
- Users require Big Data technologies supporting multiple data models and reducing data transfer.
- Users require advanced visualization techniques easily integrated with different graphical user interface (GUI), including web and mobile systems.

CEOS (http://ceos.org/) is a forum for 59 national civilian space agencies to coordinate their EO efforts. These agencies operate a combined 135 satellites orbiting the Earth and taking measurements daily. CEOS is considered to be the satellite arm of GEO.

6.2.1.4 INSPIRE Directive

In 2007, the European Commission (EC) launched the Spatial Data Infrastructure (SDI) initiative, thereby beginning the development of SDI in Europe. It was followed by a proposal for a directive of "Establishing an infrastructure for spatial information in the Community (INSPIRE)," which was adopted by the EC, the Parliament, and the Council and entered into force on

25 April 2007 (EC, 2007). From a multidisciplinary perspective to use efficiently and effectively the geographic information, the INfrastructure for SPatial InfoRmation in Europe (INSPIRE) Directive, as a legal act of the EU, sets a minimum EU standard that should be applied at the national level. According to the EU definition, a pan-European SDI means metadata, spatial datasets (as described in Annexes I, II, and III of the directive), and spatial data services; network services and technologies; agreements on sharing, access, and use; and coordination and monitoring mechanisms, processes, and procedures, established, operated, or made available in accordance with this directive. INSPIRE builds on the infrastructures for spatial information that have already been created by the member states.

The purpose of INSPIRE is to enable the formulation, implementation, monitoring activities, and evaluation of community environmental policies at European, national, and local level, and to provide publicly accessible information (EC, 2007). One of the biggest benefits of the EU SDI is an improvement in the functioning of the public administration at all levels by facilitating the administrative access to geospatial information. The departure from the existing business model and national practices based on the sales of maps and data and the lack of a complete policy framework are the reasons why these benefits would be difficult to achieve in the short term. The INSPIRE Directive Implementing Rules are relevant to the European initiatives such as e-Government and the EU interoperability framework. INSPIRE aims until 2020 to make harmonized high-quality geographic information readily available to support environmental policies at the national and European level along with policies or activities which may have an impact on the environment (see, e.g., Pashova and Bandrova, 2017).

6.2.1.5 Copernicus

The Copernicus program is one of the European flagship programs providing free and open data and information relying on satellite-based imagery, models, and in situ data. More than only data and information, the Copernicus program provides state-of-the-art models to be used for societal and environmental purposes (Copernicus Ex Ante). The Copernicus program offers information services based on satellite EO and in situ (nonspace) data. The program is coordinated and managed by the EC. It is implemented in partnership with the member states, the ESA, and the European Organisation for the Exploitation of Meteorological Satellites (EUMETSAT), the ECMWF, EU Agencies, and Mercator Océan.

Copernicus has been specifically designed to meet user requirements. Copernicus is served by a set of dedicated satellites (the Sentinel family) and contributing missions (existing commercial and public satellites). The Sentinel satellites are specifically designed to meet the needs of the Copernicus services and their users. Since the launch of Sentinel-1A in 2014, the EU set in motion a process to place a constellation of almost 20 more satellites in orbit before 2030 (Copernicus Observer, 2017). The so-called Copernicus space component consists of Sentinel-1 – land and ocean services (Sentinel-1A launched in 2014, Sentinel-1B in 2016), Sentinel-2 – land monitoring (Sentinel-2A launched in 2015, Sentinel-2B in 2017), Sentinel-3 – ocean forecasting, environmental and climate monitoring (Sentinel-3A launched in 2016, Sentinel-3B 2018), Sentinel-4 – atmospheric monitoring payload (2019), Sentinel-5 – atmospheric monitoring payload (2021) with Sentinel-5 Precursor (P) – atmospheric monitoring launched in 2017, and Sentinel-6 – oceanography and climate studies (2020).

Due to the constantly acquired imagery over the Globe, the Sentinel data are archived in an ESA rolling archive. Some of the data are archived on USGS, Google, and Amazon cloud services. For South-East Europe, there is a Collaborative Ground Segment with Sentinel archived data located in the National Observatory of Athens (NOA), Greece (HNSDMS, 2018). The mirror is powered by the EU GEANT infrastructure. With expected data volumes of 10 terabyte per day (when all Sentinel series will reach full operational capacity), data velocity highlighted by the production of global coverage with repeat time as short as 2 days for Sentinel-3, and data variety resulting from sensors in the optical and radar range at various spatial, spectral, and temporal resolutions, the Copernicus program is a game changer making EO data effectively entering the Big Data era (Soille et al., 2016).

There are around 30 existing or planned Contributing Missions for Copernicus (https://www.copernicus.eu/en/contributing-missions). They fall into the following categories: (1) synthetic-aperture radar (SAR) to observe day and night the land and the ocean; (2) optical sensors to monitor land activities and ocean dynamics; (3) altimetry systems for sea-level measurements; (4) radiometers to monitor land and ocean temperature; and (5) spectrometers for measurements of air quality. A full list of the current Copernicus Contributing Missions is provided at the ESA web site (ESA, Component Data Access, Contributing Missions, 2018a). At present, most of the Copernicus Contributing Missions data are disseminated through the ESA Third Party Mission web site, which contains a satellite data archive with the following missions: ALOS, GOSAT (GOSAT CAI, GOSAT FTS), IKONOS, JERS-1 (JERS-1 SAR Level 1 Precision Image, JERS-1 OPS Very Near Infrared Radiometer, JERS-1 SAR Level 1 Single Look Complex Image), Kompsat-2, Landsat (Landsat 5, Landsat 7, Landsat 5/7 Cloud Free, Landsat 8 NRT data), OCEANSAT-2 (Oceans-2 NRT), PLEIADES, PROBA-1 (Proba-1 CHRIS and HRC), RAPID EYE (RapidEye time series for Sentinel-2), SEASAT, SPOT (Spot 1-5; Spot 6-7), WORLDVIEW-2, and Special Collections (Image 2006 and Tropforest datasets) (ESA, Third Party Missions Online Access List, 2018b). The entire Copernicus core service datasets and the missions are accessible from ESA EOLi-SA standalone software as a Copernicus authorized user (ESA, EOLi-Sa "ESA link to Earth Observation", 2018c).

Copernicus also collects information from in situ systems (https://insitu.copernicus.eu/) such as ground stations, which deliver data acquired by a multitude of sensors on the ground, at sea, or in the air. These value adding activities are streamlined through six thematic streams of Copernicus services: Atmosphere (CAMS), Land (CLMS), Marine (CMEMS), Climate (C3S), Emergency (EMS), and Security.

The Copernicus services transform this wealth of satellite and in situ data into value-added information by processing and analyzing the data. Datasets stretching back for years and decades are made comparable and searchable, thus ensuring the monitoring of changes; patterns are examined and used to create better forecasts, for example, of the ocean and the atmosphere. Maps are created from imagery, features and anomalies are identified, and statistical information is extracted. The information services provided are freely and openly accessible to its users. The main users of Copernicus services are policymakers and public authorities who need the information to develop environmental legislation and policies or to take critical decisions in the event of an emergency, such as a natural disaster or a humanitarian crisis.

The main Copernicus data gateway for Sentinel-1, -2, and -3 is the Open Access Hub (Copernicus Open Access Hub, 2018). Sentinel data, and Copernicus service products and tools are in operation since March 2017 with a 794-terabyte data volume (rolling archive).

Recently within Copernicus different data exploitation platforms were launched employing some image analysis tools which facilitate the satellite data interpretation, i.e., CODE-DE, Sentinel data Portal and Cloud processing platform. Numerous data and exploitation platforms have been built based on Copernicus; amongst them, the most well known is Copernicus Data and Exploitation Platform Deutschland CODE-DE (https://code-de.org/en, Germany) (Reck et al., 2016).

6.2.1.6 Galileo and EGNOS

Galileo is a key component of the EC's space strategy, which focuses on fostering new services, creating business opportunities, promoting Europe's leadership in space, and maintaining Europe's strategic autonomy (EC, 2018). The high-precision Global Navigation Satellite System (GNSS) already supports emergency operations, provides more accurate navigation services, offers better time synchronization for critical infrastructures, and ensures secure services for public authorities. The Galileo Public Regulated Service (PRS) is an encrypted navigation service for government-authorized users, such as civil protection services, customs officers, and the police. This system is particularly robust and fully encrypted to provide service continuity for government users during emergencies or crises.

Putting Europe at the forefront of this strategically and economically important sector, Galileo will provide a highly accurate, guaranteed global positioning service under civilian control (https://www.gsc-europa.eu/). The Galileo services offered to users as of 2019 a continuous evolution of ground and space infrastructure, projecting Galileo towards Full Operation Capability at the end of the 2010s. Deployment of remaining ground/space infrastructure is ongoing (full system – 24 satellites, plus orbital spares to prevent interruption in service). ESA is the system architect for Galileo, managing its design, development, procurement, deployment, and validation on behalf of the EU. ESA will maintain this role, providing technical support to the European GNSS Agency, designated by the EC to run the system and provide Galileo services (https://www.gsc-europa.eu). December 2016 saw the start of Galileo Initial Services, the first step towards full operational capability. Since 2010, EGNOS has been improving accuracy and augmenting GPS, offering safety-critical applications for aviation users. Galileo is expected to spawn a wide range of applications, based on positioning and timing for transport by road, rail, air, and sea, infrastructure and public works management, agricultural and livestock management and tracking, e-banking, and e-commerce. It will be a key asset for public services, such as rescue operations and crisis management. With the new ESA Navigation Innovation and Support Programme (NAVISP), research will focus on the integration of space and terrestrial navigation and new ways to improve GNSS.

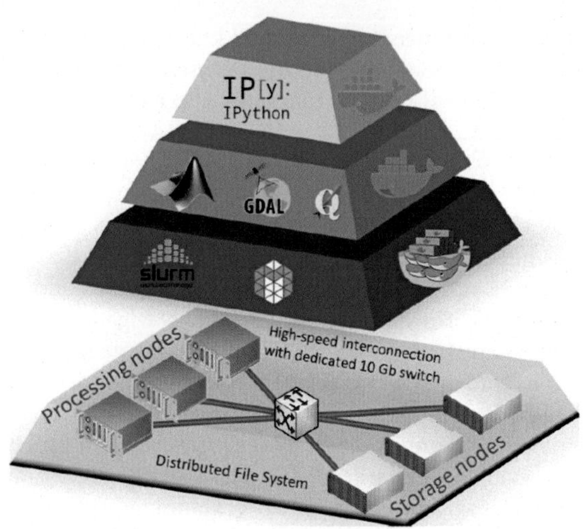

FIG. 6.4 Components of the planned JRC EO data and processing platform (JEODPP) (source: Soille et al., 2016).

6.2.1.7 JRC

Several JRC projects are exploiting EO data to achieve their goals. The scope of the JRC EO & Social Sensing Big Data (EO&SS@BD) pilot project is to propose innovative solutions addressing the needs of JRC projects (Soille et al., 2016). The envisaged architecture consists of processing servers accessing the data provided by a series of storage servers and their directly attached storage (Just a Bunch of Disks or JBODs) in a distributed file system environment (see Fig. 6.4).

6.2.1.8 ESA

Since 1975, the ESA has been conducting European space policy. ESA today has 22 member states and its leading contributors are France and Germany, represented by their respective agencies CNES and DLR. In this role, CNES guarantees Europe's independent access to space and helps prepare new-generation space systems and develop international cooperation.

ESA is one of the few space agencies in the world to combine responsibility in nearly all areas of space activity. European space operators (Arianespace/Launch services/EUMETSAT/Meteorology/Eutelsat and Inmarsat/Telecoms/SES Global/etc.) are the most successful in the world. About 85% of ESA's budget is spent on contracts with European industry. The agency is responsible for the R&D of space projects. On completion of the qualification, they are handed to outside entities for production and exploitation. Most of these entities emanated from ESA. Among the most importantEO missions are the Earth's explorers. These missions address critical and specific issues raised by the science community while demonstrating the latest observing techniques. These are GOCE (2009–2013) studying the Earth's gravity field, SMOS (2009–) studying the Earth's water cycle, CryoSat-2 (2010–) studying the Earth's ice cover, Swarm (2013–), three satellites studying the Earth's magnetic field, ADM-Aeolus (2018) studying global winds, EarthCARE (2019) studying the Earth's clouds, aerosols, and radiation (ESA/JAXA), Biomass (2021) studying the Earth's carbon cycle, and FLEX (2022) studying photosynthesis.

The most important European meteorological missions developed in cooperation with ESA's partner EUMETSAT, as Europe's contribution to the World Meteorological Organization's space-based Global Observing System. The Meteosat Second-Generation (2002, 2005, 2012, 2015–) series of four satellites provides images of the Earth from geostationary orbit. The Meteosat Third-Generation (2021–) series of six geostationary satellites provides images (four satellites) and atmospheric sounding (two satellites). The MetOp (2006, 2012, 2018) series of three satellites provides operational meteorological observations from polar orbit. The MetOp Second-Generation (2021–), two series of polar orbiters, three satellites in each series, continue and enhance meteorological, oceanographic, and climate monitoring observations from the first MetOp series. ESA operates and distributes data from Sentinel-1, -2, -3 (land), and -5P through terrestrial networks and makes data available to all users on the Sentinels Open Access Data Hub, as well as to dedicated user communities on dedicated hubs (i.e., CopHub, IntHub, and CollHub), via powerful application programming interfaces. Recently, ESA announced the tender for four Data Information and Access Services (DIAS) to provide access to the Sentinels' data through highly scalable cloud computing environments. ESA constantly monitors and reports about the performances of the access points under its responsibility. The current activity aims at establishing an independent, geographically characterized benchmark from the user perspective. This shall be achieved by accessing Sentinels data from different locations in Europe and beyond, and by systematically recording the corresponding performances. The obtained results shall be carefully analyzed concerning the statistics obtained locally at the data access points and also for analogous access experiences from notable third-party sites providing similar services (i.e., EO data discovery, view, and download). ESA organizes since 2014 annual conferences on Big Data from Space (Big Data from Space, 2017) to stimulate interactions and bring together researchers, engineers, users, and infrastructure and service providers interested in exploiting Big Data from Space (Peng et al., 2018). ESA is developing a new mode of operating in response to technological advances (e.g., cloud computing, citizen science). Starting from January 2017, ESA has designated that 25% of research funding will be oriented towards new research practices, focusing on interdisciplinary work, pairing Big Data analytics experts with Earth scientists who can interpret the results. The agency has determined that it is necessary now to invest in training existing and future scientists to use Big Data (CEOS_EOHB, 2018).

Since 2017 "Big Data from Space" conferences are coorganized by the ESA, the JRC, and the European Union Satellite Centre (SatCen). Usually, these conferences gather several hundred attendees, including researchers, engineers, users, public institutions, infrastructure suppliers, and related services people for Big Data management. These events allow to know the definition and evolution of Big Data, the new processes and applications of machine learning, policies of use, and data security and demonstrate the strength of open data sources.

Among all the topics it is important to highlight the increase in data management from the Copernicus EO platform, where different applications were revealed, including the large volume of data and the existing platforms for storage and preprocessing of the Sentinel-1 and -2 images, which revealed how in 2018 the volume of data would increase and be complemented by images from Sentinel-1B, -2B, -3A, -5P, and -3B (CGIAR-CSI, 2018). ESA has started a Climate Change Initiative, for storage, production, and assessment of essential climate data.

6.2.2 National Geospatial Catalogues and Databases

The following section describes the past and planned activities in the EO domain within the national space agencies. They are listed below alphabetically.

6.2.2.1 CNES

The Centre national d'études spatiales (CNES) (https://cnes.fr) (English: National Center for Space Studies) is the French government space agency (administratively, a "public administration with an industrial and commercial purpose"). Established under President Charles de Gaulle in 1961, its headquarters is located in central Paris and it is under the supervision of the French Ministries of Defense and Research. It operates from the Toulouse Space Center and Guiana Space Center but also has payloads launched from space centers operated by other countries. CNES is a member of the Institut au Service du Spatial, de ses Applications et Technologies. CNES is leading this initiative within a wide-ranging international partnership through which space agencies and international organizations are committing to pool data and resources to build the climate observatory together. CNES's main partners from the French scientific community – CNRS, IRD, and Meteo-France – are already on board and working towards this goal.

The PEPS platform is intended as a gateway to data from the Sentinel satellites. As such, it constitutes the first step towards closer harmonization of data dissemination platforms in Europe. With the digital revolution and the fundamental shift towards digital and new "Big Data" technologies, the issues at stake go well beyond the space sector and concern the construction of a whole new digital geoinformation ecosystem.

The Theia Land Data Center is a French national interagency organization designed to foster the use of images issued from the space observation of land surfaces (Soille et al., 2016).

6.2.2.2 CSA

In 2019–2020, the Canadian Space Agency (CSA) has planned to set up a big data center of expertise to improve the management and the use of space data by federal departments and agencies, provinces and territories, academic institutions, and Canadian industry (CSA. Departmental Plan, 2019–2020). In addition to the effort boost, EO data are used to further the socio-economic benefits with approximately 37,000 images from the RADARSAT-1 data archives, acquired in the past 17 years, and are going to be made available to the public in 2019–2020 (https://www.asc-csa.gc.ca/eng/satellites/radarsat1/default.asp).

6.2.2.3 CSIRO

The Commonwealth Scientific and Industrial Research Organisation (CSIRO) is an independent Australian federal government agency responsible for scientific research. CSIRO is one of the main actors in EO Big Data operations for Australia. Australia plays a key role in the Landsat mission as a data downlink in the Southern hemisphere and has two receiving stations that have been receiving data from the satellites since 1979. At present, the data rate for our national coverage is 80 satellite scenes every 16 days with variations due to impaired satellite function and mission priorities. The total size of the Landsat archive is therefore approximately one petabyte.

The Australian Geoscience Data Cube (http://www.datacube.org.au/) is the result of the Unlocking the Landsat Archive initiative, funded through the Australian Space Research Program, involving a collaboration between Lockheed-Martin, Geoscience Australia, the National Computational Infrastructure in Canberra, Australia (NCI), and VPAC (Lewis et al., 2014; NCI, Australian Geoscience Data Cube, 2018). The prototype Data Cube makes available, for the first time, more than three decades of satellite imagery spanning Australia's total land area at spatial resolution 25 m. The 240,000+ images show how Australia's vegetation, land use, water movements, and urban expansion have changed over the past 30 years. The Data Cube is available on both Raijin and the NCI private cloud: the former allows deep data analysis using several thousand cores; the latter provides an interactive experience for analysis, including the ability to visualize the data in situ. Australia has already signed an agreement with the EU and Sentinel data are available on Copernicus Australia and Copernicus Australasia web portals (http://www.copernicus.gov.au/). Regional data access is currently provided through both NCI's THREDDS server and the Sentinel Australasia Regional Access (SARA) interface. SARA provides intuitive map-based data search and download capability, as well as an API for advanced user interaction.

The Data Cube has been used to map observations of surface water for all of Australia between 1998 and 2012 at spatial resolution 25 m. The "Water Observations from Space" (WOfS, http://www.ga.gov.au/flood-study-web/#/water-observations) project was an information source for the National Flood Risk Information Portal. The Open Data Cube is hosted in GitHub (https://github.com/opendatacube).

6.2.2.4 DLR

The German Aerospace Center (Deutsches Zentrum für Luft- und Raumfahrt, DLR) is the national aeronautics and space research center of the Federal Republic of Germany. Its extensive research and development work in aeronautics, space, energy, transport, digitalization, and security is integrated into national and international cooperative ventures. DLR's mission comprises the exploration of Earth and the solar system and research for protecting the environment (DLR at a glance, 2018). In the cross-sectoral "Big Data Platform" project, researchers from the DLR are devising new methods for the future-oriented field of Big Data science. The interdisciplinary research project involves 21 DLR institutes from the research fields of space flight, aeronautics, transport, energy, digitalization, and security – all working together (DLR's Big Data Platform cross-sectoral project begins, 2018). DLR has two excellence centers, 10 research projects, and one accompanying research project currently receiving funding, which is managed at the DLR Project Management Agency (DLR, Big Data Research, 2018). The "Berlin Big Data Center" develops highly innovative technologies to structure huge data volumes (Berliner Kompetenzzentrum für Big Data – Berlin Big Data Center (BBDC), 2018). TELEIOS (http://www.earthobservatory.eu/) is a recent European project that addresses the need for scalable access to petabytes of EO data and the effective discovery of knowledge hidden in them. TELEIOS aims to advance the state of the art in knowledge discovery from satellite images by developing an appropriate knowledge discovery framework and applying it to SAR images obtained by the satellite TerraSAR-X of the ELEIOS partner DLR.

In Germany BigDataDays are annually organized, where DLR EO Big Data solutions are presented.

The German Satellite Data Archive (D-SDA) at the DLR has been managing large volumes of EO data in the context of EO mission payload ground segments (PGSs) for more than two decades. Hardware, data management, processing, user access, long-term preservation, and data exploitation expertise are under one roof and interact closely (Kiemle et al., 2016). The main EO scientific data portals of DLR is EOWEB GeoPortal (EGP) – the DLR's central gateway for EO data (DLR, Data Access, 2018). The TSX/TDX scientific portals (terrasar.science.dlr.de and tandemx-science.dlr.de) provide a 90 m digital elevation model (DEM) based on TerraSAR and TanDEM-X data. The DLR GeoFarm hardware organization follows the cloud-like vitalization of processing hardware (Kiemle et al., 2016).

6.2.2.5 INPE

The Brazilian National Institute for Space Research (Sp. Instituto Nacional de Pesquisas Espaciais [INPE]) has the mission to produce high quality of science and technology in space and terrestrial environment areas and to offer unique products and services for Brazilian benefit. INPE's headquarters are in Sao Jose dos Campos, but it also has facilities in other places around the country. The institute develops research in the areas of space and atmospheric sciences, engineering and Space technology, EO by satellites, meteorology, and environmental changes (INPE, 2018).

INPE distributes satellite images from the China–Brazil Earth Resources Satellite (CBERS) missions (China and Brazil), Landsat missions, and Resourcesat-1. Hence, INPE constitutes one of the most complete remote sensing catalogues in the world, with images starting in 1973 right after the launch of Landsat-1. In December 2014, INPE announced the distribution of images from the Indian Resourcesat-2 satellite. The images are received and collected at the ground station in Cuiabá, capital of the federal state of Mato Grosso, which is run by INPE. The images can be ordered for free at http://www.dgi.inpe.br/CDSR/; only a registration is required (Mühlbauer, 2014). The Amazon already has a project started for archiving CBERS data on its servers (The China–Brazil Earth Resources Satellite Mission, 2018). In the University of Münster, the Geospatial Big Data Lab for Interdisciplinary Science and Society (GLISS) lab is also established. This research lab is maintained by Gilberto Câmara as part of his activities and the CAPES-WWU Brazil Chair at the Institute for Geoinformatics at the University of Münster. Part of Dr. Camara's efforts is to establish a global land observatory which will enable the big EO data archive of INPE with geoprocessing capability. Currently proposed GLO is based on SciDB for the archive which will be geoprocessed using free and open R software with openly available geoprocessing packages (Camara, 2014). In recent years the initiative has been continued by the e-Sensing: Big Earth observation data analytics for land use and land cover change information project (http://www.esensing.org) (e-Sensing overview, 2014).

6.2.2.6 ISRO

The Indian Space Research Organisation (ISRO) holds the Indian Space Science Data Centre (ISSDC), which is the primary data center for the payload data archives of Indian space science missions. This center, located at the IDSN campus in Bangalore, is responsible for the ingesting, archiving, and dissemination of the payload data and related ancillary data for space science missions like Chandrayaan and Astrosat (https://www.issdc.gov.in/). The National Remote Sensing Centre (NRSC) is the focal point for the distribution of remote sensing satellite data products in India and its neighboring countries. NRSC has a ground receiving station at Shadnagar, about 55 km from Hyderabad.

Starting with IRS-1A in 1988, ISRO has launched many operational remote sensing satellites. ISRO is currently operating the Indian Remote Sensing (IRS) satellite system for EO applications, mainly for resource monitoring and management, as well as the Megha Tropiques mission, Oceansat, Resourcesat, Cartosat, SARAL, TES Hyperspectral, and DMSAR-1 (Overview of Indian Space sector, 2010–2020, Deloitte, 2010). Among its EO missions, ISRO also acquired data from foreign EO missions. As such the EO data volumes amassed by ISRO are among the world's largest, which also imposes the Big Data issue for the organization.

Acknowledging this trend, in 2018 under the CHAP-NET network, the Center for Soft Computing Research, Indian Statistical Institute, Kolkata has organized a Workshop on "Remotely sensed Big Data Analytics and Mining" (Remotely sensed Big Data Analytics and Mining, 2017).

6.2.2.7 JAXA

The Japan Aerospace Exploration Agency (JAXA, http://global.jaxa.jp/) was born through the merger of three institutions, namely, the Institute of Space and Astronautical Science, the National Aerospace Laboratory of Japan (NAL), and the National Space Development Agency of Japan (NASDA). It was designated as a core performance agency to support the Japanese government's overall aerospace development and utilization. JAXA, therefore, can conduct integrated operations from basic research and development to utilization. JAXA became a National Research and Development Agency in April 2015, and took a new step forward to achieve optimal R&D achievements for Japan, according to the government's purpose of establishing a national R&D agency. The JAXA and NASDA, in particular, has three strategic goals for the EO systems: contribution to the Earth science, promotion of practical use of the EO data, and advanced technology development of satellites, sensors, and ground systems (Akutsu, 2011).

To implement the strategy, a long term plan to develop a series of EO satellite missions was undertaken beginning with the launch of the NASDA Advanced Earth Observing Satellite (ADEOS) in 1996 and the ADEOS-II in 2002 then continuing with the JAXA Advance Land Observing Satellite (ALOS) satellite series with ALOS launched in 2006, ALOS-2 launched in 2014, and the ALOS-3 launch planned for 2020. For the long-term global change observation, a new concept, "Global Change Observation Mission," of a 15-year mission starting from the launch of ADEOS-II was developed. Complementing these satellites, the Tropical Rainfall Measuring Mission (TRMM) follow-on and Earth radiation missions have been developed. JAXA released "G-Portal" as a portal web site for search and deliver data of EO satellites in February 2013. G-Portal handles ten satellites' data: GPM, TRMM, Aqua, ADEOS-II, ALOS (search only), ALOS-2 (search only), MOS-1, MOS-1b, ERS-1, and JERS-1. G-Portal provides over 5.17 million products and the number of catalogues is over 14 million. Currently, G-Portal

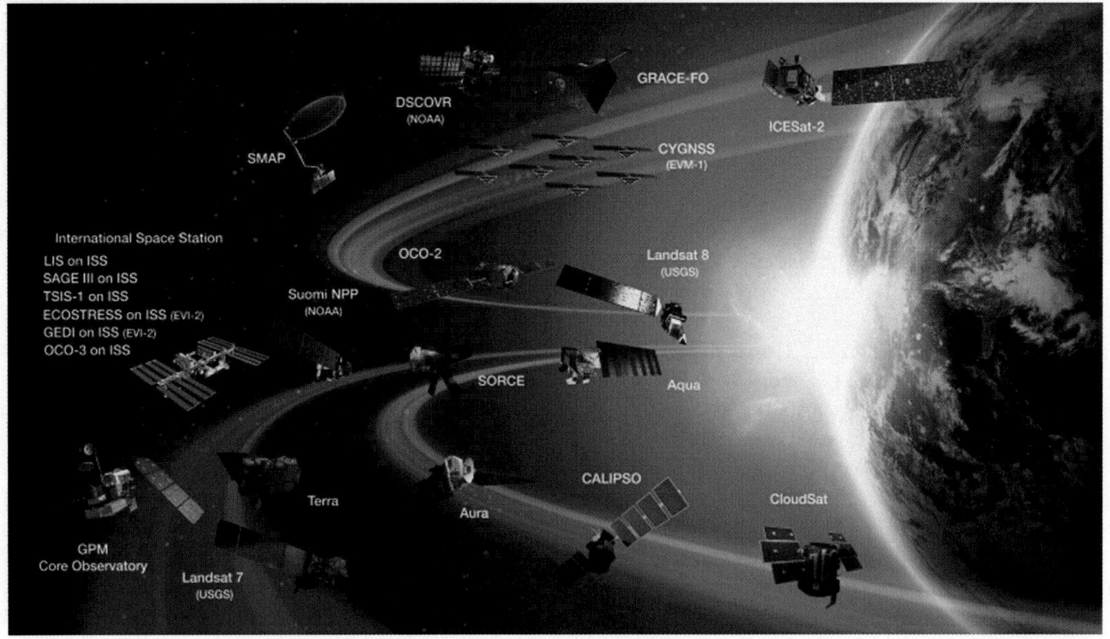

FIG. 6.5 A key component of this effort is NASA's Earth Observing System Data and Information System (EOSDIS) (source: NASA Earth Science Division Operating Missions, 2019).

receives processed satellite data without any time lag and users can search and download promptly (Ikehata, 2015; Ochiai, 2014).

6.2.2.8 NASA, USGS, NOAA

NASA is a research agency responsible for developing all civilian satellites for the US government. NASA operates many research satellites for both EO and space science application to validate the operational readiness of sensor and spacecraft platform improvements (Fig. 6.5). Access to data from all US civilian satellites is free and open. NASA data can be searched for at https://www.nasa.gov/. The EO (https://earthobservatory.nasa.gov) as a part of the EOS Project Science Office at NASA Goddard Space Flight Center serves to provide leadership in organizing open data, convening partners, and demonstrating the power of Big Data analytics through inspiring projects.

The US Geological Survey (USGS) is an operational agency that controls the Landsat satellites and provides Landsat data products in addition to numerous other in situ, aerial, and satellite data within their land remote sensing database that is accessible through the Earth Explorer web interface at https://earthexplorer.usgs.gov/.

The US NOAA is responsible for providing weather information derived from ground-based Doppler radar installations, aerial platforms, and satellite data taken over the US. NOAA controls all the polar-orbiting and geostationary weather satellites operating over the US. Access to NOAA data products can be searched for via http://www.noaa.gov/.

6.2.2.9 RADI

The Chinese Government has also paid attention to Big Data research, with the State Council releasing a report entitled "Action Plan for the Promotion of Big Data Development" in 2015, in which "Scientific Big Data" was specifically recognized (The State Council of China, 2015; Guo, 2017). In October 2017, the National GEOSS Data Sharing Platform of China was presented in the GEO Week 2017 (Peng et al., 2018). The Institute of Remote Sensing and Digital Earth (RADI) at the Chinese Academy of Sciences (CAS) (http://english.radi.cas.cn/) was founded in September 2012 through the merging of the Institute of Remote Sensing Applications and the Center for Earth Observation and Digital Earth (CEODE). RADI's main bodies of operation and research include the State Key Laboratory of Remote Sensing Science, the Center for Applied Technologies of Earth Observation, the National Engineering Center for Geoinformatics, and the CAS Laboratory of Digital Earth Sciences, the China Remote Sensing Satellite

Ground Station, and the CAS Center for Airborne Remote Sensing. The project of cloud service platform aims to establish an open international center for big Earth data. It is one of the 19 A-class strategic high-tech research projects launched by the CAS since 2011. It will integrate science and technology infrastructure with research into resources, environment, biodiversity, and ecosystems. The project strives to make breakthroughs in Earth system sciences, life sciences, and associated disciplines (China Launches Big Earth Data Project, 2018).

6.2.2.10 Roscosmos

The Roscosmos State Corporation for Space Activities (Russian: Государственная корпорация по космической деятельности “Роскосмос”), commonly known as Roscosmos, is a state corporation responsible for the space flight and cosmonautics program for the Russian Federation. The headquarters of Roscosmos is located in Moscow, while the main Mission Control space center is located in the nearby city of Korolev. The Yuri Gagarin Cosmonaut Training Center is in Star City, also in Moscow Oblast. The launch facilities used are Baikonur Cosmodrome in Kazakhstan (with most launches taking place there, both manned and unmanned), and Vostochny Cosmodrome being built in the Russian Far East in Amur Oblast. In 2015 the Russian government merged Roscosmos with the United Rocket and Space Corporation, the renationalized Russian space industry, to create the Roscosmos State Corporation (Roscosmos – Annual report 2016, 2017).

With the advancement of new technologies, the new challenges have to be met. This led to the creation of a spin-off company of Roscosmos. As it was publicly announced at the beginning of 2018, the Russian Space Systems Holding (RKS, part of Roscosmos State Corporation) announced the establishment of its subsidiary TERRA TECH (TEPPA TEX), which will offer geoinformation services to the broader market as part of the commercialization of activities in the field of EO. The operator will create geoinformation solutions based on geospatial data analysis, including remote sensing information. The first of the service line has already been tested in the "pilot" regions of Russia and demonstrated high efficiency. The work of the new commercial operator will be oriented to a broad consumer market. The company will offer fundamentally new services based on machine learning and integration with related services and technologies (navigation, geolocation, Internet of Things [IoT], Big Data, etc.) (“РОСКОСМОС” has created the commercial operator of services on the basis of the data of remote sensing of the Earth, 2018).

6.2.3 Proprietary Geospatial Databases/Catalogues/Archives

The EO industry sees six key technology drivers, i.e., Big Data analytics, cloud computing, artificial intelligence, IoT, automation, and AR/VR, transforming the way forward. Big Data and cloud continue to be the two dominant technologies driving the geospatial industry (GeoBuiz, 2018). In the 2020s, the EO data generated by private EO companies are set to escalate and outnumber the civilian EO satellites currently in orbit; see Table 6.1 (Satellite-Based Earth Observation (EO), 9th Edition, 2017). This will create an opportunity for the private sector to provide more agile EO data for various users' needs, which can complement the civilian sector (see, e.g., Towards a European AI4EO R&I Agenda, 2018). However, the Big Data issue is part of the overall picture of the future for the private sector. Some of the private companies such as Google turn to a bold action to provide free EO services based on their cloud facilities to leverage the data use of the imagery amassed for their Google Earth project.

Google Earth desktop client and Google Earth Engine are provided by Google at no cost to users. These user interface tools provide access to location-specific information about many features of interest, such as roads, businesses, photographs, and other descriptions that are overlaid on a map of the Earth which is stitched together from a series of high-resolution optical satellite images. With these tools, a user can zoom in, pan and tilt, and insert their data product overlays and annotations to convey information to themselves and others.

Amazon is already hosting the Sentinel-2 and Landsat-8 data on its servers, which are freely accessible at http://sentinel-pds.s3-website.eu-central-1.amazonaws.com and http://aws.amazon.com/public-data-sets/landsat (Soille et al., 2016). This is done with the purpose to use data and AWS processing capabilities in civilian and commercial applications. With the DigitalGlobe's Geospatial Big Data platform (GBDX) ecosystem, the DigitalGlobal company (Longmont, CO, USA) is creating building footprints quickly by leveraging machine learning in combination with DigitalGlobe's cloud-based 100-petabyte imagery library (Dan Getman, DigitalGlobe, Location Powers Big Data, September 2016). CloudEO proposes a geoinfrastructure as a service (http://www.cloudeo-ag.com). The giant in aerospace industry Airbus is having its open commercial answer for machine learning processing of the delivered satellite and low Earth orbit (LEO) UAV

TABLE 6.1
List of selected planned small satellite EO constellations (source: NSR; with modification).

Company	Country	In-orbit satellites	Planned number of satellites	Type of sensor	Full constellation deployment
BlackSky Global	USA	0	60	Optical	2019
RISAT	India	0	3	Radar	2019
Astro Digital	USA	0	16	Optical/MS	2019–2020
Hera Systems	USA	0	48	Optical	2020
IceEye	Norway	0	40	SAR	2020
AxelSpace	Japan	0	50	Optical	2020+
Planetary Resources	USA	0	10	Hyperspectral	2020+
Roscosmos	Russia	0	30	Optical/radar	2019+

imagery (Zephyr) of the company portfolio, and satellite datasets SPOT, Pleiades, DMC – Sandbox OneAtlas (https://sandbox.intelligence-airbusds.com/web/).

Other large companies such as Microsoft (Redmond, WA, USA) and Baidu (Beijing, China) are all developing their electronic maps that are supported with the remote sensing Big Data and street views' Big Data as one of the drivers. The commercial applications on Big Data are already changing the lives of the people (Peng et al., 2018). Some of the technologically agile sectors such as BFSI, smart cities, retail and logistics, and advertising have already started to harness Big Data for more targeted outreach (GeoBuiz, 2018; Liu et al., 2018).

6.3 GEOINFORMATICS

6.3.1 Definition and Subject of Geoinformatics

The scientific community believes that there is no complete equivalence between the terms geoinformatics and geomatics, as well as between geomatics and GIS. Different definitions are given to these concepts. Geoinformatics is the science and the technology which develops and uses information science infrastructure to address the problems of Earth sciences such as geography, geodesy, cartography, photogrammetry, GPS, GIS, and related branches of science and engineering. It collects and organizes the data and then analyzes them through computation and geovisualization.

Geomatics is a common scientific term encompassing geospatial technologies such as remote sensing, mapping, exploration, and global positioning systems, while storing, processing, integrating, and managing the collected geographic information through these technological tools. The International Organization for Standardization (ISO) defined geomatics (also used as geoinformatics) as a field of activity which, using a systematic approach, integrates all the means used to acquire and manage spatial data required as part of scientific, administrative, legal, and technical operations involved in the process of production and management of spatial information. These activities include, but are not limited to, cartography, control surveying, digital mapping, geodesy, geographic information systems, hydrography, land information management, land surveying, mining surveying, photogrammetry, and remote sensing.

Also, the ISO gave the following similar yet simpler definition: "Geomatics is the modern scientific term referring to the integrated approach of measurement, analysis, management, and display of spatial data." Based on the above definitions, geomatics emphasizes spatial data collection, measurement, analysis, and display of the Earth and physical objects by using various observation means to extract, manage, and apply spatial information.

6.3.2 Big Data in Geoinformatics – State of the Art and Prospects

Space agencies are not only in the process of moving their data archive holdings to cloud-based storage but are also transferring their processing and analysis software to the cloud. For example, the NASA Earth Observing System (EOS) program has undertaken a project to migrate their 20+-year archive of satellite data products to the AWS. Besides, they are implementing their processing software within cloud-based service offerings under configuration control of the Convex architectural tools. These processing software suites convert the raw data from the satellite sensors to geophysical units, apply calibration and other adjustments to correct the data, and derive higher-level outputs that identify natural and man-made changes on the Earth as maps.

When completed, the archive data products will be searchable and downloadable from the AWS S3 storage. New products that arrive daily as raw data from the satellite downlink stations will be moved there and higher-level products created and stored using cloud computing services instead of the NASA-supported servers that have been resident at all of the EOS DAACs since the inception of the EOS program.

Inherent in cloud computing services are software tools that supply data analytics and data mining capabilities for users. Examples of data analytics using the hyperspectral data from NASA's Earth Observing One (EO-1) Hyperion instrument are resident at the U.S. Geological Survey catalog at https://earthexplorer.usgs.gov/. Matsu is a cloud research platform funded by a grant from the US National Science Foundation to explore cloud capabilities for EO data.

Data analytics applied to large datasets such as the satellite hyperspectral Level 1 products provide automated capabilities to identify changes in patterns and anomalies such as hotspots from fires or volcanoes simply by setting threshold values that trigger alerts (or flags) when new data are ingested.

While raw data are delivered from the satellites in binary format, the first-level products are generated as large data files in HDF, Net-CDF, or GeoTIFF raster formats. The first-level data can be geolocated so they can be viewed on a desktop map, but these products are typically very large, ranging in size from 500 megabytes to several gigabytes. Higher-level map products are beginning to be delivered in vector-based formats (KMZ, Geo-JSON, and Shape) that are much smaller in size and show detection areas as polygons of various colors contained in map overlays. Users can view these higher-level products directly on the cloud platform (e.g., http://geojson.io/#map=2/20.0/0.0) or download them and open them on their local desktops where they can be combined with other relevant overlays on their local platforms (such as Google Earth or ArcGIS). Standard data products, as well as experimental products, can be generated in the cloud by using cloud-based processing tools or they can be generated on a user desktop and shared via the cloud.

So far, the capabilities that can be provided through cloud-based archive and processing services on data that have already been gathered have been described. However, in addition to providing these services, the cloud also enables linking together multiple satellite capabilities into what is known as a satellite "sensor web," where additional insight into potential upcoming data acquisitions across a fleet of satellites can be obtained.

Another feature that is being implemented on cloud-based platforms is known as the "data cube," (DC) where various satellites archive holdings can be ingested and multiple views of an entire area can be made comparing pixel to pixel of the same geolocation on the Earth across many sensor types and multiple years of archive holdings; see Fig. 6.6. Big Earth DC infrastructures are becoming more and more popular to provide analysis-ready data, especially for managing satellite time series (Nativi et al., 2017). Currently, there is a wider effort to adopt a common definition of the big Earth DC with the final goal of enabling and facilitating interoperability. Such "data cubes" are already in operation or under development as CEOS DC, Digital Earth Aust., Vietnam DC, Colombia DC, Swiss DC (Woodcock et al., 2018).

EarthCube is a joint initiative between the Division of Advanced Cyberinfrastructure (ACI) and the Geosciences Directorate (GEO) of the US National Science Foundation (NSF) (Soille et al., 2016).

An array database, Rasdaman (http://www.rasdaman.org) (Jacobs University Bremen and Rasdaman GmbH 2013; Baumann, 2013), empowers the EarthServer technology to integrate data/metadata retrieval, resulting in the same level of search, filtering, and extraction convenience as is typical for metadata (Array DBMS 2014). The EarthServer solution, led by the collection of requirements from scientific communities and international initiatives, provides a holistic approach that ranges from query languages and scalability up to mobile access and visualization. The result is demonstrated and validated through the development of lighthouse applications in the marine, geology, atmospheric, planetary, and cryospheric science domains (Baumann et al., 2016).

6.3.3 Challenges with 4V (Volume, Variety, Velocity, and Value) of the Geospatial and EO Big Data

Whether it is government projects, commercial applications, or academic research when characterizing Big Data, it is popular to refer to the 3Vs, i.e., remarkable growth in volume, velocity, and variety of data

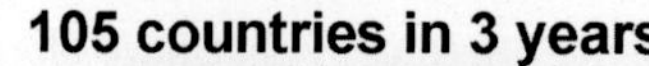

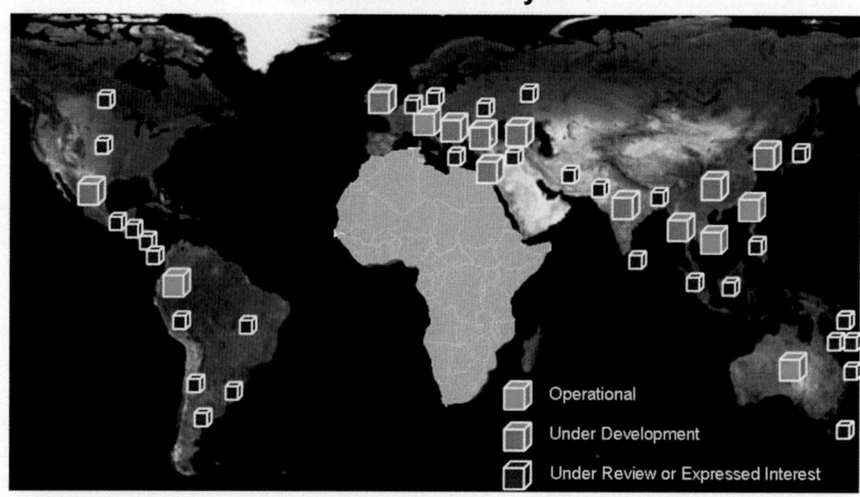

FIG. 6.6 Open Data Cube International – Road to 2020, with the permission of Bryan Killough (NASA Langley).

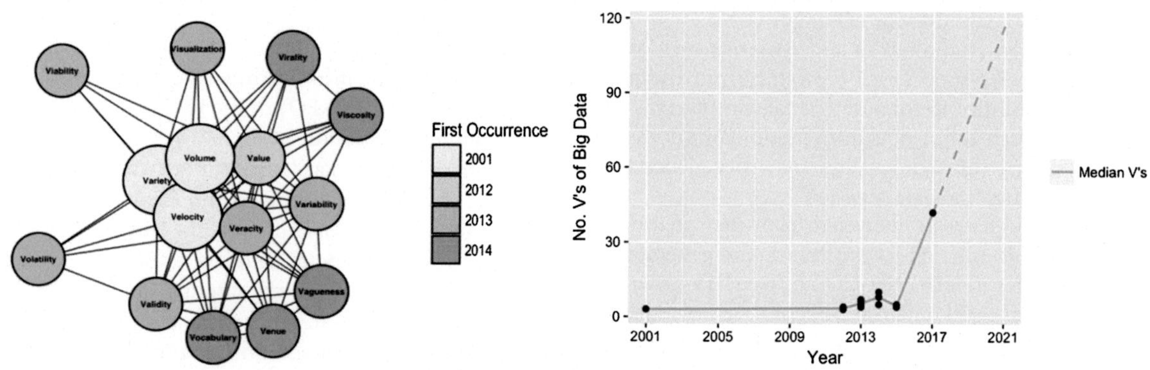

FIG. 6.7 The Vs of geospatial big-data (source: Shafer, 2017).

(Laney, 2001). For remote sensing Big Data, they could be more concretely extended to characteristics of multisource, multiscale, high-dimensional, dynamic-state, isomer, and nonlinear characteristics (Peng et al., 2018). In 2001, Gartner (perhaps) accidentally abetted an avalanche of alliteration with an article that forecast trends in the industry, gathering them under the headings Data Volume, Data Velocity, and Data Variety. Of course, inflation continues its inexorable march, and about a decade later we had the four Vs of Big Data, then 7 Vs, and then 10 Vs. In 2014, in Data Science Central, Kirk Born defined Big Data in 10 Vs, i.e., volume, variety, velocity, veracity, validity, value, variability, venue, vocabulary, and vagueness (Fig. 6.7).

As the number of Vs is increasing with time, some scholars started to mock the Vs concept by presenting the 42 Vs of Big Data and Data Science concept alluring to the Universal number 42 as the ultimate answer (Shafer, 2017).

For instance, the data volume of some of the most well-known EO satellite systems operated by public institutions, Landsat 1–6 missions, is 120 TB; the data volume of Landsat-7 and -8 is estimated from the relating metadata files provided by USGS; MODIS Terra and

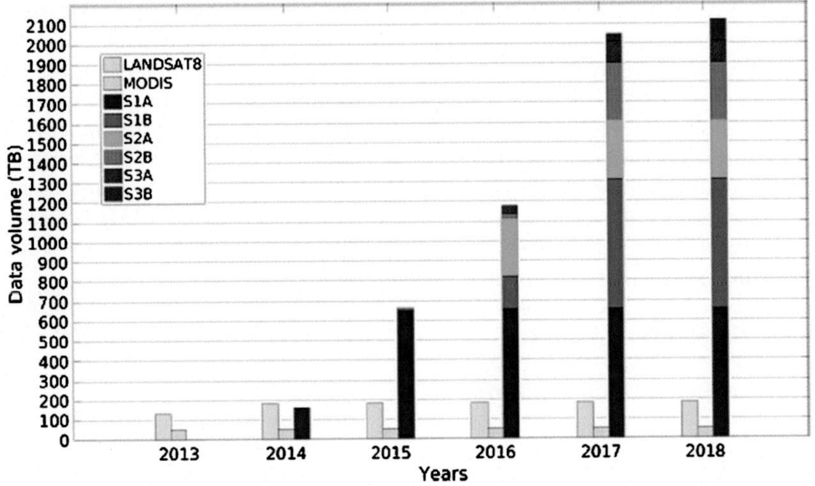

FIG. 6.8 Yearly data flow from Sentinel-1, -2, and -3 (full operational capacity) compared to MODIS and Landsat data flows, image source (source: Soille et al., 2016).

Aqua generate 70 gigabyte/day each; the data volume of Sentinel-1, -2 and -3 (see Fig. 6.8), in their twin constellations, is approximately 3.6 terabyte/day, 1.6 terabyte/day, and 0.6 terabyte/day, respectively (Soille et al., 2016).

6.3.4 HPC Computing in Geoinformatics. Simulations, Visualization, and Animations

6.3.4.1 NASA Earth Observations – NEO (https://neo.sci.gsfc.nasa.gov/)

NEO is part of the EOS Project Science Office located at NASA Goddard Space Flight Center. The NEO mission is to help the end-user to study climate and environmental changes as they occur. The web portal enables the user to browse and download the imagery of satellite data from NASA's constellation of EOS satellites. Over 50 different global datasets are represented with daily, weekly, and monthly snapshots, and images are available in various file formats such as JPEG, PNG, Google Earth, and GeoTIFF (NASA Earth Observations – NEO, 2019).

6.3.4.2 COPERNICUS Data and Information Access Services – DIAS

In 2017, the EC launched an initiative to develop the Copernicus DIAS that facilitate access to Copernicus data and information from the Copernicus services (Copernicus Observer, 2017). Today, data can be accessed through dedicated access portals set up by the Entrusted Entities for each Copernicus service, each requiring a dedicated login. National access mechanisms for Copernicus data exist in some European countries, which are called Collaborative Ground Segments and focus on the distribution of data and information (not only from Copernicus) that is of particular usefulness to national users. ESA launched a tender to establish DIAS services in 2017. Four consortia were chosen to set up DIAS computing environments under ESA management, and a fifth consortium is managed by EUMETSAT, ECMWF, and Mercator Ocean International (MOI).

The DIAS will kickstart the development of European data access and cloud processing service, open for entrepreneurs, developers, and the general public, to build and exploit their Copernicus-based services. Copernicus data and information are for the most part full, free, and open. By mandate of the EC, their production and distribution are ensured by selected institutions across Europe, the so-called Entrusted Entities. The main DIAS components are the back-office infrastructure and the DIAS interface services through which the user-established front office components can be connected to the back-office infrastructure. The DIAS-provided back office is the scalable computing environment in which users can build and operate their services based on Copernicus. The back-office will give unlimited, free, and complete access to Copernicus data and information, and any other data that may be offered by the DIAS provider. The DIAS interface services encompass tools and services that will make it easy for the users to create their applications. The environment should offer scalable computing and storage resources to the users at competitive commercial conditions. Finally, the providers of the DIAS will provide support to users and make sure their access to the DIAS data runs smoothly.

In the future, users will thus have full and free access to Copernicus data and services through the DIAS, and will, at commercial conditions to be determined by the DIAS providers, be able to process the data and information to create services for their end-users. This offering will complement existing portals that may consider installing their offering as a dedicated front office and take advantage of the DIAS themselves.

6.3.4.3 ESA Thematic Exploitation Platforms – TEPs

An important activity related to Copernicus is the thematic exploitation platforms (TEPs) of the ESA. A TEP is a collaborative, virtual work environment addressing a class of users and providing access to EO data, algorithms, and computing/networking resources required to work with them, through one coherent interface. The fundamental principle of the TEPs is to move the user to the data and tools as opposed to the traditional approach of downloading, replicating, and exploiting data "at home." Now the user community is present and visible in the platform, involved in its governance, and enabled to share and collaborate. There are currently seven TEPs addressing the following application areas: coastal, forestry, hydrology, geohazards, polar, urban themes, and food security (https://tep.eo.esa.int/).

6.4 STATE-OF-THE-ART METHODS FOR HYPERSPECTRAL IMAGE PROCESSING

6.4.1 Spectral Imaging Types

Spectral imaging is imaging that uses multiple bands across the electromagnetic spectrum. They are classified as panchromatic, multispectral, or hyperspectral, depending on the number of spectral bands, their widths, and the relative location of their centers.

- **A panchromatic image** is a single-band grayscale image with a high spatial resolution that "combines" the information from the visible R, G, and B bands. It yields a single integrated band containing no wavelength-specific information. The term pansharpening indicates the fusion of a panchromatic and a multispectral image simultaneously acquired over the same area. This can be considered as a particular problem of data fusion. Such images are produced by satellites such as Landsat with a spatial resolution of 15 m, DigitalGlobe's range of satellites, and Satellite for Observation of Earth (in French: Satellite Pour l'Observation de la Terre, or SPOT) 6/7.
- **A multispectral image** is a collection of a few bands from infrared or longer wavelengths, and X-rays and shorter wavelengths of generally different widths and spacings, which may be overlapping. Such images can be obtained, for example, from the Enhanced Thematic Mapper (ETM), which is carried onboard the Landsat 7 satellite, with a spatial resolution of 30 m.
- **A hyperspectral image** is a collection of bands which are sampled at high spectral (1–10 nm) and spatial resolutions. The spectrum is continuous, which allows detailed spectral information for applications in which the spectral characteristics change rapidly. Such images have greater noise and correlation among spectral bands. Hyperspectral images (HSIs) are generally composed of hundreds to thousands of spectral bands. Such images can be obtained, for example, from the 224-band AVIRIS, the 210-band Hyperspectral Digital Imagery Collection Experiments (HYDICE), and the HYPERION satellite sensor onboard NASA's EO-1 satellite. HSI has been increasingly used in areas such as food safety, pharmaceutical process monitoring and quality control, and other biomedical, industrial, biometric, and forensic applications. They have higher spectral resolution (i.e., larger number of bands covering the electromagnetic spectrum), but lower spatial resolution than multispectral or panchromatic images. HSI generates a three-dimensional dataset of spatial and spectral information from materials within a given scene known as a hypercube. With spatial information, the source of each spectrum can be located. HSI generally covers the light spectrum with more spectral bands (up to a few hundred) and higher spectral resolution than RGB multispectral imaging.

6.4.2 Hyperspectral Image Transformation Methods

6.4.2.1 Fourier Transform (FT)

The Fourier transform (FT) provides a way to characterize the overall regularity as well as the related concept of the frequency scale of a periodic signal. An important feature of FT is the orthogonality of the basic functions, which allows for a unique decomposition of signals. The FT is based on the use of sinusoidal basis functions and has only frequency resolution and no time resolution. A remarkable aspect of the FT is the symmetry it displays between the time domain and the frequency domain. The FT allows us to study the global regularity of the signals but does not describe and analyze the

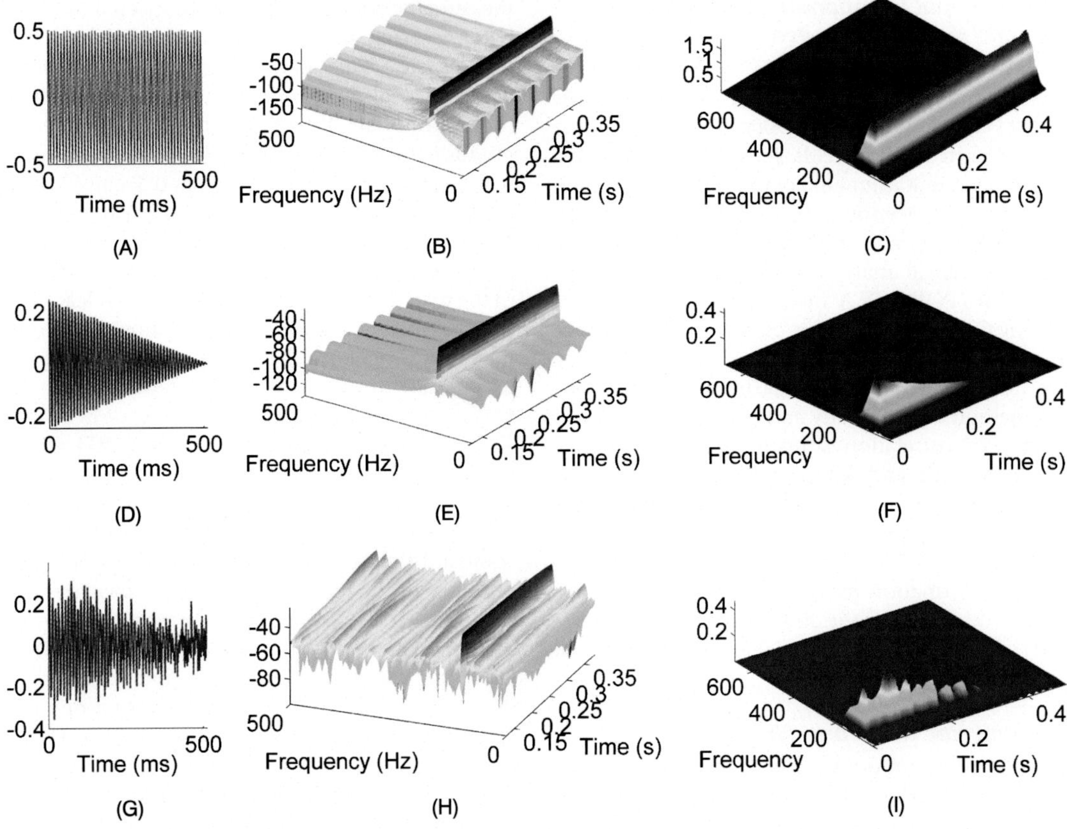

FIG. 6.9 Time-frequency representations of one-dimensional signals (time in seconds). The STFT spectrograms have an equal resolution on a linear frequency scale, while CWT spectrograms have an equal resolution on a log-frequency scale. First row: (A) A stationary signal without noise; (B) its spectrograms STFT; and (C) CWT. Middle row: (D) A nonstationary signal without noise; (E) its spectrograms STFT; and (F) CWT. Last row: (G) A nonstationary noisy signal with white Gaussian noise; (H) its spectrograms STFT; and (I) CWT.

local regularity, since considered signals can be locally regular without being globally regular. The FT does not contain information about the time evolution of the frequencies.

6.4.2.2 Short-Time Fourier Transform (STFT)

To achieve a time-dependent frequency analysis, we introduce the short-time Fourier transform (STFT) (Kaiser, 1994) shown in Fig. 6.9. This is a natural extension to the FT. It converts a time-variable signal into a signal of time and frequency variables without changing its total energy. The STFT uses modulation in the time domain to translate a frequency window. The window is shifted along with the signal; for every position, the spectrum is calculated. This leads to a time-frequency representation. The choice of window size is important. A small window gives good time resolution, but poor frequency resolution, while a big window gives good frequency, but poor time resolution. Therefore, digital filters based on the FT cannot eliminate artifacts in applications.

A simple example with three test signals (Fig. 6.9A, D, and G) shows that the STFTs shown in Fig. 6.9B, E, and H lack information on the variation of frequency. Usually, the time-frequency resolution is defined by the product of the two resolutions (time and frequency). It follows from the Heisenberg uncertainty principle (Strang and Nguyen, 1996) that it is impossible to find a scale which gives good frequency and good time resolution at the same time. Neither a window function nor a wavelet can be compactly supported in time and frequency simultaneously.

6.4.2.3 Principal Component Analysis (PCA) and Karhunen–Loève Transform (KLT)

PCA was originally introduced as a method for finding "lines and planes of closest fit to systems of points in the space" by Pearson in the early 1900s. It is known under several names in different fields, such as Hotelling transform, the discrete Karhunen–Loève transform (KLT) in the electrical engineering field, eigenvector transform in physical science, and singular value decomposition (SVD) in numerical analysis, or proper orthogonal decomposition (POD). PCA involves the calculation of the eigenvalue decomposition of the data covariance matrix or the singular value decomposition of the data matrix. It is a multivariate and maximum variance projection method used to study and explain the variance-covariance or correlation structure of a multivariate dataset to simplify and reduce dimensionality. PCA is capable of transforming the input dataset into a more compact dataset without loss of the main information. This type of transformation is performed by calculating transformation vectors (principal components, PCs), which characterize the directions of the maximum (largest) variance of the dataset in the multidimensional feature space (Jolliffe, 2002).

6.4.2.4 Wavelet Transform (WT)

A particularly important property of the wavelet transform (WT) is to reduce the dimensionality of the signal by multiscale analysis. Wavelet decomposition of hyperspectral images gives us very useful additional information such as the location and nature of high-frequency features (narrow absorption features, noisy bands, and so on) (Banskota et al., 2011; Blackburn, 2007; Hsu, 2007; Bruce et al., 2002; He et al., 2012; Pu and Gong, 2004). The discrete wavelet transform (DWT) can be used to detect amplitude variations over broad and narrow spectral regions. The results in Banskota et al. (2013) show that the forest leaf area index (LAI) estimated from the wavelet subset provided the greatest accuracy via inversion of the discrete anisotropy radiative transfer (DART) model using airborne hyperspectral data from AVIRIS. Therefore the Haar wavelet can be used as an effective tool for the construction of accurate DART model inversion. The maximum decomposition level was chosen (six levels for 184 bands, using the Haar wavelet) (Banskota et al., 2013). The first-level detail coefficients are functionally equivalent to the first derivatives of the reflectance data (Bruce et al., 2002). On the other hand, higher-level detail coefficients have reduced correlation and noise and are similar to some spectral vegetation indices (SVIs) (Darvishzadeh et al., 2008) as they tend to measure the contrast over a broad spectral interval (e.g., between green and red bands, red bands and red-edge regions, etc.). One important point is the ability of the wavelets to pick out information concerning the local properties of an HSI. The method is fast and not computationally demanding, which makes it applicable for real-time implementation. Hsu and Tseng (2000) used the WT maximum modulus for better feature extraction. The Lipschitz exponents estimated at each singularity point provide useful information about the oscillations of the hyperspectral curve for each pixel. It leads to more spectral feature extraction and distinguishes different materials. Therefore the Lipschitz exponent has been used to describe the kind of singularity.

A. Continuous Wavelet Transform (CWT) One of the reasons to consider the WT is to define a window function that is well localized in both time and frequency (Strang and Nguyen, 1996). The relationship between scale and frequency of the CWT is of fundamental importance to the representation of the local energy density. An increase of the scale improves the frequency resolution at lower frequencies and vice versa. The CWT uses scaling in the time domain and in the frequency domain, which leads to different resolutions and detailed information for the considered signal (Fig. 6.9C, F, and I). Using WTs of signals it is also possible to detect temporal interactions of their frequency components since the frequency resolution is much better.

It was shown in Stepinski (2017) that in the analysis of ultrasonic waves for nondestructive inspection of rock bolts, the Morse WT shows the influence of the dispersion effects on the chirp signal more clearly than the STFT with Hamming window. Therefore, an appropriate choice for signals with time-varying amplitude and frequency are Morse wavelets. Wavelet analysis enables spectral data to be transformed into a new representation by decomposing the original spectra into various scales (frequencies). Subsequently, the correlation between the concentration of parameters and the wavelet scales can be ascertained to identify the most sensitive spectral feature for predicting a given parameter. Alexander et al. (2014) compared CWT using pseudo-Wigner–Ville and CWT with the first-order Volterra kernels for earthquake analysis. It is shown that the Volterra CWT has a slightly broader range of high-amplitude frequency components, which means a much better CWT fit to the spectrum over the entire frequency range. CWT is used for the analysis of rocks containing cracks in earthquakes, landslides, and volcanic eruptions. It is very important to detect oriented features (e.g., segments, edges, and vector fields). In the context of wavelet analysis, one needs a directionally selective wavelet. In Rizzo et al. (2017) CWT is used to detect the abrupt change in the fracture pattern from distributed tensile microcracks to localized shear failure in a fracture network produced by triaxial deformation of a sandstone core plug.

An advanced image analysis tool based on a two-dimensional Morlet wavelet-based CWT, which is more sensitive to the direction of linear features, is good for investigating orientation changes at different scales in images of fracture patterns in faulted materials, which are anisotropic in both real and imaginary parts and also in magnitude. The Morlet wavelet is a powerful tool to quantify the scale at which the phase transition from distributed to localized deformation occurs, which can be easily applied to existing collections of thin sections to extract the critical crack lengths.

B. Discrete Wavelet Transform (DWT) When an exact reconstruction of HSI is required, orthogonal wavelets are usually used. Orthogonality is not essential in the representation of HSIs. The wavelets need not be orthogonal and in some applications, the redundancy can help to reduce the sensitivity to noise (Sheng, 1996). Additionally, to avoid introducing artifacts, the wavelets should be smooth (high number of derivatives or vanishing moments) and preferably symmetric (linear phase) (Keinert, 2003; Han, 2011). Orthogonal wavelets cannot be symmetric. A high number of vanishing moments requires a larger support width, so there is a trade-off between these two properties. To avoid pollution of the spectrum by a large number of neighboring wavelets, wavelets with short support are needed. On the other hand, the number of vanishing moments is "only" indicative for the quality of the information. The Daubechies wavelet family (D_N) has minimum support size for a given number of vanishing moments, but does not have an analytical expression. An exception is the Haar orthogonal wavelet, but it has no vanishing moments and the Haar-based DWT does not perform well when approximating smooth spectra. A compromise solution is the biorthogonal wavelets of (Keinert, 1994) and spline type (Keinert, 1995), which can be symmetric but with the loss of energy preservation. The DWT algorithm rearranges the information in HSI and emphasizes features at different scales and wavelengths.

Very useful and preferred in the wavelet analysis are orthogonal, near-symmetric Symlets and Coiflets (Monzón and Beylkin, 1999; Cooklev and Nishihara, 1999). They achieve more high-contrast HSIs where the outputs are similar to those of short filter lengths. Orthogonal Coiflet systems have been popular due to their sampling approximation property and their associated near-linear phase filter banks of even orders (Daubechies, 1992, 1993) and constructed odd-ordered ones (Tian and Wells, 1995). The DWT is very useful in HSI processing since wavelet coefficients on small scales can detect, localize, identify, and eliminate artifacts. Due to the lack of redundancy and the concentration of important data in a small number of coefficients (also known as "sparsity"), the DWT is very appropriate for signal and image compression.

The detail coefficients in the WT show promise for being able to select features that can be used in discriminating weed-free crop from crop intermixed with pitted morning glory. In Kogera et al. (2003), applying the Haar wavelet to naturally mixed hyperspectral signatures gave an overall classification accuracy of 90% to 100%, depending on pitted morning glory growth stage. This was much better than using spectral bands or PCA. The WT seems to be a promising alternative technique for selecting the most informative features from hyperspectral data. They are efficient in dimensionality reduction (Bruce et al., 2002) and classification in agricultural hyperspectral imaging (Plaza et al., 2005).

C. Undecimated Wavelet Transforms (UWT) UWT (Lang et al., 1996) is a particular WT also known as the "stationary wavelet transform." The discrete implementation can be carried out using the algorithm in Holschneider et al. (1989). Aiazzi et al. (2018) consider some applications with the two-dimensional UWT. The tests show that the algorithm preserves small spatial features in the changed regions very well, while the single geometric-mean bounded-ratio (GMBR) algorithm is more sensitive to the degradation of the spatial resolution thanks to a remarkable capability to reject the false alarms. Therefore, when application-oriented prior information is not available, the wavelet approach can detect changes at different dates.

D. Discrete Multiwavelet Transform (DMWT) Multiwavelets are a new part of wavelet theory. They are attractive in signal processing applications because in principle they can be orthogonal, symmetric, smooth, and with short support, all at the same time. Symmetry allows symmetric extension when dealing with image boundaries. Orthogonality leads to independent subimages. A higher number of vanishing moments results in a system capable of representing higher-degree polynomials with a small number of terms, which translates into better coding performance. The multiwavelet MRA is based on more channels with more than one scaling and wavelet function. Multiwavelet filter banks take into account 16 bands, where one parent gives rise to three child coefficients, compared to one child per parent for scalar wavelets. Therefore, a multiwavelet will be more effective at noise removal. Several multiwavelets have been designed (Chui and Lian, 1996; Chui et al., 2002; Rieder, 1997; Zhou, 2002). The GHM, CL, and SA4 multifilters are most commonly used orthogonal multiwavelet systems and have remarkable properties (Geronimo et al., 1994; Rieder and Nossek, 1996; Strang and Strela, 1994a, 1994b, 1995). A supercompact multiwavelet filter with compact support (0,1) and integer values are used to construct a lossless image compression system (Huang et al., 2006). Despite the poor regularity, the lossless compression methods applied to NASA AIRS ultraspectral sounder (dataNasa1) data significantly outperform methods such as CALIC (Wu and Memon, 1996), SPIHT (Said and Pearlman, 1996), JPEG2000 (Taubman and Marcellin, 2001), and the standard for onboard image compression (CCSDS-IDC) (CCSDS, 2019). Moreover, by using bias-adjusted reordering (BAR) preprocessing in Huang et al. (2004), the average compression ratio is increased from 2.25 to 2.60. The BAR scheme converts three-dimensional data into two-dimensional data by rearranging channels to place those with the highest correlation together. Hyperspectral sounder data have unique spectroscopic characteristics, which feature strong correlations between disjoint regions. This applies to both the spatial dimension, where similar absorbing gases and clouds are scattered across the geographical domain, and the spectral dimension, where these gases and clouds affect disjoint regions of the spectrum simultaneously. In this way, the wavelet or multiwavelet compression methods achieve smoother images and better results.

6.4.3 Hyperspectral Image Classification Methods

6.4.3.1 Image Fusion

A key issue in image fusion is the balance between the preservation of spectral characteristics and high spatial resolution. There exist limitations in the commonly used fusion methods, where users have no control over how much spatial detail or spectral information should be retained. In the traditional wavelet-based fusion methods considered in Donoho (1992) and Garguet-Duport et al. (1996), the low-frequency component of the high-spatial resolution image is replaced directly by that of the multispectral image, which increases the blocking effect. Good fusion methods have to guarantee the preservation of the spectral information of the multispectral image when increasing its spatial information and must allow the injection into each band of the multispectral image.

In many of the readily available fusion techniques, there are problems, such as spectral distortions or a complete change of spectral characteristics, artificial artifacts, or unnatural and artificial colors. One of the methods is wavelet-based, which retains most of the spectral characteristics but at the expense of spatial resolution. Wavelet-based fusion methods are used in González-Audícana et al. (2005), Lillo-Saavedra and Gonzalo (2006), Lei et al. (2006), Dehghani (2003), Sahu and Sahu (2014).

The method in Nunez et al. (1999) presents the proportional additive wavelet fusion (AWLP) approach, using the à trous algorithm (Lillo-Saavedra and Gonzalo, 2006). This allows the use of a dyadic wavelet to merge nondyadic panchromatic images from SPOT (10 m or 30 m resolution) with Landsat (30 m or 90 m resolution) images, respectively. The AWLP methods combine a high-resolution panchromatic image and a low-resolution multispectral image by adding some wavelet planes of the panchromatic image to the intensity component of the low-resolution image. The method is better than the intensity–hue–saturation (IHS or LHS) mergers in preserving both spectral and spatial information, but it produces additional spatial artifacts instead of spatial improvements.

A slightly better method is the wavelet-based adjustable image fusion method presented in Lei et al. (2006), which selects regions of 256×256 and 85×85 pixels, respectively, to evaluate a SPOT panchromatic image and Landsat-7 Thematic Mapper (TM) (bands 2, 3, and 4). The Landsat TM images are resampled to a 10-m resolution to obtain equal image sizes. By adjusting the parameters, the method is comparable with IHS and PCA methods and can achieve a balance between wide-range spectral characteristics and the retention of high spatial resolution.

Knowledge of both the sensor spectral response and the physical properties of the object are combined in González-Audícana et al. (2005) to obtain the WiSpeR algorithm (named by the authors to stand for window spectral response) for the merged image. The method can be understood as a generalization of AWLP methods. To estimate the spatial quality, the spatial information in each band is compared with the spatial information in each band of the original multispectral image by using a Laplacian filter and calculating the correlation between these two filtered images. High-spatial correlation coefficients indicate that many of the spatial detail information of one of the images is present in the other one.

Due to the importance of high-resolution multispectral images in many remote sensing applications, pansharpening techniques have been proposed to increase the spatial resolution of low-resolution HSIs using a high-resolution panchromatic multispectral image. In Licciardi et al. (2016), a combination of PCA and WT provides nonlinearly enhanced spatial principal components. It is shown that an accurate selection of the pansharpening method can lead to an effective improvement of the resolution of HSI and the spectral quality and spatial consistency, as well as a strong reduction in the computational time.

In Miao and Shi (2016) the authors consider a pixelwise method for spectral-spatial classification of HSIs, which combines the advantages of region-based segmentation and image fusion by using a support vector machine (SVM) classification and statistical region merging. The proposed approach is easy and efficient to implement and can be used to classify multispectral images. SVMs have superior performance in hyperspectral classification for reduction of HSIs, and when the number of training data is limited (Melgani and Bruzzone, 2004; Camps-Valls and Bruzzone, 2005).

6.4.3.2 Statistics-Based Techniques

Usually, image classifiers for HSIs include a statistics-based technique – the maximum likelihood (MaxL), artificial neural network (ANN) (Atkinson and Tatnall, 1997), and decision tree (DecT) (Goel et al., 2003) classifiers. The last two are nonparametric classifiers and depend on several factors. The spectral angle mapper (SAM) classifier is based on the theory of spectral matching, in which the spectral similarity between the reference and target spectra is used in classification.

The MaxL classifier assumes that the statistics for each class in each band are normally distributed and calculates the probability that a given pixel belongs to a specific class. Pixels are classified by using a probability threshold. ANN consists of one input layer, at least one hidden layer and one output layer. This algorithm uses standard backpropagation for supervised learning, finding a solution where traditional methods fail to give an accurate result. SAM is a physically based spectral classification that uses an n-dimensional angle to match pixels to reference spectra (Kruse et al., 1993). The algorithm determines the similarity of two spectra by calculating the angle between them.

DecT is strictly nonparametric and learns from given hyperspectral data. It uses explicit rules to classify, segment, or make predictions. It can deal with nonlinear relations and is insensitive to miss values and capable of handling numerical and categorical inputs. The technique uses calibrated reflectance data. Shafri et al. (2007) investigate a 4-hectare natural forest plot located in Forest Research Institute (FRIM) in Kepong, Selangor, Malaysia. Hyperspectral data were acquired on 27 May 2005 by the Airborne imaging spectrometer for applications (AISA) onboard a NOMAD GAF-27 aircraft. The fly-over occurred over 90 minutes between 11:05 and 12:35 h local time. The final results show the comparison of overall accuracy: the ML classifier has the highest overall accuracy (85.56%), followed by the ANN classifier (83.61%), the DecT classifier (50.67%), and a SAM classifier (48.83%).

In recent years, pattern analysis algorithms such as SVMs, kernel Fisher discriminant analysis, and kernel PCA have been developed. They can operate on general types of data and are powerful for classification and regression problems and general learning approaches. The SVM approach consists in finding the optimal hyperplane that maximizes the distance between the closest training sample and the separating hyperplane.

Kernel-based learning theory (Kwon and Nasrabadi, 2005) has been combined with the orthogonal subspace projection (OSP) algorithm, based on a linear mixture model defined in a high-dimensional feature space where data separation is expected to increase. The kernelized OSP algorithm is called kernel orthogonal subspace projection (KOSP) (Kwon and Nasrabadi, 2005). KOSP is designed to exploit higher-order correlations between the spectral bands. The results about Kernel methods in Capobianco et al. (2006) show that the selection of the best kernel algorithm is not straightforward, because it depends on the dimensionality of the training set and the spectral features of a given class (heterogeneous versus homogeneous classes, class distance to the background, class distance to classes means, textural characteristics).

6.4.3.3 Three-Dimensional Spatial-Spectral Methods

The WT using a Gabor wavelet has provided a powerful tool for texture extraction of HSIs in the two-dimensional image space, but the joint spatial-spectral structure inside a HSI cannot be exploited. This leads to the development of three-dimensional spatial-spectral methods. In Qian et al. (2013) a three-dimensional DWT is presented that is carried out by a series of one-dimensional DWT along two spatial axes and one wavelength axis in sequence.

For hyperspectral image classification various approaches have been proposed:

- three-dimensional Gabor wavelet (Jia et al., 2018);
- three-dimensional local binary pattern (3DLBP) (Jia et al., 2017);
- a multiple three-dimensional feature fusion framework (M3DF) (Zhu et al., 2018);
- hyperspectral image classification (Bioucas-Dias et al., 2013), which applies the sparse representation of the three-dimensional surface feature cube directly on the original hyperspectral image, and after that fuses three different kinds of three-dimensional features.

6.4.3.4 Learning Methods

One of the goals of classification is to separate the data into a family of clusters. The mean value of the cluster represents a statistical individual (in general virtual) representative of the cluster population if it is homogeneous (the case of a unimodal distribution of the cluster values, corresponding in general to a quasi-Gaussian shape for the distribution of the observed variables). It represents nothing if the distribution is heterogeneous (multimodal).

If the signal has been sufficiently compressed, the phase of clustering is rapid by using a classical unsupervised tool like k-means. Similar methods like the "*nuées dynamiques*" by Diday (1971) and self-organized maps are also easy to use in the case of a wide collection of data to classify. If these methods do not give results that are easy to be interpreted, SVM has the advantage to give a probabilistic explanation of the obtained clustering. The final step of deep learning can use one or more classification methods (e.g., a supervised method) following the above approaches. Other methods are hidden Markov models (HMMs), expectation-maximization (EM), and final Markov classes (FMCs), now called Markov cluster (MCL), classification expectation-maximization (CEM), PCA, and data fusion.

6.4.4 Hyperspectral Image Denoising Methods

We consider HSIs as a three-dimensional data cube data formed from two spatial and one spectral dimension, with high spatial and spectral resolution. Every pixel records light intensities at numerous wavelengths. Such an image can be modeled by a three-dimensional data cube. The variance of noise in the HSI is different in different hyperspectral bands. This makes denoising of HSIs very challenging. HSI denoising is the estimation of the original (unknown) HSI. Due to the high spectral correlation, HSI denoising is a very important step.

The HSI is assumed to be corrupted by a mixture of four types of noise sources:

- **HSI-independent noise** (SIN) is thermal or quantization noise. It is modeled by a Gaussian distribution, which is assumed to be uncorrelated spectrally, i.e., the noise covariance matrix is diagonal.
- **HSI-dependent noise** (SDN) is modeled by a Poisson distribution whose noise variance is dependent on the HSI level. Its estimation is more challenging than the HSI-independent case.
- **Sparse noise** (SpN) is impulsive noise (photons), missing pixels or/and lines, or other outliers coming from sensor problems. Impulsive noise is removed using an l_1-norm for both penalty and data fidelity terms in the minimization/optimization problem using one-dimensional and two-dimensional DWT (Rasti et al., 2018).
- **Striping noise** (StN) are random artifacts (for example, calibration errors) that may produce patterned noise.

6.4.4.1 Classical Approaches

The classical approach for the denoising of HSIs is by using WT. At different scales, coefficients at each level are retained if they are above the local noise level. This has a big advantage compared to low-pass filtering, where all high-frequency information is removed. Instead, high-frequency information that is above the noise can be retained and therefore remains in the reconstructed HSI.

The HSI can be treated as a three-dimensional data cube, and a denoising technique based on sparse analysis l_1-norm regularization and UDWT has been proposed (Rasti et al., 2013). The denoising method proposed in Chen and Qian (2011) for hyperspectral data achieves a reasonably good signal-to-noise ratio (SNR). It decorrelates the image information from the noise by using PCA and removing the noise in the low-energy PCA output channels. Most of the total energy goes into the first (about 8–10) PCA output channels, which are not denoised. The method removes the noise in each later PCA output channel by bivariate wavelet thresholding (Sendur and Selesnick, 2002).

The authors used test multispectral data from the AVIRIS in the Greater Victoria Watershed District, Canada, on 12 August 2002 and simulate hyperspectral data that were created using the United States Geological Survey and portable infrared mineral analyzer laboratory spectra in combination with fraction (abundance) maps derived from the AVIRIS data of Cuprite, Nevada, USA. The proposed denoising method performs better than the hybrid spatial-spectral noise reduction algorithm (HSSNR) (Othman and Qian, 2006), the bivariate wavelet shrinkage, the VisuShrink method, and the Wiener filter. Both the VisuShrink method and the Wiener filter remove useful features during the denoising process.

6.4.4.2 Penalty Methods

An appropriate method to preserve the spectral information in a three-dimensional data image cube is the penalized (regularized) least-squares method. It is based on the prior knowledge of the HSI, where over- or underestimating the penalty term yields either information loss or poor HSI denoising. Since the penalty term is usually selected based on the chosen model, selecting more complicated models or penalties leads to much harder optimization problems. The advantage of the penalty methods is passing of HSI denoising from the optimization with constrained to unconstrained minimization problems. Ghamisi et al. (2017) considered two-dimensional and three-dimensional denoising processing by Daubechies wavelets (D_2 and D_{10}) at five decomposition levels for spectral and spatial bases. It can be seen that three-dimensional wavelets considerably improve on conventional two-dimensional wavelets.

6.4.4.3 Linear Transformation Techniques

The probability density may have a treatment for the noise, similar to the spectral data. Linear transformation techniques are often used to eliminate noise, such as maximum noise fraction (MaxNF) (Green et al., 1988) and noise-adjusted principal components (NAPCs). These methods are adequate to eliminate noise interferences in applications such as aerial gamma-ray surveys and time series of remote sensing data. In MaxNF it can be shown that the same set of eigenvectors from the principal components transform (PCA) is obtained by procedures that maximize either the SNR or the noise fraction.

Two important properties of the MaxNF transform (not shared by principal components) are:

- it is invariant under scale changes to any band, due to its dependence on the SNR,
- it orthogonalizes the uncorrelated signal and noise components, which leads to an estimation of the covariance matrices of the uncorrelated signal.

The minimum noise fraction transform (MinNF) is an algorithm designed to determine the dimensionality of HSIs, perform denoising of the data, and reduce the computational requirements for subsequent processing. The MNF transform adopts similar arguments to the PCA to derive its components. This method is a linear transformation that uses an HSI-to-noise ratio to sort images, i.e., considering the image quality (Green et al., 1988). The MaxNF calculates an HSI-to-noise index, differently from PCA, which uses a variance-covariance matrix of the dataset. Consequently, MaxNF components will show steadily increasing image quality, unlike the usual ordering of principal components. The NAPC transform is mathematically equivalent to the MinNF transform, but the former transform can be implemented using standard principal components algorithm, without the need for matrix inversion and eigenanalysis of a nonsymmetric matrix (Lee et al., 1990). The application of the NAPC transform requires knowledge of the noise covariance matrix of the data. The conventional MaxNF is a linear matrix transform which reorders the hyperspectral data cube into an HSI space where the bands are ordered by HSI-to-noise ratio. This matrix transform is obtained from estimates of the noise and image covariances.

6.4.5 Dimensionality Reduction Methods for Hyperspectral Images

Dimensionality reduction selects spectral components with higher HSI-to-noise ratio (SNR) among neighboring bands with high correlation. Some known techniques are PCA (Jolliffe, 1986); computing KLT, which is the best data representation in the least-squares sense; SVD (Scharf, 1991), which provides the projection that best represents data in the maximum power sense; maximum noise fraction (MaxNF) (Green et al., 1988), and noise-adjusted principal components (NAPC) (Lee et al., 1990), which seeks the projection that optimizes the ratio of noise to HSI powers.

In the analysis of hyperspectral imagery, a selection of an optimal subset of bands must be performed to avoid the problems due to interband correlation. This can be achieved by employing feature extraction techniques for significant reduction of data dimensionality. Based on locally linear embedding, the Roweis and Saul approach mitigates the effects of high dimensionality on information extraction from hyperspectral imagery (Roweis and Saul, 2000), including PCA (Jolliffe, 2002), minimum noise fraction (MinNF), and linear discriminate analysis (Fukunaga, 1990). Chen and Qian (2008) consider PCA followed by a wavelet decomposition in the spatial domain. Due to a significant amount of noise after the PCA, they apply a bivariate wavelet thresholding method for denoising, which is the most efficient method to increase the peak SNR. Hence, better dimensionality reduction is achieved.

6.5 CONCLUSIVE REMARKS

This chapter provides an overview of some important current developments in EO. The achievements of the world's leading space agencies, future satellite missions, as well as opportunities and perspectives for the long-term use of these observations to achieve sustainable development are discussed. The volume of EOs is increasing at a very rapid rate and their diversity is large, which poses some challenges to their full and effective use in solving scientific and applied problems. The outlook of the big EO data international standardization efforts shows that significant efforts are made to streamline the big geospatial data storage and maintenance consistently. Depending on the availability of the resources and driven by the present-day software and hardware solutions, many national space agencies and private companies invest in cloud infrastructure, build their data cubes, or join international initiatives.

In terms of satellite image processing, future work in the Big Data processing will be the exploration of multidimensional dual-real and dual-complex wavelets, the orthogonal or biorthogonal multiwavelet transforms, and new methods for spatial domain compressed sensing. For example, WT analysis can be applied at canopy level using canopy spectra obtained from airborne and spaceborne hyperspectral data.

Another direction for research is the construction of methods for automatic determination of the number of PCA output channels. HSI analysis is an interdisciplinary field that uses and/or adapts concepts in the fields of signal and image processing, statistical inference, and machine learning theory. It should be noted that adaptations of newly developed image processing algorithms are not easy. For example, the unmixing process has the problem of developing new solutions for the separation of blind sources that cannot be interpreted in known solutions, as is noted in Bioucas-Dias et al. (2013).

Many specialized events, dealing with Big Data in general, already list geospatial Big Data among their topics. This is a trend which will increase in the years to come due to the high interest and need in the domain. However, there is still a gap, as in many other application domains, between the booming technology solutions, expertise, resources availability, and data storage capacity. This is proven by the overflow of EO data from national and international space agencies, which are already mirrored on private companies' archives. The issue is more or less acute depending on the availability of resources, but overall it is going to stay for the years to come.

The industry solutions, for the time being, seem to outpace the technology race for Big Data solutions. This is due to the international competition in the private sector as well as the agility in taking business decisions compared to the international and national space agencies and international organizations. This issue already creates a gap in the institutional capacity to handle present-day Big Data issues. The fact that climate change is one of the many global issues nowadays for which quick and sustainable solutions are needed further requires big geospatial data exploitation. However, these overarching goals, some of which are already addressed in the UN SDGs, are again in the hands of international organizations, which impose the need to work in a closer cooperation with all the stakeholders in the geospatial Big Data value chain.

6.6 LIST OF ABBREVIATIONS

AISA	Airborne imaging spectrometer for applications
ALOS	Advanced Land Observing Satellite
ANN	Artificial neural network
AVIRIS	Airborne Visible/Infrared Imaging Spectrometer
AWS	Amazon Web Service
BAR	Bias-adjusted reordering
CAS	Chinese Academy of Sciences
CEM	Classification expectation maximization
CEODE	Center for Earth Observation and Digital Earth
CEOS	Committee on Earth Observation Satellites
CNES	Centre national d'études spatiales
CSIRO	Commonwealth Scientific and Industrial Research Organisation
CWT	Continuous wavelet transform
DAAC	Distributed Active Archive Center
DART	Discrete anisotropy radiative transfer
DEM	Digital elevation model
DFT	Discrete Fourier transform
DIAS	Data Information and Access Services
DT	Decision tree
EC	European Commission
ECMWF	European Centre for Medium-Range Weather Forecasts
EM	Expectation maximization
EO	Earth observation
EOS	Earth observing systems
EOSDIS	Earth Observing System Data and Information System
ESA	European Space Agency

ETM	Enhanced Thematic Mapper
FGDC	Federal Geographic Data Committee
FMC	Final Markov class
FT	Fourier transform
GEO	Group on Earth Observations
GMBR	Geometric-mean bounded-ratio
GML	Geography Markup Language
GNSS	Global Navigation Satellite System
GUI	Graphical user interface
HDF	Hierarchical Data Format
HMM	Hidden Markov model
HSI	Hyperspectral image
HSSNR	Hybrid spatial-spectral noise reduction
HYDICE	Hyperspectral Digital Imagery Collection Experiments
IDC	International Data Corporation
INPE	Instituto Nacional de Pesquisas Espaciais
IoT	Internet of Things
IRS	Indian Remote Sensing
ISRO	Indian Space Research Organisation
ISSDC	Indian Space Science Data Centre
JAXA	Japan Aerospace Exploration Agency
JRC	Joint Research Center
KMZ	Keyhole Markup Language Zipped
KOSP	Kernel orthogonal subspace projection
KPCA	Kernel principal component analysis
LAI	Leaf area index
MaxNF	Maximum noise fraction
ML	Maximum likelihood
MinNF	Minimum noise fraction
NAPC	Noise-adjusted principal components
NASA	National Aeronautics and Space Administration
NEO	Near Earth Object
NOA	National Observatory of Athens
NOAA	National Oceanic and Atmospheric Administration
NRSC	National Remote Sensing Centre
NSF	National Science Foundation
OSP	Orthogonal subspace projection
PCA	Principal components analysis
PGS	Payload ground segments
POD	Proper orthogonal decomposition
SAM	Spectral angle mapper
SAR	Synthetic-aperture radar
SARA	Sentinel Australasia Regional Access
SDI	Spatial Data Infrastructure
SNR	Signal-to-noise ratio
STFT	Short-time Fourier transform
SVD	Singular value decomposition
SVI	Spectral vegetation indices
SVM	Support vector machine
TEP	Thematic exploitation platforms
TM	Thematic Mapper
TRMM	Tropical Rainfall Measuring Mission
UDWT	Undecimated discrete wavelet transform
UNOOSA	United Nations Office for Outer Space Affairs
UWT	Undecimated wavelet transform
WCPS	Web Coverage Processing Service
WCS	Web Coverage Service
WFS	Web Feature Service
WMS	Web Map Service
WT	Wavelet transformation

ACKNOWLEDGMENTS

The authors are gratefully acknowledged to Dr. Stefano Nativy (JRC EC), Prof. Fritz Keinert (Iowa State University), and Prof. Todor Cooklev (Indiana Purdue University) for their constructive comments for improving the quality of this chapter.

REFERENCES

Aiazzi, B., Bovolo, F., Bruzzone, L., Garzelli, A., Pirrone, D., Zoppetti, C., 2018. Change detection in multitemporal images through single- and multi-scale approaches, multi-scale CD strategy based on wavelet decomposition. In: Mathematical Models for Remote Sensing Image Processing: Models and Methods for the Analysis of 2D Satellite and Aerial Images. Springer, pp. 325–356.

Akutsu, T., 2011. JAXA's contributions to the climate change monitoring. June 7, 2011. URL: http://www.oosa.unvienna.org/pdf/pres/copuos2011/tech-20.pdf.

Alexander, N.A., Chanerley, A.A., Crewe, A.J., Bhattacharya, S., 2014. Obtaining spectrum matching time series using a Reweighted Volterra Series Algorithm (RVSA). Bulletin of the Seismological Society of America 104 (4), 1663–1673.

Atkinson, P.M., Tatnall, A.R.L., 1997. Introduction to neural networks in remote sensing. International Journal of Remote Sensing 11, 699–709.

Banskota, A., Wynne, R.H., Kayastha, N., 2011. Improving within genus tree species discrimination using the discrete wavelet transform applied to airborne hyperspectral data. International Journal of Remote Sensing 32, 3551–3563.

Banskota, A., Wynne, R.H., Thomas, V.A., Serbin, S.P., Kayastha, N., Gastellu-Etchegorry, J.P., Townsend, P.A., 2013. Investigating the utility of wavelet transforms for inverting a 3-D radiative transfer model using hyperspectral data to retrieve forest LAI. Remote Sensing 5, 2639–2659.

Bargellini, P., Cheli, S., Desmos, Y.L., Greco, B., Guidetti, V., Marchetti, P.G., Comparetto, C., Nativi, S., Sawjer, G., 2013. Big Data from Space: Event Report. European Space Agency Publication.

Baumann, P., 2013. Query language guide. URL: http://www.rasdaman.org/export/eb24d6243de082c64a898a28277cc4cb6d623f06/manuals_and_examples/manuals/doc-guides/ql-guide.pdf.

Baumann, P., Mazzetti, P., et al., 2016. Big data analytics for Earth sciences: the EarthServer approach. International Journal of Digital Earth 9 (1), 3–29. https://doi.org/10.1080/17538947.2014.1003106.

Bioucas-Dias, J.M., Plaza, A., Camps-Valls, G., Scheunders, P., Nasrabadi, N.M., Chanussot, J., 2013. Hyperspectral remote sensing data analysis and future challenges. IEEE Geoscience and Remote Sensing Magazine 1 (2), 6–36.

Blackburn, G.A., 2007. Wavelet decomposition of hyperspectral data: a novel approach to quantifying pigment concentrations in vegetation. International Journal of Remote Sensing 28, 2831–2855.

Bruce, L.M., Koger, C.H., Jiang, L., 2002. Dimensionality reduction of hyperspectral data using discrete wavelet transform feature extraction. IEEE Transactions on Geoscience and Remote Sensing 40, 2331–2338.

Camara, G., 2014. Monitoring Tropical Forests and Agriculture: the Roadmap for a Global Land Observatory. Copernicus Big EO data Workshop. Available at: https://www.slideserve.com/brad/monitoring-tropical-forests-and-agriculture-the-roadmap-for-a-global-land-observatory.

Camps-Valls, G., Bruzzone, L., 2005. Kernel-based methods for hyperspectral image classification. IEEE Transactions on Geoscience and Remote Sensing 43 (6), 1351–1362.

Capobianco, L., Carli, L., Garzelli, A., Nencini, F., 2006. Comparison of kernel-based methods for spectral signature detection and classification of hyperspectral images. Proceedings - SPIE 6365, 63650W-1–63650W-12.

Caprioli, M., Scognamiglio, A., Strisciuglio, G., Tarantino, E., 2003. Rules and standards for spatial dataquality in GIS environments. In: Proc. of the 21st International Cartographic Conference (ICC) "Cartographic Renaissance". Durban, South Africa, 10–16 August 2003. ISBN 0-958-46093-0, pp. 1740–1747.

CEOS_EOHB, 2018. Satellite Earth Observations in Support of the Sustainable Development Goals. The CEOS Earth Observation Handbook, Special 2018 Edition. Available at: http://eohandbook.com/sdg/.

Chen, G., Qian, S.-E., 2008. Simultaneous dimensionality reduction and denoising of hyperspectral imagery using bivariate wavelet shrinking and PCA. Canadian Journal of Remote Sensing 34 (5), 447–454.

Chen, G., Qian, S.-E., 2011. Denoising of hyperspectral imagery using principal component analysis and wavelet shrinkage. IEEE Transactions on Geoscience and Remote Sensing 49 (3), 973–980.

Chui, C.K., He, W., Stockler, J., 2002. Compactly supported tight and sibling frames with maximum vanishing moments. Applied and Computational Harmonic Analysis 13 (3), 224–262.

Chui, C.K., Lian, J., 1996. A study of orthonormal multiwavelet. Applied Numerical Mathematics 20 (3), 273–298.

Cooklev, T., Nishihara, A., 1999. Biorthogonal coiflets. IEEE Transactions on Signal Processing 47 (9), 2582–2588.

Craglia, M., Hradec, J., Nativi, S., Santoro, M., 2017. Exploring the depths of the global Earth observation system of systems. Big Earth Data 1 (1–2), 21–46. https://doi.org/10.1080/20964471.2017.1401284.

Darvishzadeh, R., Skidmore, A., Schlerf, M., Atzberger, C., 2008. Inversion of a radiative transfer model for estimating vegetation LAI and chlorophyll in heterogeneous grassland. Remote Sensing of Environment 112, 2592–2604.

Daubechies, I., 1992. Ten Lectures on Wavelets. Soc. Indus. Appl. Math., Philadelphia, PA.

Daubechies, I., 1993. Orthonormal bases of compactly supported wavelets II. Variations on a theme. SIAM Journal on Mathematical Analysis 24 (2), 499–519.

Dehghani, M., 2003. Wavelet based image fusion using a trous algorithm. Poster Session of the Map India Conference, 29–31 January 2003, New Delhi, India.

Diday, E., 1971. Une nouvelle méthode en classification automatique entre connaissance des formes la méthode des nuées dynamiques. Revue de Statistique Appliquée 19 (2), 19–33.

Donoho, D., 1992. Denoising by soft-thresholding. IEEE Transactions on Information Technology 41, 613–627.

Fukunaga, K., 1990. Introduction to Statistical Pattern Recognition. Acad. Press, CA.

Garguet-Duport, B., Gilrel, J., Chassery, J., Pautou, G., 1996. The use of multiresolution analysis and wavelets transform for merging SPOT panchromatic and multispectral image data. Photogrammetric Engineering and Remote Sensing 62, 1057–1066.

Geronimo, J., Hardin, D., Massopust, P., 1994. Fractal functions and wavelet expansions based on several scaling functions. Journal of Approximation Theory 78 (3), 373–401.

Ghamisi, P., Yokoya, N., Li, J., Liao, W., Liu, S., Plaza, J., Rasti, B., Plaza, A., 2017. Advances in hyperspectral image and HSI processing: a comprehensive overview of the state of the art. IEEE Geoscience and Remote Sensing Magazine 5 (4), 37–78.

Goel, P.K., Prasher, S.O., Patel, R.M., Landry, J.A., Bonnell, R.B., Viau, A.A., 2003. Classification of hyperspectral data by decision tree and artificial neural networks to identify weed stress and nitrogen status of corn. Computers and Electronics in Agriculture 39, 67–93.

González-Audícana, M., Fors, O., Núñez, J., 2005. Introduction of sensor spectral response into image fusion methods – application to wavelet-based methods. IEEE Transactions on Geoscience and Remote Sensing 43 (10), 2376–2385.

Green, A., Berman, M., Switzer, P., Craig, M.D., 1988. A transformation for ordering multispectral data in terms of image quality with implications for noise removal. IEEE Transactions on Geoscience and Remote Sensing 26 (1), 65–74.

Guo, H., 2017. Big data drives the development of Earth science. Big Earth Data 1 (1–2), 1–3. https://doi.org/10.1080/20964471.2017.1405925.

Han, B., 2011. Symmetric orthogonal filters and wavelets with linear-phase moments. Journal of Computational and Applied Mathematics 236 (4), 482–503.

He, J., Li, C., Ye, B., et al., 2012. Efficient and accurate greedy search methods for mining functional modules in protein interaction networks. BMC Bioinformatics 13, S19. https://doi.org/10.1186/1471-2105-13-S10-S19.

Holschneider, M., Kronland-Martinet, R., Morlet, J., Tchamitchian, P., 1989. A real-time algorithm for signal analysis with the help of the wavelet transform. In: Wavelets: Time-Frequency Methods and Phase Space. Springer-Verlag, Berlin, pp. 289–297.

Hsu, P.-H., 2007. Feature extraction of hyperspectral images using wavelet and matching pursuit. ISPRS Journal of Photogrammetry 62, 78–92.

Hsu, Pai-Hui, Tseng, Yi-Hsing, 2000. Wavelet-based analysis of hyperspectral data for detecting spectral features. International Archives of Photogrammetry and Remote Sensing XXXIII, Supplement B7. Amsterdam.

Huang, B., Ahuja, A., Huang, H.-L., Schmit, T.J., Heymann, R.W., 2004. Lossless compression of three-dimensional hyperspectral sounding data using context-based adaptive lossless image codec with Bias-Adjusted Reordering. Optical Engineering 43 (9), 2071–2079.

Huang, B., Sriraja, Y., Huang, H.-L., Goldberg, M., 2006. Lossless multiwavelet compression of ultraspectral sounder data. In: IEEE Inter. Symposium on Geoscience and Remote Sensing, pp. 3541–3544.

Ikehata, Y., 2015. JAXA's G-portal approach to global observation satellite big data. American Geophysical Union Fall Meeting, 14–18 December 2015, Abstract ID: IN23D-1748. URL: https://agu.confex.com/agu/fm15/webprogram/Paper59869.html.

Jia, S., Hu, J., Zhu, J., Jia, X., Li, Q., 2017. Three-dimensional local binary patterns for hyperspectral imagery classification. IEEE Transactions on Geoscience and Remote Sensing 55 (4), 2399–2413.

Jia, S., Shen, L., Zhu, J., Li, Q., 2018. A 3-D Gabor phase-based coding and matching framework for hyperspectral imagery classification. IEEE Transactions on Cybernetics 48 (4).

Jolliffe, I.T., 1986. Principal Component Analysis. Springer-Verlag, New York.

Jolliffe, T., 2002. Principal Component Analysis. Springer.

Kaiser, G., 1994. A Friendly Guide to Wavelets. Birkhäuser.

Keinert, F., 1994. Biorthogonal wavelets for fast matrix computations. Applied and Computational Harmonic Analysis 1 (2), 147–156.

Keinert, F., 1995. Numerical stability of biorthogonal wavelets. Advances in Computational Mathematics 4 (1–2), 1–26.

Keinert, F., 2003. Wavelets and Multiwavelets. Chapman & Hall/CRC Press.

Kiemle, S., et al. Big data management in Earth observation – the German satellite data archive at DLR. In: Proc. of BiDS'14, https://doi.org/10.2788/1823, pp. 45–49.

Knoop, H., 2001. ISO/TC 211 Project 19122, Geographic Information / Geomatics – Qualifications and Certification of Personnel – Status and Development, New Technology for a New Century.

Kogera, C.H., Bruceb Lori, M., Shawa David, R., Reddyc Krishna, N., 2003. Wavelet analysis of hyperspectral reflectance data for detecting pitted morningglory (Ipomoea lacunosa) in soybean (Glycine max). Remote Sensing of Environment 86, 108–119.

Kruse, F.A., Lefkoff, A.B., Boardman, J.W., Heidebrecht, K.B., Shapiro, A.T., Barloon, P.J., Goetz, A.F.H., 1993. The spectral image processing system (SIPS) – interactive visualization and analysis of imaging spectrometer data. Remote Sensing of Environment 44 (2–3), 145–163.

Kwon, H., Nasrabadi, N., 2005. Kernel orthogonal subspace projection for hyperspectral signal classification. IEEE Transactions on Geoscience and Remote Sensing 43 (12), 2952–2961.

Laney, D., 2001. 3D data management: controlling data volume, velocity and variety. Available online: https://blogs.gartner.com/doug-laney/files/2012/01/ad949-3D-Data-Management-Controlling-Data-Volume-Velocity-and-Variety.pdf. (Accessed 6 February 2018).

Lang, M., Guo, H., Odegard, J.E., Burrus, C.S., 1996. Noise reduction using an undecimated discrete wavelet transform. IEEE Signal Processing Letters 3 (1), 10–12.

Lee, J.B., Woodyatt, S., Berman, M., 1990. Enhancement of high spectral resolution remote-sensing data by a noise – adjusted principal components transform. IEEE Transactions on Geoscience and Remote Sensing 28 (3), 295–304.

Lei, D., Jing, L., Xiaobing, L., Peijun, S., 2006. A new wavelet-based image fusion method for remotely sensed data. International Journal of Remote Sensing 27 (7), 1465–1476.

Lewis, a., et al. Iterating Petabyte-Scale Earth Observation Processes in the Australian Geoscience Data Cube. In: Proc. of BiDS'14. https://doi.org/10.2788/1823..

Li, S., Dragicevic, S., Castro, F.A., Sester, M., Winter, S., Coltekin, A., Pettit, C., Jiang, B., Haworth, J., Stein, A., Cheng, T., 2016. Geospatial big data handling theory and methods: a review and research challenges. ISPRS Journal of Photogrammetry and Remote Sensing 115, 119–133.

Licciardi, G., Vivone, G., Mura, M., Restaino, R., Chanussot, J., 2016. Multiresolution analysis techniques and nonlinear PCA for hybrid pan-sharpening applications. Multidimensional Systems and Signal Processing 27 (4), 807–830.

Lillo-Saavedra, M., Gonzalo, C., 2006. Spectral or spatial quality for fused satellite imagery? A trade-off resolution using the wavelet a trous algorithm. International Journal of Remote Sensing 27 (7), 1453–1464.

Liu, P., Liping, Di, Qian, Du, Wang, Lizhe, 2018. Remote Sensing Big Data: theory, methods and applications. Remote Sensing 10 (5), 711. https://doi.org/10.3390/rs10050711.

Lynnes, C., Baynes, K., McInerney, M., 2017. Archive management of NASA Earth observation data to support cloud analysis. NASA. https://ntrs.nasa.gov/archive/nasa/casi.ntrs.nasa.gov/20170011455.pdf.

Mandl, D., Frye, S., Cappelaere, P., Handy, M., Policelli, F., Katjizeu, Mc-c., Van Langenhove, G., Aube, G., Saulnier, J.-F., Sohlberg, R., Silva, J., Kussul, N., Skakun, S., Ungar, S., Grossman, R., Szarzynski, J., 2013. Use of the Earth Observing One (EO-1) satellite for the Namibia sensorweb flood early warning pilot. IEEE Journal of Selected Topics in Applied Earth Observations and Remote Sensing 6, 298–308. https://doi.org/10.1109/JSTARS.2013.2255861.

Masser, I., 2005. The future of Spatial Data Infrastructures. https://doi.org/10.1201/9780429505904-12.

Melgani, F., Bruzzone, L., 2004. Classification of hyperspectral remote sensing images with support vector machines. IEEE Transactions on Geoscience and Remote Sensing 42 (8), 1778–1790.

Miao, Z., Shi, W., 2016. A new methodology for spectral-spatial classification of hyperspectral images. Journal of Sensors 2016, 1538973.

Monzón, L., Beylkin, G., 1999. Compactly supported wavelets based on almost interpolating and nearly linear phase filters (coiflets). Applied and Computational Harmonic Analysis 7, 184–210.

Mühlbauer, St., 2014. Free data from Indian Resourcesat-2 Satellite via INPE. http://geoawesomeness.com/free-satellite-data-from-indian-resourcesat-2-via-inpe/.

Nativi, S., 2015. GEOSS and its architecture. Boulder Colorado, USA. Retrieved from https://csdms.colorado.edu/mediawiki/images/Stefano_Nativi_CSDMS_2015_annual_meeting.pdf.

Nativi, St., Mazzetti, P., Craglia, M., 2017. A view-based model of data-cube to support big Earth data systems interoperability. Big Earth Data 1, 75–99. https://doi.org/10.1080/20964471.2017.1404232.

Nunez, J., Otazu, X., Fors, O., Prades, A., Pala, V., Arbiol, R., 1999. Multiresolution-based image fusion with additive wavelet decomposition. IEEE Transactions on Geoscience and Remote Sensing 37 (3), 1204–1211.

Ochiai, O., 2014. JAXA's Earth observation data and information system. In: Copernicus Big Data Workshop. March 13–14, 2014. European Commission, Brussels.

Othman, H., Qian, S.E., 2006. Noise reduction of hyperspectral imagery using hybrid spatial-spectral derivative-domain wavelet shrinkage. IEEE Transactions on Geoscience and Remote Sensing 44 (2), 397–408.

Pashova, L., Bandrova, T., 2017. A brief overview of current status of European spatial data infrastructures – relevant developments and perspectives for Bulgaria. Geo-spatial Information Science 20 (2), 97–108. https://doi.org/10.1080/10095020.2017.1323524.

Peng, G., et al., 2018. A conceptual enterprise framework for managing scientific data stewardship. Data Science Journal 17: 15, 1–17. https://doi.org/10.5334/dsj-2018-015.

Plaza, A., Martínez, P., Plaza, J., Pérez, R., 2005. Dimensionality reduction and classification of hyperspectral image data using sequences of extended morphological transformations. IEEE Transactions on Geoscience and Remote Sensing 43 (3), 466–479.

Pu, R., Gong, P., 2004. Wavelet transform applied to EO-1 hyperspectral data for forest LAI and crown closure mapping. Remote Sensing of Environment 91, 212–224.

Qian, Y., Ye, M., Zhou, J., 2013. Hyperspectral image classification based on structured sparse logistic regression and three-dimensional wavelet texture features. IEEE Transactions on Geoscience and Remote Sensing 51 (4), 2276–2291.

Rajabifard, A., Feeney, M.E., Williamson, I., Masser, I., 2003. National SDI-initiatives. (Chapter 6). In: Williamson, I., Rajabifard, A., Feeney, M.E. (Eds.), Developing Spatial Data Infrastructures: From Concept to Reality. Taylor & Francis, pp. 95–109.

Rasti, B., Scheunders, P., Ghamisi, P., Licciardi, G., Chanussot, J., 2018. Noise reduction in hyperspectral imagery: overview and application. Remote Sensing 10 (3), 482.

Rasti, B., Sveinsson, J.R., Ulfarsson, M.O., Benediktsson, J.A., 2013. Hyperspectral image restoration using wavelets. Proceedings - SPIE 8892, 207-1–207-9.

Reck, C., Campuzano, G., Dengler, K., Heinen, T., Winkler, M., 2016. In: German Copernicus Data Access and Exploitation Platform. BiDS'16, Teneriffa, Spain.

Reinsel, D., Gantz, J., Rydning, J., 2017. Data Age 2025: The Evolution of Data to Life-Critical Don't Focus on Big Data. IDC Analyze the Future, Framingham.

Rieder, P., 1997. Parameterization of symmetric multiwavelets. In: Proc. Int. Conf. Acoust., Speech, Signal Processing, Vol. 3, pp. 2461–2465.

Rieder, P., Nossek, J., 1996. Smooth multiwavelets based on two scaling functions. In: Proc. of the IEEE Inter. Sym. on Time-Frequency and Time-Scale Analysis, pp. 309–312.

Rizzo, R.E., Healy, D., Farrell, N.J., Heap, M.J., 2017. Riding the right wavelet: quantifying scale transitions in fractured rocks. Geophysical Research Letters 44 (11), 808–811. 815.

Roweis, S.T., Saul, L.K., 2000. Nonlinear dimensionality reduction by locally linear embedding. Science 290, 2323–2326.

Sahu, V., Sahu, D., 2014. Image fusion using wavelet transform: a review. Global Journal of Computer Science and Technology: F, Graphics & Vision 14 (5).

Said, A., Pearlman, W.A., 1996. A new, fast, and efficient image codec based on set partitioning in hierarchical trees. IEEE Transactions on Circuits and Systems for Video Technology 6, 243–250.

Scharf, L.L., 1991. Statistical HSI Processing, Detection Estimation and Time Series Analysis. Addison–Wesley, Reading, MA.

Sendur, L., Selesnick, I., 2002. Bivariate shrinkage with local variance estimation. IEEE Signal Processing Letters 9 (12), 438–441.

Shafer, T., 2017. The 42 V's of Big Data and Data Science. https://www.kdnuggets.com/2017/04/42-vs-big-data-data-science.html. (Accessed 6 February 2018).

Shafri, H.Z.M., Suhaili, A., Mansor, S., 2007. The performance of maximum likelihood, spectral angle mapper, neural network and decision tree classifiers in hyperspectral image analysis. Journal of Computer Science 3 (6), 419–423.

Sheng, Y., 1996. Wavelet transform. In: Poularikas, A.D. (Ed.), The Transforms and Applications Handbook. CRC Press, pp. 747–827.

Soille, P., Burger, A., Rodriguez, D., Syrris, V., Vasilev, V., 2016. Towards a JRC Earth observation data and processing platform. In: Proceedings of the Conference on Big Data From Space. BiDS'16, Santa Cruz de Tenerife, pp. 15–17.

Stepinski, T., 2017. Time-frequency analysis of guided ultrasonic waves used for assessing integrity of rock bolts. In: Smart Materials and Nondestructive Evaluation for Energy Systems. Proc. of SPIE, vol. 10171, pp. 101710J-1–101710J-10.

Strang, G., Nguyen, T., 1996. Wavelets and Filter Banks. Wellesley–Cambridge Press, Welleseley MA, USA.

Strang, G., Strela, V., 1994a. Finite element multiwavelet. In: Proc. of Inter. Sym. on Time- Frequency and Time-Scale Analysis, pp. 32–35.

Strang, G., Strela, V., 1994b. Orthogonal multiwavelets with vanishing moments. Journal of Optical Engineering 33 (7), 2104–2107.

Strang, G., Strela, V., 1995. Short wavelets and matrix dilation equations. IEEE Transactions on Signal Processing 45 (1), 108–115.

Taubman, D., Marcellin, M.W., 2001. JPEG2000: Image Compression Fundamentals, Standards and Practice. Springer, USA.

Tian, J., Wells, R., 1995. Vanishing moments and wavelet approximation. Tech. Rep. CML TR95-01. Computational Mathematics Laboratory, Rice University, Houston, TX.

Woodcock, R., Paget, M., Taib, R., Wang, P., Held, A., 2018. CEOS Open Data Cube and Earth Analytics Industry Innovation at CSIRO, What on Earth Colloquium, New Zealand 2018. URL: https://static1.squarespace.com/static/59f0e98ffe54efc487ec7446/t/5aa71e378165f5b3869dd0ee/1520901860649/Robert+Woodcock+Open+Data+Cube.pdf.

Wu, X., Memon, N., 1996. CALIC – a context based adaptive lossless image codec. IEEE ICASSP Conference Proceedings 4, 1890–1893.

Zhou, D.X., 2002. Interpolatory orthogonal multiwavelets and refinable functions. IEEE Transactions on Signal Processing 50 (3), 520–527.

Zhu, J., Hu, J., Jia, S., Jia, X., Li, Q., 2018. Multiple 3-D feature fusion framework for hyperspectral image classification. IEEE Transactions on Geoscience and Remote Sensing 56 (4).

Internet Sources

Big Data from Space, 2017. URL: http://www.bigdatafromspace2017.org/. https://directory.eoportal.org/web/eoportal/satellite-missions/c-missions/cbers-1-2, 2017.

CGIAR-CSI, 2018. A Note from the ESA Big Data from Space Conference 2017 (BiDS'17). Available at: https://cgiarcsi.community/2018/02/13/a-note-from-the-esa-big-data-from-space-conference-2017-bids17/.

China Launches Big Earth Data Project. http://english.radi.cas.cn/Research/RP/201803/t20180313_190698.html, 2018.

CSA, Departmental Plan 2019–2020. https://www.asc-csa.gc.ca/eng/publications/dp-2019-2020.asp.

Consultative Committee on Space Data Systems (CCSDS), 2019. https://public.ccsds.org/default.aspx.

Copernicus Open Access Hub, 2018. URL: https://scihub.copernicus.eu.

Copernicus Observer, 2017. Four consortia selected to set up Copernicus cloud-based platforms for Data and Information Access Services (DIAS). https://www.copernicus.eu/en/copernicus-dias-contracts-signed.

(dataNasa1). ftp://ftp.ssec.wisc.edu/pub/bormin/Count/.

DLR, Data Access, 2018. URL: https://www.dlr.de/eoc/en/desktopdefault.aspx/tabid-8799/.

DLR at a glance, 2018. URL: https://www.dlr.de/EN/organisation-dlr/dlr/dlr-at-a-glance.html.

DLR, Big Data Research, 2018. URL: https://www.dlr.de/pt/en/desktopdefault.aspx/tabid-9928/16990_read-41089/.

DLR's Big Data Platform cross-sectoral project begins, Big Data Science – using large scientific datasets. Date published online: 18 July 2018. URL: https://www.dlr.de/content/en/articles/news/2018/3/20180718_dlr-s-big-data-platform-cross-sectoral-project-begins_28966.html.

EC, 2007. Commission of the European Communities, Directive 2007/2/EC of the European Parliament and of the Council of 14 March 2007 Establishing an Infrastructure for Spatial Information in the European Community (INSPIRE). Official Journal of the European Union L108, 1–14. https://eur-lex.europa.eu/legal-content/EN/TXT/?qid=1585260681787&uri=CELEX:32007L0002.

EC, 2018. Space policy: Galileo Security Monitoring Centre back-up site moves to Spain. http://europa.eu/rapid/press-release_IP-18-389_en.htm.

ESA, 2018a. Space component data access, contributing missions. https://spacedata.copernicus.eu/web/cscda/missions. (Accessed 10 March 2018).

ESA, 2018b. Third Party Missions Online Access List. https://tpm-ds.eo.esa.int/collections/. (Accessed 14 March 2018).

ESA, 2018c. EOLi-Sa "ESA link to Earth Observation". https://earth.esa.int/web/guest/eoli. (Accessed 14 March 2018).

e-Sensing overview, 2014. URL: http://www.esensing.org/docs/e-sensing_overview.pdf.

EO4SDG, 2018. URL: eo4sdg.org.

GeoBuiz, 2018. Geospatial Industry Outlook & Readiness Index. Geospatial Media and Communications, 116.

GEO, 2019. URL: www.earthobservations.org.

GEOSS Portal. URL: www.geoportal.org.

HNSDMS, 2018. Hellenic National Sentinel Data Mirror Site. https://sentinels.space.noa.gr/. (Accessed 14 March 2018).

INPE, 2018. National Institute for Space Research, Ministry of Science, Technology, Innovation and Communications, São José dos Campos, São Paulo, Brazil. http://www.inpe.br/. (Accessed 14 December 2018).

NASEM, 2018. National Academies of Sciences, Engineering and Medicine: Thriving on Our Changing Planet: A Decadal Strategy for Earth Observation From Space. The National Academies Press, Washington DC. https://doi.org/10.17226/24938.

NASA Earth Observations – NEO, 2019. About. URL: https://neo.sci.gsfc.nasa.gov/about/.

NASA Earth Science Division Operating Missions, 2019. URL: https://eospso.nasa.gov/content/nasas-earth-observing-system-project-science-office (Date online: 10 October 2019).

NCI, Australian Geoscience Data Cube, 2018. URL: https://nci.org.au/research/publications/research-articles/australian-geoscience-data-cube-foundations-and-lessons.

NOAA, Big Data Project, 2018. URL: http://www.noaa.gov/big-data-project.

OGC, 2017. http://docs.opengeospatial.org/wp/16-131r2/16-131r2.html.

Overview of Indian Space sector, 2010–2020, Deloitte, p. 84.

Remotely sensed Big Data Analytics and Mining, 2017. URL: http://sites.ieee.org/kolkata-grss/2017/12/21/rsbdam/.

"РОСКОСМОС" has created the commercial operator of services on the basis of the data of remote sensing of the Earth. URL: http://russianspacesystems.ru/2018/02/21/roskosmos-sozdal-kommercheskogo-operatora-dzz/, 2018 (in Russian).

Roscosmos - Annual report 2016, Moscow. URL: https://www.roscosmos.ru/media/img/docs/Reports/otcet.2016.pdf, 2017 (in Russian).

The China–Brazil Earth Resources Satellite Mission, AWS Government, Education, & Nonprofits Blog. URL: https://aws.amazon.com/blogs/publicsector/the-china-brazil-earth-resources-satellite-mission/, 2018.

The State Council of China, 2015. Action plan for the promotion of big data development. URL: http://www.gov.cn/zhengce/content/2015-09/05/content_10137.htm (in Chinese).

Towards a European AI4EO R&I Agenda, v.1.0. AI4EO R&D community consultation, ESRIN, 26-27 March 2018. URL: https://eo4society.esa.int/wp-content/uploads/2018/09/ai4eo_v1.0.pdf, 2018, pp. 26–27.

UN-GGIM. http://ggim.un.org/UNGGIM-wg6/.

CHAPTER 7

High-Performance Techniques for Big Data Processing

PHILIPP NEUMANN, PROF, DR • JULIAN KUNKEL, DR

7.1 INTRODUCTION

Supercomputing and Big Data Supercomputers deliver a hundred- to millionfold performance compared to regular desktop PCs. This allows to tackle problems which could not be solved otherwise in a reasonable amount of time. At the same time, the massive amount of data generated and collected through sensors, experiments, etc., is constantly increasing, requiring new and analogously evolving approaches at hardware and software level for data processing and analysis.

Both *Big Data processing* and *high-performance computing (HPC)*, as well as their convergence, lead to new observations and understandings of phenomena which would otherwise be too fast or too slow to be comprehended in vitro, or, simply, not accessible (Kaufmann and Smarr, 1992; Chen and Zhang, 2014). A prerequisite to exploit supercomputers is, however, a thorough understanding of hardware and software design which is vital for providing the necessary computing power for scientists. The same holds for data processing in supercomputing environments, clouds, or streaming-based systems.

Outline: How to Read This Chapter In the following, we shed light on recent hardware trends, programming paradigms, and software in the scope of HPC and Big Data processing. In Section 7.2, we provide an overview of computing architectures that are currently in use and point out recent trends, interwoven with selected programming examples. Afterwards, we discuss how these architectures are leveraged in the context of distributed computing in Section 7.3. We focus on cluster architectures and, thus, supercomputing aspects, and further discuss aspects of cloud computing in this section. Section 7.4 is divided into two subsections, addressing storage and data management concepts for both (super)computing and Big Data processing. Having finished the discussion on rather hardware-aware aspects at this point, focus is shifted to performance metrics, benchmarking, and performance modeling and analysis for computing and data handling in Section 7.5. This will equip the reader with knowledge on metrics and approaches to evaluate performance of both hardware and applications running on it. Section 7.6 is dedicated to high-performance data analytics. In particular, this section is meant to point out the intersection of high-performance computing and (big) data analytics at the example of numerical linear algebra, including a list of available software packages and libraries. We close the discussion on high-performance techniques for Big Data processing in Section 7.7 with concluding remarks on HPC, Big Data, and the convergence of both fields.

The interested reader is further referred to Ma et al. (2015) for a review on remote sensing Big Data computing, with a bigger focus on current challenges to be overcome in this field.

Expected Learning Outcomes After reading the chapter, the reader has gained overviews on:

- basic concepts of current compute architectures and storage, including first insights into programming these devices,
- metrics and methods to evaluate performance of particular hardware,
- similarities, differences, and convergence of HPC and Big Data processing.

This shall support developers and users of particular HPC and Big Data software in determining their optimal processing platforms, evaluate performance of their application, and design and improve their application, embracing current hardware architecture styles and hardware trends.

7.2 COMPUTE ARCHITECTURES

7.2.1 Cache-Based Systems

Typically, processing of data in a computer follows the *von Neumann architecture* (von Neumann, 1993): A *central processing unit* (CPU) executes one instruction after the other. Each instruction causes one of the available processing units to perform modifications of the data stored in a memory system. In fact, most processors

Knowledge Discovery in Big Data from Astronomy and Earth Observation. https://doi.org/10.1016/B978-0-12-819154-5.00017-5

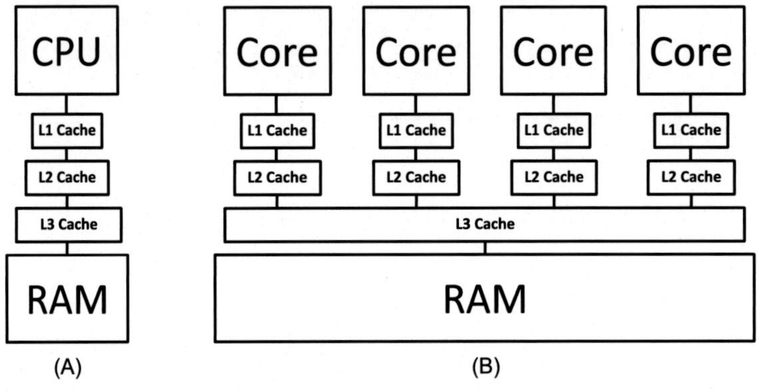

FIG. 7.1 (A) Cache architecture and (B) multicore architecture with shared L3 cache.

TABLE 7.1
Overview of commodity hardware, instruction sets, and vectorization properties. Architecture: codename of underlying hardware. Instruction set: supported vectorization method. SSE, Streaming SIMD Extensions; AVX, Advanced Vector Extensions; Number of registers, number of registers per compute core; Register width, size of a register (bits).

Architecture	Vendor	Instruction set	Number of registers	Register width
Nehalem	Intel	SSE	8/16	128 bit
Sandy Bridge, Ivy Bridge	Intel	AVX	16	256 bit
Bulldozer, Piledriver, Jaguar	AMD	AVX	16	256 bit
Haswell, Broadwell, Kaby Lake	Intel	AVX2	16	256 bit
Excavator, Zen	AMD	AVX2	16	256 bit
Skylake-X	Intel	AVX-512	32	512 bit
Knights Landing (Xeon Phi)	Intel	AVX-512	32	512 bit

(e.g., standard processors for desktop PCs such as the Intel Xeon family) implement a *memory hierarchy*: to speed up search for and access of data, the slow, but large, main memory is extended by a hierarchy of smaller memory segments, so-called *caches*; see Fig. 7.1A. Depending on the underlying strategy for reading and writing data, data are only fetched from/written to main memory/higher-level cache if they are not found in a lower-level cache. Operations (such as floating point arithmetics) are performed in the smallest and thus fastest memory – the so-called registers, which are embedded in the processing unit itself and typically hold between 4 and 16 floating point numbers. Floating point operations, that is, additions and multiplications, can be carried out *simultaneously for all numbers within the considered registers (vectorization)*, analogously to component-wise additions and multiplications on vectors in linear algebra. Making efficient use of this vectorization requires, however, data to be *aligned in memory*, so that they are loaded contiguously into a register for processing. Vector widths have been increasing from 128 to 512 bits over the last years, implying an increase from 4/2 to 16/8 float/double values that can be processed at a time; see Table 7.1 for an overview of hardware, corresponding register widths, and supported vector instruction sets.

Modern processors provide multiple functional units, which can operate simultaneously to manipulate data. An example is given by the latest Intel Skylake-X compute cores, featuring two AVX-512 fused multiply-add units. Thus, a CPU can *execute multiple instructions on multiple scalar values concurrently*. Therefore, a CPU (or the compiler for the system) keeps track of data dependencies to ensure that the computation result is identical to a sequential execution. In the case of a data-intensive application, the use of cache-based architectures implies that if data in a small cache or in the registers can be reused, less time has to be spent for looking them up in memory – data access and processing are accelerated, respectively.

7.2.2 Multicore Systems

Multicore Architecture In the past, application performance was improved by packing more functionality into a single chip and by incrementing the clock frequency of the processor. However, power consumption of a processor is proportional to the square of the clock frequency. Thus, combined with the steady miniaturization, at some point chips would have been designed that produce more heat than a nuclear power plant or even the sun (Ronen et al., 2001). As a result of this observation, a change in strategy was implemented. In order to utilize available transistors, multiple processing units are replicated on a single chip and clocked with a lower frequency. With this approach the aggregated performance of all cores is higher than by increasing the complexity of a single core. Nowadays, all commodity chips consist of multiple cores (*multicore chips*) and operate on a shared main memory (*shared memory architecture*); see Fig. 7.1B.

Programming Models for Multicore Systems Various programming models have evolved to leverage multicore performance, including OpenMP (OpenMP web site, Accessed: 2018-06-20), Intel Threading Building Blocks (Intel Threading Building Blocks web site, Accessed: 2018-06-20), or Cilk (Cilk web site, Accessed: 2018-06-20). *Open Multiprocessing (OpenMP)* is the most commonly used programming model and makes use of a master–worker concept. First, the programmer instruments the compute-/data-intensive regions of the program with precompiler-based OpenMP directives (so-called pragmas). These regions are called *parallel regions*. Launching the program after compiling, a master thread[1] is created and runs through the program instructions. If it encounters a parallel region, it forks worker threads. Every worker thread processes only a part of the parallel region. Leaving the parallel region, the worker threads are discarded (or set asleep) and the master thread continues execution. Listing 7.1 depicts an OpenMP parallelization of a vector–vector addition in C.

Work sharing constructs define how the work inside the parallel block is distributed across available threads. The `for` clause in Listing 7.1 distributes the values of the loop variable into as many chunks as threads are

Listing 7.1: OpenMP implementation of a vector addition. The pragma statement generates N threads, creates N chunks of vector data, and makes each thread execute the loop body on one of these chunks.

```
#pragma omp parallel for
for (i=0; i<100; i++){
  a[i]=b[i]+c[i];
}
```

available; in this example, blocks of `i` between 0 and 99 are distributed among the threads. Every thread carries out the addition for its chunk of data. Using four threads, for example, results in processing the values 0–24 (thread 0), 25–49 (thread 1), 50–74 (thread 2), 75–99 (thread 3).

OpenMP further provides clauses for thread synchronization, thread scheduling, or indication of data attributes; for details, see Chapman et al. (2007), Van Der Pas et al. (2017).

7.2.3 Manycore Systems

Manycore Architecture and Accelerators Similar to multicore systems, *manycore architectures* feature an even bigger set of compute cores per device. Meant and optimized for a high degree of parallelism and HPC, the devices feature up to O(5000) cores,[2] high throughput, and, compared to multicore systems, relatively low energy consumption. For these reasons, these architectures have evolved well into the field of HPC and (big) data analysis, with 7 out of the top 10 supercomputers in the world featuring GPUs (TOP500 web site, Accessed: 2018-06-20).

Manycore systems such as GPUs are typically not equipped with an operating system and do not execute corresponding routines, such as for memory management. Hence, they are normally attached to a traditional multicore processor as a host system. Therefore, in a typical setup, GPUs are deployed on a server system connected via the PCI express bus and are used to speed up highly parallel workloads; this coined the term *accelerators*. The separation of GPU and host memory increases the programming complexity. To reduce the burden for the programmer, modern systems such as Nvidia's Unified Memory provide a virtual address space that manages memory and data accesses on host and GPU – copying data between them if needed.

Lately, to deliver even more performance on machine learning problems, half-precision support (i.e., 16-bit floating point) has been stepping into accelerator architectures, with one example given by Nvidia

[1] A thread is a single stream of instructions that is processed sequentially.

[2] For example, the Nvidia Tesla V100.

Tesla P100 GPUs. Other examples for accelerators comprise the Intel Xeon Phi architecture, including the deep learning-specialized Knights Mill and the rather supercomputing-driven Knights Landing architecture,[3] as well as the Sunway 26010 RISC processor (containing 256 cores), which is currently used in Sunway TaihuLight (TOP500 web site, Accessed: 2018-06-20), the fastest supercomputer in the world at the time of writing.

Programming Models for Accelerators Accelerator programming is still restricted to particular languages or language extensions, with the most prominent ones given by CUDA, a programming interface from Nvidia for GPUs, and OpenACC, which has been developed by Cray, CAPS, Nvidia, and PGI. An exception is given by Intel's Xeon Phi systems, which are able to execute code written for "host architectures"; however, this does not guarantee optimal performance and additional code optimization and tuning is often still required in this case to account for the architectural differences between host and accelerator.

CUDA, originally named *Compute Unified Device Architecture*, enables embedding GPU-related instructions in a slightly extended programming language in C. These instructions allow to transfer data between CPU host and GPU and the execution of C-written kernels on the GPU, including a corresponding configuration of thread usage and block-based distribution within the GPU. Listing 7.2 depicts a CUDA implementation of a vector addition in CUDA C.

7.2.4 Other Architectures

Besides, other architectures are currently entering the HPC market and appear to be very promising. *ARM processors* have lately been integrated in supercomputers, comprising the ThunderX2 supercomputer architecture by CRAY (Hemsoth, 2017) and the post-K supercomputer currently developed by Fujitsu for the Japanese research and supercomputing institution RIKEN (Feldman, 2016). ARM processors belong to the RISC[4] family which, compared to the CISC[5] concept, which, e.g., Intel processors are based on, make use of reduced instruction sets. On the one side, this yields lower performance and thus slower execution times, as well as less flexibility with regard to the operating systems that can be used on ARM systems. On the other side, due to the reduced instruction sets, a smaller amount of transistors is required to store the instructions, resulting in smaller and, thus, more energy-efficient chips. Particularly the latter aspect becomes increasingly important for peta- and exascale[6] supercomputers.

Field-programmable gate arrays (FPGAs) have also gained wider interest in HPC. FPGAs are unique in the sense that the underlying hardware remains configurable and thus tunable per application. Indeed an application may program the hardware logic according to its needs. This can enable significant speedups for various applications, but comes to the authors' knowledge with significant efforts for software porting. Similarly, *Application-specific integrated circuits* (ASICs) are customized circuits for a particular use case offering higher clock rate (and hence performance) but the logic cannot be reconfigured. Design of an ASIC costs a million dollars. ASICs are only used in HPC for the features that are already an integral part of modern CPUs, like data compression and encryption. Aside from HPC, ASICs are widely used in blockchain technology like for Bitcoin mining as the algorithm is fixed.

7.3 DISTRIBUTED SYSTEMS

7.3.1 Clusters

Cluster Architecture Cluster architectures comprise several *compute nodes* that are combined and connected through a high-performance network; see Fig. 7.2. Every node is an independent component with at least one CPU (and, potentially, one or more accelerators) and has its own main memory. Due to the latter, clusters belong to the class of *distributed memory systems*. Data of one node can be transferred over the network to the memory of another node. Besides internode connectivity, a storage system is connected to the nodes.

Today's supercomputers, such as the supercomputer Mistral hosted at the German Climate Computing Center (cf. Fig. 7.3), are typically of the cluster type. The tight, that is, fast, interconnect between the compute nodes allows rapid data exchange (compared to, e.g., cloud computing) and, thus, allows independent nodes to collaborate to *rapidly compute solutions for large-scale tightly coupled problems*. For example, computing a weather forecast involves the computation of wind speeds; although these wind speeds regionally differ and it thus makes sense to compute these speeds within different subregions on different compute nodes,[7] the

[3] A successor Knights Hill of Knights Landing has been canceled in November 2017.

[4] Reduced instruction set computer.

[5] Complex instruction set computer.

[6] A petascale/exascale computer performs $10^{15}/10^{18}$ floating point operations per second.

[7] This technique is referred to as *domain decomposition* in numerical simulation.

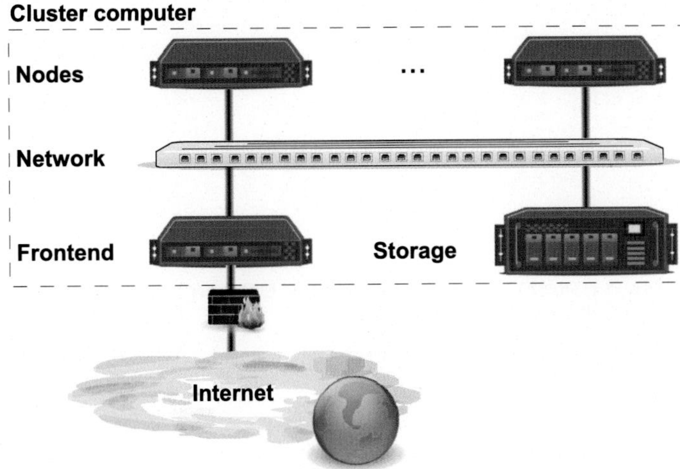

FIG. 7.2 Schematic view of a cluster computer. A scientist can connect to the front-end and can either work interactively on a reserved number of nodes, or submit compute jobs that will be distributed and scheduled by a preinstalled scheduling software.

FIG. 7.3 Supercomputer Mistral hosted at the German Climate Computing Center. Mistral features ca. 99,000 compute cores and a maximum performance of three petaflops.

Listing 7.2: CUDA implementation of a vector addition.

```
__global__
void addab(int n, float* a, float *b, float *c)
{
  int i = blockIdx.x*blockDim.x + threadIdx.x;
  if (i < n) a[i] = b[i] + c[i];
}

int main(void)
{
  ...
  /* Variable declaration */
  float *a, *b, *c, *d_a, *d_b, *d_c;

  /* Memory for all vectors (a, b, c) is allocated on both CPU host
   * (\texttt{malloc()}) and GPU (\texttt{cudaMalloc()). */

  a = (float*)malloc(N*sizeof(float));
  b = (float*)malloc(N*sizeof(float));
  c = (float*)malloc(N*sizeof(float));

  cudaMalloc(&d_a, N*sizeof(float));
  cudaMalloc(&d_b, N*sizeof(float));
  cudaMalloc(&d_c, N*sizeof(float));

  /* Initialize data on the host (CPU) system */
  for (int i = 0; i < N; i++) {
    a[i] = 0.0; b[i] = 1.0; c[i] = 2.0;
  }


  /* Input vectors b, c are transferred from host to GPU .
   * Note this manual copy is not needed with Unified Memory. */
  cudaMemcpy(d_b, b, N*sizeof(float), cudaMemcpyHostToDevice);
  cudaMemcpy(d_c, c, N*sizeof(float), cudaMemcpyHostToDevice);

  /* The addition is carried out on the GPU, i.e., spawn the kernel
   * addab with a specific parallelism. The numbers in the chevrons
   * denote the number of 1D blocks (N+255/256) and threads per
   * block (256) that CUDA will use for this operation.
   * This information is used in the kernel addab to check for
   * out-of-bounds situations of thread-based operations. */
  addab<<<(N+255)/256, 256>>>(N, d_a, d_b,d_c);

  /* Transfer back the result from GPU to host. */
  cudaMemcpy(a, d_a, N*sizeof(float), cudaMemcpyDeviceToHost);
  ...
  cudaFree(d_a); cudaFree(d_b); cudaFree(d_c);
  free(a); free(b); free(c);
}
```

speeds are still tightly coupled at the interfaces of neighboring subdomains; see Fig. 7.4 for a global high-resolution weather simulation. A rapid exchange and synchronization of the wind speeds at these boundaries is therefore essential for efficient weather forecasting.

The node count in clusters can be moderate (e.g., 10–100 in small-sized clusters), or it can reach O(10,000) or even more in the case of supercomputers. As an example, Sunway Taihulight, the fastest machine on the TOP500 list (TOP500 web site, Accessed: 2018-06-20), currently features 40,960 compute nodes, each equipped with 260 cores, yielding a total of 10,649,600 cores and a performance of up to 93 petaflops, that is, $93 \cdot 10^{15}$ floating point operations per second.

Due to the vast number of nodes, not every node can be directly connected to every other node with a physical connection. Network topologies establish communication between nodes via intermediate links (switches or other nodes). Several network topologies have been found reasonable as they provide a high bisection bandwidth and a low number of indirections and, hence,

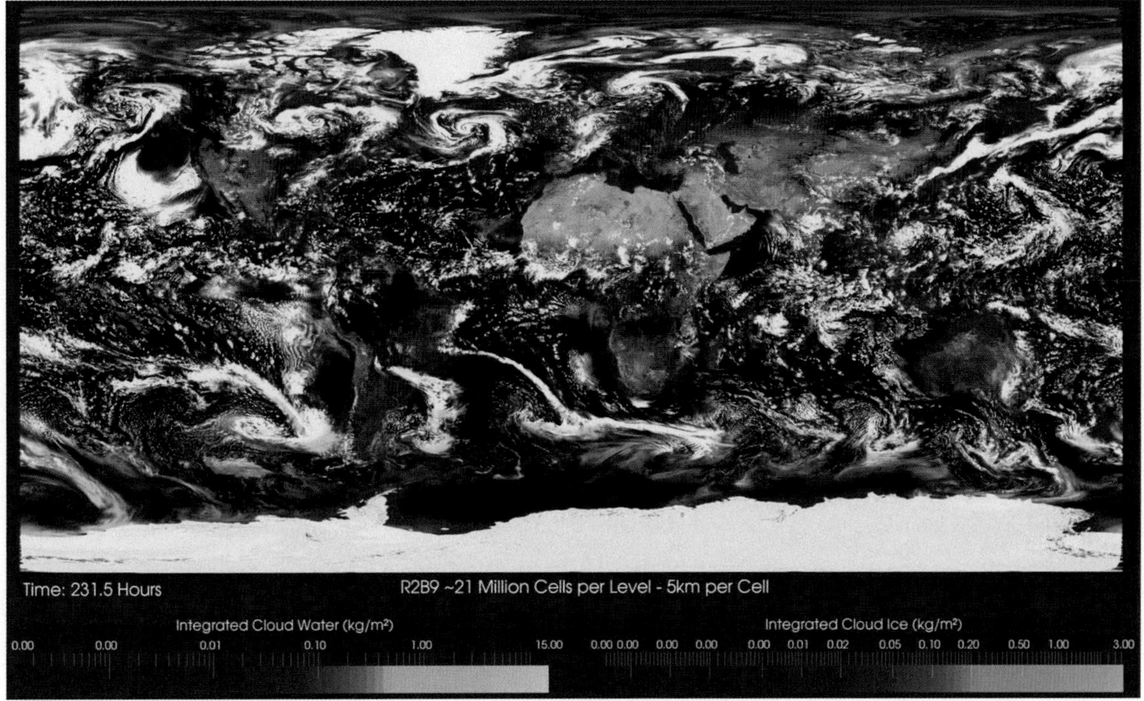

FIG. 7.4 Global 5-km resolution weather simulation based on the Icosahedral Nonhydrostatic (ICON) model.

latency. Examples are fat tree, torus, and dragon fly topologies (Dally and Towles, 2003; Lee and Kalb, 2008; Kim et al., 2008); see Fig. 7.5 for a 2D-torus topology example. Similar to the cache hierarchy paradigm within a CPU, data access becomes slower (a) the more data needs to be sent through a network and (b) the bigger the distance between two communicating nodes in the network is.

Design and performance of a network depend on the technology deployed and the layout of the interconnect between the network components. Gigabit Ethernet (Ethernet, 2001) is a commodity technology provided in consumer hardware which is often used in Big Data environments. In contrast to the cost-effective Ethernet, the InfiniBand technology (Mellanox Technologies, 2000), for instance, is specifically designed for HPC.

Data management in bigger clusters is performed by distributed file systems, which scale with the demands of the users. Usually, multiple file systems are deployed to deal with disjoint requirements.

In a typical setup two file systems are provided: a fast and large scratch space to store temporary results, and a slower, but highly available volume to store user data (e.g., for home directories). Looking at a particular distributed file system, a set of storage servers provides a high-level interface to manipulate objects of the namespace. The *namespace* is the logical folder and file structure the user can interact with; typically it is structured in a hierarchy. Files are split into parts which are distributed on multiple resources and which can be accessed concurrently to circumvent the bottleneck of a single resource. Replication of a part on multiple servers increases availability in the case of server failure. Truly *parallel file systems* support concurrent access to disjoint parts of a file; in contrast, conventional file systems serialize I/O to some extent.

An example of the *hierarchical namespace* and its mapping to servers is given in Fig. 7.6. The directory "home" links to two subfolders which contain some files. In this case, directories are mapped to exactly one server each, while the data of logical files is maintained on all three servers. Data of "*myFile.xyz*" are split into ranges of equal size and these blocks are distributed round-robin among the data servers. Each server holds three fragments of the file.

Storage devices are required to maintain the state of the file system in a persistent and consistent way. A stor-

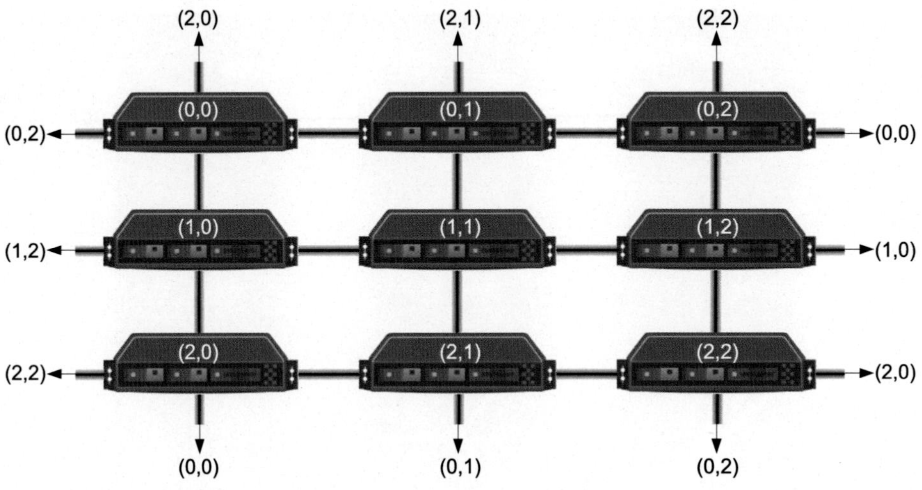

FIG. 7.5 Network topology of a 2D-torus. Leftmost nodes (X,0) are interconnected with the rightmost nodes (X,2). Top nodes (0,X) are connected with the bottom nodes (2,X).

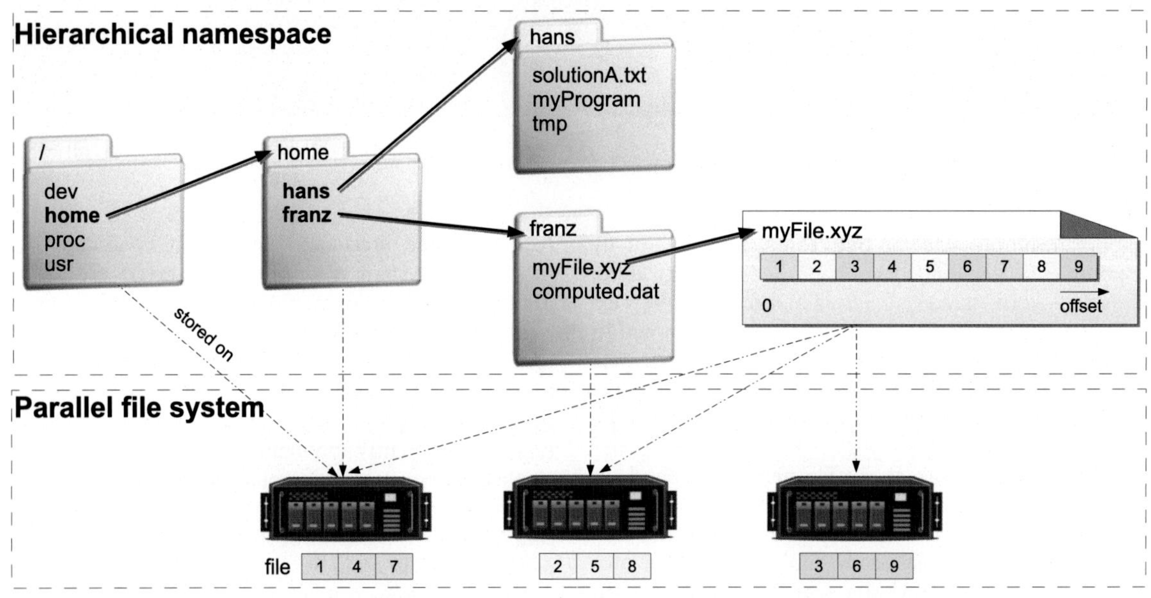

FIG. 7.6 Example of a hierarchical namespace and mapping of the objects to servers of a parallel file system. Here, metadata of a single logical object belong to exactly one server, while file data are distributed across all servers.

age device can be either directly attached to a server (e.g., a built-in hard disk), or the server controls devices attached to the network. In contrast to the interface provided by file systems, the interface to storage devices is low-level. At the block level, data can be accessed with a granularity of full blocks by specifying the block number and *access type* (read or write).

In a *storage area network* (SAN), the block device seems to be connected directly to the server. To communicate with the remote block storage device, servers use the *small computer system interface* (SCSI) command protocol. A SAN could share the communication infrastructure, or another network just for I/O can be deployed. Fibre Channel and Ethernet are common network technologies with which to build a SAN. In the latter case, SCSI commands are encapsulated into the IP protocol, that is, the so-called *Internet SCSI* (iSCSI). Therefore, the existing communication infrastructure can be used for I/O, too. For details on high-performance storage and I/O, see Section 7.4.1.

Programming Models for Distributed Memory Systems Two models used to program the distributed memory architecture of cluster machines, i.e., to exchange data between the nodes, are MPI and PGAS. The most widely used *Message Passing Interface* (MPI) (Message Passing Interface Forum, 2015) typically follows the single-program multiple-data (SPMD) approach: a program is extended by MPI routines that trigger the sending and receiving of messages. Several *processes* of the same program are then launched on various compute resources (that is, compute cores on compute nodes) and MPI orchestrates communication between these processes. Every process takes care of its own memory resources. Alternatively, multiple program executables may be launched and connected via MPI. This approach is known as multiple-program multiple-data paradigm (MPMD). MPI also offers routines for parallel input and output of data.

Listing 7.3 shows two variants to sum up entries in a vector which is distributed among processes using MPI. With the *partitioned global address space* (PGAS) programming model, a process can access data that are stored in remote memory (that is, in the memory of another node) by using the syntax of the underlying programming language, such as array access. PGAS requires a run-time environment which ensures that, if required, data are automatically transferred between the nodes' memory. However, the data partitioning is still encoded by the programmer. Example language extensions which enable remote memory access in C and Fortran are the *Unified Parallel C* (UPC) and *Coarray Fortran* (CAF), respectively.

On the Use of Supercomputers The productive usage of supercomputers requires to request resources and to program efficient parallel applications with a suitable programming model. Traditionally, a supercomputer is built rather independently from the applications, which are to be executed later on in the system. Thus, the demands of the software and the system capabilities may not match optimally. Recent studies show the importance of software-hardware *codesign* to cope with the challenges of applications on large-scale systems (Joubert et al., 2009; Sarkar et al., 2009). With this approach, software aspects are kept in mind as the hardware is designed, and the hardware design also influences the programming model of the software.

To access a supercomputer, a user logs into a cluster front-end that is connected to all nodes and the storage (see Fig. 7.2). Typically, the storage and file systems can be accessed from the front-end allowing access to all data. To ensure best performance of the tightly coupled parallel applications, nodes are allocated exclusively to user jobs. The user has to request the resources via a workload manager, which manages the queue of *batch jobs* that are submitted by various users. The jobs are created through every user separately by writing a short instruction script and submitting it for execution by the workload manager. The workload manager assigns resources for the jobs, runs scheduling algorithms, and dispatches jobs according to defined priority schemes. Various scheduling environments are in use nowadays, with the most common ones comprising SLURM Workload Manager, LoadLeveler, and Portable Batch System. Listing 7.4 shows an exemplary SLURM script for job submission.

Besides *batch processing* (i.e., job scheduling), *interactive mode* enables the user to allocate a number of nodes and interactively run his/her executables on these resources. This is particularly useful for program debugging. Interactive sessions are typically restricted to rather limited node counts. They are also managed through the scheduling environment.

7.3.2 Cloud Computing

In contrast to clusters that consist of tightly coupled compute nodes, *cloud computing* makes use of pools of compute units that are potentially geographically distributed. These units are interconnected and virtualized so that they can be used as single or unified compute resources. Most clouds further come with provisioned IT- and service infrastructure, such as the Amazon Cloud EC2.

Cloud computing has several advantages for users. In contrast to hosting a cluster on one's own, no staff is required to administrate the hardware. The elasticity of the cloud allows to scale the need for compute and storage according to the own needs – one has to pay only for as much resources as were used. This allows for a reliable planning of costs, particularly for startups. Software cost is reduced (as long as it comes with the service infrastructure of the cloud). Data are reliably backed up,

Listing 7.3: Summation of vector entries using MPI. Variant 1 sequentially sends partial sums from one rank to the next (typically slow due sequentialism in sending/receiving). Variant 2 uses an MPI reduction to accumulate all local partial sums on rank 0.

```
int main(int argc,char* argv[]){
  MPI_Init(&argc,&argv);
  // global size of vector
  int N = 10;

  // determine total number of ranks and the rank of this process
  int rank=-1; int size=-1;
  MPI_Comm_rank(MPI_COMM_WORLD, &rank);
  MPI_Comm_size(MPI_COMM_WORLD, &size);
  // determine local vector size: every rank receives
  // equal portion of the vector; last rank receives
  // same portion and potential rest of global vector
  int localN= (rank<size-1) ? N/size : N/size+N%size;
  printf("Rank: %d, localN=%d\n", rank,localN);

  // allocate and initialize local portion of vector
  int *a=(int*) malloc(sizeof(int)*localN);
  for (int i=0;i<localN;i++) a[i]=rank*(N/size)+i;
  // local summation; store result in sum
  int sum=0; int buf=0;
  for (int i=0;i<localN;i++) sum+=a[i];

  // Variant 1 - sequential summation
  buf=sum;
  for (int r=size-1; r>0; r--){
    // receive latest partial from rank r on rank r-1,
    // store it in buf and add local sum
    if (rank==r-1){
      MPI_Recv(&buf,1,MPI_INT,r,0,MPI_COMM_WORLD, MPI_STATUS_IGNORE);
      buf=buf+sum;
    }
    // send buf from rank r to rank-1. Called after receive
    // due to blocking nature of both commands
    if (rank==r) MPI_Send(&buf,1,MPI_INT,r-1,0,MPI_COMM_WORLD);
  }
  if (rank==0) printf("Sum via sequential summation: %d\n", buf);

  // Variant 2 (more efficient) - summation via MPI reduction
  buf=0;
  MPI_Reduce(&sum,&buf,1,MPI_INT,MPI_SUM,0,MPI_COMM_WORLD);
  if (rank==0) printf("Sum via reduction: %d\n", buf);

  free(a);
  MPI_Finalize();
  return 0;
}
```

due to the virtualization within the cloud and the geographical distribution of resources.

Cloud computing also has some potential drawbacks. Firstly, one may become dependent on the cloud provider and, potentially, dominating client institutions that may have a greater influence on service and IT infrastructure design and development in the cloud. Security is another hot topic: how safe are the data in such a distributed environment? Which ownership and rights apply, for example in a multinational cloud? Can data be accessed if the cloud is (temporarily) out of service at some point? For HPC, particularly the lower and less predictable performance of the virtual machines was an issue for many years. However, cloud providers have integrated HPC capabilities into their infrastructure over the last few years – still public data centers are a cheaper option for scientists.

Concluding, clouds are an interesting alternative for weakly coupled distributed applications and if there is no need for rapid exchange of data within an application, for example when carrying out many (potentially different) evaluations on a set of big, yet lim-

Listing 7.4: SLURM bash script to run the executable *burgers* on four nodes, using eight MPI processes and 36 OpenMP threads per MPI process. Standard and error output are written into the file out.log.

```
#!/bin/bash
#This is a comment, lines starting with the keyword SBATCH are
#interpreted by the workload manager like an argument.
#
#SBATCH —nodes=4
#SBATCH —time=00:15:00
#SBATCH —output=/home/user/burgers/out.log
#SBATCH —error=/home/user/burgers/out.log
#SBATCH —exclusive

cd /home/user/burgers
srun —ntasks=8 —ntasks-per-node=2 —cpus-per-task=36 ./burgers
```

ited, amounts of data. In the latter scope, the cloud is basically considered as an extension for compute-intensive desktop applications (Armbrust et al., 2009). See Sadashiv and Kumar (2011) for a more detailed comparison between cloud and cluster computing and Armbrust et al. (2009) for more information on cloud computing, use cases, and services.

7.4 STORAGE AND DATA MANAGEMENT

Data management is a crucial aspect of scientific discovery, particularly for data-intensive science fields like climate research and astronomy. As climate researchers run numerical simulations on supercomputers to explore future climate, they need to analyze and preserve the data for further reference. Similarly, the rate of generating observational data can be extreme. When the radio telescope Square Kilometre Array (SKA) will be completed by 2023, it will generate science-ready data for analysis at a rate of 3 petabyte/day (Rosolowsky et al., 2016). Therefore, data centers supporting these scientific domains have a stronger focus on storage capacity compared to those primarily serving compute-intense sciences. For example, the German Climate Computing Center provides 53 petabyte of storage capacity on a parallel file system and a tape archive with 67,000 tapes.

7.4.1 High-Performance Storage and I/O

Understanding the difference between HPC I/O and Big Data requires us to describe the hardware and software stack used in HPC first; see Fig. 7.7B. High-performance storage provides the hardware and software technology to query and store large data volumes at high velocity of input/output while ensuring data consistency. On upcoming exascale systems, workflows will harness hundreds of thousands of processors with millions of threads while producing or reading petabytes to exabytes of data.

Storage Media and Hardware In the 2010s, the technology of storage media evolved quickly, enriching the storage landscape with alternative technology that has different characteristics in terms of power, latency, throughput, and resilience. Low-level storage media are considered a block device offering only read/write to individual and coarse-grained blocks of data (like 4-kB blocks).

Around 2010, data centers provided typically a tiered storage consisting of an online storage with some 1000 hard disk drives (HDDs) and a near-line storage in the form of a tape archive. The former enables interactive I/O while the latter implies request times in the order of minutes. With the advent of solid-state disks (SSDs), an additional storage tier was introduced, enabling fast random I/O. Over time, the capacity price advantage of HDDs decreased, and some centers started to replace their HDD environment completely.

Recent advances in new storage technologies such as Intel Optane (Tallis, 2017) with 3DXPoint (Malventano, 2017), Seagate Nytro (Seagate, 2017), KoveXPD (Kove, 2017), and NVME (Strass, 2016) promise to bring high-capacity nonvolative memory (NVM) with performance characteristics (latency/bandwidth/energy consumption) that bridge the gap between DDR[8] memory and SSD/HDD. Certain storage like NVM allows byte addressing instead of the block access of traditional media. Such storage is called *storage class memory* (SCM) since it can be treated similar to memory, potentially even replacing memory DIMMs[9] with alternative storage DIMMs.

[8] Double data rate.

[9] Dual Inline Memory Module.

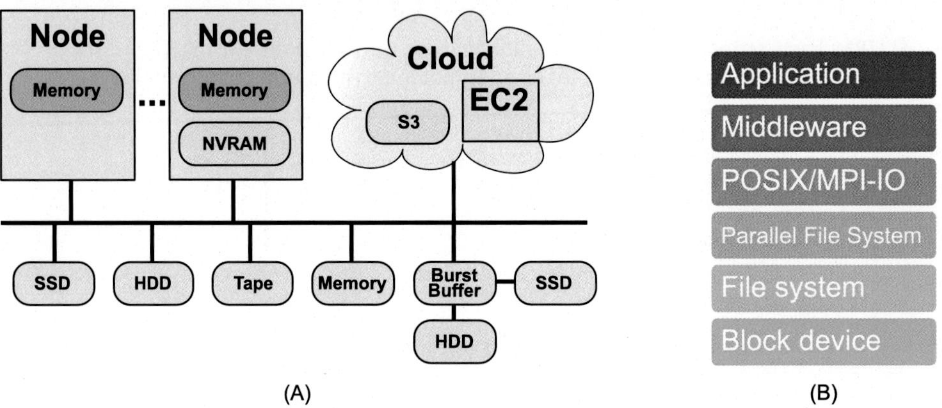

FIG. 7.7 Typical HPC environment.

However, due to high cost, these technologies will be used mostly in the upper layers of the I/O stack (see Fig. 7.7B). At the same time, they require a careful integration into the existing I/O stack or demand the development of next-generation storage systems. For large quantities of data, e.g., for archival, near-line storage such as tape is expected to remain relevant for a prolonged time period, as it is reliable and very affordable, though I/O latency is poor, and individual drive bandwidth is relatively low. Massive array of idle disks (MAID) has been considered as an alternative; this architecture relies on HDDs as storage media. In an archive of many HDDs, only a subset of them can be connected and/or powered concurrently. In contrast to a MAID, a tape archive has still a cost advantage, and as tape can be considered a three-dimensional magnetic storage compared to a two-dimensional disk, there is still room for increasing capacity significantly.

Storage media are often provided in optimized storage appliances consisting of servers attached or connected to the storage media and network interconnects.

All this implies that the I/O storage stack will gain more layers, increasing its complexity and making overall HPC scientific workflows harder to be optimized, even for experts. Fig. 7.7A illustrates the compute, storage, and network infrastructure of a potential future data center. In the figure, we do not distinguish between the storage interface (e.g., NoSQL, NVMEoF, file, object), but we list potential storage technology. Network attached (nonvolatile) memory and SSDs are the fastest technology providing reliable access times while HDDs and tapes provide capacity for latency-insensitive data. Some storage systems may manage internal storage tiers explicitly or by software, for example, the so-called burst buffers. A burst buffer is a set of servers that can absorb write bursts extremely well, i.e., it has a fast network connection to client nodes and may use faster storage technology. Note that such a system may potentially hide the internal storage tiers from access.

Software Stack A typical I/O stack is shown in Fig. 7.7B. Traditionally, a parallel application uses an *I/O middleware* to access data which provides a data access and manipulation API[10] for higher-level objects such as variables. High-level object-oriented data models such as HDF5 (The HDF group, 2015), NetCDF (Rew et al., 2008), and ADIOS (Podhorszki et al., 2016) have proven to be an intuitive and easy way to organize scientific data and thus have been adopted by multiple scientific communities. Internally, the data are stored in machine-independent and self-describing objects, which are organized in a flat or hierarchical namespace. Naturally, these interfaces provide operations for accessing and manipulating data that are tailored to the needs of users. Traditionally, such middleware is also responsible for conversion of the user data into the file, which is still a contiguous byte array and is suboptimal for parallel or concurrent access. In the 2010s, data centers realized that existing I/O middleware is unable to exploit the deployed parallel file systems in HPC systems for various reasons. As a consequence, data centers started to develop new middleware such as PNetCDF (Li et al., 2003), SIONlib (Frings et al., 2009), GLEAN (Vishwanath et al., 2011), ADIOS (Podhorszki et al., 2016), PLFS (Bent et al., 2009), and data clay (Cortes et al., 2015). Various of these interfaces are now used in applications that run on large-scale machines.

[10]Application programming interface.

Additional data services (not shown in Fig. 7.7B) such as SILO (Miller, 2015), MARS (Raoult, 1997) or iCHIP (Lawerenz et al., 2007) are built on top of these middleware APIs and offer convenient access for certain domains; examples are meshes and fields for SILO, or meteorological data for MARS.

The middleware uses a widely available *file system interface* like POSIX or MPI-IO. A file system organizes data as files into a hierarchical namespace consisting of directories and provides interfaces to deal with these objects.

In contrast to a local or distributed file system, a *parallel file system* is explicitly designed for achieving optimal and concurrent access to files in massively parallel environments, as employed in HPC systems. To achieve this performance, internal data of a file are physically scattered among a subset of the available servers and their I/O subsystems. This enables these servers to participate in one I/O operation, thus bundling their hardware resources to achieve higher aggregated performance.

An alternative to a file system is *object storage*. Amazon's Simple Storage Service (S3) interface[11] is widely spread in industry and cloud computing, yet rarely used in HPC. Object storage provides simpler APIs and naming schemes, e.g., an object ID instead of a hierarchical view, and may allow read/writes of whole objects only instead of the byte-level granular access offered by file systems. That means, updating an object requires to completely rewrite the object on the object storage.

Finally, the underlying storage media is dealt with as *block device*, i.e., access is performed in coarse-grained blocks of kilobytes of data at least. With SCM and NVM, a new technology is pushed into the software stack which enables byte-addressable access that we are used to from main memory allowing processors to access storage with normal load and store instructions. Ultimately, this blurs the line between storage and memory unifying both concepts to some extent. Programming SCM is still subject to research as additional interfaces are necessary to efficiently program the systems. A prominent interface under development is persistent memory I/O (pmem.io) (pmem.io - Persistent Memory Programming, Accessed: 2017-08-25). SCM has not been found in productive use in HPC workflows yet.

Hierarchical Storage Until recently, state-of-the-art I/O stacks consisted of two I/O layers: online and nearline storage. The performance gap between the highest and lowest layers is bridged by intermediate layers and the use of caching. Some are now supplemented by burst buffers and special purpose hardware – particularly random I/O accelerators (often an all-flash storage system). Traditionally, utilizing these storage tiers was accomplished manually via system-wide policies as provided by HPSS (Jes, 2017), or by user-provided scripts that migrate data in a separate step. Whether or not the selected strategy was beneficial for the access pattern was not always known. All existing middlewares, e.g., MPI, HDF5, NetCDF, implicitly assume such a manual workflow, and do not address I/O layer management.

With the greater variety of performance characteristics associated with new storage systems, the naive approach becomes inefficient, because manual management would be too complicated and explicit data migration is likely to waste system performance. Burst buffers such as DDN IME (DDN Storage, 2015) or Cray DataWarp (Cray Inc., 2015) aim to achieve transparency by organizing I/O layers as a kind of cache hierarchy, in which I/O requests are propagated through the I/O layers.

7.4.2 Big Data Storage

In Big Data storage systems, *commodity-of-the-shelf* (COTS) technology is used. Servers are typically equipped with Gigabit Ethernet and local storage like a RAID[12] of HDDs or SSDs.

Big Data tools exploit this local storage and compute by shipping the application code to the data. A new era of Big Data tools was initiated when Google published a paper on the Map/Reduce programming paradigm (Dean and Ghemawat, 2004) that used the Google File System (GFS).

An example program for Map/Reduce is given in pseudocode in Listing 7.5. The code for the mapper is executed once for each record and may emit multiple results in the form of key/value pairs – we assume any data type is acceptable. This allows to filter, manipulate, and split the record data accordingly. After all Map operations have finished, data are sorted by the key and sent to Reduce functions. The Reduce function is called for each key together with all values and is able to aggregate across the values for each key and output a list of results for the given key. Therewith, Reduce allows to filter and manipulate data once more.

In the example, we summarize the income of people living in Germany per city and output cities that have at least 10 k income. After running such a program, the output of the Reduce stage is found in files.

[11] https://aws.amazon.com/s3/.

[12] Redundant Array of Independent Disks.

Listing 7.5: Pseudocode of a Map/Reduce program that summarizes income of people per city.

```
list<Key,Value> map(Record person){
        // We assume the record contains the data of a person

        // filter people living not in the EU
        if ( person.country == "Germany" )
                list.append(person.city, person.income)
}

list<Value> reduce(Key k, list<Value> data){
        // The key is now a city
        // Data is the list of peoples' incomes
        float sum = 0;
        for each income in data{
                sum += income;
        }
        // Check if at least 10k is earned, then output the value
        if( sum > 10000 ){
                list.append(sum)
        }
}
```

The Map/Reduce execution paradigm ships code that performs the Map task to the servers that hold fragments of the file allowing for massive concurrent execution. This is actually necessary, as Big Data networks are made of commodity hardware that is slower than HPC networks. The splitting of a file into individual records must be understood to identify how work can be distributed as processing of complete records is expected by the user. This embeds the knowledge about file formats into the storage framework and provides means to abstract from the specific file format. To conclude, GFS with Map/Reduce fuses computation and storage paradigm into one efficient solution.

The ideas were quickly adopted by the community and implemented in the open source variant Apache Hadoop (Apache Hadoop web site, Accessed: 2018-06-27). Hadoop provides the Hadoop Distributed File System (HDFS) and implements the Map/Reduce data processing paradigm.

Nowadays, a huge software landscape and industry are built on top of HDFS and Map/Reduce that increased the variety of programming models and services. While these systems imply certain limitations over traditional data management and file systems, they serve their purpose well to handle data that have never been modified (immutable) but only appended. For example, with Hive (Thusoo et al., 2009), SQL queries can be executed on files stored on HDFS. Practitioners quickly realized that these distributed systems bring cost, performance, and usability benefits over centralized relational database management systems that were widely used.

Additional NoSQL[13] storage solutions emerged quickly that define their own high-level data model to deal with semistructured data. Examples are Cassandra, CouchDB, BigTable, MongoDB, and Neo4J. While data warehouses aim to stretch database technology to cope with business intelligence data and Big Data demands, NoSQL solutions abandon the atomicity, consistency, isolation, and durability (ACID) consistency model for transactions in favor of relaxed consistency models such as basically available, soft state, eventual consistency (BASE).

Nowadays, the Map/Reduce bulk data processing model is abandoned since it requires a synchronization after the Map step; this barrier slows down the processing of workflows involving several Map/Reduce steps. Next-generation Big Data tools such as Spark and Flink perform operations on large datasets and manage compute and storage resources (Madden, 2012). The basic building blocks of Flink programs are streams and transformations. Conceptually, a stream is a (potentially never-ending) flow of data records, and a transformation is an operation that takes one or more streams as input and produces one or more output streams as a result. When executed, Flink programs are mapped to streaming data flows, consisting of streams and transformation operators. A data flow starts with one or more sources and ends in sinks. Data flows typically resemble arbitrary directed acyclic graphs (DAGs).

In order to speed up data processing in-memory, databases and processing engines load data into mem-

[13]Not only SQL.

ory. The problem here is that a database can neither do performance optimization directly on storage nor load data to some other location, e.g., remote memory.

7.5 ASSESSING PERFORMANCE

The complexity of a supercomputer architecture – communicating nodes, each node comprising one or several CPUs with multiple compute cores with each core executing (multiple) instructions, and storage – renders the process of understanding the performance of a program executed on such a platform a challenging task. High performance of a program is not only essential to have a short "time-to-solution" for a certain compute problem, but also to efficiently exploit energy, with every supercomputer in the top 10 (TOP500 web site, Accessed: 2018-06-20) consuming more than 1 MW of power. Data transfer is typically the most expensive operation and should therefore be avoided at all levels: as a rule of thumb, transfer through the network is slow (compared to memory and caches) and very expensive, data transfers between CPU and memory are still expensive, and floating point operations are basically for free; see Table 7.2 for an overview of latency times to access data from different locations (Huang et al., 2014; Lüttgau et al., 2018).

TABLE 7.2
Latencies for accessing data in different cache levels (L1, L2, L3), main memory (RAM), over networks (InfiniBand, Ethernet) or from (local) disks (SSD, HDD) (Lüttgau et al., 2018).

Level	Latency
L1 cache	$\approx$ 1 ns
L2 cache	$\approx$ 5 ns
L3 cache	$\approx$ 10 ns
RAM	$\approx$ 100 ns
InfiniBand	$\approx$ 500 ns
Ethernet	$\approx$ 100,000 ns
SSD	$\approx$ 100,000 ns
HDD	$\approx$ 10,000,000 ns

This induces several guidelines for HPC software development:

- Data alignment within a node is necessary to feed whole *cache lines* to the CPU registers and process them efficiently. This may contradict well-established software design patterns from object-oriented programming (OOP). For example, in OOP, one might model a globe covering grid for atmospheric motion by defining a two-dimensional array of grid cells. Each grid cell is modeled to hold three floats for flow velocity components, a float for the pressure, and a float for the temperature. If an operation should be carried out on only one of the values (e.g., the pressure), this OOP model will fail in terms of performance as the pressure values are not aligned in memory and cannot be loaded "en-bloc" into registers (e.g., four pressure values need to be fetched from four distant memory locations, separated by four float values each). Therefore, this "array-of-structures" (AoS) approach is often exchanged for a "structure-of-arrays" (SoA) layout which, in our example, arranges every velocity component, pressure, and temperature value in a separate long array and bundles these arrays in a grid class. This yields data alignment for every variable field.
- Synchronization points within a program should be avoided. Synchronizing steps within a program imply that some instances, i.e., nodes/ cores/threads have to wait for others. This introduces not only overhead which typically increases with the number of waiting instances but also blocks until the slowest task reaches the synchronization point.
- Parallel execution needs to be fostered at all levels, and sequential program parts have to be minimized. The *speedup* of a program is defined as the ratio of execution time on one versus N instances (that is, cores or nodes). In the ideal case of a perfectly parallelized program, running a program on N instances delivers a speedup of N. However, most programs feature some sequential parts, due to the used algorithms, I/O operations, etc. Let t denote the total execution time of a program running on a single instance and t_S its sequential (not parallelized) part. According to *Amdahl's law* (Amdahl, 1967), the maximum speedup arises as t/t_S. That is, even if 99% of a program can be executed embarrassingly parallel, the program cannot be accelerated beyond a speedup of 100.
- Data duplication and data transfers between nodes should be avoided. This means that data should be partitioned between the processes instead of being replicated; for example, each process should hold a fraction of data. A widely used technique in numerical analysis is given by domain decomposition methods that provide means to split the domain (and

data) into subdomains that can be solved independently – but require data exchange on the domain boundaries.

All these points immediately translate into prerequisites for efficient algorithms. Still, it is often not trivial to deduce what "high performance" for a particular application program means.

7.5.1 Compute Metrics: Moore's Law, Floating Point Performance, and Bandwidth

Various metrics have been proposed to categorize and "measure" HPC systems (i.e., supercomputers), with floating point performance being the most prominent one. Between 1993 and 2011, the performance of the fastest listed supercomputer improved from 60 gigaflops to about 10 petaflops, which is a factor of roughly 2^{17}. This exponential growth in computing performance, i.e., doubling performance every 18–24 months, is proportional to "Moore's law" (Thompson and Parthasarathy, 2006): Gordon E. Moore observed that the number of transistors which can be placed on an integrated circuit roughly doubled every year between 1958 and 1965 and predicted that this trend would continue in the future. On the one hand, the factor correlates to the improvement of chip design and miniaturization that has been achieved in these 18 years and is analogous to Moore's law. On the other hand, it is a result of the improved clustering of independent chips into a large supercomputer and of increased investments. Lately, Moore's law started to stagnate at supercomputing level, and it is not clear whether and how performance boosts will occur in the future. In particular, while obstacles to achieve petascale performance have been overcome, developing machines and parallel algorithms for exascale supercomputers is a task which must still be solved.

Another important metrics is given by memory throughput or *bandwidth*, that is, the amount of data that can be transferred from (main) memory to the CPU. It is typically given in byte/s and, together with floating point performance, enables effective performance modeling considerations; see for example the roofline model in Section 7.5.4.

7.5.2 Data Metrics: the Five Vs

Big Data processing is typically defined and characterized through the five *Vs*. The *volume* of the data, measured in bytes, defines the amount of data produced or processed. The *velocity* at which data are generated and processed (e.g., bytes per second) corresponds to another characteristic. Both data volume and velocity also play a role for computing and have similar metrics in this regard, with data velocity rather defined by throughput. The *variety* gives information on the diversity of data that are collected. This covers data format and structure (structured like a database, unstructured like human-generated text/ speech, or semistructured like HTML). Data can be rather similar (e.g., when collecting measurements through the same apparatuses from different sources), or extremely different, without any kind of obvious relation, with the latter turning out to exist and being essential after some processing of the data. Besides these three Vs (Pettey and Goasduff, 2011; Laney, 2001), two more characteristics have evolved that are frequently referred to in the Big Data context. The *validity* denotes the quality or actual trustworthiness in the data. For example, damaged data or incorrect values due to wrong data measurements may deteriorate a dataset. Finally, the *value* of data corresponds to the actual meaning of the data in a prescribed context. For example, data on customer satisfaction are very valuable for a company.

Big Data implies typically high volume, nonstatic, and frequently updated (velocity) data that involve a variety of data formats and particularly unstructured data, potentially wrong data (validity), and low value in its origin. Refinement is hence required to add value. Overviews on the Vs for Earth observation data can be found in Guo et al. (2015), Nativi et al. (2015). In particular, Nativi et al. (2015) conclude amongst others that heterogeneity "is really perceived as the most important challenge for Earth Sciences and Observation infrastructures."

7.5.3 Benchmarking

Benchmarking refers to the measurement of hardware performance via standardized problem settings. Running benchmark suites on an HPC system yields insights into the actual capabilities of the hardware in terms of floating point, memory, network, and storage performance and thus lays a practical foundation for application-specific performance investigation. In the following, we give examples of typical benchmarks used in HPC.

The *LINPACK* benchmarks make use of linear algebra routines to solve large, dense linear systems of equations. Such linear algebra problems are very compute-intensive; LINPACK hence extends the performance view on an HPC system from the *theoretical peak performance* to actual, achievable floating-point performance. LINPACK is used to rank the supercomputers in the TOP500 list. Also relying on LINPACK, the Green500 list comprises the most energy-efficient supercomputing systems, measured through LINPACK performance in terms of (flop/W).

The *STREAM* benchmark (STREAM benchmark web site, Accessed: 2018-06-20) consists of four vector-type operations (copy, scale, triad, add) and is used to determine the actual memory bandwidth (in byte/s) of a system. Analyzing I/O with benchmarks is highly complex, as it depends on all other hardware components and achieved performance can vary across orders of magnitude. Also, raw I/O throughput must be distinguished from the metadata performance when manipulating the storage namespace (like the hierarchical directory tree). Example benchmarks are IOR and MDTest,[14] which are widely used to evaluate I/O and metadata throughput of the typical bulk-synchronous access patterns, respectively. During the 2010s, various new lists complemented the TOP500: the Green500, Graph 500,[15] and IO-500,[16] each aiming to track alternative metrics and to allow long-term analysis.

7.5.4 Performance Modeling

Performance modeling can be used to investigate (parts of) a program and derive a mathematical model to predict the performance on a certain hardware, given an optimal implementation. At core level, the analytical *roofline model* (Williams et al., 2009) provides good estimates for a wide range of problems. The underlying idea is that the performance of a program is either limited by clock speed P_{max} (in flops) or by data transfer along the slowest necessary path in terms of memory bandwidth b_s (in byte/s). Measuring b_s in a system, e.g., through the STREAM benchmark, and determining the *computational intensity* I (in flop/byte) of the program (i.e., counting the number of computations, data loads, and stores in the source code), deliver a simple upper bound for the expected performance P (in flops):

$$P = \min(P_{max}, b_s \cdot I). \tag{7.1}$$

Fig. 7.8 illustrates the approach: the continuous red/-green lines dictate the upper bound, the so-called "roofline," of the expected performance. Several extensions to the roofline model exist, for example to take into account cache hierarchies (Ilic et al., 2014). For multicore systems, the analytical execution-cache-memory model performance model has evolved in the last years (Hofmann et al., 2015). Besides fully analytical models, automatic performance model generation,

[14]https://github.com/hpc/ior.
[15]https://graph500.org/.
[16]http://www.vi4io.org/io500/start.

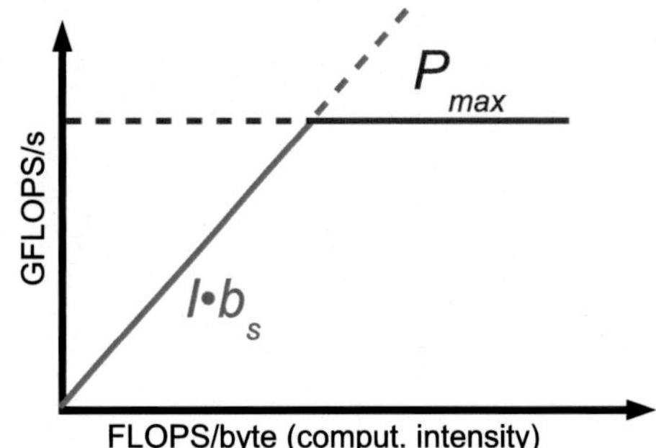

FIG. 7.8 Illustration of the roofline model.

e.g., to determine scalability bugs or parallelization bottlenecks, has become available. An example is given by the software Extra-P (Calotoiu et al., 2013).

7.5.5 Performance Analysis

To measure the performance of a program, several analysis software packages have become well-established in HPC. The software Gprof (Graham et al., 1982) is open source and easily lists the run-times of the (most costly) function calls within a program. Valgrind (Valgrind web site, Accessed: 2018-06-20; Nethercote and Seward, 2007) provides a set of tools to detect memory and threading issues within an application. Scalasca is an open source solution for measuring and analyzing program run-times with particular regard to communication analysis (Scalasca web site, Accessed: 2018-06-20; Geimer et al., 2010). Score-P (Score-P web site, Accessed: 2018-06-20; Knüpfer et al., 2012) and Vampir (Vampir web site, Accessed: 2018-06-20) similarly enable source code instrumentation and program performance analysis. Score-P further integrates the Extra-P tool. Commercial software packages for program profiling and analysis are given by Intel VTune Amplifier (Intel VTune Amplifier web site, Accessed: 2018-06-20) and ARM Tools (ARM Tools web site, Accessed: 2018-06-20).

7.6 HIGH-PERFORMANCE DATA ANALYTICS

Having discussed compute and storage hardware as well as performance metrics and measurement approaches, this section shall point at one particular link between HPC and Big Data processing in a "bottom-up" sense, considering the integration of numerical linear algebra in data analytics. Data analytics is not only important for commercial Big Data applications, but also for scientific workflows executed on HPC systems. In the following, Big Data science that makes use of HPC systems is referred to as *high-performance data analysis* (HPDA).

According to Tracy (2014), there are many reasons why HPDA has become a major pillar in R&D and industry:

- *"the ability of increasingly powerful HPC systems to run data-intensive modeling and simulation problems at larger scale, at higher resolution, and with more elements (e.g., inclusion of the carbon cycle in climate ensemble models)*
- *proliferation of larger, more complex scientific instruments and sensor networks, from 'smart' power grids to the Large Hadron Collider and Square Kilometer Array*
- *the increasing transformation of certain disciplines into data-driven sciences...*
- *availability of newer advanced analytics methods and tools like MapReduce/Hadoop, graph analytics, semantic analysis, knowledge discovery algorithms and others*
- *the escalating need to perform advanced analytics in near-real-time..."*

These increasing needs go hand in hand with architectural hardware and software challenges and changes, yielding a significant adoption of hardware to data analytics needs, in particular to machine learning (cf. Section 7.2.3). Depending on the use case, the optimal hardware for a Big Data analytics problem may consist of a number of fat nodes with large memory and high bandwidth, or a big number of nodes with smaller memory (but providing a higher level of parallelism and compute power) (Appuswamy et al., 2013; Xenopoulos et al., 2016).

Besides – as the cost factor plays a more and more essential role – cloud computing may also be a cost-effective approach. Xenopoulos et al. conclude that *"cloud computing is more cost effective for small or experimental data workloads, but HPC is more cost effective at scale"* (Xenopoulos et al., 2016).

Hence, Big Data and HPC technology converge, amongst others in the field of HPDA. There are two directions for technology transfer between Big Data and HPC communities. On the one hand, HPDA aims, amongst others, to harness HPC infrastructure to enable extremely performant analysis of scientific workflows. Also, HPC researchers explore the potential usage and benefit of Big Data technology. On the other hand, HPC infrastructures are able to execute Big Data software stacks. Often, vendors provide optimized versions of Big Data software, for example, changes to parallel file systems to support HDFS efficiently. Besides, various projects modify Big Data tools to make scientific data formats such as NetCDF available to software such as Hadoop or Spark. The latter enables the application of Big Data toolchains to scientific data.

Numerical Linear Algebra and Its Applications in Data Science Data analytics comprises various (subfields of) methods, including clustering analyses, classification, outlier detection, machine learning approaches, regression, or optimization.

The backbone in most of these approaches is given by building blocks from numerical linear algebra. *General matrix to matrix multiplication* (GEMM) is particularly relevant in (amongst others) deep learning (Warden, 2015); deep learning methods are employed amongst others for Earth observation classifications (Marmanis et al., 2016); *Sparse matrix-matrix, matrix-vector, and related operations* are essential amongst others for processing unstructured data (Buono et al., 2016), e.g., in graph-based data mining or machine learning (Peng et al., 2016). Depending on the sparsity of the underlying system and operation, an efficient implementation thereof may be compute- or memory-bound, with a high level of sparsity typically resulting in a memory-bound algorithm. *Eigenvalue and eigenvector problems* correspond to the algorithmic ingredient of many data analyses. The *singular value decomposition* (SVD), for example, represents the actual ingredient of the proper orthogonal decomposition/principal component analysis; see amongst others (Anzt et al., 2017) for an overview of relevant solvers and algorithms for SVD as well as for a discussion of SVD performance on hybrid CPU/GPU systems. Similarly, solving eigenvalue problems such as determining largest/smallest eigenvalues of a system enables the simplification and reduction of a complex system to a small set of relevant modes (i.e., eigenvalues and eigenvectors). In statistics, determining the largest eigenvalues of a covariance matrix corresponds to finding the most influential factors to variance. Finally, *linear solvers* are used at the heart of various data analytics algorithms. Conjugate gradient methods and variants thereof may represent an efficient approach to solve ordinary least-squares problems, or, for example, Lasso regression problems (Mirone and Paleo, 2017). Algebraic multigrid methods have been successfully applied to data analysis problems such as clustering, segmentation, or quantization (Galun et al., 2015).

Numerical Linear Algebra and Data Analytics Software With this variety in numerical linear algebra routines used in data analytics, various long-standing, as well as rather new, implementations thereof have found their way into data analytics software. In the following, most common numerical linear algebra programs and libraries are discussed. They are linked to software for data analytics that uses them.

The *Basic Linear Algebra Subprograms* (BLAS) are a specification for common linear algebra routines, originally addressing specifically dense systems. Most common open source implementations thereof are given by OpenBLAS and ATLAS. The BLAS implementations are in use amongst others in the statistics software *R* (The R Project for Statistical Computing web site, Accessed: 2018-06-20) or in *Caffe* (Caffe web site, Accessed: 2018-06-20; Jia et al., 2014), a C++-written library for deep learning (Intel MKL [see below], OpenBLAS, or ATLAS).

The *Linear Algebra Package (LAPACK)* consists of linear solvers, eigenvalue and SVD solvers. *ScaLAPACK* includes parts of LAPACK and is specifically designed for distributed memory architectures. Intel's *math kernel library (MKL)* comprises BLAS, LAPACK, and ScaLAPACK functionality as well as solvers for sparse systems. MLlib (MLlib web site, Accessed: 2018-06-20), Apache's scalable machine learning library which is developed as part of Apache Spark, relies on BLAS (through the Java library Jblas) and LAPACK.

Eigen is a C++ template library, providing routines for matrix and vector operations and several numerical solvers, and for linear algebra: matrices, vectors, numerical solvers, etc. TensorFlow (TensorFlow web site, Accessed: 2018-06-20), an open source Google development for data flow programming and machine learning, uses Eigen and Intel MKL.

Apache Mahout (Apache Mahout web site, Accessed: 2018-06-20) is a "distributed linear algebra framework and mathematically expressive scala DSL" (Apache Mahout web site, Accessed: 2018-06-20) (domain-specific language), with focus on data science/statistics. Mahout traditionally used Apache Spark as a back-end for cluster computing, and has opened up to support other back-ends, including H2O and Flink, since its release 0.10.0.

Libraries, that have been specifically dedicated to solving (selected) eigenvalue problems comprise *FEAST*, *ELPA* (direct solver for symmetric matrices), and *SPECTRA* (computation of a specified number of eigenvalues/eigenvectors), with the latter built on top of Eigen.

Dedicated solutions for GPUs are provided through Nvidia's *cuBLAS* (CUDA-based BLAS functionality), *cuSPARSE* (sparse matrix support, etc.), and *cuSOLVER* (dense and sparse direct solvers for linear systems of equations, eigenvalue computation), with the latter based on cuBLAS and cuSPARSE. cuBLAS, cuSPARSE, and cuDNN (building blocks for neural network computations) are in use within Nvidia's Deep Learning SDK (NVidia Deep Learning SDK, Accessed: 2018-06-20) which provides tools and libraries for deep learning. *Matrix algebra for GPU and multicore architectures* (MAGMA) is another library for dense linear algebra, addressing heterogeneous architectures such as CPU/GPU systems.

7.7 CONCLUDING REMARKS ON HPC, BIG DATA, AND THEIR CONVERGENCE

Throughout this chapter – in particular in the last section – we have pointed out several intersections of HPC and Big Data handling. At a high level, throughput-oriented hardware becomes more and more essential in several memory-bound HPC applications and is therefore not exclusively essential for Big Data problems. The determination of optimal parallel processing of computation and/or data is essential for both HPC and Big Data. Similarly, despite several differences and challenges from the underlying application perspectives and actual hardware configurations, the need for high-performance storage and I/O has greatly increased in both disciplines. In terms of data processing, as pointed out in the last section, high-performance data analytics and machine learning currently are right at the intersection of both HPC and Big Data. This even holds for very basic algorithmic components, such as numerical linear algebra, which has historically been at the heart of many HPC applications. Yet, every application exposes specific needs to compute and storage hardware, processing speeds, data access, and reliability; Table 7.3 summarizes and compares several hardware and software characteristics for both HPC and Big Data. Several challenges need to be overcome in the future to leverage both HPC and Big Data paradigms more efficiently when it comes to processing Earth observation data. Ma et al. (2015) name, amongst others, parallel I/O performance, optimal scheduling of data-dependent tasks, and efficient and productive programming of remote sensor applications on hierarchical cluster-based parallel systems.

ACKNOWLEDGMENT

Parts of this work have been carried out under the umbrella of the *Centre of Excellence in Simulation of Weather and Climate in Europe* (ESiWACE), which has received funding from the European Union's Horizon 2020 research and innovation program under grant no. 675191; this material reflects only the author's view and the European Commission is not responsible for any use that may be made of the information it contains.

TABLE 7.3
Characteristics of HPC and Big Data in hardware and software.

(a) Hardware		
Characteristics	**Big Data**	**HPC**
Type	Commodity-of-the-shelf	Cutting-edge
Price	Cheap	Expensive
Performance	Average	High end
Reliability	Average	Very high

(b) Software		
Characteristics	**Big Data**	**HPC**
Shared storage	Yes	Yes
POSIX file system	No	Yes
Local data access	Yes	No
File replication	Yes	No
Programming model	bulk, streaming	MPI, OpenMP, OpenACC, CUDA

REFERENCES

Amdahl, G.M., 1967. Validity of the single processor approach to achieving large scale computing capabilities. In: Proceedings of the April 18–20, 1967, Spring Joint Computer Conference, AFIPS '67 (Spring). ACM, pp. 483–485.

Anzt, H., Dongarra, J., Gates, M., Kurzak, J., Luszczek, P., Tomov, S., Yamazaki, I., 2017. Bringing High Performance Computing to Big Data Algorithms. Springer International Publishing, pp. 777–806.

Apache Hadoop web site, Accessed: 2018-06-27. http://hadoop.apache.org.

Apache Mahout web site, Accessed: 2018-06-20. https://mahout.apache.org/.

Appuswamy, R., Gkantsidis, C., Narayanan, D., Hodson, O., Rowstron, A., 2013. Scale-up vs scale-out for Hadoop: time to rethink? In: Proceedings of the 4th Annual Symposium on Cloud Computing, SOCC '13. ACM, pp. 20:1–20:13.

ARM Tools web site, Accessed: 2018-06-20. https://www.arm.com/products/development-tools.

Armbrust, M., Fox, A., Griffith, R., Joseph, A.D., Katz, R.H., Konwinski, A., Lee, G., Patterson, D.A., Rabkin, A., Zaharia, M., 2009. Above the clouds: a Berkeley view of cloud computing. Technical Report.

Bent, J., Gibson, G., Grider, G., McClelland, B., Nowoczynski, P., Nunez, J., Polte, M., Wingate, M., 2009. PLFS: a checkpoint filesystem for parallel applications. In: Proceedings of the Conference on High Performance Computing Networking, Storage and Analysis, SC '09. ACM, New York, NY, USA, pp. 21:1–21:12.

Buono, D., Petrini, F., Checconi, F., Liu, X., Que, X., Long, C., Tuan, T.-C., 2016. Optimizing sparse matrix-vector multiplication for large-scale data analytics. In: Proceedings of the 2016 International Conference on Supercomputing, ICS '16. ACM, pp. 37:1–37:12.

Caffe web site, Accessed: 2018-06-20. http://caffe.berkeleyvision.org/.

Calotoiu, A., Hoefler, T., Poke, M., Wolf, F., 2013. Using automated performance modeling to find scalability bugs in complex codes. In: Proceedings of the Conference on High Performance Computing Networking, Storage and Analysis, SC '13, pp. 1–12.

Chapman, B., Jost, G., Van Der Pas, R., 2007. Using OpenMP: Portable Shared Memory Parallel Programming. MIT Press.

Chen, C., Zhang, C.-Y., 2014. Data-intensive applications, challenges, techniques and technologies: a survey on big data. Information Sciences 275, 314–347.

Cilk web site, Accessed: 2018-06-20. https://www.cilkplus.org/.

Cortes, T., Queralt, Anna, Martí, J., Labarta, J., 2015. dataClay: towards usable and shareable storage. http://www.exascale.org/bdec/sites/www.exascale.org.bdec/files/whitepapers/dataClay%20at%20BDEC%20Barcelona.pdf. (Accessed 12 July 2017).

Cray Inc., 2015. Cray XC40 DataWarp's applications I/O accelerator, Cray Inc., Cray Inc. 901 Fifth Avenue, Suite 1000 Seattle, WA 98164.

Dally, W., Towles, B., 2003. Principles and Practices of Interconnection Networks. Morgan Kaufmann Series in Computer Architecture and Design.

DDN Storage, 2015. Burst Buffer & Beyond; I/O & Application Acceleration Technology. DDN Storage.

Dean, J., Ghemawat, S., 2004. MapReduce: simplified data processing on large clusters. In: Sixth Symposium on Operating System Design and Implementation, OSDI '04, pp. 137–150.

Feldman, M., 2016. Fujitsu switches horses for Post-K supercomputer, will ride ARM into exascale. https://www.top500.org/news/fujitsu-switches-horses-for-post-k-supercomputer-will-ride-arm-into-exascale/.

Frings, W., Wolf, F., Petkov, V., 2009. Scalable massively parallel I/O to task-local files. In: Proceedings of the Conference on High Performance Computing Networking, Storage and Analysis, SC '09. ACM, pp. 17:1–17:11.

Galun, M., Basri, R., Yavneh, I., 2015. Review of methods inspired by algebraic-multigrid for data and image analysis applications. Numerical Mathematics: Theory, Methods and Applications 8 (2), 283–312.

Geimer, M., Wolf, F., Wylie, B.J.N., Ábrahám, E., Becker, D., Mohr, B., 2010. The Scalasca performance toolset architecture. Concurrency and Computation: Practice and Experience 22 (6), 702–719.

Gigabit Ethernet - Technology and Solutions, 2001. http://www.intel.com/network/connectivity/resources/doc_library/white_papers/gigabit_ethernet/gigabit_ethernet.pdf.

Graham, S.L., Kessler, P.B., Mckusick, M.K., 1982. Gprof: a call graph execution profiler. SIGPLAN Notices 17 (6), 120–126.

Guo, H.-D., Zhang, L., Zhu, L.-W., 2015. Earth observation big data for climate change research. Advances in Climate Change Research 6 (2), 108–117.

Hemsoth, N., 2017. Cray ARMs highest end supercomputer with ThunderX2. https://www.nextplatform.com/2017/11/13/cray-arms-highest-end-supercomputer-thunderx2/. (Accessed 20 June 2018).

Hofmann, J., Eitzinger, J., Fey, D., 2015. Execution-cache-memory performance model: introduction and validation. Computing Research Repository. arXiv:1509.03118 [abs].

Huang, Jian, Schwan, Karsten, Qureshi, Moinuddin K., 2014. NVRAM-aware logging in transaction systems. Proceedings of the VLDB Endowment 8 (4), 389–400.

Ilic, A., Pratas, F., Sousa, L., 2014. Cache-aware roofline model: upgrading the loft. IEEE Computer Architecture Letters 13 (1), 21–24.

Intel Threading Building Blocks web site, Accessed: 2018-06-20. https://www.threadingbuildingblocks.org/.

Intel VTune Amplifier web site, Accessed: 2018-06-20. https://software.intel.com/en-us/intel-vtune-amplifier-xe.

Jes, 2017. High-Performance Storage Systems Answering the Data Explosion with Massive Scale and Compelling Economics.

Jia, Y., Shelhamer, E., Donahue, J., Karayev, S., Long, J., Girshick, R., Guadarrama, S., Darrell, T., 2014. Caffe: convolutional architecture for fast feature embedding. Computing Research Repository. arXiv:1408.5093 [abs].

Joubert, W., Kothe, D., Nam, H.A., 2009. PREPARING FOR EXASCALE: Application Requirements and Strategy. Technical Report. ORNL Leadership Computing Facility.

Kaufmann, W.J., Smarr, L.L., 1992. Supercomputing and the Transformation of Science. W. H. Freeman & Co.

Kim, J., Dally, W.J., Scott, S., Abts, D., 2008. Technology-driven, highly-scalable dragonfly topology. In: 2008 International Symposium on Computer Architecture, pp. 77–88.

Knüpfer, A., Rössel, C., Mey, D.a., Biersdorff, S., Diethelm, K., Eschweiler, D., Geimer, M., Gerndt, M., Lorenz, D., Malony, A., Nagel, W.E., Oleynik, Y., Philippen, P., Saviankou, P., Schmidl, D., Shende, S., Tschüter, R., Wagner, M., Wesarg, B., Wolf, F., 2012. Score-P: a joint performance measurement run-time infrastructure for Periscope, Scalasca, TAU, and Vampir. In: Brunst, H., Müller, M.S., Nagel, W.E., Resch, M.M. (Eds.), Tools for High Performance Computing 2011. Springer Berlin Heidelberg, pp. 79–91.

Kove, 2017. Kove XPD. http://kove.net/downloads/Kove-XPD-L3-datasheet.pdf. (Accessed 24 August 2017).

Laney, D., 2001. 3D data management: controlling data volume, velocity, and variety. Technical Report. META Group, Application Delivery Strategies.

Lawerenz, C., Eils, J., Eils, R., 2007. iCHIP: Plattform für NGFN Datenintegration. GenomXPress 2, 7–10.

Lee, D.S., Kalb, J.L., 2008. Network topology analysis. Technical Report.

Li, J., Liao, W.-K., Choudhary, A., Ross, R., Thakur, R., Gropp, W., Latham, R., Siegel, A., Gallagher, B., Zingale, M., 2003. Parallel netCDF: a high-performance scientific I/O interface. In: Proceedings of the 2003 ACM/IEEE Conference on Supercomputing, SC '03. ACM, p. 39.

Lüttgau, J., Kuhn, M., Duwe, K., Alforov, Y., Betke, E., Kunkel, J., Ludwig, T., 2018. Survey of storage systems for high-performance computing. Supercomputing Frontiers and Innovations 5 (1).

Ma, Y., Wu, H., Wang, L., Huang, B., Ranjan, R., Zomaya, A., Jie, W., 2015. Remote sensing big data computing: challenges and opportunities. Future Generation Computer Systems 51, 47–60.

Madden, S., 2012. From databases to big data. IEEE Internet Computing 16 (3), 4–6.

Malventano, A., 2017. How 3D XPoint phase-change memory works. https://www.pcper.com/reviews/Editorial/How-3D-XPoint-Phase-Change-Memory-Works. (Accessed 12 July 2017).

Marmanis, D., Datcu, M., Esch, T., Stilla, U., 2016. Deep learning Earth observation classification using ImageNet pretrained networks. IEEE Geoscience and Remote Sensing Letters 13 (1), 105–109.

Mellanox Technologies, 2000. Introduction to InfiniBand, document number 2003WP. http://www.mellanox.com/pdf/whitepapers/IB_Intro_WP_190.pdf.

Message Passing Interface Forum, 2015. MPI: A Message-Passing Interface Standard – Version 3.1. Technical Report.

Miller, M., 2015. Silo/HDF5 and Portable, Scalable, Parallel I/O. Technical Report. Lawrence Livermore National Laboratory (LLNL), Livermore, CA.

Mirone, A., Paleo, P., 2017. A conjugate subgradient algorithm with adaptive preconditioning for the least absolute shrinkage and selection operator minimization. Computational Mathematics and Mathematical Physics 57 (4), 739–748.

MLlib web site, Accessed: 2018-06-20. https://spark.apache.org/mllib/.

Nativi, S., Mazzetti, P., Santoro, M., Papeschi, F., Craglia, M., Ochiai, O., 2015. Big Data challenges in building the Global Earth Observation System of Systems. Environmental Modelling & Software 68, 1–26.

Nethercote, N., Seward, J., 2007. Valgrind: a framework for heavyweight dynamic binary instrumentation. In: Proceedings of the 28th ACM SIGPLAN Conference on Programming Language Design and Implementation, PLDI '07. ACM, pp. 89–100.

NVidia Deep Learning SDK, Accessed: 2018-06-20. https://developer.nvidia.com/deep-learning-software.

OpenMP web site, Accessed: 2018-06-20. https://www.openmp.org/.

Peng, J., Xiao, Z., Chen, C., Yang, W., 2016. Iterative sparse matrix-vector multiplication on in-memory cluster computing accelerated by GPUs for big data. In: 2016 12th International Conference on Natural Computation, Fuzzy Systems and Knowledge Discovery (ICNC-FSKD), pp. 1454–1460.

Pettey, C., Goasduff, L., 2011. Gartner says solving 'Big Data' challenge involves more than just managing volumes of data. https://www.gartner.com/newsroom/id/1731916. (Accessed 27 June 2018).

pmem.io - Persistent Memory Programming, Accessed: 2017-08-25. http://pmem.io.

Podhorszki, N., Liu, Q., Logan, J., Mu, J., Abbasi, H., Choi, J.-Y., Klasky, S.A., 2016. ADIOS 1.11 User's Manual. Oak Ridge National Laboratory.

Raoult, B., 1997. The architecture of the new MARS server. In: Sixth Workshop on Meteorological Operational Systems, pp. 90–100.

Rew, R., Davis, G., Emmerson, S., Davies, H., Hartne, E., 2008. ADIOS 1.11 User's Manual. version 4.0-snapshot2008122406.1-beta2 edn. Unidata Program Center.

Ronen, R., Member, S., Mendelson, A., Lai, K., Lu, S.-L., Pollack, F., Shen, J.P., 2001. Coming Challenges in Microarchitecture and Architecture, Vol. 89. IEEE, pp. 325–340.

Rosolowsky, E., Gaensler, B., Baum, S., Spekkens, K., Stil, J., 2016. The square kilometre array and data intensive radio astronomy. https://www.computecanada.ca/wp-content/uploads/2016/03/SPARC2-2016-White-Paper-SKA-CASCA.pdf.

Sadashiv, N., Kumar, S.M.D., 2011. Cluster, grid and cloud computing: a detailed comparison. In: 6th International Conference on Computer Science Education, ICCSE '11, pp. 477–482.

Sarkar, V., Harrod, W., Snavely, A.E., 2009. Software challenges in extreme scale systems. Journal of Physics: Conference Series 180 (1), 012045.

Scalasca web site, Accessed: 2018-06-20. http://www.scalasca.org/.

Score-P web site, Accessed: 2018-06-20. http://www.vi-hps.org/projects/score-p/.

Seagate, 2017. Flash acceleration on HPC storage. http://files.gpfsug.org/presentations/2017/Manchester/SEAGATE_Flash_Acceleration_of_HPC_Storage.pdf. (Accessed 24 August 2017).

Strass, H., 2016. An introduction to NVMe. http://www.seagate.com/files/www-content/product-content/ssd-fam/nvme-ssd/nytro-xf1440-ssd/_shared/docs/an-introduction-to-nvme-tp690-1-1605us.pdf. (Accessed 12 July 2017).

STREAM benchmark web site, Accessed: 2018-06-20. https://www.cs.virginia.edu/stream/.

Tallis, B., 2017. The Intel Optane SSD DC P4800X (375GB) review: testing 3D XPoint performance. http://www.anandtech.com/show/11209/intel-optane-ssd-dc-p4800x-review-a-deep-dive-into-3d-xpoint-enterprise-performance/5. (Accessed 12 July 2017).

TensorFlow web site, Accessed: 2018-06-20. https://www.tensorflow.org/.

The HDF group, 2015. HDF5 User's Guide. 1.11.x edn. The HDF Group.

The R Project for Statistical Computing web site, Accessed: 2018-06-20. https://www.r-project.org/.

Thompson, S.E., Parthasarathy, S., 2006. Moore's law: the future of Si microelectronics. Materials Today 9 (6), 20–25.

Thusoo, A., Sarma, J.S., Jain, N., Shao, Z., Chakka, P., Anthony, S., Liu, H., Wyckoff, P., Murthy, R., 2009. Hive: a warehousing solution over a map-reduce framework. Proceedings of the VLDB Endowment 2 (2), 1626–1629.

TOP500 web site, Accessed: 2018-06-20. top500.org. List - November 2017.

Tracy, S., 2014. Big Data meets HPC. https://www.scientificcomputing.com/article/2014/03/big-data-meets-hpc. (Accessed 20 June 2018).

Valgrind web site, Accessed: 2018-06-20. http://valgrind.org/.

Vampir web site, Accessed: 2018-06-20. https://www.vampir.eu/.

Van Der Pas, R., Stotzer, E., Terboven, C., 2017. Using OpenMP – The Next Step. MIT Press.

Vishwanath, V., Hereld, M., Morozov, V., Papka, M.E., 2011. Topology-aware data movement and staging for I/O acceleration on Blue Gene/P supercomputing systems. In: Proceedings of the Conference on High Performance Computing Networking, Storage and Analysis, SC '11. ACM, New York, NY, USA, pp. 19:1–19:11.

von Neumann, J., 1993. First draft of a report on the EDVAC. IEEE Annals of the History of Computing 15, 27–75.

Warden, P., 2015. Why GEMM is at the heart of deep learning. https://petewarden.com/2015/04/20/why-gemm-is-at-the-heart-of-deep-learning/. (Accessed 20 June 2018).

Williams, S., Waterman, A., Patterson, D., 2009. Roofline: an insightful visual performance model for multicore architectures. Communications of the ACM 52 (4), 65–76.

Xenopoulos, P., Daniel, J., Matheson, M., Sukumar, S., 2016. Big data analytics on HPC architectures: performance and cost. In: 2016 IEEE International Conference on Big Data (Big Data), pp. 2286–2295.

CHAPTER 8

Query Processing and Access Methods for Big Astro and Geo Databases

KARINE ZEITOUNI, PROF, PHD • MARIEM BRAHEM, PHD • LAURENT YEH, ASSOC PROF, PHD • ATANAS HRISTOV, PHD

8.1 BIG DATA MANAGEMENT

Accelerating the development and deployment of advanced computing technologies and complex databases will require a comprehensive strategy integrating efforts from invention to deployment. High-performance computing (HPC) has become an essential tool in numerous social and natural sciences. The modern HPC systems are composed of hundreds of thousands of computational nodes, as well as complex interconnected topologies. Existing high-performance algorithms and tools already require intensive programming and optimization efforts to achieve high efficiency on current supercomputers. On the other hand, these efforts are platform-specific and nonportable.

A core challenge in petabyte-scale data collected from astronomical surveys or Earth observation is the need to process these data with highly effective algorithms and tools where the data processing costs grow exponentially. The large amount of data that are being generated by the Earth and universe observation will not only benefit from, but require high-performance algorithms and tools due to the computational volume and the complexity of the existing models.

The vast availability of unstructured datasets presents Earth observation analytics and astronomy with new challenges. It is clear that the complexity of the astronomical data is so high that the solution can only be found in applying the computational resources of the most powerful high-performance systems and by developing novel solutions for Big Data.

Big data management usually refers to efficient and advanced data management techniques that can be applied on large volume of data. Recently, the Big Data management has become a "hot" topic in computer science and the number of research articles that address this topic is continuously growing. Taking into account the current work in this field, many new and effective query paradigms, architectures, and systems have been proposed.

Map/Reduce Map/Reduce is a data processing paradigm for large-scale data processing. This paradigm is composed of two functions: *Map* and *Reduce*. The *Map* function performs partitioning. Basically, it maps a set of values into an intermediate result which is another set of key/value pairs. All key/value pairs sharing the same key define a partition. On the second stage, the shuffling process transfers the map tasks output results to the reducer tasks. The reduce function executed in every reducer task use combining strategies to reduce the intermediate results having same key. One of the most used implementation of Map/Reduce model is Apache Hadoop. The Hadoop framework is designed to deal with the classical distributed computing issues and to detect and handle failures in order to deliver highly available services. On the other hand, it takes care about data locality by sending operations on the nodes that have a chunk data as much as possible. This approach relieves programmers from having to deal with the abovementioned issues. Besides the Map/Reduce module, the Apache Hadoop framework is composed of several additional parts, including Hadoop Common for the implementation of user Map/Reduce jobs; Hadoop Distributed File System (HDFS); and Hadoop YARN as the resources manager.

Directed Acyclic Graph Directed acyclic graph (DAG) is another data processing paradigm for effective Big Data management. A DAG is a finite directed graph composed of a finite set of edges and vertices. In DAG each edge is directed from one vertex to another, without cycles. Due to possibilities to model many different types of data, it can be easily implemented on distributed computing systems. DAG is very useful for applications with task dependencies, meaning the input and the output of the application depends on other tasks. There are two types of dependencies in DAG paradigm:

- explicit, where the dependencies are specified by the developer through explicit instructions,

Knowledge Discovery in Big Data from Astronomy and Earth Observation. https://doi.org/10.1016/B978-0-12-819154-5.00018-7

- implicit, where the dependencies are detected by the system through analyzing of the input, the output, and the tasks of the program.

One of the most popular implementations of the DAG paradigm is Apache Spark, but Spark does not implement implicit dependencies. The Spark framework was developed in 2012, in order to avoid the main bottleneck of the Hadoop Map/Reduce implementation.

The main difference between Hadoop and Spark is that all Map/Reduce frameworks keeps the intermediate results on the distributed file system. On the other hand, Spark keeps the data on the local memory and queries it repeatedly in order to achieve better performance. Usually, Spark obtains better performance for applications that can be viewed by the system as a pool of independent stages working on different computational nodes. On each stage, the same code is executed with different input parameters.

NoSQL Recently, the amount of data that need to be stored on distributed computing systems and databases has been growing exponentially. The possibilities to hit the scalability limitations are very high, which will lead to significant reduction in the efficiency of querying and analysis. The main problem of the traditional relational databases is that they are not able to efficiently scale horizontally over distributed computing systems.

One of the biggest challenges is to find an effective way to store and manage the large amounts of data that are generated by the astro and geo applications. In order to enable horizontal scalability over distributed systems, one of the recent approaches as an alternative to traditional transactional relational database systems is the NoSQL approach.

The NoSQL movement began in San Francisco in June 2009, and is growing rapidly. They define some distinguish characteristics, such as (i) it can scale horizontally, (ii) it replicates/distributes data over many servers, (iii) it is mostly query-oriented and has few updates, and updates and inserts are asynchronous, (iv) it is schema-free or has flexible schemas; the system can work without a schema that describe the data.

NoSQL solves most of the common issues that arise from storing and managing Big Data, but it still has some issues for the analysis of Big Data. In order to override the issue with Big Data analysis, many Map/Reduce solutions for querying and analyzing have been proposed. Even though Apache Hadoop reduces querying time and solves scalability issues, it requires highly professional developer skills to use it. On the other hand, it has very low level of abstraction and the Map/Reduce approach needs to be used for all applications, even for very simple tasks. This method is very time consuming, and it is not suitable for the companies. Thus, some companies go a step further and develop distributed systems for simple data analysis based on SQL-like or command language. Apache Pig translates a command language to a sequence of Map/Reduce jobs, and Apache Hive does the same but from an SQL-like query. Both take the benefit of Map/Reduce to process very large datasets. Recent extensions of Pig and Hive allow to execute with main memory engine instead of Hadoop Map/Reduce. This allows more interactivity when data analysts explore data.

The NoSQL data model addresses several issues that the relational model is not designed to address, for instance, applications that require process graphs (e.g., social network), XML, or time sequences.

8.2 QUERY PROCESSING STEPS

Query processing is translating a user query into an executable plan for a system and executing it. For queries which are formulated in procedural language (Eldawy and Mokbel, 2015; Apache Pig) or by simple API (Nishimura et al., 2011; Tang et al., 2016; Kini and Emanuele, n.d.), this requires programming skills. In this case, it is the responsibility of the user to produce an optimal query plan.

Besides, geospatial or astronomical SQL-like query language is broadly successful, because SQL is user-friendly and is a declarative language – the user declares what he/she wants, without telling the SQL engine how to solve the query. However, common Big Data systems with SQL language like Spark SQL and Hive (Apache Hive) are unable to meet the spatial queries. At best, they require query processing extensions. In this subsection we describe the key points of query processing in representative systems that support geospatial or astronomical queries.

A typical SQL query processing workflow is as follows:

- The query in the form of an SQL statement is parsed by an SQL parser to build up a syntax tree, in which keywords, expressions, variables, etc., are all identified.
- Then the evaluation step is started. Each SQL statement that refers to resources is bound to concrete objects in the spatial data schema (columns, tables, views, etc.).
- After the query optimizer provides a series of candidates (logical execution plans), it chooses an optimal plan according to statistics (e.g., the histogram of the

value distributions in each column, heuristics to reduce disk I/O, CPU consumption, or network costs). The physical plan is generated through binding data sources, and choosing the appropriate algorithms in order to improve query performance.

- Finally, the physical query plan is put into the executing queue of the query execution engine.

Most Big Data processing systems do not provide support for spatial data and operations, even though they offer a partial implementation of the SQL standard. Indeed, several problems should be addressed to integrate a spatial or astronomical data type as a basic domain of values and to be able to query it in the SQL language.

Parsing Stage The primary responsibility of the parser is to determine the grammatical structure of its input and, if well-formed, construct a data structure containing an abstract representation of that input (i.e., parse tree). A well-formed SQL query always start by the `SELECT` clause and follows by the FROM clause.

Many big data systems with an SQL interface also provide user definition function (UDF) utility, that is, users can customize the required UDF to process spatial data. Data types and operators for spatial UDF should follow as well as possible the standard. For example, this is how Aji et al. (2013) extends Hive SQL, and Huang et al. (2017) extends Spark SQL with spatial features. Therefore, using the UDF facility allows taking full advantage of built-in parsers like Spark SQL parsing as in Xie et al. (2016) and Huang et al. (2017) to achieve the support of SQL Spatial.

The aforementioned prototypes addressed geospatial applications. The standard used query language in astronomy is ADQL (GAVO Data Center, 2008), promoted by the International Virtual Observatory Alliance (IVOA), which is an SQL-like language and is widely used by the community of astronomers. However, it is not completely compliant with the SQL standard. For this reason, the existing SQL built-in parser for SQL cannot process ADQL queries, and a custom parser is required. To take full advantage of the built-in query processor of the existing system, another way is to use an ADQL parser to translate it to an SQL-compliant statement with spatial UDF. ADQL Library provides such parser that translates ADQL to Postgresql/Q3C or Postgres/pgSphere. AstroSpark (Brahem et al., 2016) adapts this parser to translate ADQL into Spark SQL with UDF.

Evaluation Stage The evaluation step (Gardarin and Valduriez, 1989) is to bound SQL statements to concrete resources in data schemas (i.e., columns, tables, views, etc.).

When the data schema refers to a large data volume and no index exists, which is the case for most Big Data systems, a commonly used approach is to split spatial data into partitions. To save useless disk I/O, the idea is to bind only relevant partitions instead of the whole dataset. In Wang et al. (2011), data are split in chunks according a spatial grid. In Brahem et al. (2016), data are split according to the HEALPix order. Every data element (i.e., representing a star) is tagged by the HEALPix index of the cell where it is located. Then, all the data are sorted according to the tagged HEALPix index, and the dataset is divided into equal length size partitions. A partition could be virtual, as shown in Yu et al. (2015). The most common approach is to split the sky according to a grid. All objects located in a given cell of the grid are virtually in a same partition. For these objects in the cell, they are tagged with the same partition ID.

To exploit this feature at this stage, the idea as shown in Wang et al. (2011) is to bind to relevant partitions instead of the entire dataset. For that, the process is to rewrite the initial query (i.e., the logical plan) to a set of subqueries where each subquery addresses a partition. Then, the initial query is transformed into a union of subqueries, where each subquery is sent to the processing node which holds the partition. In this process, an additional subquery that merges the results from each node is also added to the transformed query.

To achieve this, as most spatial queries are applied on a limited area of the space, the first step is to extract from the user query the minimal spatial bound (MSB), which is a geometric object (e.g., rectangle or circle). If an object in a partition for which the shape intersects the MSB, the partition is relevant, and a subquery is generated to address it.

In existing systems that offer partitioning capability as in Apache Hive or Apache Spark, this task is usually done at the optimization stage, where the optimizer determines relevant partitions according to metadata which describe how the data are partitioned. In the case of spatial data, as illustrated in Brahem et al. (2016), the optimizer requires a new rule that allows to exploit this feature. For that, the optimizer extracts the MSB from the user query which allows to determine relevant Healpix cells. With these cells and facing it with the partitioning metadata description, the optimizer could deduce all relevant partitions. The new rule in the optimizer will add a constraint on relevant partitions by extending the WHERE clause. The constraint is used at the execution step for loading in memory only relevant data.

Optimization Stage At this stage, the query optimizer provides several candidate execution plans according to heuristics, and it produces an optimal plan according to the cost-based function (Xie et al., 2016) which integrates running statistics, the existing index, disk I/O cost, and so on. The optimization differs if the engine is memory-oriented or disk-oriented. Systems build on an existing query optimizer could take advantage of implemented heuristics for reducing I/O costs for disk-oriented processing, or CPU costs for memory-oriented processing. Heuristics can change the order of join operators and add projection operators to prune nonnecessary attributes during query execution (Gardarin and Valduriez, 1989).

However, traditional query optimizers are not able to optimize queries with complex UDF (i.e., spatial UDF) because it appears as a black box. This often leads to very expensive query evaluation plans. For example, when Spark SQL implements a spatial distance join via UDF, it has to use the expensive Cartesian product approach, which is not scalable for large data. To avoid this, it is beneficial to replace it by an optimized algorithm for this purpose (Xie et al., 2016; Huang et al., 2017). Many Big Data systems provide an extensibility in the built-in optimizer as in Catalyst. It permits to modify the optimizer using mechanisms like rules or overload existing rules; otherwise, the user must rewrite the optimizer as in Xie et al. (2016). After the logical plan is optimized, a series of candidate physical plans are generated through binding data, physical operators, and index use when relevant, and/or data organization like partitioning that allows speedup of data access. Then, the optimizer chooses an optimal plan to put into the executing queue.

For systems based on the Map/Reduce paradigm, the query optimizer proceeds differently. The Map/Reduce paradigm is based on *job*, which is a unit of execution. If we consider that each algebraic operator in a query plan is translated into a job, and each job (i.e., operator) reads data from the disk and stores the result back to the disk, the query cost could be prohibitive. To minimize the disk I/O in the case of SQL query, the optimizer of a Map/Reduce framework decomposes the logical plan into a sequence of jobs, where each job could contain more than one algebraic operator (i.e., join, project, select, etc.) (Apache Hive). For procedural query language where each instruction could be an algebraic operator, a system like PIG introduces the lazy query evaluation. The idea is to record the sequence of instructions (i.e., the program) until a materialization instruction (e.g., STORE, DUMP in PigLatin) is reached. With the records of instructions, the query optimizer could reorder and group instructions to provide a sequence of optimal jobs.

Execution Stage The last stage is the execution on an engine which could evaluate the query plan. Almost all existing spatial Big Data are based on the existing NoSQL Databases, of which Spark and Hadoop Map/Reduce are a part. Using such a system provides properties like parallelism, scalability, reliability, and fault tolerance, which are valuable for processing large datasets. The Map/Reduce framework is designed for batch processing that is efficient when the program can be done in one job. On the other side, memory-oriented frameworks like Spark or Flink are interesting when interactivity with users is required because it avoids data reloading between operators of an algebraic query plan, in contrast to the Hadoop Map/Reduce framework. This last point is a big advantage because Spark applies the lazy evaluation (i.e., some operators do not materialize intermediate results).

In contrast to traditional distributed database systems, the benefit of using NoSQL systems is that the query optimizer does not have to take care of an optimal distribution plan for sending task to processing nodes. This is usually dynamically done by the resources manager (Yarn) in the case of the Map/Reduce framework or by the low level of the system (e.g., Spark). The resources manager is able to load balancing the execution even in the presence of nodes with resource gap, and in presence of data skew in the case of Map/Reduce.

However, using an existing NoSQL system for execution of a query plan is not straightforward because there is no spatial support. If the framework provides extensibility to implement new value domains, it is used to define a spatial value domain with ad hoc operators. This was the case in Yu et al. (2015) where the authors extended the resilient distributed datasets (RDDs) of APACHE SPARK to the spatial domain (named sRDDs) and added new spatial algorithms like kNN search. Otherwise, it requires rewriting part of the framework code to achieve the spatial data processing, as in SpatialHadoop (Xie et al., 2016; Eldawy and Mokbel, 2015). This approach offers more flexibility but it has the drawback that at each update of the framework, the implemented code could be invalidated.

The spatial index is often implemented in spatial systems (Tang et al., 2016; Nishimura et al., 2011; Xie et al., 2016) because they could drastically save the required processing resources. It could save the number of disk I/O (Eldawy and Mokbel, 2015), or reduce memory access (Huang et al., 2017). However, Map/Reduce, unlike DBMS, does not have the notion of index. Each job works with the whole dataset in input. To optimize this, many works have implemented a Global and Local index in the Map/Reduce framework as in Spatial Hadoop (Eldawy and Mokbel, 2015). At the execution stage, systems use *global index* to determine the subspace to be processes and also to determine for each worker the part of the subdataset to process. The *local index* is to speed up the processing inside of nodes, which is useful for complex spatial processing.

8.3 ACCESS METHODS

Physical data organization has a primary role in query optimization, whatever the data management technology. Each data management system implements various techniques, including internal data structures (e.g., B-tree index) and algorithms to optimize the data access. In conventional databases, the so-called database physical design is an important step, which is concerned with setting the access methods according to the database characteristics, the underlying hardware, and the expected query load. At the query time, the optimizer chooses the best access path among the existing access methods, and combines them to generate the physical query plan.

Astronomical and Geospatial Data Access The access methods are even more crucial in astronomical and geospatial Big Data management. The reasons for this are manifold:

(i) Spatial queries, i.e., involving spatial criteria, are frequent, and spatial data typically constitute larger amounts of data than conventional alphanumeric data. It is therefore crucial to reduce the cost associated to the data access as much as possible, and avoid scanning the whole dataset by using spatial-aware access methods.
(ii) These queries are complex and costly, since they involve geometrical computation. Hence, beyond reducing the I/O costs, access methods also save the CPU costs.
(iii) Big Data make use of distributed systems, with horizontal partitioning as a technique to spread the data over multiple cluster nodes. The way to partition the data widely impacts the performances of the system. In particular, favoring spatial locality within partitions is a desirable feature which limits the communication costs.

In this section, we focus on spatial access methods (SAM) (Gaede and Günther, 1998; Manolopoulos et al., 2005a) and their adaptation to the context of Big Data in astronomy and geospatial applications. The reader interested in the nonspatial queries can refer to this study in the context of astronomy (Mesmoudi et al., 2016). The main difference with the access to scalar data is the complexity of the spatial predicates (e.g., geometric intersection or inclusion) that are not limited to exact or interval search on one-dimensional attribute values.

Spatial Indexing A common technique to avoid geometrical computation on complex shapes is to first approximate them with a minimum bounding rectangle (MBR) (as illustrated in Figs. 8.1 and 8.2), and then to build an index structure based on the MBRs, employed as index key associated to the spatial data that it represents. The index aims at reducing the search space by filtering the candidates. A refinement step is necessary to get the exact result. This comes down to building a secondary data structure suitable for n-dimensional rectangles (where n is mostly two or three). Note that even for point data, spatial indexing is commonly used to improve multidimensional range queries. In this particular case, the spatial feature and its MBR are identical, and then, the refinement step is useless.

In their survey, Gaede and Günther (1998) categorize spatial access methods in three classes: the overlapping methods, the clipping methods, and those that transform data. An example of overlapping SAM is R-tree (standing for rectangle tree) and R*-tree, whereas R+-tree adopts clipping, and the space filling curves approach is representative of the transformation-based SAM.

We describe the main SAM hereafter, and highlight those proposed for astronomical applications. We then discuss their adaptation to the Big Data context, and summarize some existing approaches.

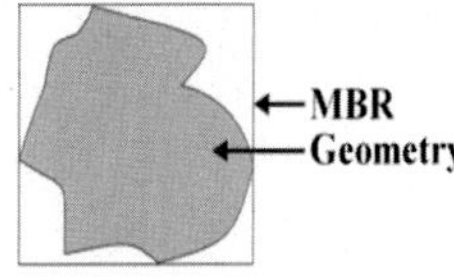

FIG. 8.1 Minimum bounding rectangle of a spatial object.

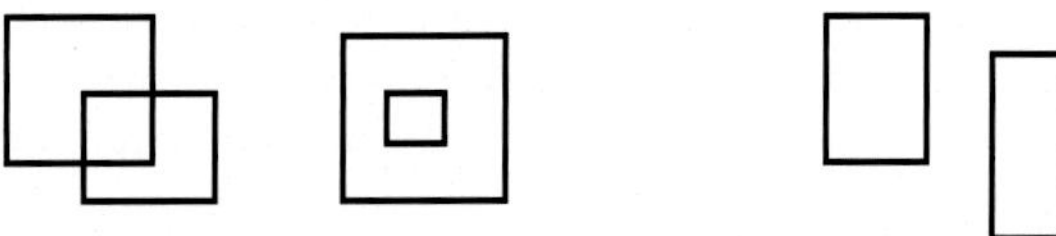

FIG. 8.2 MBR-based filtering: Objects having disjoint MBRs cannot intersect and are pruned without geometrical computation (right); others are candidates (the two left).

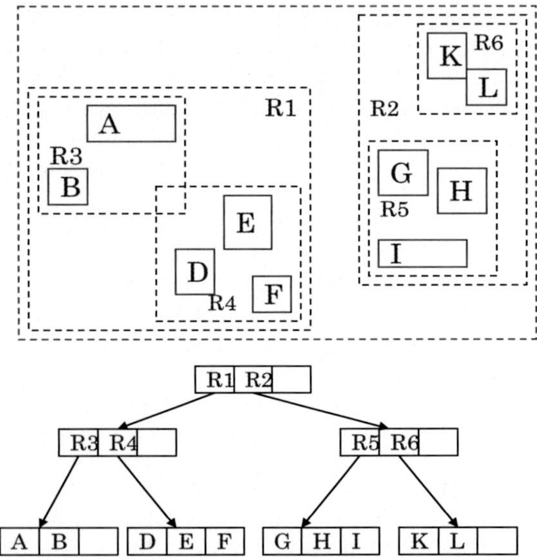

FIG. 8.3 An example of R-tree.

R-tree is an early index structure inspired by B+-tree, which has been proposed by Guttman (1984). It indexes a collection of rectangles, in a tree where each node (or leave in the lower level) is assigned its MBR, and a parent node contains the MBRs of its children nodes (see Fig. 8.3). As in B+-tree, the number of entries per node is bounded, which sometimes entails node splitting during the insertion process or node merging after several deletions. Note that this process may lead to overlapping MBRs within the same level of the tree.

Formally, an R-tree is defined as fellows (@ denotes a pointer):

- a leaf node contains a sequence of (RECT), where RECT is a rectangle to index;
- other nodes contain a sequence of (MBR, @NODE) where MBR is the minimum bounding rectangle covering all the rectangles of the referenced child node;
- the number of entries in a node, except in the root node, is between a lower and an upper bound.

The disadvantage of the overlaps is that the search may need to traverse several paths of the tree when the query falls in the intersection of several MBRs of nodes, and this increases when the construction does not minimize the dead space (i.e., the space covered by a node's MBR but not by its children nodes). This has motivated the proposal of, for example, R*-tree (Beckmann et al., 1990), which builds and maintains an R-tree while limiting the overlaps.

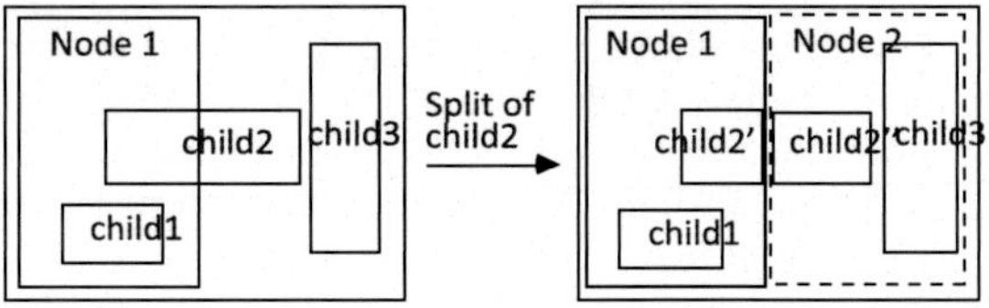

FIG. 8.4 Propagation process of a node split in R+-tree index.

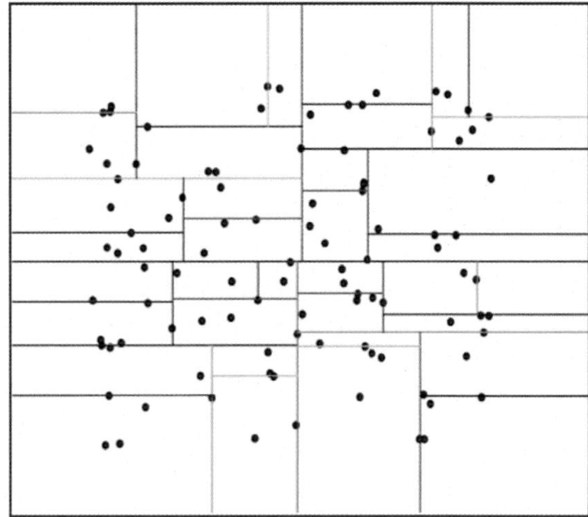

FIG. 8.5 The process of kd-tree binary space partitioning.

Another variant of R-tree is R+-tree, proposed by Sellis et al. (1987), which belongs to the category of clipping methods. In this data structure, the MBRs of the nodes of the same level are disjoints. At this end, the creation and maintenance process were modified so that (i) the original rectangles can be duplicated in each leaf which MBRs intersect; and (ii) a node split is propagated to the lower levels of the tree so that nodes cannot overlap. This is illustrated in Fig. 8.4. Other SAMs in the clipping category can be mentioned, including grid files, quad-trees, and kd-trees (illustrated in Fig. 8.5), which apply various spatial partitioning types.

Lastly, a transformation-based SAM consists of embedding the original space in an alternative representation that could be dealt with more easily. The most used transformation approach is space ordering, also called linearization by means of space filling curves. In fact, spatial queries can be viewed as multidimensional range queries. Ranges are well supported by traditional (nonspatial) access methods, such as B-trees, that employ the total order of the indexed key. However, there is no obvious order in n-dimensional space. To cope with this, the idea is to divide the space into grid cells and

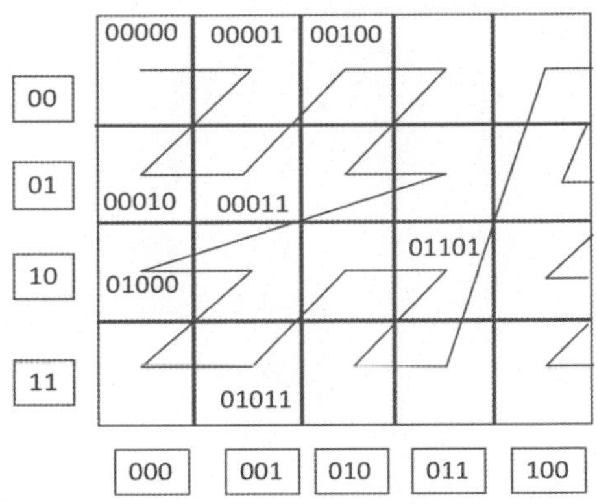

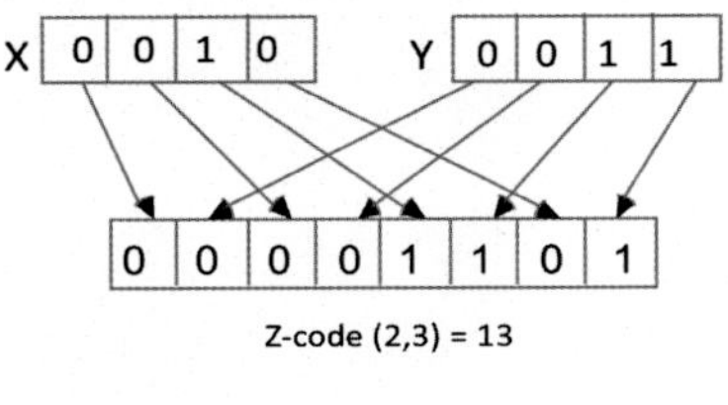

FIG. 8.6 Z-order space filling curve.

order the cells close to each other. This results in cell indices that follow a space filling curve so that close cells in space get close indices with a high probability (Moon et al., 2001). Indexed data are assigned the cell indices where they are located. A query window is also transformed to a list of indices of the cells (mostly consecutive thanks to the locality property), and can be answered by using a simple, yet efficient index like a B+-tree. There exist variants of transformations with filling curves, among which Z-order[1] (see Fig. 8.6) and Hilbert are the most common.

Spatial Indexing for Astronomical Data The majority of SAMs assume planar Cartesian coordinates. Astronomical reference systems are, on the contrary, based on spherical coordinates. Specific SAMs have been proposed for this purpose. Especially HTM (Kunszt et al., 2000) in the context of the Sloan Digital Sky Survey (SDSS) applies a hierarchical triangular tessellation of a sphere associated with a linearization. On the other hand, HEALPix (Gorski et al., 2005), standing for Hierarchical Equal Area iso-Latitude Pixelization, is another widely used spherical indexing scheme for efficient astronomical numerical analysis, including spherical harmonic and multiresolution analysis. A parameter, called NSIDE, governs the level to consider in the hierarchy of this index, and so the resolution, as illustrated in Fig. 8.7. Thanks to its geometrical properties, HEALPix supports two different ordering schemes: per isolatitude ring, or nested, similar to Z-order.

Access Methods for Big Spatial Data The question is: *How to adapt SAMs to the Big Data context?* In fact, it is not straightforward to apply the existing data structures and the corresponding algorithms to optimize a big geospatial or astronomical database. The main difference is the granularity of data management, which is no longer observation (or a tuple), but larger splits that are processed by separated worker nodes. Therefore, a unique index is unsuitable.

The general idea proposed in the literature (Eldawy and Mokbel, 2015; Aji et al., 2013) is to define a global and a local index. The global index applies to the splits, and contributes in the organization of partitions, and the limitation of the internode communication. The local index limits the access and computation at the level of one node. The implementation of this principle differs however from one system to another. SIMBA (Xie et al., 2016) and SpatialHadoop both use R-trees for global and local indexing (SpatialHadoop also proposes a global grid index as an alternative) and a local index. ASTROIDE adopts a linearization technique according to HEALPix indices, and astutely leverages the built-in access methods such as range partitioning to optimize the data access and filtering.

A recent study in Pandey et al. (2018) has surveyed some of the available big spatial data analytics systems, and compares five of them which are based on the Spark framework. This indexing scheme is reported as well as its cost in term of memory consumption.

8.4 QUERY OPTIMIZATION

Query optimization is a crucial and difficult part of the overall query processing; it is the activity of producing an efficient query execution plan (QEP). The task of the query optimizer is to compute the requested spatial query efficiently and to select the optimal plan with

[1]The Z-code value is calculated by interleaving the binary representations of column and row numbers.

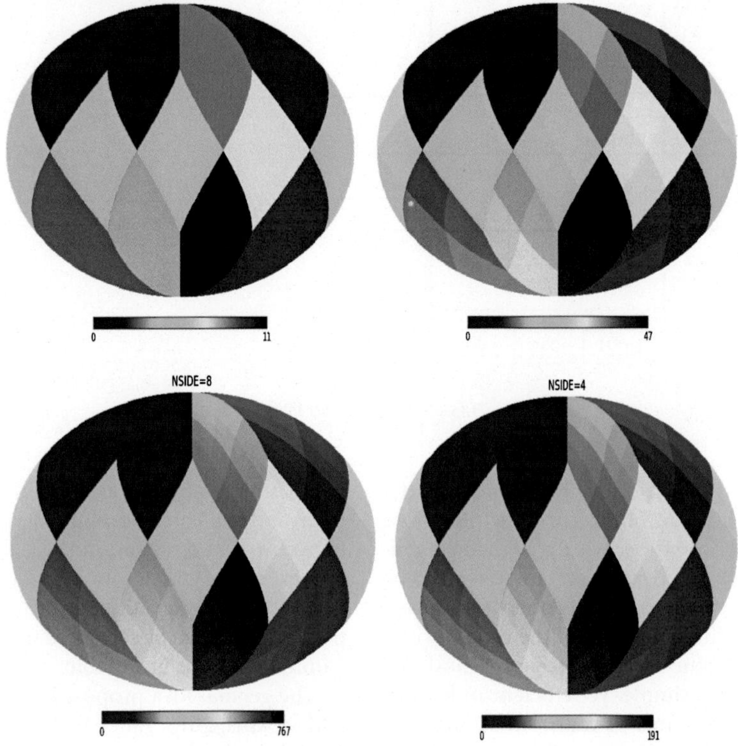

FIG. 8.7 HEALPix partition of the sphere (NSIDE = 1, 2, 4, 8).

minimum cost among several alternatives for processing the query. All plans are equivalent in terms of final output but may be widely different in their costs, i.e., the query execution times. The difference between the two plans can be enormous. Therefore, the query optimizer has an important role in modern spatial distributed database systems.

There are two other major elements which are necessary to completely determine all aspects of query optimization. On the one hand, we need to describe how to search through the set of all possible QEPs. On the other hand, we need to compare different QEPs and decide which one is the best. In general, the decision is based on the costs of various resources, such as CPU and disk I/O.

The query optimizer is represented by three components, as shown in Fig. 8.8: search space, cost model, and search strategy.

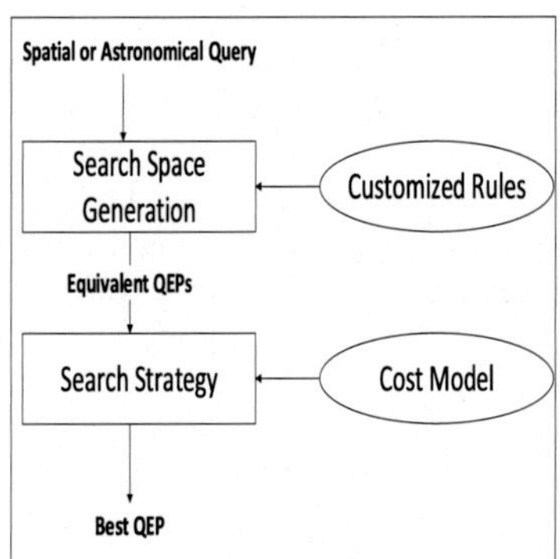

FIG. 8.8 Query optimizer architecture.

Search Space The search space is defined as a set of equivalent QEPs for a given query that can be generated using transformation rules. The objective is to reduce the search space by applying heuristics (e.g., push the projection operator as close as possible to the data source).

Many different approaches for query optimization were proposed in traditional database systems. Typical examples are predicate pushdown, in which predicates are applied as early as possible in the query, commutative (e.g., $R \bowtie S = S \bowtie R$) and associative laws (e.g., $(R \bowtie S) \bowtie T = R \bowtie (S \bowtie T)$), and duplicate elimination. Such rules can significantly improve query execution time. However, they do not support spatial and astronomical data due to data skewness and query complexity.

Spatial queries such as nearest neighbors or distance joins are expensive to process and may lead to computation skew in some nodes of a cluster. Processing such queries in distributed systems requires some data partitioning techniques that divide the dataset into approximately equal-size partitions to ensure load balancing and avoid data skewness. Rule-based optimization in this context exploits spatial partitioning to access the smallest possible number of partitions, avoiding Cartesian product or performing projection on spatial indices as early as possible.

For illustration, let us consider two large astronomical datasets R, S and a cross-matching query that computes all pairs of points $(r, s) \in R \times S$ where the spherical distance between r and s is lower than ε. Cross-matching is crucial in astronomy; it allows correlating two catalogues, matching newly extracted sets of objects with an existing catalogue, or even tracking transients from a series of observations. This query, in a nutshell, is a join on a distance condition. But, it could be extremely slow when applied to large catalogues, due to the pairwise distance computation costs, along with the communication and the I/O costs incurred. This issue has been at the root of several works, among which recent algorithms in the context of parallel data processing. HEALPix is commonly used to limit to target only the objects that are in the same area of the sky (Pineau et al., 2011). Thus, Nadvornik (2015) assigns all objects in one (large) pixel to one node, assuming random objects distribution on the sphere. Brahem et al. (2017) present a new cross-matching algorithm called HX-MATCH. It astutely combines HEALPix based indexing and adaptive partitioning, to achieve an efficient pruning and to remedy the potential skew of celestial objects. HX-MATCH maps the original predicate into an equi-join between R and S on HEALPix indices (which is much faster than a distance join) in order to generate candidate pairs sharing the same cell. For example, a cross-matching query between two catalogs `gaia` and `igsl`, with a radius of 2 arc-seconds is expressed using the following ADQL expression (GAVO Data Center, 2008 is defined by the International Virtual Observatory Alliance (IVOA) as a standard to query astronomical data):

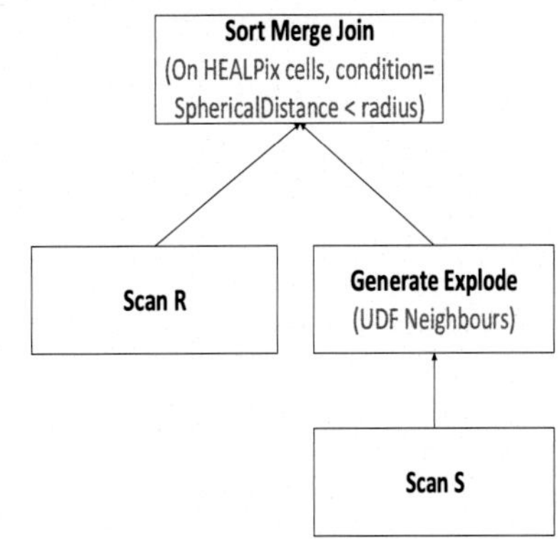

FIG. 8.9 Cross-matching QEP in ASTROIDE.

```
SELECT * FROM gaia R JOIN igsl S
ON 1=CONTAINS(POINT('ICRS', R.ra, R.dec),
   CIRCLE('ICRS', S.ra, S.dec, 2/3600));
```

To execute this query using HX-MATCH, ASTROIDE (Brahem et al., 2016, 2018a, 2018b) starts by scanning the two input files. Then, it applies the *Generate* logical operator to calculate the list N of neighboring cells for each row in S and generate a new row for each element in N. ASTROIDE duplicates all objects of a reference catalogue (here S) in neighboring cells, runs a sort-merge-join algorithm on HEALPix indices, and, finally, filters the output result according to the spherical distance predicate (see the physical plan in Fig. 8.9).

Other spatial queries can follow a similar process. The spatial selection query is a simple query type to find objects that geometrically intersect a given selection region. For example, users may be interested in spatial features located in a given rectangle:

```
SELECT  c.name, c.loc
FROM   county c
WHERE ST_Intersects (c.loc, given_shape_WKT)
```

For spatial selection queries, the query optimizer injects transformation rules to avoid scanning all records. It uses indices to prune out partitions that do not overlap the query area and scans only relevant partitions.

Other transformation rules according to Silva et al. (2013) can be used, as is the case when we combine spatial queries with relational predicates, e.g., applying the relational predicate after computing the kNN join (the kNN join query combines each object in a dataset A with the k closest objects in another dataset B), i.e., $(\sigma_A(A \bowtie_{kNN} B))$ is equivalent to applying the predicate to the outer relation before computing the kNN join, i.e., $(\sigma_A(A) \bowtie_{kNN} B)$.

To implement a spatial query optimizer, two approaches are possible:

- The first approach is to develop a new optimizer over a distributed system.
- The second approach is to extend the functionalities of an existing optimizer such as Catalyst or Calcite by adding new rules to support spatial queries. The advantage of this strategy is to leverage all the benefits of such optimizers. Here, relational-based operators are processed by the backbone optimizer. For example, Hadoop GIS (Aji et al., 2013) proposes a rule-based optimizer for spatial data. If a query does not contain any spatial operations, the resulting QEP is exactly the same as the one generated from Hive. However, if the query contains spatial operations, the QEP is regenerated with customized handling for spatial operators.

Search Strategy The search strategy or enumeration algorithm determines the algorithms applied to explore the search space. The most popular strategy is dynamic optimization which works in a bottom-up way and builds execution plans starting from base relations and joining one more relation at each step until the complete plans are obtained. Other alternative enumeration algorithms have been proposed. Steinbrunn et al. (1997) present an overview of these algorithms and Kossmann (2000) describes the most important approaches in the literature.

Cost Model This module requires analytical formulas to estimate the size of intermediate spatial data and cost functions to predict the cost of operators. It assigns an estimated cost to each possible QEP.

This cost model relies on:

- statistics on the relations which include various metadata (e.g., the number of rows, the number of disk pages of the relations, histogram of distributions of column values) and indexes,
- formulas to estimate the selectivity of various predicates and the sizes of the output for each operator in the query plan,
- formulas to estimate the CPU and I/O costs for every operator in the query plan.

In distributed systems, cost functions can be expressed with respect to local processing time (CPU time, I/O time) and communication time (time to initiate a transmission, time to transmit the data). The communication cost is a key factor of distributed systems performances; it is a linear function of the amount of data to be transmitted. Thus, an optimizer has to reduce the amount of data transmitted. In the spatial context, we can take into account partitioning approaches to retrieve only relevant partitions. This allows to reduce the amount of data transmitted. Thus, accessing unnecessary partitions incurs useless communication costs. The role of the cost model is to make the best use of existing indices and statistics in order to select the most efficient QEP. The objective in astronomical systems is to leverage the index and partitioning support.

Regarding Big Data technologies, Hive introduces a cost model using Calcite. Calcite is currently the most widely adopted optimizer for Big Data analytics in the Hadoop ecosystem. Calcite is adopted by Hive, Drill, Storm, and many other query processing engines, providing them with advanced query optimizations. Calcite applies various optimizations such as query rewrite, join reordering, and join algorithm selection. Calcite has a plan pruner that can choose the cheapest query plan. The chosen logical plan is then converted by Hive into a physical operator tree, optimized and converted into jobs, and then executed on the Hadoop cluster. Regarding the Spark optimizer, Catalyst introduces rule- and cost-based optimization but very little work has been devoted to the cost model; it uses a simple cost model. Cost-based optimization was added in the recent Spark version 2.0 and focused on cardinality estimation, broadcast versus shuffled join, and join reordering. Thus, future Spark versions are continuously evolving to finish with a robust cost model. In Brahem (2019), ASTROIDE (a spark extension to execute astronomical queries) reuses the cost model integrated in the Catalyst optimizer.

In the literature, the development of a cost model for a spatial query has been identified as a complex search problem. For example, using R-tree, the number of disk I/Os can be estimated using the number of nodes visited during R-tree traversal. In Manolopoulos et al. (2005b), authors present different cost functions for spatial selection and join queries using R-trees.

The cost of QEPs depends also on the estimated size of intermediate spatial data that are generated during query execution. This estimation is carried out using statistics and formulas to predict the cardinalities of spatial operations. Simba (Xie et al., 2016) applies a cost model that takes into consideration indices and statistics to choose the most efficient physical plans. Specifically, it defines the selectivity of a predicate for a partition as the percentage of records in the partition that satisfy the predicate. If the selectivity is greater than a certain threshold, SIMBA will scan all the partitions rather than using indices. SIMBA uses also a *broadcast join* to optimize spatial join queries when one of the input dataset size is below a given threshold. SIMBA employs local joins by skipping the data partition phase on the small table and broadcasts it to every partition of the large table.

8.5 FROM BIG DATA MANAGEMENT TO BIG DATA ANALYTICS AND VICE VERSA

The concept of Big Data management also refers to dealing with the problems of the key features of big data: volume, velocity, variety, and veracity. Effective and efficient Big Data management should meet ongoing data lifecycle needs that include these features. The data lifecycle usually needs to follow the decision making processes. Good decisions depend on data analysis results. The results can vary according to the different types of Big Data analytics in data management. The first level of Big Data analysis gives a description of the data. At this level answer of the question "What is done?" can be found. The second level or type of Big Data analytics refers to diagnoses and gives an understanding of "why it happens." The third type of Big Data analytics, known as predictive, uses data to discover "what could happen." The last, perspective Big Data analytics, uses the latest technologies such as deep learning to understand what the impact is on future decisions. With prescriptive analytics, it becomes possible to understand and grasp future opportunities or mitigate future risks as predictions, continuously updated with new data coming in. Analysis can be viewed as the final and most difficult to realize step in the data management process. Here it should be mentioned that Big Data management depends on the quality of data analytics. Cleansing, profiling, mapping, and visualization are some of the crucial quality factors that influence all aspects of astro and geo data. The relation of Big Data management and Big Data analytics of geo and astro data is in two directions: scale and scope of data analytics, and design considerations (hardware and software systems).

8.5.1 Open Issues in the Intersection of Big Data Management and Big Data Analytics of Geo and Astro Data

In spite of their development in different communities, either astroinformatics or geomatics, data management and analytics of astronomical and geospatial data share the same characteristics, and raise the same challenges when it comes to access, query, or analysis of the spatial features over Big Data. The very first challenge is to deal with the data volume, which is tremendous in many geo and astro datasets. Even the data ingestion and preprocessing (such as cleansing) become challenging. In fact, it becomes difficult to judge data quality within a reasonable amount of time.

Another challenge of management of geo and astro data is a difficulty of data integration due to the diversity of data sources. Furthermore, the nature of geo and astro data, described by complex data types using specific data structures, makes it more difficult to match and integrate than the conventional tabular data. It is worth noting the importance in the integration process of the metadata, such as the precision and the spatial reference system, which characterize both geo and astro datasets.

"Timeliness" of the data poses another challenge in the intersection of Big Data management and Big Data analytics of geo and astro data. Timeliness leads to higher requirements for processing technologies. Nowadays, a lot of attention is paid to it. Actually, Big Data management is enabled by recent advances in technologies.

Other issues are opened in the intersection of Big Data management and Big Data analytics in general. Without further details, here several of them should be mentioned that are characteristic for geo and astro data: availability (accessibility, timeliness), usability (credibility), reliability (accuracy, consistency, integrity, and completeness), and relevance (fitness) and presentation quality (readability).

8.6 DISCUSSION AND OUTLOOK

We have broadly introduced the topic of query processing in big geospatial and astronomical data. We have highlighted their main specificity and outlined the main steps, the related methods, and the more recent work in this area. In spite of their characteristics, the approach of data management is analogous to that of traditional data, or general purpose Big Data.

This chapter has focused on vector data. Recent work also addresses raster data, which constitute a wide part of Earth observation, and the sources in sky surveying (Soille et al., 2018; Li et al., 2017). Here again, the data organization, big arrays, pyramidal stricture, etc., have great importance in the support of the data access and analysis. Providing a high-level and efficient query engine in this context would be an interesting perspective. Also, the joint management of raster and vector data is useful, and, at the best of our knowledge, missing in the context of Big Data.

REFERENCES

ADQL Library. URL: http://cdsportal.u-strasbg.fr/adqltuto/. (Accessed 6 July 2018).

Aji, Ablimit, et al., 2013. Hadoop GIS: a high performance spatial data warehousing system over MapReduce. Proceedings of the VLDB Endowment 6 (11), 1009–1020.

Apache Hive. https://hive.apache.org/. (Accessed 6 July 2018).

Apache Pig. https://pig.apache.org/. (Accessed 6 July 2018).

Apache Spark. https://spark.apache.org/. (Accessed 6 July 2018).

Beckmann, Norbert, et al., 1990. The R*-tree: an efficient and robust access method for points and rectangles. ACM SIGMOD Record 19 (2), 322–331.

Brahem, Mariem, Lopes, Stephane, et al., 2016. AstroSpark: towards a distributed data server for Big Data in astronomy. In: Proceedings of the 3rd ACM SIGSPATIAL PhD Symposium. SIGSPATIAL PhD '16. ACM, Burlingame, California. ISBN 978-1-4503-4584-2, pp. 3:1–3:4. URL: http://doi.acm.org/10.1145/3003819.3003823.

Brahem, Mariem, Zeitouni, Karine, Yeh, Laurent, 2017. HX-MATCH: in-memory cross-matching algorithm for astronomical big data. In: International Symposium on Spatial and Temporal Databases. Springer, pp. 411–415.

Brahem, Mariem, Yeh, Laurent, Zeitouni, Karine, 2018a. ASTROIDE: a unified astronomical big data processing engine over Spark. IEEE Transactions on Big Data (ISSN 2372-2096). https://doi.org/10.1109/TBDATA.2018.2873749.

Brahem, Mariem, Yeh, Laurent, Zeitouni, Karine, 2018b. Efficient astronomical query processing using Spark. In: 26th ACM SIGSPATIAL International Conference on Advances in Geographic Information Systems. SIGSPATIAL '18. ACM, Seattle, WA, USA.

Brahem, Mariem, 2019. Spatial Query Optimization and Distributed Data Server - Application in the Management of Big Astronomical Surveys. PhD thesis. University of Paris-Saclay, France. URL: https://tel.archives-ouvertes.fr/tel-02100861.

Calcite. https://calcite.apache.org. (Accessed 6 July 2018).

Catalyst. https://databricks.com/session/a-deep-dive-into-spark-sqls-catalyst-optimizer. (Accessed 20 December 2019).

Eldawy, Ahmed, Mokbel, Mohamed F., 2015. SpatialHadoop: a MapReduce framework for spatial data. In: 2015 IEEE 31st International Conference on Data Engineering, ICDE 2015. Vol. 2015-May. IEEE Computer Society, pp. 1352–1363. English (US).

Gaede, Volker, Günther, Oliver, 1998. Multidimensional access methods. ACM Computing Surveys (CSUR) 30 (2), 170–231.

Gardarin, G., Valduriez, P., 1989. Relational Databases and Knowledge Bases. Addison-Wesley. ISBN 9780201099553. URL: https://books.google.fr/books?id=gagmAAAAMAAJ.

GAVO Data Center, 2008. ADQL Query. VO resource provided by the GAVO Data Center. URL: http://dc.zah.uni-heidelberg.de/__system__/adql/query/info.

Gorski, Krzysztof M., et al., 2005. HEALPix: a framework for high-resolution discretization and fast analysis of data distributed on the sphere. The Astrophysical Journal 622 (2), 759.

Guttman, Antonin, 1984. R-trees: a dynamic index structure for spatial searching. In: ACM SIGMOD, International Conference on Management of Data. ACM, pp. 47–54.

Huang, Zhou, et al., 2017. GeoSpark SQL: an effective framework enabling spatial queries on Spark. ISPRS International Journal of Geo-Information (ISSN 2220-9964) 6 (9), 2220–9964. https://doi.org/10.3390/ijgi6090285. URL: http://www.mdpi.com/2220-9964/6/9/285.

Kini, A., Emanuele, R., n.d. GeoTrellis: adding geospatial capabilities to Spark. http://spark-summit.org/2014/talk/geotrellis-adding-geospatial-capabilities-to-spark.

Kossmann, Donald, 2000. The state of the art in distributed query processing. ACM Computing Surveys (CSUR) 32 (4), 422–469.

Kunszt, P.Z., et al., 2000. The indexing of the SDSS science archive. In: Astronomical Data Analysis Software and Systems IX. Vol. 216, p. 141.

Li, Zhenlong, et al., 2017. A spatiotemporal indexing approach for efficient processing of big array-based climate data with MapReduce. International Journal of Geographical Information Science 31 (1), 17–35.

Manolopoulos, Yannis, et al., 2005a. R-Trees: Theory and Applications. Springer-Verlag New York, Inc., Secaucus, NJ.

Manolopoulos, Yannis, et al., 2005b. R-Trees: Theory and Applications (Advanced Information and Knowledge Processing). Springer-Verlag New York, Inc., Secaucus, NJ.

Mesmoudi, Amin, Hacid, Mohand-Said, Toumani, Farouk, 2016. Benchmarking SQL on MapReduce systems using large astronomy databases. Distributed and Parallel Databases 34 (3), 347–378.

Moon, Bongki, et al., 2001. Analysis of the clustering properties of the Hilbert space-filling curve. IEEE Transactions on Knowledge and Data Engineering 13 (1), 124–141.

Nadvornik, Jiri, 2015. Cross-matching Engine for Incremental Photometric Sky Survey. Master thesis. Faculty of Information Technology, Czech Technical University in Prague. URL: http://arxiv.org/abs/1506.07208.

Nishimura, Shoji, et al., 2011. MD-HBase: a scalable multi-dimensional data infrastructure for location aware services. In: Mobile Data Management (MDM). IEEE Computer Society.

NoSQL Databases. http://nosql-database.org/. (Accessed 29 May 2018).

Pandey, Varun, et al., 2018. How good are modern spatial analytics systems? Proceedings of the VLDB Endowment 11 (11).

Pineau, F.-X., Boch, Thomas, Derriere, Sebastien, 2011. Efficient and scalable cross-matching of (very) large catalogs. In: Astronomical Data Analysis Software and Systems XX. Vol. 442, p. 85.

Sellis, Timos K., Roussopoulos, Nick, Faloutsos, Christos, 1987. The R+-tree: a dynamic index for multi-dimensional objects. In: Proceedings of the 13th International Conference on Very Large Data Bases. VLDB 87. Morgan Kaufmann Publishers Inc. ISBN 0-934613-46-X, pp. 507–518. URL: http://dl.acm.org/citation.cfm?id=645914.671636.

Silva, Yasin N., et al., 2013. Similarity queries: their conceptual evaluation, transformations, and processing. The VLDB Journal—The International Journal on Very Large Data Bases 22 (3), 395–420.

Soille, P., et al., 2018. A versatile data-intensive computing platform for information retrieval from big geospatial data. Future Generations Computer Systems 81, 30–40.

Steinbrunn, Michael, Moerkotte, Guido, Kemper, Alfons, 1997. Heuristic and randomized optimization for the join ordering problem. The VLDB Journal—The International Journal on Very Large Data Bases 6 (3), 191–208.

Tang, Mingjie, et al., 2016. LocationSpark: a distributed in-memory data management system for big spatial data. Proceedings of the VLDB Endowment (ISSN 2150-8097) 9 (13), 1565–1568. https://doi.org/10.14778/3007263.3007310.

Wang, Daniel L., et al., 2011. Qserv: a distributed shared-nothing database for the LSST catalog. In: State of the Practice Reports. SC '11. ACM, Seattle, Washington. ISBN 978-1-4503-1139-7, pp. 12:1–12:11. URL: http://doi.acm.org/10.1145/2063348.2063364.

Xie, Dong, et al., 2016. Simba: efficient in-memory spatial analytics. In: Proceedings of the 2016 International Conference on Management of Data. SIGMOD '16. ACM, San Francisco, California, USA. ISBN 978-1-4503-3531-7, pp. 1071–1085. URL: http://doi.acm.org/10.1145/2882903.2915237.

Yarn. https://hadoop.apache.org/docs/current/hadoop-yarn/hadoop-yarn-site/YARN.html. (Accessed 6 July 2018).

Yu, Jia, Wu, Jinxuan, Sarwat, Mohamed, 2015. GeoSpark: a cluster computing framework for processing large-scale spatial data. In: Proceedings of the 23rd SIGSPATIAL International Conference on Advances in Geographic Information Systems. SIGSPATIAL '15. ACM, Seattle, Washington. ISBN 978-1-4503-3967-4, pp. 70:1–70:4. URL: http://doi.acm.org/10.1145/2820783.2820860.

CHAPTER 9

Real-Time Stream Processing in Astronomy

VELJKO VUJČIĆ • DARKO JEVREMOVIĆ, DR

9.1 INTRODUCTION

One of the obstacles in "Big Data"-related use cases is when large quantities of data have to be analyzed as they happen. Systems for algorithmic trading, social networks analytics, numerous kinds of monitoring applications, intelligent data transfer and integration architectures, they all employ stream analysis in distributed systems on real-time Big Data with success. Even probabilistic and predictive scenarios using complex or uncertain data can be tackled with event processing technologies. Although classification algorithms and knowledge extraction are not new to the astronomical science, they were not commonly employed "online"[1] and over Big Data. In the next few years, (near-)real-time processing will become more relevant and the field should prepare for the advent of new instruments and techniques.

The nature of most astronomical objects is that they change slowly and sometimes these changes are hard to detect even over the course of a human lifespan. On the other side there are phenomena which in fact are changing very rapidly and *those* present a challenge in terms of stream processing and astronomical event classification. Time scales of changes for physical properties of celestial objects range from milliseconds (i.e., pulsars) to millions of years. Variable and transient objects are rare in comparison to slow-changing "static" sky, but one of the big science goals of the 2020s will be the "hunt for the rarest of the rare" (Narayan et al., 2015).

Evidence about observations of variable stars may go all the way to ancient Babylon and Egypt, although there are some disputes. Old Egyptians recorded a changes in brightness of an object with a period of 2850 days, which might correspond to what the period of Algol would have been around 3000 years ago. Algol and its changes in brightness can be easily followed with the naked eye. In modern times the first discovered variable star was o Ceti, better known as Mira, in the beginning of the 16th century (Algol followed soon afterwards).

Other important observations are so-called historical supernovae in the Milky Way. Chinese astronomers recorded a total of 20 candidates (stars that suddenly explode, emit spectacular luminosity, and then gradually fade away) and some of them were simultaneously observed by Japanese, Arabic, and European astronomers. The first universally confirmed (Green, 2002) supernova, SN1006, was seen in the 11th century by observers around the world as it was an extremely bright event. Its authenticity was confirmed by contemporary research of supernova remnants (Gardner and Milne, 1965). The last of supernovae in the era before telescopes was observed in 1604. Since then we did not record a supernova in our galaxy. In the 16th century, Tycho Brahe noticed parallax[2] difference when comparing a supernova and comets (for a supernova, parallax is significantly smaller). This proved that stars are situated far away from us, outside the solar system.

After the discovery of the telescope, and especially photography, the number of detected variable objects in the sky has risen dramatically. Furthermore, the recent advent of digital photography which uses charge-coupled devices allows a cheap and simplified workflow (no development of photo plates, physical properties are measured numerically, automatized software pipelines, etc.).

Historically, there were long-term observational programs where specific, more "interesting" objects, which showed unusual changes, were followed for many months or years (i.e., activities of active galactic nuclei, long-term periodic variable stars, etc.). Also, there were short-term observational campaigns which focused on specific objects, trying to understand their nature. Several groups of scientists with instruments located at different longitudes were doing (almost) continuous monitoring of chosen objects for several days/weeks (i.e., MUSICOS campaigns during late 1980s and early 1990s, Catala et al., 1993; Foing et al., 1994).

[1] *Online* in a sense of immediate continuous processing, as opposed to *offline* or batch processing.

[2] Parallax is an ostensible change of position of an object relative to the background, due to the change of position of the observer.

Knowledge Discovery in Big Data from Astronomy and Earth Observation. https://doi.org/10.1016/B978-0-12-819154-5.00019-9

There are also joint campaigns for tracking of LIGO/Virgo gravitational wave events which joined several dozens of instruments. These events are time-critical, and thus specific services and tools are developed for the triggering follow-up[3] observations (Abadie et al., 2012; Abbott et al., 2017). These led to very interesting multimessenger observations of gravitational wave events and have a potential for furthering our understanding of violent events connected to gravitational wave production.

The field of time-domain astronomy is in its very early development phase. We did so far follow a very small number of objects in time and mainly based largely on chance detection of changes and done mainly on a short time scale. With the advent of ZTF[4] and later LSST,[5] we will be able to follow changes in millions of object and have to develop "smart" ways of choosing the candidates for intensive photometric and/or spectroscopic follow-up observations.

The Zwicky Transient Facility (ZTF) is a new time-domain survey that started operations at the Palomar Observatory in the beginning of 2018 and is planned to operate as a short-term project. ZTF searches and catalogues rare and exotic transients. Its data products will drive novel studies of supernovae, variable stars, binaries, AGNs, and asteroids. ZTF started to publicly disseminate part of their detected events in mid-2018 (Masci et al., 2018). ZTF can be perceived as a precursor of Large Synoptic Survey (LSST, which will come online 2022/2023, Ivezic et al., 2008): their cadence (the way a telescope chooses target fields for observation) is similar, but the depth of the ZTF apparatus is significantly shallower – LSST will be able to fetch much fainter objects, resulting in a population larger by one to two orders of magnitude. Despite differences in design, responsiveness, and data volumes, ZTF presents an opportunity to get prepared for the LSST-era of big volumes of data and real-time processing of streams. Also, we will have a chance to learn about necessary infrastructure for follow-ups of important and interesting events in order to extract as much science as possible from these surveys.

This chapter deals with technologies and concepts which will possibly allow astronomers/observatories to easily handle vast quantities of data coming from these new facilities/surveys and extract useful information.

[3]A *follow-up* is a common term in astronomy for additional and timely observations of a phenomenon of particular interest. Follow-ups are often done with different telescopes from the one that originally spotted the event, in order to extract specific physical features of the astronomical object.

[4]ZTF homepage: https://www.ztf.caltech.edu/.

[5]LSST homepage: https://www.lsst.org/.

9.2 EVENT PROCESSING CONCEPTS

Event processing is an umbrella term for technologies that are conceptually centered around ingestion, manipulation, and dissemination of events. An event could be any discrete occurrence, depending on the system ontology and granularity of its concepts. Event processing enables real-time pattern application, suitable for large input streams where automatized reaction is crucial.

Systems based around events perceive events as entities that arrive from the outer world, pertinent to the relevant domain. An event is a discrete occurrence that the system is capable of detecting. It reflects something that positively happened and cannot be undone. Thus, the system is defined by its domain, which narrows down the area of interest, and its detection potentials.

The main difference between event processing applications and common transactional (or request–response) systems is their query paradigm. While traditional applications make queries and serve previously stored data per request, event processing applications run continuous queries through streams of new data and deliver discoveries if and when they happen.

Although notion of event handling is not new in computer science, systems entirely based around events started to emerge at the beginning of the 2000s.[6] Some of the use cases for event processing are:

- real-time analytics,
- monitoring, fault detection,
- diagnostics made in conjunction with historical data,
- data are incoming continuously, organized in streams,
- huge loads of data, which are a challenge to process with traditional applications.

Events can be produced by a wide range of devices, ranging from simple or combined physical sensors (i.e., Internet of Things) to enterprise software systems. They could be a purely simulated product or initiated by human participation. Devices such as RFID reader, seismometer, or GPS locator are usually called "hardware" producers while sensor simulators, application monitors, or stream aggregators are labeled as "software" producers – although these categories often overlap. Such entrant events are called "simple" or "raw" and they present a basis for further manipulation. Correlation of multiple "raw" events may lead to detection of "complex" events, which is the ultimate goal of complex event processing mechanisms. A complex event "summarizes, represents, or denotes a set of other events" (Luckham, 2011).

[6]Main concepts of event-driven architecture were defined in the paper Michelson (2006), and later thoroughly elaborated in the book *Event Processing in Action* (Etzion and Niblett, 2011).

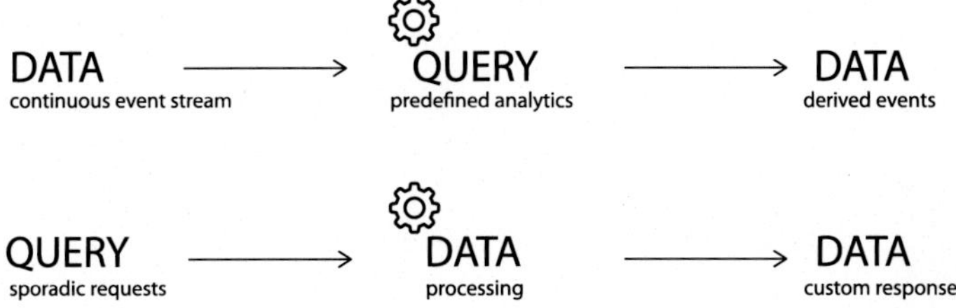

FIG. 9.1 Comparison of query paradigms in event processing and common transactional applications.

One domain's simple event can be another domain's complex event – e.g., a successful airplane landing can be a simple event in air traffic control system, but a derived, complex event in an internal aircraft monitoring system.

In a document Tsimelzon (2006) published by the now defunct event processing brand *Coral8*[7] in the mid 2000s, there is a definition of ten design patterns of complex event processing. They are stated here, with short comments.

Filtering: Simple restriction, as in the SQL WHERE clause, where single events (objects, rows) are processed. Common use of filtering is reduction of the initial data load, where filtered data reflect areas of interest more precisely. It can exist in more complex form, as an evaluation of an external function.

Caching: Scenario when events are cached in memory for joint processing. The concept of *windows* is here introduced, as autochthonous and inherent CEP mechanism. Windows are the basis for aggregation and most of the CEP patterns. Caching applies to data streams as well as to "static" from traditional databases (which contain historical or referent data).

Aggregation: Group statistics over cached data (min/max, count, average, sum, standard deviation, etc.).

Database lookups: Access to historical or referent data, which is combined with the entry stream event data. This is also called "enrichment."

Database writes: Writes of events or periodical reports.

Correlation: Joining and comparing multiple streams/sources, as in the SQL JOIN operation.

Patterns: Pattern matching is a key characteristic of CEP systems. Patterns shape multiple simple events into a single complex event. Pattern formulation is commonly given through the definition of presence or absence of events, in a certain temporal interdependence.

Finite automata: Underlying mathematical model on top of which CEP mechanisms are usually built.

Hierarchical events: When events have a nested structure.

Dynamic queries: Besides the continuous nature of stream processing, the user can pose an ad hoc query.

Based on graphical terminology established in *Event Processing in Action* (Etzion and Niblett, 2011), Fig. 9.2 shows a network of astronomical software entities which communicate using the common VOEvent format:

- a stream producer (LSST),
- a simple filtering agent,
- a pattern detection agent connected to the external databases and services for data enrichment,
- channels which disseminate processed events to subscribers or to a pool for follow-up observations.[8]

9.2.1 Tools and Differences

In the last couple of years the event processing market has been fairly stable, with well-established products and packages. Widely used open source frameworks include Esper (EsperTech)[9] and Drools Fusion (JBoss),[10] with optional commercial versions focused mainly on infrastructural improvements.

Here is a list of event processing tools for 2019. It includes complex event processing tools as well as event

[7]The business case of Coral8 well illustrates the instability of the CEP tools market. It was merged with Aleri in 2009; the merger was bought in 2010 by Sybase, which was sold to SAP in the same year. Sybase was fully integrated to SAP in 2014 and ceased to exist under its own name.

[8]There are several efforts to develop Target Observation Managers (led by Las Cumbres Observatories) and automatic schedulers for possible targets of opportunity which also includes apart LCO, NOAO/SOAR, and Gemini telescopes.

[9]http://www.espertech.com/esper/.

[10]http://drools.jboss.org/drools-fusion.html.

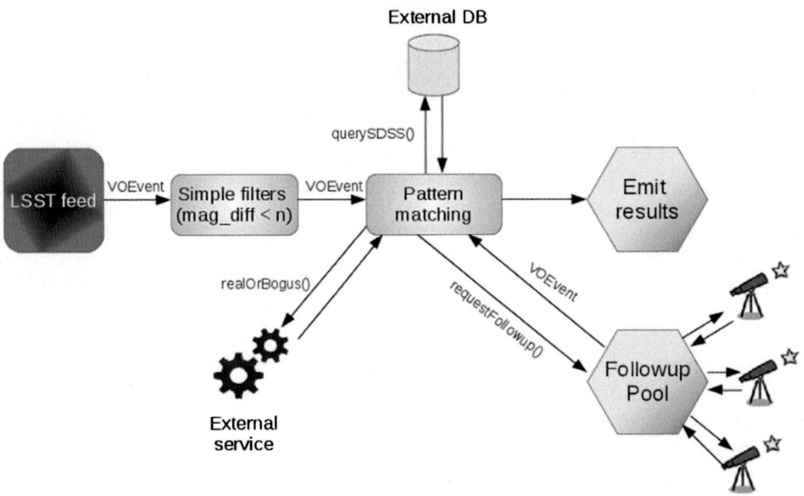

FIG. 9.2 A proposed event processing network for an astronomical data stream.

stream processing (ESP) and distributed stream computing platforms (DSCPs):

- Amazon **Kinesis**
- Apache **Apex**
- Apache **Flink**
- Apache **Heron**
- Apache **Kafka**
- Apache **Samza**
- Apache **Spark**
- Apache **Storm**
- EsperTech **Esper, NEsper**
- Evam **Evam Streaming Analytics**
- DataTorrent **RTS (Real-time Streaming)**
- **FeedZai**
- Fujitsu **Interstage Big Data Complex Event Processing Server**
- Hitachi **Streaming Data Platform (HSDP)**
- IBM **Streaming Analytics**
- IBM **Operational Decision Manager (ODM)**
- Informatica **RulePoint CEP**
- Google **DataFlow**
- LG CNS **EventPro**
- Microsoft **Azure Stream Analytics**
- Microsoft **StreamInsight**
- Microsoft **Trill**
- OneMarketData **OneTick CEP**
- Oracle **Event Processor**
- RedHat **Drools Fusion**
- SAP **Event Stream Processor**
- SAS **DataFlux**
- **ScaleOut Software**
- Software AG **Apama Event Processing Platform**
- SQLStream **s-Server**
- Striim **Striim**
- Tibco **BusinessEvents**
- Tibco **StreamBase**
- Vitria **Operational Intelligence Platform**
- WS02 **Complex Event Processor**
- WS02 **Siddhi**

Many of these tools are purely commercial and integrated into larger software packages. Some of them do not implement time domain inference mechanism, but focus on infrastructure and Big Data performance – messaging, scalability, resource management, fault tolerance, or parallelization. It is possible to combine some of the aforementioned tools – e.g., Esper inside an Apache Storm bolt (Dudziak, 2011), where Storm is used for data stream manipulation and communication with external APIs (such as Twitter in the cited example) and Esper for data processing, i.e., tweet count and maximum number of retweets. Commercial version of Esper supports horizontal scaling which again relies on Apache Kafka and Apache Zookeeper architecture.

Event processing tools could be categorized on the basis of how they treat the event state (Zapletal, 2015). ESP events are usually *stateless*, in a sense that they are not kept in the system, but used for group analytics and trend monitoring. In CEP, certain single discrete events might be kept for periods of time (*stateful*) and topologies are applied upon them. As usual, these categories are imperfect because CEP tools offer periodical, stateless processing while certain ESP tools like Storm offer language enhancements (Trident[11]) for defining topologies in a stateful manner.

A detailed analysis of multiple Apache stream processing environments (Zapletal, 2016) states key characteristics of event processing tools, some of which are:

- **Streaming model**: *native* – continuous model which processes individual events as they arrive – or *micro-batch* – which is basically an offline algorithm adjusted to imitate stream processing,
- **API**, or event processing language: compositional or declarative,
- **State management** (explained in the paragraph above),
- **Fault tolerance** + system recovery,
- **Performance**: latency, throughput, scalability,
- **Maturity** of the platform.

Under these criteria we could categorize Esper as a platform with a continuous streaming model which implements a declarative event processing language and can work in a stateful as well as stateless mode, with a medium/high throughput (with a tuned JVM, 500K+ events per second on a common dual 2 GHz processor[12]) and medium low latency (<3 us, <10 us with 99% predictability[12]).

9.2.2 Esper Versus Drools

As was already stated, Esper and Drools Fusion are the only mature open source tools available on the CEP market and as such might be of particular interest to astronomers willing to develop full-blooded event "brokers" or just play around with available astronomical alert streams. Although they are both complex event processing Java platforms, Esper and Drools are conceptually quite different. Esper is built around an SQL-enhanced language for event processing, while Drools Fusion is an adaptation of a rule-based engine which was adapted to work in a reactive mode. Looking at their surface, Esper resembles transactional applications while Drools inherits from Expert systems or decision support systems.

Esper EPL Syntax Esper EPL syntax is an extension of SQL with support for time-based reasoning.

```
select ...
from ... (pattern / match_recognize)
[where ...]
[group by ...]
[having ...]
```

[11] http://storm.apache.org/releases/current/Trident-state.html.
[12] http://esper.espertech.com/release-7.1.0/esper-reference/html/performance.html.

Optionally, we can define patterns inside the FROM clause. There are two different approaches for pattern definition, i.e., via the PATTERN operator, or via the MATCH_RECOGNIZE operator. Data "windows" can be specified by adjoining stream definition in the FROM clause:

```
select avg(delta_mag)
from LsstStream.win:time(30 sec)
```

Esper syntax is covered with more detail in Section 9.3.1.

Drools Fusion Drools Fusion is a rule-based declarative language.

```
rule "rulename"
when
// this pattern occurs
then
// do something
end
```

Drools Fusion supports temporal operators of interval algebra:

```
$eventA : EventA( this after[3m30s, 4m] $eventB )
```

as well as contextual windows:

```
LsstStream() over window:time( 2m )
```

Esper Versus Drools Fusion We choose to present CEP concepts (in the next section) using Esper EPL. Simply put, Esper is faster, somewhat more powerful, and easier to learn. Here is a more comprehensive list of strengths and weaknesses of both tools:

- Both are well-documented Java-based frameworks.
- They are easy to integrate with other Java tools, e.g., for visualization. Drools Fusion is a part of the jBPM workflow engine.
- Drools does not have syntax for grouping enabled and requires workarounds.
- Esper EPL statements are represented as strings, thus lack syntax checking and are hard to debug.
- Drools has richer support for interval algebra.
- Esper has more expressive power, partly because SQL itself is quite flexible, partly due to language extensions.
- Esper can ingest events serialized as Avro or XML (along with schema recognition and XPATH queries).
- Drools enables "hot deployment" (without restarting service) of rules. Esper does it only in the Enterprise edition.
- Esper is several times quicker in general cases and scales more reliably.

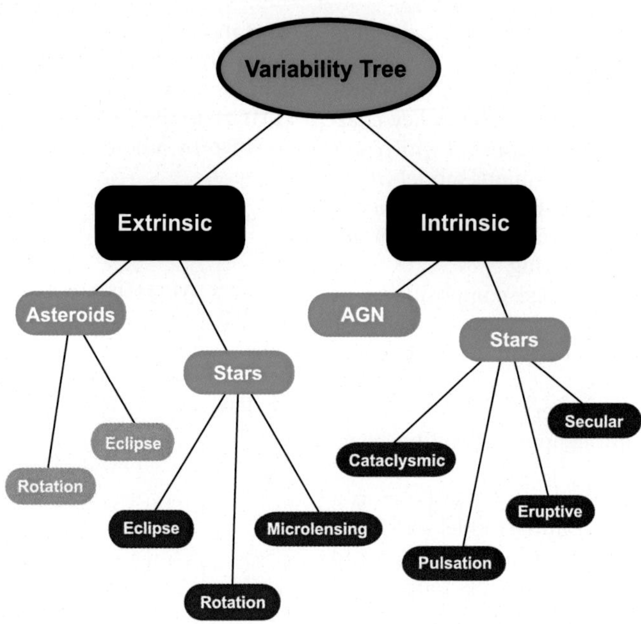

FIG. 9.3 Main categories of variable objects, based on Eyer and Mowlavi (2008).

9.3 APPLICATION OF EVENT PROCESSING TO ASTRONOMY

As we mentioned in the introduction, every astronomical object is continuously changing – whether we observe change in position, physical transformation, change in radiative flux, or some other property. However, we can make a distinction between "static" sky (no significant change in properties over long time) and variable sky.[13]

In Fig. 9.3, variable objects are divided into categories based on the cause of the variability. Intrinsic objects change physically, whether reversibly (as in stars pulsation) or irreversibly (cataclysmic events). Movement causes various extrinsic phenomena, such as gravitational lensing, eclipsing, etc. These categories are only top-level branches of the variability tree – some of them have tens of leaves, such as pulsating stars.

The classification depends on the detectability of a certain kind of variability and on the science goal behind certain observational scenarios. For example, the Kepler mission (Koch et al., 2010) is specialized for small planet transits and lightcurve analysis of Milky way objects.

In a vastly heterogeneous ecosystem of numerous telescopes, data centers, and astronomical software applications, there are efforts to unify formats and protocols for communication between different entities. *Virtual Observatory* is an international initiative and a virtual space *in vivo* which employs interoperability of various physically distributed astronomical subjects. It is federated by IVOA,[14] which consists of different working groups engaged on standards and recommendations for data providers, scientists, and developers.

Data Formats for Interoperability Description of astronomic events is standardized in an XML format called *VOEvent*. A VOEvent represents a celestial event and its measured properties; it embraces other IVOA standards (such as unique "IVORN" identifiers of astronomical entities or Unified Content Descriptors); it is designed to support different data representations, with sections dedicated to provenance, scientific hypothesis, and others (from the official document Seaman et al., 2011):

- who – identification of scientifically responsible author,
- what – event characterization modeled by the author,
- where and when – space–time coordinates of the event,
- how – instrument configuration,
- why – initial scientific assessment,

[13] This excludes our Sun and the planets because of their proximity.

[14] International Virtual Observatory Alliance, ivoa.net.

- citations – follow-up observations,
- description – human-oriented content,
- reference – external content.

At this moment, VOEvent is represented in XML[15] and there are several projects which provide streams to the public, such as GCN, 4 Pi Sky, LOFAR, etc. (for a detailed list see the IvoaVOEvent wiki page[16]). These streams can be accessed with Comet[17] or Dakota[18] open source tools, colloquially called "brokers," which can receive, distribute, and filter events.

In June 2018, ZTF started distributing nightly tarballs of events in AVRO format,[19] which is an Apache data serialization system based on JSON, with schema definition and data compression. ZTF events reflect the internal data model, and do not comply with the VOEvent standard. ZTF will offer real-time streaming for authorized subscribers, which will be a precursor to the LSST stream. As of now, the format of the LSST stream is still unknown and may or may not comply with the IVOA standard at the time data production starts.

9.3.1 Esper EPL

Esper EPL is a specialized language for event processing based on SQL query language. The Esper web site[20] states that EPL is a "declarative language for dealing with high frequency time-based event data" and that it "is compliant to the SQL-92 standard and extended for analyzing series of events and in respect to time."

We chose to present CEP concepts with examples coded in EPL. We will begin with simple statements and gradually increase in complexity.

The Basics We create a (vanilla) stream schema,

```
create schema LsstStream(diaObjectId int,
    diaSourceId int, ssObjectId int,
    filterName string);
```

where *diaObjectId* is the unique identifier of a celestial object, *diaSourceId* and *ssObjectId* are unique identifiers of a single occurrence of that object (for outside and inside the solar system, respectively; they mutually exclude each other), and *filterName* is the optical filter used in capturing that occurrence. Objects which were spotted for the first time do not have a *diaObjectId*.

Projection (SELECT clause) is done as in SQL. We count objects which had previous association as *oldcount* and new objects as *newcount*

```
select count(diaObjectId) as oldcount,
        sum(case
            when diaObjectId is null then 1
        end) as newcount,
    filterName
from LsstStream
```

And so is selection (WHERE clause) – we choose only objects that do not belong to the solar system,

```
where diaSourceId is not null
```

grouping (by optical filter),

```
group by filterName
```

and group condition (when newcount passes a threshold). As in the SQL standard, alias from select clause cannot be used in having clause.

```
having sum(case
            when diaObjectId is null then 1
          end) > 100
```

Windows As was stated earlier, the time domain is inherent to the event processing. The concept of *(data) windows* – a "lookup" into the future – is one of the essential features of CEP. Data windows allow the use of inference mechanisms over selected objects in a certain period of time.

Esper implements time and length windows with basic and complex variations. Here we will cover some of them (for more details, see Chapter 13 in Esper documentation).[21]

If we want to place our report into a time window, we might add

```
from LsstStream.win:time(1 hour)
```

where we would get a sliding count of objects for the last hour, refreshed with each new occurrence of an adequate object. We call this a *sliding time window.*

To rely on external timestamps instead of on system time, we would declare

```
from LsstStream.win:ext_time(1 hour)
```

Static time window reminds of periodical, batch query. Such formulation will issue a single report after each hour.

---

[15] VOEvent XML schema definition: www.ivoa.net/xml/VOEvent/VOEvent-v2.0.xsd.

[16] IVOA VOEvent wiki page: http://wiki.ivoa.net/twiki/bin/view/IVOA/IvoaVOEvent.

[17] https://comet.readthedocs.io/en/stable/.

[18] http://voevent.dc3.com/.

[19] Nightly archive of ZTF events in AVRO format: https://ztf.uw.edu/alerts/public/.

[20] http://www.espertech.com/esper/.

[21] http://esper.espertech.com/release-7.1.0/esper-reference/html/epl-views.html.

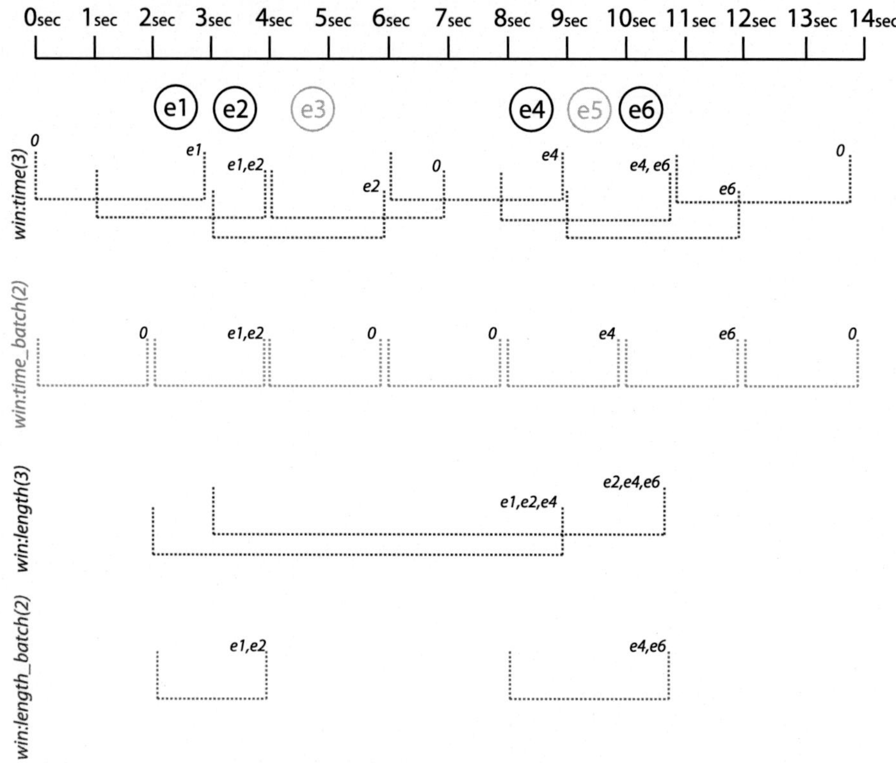

FIG. 9.4 Four basic flavors of event processing "windows."

```
from LsstStream.win:time_batch(1 hour)
```

Sliding length window relates to a report which contains *n* last events which comply to the restriction. It refreshes with each new entrant event.

```
from LsstStream.win:length(1000)
```

Static length window processes a group of *n* events, to proceed to the next group of *n* events, without overlapping.

```
from LsstStream.win:length_batch(1000)
```

Fig. 9.4 shows comparison of four fundamental types of CEP windows: time and length window, each of which can be continuous (sliding) and static (batch). The X-axis shows a continuous time scale in seconds. Entrant events are marked as e_n. Events that do not comply with an imaginary constraint, and thus do not enter the window pool, are painted light gray. Windows are represented with dashed lines, with beginnings and ends marked. They work as half-open $[a, b)$ intervals. On the left side, the EPL expression is written vertically.

Window pool, or lineup – which consists of individual events that entered the particular window – is stated on the exiting edge of that window. Sliding time window will not report only when there is a status quo, i.e., there is no positive or negative change in the lineup of events from the previous report. Other window types are conceptually simpler and they were already described in the examples above.

Context Context is introduced when data need to be partitioned (fragmented). Contexts can be reused, which reduces code redundancy.

```
create context RaDec partition by ra, dec
from LsstStream
```

This context expression separates groups of events according to their position in the equatorial coordinate system (right ascension and declination) – or to put it more simply, it associates all events that happened at the same position in the sky (to the same celestial object, in the perfect case).

Context definition takes over the group by clause and associates events as a basis for pattern formulation. Combination of context and group by clause would result in nested grouping. The expression

```
context RaDec
select count(*) from LsstStream;
```

replaces

```
select count(*) from LsstStream;
group by ra, dec
```

Joined Streams Joint streams are enabled via the INSERT INTO clause. It is possible to join data from multiple streams using the JOIN operation. For a flexible usage of multiple heterogeneous streams, we can use "Variant Stream". Here, heterogeneous events are treated as equals in the same FROM clause and all the attributes are treated as optional. Variant stream is defined in an XML configuration file:

```
<variant-stream name="VariantStream">
    <variant-event-type name="LsstStream"/>
    <variant-event-type name="ZtfStream"/>
</variant-stream>
```

This kind of "ambiguous" stream is particularly interesting for astronomy – of the sky – resulting in vastly different sets of attributes. The things which associates them are key attributes, which again can internally carry different names but the same Unified Content Descriptor.[22] Variant streams simplify application of patterns with complex interdependencies of events, from multiple data sources. Possible interaction of events from two streams, with complexity of n^2

```
pattern
[(LsstStream->ZtfStream) or
(ZtfStream->LsstStream) or
(LsstStream->LsstStream) or
(ZtfStream->ZtfStream)]
```

becomes

```
pattern
[VariantStream->VariantStream]
```

Patterns As we already stated, pattern matching is in the core of complex event processing. It usually involves temporal relationship between two or more events (or absence of events!).

Here is the mathematical model of a "mag per hr" pattern, given in pseudocode, as suggested by Željko Ivezić (University of Washington, LSST). It detects changes in brightness of an object and triggers an alert if it increases at a rate of 1 magnitude per hour or quicker (with a constraint that the object was previously undetected). The pattern may be used for GRB detection, or with modified constraints for detection of some other phenomena.

```
dmdt = (m1-m2) / (t2-t1)
if (dmdt >= 1 and object was previously
    undetected)
then alert
```

Now we will see how the pseudocode translates to EPL code.

We extend schemas from previous examples and define the *zeroPoint* variable for flux to magnitude transformation. Current system time is taken as detection time (just for show):

```
create schema LsstStream(ra double, decl double,
    flux double, diaObject boolean);

create variable double zeroPoint =
    -2.5*Math.log10(3631);

insert into LsstStreamTransformed
select ra,
      decl,
      diaObject,
      -2.5*Math.log10(flux) - zeroPoint as mag,
      current_timestamp() as detectionTime
from LsstStream;
```

Here MATCH_RECOGNIZE expression comes into play:

```
select * from LsstStreamTransformed
  match_recognize (
    partition by ra, decl
    measures A.ra as ra, A.decl as decl,
        A.detectionTime as firstDetection,
        B.detectionTime as secondDetection,
        A.mag as firstMag, B.mag as secondMag,
        Math.abs(B.mag - A.mag)/(
        (B.detectionTime-A.detectionTime)/3600000
        ) as ratePerHour
    pattern (A B)
    define
      A as A.diaObject=false,
      B as Math.abs(B.mag - A.mag)/(
      (B.detectionTime-A.detectionTime)/3600000
      ) >=1
)
```

[22]The Unified Content Descriptor (UCD) is a formal vocabulary by IVOA which describes astronomical concepts and attempts to provide interoperability between heterogenous data sets.

9.4 CONCLUSION

Astronomy is traditionally a slow-paced discipline, as most phenomena are happening on long time scales. Nonetheless, occasional transient events can, aside from being fascinating in their nature, tell us about other aspects of science such as typology or cosmology. "Time-domain astronomy" covers astronomical research where time is a fundamental component. In a general sense, its aim is discovery of transient objects after several successive observations. In the new data-rich environment, which will start during the early 2020s, one can expect that machine learning and some form of "artificial intelligence" will be necessary to sift through the millions of alerts (events) which will be issued every night by facilities such as LSST. We described one possible direction in achieving these using event (stream) processing techniques.

ACKNOWLEDGMENT

The gear icon for Fig. 9.1 was made by https://www.flaticon.com/authors/gregor-cresnar.

REFERENCES

Abadie, J., et al., 2012. First low-latency LIGO+Virgo search for binary inspirals and their electromagnetic counterparts. Astronomy & Astrophysics 541, A155. https://doi.org/10.1051/0004-6361/201218860. arXiv:1112.6005.

Abbott, B.P., et al., 2017. Multi-messenger observations of a binary neutron star merger. The Astrophysical Journal 848, L12. https://doi.org/10.3847/2041-8213/aa91c9. arXiv:1710.05833 [astro-ph.HE].

Catala, C., Foing, B.H., et al., 1993. Multi-site continuous spectroscopy – part one – Overview of the MUSICOS 1989 campaign organization. Astronomy and Astrophysics.

Dudziak, Thomas, 2011. Storm & Esper. [Online]. URL: https://tomdzk.wordpress.com/2011/09/28/storm-esper/. (Accessed May 2014).

Etzion, Opher, Niblett, Peter, 2011. Event Processing in Action. Manning Greenwich.

Eyer, Laurent, Mowlavi, Nami, 2008. Variable stars across the observational HR diagram. Journal of Physics. Conference Series 118 (1), 012010.

Foing, B.H., et al., 1994. Multi-site continuous spectroscopy. 2: Spectrophotometry and energy budget of exceptional white-light flares on HR 1099 from the MUSICOS 89 campaign. Astronomy and Astrophysics.

Gardner, F.F., Milne, D.K., 1965. The supernova of AD 1006. The Astronomical Journal 70, 754.

Green, D.A., 2002. Historical supernovae and their remnants. Highlights of Astronomy 12, 350–353.

Ivezic, Z., et al., 2008. LSST: from science drivers to reference design and anticipated data products. ArXiv e-prints arXiv:0805.2366.

Koch, David G., et al., 2010. Kepler mission design, realized photometric performance, and early science. The Astrophysical Journal Letters 713 (2), L79.

Luckham, David C., 2011. Event Processing for Business: Organizing the Real-Time Enterprise. John Wiley & Sons.

Masci, F., et al., 2018. The Zwicky Transient Facility public alert stream. The Astronomer's Telegram, 11685.

Michelson, Brenda M., 2006. Event-Driven Architecture Overview. Patricia Seybold Group.

Narayan, G., et al., 2015. ANTARES: hunting the "rarest of the rare" in the time-domain. IAU General Assembly 22, 2258269.

Seaman, R., et al., 2011. Sky Event Reporting Metadata Version 2.0. Ed. by R. Seaman and R. Williams. IVOA Recommendation 11 July 2011. arXiv:1110.0523 [astro-ph.IM].

Tsimelzon, M., 2006. Complex event processing: ten design patterns. [Online]. URL: http://complexevents.com/wp-content/uploads/2007/04/Coral8DesignPatterns.pdf. (Accessed March 2018).

Zapletal, Petr, 2015. Introduction into distributed real-time stream processing. [Online]. URL: https://www.cakesolutions.net/teamblogs/introduction-into-distributed-real-time-stream-processing. (Accessed March 2018).

Zapletal, Petr, 2016. Comparison of Apache stream processing frameworks: part 1. [Online]. URL: https://www.cakesolutions.net/teamblogs/comparison-of-apache-stream-processing-frameworks-part-1. (Accessed March 2018).

CHAPTER 10

Time Series

ASHISH MAHABAL, PHD

10.1 INTRODUCTION

Time series are ubiquitous. We find them everywhere in nature, and in man-made things. Temperatures and day lengths throughout a year, rising tempers, and time spent working or working out over a year. One can obtain time series for temperature at equi-spaced time intervals, giving rise to what are called regular time series. On the other hand, owing to weekends and other sporadic holidays, not to mention sick days and vacations, the time series of arriving at one's office over a year is gappy and hence an irregular time series. In the very last part of this chapter we will touch upon some time series related to Earth science, but otherwise we will mostly concern ourselves with astronomical time series. Astronomical time series refer to the variation in the amount of electromagnetic radiation received from various sources as a function of time, and are generally referred to as light curves.

The study of time series in astronomy has led to an entire subfield known as time-domain astronomy (TDA). It is the modern large synoptic sky surveys generating hundreds of millions of light curves that are fueling time-domain astronomy and helping explore rich phenomenology from asteroids to pulsating stars, and from supernovae to black holes.

In this chapter we will look at basic properties of time series, how time series are analyzed, and how their study reveals a very dynamic universe.

10.2 BASICS OF TIME SERIES

For anthropomorphic reasons, the response of our eyes is logarithmic, and hence the system that developed for measuring light involves taking the logarithm of flux from astronomical sources, and we will use the resulting scale – called magnitudes – throughout the chapter. In certain cases it is advantageous to use the raw flux, and if so, we will point that out. Magnitude, $m = -2.5 \times \log(f) + c$, where c is fixed such that the magnitude of the star Vega is 0 (in one standardized system – unfortunately there are many; going into those details may take us too afar). Since this magnitude is based on light that reaches us, and has nothing to do with how far the object is, it is called apparent magnitude. Owing to the negative sign, brighter objects have numerically smaller magnitudes. The sun has a magnitude of about -26.5, the moon has -12.5, and the faintest stars one can see in the night sky are about magnitude 6. The magnitudes for the sun and the moon will of course change though the year and in the case of the moon daily as different fractions of reflected light reach us.

Some of the oldest astronomical time series (besides cataloging the sun and the moon), are those of the planets as they wandered and waxed and waned. Galileo also kept track of the Jovian moons. A majority of the astronomical time series come from surveys large and small. At times one obtains high-cadence observations of a small set of objects of interest (e.g., to study transients, microlensing, or eclipsing binaries – see Fig. 10.1 for instance), but a majority – hundreds of millions of them – come from large sky surveys, e.g., the Digitized Palomar Observatory Sky Survey (DPOSS; Weir, 1995), Catalina Sky Survey (CSS; Djorgovski et al., 2008; Drake et al., 2009; Mahabal et al., 2011), Palomar Transient Factory (PTF; Law et al., 2009), Zwicky Transient Facility (ZTF), and so on (more details can be found in the chapter on astronomy surveys). Most of these have one or more scientific aims and the science determines the cadence of the survey, e.g., the focus of CSS is to look for near-Earth asteroids and four observations are separated by ten minutes to spot the movement of the asteroids. But thanks to the number of observations available, and improving time-series methods, a wider variety of science can still be done with most surveys. In this chapter we will look at a few datasets, methods being applied to them, and the future, all in brief.

Knowledge Discovery in Big Data from Astronomy and Earth Observation. https://doi.org/10.1016/B978-0-12-819154-5.00021-7

FIG. 10.1 A transient detected by the CRTS survey. Note the differences in the central rectangles of the two images taken at different times. Sources in the rest of the image do not seem to have changed much between the two epochs.

10.3 EARLY AGNOSTIC TIME-DOMAIN STUDIES

One of the earliest astronomical time-domain studies where we did not start with the aim of studying a particular type of object was with DPOSS. DPOSS observations were carried out using photographic plates. This was the last large pre-CCD survey. The entire Northern sky was covered using a thousand 6.5 deg × 6.5 deg plates with the centers separated by 5 deg. This provided a healthy overlap of 1.5 degree wide strips. Three filters (emulsions) were used i.e., J, F, and N, each with roughly one hour long exposures. That provided up to six points in time – two each in the three filters. Adjacent plates were taken one hour apart, but scheduling constraints meant they could be separated by up to 15 years, the approximate length of the survey. Thus these time-"series" were not only very sparse – consisting of just a few points – but also extremely irregular. There are a limited number of things one can do with such data. One can ask for certain basic statistics like change in magnitude per filter (called amplitude), median magnitude, standard deviation, etc. These numbers can help separate the outliers – objects that behave differently from the bulk. Objects that vary discernibly are a small fraction, and that is what we could glean from such an exercise.

Thus, the basic input that one has is the time series in the form of magnitudes as a function of time. In general the magnitude (called *mag* hereafter) is accompanied by an error term (called *magerr* hereafter). The data can be neatly arranged in three columns: time, *mag*, and *magerr*. Many methods use the derived statistical features as a proxy for the original time series. We will look at some such methods and also see how they are sometimes more usable than the time series themselves.

If one wishes to combine time series in different filters, there are additional considerations (e.g., the DPOSS data should really be treated as three related time-series of two points each). VanderPlas and Ivezić (2015) show how to combine data in multiple filters to look for periodicity. We will briefly bring that up again later, but we will largely concern ourselves with time series obtained through single passbands.

10.4 LONGER TIME SERIES WITH UNIFORM PASSBANDS

Longer astronomical time series tend to be irregular. There are many reasons for this, e.g., in optical astronomy one cannot carry out observations during the day, and similarly for roughly half the year objects are in the direction of the sun – due to the Earth's annual motion – and cannot be observed. Then there are interruptions due to unfavorable weather, and other factors like proximity to the moon (in projection on the sky), varying priorities of observations, etc. Besides resulting in irregular light curves, this leads to different objects getting observed a different number of times. This can be problematic when comparing light curves. For this reason too, extracting statistical features from light curves becomes useful. For each light curve one can derive the same number of features, and thus match their dimensionality for comparisons.

We already mentioned a few sky surveys. Some others are: Palomar-QUEST (Djorgovski et al., 2008), Kepler, Gaia (Gaia Collaboration, 2016), LSST (Ivezić et al., 2008), etc., just in the optical wavebands. Then there are radio surveys (e.g., to search for fast radio bursts [FRBs]), to obtain continuous gravitational wave data, etc. A more comprehensive list can be found in Chapter 5.

10.5 FEATURES

A vast majority of features[1] (see Fig. 10.2) can be computed from the light curves; see for examples Richards et al. (2011) and Faraway et al. (2016). There is also a service available from Harvard, called FATS,[2] that returns a set of features when one uploads a light curve. Some of the features can be simple ones like amplitude, standard deviation, mean, median, skew, kurtosis, etc. This class of features are agnostic to the nature of the source. If one is aware of the types of sources

[1] An alternate word used for feature is metric, especially when it is computed. Another alternate word used is statistic.

[2] http://isadoranun.github.io/tsfeat/FeaturesDocumentation.html.

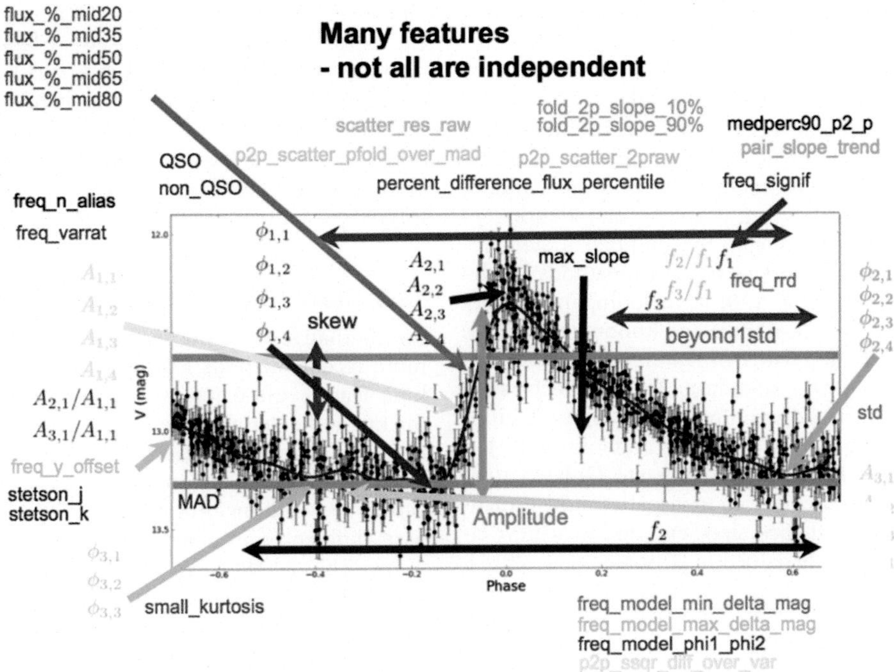

FIG. 10.2 Various features are shown superimposed on a phased light curve of an RR Lyrae star. The features are not all independent, and dimensionality reduction has to follow feature extraction before other processing. Figure courtesy Adam Miller, features based on Richards et al. (2012). A few of the feature names may not be legible as they overlap with other text or marks, but that is not critical since the features shown here are indicative, and not exhaustive.

in the sample one can design features to help separate the different classes. Examples of this type include P, the peakiness metric, for distinguishing between supernovae (typically single broad peaks), flaring M-dwarfs (typically single sharp peak), cataclysmic variables (sporadic peaks due to stochasticity), etc., where $P = \frac{1}{t_{span}}\sqrt{\frac{1}{N}\Sigma_i w_i (p_i - p_m)^2}$, and multiple peaks ($p_i$) are weighted by their extent.

In addition to the whole-curve features, there are residual features. Examples are the M and Q metrics (Cody et al., 2014). Q measures the degree of periodicity by computing how close the light curve points are to the systematic noise floor before and after the phased trend is subtracted from the light curve $Q = \frac{RMS^2_{resid} - \sigma^2}{RMS^2_{raw} - \sigma^2}$. Here RMS_{resid} is the RMS after subtracting the phased light curve. Similarly, M measures the extent of flux asymmetry, $M = (< d_{10\%} > - d_{med})/\sigma_d$, where the median is first subtracted from $< d_{10\%} >$, the mean of all data at the top and bottom decile of the light curve.

One can also derive cluster-based features. For example, CSS observes a sky position four times within 30 minutes, and then returns to the same position irregularly, with the next visit being a day, or a week, or even a month later. In such a case, it can be useful to combine the four observations and treat them as a single point. That ensures that one does not have to handle short time scales (ten minutes) and long time scales (month-long) simultaneously. A variety of such features are discussed by Faraway et al. (2016).

One can thus produce several tens of such features. Light curves of nonuniform length get converted to a uniform set of numbers. In this high-dimensional space one can cluster similar types of objects, look for newer subclasses, and so on. But working in high-dimensional space comes with its own problems. We will see next how that can be mitigated.

10.5.1 Dimensionality Reduction

Working in high dimensions has several problems, the biggest is that most methods require exponentially increasing computational time as dimensions increase. To begin with, not all features that are computed are orthogonal to each other. Traditionally methods like principal component analysis (PCA) have been used for dimensionality reduction. However, PCA provides new dimensions that are a combination of input dimensions, and often difficult to interpret physically.

Moreover, different features are more useful in helping discriminate different classes of objects. Thus one can approach dimensionality reduction by building specific features for a given problem, or use more agnostic methods for reducing the number of standard features in which to work. A few such methods are described by Donalek et al. (2013), including the Fast Relief Algorithm (Robnik-Šikonja and Kononenko, 2003), which relies on the quality of features, the Fisher discriminant ratio (FDR; Fukunaga, 1990), which ranks features by their class discriminatory power, correlation-based feature selection (CFS; Kohavi and John, 1997) to select features with low redundancy, etc. Of late virtual reality (VR) is also being used to visualize more dimensions using colors, shapes, sizes, etc., and to achieve effective dimensionality reduction (Donalek et al., 2014).

10.6 PERIOD FINDING

We can associate a period with a fraction of the stars that are observed, even with the sparse light curves we often have. These could be binary stars, pulsating stars of different types, stars with spots, etc. A small fraction is also large period variable stars. Fourier techniques, along with the Lomb–Scargle periodogram (Deeming, 1975; Lomb, 1976; Scargle, 1982), are used extensively and sometimes blindly without regard to possible aliasing resulting from regular observations (e.g., similar time every night, or similar time of the year for a few years). Graham et al. (2013) compare different period finding methods as applicable to light curve data. These methods have turned out to be very useful in looking for exoplanets (planets orbiting stars other than the sun) owing to the eclipsing they effect. The effectiveness of the methods with sparse data is evident from a recent discovery of two white dwarfs orbiting each other in under seven minutes (Burdge et al., 2019).

10.7 EARLY CHARACTERIZATION/ CLASSIFICATION

The features are key to do early characterization, especially if one is after hierarchical classification. The procedures are often implemented so as to do the broadest separations first, followed by finer and finer subclassifications. An alternative is to do binary classifications. Something to be careful about is that the classifier does not get stumped by an object that does not belong to either class. To ensure that, one can do a series of steps, e.g., Class A versus everything else, Class B versus everything else, and then Class A versus Class B versus everything else. Doing this in a problematically consistent manner ensures that outliers are correctly captured. Any number of standard traditional machine learning classifiers can be used for the classification, e.g., random forests and support vector machines.

10.8 CLASSIFICATION USING CNNS

An alternative to doing feature classification is using more recent methods like deep learning which combine feature extraction with classification. By letting the method determine the features one can ensure that the features are unbiased (Mahabal et al., 2017). Since CNNs work best with images, light curves were converted to two-dimensional images by obtaining pairwise differences in time and brightness between all points. Thus a light curve with n points results in a density plot – a kind of structure function – with $o(n^2)$ points. One great advantage is that light curves with uneven lengths get converted into two-dimensional images with identical dimensions, and are thus more suitable for direct comparison. One drawback of this and other deep learning methods is lack of interpretability, and a lot of work is going on to surmount that problem.

10.9 RNNS, LSTMS, ETC.

Though by using CNNs with image representation of time series circumvents the possibly subjective step of extracting features, the representation itself involves uneven binning and can be subjective. While there would be ways of selecting optimal bins for given light curve sets, that still needs to be researched into. Recurrent neural networks (RNNs) are a different type of neural network that can directly operate on light curves. The long short-term memory (LSTM) variant retains short-term memory for longer durations by gating parts of information pieces that can be selectively retained and executed much like logic circuits can compute solutions to complex problems based on its inputs. Large and irregular gaps in astronomical time series are still an issue for these networks that have seen great success with sequence to sequence (seq2seq) and other natural language processing (NLP) applications. Other variants, like gated recurrent units (GRUs) and dilated convolutional RNNs, are also being explored.

One of the early applications came from Charnock and Moss (2017), and more recently Naul et al. (2018) used light curves in an encoder-decoder mode to create

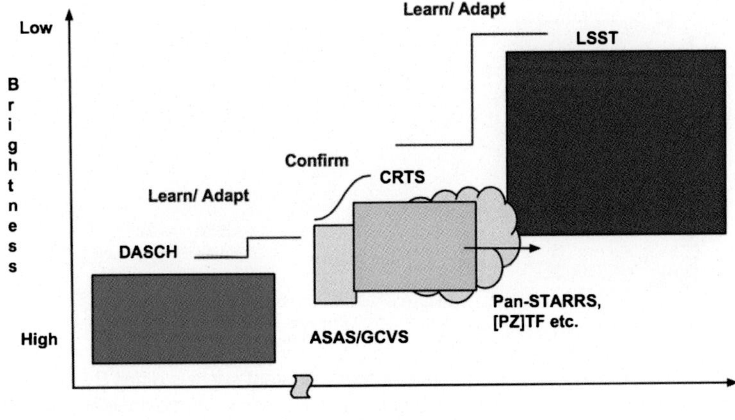

FIG. 10.3 Surveys differ in many aspects. That is why combining them becomes a powerful way to expand their capabilities. The depths, the time baselines, the filters, the area covered, the resolution and so on and on.

same-length versions from uneven lengths and effected state-of-the-art classifications.

10.10 AVAILABILITY OF LIBRARIES

This is a great time to get into the analysis of time series data. Thanks to many surveys light curves are available in abundance, there are excellent machine learning libraries in languages like Python and R, and frameworks to build software stacks with reproducible results are available. Besides FATS, which we already mentioned, there are other libraries, like Vartools,[3] providing useful functionality.

10.11 REAL TIME ASPECTS

Long back imaging the entire sky was a Herculean task. Thanks to – among other things – better electronics and bigger storage space available, it is not only possible to image the same part of the sky multiple times, but also to compare newer images with deeper coadded archival images in real-time to look for transients – objects that vary substantially in a relatively short amount of time. Not only have thousands of supernovae been thus found, they are now routinely found well before they peak in brightness, thus opening up new aspects of the time domain. We have been moving from a day old transients to those that are just hours and minutes old. Multimessenger astronomy involving simultaneous observations with telescopes providing electromagnetic signals, combined with gravitational wave data, and neutrinos from experiments like IceCube are becoming commonplace. Some aspects and details of these can be found in the chapter on real-time processing.

We will not be able to go into details of some other allied topics like transfer learning and domain adaptation, where data collected from archives are used for training with limited inference possible on new datasets, thus giving new surveys a running start (see Figs. 10.3 and 10.4).

Surveys like ZTF publish transients found each night publicly along with metadata for other groups to carry out follow-up observations of interesting transients. ZTF finds of the order of 100,000 transients every night. In preparation for LSST, which will produce an order of magnitude more transients, many brokers – frameworks to classify these alerts – are developing, and often sharing, code, e.g., ALeRCE, Antares, Lasair, etc.

10.12 TOPICS NOT ADEQUATELY COVERED

While we have touched upon many aspects of time series, especially from an optical perspective, this is a vast topic, and we have left unmentioned far more topics, many of them critical to the field, e.g., various autocorrelations (ARMA, CARMA, CARFIMA, etc.). A few statistical and somewhat complementary methods are discussed in another chapter (Chapter 11 by Andronov).

We will end the chapter by mentioning parallels with other fields. Time series of water data are an example. Just like we have different wavelengths at which observations can be carried out in astronomy, and once

[3] https://www.astro.princeton.edu/~jhartman/vartools.html.

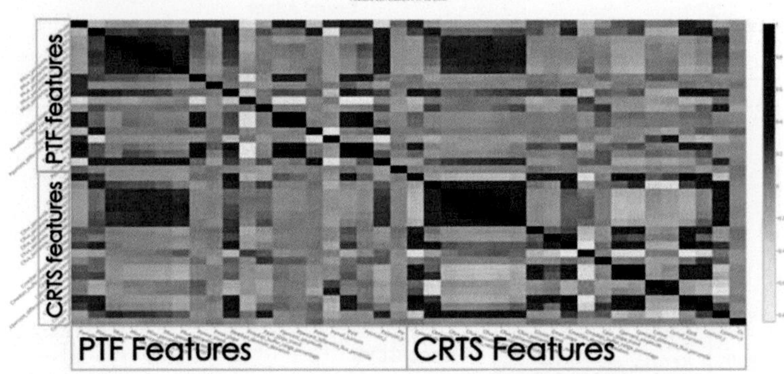

FIG. 10.4 Mathematically identical features derived from different surveys differ. By mapping the features spaces they occupy from one to the other, learning in a newer survey can be accelerated. Shown here are features for CRTS and PTF. The features shown are a subset of the ones mentioned elsewhere in the chapter. Again, the specific features shown here are not critical, but indicative.

combined they provide a more complete picture, similarly one can obtain observations related to overground and underground water, snow, moisture, etc., and combining them provides a more complete picture. This combining generally involves different scales and resolutions in the different layers making it necessary to do intelligent interpolations and approximations. For example, to improve the accuracy of modeled groundwater estimates and allow the representation of WTD at finer spatial scales, Stampoulis et al. (2019) implemented a novel approach to integrate data from the Gravity Recovery and Climate Experiment (GRACE), by augmenting the Variable Infiltration Capacity (VIC) hydrologic model. Validation is required whenever different datasets are being integrated, and often new observations from independent instruments providing the stamp of approval. For example NASA's ultra-high-resolution Airborne Snow Observatory (ASO), offers an opportunity to improve model estimates for measuring snow water equivalent (SWE) by providing a high-quality validation target (Oaida et al., 2019). There are sensors in US rivers which collect near-continuous time-series of water flow. These allow one to build models of the flow, and computing where excess water will go when rain falls becomes simpler.

Another example is that of earthquakes and their aftershocks. It is now possible to predict possible aftershocks with probability intervals after a major earthquake. Since the geographic locations of the aftershocks differ a bit, that adds to the complexity of the problem. Using accelerometers in smartphones an early-warning system has been developed and is being currently tested. More geoscience examples can be found in another chapter (Chapter 14 by Kern et al.)

Yet another example is longitudinal data for patients where progression of some feature like a tumor is being investigated. While the dataset is very different, once abstracted methodologies very similar to those used in astronomy can be applied to these time series as well.

Keep an eye out for time series and you will find them everywhere.

ABBREVIATIONS

CNN	Convolutional neural network
CSS	Catalina Sky Survey
DPOSS	Digitized Palomar Observatory Sky Survey
GRU	Gated recurrent unit
LSTM	Large short-term memory
PTF	Palomar Transient Factory
RNN	Recurrent neural network
ZTF	Zwicky Transient Facility

ACKNOWLEDGMENTS

We acknowledge that this rather short introduction is based on a lot of work done across surveys with many collaborators, but in particular S George Djorgovski, Andrew Drake, and Matthew Graham. We also acknowledge support from NSF (1640818, AST-1815034) and IUSSTF (JC-001/2017).

REFERENCES

Burdge, Kevin B., Coughlin, Michael W., Fuller, Jim, Kupfer, Thomas, Bellm, Eric C., Bildsten, Lars, Graham, Matthew J., Kaplan, David L., van Roestel, Jan, Dekany, Richard G., Duev, Dmitry A., Feeney, Michael, Giomi, Matteo,

Helou, George, Kaye, Stephen, Laher, Russ R., Mahabal, Ashish A., Masci, Frank J., Riddle, Reed, Shupe, David L., Soumagnac, Maayane T., Smith, Roger M., Szkody, Paula, Walters, Richard, Kulkarni, S.R., Prince, Thomas A., 2019. General relativistic orbital decay in a seven-minute-orbital-period eclipsing binary system. Nature 571 (7766), 528–531. https://doi.org/10.1038/s41586-019-1403-0.

Charnock, T., Moss, A., 2017. Deep recurrent neural networks for supernovae classification. The Astrophysical Journal Letters 837, L28. https://doi.org/10.3847/2041-8213/aa603d.

Cody, Ann Marie, Stauffer, John, Baglin, Annie, Micela, Giuseppina, Rebull, Luisa M., Flaccomio, Ettore, Morales-Calderón, María, Aigrain, Suzanne, Bouvier, Jérôme, Hillenbrand, Lynne A., 2014. CSI 2264: simultaneous optical and infrared light curves of young disk-bearing stars in NGC 2264 with CoRoT and Spitzer—evidence for multiple origins of variability. The Astronomical Journal 147 (4), 82. https://doi.org/10.1088/0004-6256/147/4/82.

Deeming, T.J., 1975. Fourier analysis with unequally-spaced data. Astrophysics and Space Science 36, 137–158. https://doi.org/10.1007/BF00681947.

Djorgovski, S.G., Baltay, C., Mahabal, A.A., Drake, A.J., Williams, R., Rabinowitz, D., Graham, M.J., Donalek, C., Glikman, E., Bauer, A., Scalzo, R., Ellman, N., 2008. The Palomar-Quest digital synoptic sky survey. Astronomische Nachrichten 329, 263. https://doi.org/10.1002/asna.200710948.

Donalek, Ciro, Arun Kumar, A., Djorgovski, S.G., Mahabal, Ashish A., Graham, Matthew J., Fuchs, Thomas J., Turmon, Michael J., Sajeeth Philip, N. , Yang, Michael Ting-Chang, Longo, Giuseppe, 2013. Feature selection strategies for classifying high dimensional astronomical data sets. ArXiv e-prints arXiv:1310.1976.

Donalek, Ciro, Djorgovski, S.G., Davidoff, Scott, Cioc, Alex, Wang, Anwell, Longo, Giuseppe, Norris, Jeffrey S., Zhang, Jerry, Lawler, Elizabeth, Yeh, Stacy, Mahabal, Ashish, Graham, Matthew, Drake, Andrew, 2014. Immersive and collaborative data visualization using virtual reality platforms. ArXiv e-prints arXiv:1410.7670.

Drake, A.J., Djorgovski, S.G., Mahabal, A., Beshore, E., Larson, S., Graham, M.J., Williams, R., Christensen, E., Catelan, M., Boattini, A., Gibbs, A., Hill, R., Kowalski, R., 2009. First results from the Catalina Real-Time Transient Survey. The Astrophysical Journal 696, 870884. https://doi.org/10.1088/0004-637X/696/1/870.

Faraway, J., Mahabal, A., Sun, J., Wang, X.-F., Wang, Y.G., Zhang, L., 2016. Modeling light curves for improved classification. Statistical Analysis and Data Mining: The ASA Data Science Journal 9 (1), 1–11. https://doi.org/10.1002/sam.11305.

Fukunaga, Keinosuke, 1990. Introduction to Statistical Pattern Recognition, 2nd ed. Academic Press Professional, Inc., San Diego, CA, USA.

Gaia Collaboration, Brown, A.G.A., Vallenari, A., Prusti, T., de Bruijne, J.H.J., Mignard, F., Drimmel, R., Babusiaux, C., Bailer-Jones, C.A.L., Bastian, U., et al., 2016. Gaia Data Release 1. Summary of the astrometric, photometric, and survey properties. Astronomy & Astrophysics 595, A2. https://doi.org/10.1051/0004-6361/201629512.

Graham, M.J., Drake, A.J., Djorgovski, S.G., Mahabal, A.A., Donalek, C., Duan, V., Maker, A., 2013. A comparison of period finding algorithms. Monthly Notices of the Royal Astronomical Society 434, 3423–3444. https://doi.org/10.1093/mnras/stt1264.

Ivezić, Ž., Kahn, S.M., Tyson, J.A., Abel, B., Acosta, E., Allsman, R., Alonso, D., AlSayyad, Y., Anderson, S.F., Andrew, J., et al., 2008. LSST: from science drivers to reference design and anticipated data products. ArXiv e-prints arXiv:0805.2366.

Kohavi, Ron, John, George H., 1997. Wrappers for feature subset selection. Artificial Intelligence 97 (1–2), 273–324. https://doi.org/10.1016/S0004-3702(97)00043-X.

Law, N.M., Kulkarni, S.R., Dekany, R.G., Ofek, E.O., Quimby, R.M., Nugent, P.E., Surace, J., Grillmair, C.C., Bloom, J.S., Kasliwal, M.M., Bildsten, L., Brown, T., Cenko, S.B., Ciardi, D., Croner, E., Djorgovski, S.G., van Eyken, J., Filippenko, A.V., Fox, D.B., Gal-Yam, A., Hale, D., Hamam, N., Helou, G., Henning, J., Howell, D.A., Jacobsen, J., Laher, R., Mattingly, S., McKenna, D., Pickles, A., Poznanski, D., Rahmer, G., Rau, A., Rosing, W., Shara, M., Smith, R., Starr, D., Sullivan, M., Velur, V., Walters, R., Zolkower, J., 2009. The Palomar Transient Factory: system overview, performance, and first results. Publications of the Astronomical Society of the Pacific 121, 1395–1408.

Lomb, N.R., 1976. Least-squares frequency analysis of unequally spaced data. Astrophysics and Space Science 39, 447–462. https://doi.org/10.1007/BF00648343.

Mahabal, A.A., Djorgovski, S.G., Drake, A.J., Donalek, C., Graham, M.J., Williams, R.D., Chen, Y., Moghaddam, B., Turmon, M., Beshore, E., Larson, S., 2011. Discovery, classification, and scientific exploration of transient events from the Catalina Real-time Transient Survey. Bulletin of the Astronomical Society of India 39, 387–408.

Mahabal, A., Sheth, K., Gieseke, F., Pai, A., Djorgovski, S.G., Drake, A.J., Graham, M.J., 2017. Deep-learnt classification of light curves. In: 2017 IEEE Symposium Series on Computational Intelligence (SSCI), pp. 2757–2764.

Naul, Brett, Bloom, Joshua S., Pérez, Fernando, van der Walt, Stéfan, 2018. A recurrent neural network for classification of unevenly sampled variable stars. Nature Astronomy 2, 151–155. https://doi.org/10.1038/s41550-017-0321-z.

Oaida, Catalina M., Reager, John T., Andreadis, Konstantinos M., David, Cédric H., Levoe, Steve R., Painter, Thomas H., Bormann, Kat J., Trangsrud, Amy R., Girotto, Manuela, Famiglietti, James S., 2019. A high-resolution data assimilation framework for snow water equivalent estimation across the western United States and validation with the Airborne Snow Observatory. Journal of Hydrometeorology 20 (3), 357–378. https://doi.org/10.1175/JHM-D-18-0009.1.

Richards, J.W., Starr, D.L., Butler, N.R., Bloom, J.S., Brewer, J.M., Crellin-Quick, A., Higgins, J., Kennedy, R., Rischard, M., 2011. On machine-learned classification of variable stars with sparse and noisy time-series data. The Astronomical Journal 733, 10. https://doi.org/10.1088/0004-637X/733/1/10.

Richards, Joseph W., Starr, Dan L., Miller, Adam A., Bloom, Joshua S., Butler, Nathaniel R., Brink, Henrik, Crellin-Quick, Arien, 2012. Construction of a calibrated probabilistic classification catalog: application to 50k variable sources in the all-sky automated survey. The Astrophysical Journal. Supplement Series 203 (2), 32. https://doi.org/10.1088/0067-0049/203/2/32.

Robnik-Šikonja, Marko, Kononenko, Igor, 2003. Theoretical and empirical analysis of ReliefF and RReliefF. Machine Learning 53 (1), 23–69. https://doi.org/10.1023/A:1025667309714.

Scargle, J.D., 1982. Studies in astronomical time series analysis. II. Statistical aspects of spectral analysis of unevenly spaced data. The Astrophysical Journal 263, 835–853. https://doi.org/10.1086/160554.

Stampoulis, Dimitrios, Reager, John T., David, Cédric H., Andreadis, Konstantinos M., Famiglietti, James S., Farr, Tom G., Trangsrud, Amy R., Basilio, Ralph R., Sabo, John L., Osterman, Gregory B., Lundgren, Paul R., Liu, Zhen, 2019. Model-data fusion of hydrologic simulations and GRACE terrestrial water storage observations to estimate changes in water table depth. Advances in Water Resources 128, 13–27. https://doi.org/10.1016/j.advwatres.2019.04.004.

VanderPlas, Jacob T., Ivezić, Željko, 2015. Periodograms for multiband astronomical time series. The Astrophysical Journal 812 (1), 18. https://doi.org/10.1088/0004-637X/812/1/18.

Weir, N., 1995. Automated Analysis of the Digitized Second Palomar Sky Survey: System Design, Implementation, and Initial Results. Dissertation. California Institute of Technology.

CHAPTER 11

Advanced Time Series Analysis of Generally Irregularly Spaced Signals: Beyond the Oversimplified Methods

IVAN L. ANDRONOV, DSC, PROF

11.1 INTRODUCTION

Time series analysis is the tool to study the dynamics and evolution of processes and objects of any nature. Its main aim is to extract information with a minimum number of parameters needed for approximating the data within a given interval, or to forecast it outside, with a required (or best) accuracy. In other words, the famous "Occam's Razor" principle is "*Entia non sunt multiplicanda praeter necessitatem*," or "Entities are not to be multiplied beyond necessity" (Schaffer, 2015).

Many of the well-known methods were inspired by the data being used in astro- and geo-informatics. "Ideally," as in mathematical analysis, one has to analyze an infinite continuous function with continuous derivatives of any order. "Back to reality," and the "ideal" is a finite number n of data points $(t_k, x_k,\ k = 1..n)$ with the additional "bonus" of a regular spacing of the arguments: $t_k = t_0 + \delta \cdot k$, where δ is called "time resolution," "time step," etc. The regularity is a necessary condition for the Fourier transform (FT), fast Fourier transform (FFT), and analysis using the autocorrelation function (ACF) and cross-correlation function (CCF) (Fisher, 1954; Scheffe, 1959; Anderson, 1958, 2003; Press et al., 2007). A review of the history of the methods was presented by Wermuth (2011). Combinations of methods are used in neural networks and machine learning (Haykin, 1999).

The data are often irregularly spaced, having gaps and irregular argument distributions either in time, or (if periodic) in the phase domain. Anyway, often methods have been applied, which contain a sequence of few simple steps and simplified formulas. Even for test data with well-defined properties, this may lead to significant differences between the parameters of the test model and the values obtained using simplified methods. This type of partial modeling is often called "de-trending" (or "trend removal") and "pre-whitening." Mikulášek (2007a) called such type of modeling "Matrix-Phoebia," as, mathematically, this is similar to ignoring the nondiagonal values of the matrix of normal equations in the least-squares (LS) method.

Below we present references to the correct ("non-simplified") methods and show in which cases they are significantly better than the common "simplified" ones. Obviously, "simplified" and "non-simplified" methods should produce the same corresponding parameters and approximations, if the conditions used for the "simplified" methods are satisfied. If not, an improved method should be used instead of one of the simplified ones.

The variety of methods reflects the variety of types of variability. If we could have an infinite number of observations, theoretically one could get an infinite number of values of the functions describing the FT. Practically, in astro- and geo-time series, the data are not only finite in the number of observations, but also often have gaps between single observations (e.g., photographs from Harvard, Sonneberg, Odessa, Moscow, Asiago plate collections and others, CCD photometric surveys [NSVS, Wozniak et al., 2004; ASAS, Pojmanski, 2002; OGLE, Paczynski, 1986; Udalski et al., 1997; CRTS, Drake et al., 2009; MASTER, Lipunov et al., 2010; WASP, Butters et al., 2010; Street et al., 2003, ZTF; etc.], or space observations [Hipparcos/Tycho, Høg et al., 2000; KEPLER, 2019; TESS, 2019; WISE, Wright et al., 2010; GAIA, 2019]).

Many of these signals contain different contributions, for the extraction of which one has to use complicated models, rather than a sequence of simple ones. Such an approach significantly improves the accuracy, which allows us not only to determine the model parameters, but also to more surely estimate their statistical significance and to avoid wrong detection/discovery and thus fake interpretations.

The elaboration of advanced algorithms and programs was inspired by long-term collaboration with

Knowledge Discovery in Big Data from Astronomy and Earth Observation. https://doi.org/10.1016/B978-0-12-819154-5.00022-9

the "Inter-Longitude Astronomy" (ILA) team. This is a joint name of a series of smaller temporarily projects on observation and interpretation of concrete variable stars or groups of stars. It has no special funds; professional and amateur astronomers take part based on their own resources. This "ILA" project is in some way similar and complementary to other projects like WET, CBA, VSOLJ, BRNO, MEDUZA, and currently UkrVO (Vavilova et al., 2012) and Astroinformatics (Vavilova et al., 2017), where many of us take part. Previous reviews on the ILA project were published by Andronov et al. (2003a, 2014, 2017a).

11.2 STATISTICAL PROPERTIES OF THE FUNCTIONS OF (CORRELATED) PARAMETERS OF THE LS FITS

11.2.1 Least-Squares Method: Test Functions

The most common method for the determination of the statistically optimal approximation with a corresponding set of parameters is called the least-squares (LS) method and was proposed about two centuries ago by Carl Friedrich Gauss (1777–1855). It has been described in thousands of textbooks and monographs (e.g., Anderson, 1958, 2003; Forsythe et al., 1977; Press et al., 2007). LS problems with restrictions were discussed by Lawson and Hanson (1974). For a more complete description, we present an extended set of formulas, which are typically omitted.

Typically, it may be written in a form, with minimizing the weighted sum of the residuals of the observational points x_k from the calculated values x_{Ck} at the arguments t_k, i.e.,

$$\Phi = \sum_{k=1}^{n} (x_k - x_{Ck})^2. \tag{11.1}$$

However, in many cases, one should take into account the weights of the observations $w_k = \sigma_0^2/\sigma_k^2$, where σ_k is the standard error of the observation x_k and σ_0 is called "the unit weight error." Then we have

$$\Phi = \sum_{k=1}^{n} w_k \cdot (x_k - x_{Ck})^2. \tag{11.2}$$

For the pure Gaussian noise, the random value $U = \Phi/\sigma_0^2$ is distributed according to the χ^2 distribution with $(n - m)$ degrees of freedom, where m is the number of the independent parameters C_α $(\alpha = 1..m)$ used for computation of the approximation $x_C(t)$.

More complicated, but statistically justified, is the use of the weight matrix w_{jk}, thus leading to

$$\Phi = \sum_{kj=1}^{n} w_{kj} \cdot (x_k - x_{Ck}) \cdot (x_j - x_{Cj}). \tag{11.3}$$

For the "correct" evaluation of the matrix w_{ij}, one should use

$$w_{ij} = \sigma_0^2 \mu_{ij}^{-1}, \tag{11.4}$$

where μ_{ij} is the covariation matrix of the errors of the observations, μ_{ij}^{-1} is an inverse matrix, and σ_0 is any positive constant. Generally, we do not mention the precise values of μ_{ij}, but, by setting the coefficients w_{ij}, we automatically assume the matrix $\mu_{ij} = \sigma_0^2 \cdot w_{ij}^{-1}$.

This expression resembles the metrics in the tensor analysis and shows a squared "distance" between the observations and the approximation.

Next improvement contains the filter (weight) function $p(z_k, z_j)$, $z_k = (t_k - t_0)/\Delta t$, which is dependent on the times of observations t_k, t_j and on the "shift" t_0 and "scale" Δt, as typically defined in the wavelet analysis. The general expressions for this case are presented by Andronov (1997). For practical purposes, the replacement of w_k in Eq. (11.2) by $p(z_k) \cdot w_k$ may be recommended, so neglecting possible correlations between the statistical deviations of the data from the "true values."

All these four approaches may be written using a generalized form of the scalar product of vectors, i.e.,

$$(\vec{a} \cdot \vec{b}) = \sum_{kj=1}^{n} p(z_k, z_j) \cdot w_{kj} \cdot a_k \cdot b_j. \tag{11.5}$$

The vectors $\vec{a}$ and $\vec{b}$ are called "orthogonal" if $(\vec{a} \cdot \vec{b}) = 0$. The "squared" vector may be defined in a usual way:

$$\vec{a}^2 = (\vec{a} \cdot \vec{a}) = \sum_{kj=1}^{n} p(z_k, z_j) \cdot w_{kj} \cdot a_k \cdot a_j. \tag{11.6}$$

Then the test function may be generally written as

$$\begin{aligned} \Phi &= (\vec{x} - \vec{x}_C)^2 \\ &= \sum_{kj=1}^{n} p(z_k, z_j) \cdot w_{kj} \cdot (x_k - x_{Ck}) \cdot (x_j - x_{Cj}) \end{aligned} \tag{11.7}$$

and one should determine the values of the coefficients (often called "parameters") C_α, $\alpha = 1..m$, of the approximation $x_C(t, C_\alpha)$.

In some cases, the coefficients are determined using the principle of the maximum likelihood. However, under a common assumption of normal probability distribution of observational errors, the parameters of the maximum likelihood are exactly at the minimum of the function Φ.

11.2.2 Linear Least Squares

The simplest case of the approximation $x_C(t, C_\alpha)$ is a linear combination of so-called basic functions $f_\alpha(t)$, i.e.,

$$x_C(t, C_\alpha) = \sum_{\alpha=1}^{m} C_\alpha f_\alpha(t). \tag{11.8}$$

In this case, the minimization of the test function is reduced to the solution of the following system of normal equations:

$$\sum_{\alpha=1}^{m} A_{\alpha\beta} C_\alpha = B_\beta, \tag{11.9}$$

$$C_\alpha = \sum_{\alpha=1}^{m} A^{-1}_{\alpha\beta} B_\beta, \tag{11.10}$$

where $A_{\alpha\beta} = (\vec{f}_\alpha \cdot \vec{f}_\beta)$, $B_\beta = (\vec{x} \cdot \vec{f}_\beta)$, and $A^{-1}_{\alpha\beta}$ is the matrix, inverse to $A_{\alpha\beta}$. The matrix $A_{\alpha\beta} = A_{\beta\alpha}$ is symmetrical.

If the set of the basic functions is orthogonal ($A_{\alpha\beta} = (\vec{f}_\alpha \cdot \vec{f}_\beta) = A_{\alpha\alpha}\delta_{\alpha\beta}$), the inverse matrix is diagonal ($A^{-1}_{\alpha\beta} = (1/A_{\alpha\alpha}) \cdot \delta_{\alpha\beta}$), and $C_\alpha = B_\alpha / A_{\alpha\alpha}$.

The following relations are valid: $((\vec{x} - \vec{x}_c) \cdot \vec{f}_\alpha) = 0$, $((\vec{x} - \vec{x}_c) \cdot \vec{x}_c) = 0$,

$$((\vec{x} - \vec{x}_c) \cdot (\vec{x} - \vec{x}_c)) = (\vec{x} \cdot \vec{x}) - (\vec{x}_c \cdot \vec{x}_c). \tag{11.11}$$

Introducing the "reference" vector

$$\vec{x}_0 = \sum_{\alpha=1}^{m} C_{0\alpha} \vec{f}_\alpha, \tag{11.12}$$

$$(\vec{x} - \vec{x}_C)^2 = (\vec{x} - \vec{x}_0)^2 - (\vec{x}_C - \vec{x}_0)^2. \tag{11.13}$$

This extended relation is valid for any set of constant coefficients $C_{0\alpha}$.

Typically, if it is used, an abbreviated form $\vec{x}_0 = \vec{x}_1 = (C_{11}, C_{11}, ...C_{11})$ is applied, where C_{11} is a solution of the one-parameter fit $x_c(t) = C_{11}$, (so $f_1(t) = 1$), i.e., a weighted sample mean of x_k. In a general case of nonorthogonal basic functions, the coefficients C_α are dependent on m, so it might be recommended to write a complete form $C_{m\alpha}$ (as we did for C_{11}). So, generally, $C_{m\alpha} \neq C_{L\alpha}$ for different number of parameters. However, after this remark, we will still use C_α as a short designation of $C_{m\alpha}$ for current m.

11.2.3 Influence of Deviations of Coefficients

In the "linear combination" case (Eq. (11.8)), the test function for any coefficients $\tilde{C}_\alpha = C_\alpha + D_\alpha$ may be rewritten as

$$\Phi(\tilde{C}_\alpha) = \Phi_m + \sum_{\alpha,\beta=1}^{m} A_{\alpha\beta} \cdot (\tilde{C}_\alpha - C_\alpha) \cdot (\tilde{C}_\beta - C_\beta),$$

$$\Phi_m = \Phi(C_\alpha) = \Phi_0 - \sum_{\alpha,\beta=1}^{m} A_{\alpha\beta} \cdot C_\alpha \cdot C_\beta, \tag{11.14}$$

$$\Phi_0 = (\vec{x} \cdot \vec{x}).$$

Here D_α are some coefficients. Let

$$x_D(t) = \sum_{\alpha=1}^{m} D_\alpha \cdot f_\alpha(t). \tag{11.15}$$

Similarly to common designations $\vec{x} = \vec{O}$ (observed), $\vec{x}_C = \vec{C}$ (calculated), let $\vec{x}_D = \vec{D}$ (deviation). Then Eq. (11.14) may be rewritten for the squares (scalar products) of vectors, and we have

$$\begin{aligned}
&(\vec{O} - \vec{C} - \vec{D})^2 \\
&= (\vec{O} - \vec{C})^2 + \vec{D}^2 = \vec{O}^2 - \vec{C}^2 + \vec{D}^2 = \Phi(C_\alpha + D_\alpha) \\
&= \Phi_0 - \sum_{\alpha,\beta=1}^{m} A_{\alpha\beta} \cdot C_\alpha \cdot C_\beta + \sum_{\alpha,\beta=1}^{m} A_{\alpha\beta} \cdot D_\alpha \cdot D_\beta.
\end{aligned} \tag{11.16}$$

A two-dimensional simple geometrical interpretation of these equations is shown in Fig. 11.1. As the vectors $\vec{C}$ and $\vec{O} - \vec{C}$ are orthogonal, the squares of their length are related by the Pythagoras theorem.

For the model (11.8), the test function $\Phi(\tilde{C}_\alpha)$ is an m-dimensional paraboloid with a single minimum at $\tilde{C}_\alpha = C_\alpha$. This relatively rare equation may be useful either to see the influence of the rounding of the determined coefficients on the test function, or to determine "confidence intervals" for the parameters (cf. Cherepashchuk, 1993).

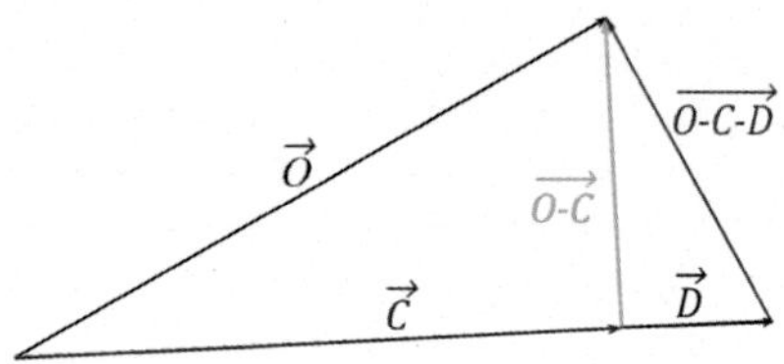

FIG. 11.1 Two-dimensional illustration of the m-dimensional equation (11.16).

11.2.4 Linear Approximation

The mostly used formula is the linear approximation

$$x_C(t) = C_1 + C_2 \cdot t = \tag{11.17}$$

$$= \bar{x} + C_2 \cdot (t - \bar{t}), \tag{11.18}$$

$$\sigma^2[x_C(t)] = \sigma^2_{\bar{x}} \left(1 + \frac{(t-\bar{t})^2}{\sigma^2_t}\right) = \sigma^2_{\bar{x}} \left(1 + \frac{t^2 - 2t\bar{t} + \bar{t}^2}{\sigma^2_t}\right), \tag{11.19}$$

$$\sigma^2_{\text{wrong}}[x_C(t)] = \sigma^2[C_1] + \sigma^2[C_2] \cdot t^2 = \sigma^2_{\bar{x}} \left(1 + \frac{\bar{t}^2 + t^2}{\sigma^2_t}\right). \tag{11.20}$$

Here the mean values $\bar{x} = B_1/A_{11}$, $\bar{t} = A_{12}/A_{11}$, the variance $\sigma^2_{\bar{x}} = \sigma^2_0/n$, and $\sigma_{\text{wrong}} = \sigma_{\text{w}}$ is "wrong" (or "oversimplified") estimate corresponding to Eq. (11.20). The mean values are here defined in a general case, not restricting to the case of equal weights (i.e., assuming that $\sigma_k = \text{const}$). These expressions are valid for any of the forms of the matrix w_{ij} with a definition of the sample mean value of any function as

$$\sigma^2_{\text{wrong}}[x_C(t)] = \sum_{\alpha}^{m} R_{\alpha\alpha} \cdot (f_\alpha(t))^2 = \sum_{\alpha}^{m} (f_\alpha(t) \cdot \sigma[C_\alpha])^2. \tag{11.21}$$

The difference in the error estimates using the correct formula and the incomplete one is shown in Fig. 11.2.

For the uniform distribution of times, starting from time zero, and large n, the accuracy of the zero point is $\sigma[C_1] = 2\sigma_{\bar{x}}$, and this is exactly twice larger than the accuracy of the zero point $\sigma_{\bar{x}}$ in Eq. (11.20).

In Fig. 11.2, the illustrative dependence of $x = \Delta(R - I)$ on $t = B - V$ is shown based on Table 2 from Kim et al. (2004). The best fit line is $x_c(t)$. The difference between the error estimates is due to nonorthogonality of the set of basic functions $f_\alpha(t_k)$, which leads to an incorrect formula. Thus it should be recommended to use a slightly complicated expression $x_C(t) = \bar{x} + C_2 \cdot (t - \bar{t})$ with an error estimate.

Often the data are regular and the first moment is set to zero, i.e., the time is converted to $t_k = t_{0k} - t_{01}$. Here t_{0k} are times with an arbitrary zero point in this case $\bar{t} = (n-1)/2\delta$, $\sigma_t = ((n^2-1)/12)^{1/2}\delta$. Thus for any n, $\sigma[x_C(0)] = \sigma[C]_1 = 2\sigma_{\bar{x}}(1 - 1.5/(n+1))^{1/2}$. For large n, $\sigma[C_1] = 2\sigma_{\bar{x}}$. Using a simplified formula (11.21), one gets $\sigma[x_C(\bar{t})] = \sqrt{7}\sigma_{\bar{x}}$ instead of the correct value of $\sigma_{\bar{x}}$.

Besides the linear regression (Eq. (11.17)) with coefficients C_2 determined using the LS, more often are used other lines with different slopes. They all have

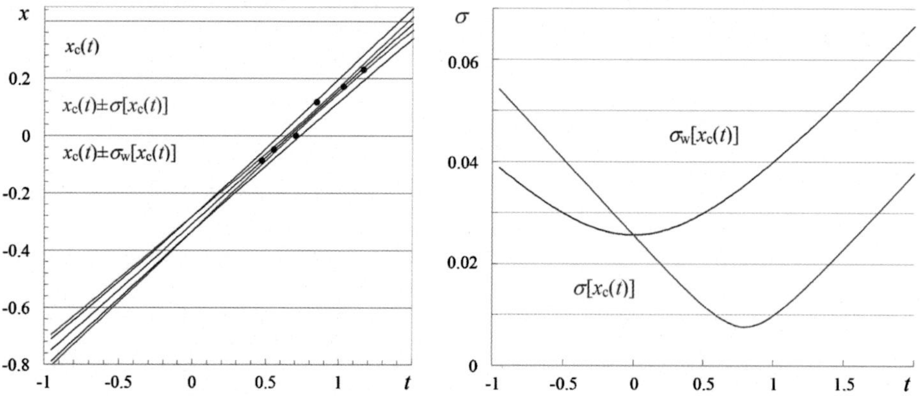

FIG. 11.2 The dependence of the smoothing linear function $x_C(t)$ and its "$\pm 1\sigma$" error corridors for the correct equation (11.19) and wrong (simplified) equation (11.20). For the illustration, we have used the dependence of $\Delta(V - R)$ on the color index $(B - V)$ (Table 2 from Kim et al. (2004)).

a cross-point at $(\bar{t}, \bar{x})$, so $x_\epsilon = \bar{x} + b_\epsilon \cdot (t - \bar{t})$. Introducing the second-order moments $\mu_{xx} = \bar{x}^2/A_{11} - \bar{x}^2$, $\mu_{tx} = B_2/A_{11} - \bar{t}\bar{x}$, $\mu_{tt} = A_{22}/A_{11} - \bar{t}^2$, the coefficients are $b_1 = C_2$ (the best fit approximation for $x_C(t)$ as a function of t, i.e., minimizing $(\vec{x} - \vec{x}_C)^2$); $b_2 = \mu_{tx}/\mu_{xx}$ for minimizing $(\vec{t} - \vec{t}_C)^2$; $b_3 = \mu_{tx}/\mu_{xx}$ for minimizing $(\vec{t} - \vec{t}_C)^2$; $b_3 = -q + (q^2+1)^{1/2}$, $b_4 = -q - (q^2+1)^{1/2}$, $q = (\mu_{tt} - \mu_{xx})/(2\mu_{tx})$ are slopes of the "orthogonal" regression (the coefficient of the main line has the same sign as μ_{tx}). Two-dimensional orthogonal regression is a kind of principal component analysis (PCA), which is discussed below. One may also apply PCA to normalized data, with a corresponding slope $b_5 = \mathrm{sign}(\mu_{tx}) \cdot (\mu_{xx}/\mu_{tt})^{1/2}$, which is dependent on the sign of μ_{tx}, but not on its value.

The correlation coefficient $r = \mu_{tx}/(\mu_{xx}\mu_{tt})^{1/2}$. Its accuracy $\sigma[r] = (1 - r^2)/\sqrt{n-1}$, but, for the significance test, the value of $t = r/\tilde{\sigma}[r]$ is used, where $\tilde{\sigma}[r] = \sqrt{1-r^2}/\sqrt{n-1}$. Another test of the "null hypothesis" is for $t = \sqrt{n-3} \cdot \ln\sqrt{(1+r)/(1-r)}$ (Fisher, 1954). More details on application of correlation analysis are discussed by Isobe et al. (1990). The case of unequal independent statistical errors in *both* coordinates was reviewed by Press et al. (2007).

11.2.5 Linearization

There are also some approximations, the parameters of which may be more easily estimated by replacing initial functions with other ones. For example, for a power dependence model $x = a \cdot t^b$, $\log x = \log a + b \cdot \log t$, or $\tilde{x} = \tilde{a} + b \cdot \tilde{t}$. Obviously, it is acceptable only for positive values of all x, t. Similarly, for an often exponential model $x = a \cdot \exp(bt)$, $\log x = \log a + b \cdot t$, or, again, $\tilde{x} = \tilde{a} + b \cdot t$. In this case, the expressions become linear, and the parameters are determined without using slightly more complicated nonlinear optimization. However, the coefficients obtained using "linearized" and initial equations are generally different. They coincide only if the residuals are zero, containing no observational errors. For small $\sigma_k \ll x_k$, $\sigma[\log_z(x_k)] \approx \sigma_k/x_k/\ln z$. So it should be recommended to use the parameters from the "linearized" model only as initial values for further iterations in a nonlinear model.

11.2.6 Statistical Properties of Functions of Coefficients

In the general case of nonorthogonal basic functions, the statistical errors are highly correlated. Thus it is important to define a complete covariance matrix for the errors of the coefficients $R_{\alpha\beta} = \langle C_{d\alpha} C_{d\beta}\rangle$. Generally,

$$R_{\alpha\beta} = \sum_{\gamma\epsilon=1}^{m} \sum_{i,j,k,L=1}^{n} A^{-1}_{\alpha\gamma} A^{-1}_{\epsilon\beta}\, p(z_i, z_j) \cdot w_{ij}\, p(z_k, z_L) \cdot w_{kL} f_\gamma(t_i) f_\epsilon(t_k) \mu_{jL}. \tag{11.22}$$

This is a statistically correct full version.

Under two assumptions ($p(z_i, z_j) = 1$ and Eq. (11.4)), this long equation significantly shortens to

$$R_{\alpha\beta} = \sigma_0^2 \cdot A^{-1}_{\alpha\beta}. \tag{11.23}$$

The statistical errors (accuracy) of the coefficients are

$$\sigma[C_\alpha] = \sqrt{R_{\alpha\alpha}}. \tag{11.24}$$

The variance (the squared error estimate of $\sigma[G]$) of the general function $G(C_\alpha)$ of coefficients is

$$\sigma^2[G] = \sum_{\alpha\beta=1}^{m} R_{\alpha\beta} \cdot \frac{\partial G}{\partial C_\alpha} \cdot \frac{\partial G}{\partial C_\beta}. \tag{11.25}$$

One may note a common "simplified" (generally wrong) relation

$$\sigma^2[G] = \sum_{\alpha}^{m} R_{\alpha\alpha} \cdot \left(\frac{\partial G}{\partial C_\alpha}\right)^2 \tag{11.26}$$

$$= \sum_{\alpha}^{m} \left(\frac{\partial G}{\partial C_\alpha} \cdot \sigma[C_\alpha]\right)^2. \tag{11.27}$$

This is what Mikulášek (2007a) called the "Matrix-Phoebia." Both relations coincide only if the matrix $R_{\alpha\beta}$ (and so $A_{\alpha\beta}$) are diagonal. For a simplest comparison, this is like using $a^2 + b^2$ for a result of $(a+b)^2 = a^2 + 2ab + b^2$.

Particularly, for the smoothing function (Eq. (11.8)) itself,

$$\sigma^2[x_C(t)] = \sum_{\alpha\beta=1}^{m} R_{\alpha\beta} \cdot f_\alpha(t) \cdot f_\beta(t), \tag{11.28}$$

with a "simplified" expression

$$\sigma^2[x_C(t)] = \sum_{\alpha}^{m} R_{\alpha\alpha} \cdot (f_\alpha(t))^2 = \sum_{\alpha}^{m} \cdot (f_\alpha(t) \cdot \sigma[C_\alpha])^2. \tag{11.29}$$

This equation is valid *only* if the matrix $R_{\alpha\beta}$ is diagonal, so the matrix of normal equations $A_{\alpha\beta}$ is diagonal.

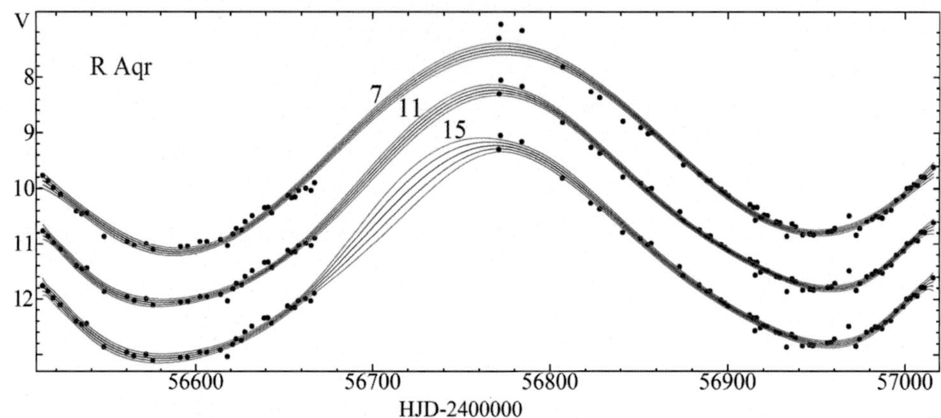

FIG. 11.3 Part of the light curve of R Aqr, the symbiotic binary star with a pulsating component. The observations were taken from the AAVSO international database in the filter V. The approximations $x_c(t)$ and the error corridors $x_c(t) \pm \sigma[x_c(t)]$ and $x_c(t) \pm 2\sigma[x_c(t)]$ for different numbers of parameters m correspond to different criteria for the determination of m.

In other words, the system of basic vectors (vectors of values of the basic functions) is orthogonal.

The "Matrix-Phoebia" (Mikulášek, 2007a) in using this equation (11.29) is in replacing the matrix $R_{\alpha\beta}$ by a diagonal matrix, arbitrarily setting nondiagonal values to zero.

In the popular electronic tables (Microsoft Excel, Open/LibreOffice Calc, GNUmeric), there is a possibility of approximation the data using polynomial and some linearized approximations. However, no error estimates of the parameters and smoothing function are available, as well as the use of the weights. With the function *LINEST*, it is possible to determine parameters and their estimates for the simplest case $w_{ij} = \text{const} \cdot \delta_{ij}$. Even the diagonal form $w_{ij}\text{const} \cdot \delta_{ij}$ and the covariation matrix $R_{\alpha\beta}$ are not available.

An illustration of the approximation of the real data is shown in Fig. 11.3. The methods to determine an optimal number of parameters is discussed below.

11.2.7 Accuracy of the Derivative and Moments of Crossings

Similarly, for the derivative of degree q of the smoothing function $x_C^{(q)}$, $f_\alpha^{(q)}(t) = \mathrm{d}^q f_\alpha(t)/\mathrm{d}t^q$,

$$\sigma^2[x_C^{(q)}(t)] = \sum_{\alpha\beta=1}^{m} R_{\alpha\beta} \cdot f_\alpha^{(q)}(t) \cdot f_\beta^{(q)}(t). \quad (11.30)$$

For "running approximations" with $p(z_k, z_j) \neq$ const, when an approximation consists of mid-points of the numerous smoothing functions, the expressions are more complicated, as presented by Andronov (1997).

The moment of crossing t_{cross} of the approximation of the constant level x_{cross} (e.g., the gamma velocity or some constant brightness) is determined as the root of equation $x_C(t_{cross}) = x_{cross}$ and has a standard error of

$$\sigma[t_{cross}] = \frac{\sigma[x_C(t_{cross})]}{|\dot{x}_C(t_{cross})|}, \quad (11.31)$$

where $\dot{x}_C(t_{cross})$ is the derivative dx/dt at the moment t_{cross}.

Alternately, Andronov et al. (2008) and Andronov and Andrych (2014) used the inverse approximation $t_c(x)$, where results may be expressed as $t_{cross} = t_c(x_{cross})$, so $\sigma[t_{cross}] = \sigma[t_C(x_{cross})]$.

Such method is preferred when the duration of the ascending or descending branches is much smaller than the duration of the "outburst" (or "eclipse"). Some algorithms that still use $x_C(t)$ approximations are implemented in the software MAVKA (Andrych and Andronov, 2019).

11.3 STATISTICALLY OPTIMAL NUMBER OF PARAMETERS

11.3.1 "Esthetic" (User-Defined)

There are few methods to determine a number of parameters. Probably the most common one may be called an "esthetic," as the user chooses himself, looking at the approximation and visually estimating its quality. It is realized in the electronic tables like Microsoft Excel,

GNUmeric, Libre Office, Open Office, Kingsoft Office, etc. The default approximation is a polynomial one, and the user chooses the degree of the polynomial. An example is shown in Fig. 11.3. A similar approach is realized in the number of principal components used for filtering multichannel signals (Golyandina et al., 2001).

11.3.2 Analysis of Variance (ANOVA)

The comparison between the values of Φ_m may be done in a few ways. In many cases, there is a comparison between the smoothing function obtained with $m-q$ and m parameters. Very often, the parameter q is set to 1 (e.g., for algebraic polynomials) or 2 (for trigonometrical polynomials). The difference between them may be scaled easily as $B=(\Phi_{m-q}-\Phi_m)/\Phi_{m-q}$. Assuming that the residual signal $x_k - x_{cm}(t_k)$ is a random Gaussian noise with a theoretical covariation matrix μ_{jk}, the random values of Φ_m/σ_0^2 are expected to obey the χ^2_{n-m} random distribution. As the value of σ_0 is unknown, one uses the sample value of σ_{0m}. The ratio σ^2_{0m}/σ_0^2 obeys a "reduced" χ^2_{n-m} random distribution.

Similarly, the value

$$B=\frac{\Phi_{m-q}-\Phi_m}{\Phi_{m-q}} \tag{11.32}$$

has the B (Beta) distribution with the parameters $q/2$ and $(n-m)/2$. The FAP may be computed in many programs, also in the electronic tables, as the function *BETADIST*.

Traditionally, another related value,

$$F=\frac{n-m}{q}\cdot\frac{\Phi_{m-q}-\Phi_m}{\Phi_m}=\frac{n-m}{q}\cdot\frac{B}{1-B}, \tag{11.33}$$

is used, instead of the value B. If the observational errors obey the normal (Gaussian) probability distribution, the value of F is a random variable corresponding to the Fisher F distribution with q and $n-m$ degrees of freedom (Fisher, 1954). It also may be computed using the electronic tables. Obviously, the estimate of FAP will be the same, if using either B or F. The inverse relation is $B=qF/(n-m+qF)$.

In the particular case $q=1$, one may determine the same value of FAP using the Student T distribution for the value $T=C_m/\sigma[C_m]$ for the last coefficient of the approximation ($\alpha=m$).

As the system of basic functions is generally not an orthogonal one, the T distribution is valid only for the last coefficient C_m. For fast estimates, people use the "3σ" criterion, i.e., the coefficient is decided to be statistically significant if $|T|=|C_m/\sigma[C_m]|\geq 3$. The FAP may be computed in the electronic tables using the function $2\cdot\mathrm{TDIST}(|T|, n-m)$.

In a frequent case $f_1(t)=1$, the quality of the fit is measured as the square of the correlation coefficient r between the observed and calculated data,

$$S=r^2=1-\frac{\Phi_m}{\Phi_1}. \tag{11.34}$$

Here S is the same as B (Eq. (11.32)) with special parameter $q=m-1$.

11.3.3 "Best Accuracy" and Related Estimates

Another method was proposed by Andronov (1994a) (see also Andronov, 2003; Andronov and Marsakova, 2006), in which the number of parameters is determined to get the best accuracy estimate of the chosen phenomenological parameter, e.g., the r.m.s. accuracy of the smoothing function $\sigma_m[x_c]$ at the times of observations,

$$\begin{aligned}\sigma_m^2[x_c]&=\frac{1}{N}\sum_{k=1}^{n}\sigma^2[x_c(t_k)]\\&=\frac{\sigma_{0m}^2}{N}\sum_{k=1}^{n}\sum_{\alpha\beta=1}^{m}A^{-1}_{\alpha\beta}f_\alpha(t_k)f_\beta(t_k).\end{aligned} \tag{11.35}$$

Only in the case $w_{jk}=\mathrm{const}\cdot\delta_{jk}$, this expression is simplified to

$$\sigma_m^2[x_c]=\frac{m}{N}\sigma_{0m}^2=\frac{m}{N}\cdot\frac{\Phi_m}{n-m}. \tag{11.36}$$

For the more general case $w_{jk}=w_k\cdot\delta_{jk}$, one may introduce a weighted version

$$\sigma_m^2[x_c]=\frac{1}{W}\sum_{k=1}^{n}w_k\sigma^2[x_c(t_k)]=\frac{m}{W}\sigma_{0m}^2=\frac{m}{W}\cdot\frac{\Phi_m}{n-m}, \tag{11.37}$$

where $W=\sum_{k=1}^n w_k$.

Similarly, one may introduce "continuous" versions of the root mean squared accuracy,

$$\sigma^2_{mxc}=\frac{1}{W_I}\int_{t_{\min}}^{t_{\max}}w_I(t)\sigma^2[x_c(t)]\,dt, \tag{11.38}$$

$$W_I=\int_{t_{\min}}^{t_{\max}}w_I(t)\,dt, \tag{11.39}$$

where $w(t)$ is a (user-defined) weight function, which is nonnegative in the interval $[t_{\min}, t_{\max}]$.

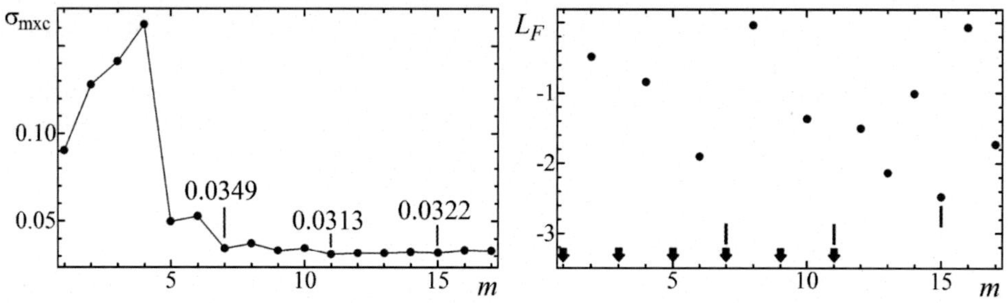

FIG. 11.4 *Left:* Dependence of r.m.s. accuracy of the smoothing function σ_{mxc} at the arguments of observations on the number of parameters m, for the part of the light curve of R Aqr shown in Fig. 11.3. The numbers correspond to the values of σ_{mxc}. *Right:* Dependence of $L_F = \lg$ FAP on m. The vertical bars correspond to the same three values of m.

Particularly, the function $W_I(t)$ may be a constant in this interval, so making the best r.m.s. accuracy estimate for the interval. Otherwise, it may be a Dirac's $W_I(t) = \delta(t - \tilde{t})$, producing the value for a chosen time $\tilde{t}$.

Generally, σ_{mxc} initially decreases with increasing m, as Φ_m increases, then increases, as σ_{0m} reaches a "standstill," and the multiplier m becomes more important. The value of m which corresponds to the minimum of σ_{mxc}, may be recommended to be a statistically optimal one. It may be significantly smaller than the value determined from the FAP.

An additional algorithm to determine the statistically optimal value of m is based on the signal-to-noise ratio (SNR), which is usually defined as SNR $= P_{\rm signal}/P_{\rm noise}$, where P is power, which is proportional to corresponding variances: SNR $= \sigma^2_{\rm signal}/\sigma^2_{\rm noise}$. Sometimes, an "amplitude" SNR is used, i.e., SNR $= \sigma_{\rm signal}/\sigma_{\rm noise}$.

In Fig. 11.4, the dependence of the r.m.s. σ_{mxc} on m is shown. The smallest marked value of $m = 5$ corresponds to the beginning of the wide minimum, which resembles a "standstill." The largest value $m = 15$ corresponds to the 3σ criterion with a corresponding FAP $= 0.0034$. The middle value $m = 11$ corresponds to the minimal σ_{mxc}. Current data are of a good CCD accuracy, thus small systematic variations lead to larger optimal values of m than in the case of more noisy (e.g., visual) observations.

There are many weight (often called "window") functions $p(z)$. The classical "finite-length" (if $-1 \le z \le 1$, zero outside) ones are the "rectangular" one $p(z) = 1$, the Bartlett function $p(z) = 1 - |z|$, the "const+cosine" function $p(z) = a_0 + (1 - a_0)\cos(\pi z)$ (for $a_0 = 0.5$ and $a_0 = 0.53836$, they are called the Hann and Hamming filter, respectively), and the infinite Gaussian $p(z) = \exp(-z^2/2)$. There are many "hybrid" window functions, which are often multiplicative or additive combinations of other window functions (e.g., Prabhu, 2014). Andronov (1997) proposed a simpler function $p(z) = (1 - z^2)^2$, which does not need computation of exponents or trigonometric functions, and so is time consuming for computations.

11.4 NONLINEAR LS METHOD AND DIFFERENTIAL CORRECTIONS

Generally, it is suitable to distinguish between "linear" and "nonlinear" parameters in the approximation. The "linear" parameters are those for which the approximation depends on the corresponding parameter linearly (as in Eq. (11.8)). Instead, "nonlinear" parameters are included in the "nonlinear" basic functions f_α.

The most common example is a sine approximation

$$x_C[t] = C_1 + C_2 \cdot \cos(C_4 \cdot (t - t_0)) + C_3 \cdot \sin(C_4 \cdot (t - t_0)), \qquad (11.40)$$

where C_1, C_2, C_3 are "linear" parameters and $C_4 = 2\pi/P = 2\pi f$ is a "nonlinear" one. Here $P = 2\pi/C_4 = 1/f$ is the period and f is frequency. The parameter C_4 is "nonlinear," as the value of the smoothing function $x_C(t)$ changes with C_4 nonlinearly. The parameter t_0 may be arbitrary. It influences the coefficients C_2 and C_3, but not the sum (Eq. (11.40)).

To determine its value, one may compute a sequence of values of the test function for a set of "nonlinear" parameters, determining the "linear" parameters using the LS described above. Then the set corresponding to the minimum is determined, which corresponds to the minimum of the test function Φ. Then we either decrease the step for the grid of parameters and determine the value of C_4, or use iterations to determine a more precise position of the minimum. The parameter t_0 is often set to zero, or to the beginning of observations. However, for faster convergence of the iterations, it is recommended to set it to a sample mean.

For this purpose, there are many methods; among them the most popular are (e.g., Press et al., 2007):

- Gauss–Newton algorithm,
- gradient descent algorithm,
- Levenberg–Marquardt algorithm,
- conjugate gradients algorithm,
- simplex method,
- coordinate descent,
- Monte Carlo (random arguments).

All these methods achieve more accurate determination of the parameters. But, to estimate statistical properties, the method of differential corrections (Gauss–Newton algorithm) should be finally used to allow estimating the matrix $R_{\alpha\beta}$ and further estimating the accuracy of coefficients and approximation itself. The main idea is to calculate the corrections to the nonlinear coefficients in such a way that the test function Φ should reach its deeper minimum. Let us enumerate the nonlinear parameters from $m+1$ to $L=m+q$. Then the LS method is applied twice, separately for m initial equations with some input values of $C_{m+1}..C_L$, and later to the system of L equations for the residuals, i.e.,

$$\sum_{\alpha=1}^{L} A_{\alpha\beta}\cdot\Delta C_\alpha = \Delta B_\beta, \tag{11.41}$$

$$\Delta B_\beta = ((\vec{x}-\vec{x_C})\cdot\vec{f_\beta}). \tag{11.42}$$

Here the basic functions $f_\beta(t)=\partial x_C(t)/\partial C_\beta$. For $\beta = 1..m$, they contain only "nonlinear" coefficients. However, additional basic functions with $\beta = m+1..L$ contain both "linear" and "nonlinear" coefficients:

$$f_\beta(t)=\sum_{\gamma=1}^{m} C_\gamma\cdot\frac{\partial f_\gamma(t)}{\partial C_\beta}. \tag{11.43}$$

One may note that the inner part of the matrix $A_{\alpha\beta}$, $\alpha,\beta=1..m$, is the same for both systems (of m and L equations), and $\Delta B_\beta=0$ (within rounding errors) for $\beta=1..m$.

The next step is to add differential corrections ΔC_α to the input values of C_α and replace them: $C_\alpha+\Delta C_\alpha$. These iterations are repeated while all $|\Delta C_\alpha|$ will decrease below some limiting accuracy ε, or the number of iterations will not exceed some limiting value (e.g., 30). If the initial "guess" of the parameters is good, only few iterations are needed to reach the "ε" limit. If not, the Levenberg–Marquardt method is used. It is based on adding to the diagonal elements of the matrix $A_{\alpha\beta}$ values of $\lambda>0$ (Levenberg, 1944) or $\lambda\cdot A_{\alpha\alpha}$ (Marquardt, 1963). That is, only the diagonal elements of the matrix $A_{\alpha\alpha}$ are multiplied by a factor of $(1+\lambda)$, so the modified LS equations

$$\sum_{\alpha=1}^{L}(A_{\alpha\beta}+\lambda\cdot\delta_{\alpha\beta})\cdot\Delta C_\alpha=\Delta B_\beta \tag{11.44}$$

are solved (Marquardt, 1963). After moving of the iterations from the "risk zone" with large λ, its value should be decreased to a final value of zero to allow correct values of the matrices $A_{\alpha\beta}$ and $R_{\alpha\beta}$. These methods are similar to the "Tikhonov regularization" (Tikhonov, 1963). A very important point is to choose a correct initial point, as it may lead to, instead of a global minimum, a local minimum, or even a maximum. If the initial values are close to the solution, the iterations converge rapidly. Sometimes, after an iteration, the value of the test function may become larger. In this case, one may use smaller steps and move to a closer point $C_\alpha+\lambda_{\text{step}}\cdot\Delta C_\alpha$, $0<\lambda_{\text{step}}\le 1$. This will slow down convergence of the iterations, but may make the interval of convergence wider. In a simpler (but generally slower) method of "steepest descent," no inverse matrix is needed, and one may just estimate $\Delta C_\alpha=\Delta B_\alpha/A_{\alpha\alpha}$. Asymptotically, for large $\lambda\gg A_{\alpha\alpha}$ for all α, the direction of the vector of differential corrections changes from that for "differential corrections" to that of the "steepest descent" with $\lambda_{\text{step}}\approx 1/(1+\lambda)\ll 1$.

The Monte Carlo method needs ranges for all available parameters instead of the initial point of iterations. Its realization is the easiest in the computer program, but needs too many test samples N_C to get accuracy δ_C: $N_C\approx\delta_C^{-q/2}$, where q is the number of (naturally, "nonlinear") parameters, for which the Monte Carlo search is applied (Andronov and Tkachenko, 2013).

11.5 NONUNIQUE MINIMUM OF THE TEST FUNCTION

Sometimes the data do not allow to determine all physical parameters needed, as the information is not sufficient. For example, a visible magnitude m is related to the absolute magnitude M and the distance r in parsec: $m=M-5+5\cdot\lg r$. The observational parameter is m, and, from one equation, it is not possible to get two unknown parameters M and r. However, to determine the distance r from other measurements, e.g., parallax from ground-based or space

(HIPPARCOS, GAIA) observatories, one may determine an absolute magnitude separately. Similarly, the half duration of the eclipse is related to radii of both stars (assumed to be spherically symmetrical), the distance between them, and the inclination i (Shul'Berg, 1971; Andronov and Tkachenko, 2013). Thus one may determine less phenomenological parameters than the physical ones. Or, for the same values of the phenomenological parameters, one may get a region of physical parameters, which are thus poorly defined separately.

Another problem may appear if there are few minima of the test function for different sets of parameters. Sometimes, these minima are of comparable depth, and occasionally the deepest minimum may correspond to a wrong set. This is why the results from discovery papers should be checked and corrected with better accuracy, using additional further observations. These simple examples show necessity of complementary methods.

11.5.1 Bootstrap Method

An alternate method for determination of the accuracy of the model parameters is the so-called "bootstrap" method (Efron, 1979; Efron and Tibshirani, 1993; Shao and Tu, 1996). The parameters are determined for the initial data, and used as a solution. Then new datasets are generated using random numbers $j = \text{int}(\text{random} \cdot n) + 1$, where "random" is a pseudorandom number, which is uniformly distributed in a range $(0, 1)$. So some initial data points are missing, and other ones may be used one or a few times.

Andrych et al. (2020) and Andronov and Kulynska (2020) discussed statistical properties of the approximations of the "bootstrap-generated" data sets in more detail.

Generally, the sample distribution may be asymmetrical. The sample mean of a given parameter may differ from the initial value, so one may use, as an accuracy estimate, the r.m.s. deviation of the generated value from the initial one. Sometimes, instead of single σ, there are asymmetrical positive (σ_+) and negative (σ_-) errors corresponding to, e.g., the 95% confidence interval.

This challenges an usual assumption on Gaussian distribution of statistical errors. Moreover, this type of the confidence interval is not consistent with the definition of weights. For the normal distribution, this interval is 1.96 times larger than the standard error. Approximately, for a sample value, one may just divide by this factor. This factor is larger for the Student distribution for a smaller number of degrees of freedom. Another disadvantage of the bootstrap method is that, due to a decrease of the number of different arguments, and hence much larger gaps, the "best" approximation may be unrealistically shifted as compared to that for the real sample.

11.5.2 Determination of Times of Minima/Maxima (ToM)

There is a special kind of analysis of the period and its possible changes, based on the "times of minima/maxima" (ToM) (AAVSO) or the "moments of characteristic events" (Tsesevich, 1973; Dumont et al., 1978; Andronov, 1988). Then, from many "near-extremum" observations, the only information that is extracted is the moment of time, which corresponds to a minimum or maximum of the approximation. Some of the methods were discussed by Andronov (2005).

For the moment of extremum t_e (maximum or minimum, what is commonly used to compile international databases) is determined as the root of equation $\dot{x}_C(t_e) = 0$ and has a standard error of

$$\sigma[t_e] = \frac{\sigma[\dot{x}_C(t_e)]}{|\ddot{x}_C(t_e)|}. \tag{11.45}$$

For example, for a simplest parabolic fit

$$x_C[t] = C_1 + C_2 \cdot (t - t_0) + C_3 \cdot (t - t_0)^2, \tag{11.46}$$

$t_e = t_0 - C_2/(2C_3)$ and

$$\sigma^2[t_e] = \frac{R_{22}C_3^2 - 2R_{23}C_2C_3 + R_{33}C_2^2}{4C_3^4}. \tag{11.47}$$

For the polynomial of arbitrary order $s = m - 1$

$$x_C[t] = C_1 + C_2 \cdot (t - t_0) + C_3 \cdot (t - t_0)^2 + \ldots + C_m \cdot (t - t_0)^{m-1}, \tag{11.48}$$

the position of the extremum is determined numerically by solving the equation $\dot{x}_C[t_e] = 0$ using the Newton–Raphson method of iterations $t_e := t_e + \delta t_e$, where $\delta t_e = -\dot{x}_C[t_e]/\ddot{x}_C[t_e]$, until $|\delta t_e| \le \epsilon$, where ϵ may be set to the desired accuracy. In practice, this may be a computer accuracy, when $t_e + \delta t_e - t_e$ is equal to zero because of the rounding errors. The starting point is determined on a regular grid within the given data interval. The type of the extremum corresponds to that from the parabolic approximation ($s = 2$).

The statistical error is estimated similarly to Eq. (11.31):

$$\sigma[t_e] = \frac{\sigma[\dot{x}_C(t_e)]}{|\ddot{x}_C(t_e)|}. \tag{11.49}$$

As the polynomials are most common functions used for approximations (e.g., in the electronic tables),

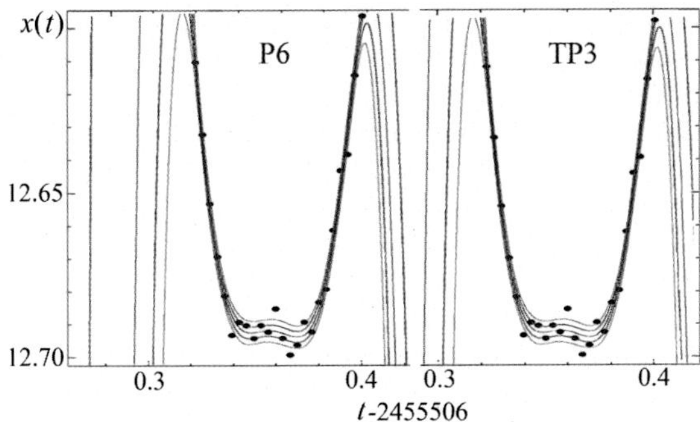

FIG. 11.5 Approximations of the eclipse of the binary system GSC 3692-00624 = 2MASS J01560160+5744488 using a short data sample published by Devlen (2015). *Left:* Algebraic polynomial of degree $m-1=6$. *Right:* Trigonometric polynomial of degree $(m-1)/2=3$. The $\pm 1\sigma$ and $\pm 2\sigma$ "error corridors" are shown.

they were often used for the ToM determination. For example, the catalogue of 6509 extrema of semiregular variables (Chinarova and Andronov, 2000) was used using this method. The number of the parameters $m = s+1$ was determined separately for each interval of data near extremum. The same algorithm was implemented in other software (Breus, 2007; Andrych et al., 2015). In practice, we test orders s up to eight only. In the popular commercial software PERANSO (Paunzen and Vanmunster, 2016), the degree of the polynomial s is user-defined, so the "esthetic" method is applied.

In Fig. 11.5, the approximations of the short observational run near the minimum are shown, using the same number of parameters $m=7$ for the sixth-order algebraic polynomial and the third-order trigonometrical polynomial. Remarkable coincidence of both approximations in the interval of observations is observed, but a drastic difference outside the interval. These "common" approximations show a bad approximation at the bottom of eclipse, where a statistical error of one point "makes" the minimum "split."

To avoid this physically unreal behavior, we have proposed other phenomenological approximations, which were applied to the same data by Andrych et al. (2017). They are shortly described below.

For theoretically symmetrical signals (e.g., eclipses), it is natural to use symmetrical functions, e.g., polynomials of order $s=2(m-1)$, i.e.,

$$x_C[t] = C_1 + C_2 \cdot (t-t_e)^2 + C_3 \cdot (t-t_e)^4 + \ldots + C_m \cdot (t-t_0)^{2(m-1)}. \tag{11.50}$$

Here $t_e = C_{m+1} = C_L$ is the symmetry point which corresponds to the extremum and is determined numerically using the Newton–Raphson method. Obviously, for $m=3$, an either "ordinary" or "symmetrical" polynomial is a parabola, and there is a difference in the form of two functions, but not their values or properties. For polynomials of higher orders, the absence of terms with odd power indices keeps the function symmetrical.

Even for a "complete" polynomial with both odd and even power indices, the form

$$x_C[t] = C_1 + C_2 \cdot (t-t_e)^2 + C_3 \cdot (t-t_e)^3 + \ldots + C_m \cdot (t-t_0)^m \tag{11.51}$$

is preferable, as t_e is the moment of extremum, even if there may be up to $(m-1)$ real roots of the equation $\dot{x}_C[t_e]=0$.

Although polynomials are commonly used for approximations, they are not the best functions because of the Gibbs effect and apparent waves at the approximation. Smaller amplitudes of waves show cubic splines (Andronov, 1987). Marsakova and Andronov (1996) have proposed a spline with varying degree and subintervals - the "asymptotic parabola," which consists of two straight lines connected with a parabola. This is effective for asymmetric extrema, e.g., in the light curves of pulsating variables of the types RR Lyrae, δ Cephei, and Mira et al.

It is suitable to write the approximation in the form

$$x_C[t] = C_1 + C_2 \cdot G((t-C_3), C_4, \ldots, C_{mp}) \tag{11.52}$$

to determine two "linear" parameters C_1, C_2 and $L-2$ "nonlinear" parameters $C_3, \ldots, C_L$. If choosing the functions G in such a way that $G=0$ at $t=C_3$, the explanation of the "linear" parameters is simple: C_1 is the smoothed signal value at the extremum. With an additional condition $G \to 1$ far away from the extremum, C_2 is equal to the amplitude, i.e., the difference between the "extremal" and "quiet" values of the approximation.

If the "quiet" part is present, one may use another definition, $x_C[t] = \tilde{C}_1 + \tilde{C}_2 \cdot \tilde{G}((t-C_3), C_4, \ldots, C_{mp})$, with obvious relations between the coefficients $\tilde{C}_1 = C_1 - C_2$, $\tilde{C}_2 = -C_2$ with the same values of other parameters.

For the typical near-extremum observations, there is no "quiet" part, so C_2 may be unrealistically large, even formally "infinite."

In this case, one should use a "restricted" model - e.g. setting at least one of the "nonlinear" parameters to some limiting value and recomputing the fit for a smaller number of parameters L. In some cases, the "inner interval" is a wide as the observations. Then the model simplifies to a "singular interval" one, and may be an ordinary parabola, as, e.g. in the method of "asymptotic parabola" (Andrych et al., 2015) and "parabolic spline" (Andrych et al., 2020).

In some cases, we can just use power series (sometimes even noninteger power indices; e.g., Andronov et al., 2017b).

For the "global" approximation with a single analytical function, Andronov (2005) proposed

$$\tilde{G}((t-C_3),C_4,C_5) = \frac{2}{\exp(C_4\cdot(t-C_3))+\exp(-C_5\cdot(t-C_3))}. \quad (11.53)$$

This an extension of the classical hyperbolic secant function $\mathrm{sech}(z)=2/(e^z+e^{-z})$ for the case of asymmetrical ascending and descending branches. The connection to physics of the process is that $\tau_- = 1/C_4$ and $\tau_+ = 1/C_5$ are characteristic times of exponential rise/decay at the beginning/end. The accuracy estimate is $\sigma[\tau_-] = \sigma[C_4]/C_4^2$ and, similarly, $\sigma[\tau_+]=\sigma[C_5]/C_5^2$. The position of the extremum is shifted; we now have

$$t_e = C_3 + \ln(C_5/C_4)/(C_4+C_5). \quad (11.54)$$

The accuracy of this function of three coefficients C_3, C_4, and C_5 is estimated using Eq. (11.25).

The following analytical approximation, which is similar to a probability distribution, was proposed by Bódi et al. (2016):

$$\tilde{G}((t-C_3),C_4,C_5) = \exp(-\ln 2\cdot C_5\cdot(\ln(C_4\cdot(t-C_3)+1))^2). \quad (11.55)$$

These two functions are time consuming, because of numerous computations of exponents and logarithms during a "brute force" determination of three parameters before using the differential corrections. Also, during the iterations, the values should be checked for being in a reasonable interval. For example, for the latter function, $C_4 \to 0$ for an exactly symmetrical signal, which causes $C_2\cdot C_5 \to \infty$. In this case, the function should be changed to a symmetrical polynomial, or to a Gaussian

$$\tilde{G}((t-C_3),C_4) = \exp(-C_4\cdot(t-C_3)^2). \quad (11.56)$$

This method is also widely applied, as some software uses it to fit spectral lines.

However, for symmetric signals similar to the eclipses of the eclipsing binaries, it is useless, as the eclipses are of finite length. Moreover, the eclipse duration is an optional parameter to be listed in the "General Catalogue of Variable Stars" (Samus et al., 2017). Thus one should use approximations of a finite length. Andronov (2012) proposed a "New Algol Variable" (NAV) function

$$\tilde{G}((t-C_3),C_4) = (1-(|t-C_3|/C_4)^{C_5})^{1.5}, \quad (11.57)$$

where the parameter C_5 determines the "flatness" of the shape near the mid-eclipse, i.e., $C_5=1$ corresponds to a lower physical limit (when the stars have the same size, and the total eclipse is very short as compared to the eclipse duration); $C_5=2$ corresponds to the "classical" mathematical function, which has a nonzero second derivative at the extremum. With increasing C_5, the shape becomes flatter, and very large values may correspond to short ascending (or descending) branch, which is typical for exoplanet transits. The power 1.5 asymptotically describes the shape of the eclipse of spherical (or even ellipsoidal) stars close to the outer contact. Added to a trigonometrical polynomial, the NAV function is effective not only for the Algols, but also for EB and EW - eclipsing systems with more smooth variations than in EA (Tkachenko et al., 2016).

Mikulášek (2015) introduced few functions improving the Gaussian. At first, the simple parabola in the argument was replaced by a hyperbolic cosine, i.e.,

$$\tilde{G}((t-C_3),C_4) = \exp(1-\cosh((t-C_3)/C_4). \quad (11.58)$$

This function tends to zero faster than a Gaussian, but still has a parabolic shape close to mid-eclipse. Thus also a "noninteger" power shape was introduced, i.e.,

$$\tilde{G}((t-C_3),C_4) = 1-(1-\exp(1-\cosh((t-C_3)/C_4)))^{C_5}. \quad (11.59)$$

Andrych et al. (2017) introduced a series of symmetrical functions, where the interval of observations is split into three subintervals (physically, the beginning/ending and middle branches are described using different formulas). They called these functions "wall-supported" (WS) ones. The WS parabola seems the best for describing the exoplanet transits. The WS line is good for total eclipses of stars of comparable sizes. WS "asymptotic parabola" is good for the intermediate cases. For an extreme case of very wide interval beginning close to the previous extremum and ending close to the next one, we have introduced the quadratic spline. In all these functions, the positions of the points splitting the interval are "nonlinear" parameters, which are determined to get the best fit.

For wider intervals, which contain completely the ascending and descending branches, Andronov et al. (2017b) tested almost a half hundred modifications of the shapes ("patterns").

Obviously, a wide variety of functions needs numerical criteria to choose the best one. Currently, all these 21 methods are implemented in the software MAVKA, which has an option for automatic determination of the method with best accuracy estimate (Andrych and Andronov, 2019) from a list of chosen function(s).

11.6 PERIODOGRAM ANALYSIS: PARAMETRIC VERSUS NONPARAMETRIC METHODS

11.6.1 From Time to Phase

The truly periodic signal satisfies the condition $x(t + k \cdot P) = x(t)$ for any integer k, and P is the period. According to this definition, the values $2P$, $3P$, etc., are also "periods." So it is usually adopted to use the minimal positive value as the "period."

In some cases, the physical period is $2P$, e.g., for eclipsing binary stars of the EW-type or for the elliptical variables. Their light curves have two similar waves, which are mirror (reflection) symmetrical in respect to the primary (deeper) or secondary minimum. The difference between these "reflected" parts is usually within observational errors. In this case, the period P is called the "photometric" period, or the "formal" period, while $2P$ is called the "true," "physical," or (for binary stars) the "orbital" period.

The main idea for the periodic functions is to "pack" all the data into a single interval. Typically, the time is thus shifted by an integer number of cycles, so an age may be introduced, $\tau_P = t - T_0 - P \cdot E$, where T_0 is called the initial epoch and $E = \mathrm{INT}((t - T_0)/P)$ is the cycle number. Here the function INT is defined as the largest integer which does not exceed unity. For example, int(−2.7) = −3. In this case, $E = -3$, $\phi = +0.3$. In some computer languages, the value of this function is set to −2, so a correction is needed. This should be checked in concrete programming environments.

The "age" is typical for an everyday life situation, e.g., time measured in 12- or 24-hour format (or a decimal part of the Julian Day). It is measured in units of time. To scale the signals with different periods, the dimensionless "phase" $\phi = \tau_P/P$. So $t = T_0 + P \cdot E + \tau_P$, $= T_0 + P \cdot (E + \phi)$.

The astronomical definition of the phase is different from an usual definition in physics and mathematics $\varphi = 2\pi\phi$, i.e., in radians.

So $0 \le \tau_P < P$, $0 \le \phi < 1$. However, this range may be extended for some methods and for better illustration. For example, the phase $\phi = 0.99$ is the same as $\phi = -0.01$. In Fig. 11.6, the computed nonsinusoidal signal is shown for a random distribution of time. The points should be shifted by an integer number of periods to "be moved" to the main interval.

11.6.2 "Parametric" ("Point-Curve") Methods

The periodogram analysis is based on estimate of the "quality" of the phase light curve by some parameter, which is called "the test function." This test function is computed for different trial periods P or, alternately, frequencies $f = 1/P$. Traditionally, the periodogram analysis is divided into two large groups, which are called parametric (or "point-curve") or nonparametric ("point-point"). In the first group, the phase curve is compared to some approximation (smoothing curve). The test functions are used similarly to Φ in Eq. (11.1), so the position of its minimum (as a function of the period P, rarely also of the initial epoch T_0) is to be found. However, in many cases, the amplitude of the signal is small as compared to the noise, so the relative amplitude of the test function is not too large, so it is far enough from the zero level. So it may be suitable to introduce the test function

$$S(f) = r^2 = 1 - \Phi_{m+q+1}/\Phi_{q+1}. \tag{11.60}$$

Typically, the value $q = 0$ is used, i.e., the approximation by a constant. However, this definition of the periodogram was used by Andronov and Baklanov (2004) for the periodogram analysis with a trend approximated by the polynomial of order q.

The trigonometric polynomial is most often used for the periodogram analysis. However, other periodic functions may be used, e.g., "piecewise constant" splines (Jurkevich, 1971). This method was improved by Stellingwerf (1978), who partially removed the dependence of the test function on the initial epoch. This algorithm is called "phase dispersion minimization" (PDM) and is often used. The test functions are dependent on the number of intervals (bins) m, which is a free parameter. We recommend to use at least $m \ge 3$ for expected near-sinusoidal variability and $m \ge 5$ for double-peaked curves of EW-type stars. However, for light curves with sharp parts (eclipsing binaries, RR Lyrae-type stars), the number m should be increased so that at least two subintervals cover such sharp parts. Generally, according to the Sturges (1926) rule, the number of bins should be $m \approx 1 + 3.32 \cdot \lg n \approx 1 + \log_2 n$.

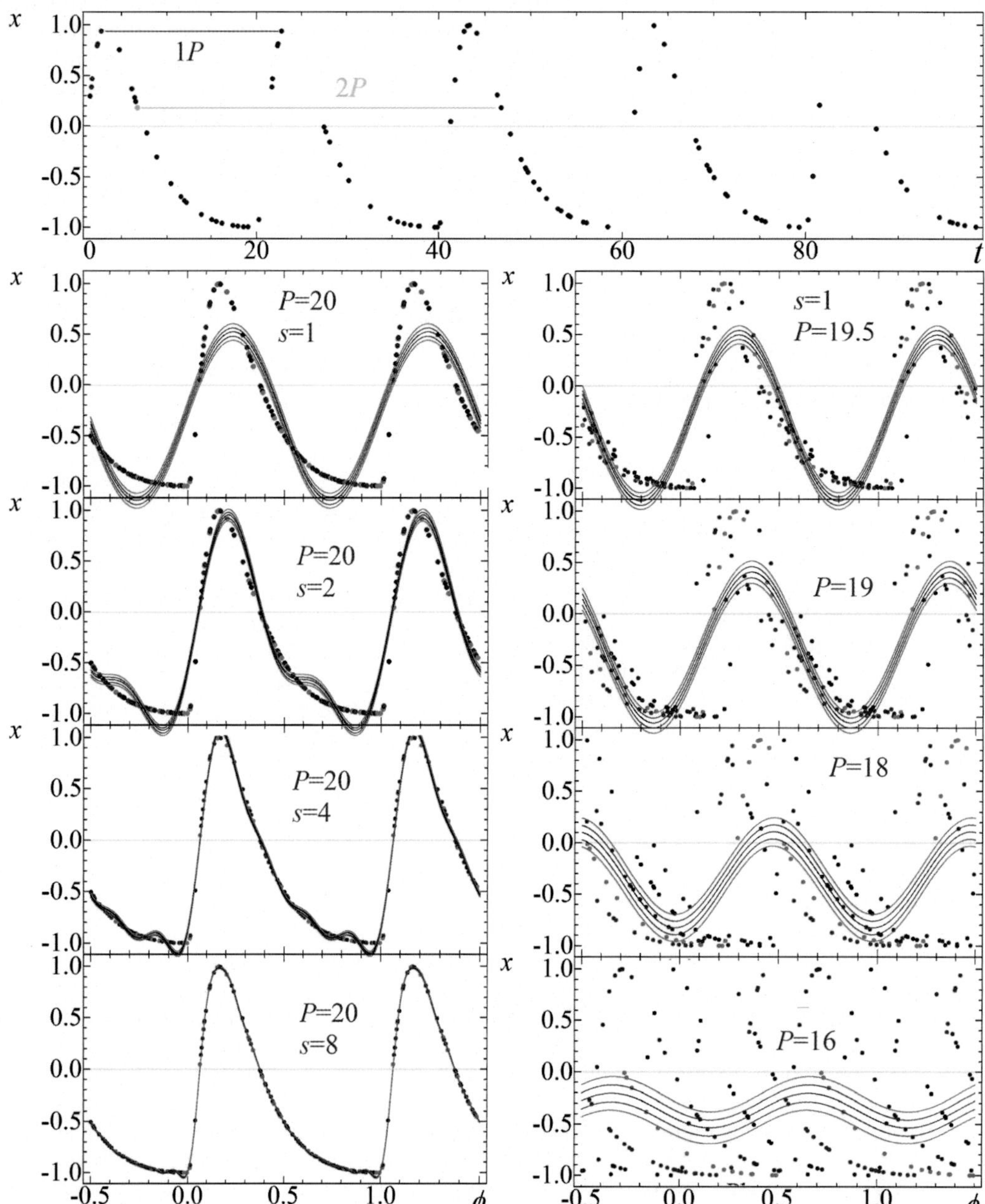

FIG. 11.6 *Top:* The simulated truly periodic signal with a period $P = 20$ with a highly asymmetrical shape. Such shape is similar to periodic flashes (or, in astronomy, to RR Lyr-type stars). The points are shifted by an integer number of periods P inside the preferred interval. *Left:* Trigonometric polynomial approximations of orders $s = 1, 2, 4, 8$ for the true period. *Right:* Phase curves for trial periods, which deviate from the true period, with a cosine (TP1) approximation. Points are shown in different colors, to see that different "seasons" show good curves, but they are shifted between each other. The $\pm 1\sigma$ and $\pm 2\sigma$ "error corridors" are shown. The approximations were made using MCV software (Andronov and Baklanov, 2004).

Andronov (1987) proposed to use more smooth cubic splines, and, additionally, to remove dependence on T_0 either by smoothing the cubic splines, or to determine the best phase shift corresponding to a minimum of the test function.

11.6.3 "Nonparametric" ("Point-Point") Methods

In another group of methods, which do not assume the basic functions, the distance between the points (different in different methods) is taken into account. The most famous method was published by Lafler and Kinman (1965), typically abbreviated to LK.

The test function is

$$\Theta = C \sum_{k=1}^{n} \theta(\Delta x_k, \Delta\phi_k), \quad (11.61)$$

where $\Delta x_k = x_k - x_{k-1}$, $\Delta\phi_k = \phi_{k-1}$, and, "formally," $x_0 = x_n$, $\phi_0 = \phi_n - 1$. Its minimum corresponds to the best period. Here the data are sorted according to the phases ϕ_k for each trial period. The summand $\theta(\Delta x_k, \Delta\phi_k)$ is some kind of "distance" between the two points, which are subsequent in phase. The scaling parameter $C > 0$ is arbitrary. It may be set, e.g., to unity, or to make the mathematical expectation for a pure noise to 1, 2, or any positive value.

In the original LK method, $\theta(\Delta x, \Delta\phi) = (\Delta x)^2$, and the scaling factor $C = \sigma_x^{-2}/n$ is inversely proportional to the variance of the data σ_x^2.

There were numerous modifications of the method, e.g., $\theta = |\Delta x|$ (see the appendix by Deeming to Bopp et al., 1970), $|\Delta x|^\gamma$ (Pelt, 1975), $(\Delta x)^2/((\Delta\phi)^2 + \epsilon^2)$, $|\Delta x|/(\Delta\phi + \epsilon)$ (Renson, 1978), $\sqrt{((\Delta x)/\epsilon)^2 + (\Delta\phi)^2}$ (Dworetsky, 1983), etc.

The comparative study of these methods was presented by Andronov and Chinarova (1997) with recommendations on values of ϵ.

Pelt (1975) proposed a more general relation, taking into account the distance measure not between the pairs of subsequent in phase points, but with neighbors in some interval of phases, i.e.,

$$\Theta = C \sum_{k=1}^{n} \sum_{j=1}^{k-1} (x_k - x_j)^2 \cdot \tilde{\theta}(|\phi_k - \phi_j|), \quad (11.62)$$

where the simplest weight function $\tilde{\theta} = 1$ if $|\phi_k - \phi_j| \le \Delta\phi_{max}/2 << 1/2$.

The full width of this interval $\Delta\phi_{max}$ is an additional free parameter. If it is narrow, the fluctuations of the periodogram are large.

There is no single best method among these; otherwise the others could "go to history." Some modifications called "string/rope length methods" were discussed by Clarke (2002). In practice, the LK method is the most popular one among "point-point" the ones, and it has been implemented in many computer programs, e.g., VSCalc (Breus, 2007), Peranso (Paunzen and Vanmunster, 2016), etc. According to the ADS, this paper has been cited 652 times already in 2019.

Periodograms are discontinuous, so it is often possible to get a local minimum, which is shifted from the "true" position.

The periodograms have the deepest minimum at the main period, whereas there are minima at $2P$, $3P$, kP, etc. The depth gradually decreases with the multiplier k, as the same number of points are distributed less densely at the main wave, so the systematic differences between the points become relatively large. The mathematical expectation of the normalized function Θ for a pure noise (no signal) in the LK method is 2.

For a better apparent contrast of the minimum of the test function Θ to the "noisy continuum" at the periodogram, it should be recommended to use $\lg\Theta$ instead of Θ itself.

This behavior is opposite to that of the one based at the sinusoidal approximation, where the possible peaks appear at multiple frequencies, rather than periods. The width of the peaks is nearly constant.

The periodogram for the two-point distance is discontinuous. So the minimal value of the test function on a grid of frequencies or period s may be shifted from an expected one, even if the signal is an accurate periodic function without noise.

The recommended frequency step is $\Delta f = \Delta\phi/(t_n - t_1)$, where $\Delta\phi$ is the change of the phase difference between the first and the last observations.

For the sinusoidal signals, a value of $\Delta\phi$ between 0.04 and 0.06 is recommended, but not worse than 0.1. For the trigonometric polynomial of order s, the recommendation is $\Delta f = \Delta\phi/(t_n - t_1)/s$. In the FT (see next section), $\Delta\phi = 1 - 1/n$. But there the period is exactly fixed to the duration of the observations $n\delta = (t_n - t_1)/(n-1) \cdot n = jP$. The periodograms of different kind are shown in Fig. 11.7.

11.7 WHAT IS THE "ORTHODOX" FOURIER TRANSFORM FOR DISCRETE DATA?

The FT is one of the most popular methods for data analysis, as well as for solving tasks in mathematical physics and other directions of mathematics.

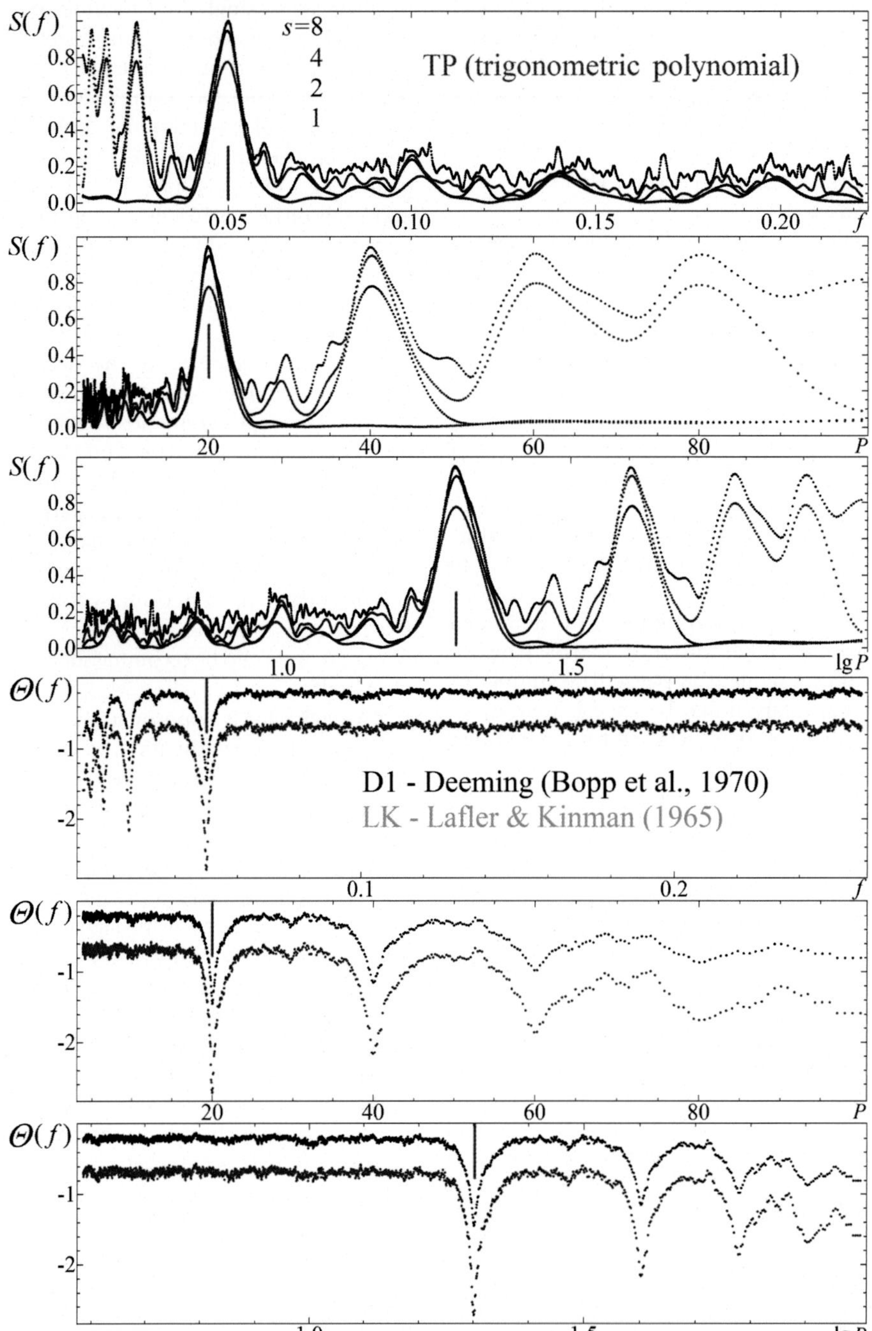

FIG. 11.7 *Top:* The simulated truly periodic signal with a period $P = 20$ with a highly asymmetrical shape. Such a shape is similar to periodic flashes (or, in astronomy, to RR Lyr-type stars). Periodograms for "parametric" or "nonparametric" are computed using the TP, LK, and D1 methods and are represented as functions of $f = 1/P$, P, and $\lg P$.

A Google search shows almost 38 million links to publications for the "Fourier Transform," among them thousands of books containing the description (e.g., Anderson, 1958; Bendat and Piersol, 2010; Press et al., 2007). Excellent reviews on FT and the discrete Fourier transform (DFT) are posted on Wikipedia. There are many versions with different designations.

The original work was published by Fourier (1822) almost two centuries ago. One of the classical monographs dedicated solely to FT was published by Tolstov (2012).

The classical FT is typically defined as

$$\hat{x}(f)=\int_{-\infty}^{+\infty} x(t)\cdot e^{-i2\pi ft}dt, \qquad (11.63)$$

with an inverse transform

$$x(t)=\int_{-\infty}^{+\infty} \hat{x}(f)\cdot e^{i2\pi ft}df. \qquad (11.64)$$

Here t is time and $f=1/P$ is frequency, P is period, and $i^2=-1$. As the integrals should be limited, the integral of the function $|x(t)|$ should exist.

The FT is a very powerful tool used in numerous analytical studies, e.g., mathematical physics, statistics, etc. In reality, it has strong limitations due to the absence of infinite information from the observed signals. The most common adaptation of the method to discrete data may be called DFT, which also has some versions.

It is assumed that the signal is defined at a set of discrete points $t_k=t_0+k\delta$, $k=0..n_1$, where $n_1=n-1$. The inverse FT becomes a sum:

$$\begin{aligned} x_c(t)&=C_1+\sum_{j=1}^{s}(C_{2j}\cos(2\pi jft)+C_{2j+1}\sin(2\pi jft))\\ &=C_1+\sum_{j=1}^{s}R_j\cdot\cos(2\pi jf(t-T_{0j})), \end{aligned} \qquad (11.65)$$

which is called the "trigonometrical polynomial" of order s.

The relations are listed as

$$\begin{aligned} C_{2j}&=R_j\cdot\cos(2\pi jfT_{0j}),\\ C_{2j+1}&=R_j\cdot\sin(2\pi jfT_{0j}),\\ R_j&=\sqrt{C_{2j}^2+C_{2j+1}^2},\\ T_{0j}&=\mathrm{atan2}(C_{2j+1},C_{2j}/(2\pi)+k)\cdot P/j, \end{aligned} \qquad (11.66)$$

where R_j is called "semiamplitude," i.e., the difference between the maximum deviation of the wave from its mean value, so $2\cdot R_j$ is a full amplitude between the maximum and minimum of the corresponding wave, and T_{0j} is called the initial epoch (moment of time, which corresponds to the maximum of the wave).

One may choose any integer value of k in this equation – e.g., if we say that the Sun is highest close to noon, "noon" is the initial epoch. Of which day? Of each. Many authors use the earliest initial epoch occurring during the observations. However, the best results for the matrix of normal equations will be obtained choosing T_{0j} closest to the sample (weighted) mean value of times of the observations.

It is also important to note that the brightness in astronomy is measured in "stellar magnitudes," so the minimum of brightness corresponds to a minimum of the stellar magnitude. This should be clearly written in the papers to avoid misinterpretation by other authors.

Terminologically, the wave j is called the $(j-1)$-th harmonic of the main period ($j=1$). However, some authors use the j-th wave as the j-th harmonic. It is some type of scientific slang, as $j=1$ is a main wave, and not its harmonic. This also could lead to misinterpretation.

Here it is suggested that the signal repeats from $-\infty$ to $+\infty$ with a period $P=n\delta/j$. Obviously, it is not possible to determine more parameters than there are observations, thus the number of frequencies is limited to $s_{max}=\mathrm{int}(n/2)$. The set of the basic functions is $f_1=1$, $f_{2j}=\cos(2\pi jft)$, $f_{2j+1}=\sin(2\pi jft)$.

The set of frequencies for the classical FT is discrete ($f_j=2\pi j/n$), and the coefficients may be easily determined as

$$\begin{aligned} C_1&=\frac{1}{n}\sum_{k=1}^{n}x_k,\\ C_{2j}&=\frac{2}{n}\sum_{k=1}^{n}x_k\cos(2\pi kj/n),\\ C_{2j+1}&=\frac{2}{n}\sum_{k=1}^{n}x_k\sin(2\pi kj/n). \end{aligned} \qquad (11.67)$$

If n is even, then $j_{max}=s=n/2$, and then the coefficient C_n should be twice smaller than in the equation above, and C_{n+1} is not used, or is set to zero.

An example of inadequate use of the FT is shown in Fig. 11.8. The sums Eq. (11.65) converge to a function, which has a discontinuity. So the point at the border is a mean from the values left and right from this point. There are apparent waves (the Gibbs phenomenon) of

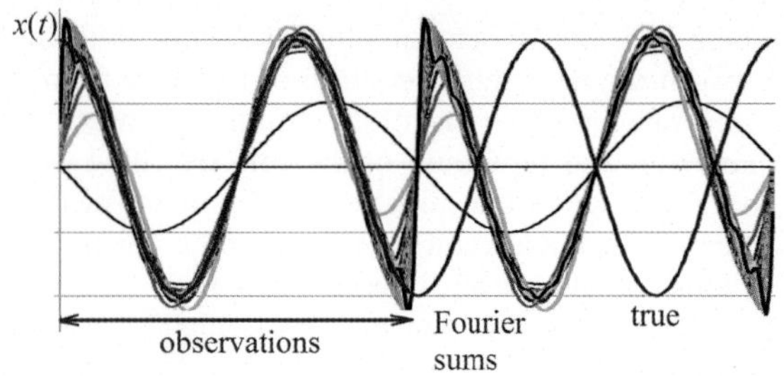

FIG. 11.8 Approximations of a cosine function "observed" during $0.75P$ by Fourier sums (trigonometrical polynomials). For the discrete Fourier transform (DFT), the "period" is assigned to the duration of the data $n\delta$, and not to the true period. The Fourier sums smooth jumps, the slope of which increases with increasing degree of the trigonometrical polynomial s.

no physical meaning, and an abrupt switch between the levels of discontinuous function.

This problem does not arise, if to determine a statistically optimal period from the data, not from the duration of observations.

For irregularly spaced data, the statistically correct decision is to use LS. However, there are some methods, the authors of which still call "Fourier transform," even though the conditions are not satisfied.

The simplest formula was used by Deeming (1975). He just used

$$C_1 = \bar{x},$$
$$C_2 = \frac{2}{n}\sum_{k=1}^{n} x_k \cos(2\pi f t_k) = C(f), \qquad (11.68)$$
$$C_3 = \frac{2}{n}\sum_{k=1}^{n} x_k \sin(2\pi f t_k) = S(f)$$

as the approximation

$$x_C(t) = C_1 + C_2 \cos(2\pi f t) + C_3 \sin(2\pi f t). \qquad (11.69)$$

This coincides with the Fourier coefficients only under conditions of orthogonality of the basic functions, which is generally not fulfilled. Moreover, the coefficients C_2 and C_3 are dependent on the zero point, so later it was recommended to use $(x_k - \bar{x})$ instead of x_k in the corresponding equations.

Lomb (1976) proposed a partial improvement, making a model as in equation (11.69), but fixing $C_1 = \bar{x}$, and using C_2 and C_3 from an LS approximation with $m = 2$ parameters. Scargle (1982) got the same periodogram, but shifting the initial phase for each trial frequency to make orthogonal the basic functions $\vec{f}_2$ and $\vec{f}_3$. He also studied statistical properties of such a periodogram if the signal is pure noise and the accuracy of the data is known. The method was further referred to as the "Lomb–Scargle" method and is the most popular one, having 3851 citations in the papers listed in the ADS in 2019.

The problem in this method is neglecting nonorthogonality of the basic function $\vec{f}_1$ with two others. It is not so important if the observations cover the phase more or less homogeneously. However, many stars have periods, which are ~ 2 times longer than the typical duration of observations during the night.

In Fig. 11.9, there are examples of approximation of short runs using this most popular method of period search. For the near-extremum symmetrical data, the apparent period is equal to the duration of the data (in this sample, $0.5P$). For the descending branch, if we wish to remove the linear trend, the periodogram shows a peak at $0.36P$. At the same time, the complete three-parameter LS fit (Andronov, 1994a) produces an exact approximation, so a correct value of the period,

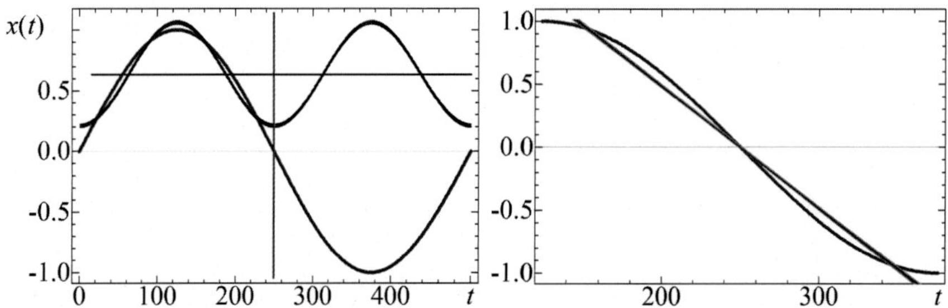

FIG. 11.9 Approximations of a sine function "observed" during $0.5P$ by the Lomb–Scargle method. *Left:* Approximation using the first half of the period. Because the mean value is shifted, the formal period in this method is $0.5P$, and the approximation has two apparent waves during one true period. *Right:* The middle half and the linear trend. The detrended data show an apparent period of $0.36P$.

amplitude, phase, and a zero point. The "sparing" of computer time was important in 1976, but now the computers are fast enough to determine parameters of complete models.

From the non-LS methods, there are "CLEAN" (Roberts et al., 1987) and "CLEANEST" (Foster, 1995). The similarity law in spectral estimation of time series was analyzed by Terebizh (1998).

11.7.1 LS-Based DFT

For a connection between the continuous and discrete FTs, one may refer to Wehlau and Leung (1964), where the discrete signal may be written as a multiplication of functions describing different types of irregularities of the distribution of the arguments of the signal. Among them, there is a smoothing of the data during the exposure time, distribution of data in time.

As the FT of the multiplication of functions is a convolution of their FTs, in the resulting spectrum, there will be biases, i.e., the peaks at the beat frequencies. This is clearly seen in ground-based photometrical surveys. The observations are carried out during a few months in the season. Thus the interval of phases slightly shifts from season to season. An apparent light curve shows an apparent longer period, which does not exist in a reality. Similarly, there may be problems if there are larger gaps between the nights even if a complete light curve is observed.

An example is shown in Fig. 11.10. The model data are a pure sinusoid with a true period ($P_0 = 1.1$), with a period of sampling $P_s = 1$. In the top left figure, the data show a larger "beat" period $P_{beat} = P_s \cdot P_0/|P - P_s|$ (11, in our example) is present. For this data sample, the periodogram is the dependence of the test function $S(f) = S(|f + j \cdot f_s|)$ on trial frequency, and thus shows equal peaks, which correspond not only to the correct frequency $f_0 (= 1/P_0 = 0.90909)$, but for *any* frequency $f = |f_0 + j \cdot f_s|$, where j is any integer. Moreover, the periodogram is reflection symmetrical around frequencies $j \cdot f_s/2$. Thus it is recommended to use frequencies in the range $(0, f_s/2]$ if the data are equidistant. The value $f_s/2$ is called the "Nyquist" frequency f_N. In other words, one should have at least two points per period. This "main range" of frequencies does not prevent the periods to be shorter (as in this case). However, one may not distinguish between the peaks of equal height to choose the correct period.

The situation becomes better if there are observations shifted from the "main periodicity." In Fig. 11.10, the "complete" dataset is with the time interval $\delta = 0.1$, so 10 points per unit time interval. The intermediate values of subsequent points are 2 and 5. It is clearly seen that the height of the "bias" peaks strongly decreases with increasing number of points per P_s. Obviously, "the best" is the case of "no gaps." Also, we have compared the periodograms for the same *number of data per P_s* for equidistant and "random" distribution of the arguments. The random distribution shows very low peaks as compared to the main peak with height $S(f_0) = 1 = 100\%$.

In these samples, we illustrated the influence of the distribution of data in time, with an exact sinusoid in the signal. Naturally, the observational noise of the signal x_k will add noise and spurious peaks at the periodogram.

To describe the peaks, sometimes it is recommended to show a "spectral window" of the observations. The spectral window may be defined as a complex function

$$\begin{aligned} W(f) &= \frac{1}{n}\sum_{k=1}^{n} e^{2\pi i f t_k} \\ &= \frac{1}{n}\sum_{k=1}^{n} (\cos(2\pi f t_k) + i \cdot \sin(2\pi f t_k)) = \tilde{C} + i\tilde{S}. \end{aligned} \tag{11.70}$$

For equidistant points $t_k = t_0 + k \cdot \delta$,

$$W(f) = e^{i\pi \cdot f \cdot (2t_0 + (n-1)\delta)} \cdot \frac{\sin(\pi f \delta n)}{n \sin(\pi f \delta)}. \tag{11.71}$$

The first multiplier has an unit absolute value, and the function

$$\tilde{W}(f) = \frac{\sin(\pi f \delta n)}{n \sin(\pi f \delta)} \tag{11.72}$$

is real, symmetric, and periodic with a period $1/\delta$. Some properties are $\tilde{W}(j/\delta) = 1$, $\tilde{W}(j/(n\delta)) = 0$ for noninteger ratios j/n. However, the values between these points are nonzero, biasing the periodogram.

For nonequidistant points, C and S do not cross zero at the same frequency, but one may determine the smallest value of f, which corresponds to the minimum of $\tilde{W}(f)| = (C^2 + S^2)^{1/2}$. This value may be named Δ_{f_0} and may be used to estimate the "effective number of frequencies" N_{eff} in the periodogram analysis to estimate a false alarm probability (FAP) of a given peak (see Andronov, 1994a for more details).

To decrease the height of the peaks, it should be recommended, whenever possible, to break periodicity of the signal. For ground-based observations, it may be recommended to use telescopes at different longitudes.

The width of the peaks in the periodogram is constant for frequency and is proportional to P^2 for each peak, including the aliases!

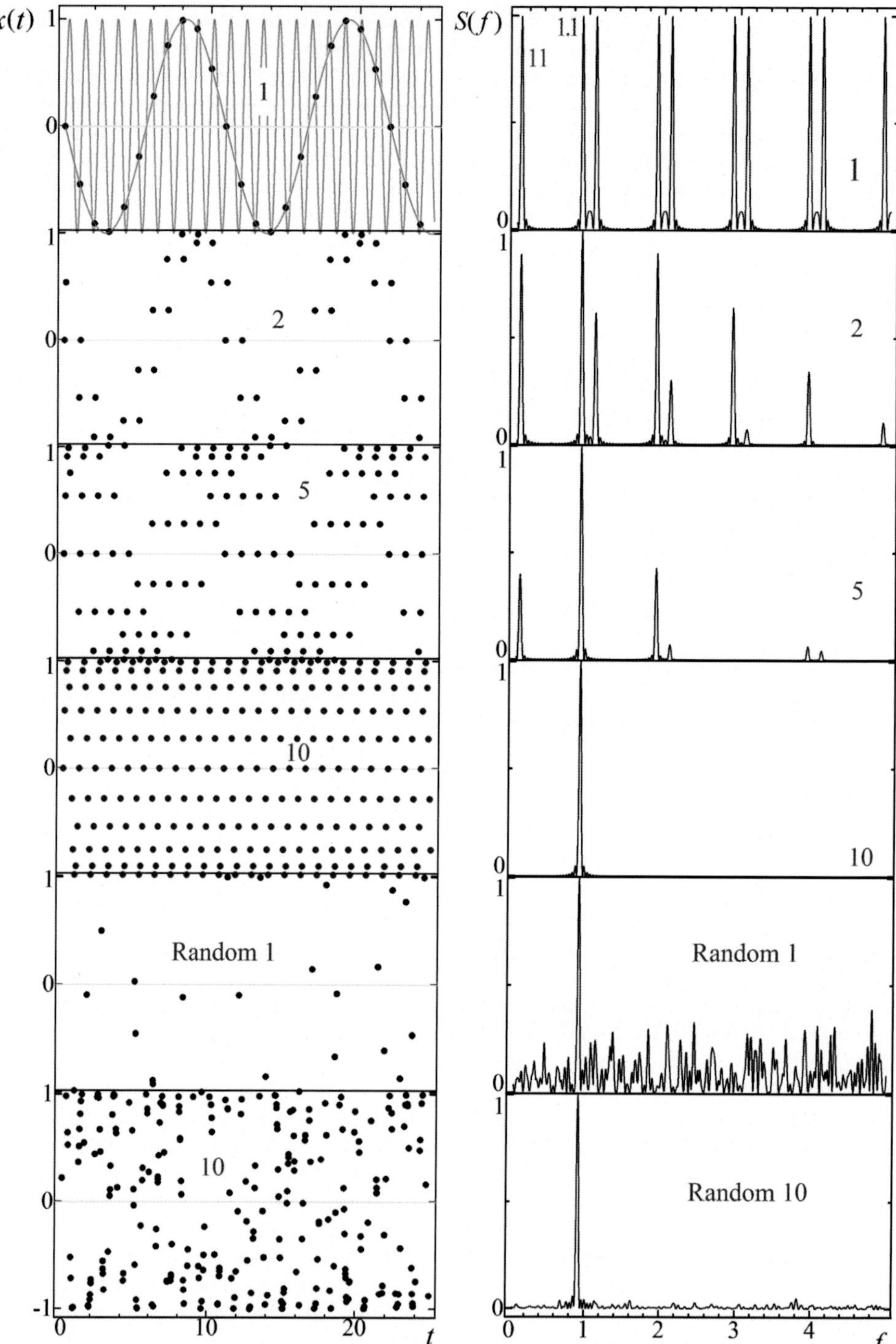

FIG. 11.10 *Left:* The simulated sinusoidal signal with a period $P = 1.1$ with $(1, 2, 5, 10)$ sampling points per unit time. At the top figure, the sampling is 1, for lower figures, the points are added every 0.1. *Right:* The periodograms are computed using the TP1 methods and represented as functions of $f = 1/P$.

11.8 PERIODOGRAM ANALYSIS OF SIGNALS WITH APERIODIC OR PERIODIC TRENDS: WHEN DETRENDING AND PREWHITENING LEAD TO GENERALLY WRONG RESULTS

Often the astro and geosignals contain different contributions with different properties, shape, timescale, etc. These contributions are to be taken into account simultaneously to get statistically optimal results.

The simplest model for a periodic signal with a linear trend is

$$x_c(t) = C_1 + C_2 \cdot \tilde{t} + C_3 \cos(2\pi f \tilde{t}) + C_4 \sin(2\pi f \tilde{t}), \tag{11.73}$$

where $\tilde{t} = t - \bar{t}$. For the fixed value of f, the term $\bar{t}$ is usually omitted (i.e., set to zero). However, for differential corrections, it is recommended to use a sample mean value of the times of observations. This improves the quality of the matrix of normal equations and thus significantly speeds up convergence of iterations (Andronov, 1994a, 2003). In the software MCV, the user may choose to remove the mean value from times of observations before the analysis.

The periodogram may be defined as $S(f) = 1 - \Phi_4/\Phi_2$, where m in Φ_m shows the number of parameters.

Recently, Olspert et al. (2018) used the same model and called it "generalized Lomb–Scargle periodogram with trend" and applied Bayesian methods for the period estimates.

For example, for a multiharmonic multiperiodic signal superimposed onto a trend approximated with an algebraic polynomial, one may write a model

$$x_c(t) = C_1 + \sum_{\alpha=1}^{s_0} C_{\alpha+1}\tilde{t}^{\alpha} + \\ + \sum_{\beta=1}^{s_P}\sum_{\alpha=1}^{s_\beta} (C_j \cos(2\pi\alpha f_\beta \tilde{t}) + C_{j+1} \sin(2\pi\alpha f_\beta \tilde{t})), \tag{11.74}$$

$$j = 2\alpha + s_0 + 2\sum_{\gamma=1}^{\beta-1} s_\gamma.$$

Here s_β are degrees of the trigonometrical polynomial corresponding to period $P_\beta = 1/f_\beta$, and s_P is the number of periods.

Such an approximation is used in the software MCV (Andronov and Baklanov, 2004) for up to $s_P = 3$ and a total number of parameters $m = 21$ in the standard version.

There is a possibility to improve the initial values of the period using differential corrections. This should be done with care, as sometimes the periods taken from a periodogram are biases, so the matrix of normal equations is (nearly) degenerate, and the iterations do not converge to the statistically optimal solution. One may recommend to make corrections to only one period, keeping others fixed.

In many monographs and textbooks, there are recommendations to remove slow trends ("detrending") or remove periodic components ("prewhitening") before further time series analysis. Although we prefer to use complete models using neither detrending nor prewhitening, in the MCV, we have often included this option.

11.9 ANALYSIS OF MULTIPERIODIC, MULTIHARMONIC, AND MULTISHIFT SIGNALS

Similarly to previous expressions, one may write an unsimplified expression taking into account all the following components of variability: multiple periods, nonsinusoidal shape, and possible shifts between the observations of different observers or different runs.

For the periodogram analysis of multiharmonic signals with an algebraic polynomial trend, in the MCV, we propose to use

$$x_c(t) = C_1 + \sum_{\alpha=1}^{s_0} C_{\alpha+1}\tilde{t}^{\alpha} + \\ + \sum_{\alpha=1}^{s} (C_j \cos(2\pi\alpha f \tilde{t}) + C_{j+1} \sin(2\pi\alpha f \tilde{t})), \tag{11.75}$$

$$j = 2\alpha + s_0.$$

Generally, if there are shifts, a special study is needed to explain their nature. Examples are during multitelescope campaigns, when there are different instrumental systems. Assuming that these differences are not important, one may determine these shifts from this model and then maybe subtract them from the original data to make a joint dataset. In common programs, this option is not available, thus sometimes one may subtract a shift "by eye." In other programs, a sample mean value is subtracted. Before we discussed how such detrending may affect results of the periodogram analysis. Thus we recommend to use a complete expression, which includes a separate shift C_α for each filter/channel α, i.e.,

$$x_{C\alpha}(t) = C_\alpha + \sum_{j=1}^{s} \left(C_{s_0-1+2j}\cos(2\pi j f\tilde{t}) + C_{s_0+2j}\sin(2\pi j f\tilde{t})\right) \quad (11.76)$$

11.10 RUNNING APPROXIMATIONS

11.10.1 General Expressions

They are based on the local approximation $x_C(t, t_0, \Delta t)$ in the interval $(t_0 - \Delta t, t_0 + \Delta t)$ and only a central point $t = t_0$ is taken into account, so the smoothing function $\tilde{x}_C(t_0, \Delta t) = x_C(t_0, t_0, \Delta t)$. "Running means" or (a synonym) "moving averages" are most famous local approximations when the points have an equal weight and are approximated in this "running" interval by a sample mean,

$$x_C(t, t_0, \Delta t) = \frac{1}{n(t_0, \Delta t)} \sum_{|t_k - t_0| \leq \Delta t} x_k, \quad (11.77)$$

where $n(t)$ is the number of points inside this interval. This may be easily generalized to a case of different weights of observations w_k, i.e.,

$$x_C(t, t_0, \Delta t) = \frac{1}{n(t_0, \Delta t)} \sum_{|t_k - t_0| \leq \Delta t} w_k \cdot x_k, \quad (11.78)$$

$$n(t_0, \Delta t) = \sum_{|t_k - t_0| \leq \Delta t} w_k. \quad (11.79)$$

Generally, $\tilde{x}_C(t_0, \Delta t)$ is a piecewise constant function, i.e., is a spline of degree 0 and defect 1. Where the times are distributed regularly and all $w_k = 1$, the running approximations are usually defined at the arguments of observations t_k, i.e.,

$$\tilde{x}_C(t_k, \Delta t) = x_C(t_k, t_k, \Delta t) = \sum_i h_i(\Delta t) \cdot x_{k+i}. \quad (11.80)$$

The final value of Δt should be fixed, so the coefficients h_i of the filter are fixed. The summation takes place for all i with $h_i \neq 0$. The statistical accuracy is

$$\sigma[\tilde{x}_C(t_k, \Delta t)] = \sigma \cdot \sqrt{\sum_i h_i^2(\Delta t)}, \quad (11.81)$$

assuming that all $\sigma_k = \sigma$. The filters may be asymmetrical, $h_{-i} = -h_i$, symmetrical, $h_{-i} = h_i$, or "general." There are also the "differentiating filters," e.g., $h_i = (-\frac{1}{2}, 0, +\frac{1}{2})$ for the numerical value of the derivative, or "integrating" ones, e.g., $h_i = (\frac{1}{3}, \frac{1}{3}, \frac{1}{3})$ for the "running mean," $h_i = (\frac{1}{3}, \frac{1}{3}, \frac{1}{3})$ for the "running mean," $h_i = (\frac{1}{4}, \frac{2}{4}, \frac{1}{4})$ for the central point in the linear approximation and for integration using the trapezium method, and $h_i = (\frac{1}{6}, \frac{4}{6}, \frac{1}{6})$ for the integration using the three-point parabolic approximation for the function. There is a common problem with "border intervals," as there may be missing points needed for this sum. Practically, there may be few approaches. The correct one is to recompute the values of h_i for the asymmetrical case. Simple approaches are to set missing values to "zero" (or the sample mean of the whole series); to set to the nearest "normal" value; or to assume that the series are periodic, so $x_{k+j \cdot n} = x_k$ for all integer j. For a running mean, one may not use missing values and compute a sample mean for a smaller number of points near the border. Also, one may divide the sum by a sum of used coefficients h_i. All these approximations lead to "biased" values close to the borders, so we prefer to use the sets of h_i recomputed for a smaller number of points using the same basic functions and the window function. General principles of design of the digital filters were presented in monographs (e.g., Hamming, 1997).

11.10.2 Running Approximations and Scalegram Analysis for Irregularly Spaced Data

Obviously, if the variations are periodic, the global cosine approximations are preferred. However, many signals are not harmonic; the individual oscillations may vary in amplitude, shape, cycle length, etc. For such "quasiperiodic" oscillations, local approximations at "running" intervals are more effective. The main idea is to compare approximations with observations at different timescales. For small Δt, the approximation will be dominated by observational errors, and will have apparent waves of low statistical significance. For very large Δt, the approximation will be bad because of large systematic difference from the observations.

Andronov (1997) has studied a general case of running approximations with arbitrary test functions using additional weight functions (= windowing functions = filter functions). In this case, Eq. (11.5) should be replaced by

$$(\vec{a} \cdot \vec{b}) = \sum_{k=1}^{n} p(z_k) \cdot w_k \cdot a_k \cdot b_k. \quad (11.82)$$

Here $p(z)$ is the weight function dependent on the (dimensionless) parameter $z = (t - t_0)/\Delta t$. The localized function $p(z)$ should be zero outside the interval of smoothing (i.e., for $|z| > 1$). For smoothing, one may recommend symmetrical functions $p(-z) = p(z)$; otherwise the "forward" and "backward" approximations will not coincide. The simplest weight function is the "rectangular" one, $p(z) = 1$. The disadvantage of this method is that the smoothing function does not smoothly vary with time t_0 at the mid-interval, and has discontinuity when a single point enters/leaves the interval of smoothing. Thus one may recommend to use additional restrictions for derivative $p'(\pm 1) = 0$ and $p(z) \leq 0$. Obviously, there may be an infinite number of such functions, with among them the Hann function $p(z) = \cos^2(\pi z/2) = (1 + \cos(\pi z))/2$ (Hamming, 1997). Andronov (1997) proposes the much simpler function (and thus faster for computations) $p(z) = (1 - z^2)^2$. It has an additional advantage: as the derivative $p'(\pm 1) = 0$, the derivative of the smoothing function $\tilde{x}_C(t_k, \Delta t)$ is continuous everywhere. A global approximations using the cubic splines (Andronov, 1987) has a continuous second derivative. Because the real time series are always limited, and often have irregular gaps, there is no real advantage of using infinite functions like a Gaussian $p(z) = \exp(-c \cdot z^2)$, an exponent $p(z) = \exp(-c \cdot |z|)$, etc. The next user's choice is to define a set of the basic functions $f_\alpha(z)$. Following the tradition of using algebraic polynomials, the parabola is the next case after the constant. Obviously, there may be lines in between. But they may be skipped in a "good" case of equidistant observations, as they produce the same values $\tilde{x}_C(t_k, \Delta t)$ at the moments t_k. In other words, the polynomials of orders (0,1), (2,3), etc., are different for the same data, but the central point is the same. Thus the "running parabola" will exactly fit the cubic polynomial at times t_k. Moreover, the approximation has much better properties as compared with either the "running weighted mean" with this $p(z) = (1 - z^2)^2$, or the "unweighted" ($p(z) = 1$) parabola. The only parameter which remained to be determined, is the filter half-width Δt. For this purpose, one has to compute "scalegrams," i.e., the dependence of the test functions on Δt. Andronov (1987) used the following characteristics: σ_{O-C}, the unbiased estimate of the r.m.s. deviation of the observations from the fit; σ_{xC}, r.m.s. accuracy of the smoothing function at the moments of observations; and the SNR. These dependencies are shown in Fig. 11.11. For an illustration, we have used a relatively short part of the light curve of R Aqr in the filter V from the AAVSO database. It was used above for the polynomial approximation in Fig. 11.3.

Briefly, the dependence σ_{O-C} should have two "standstills." At small Δt, which corresponds to statistical noise (for this sample, the minimum corresponds to 0.070^m), the function has a transition to the "standstill"

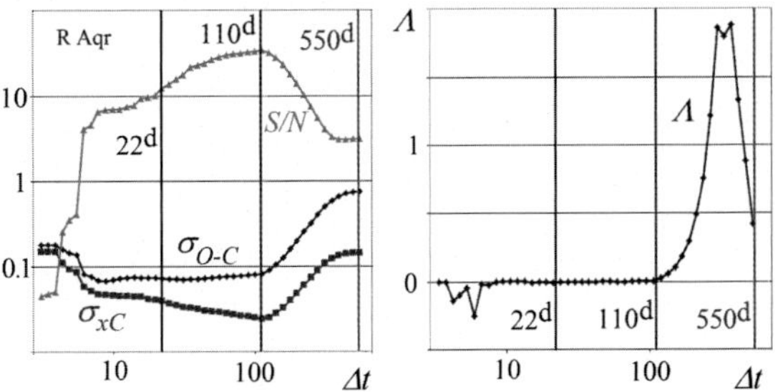

FIG. 11.11 Scalegrams σ_{O-C}, σ_{xC}, and SNR as functions of Δt (left) and the Λ scalegram (right) for the part of the light curve of R Aqr (same as for the polynomial approximation at Fig. 11.12). The vertical lines correspond to the optimal value $\Delta t = 110^d$ and five times larger and smaller for comparison.

at large Δt due to increasing influence of the systematic deviations of the approximation from the observations.

From the position of the transition, it is possible to estimate the "period," and the levels of the standstill allow to estimate the effective semiamplitude of the variations. Andronov and Chinarova (2003) introduced these new effective characteristics of quasiperiodic signals, namely, the effective amplitudes, periods (timescales), and slopes of the scalegram. They have been determined for 173 semiregular variables. Five stars are characterized by outstanding values of at least one of the parameters and were chosen for additional observations to check their peculiar behavior.

The dependence of σ_{xC} is more complicated, and has a single minimum of 0.024^m at $\Delta t = 110^d$. This is one of the criteria to choose Δt for a final approximation. The "amplitude" SNR is 34. For very noisy observations, the maximum SNR is typically shifted towards smaller Δt as compared to a minimum of σ_{xC}. Andronov (2003) introduced the "Λ scalegram," which more resembles different periodograms, i.e., showing "peaks" instead of "transitions." From the position and height of the peak, it is possible to determine the effective period P_Λ and semiamplitude R_Λ. For the data on R Aqr, $P_\Lambda = 345^d$. The period estimate is smaller than $360^d \pm 2^d$ from the cosine fit, with the initial epoch for the maximum brightness (minimum magnitude) $T_0 = 2456769.8 \pm 1.1^d$ and semiamplitude $R = 1.74^m \pm 0.05^m$. Andronov (1997) has shown that the optimal $\Delta t \approx 0.511P$ for a pure sinusoid with many (nearly regular) observations during a period. A smaller value of P_Λ is explained by the presence of harmonics of the main oscillation. In cases where the variability is present at many timescales with variable cycle lengths, the "transitions" at σ_{O-C} occur at different Δt. Using the σ scalegram, Andronov et al. (1997) discovered a

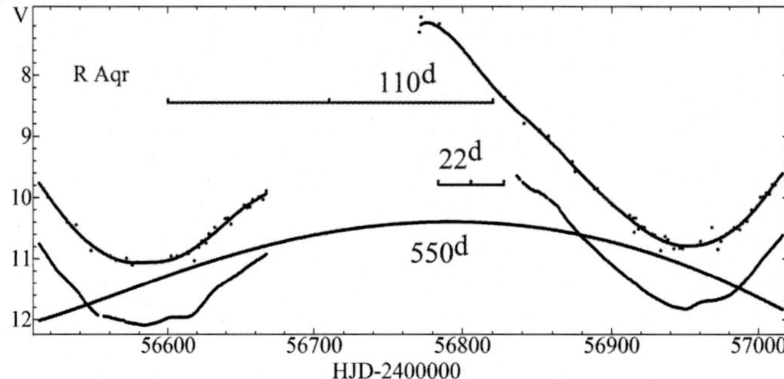

FIG. 11.12 Part of the light curve of R Aqr (same as for the polynomial approximation in Fig. 11.3). The approximations $x_c(t)$ using "running parabola" fit with $\Delta t = 110^d$ (optimal) and five times larger and smaller (the curves are shifted by 1^m. Horizontal bars and numbers show the values of the filter half-width Δt).

fractal type of variability of the magnetic cataclysmic variable AM Her at a very wide range of timescales from 3 seconds to 30 years. The unbiased scatter estimate increases with the filter half-width according to a power law $\sigma_{O-C} \propto (\Delta t)^{0.180}$ from 10^{-4} to 3000 days. The estimate of the fractal dimension is $D = 0.32$.

11.10.3 Running Sines

For nearly sinusoidal signals, it is expected to use a "running approximation"

$$\tilde{x}_C(t, \Delta t) = C_1 + C_2 \cos(\omega \cdot (t - T_0)) + C_3 \sin(\omega \cdot (t - T_0)), \quad (11.83)$$

$$\tilde{x}_C(t, \Delta t) = C_1 - R \cos(\omega \cdot (t - T_M)), \quad (11.84)$$

where $\omega = 2\pi/P$, P is the period, and T_0 is the initial epoch. The coefficients C_α are functions of t_0 and Δt, as only the data in the interval $[t_0 - \Delta t, t_0 + \Delta t]$ are used. The following parameters are determined: C_1, the mean (over the period) value; R, the semiamplitude; T_M, the moment of maximum brightness (minimum stellar magnitude); and brightness at maximum $x_{max} = C_1 - R$ and minimum $x_{max} = C_1 + R$. Obviously, T_M may be easily converted to a phase, using T_0 and P. The recommended value $\Delta t = P/2$, but, for large gaps in the observations, it may be enlarged, e.g., to $\Delta t = P$. This method is effective for studies of many types of stars - semiregular, symbiotic, Mira-type, intermediate polars, RR Lyrae-type with the Blazhko effect. This approximation is close to that used in physics for studies of amplitude and phase modulations. Another example may be temperature changes during a day with a yearly wave. A review of the method was presented by Andronov and Chinarova (2013). An extension of this method may be a shift (in C_1, T_M) and scaling (in R) of nonsinusoidal shapes ("patterns," "templates").

11.10.4 Wavelet Analysis

Wavelet analysis is another extension of the Fourier analysis and of running approximations. It is also related to dynamic spectra, when the Fourier analysis is carried out using the window function. In astronomy, the "seasonal" periodograms are often used, as there are natural borders of the series of more or less dense observations. A similar situation is for time series obtained during a night with gaps between them. In this case, there is no sense for a constant Δt, and the complete data are binned to (unequal in length) subintervals. If the signal has no gaps, these intervals may be chosen also as subsequent nonoverlapping ones. In this case of the nonoverlapping subintervals, the values at the periodogram (and other parameters) are statistically independent.

For overlapping intervals, the "morphing" of the parameters is more smooth. Similarly to the running approximations, one may use not only a rectangular window, but also smooth ones.

If Δt is not dependent on the trial period, the terminology is the "dynamic spectra." For the wavelet analysis, it is assumed that $\Delta t = \tilde{c} \cdot P$, where $\tilde{c}$ is some constant. For $\tilde{c} \to \infty$, the approximation becomes "global" rather than "local."

Andronov (1998, 1999) reviewed the statistical properties of the test functions of the weighted LS improvement of the Morlet-type wavelet. Beyond the typical "wavelet maps," the additional features are introduced, e.g., the (weighted mean) wavelet periodogram, the wavelet skeleton, and the wavelet approximation.

For irregularly spaced signals, the "noise" at the wavelet periodogram may be decreased by a few dozen percent, and, in some extreme cases, even by a factor of a few times.

For the wavelet periodogram, one may use four different functions, which have different asymptotic behavior at large and small trial periods. They were compared to the scalegrams (e.g., Andronov and Chinarova, 2001).

The analysis of citation of these papers at the site Researchgate.net shows numerous applications of the method not only for astro- and geodata, but even in medicine (cardiograms).

An example of dependence of the wavelet periodogram on the effective width of the window function and a comparison to the "running sine" analysis was presented by Chinarova (2010). The drastic variations of the semiamplitude of pulsations of the semiregular variable RU And from nearly constant (0.027^m) to 1.204^m (Mira-type pulsating variable) were detected.

Andronov and Kulynska (2020) discussed a modification of the Morlet-type wavelet by using a compact weight function $p(z) = (1 - z^2)^2$ (initially proposed by Andronov, 1997 and discussed by Andronov, 1999), and have tested statistical properties of the parameters using the traditional matrix form, as well as the "bootstrap"- generated data.

Mann and Haykin (1991) introduced a generalization of Gabor's Logon Transform, which they called "the chirplet transform" it is pointed to analysis of signals with frequency variations.

11.11 MOMENTS OF CHARACTERISTIC POINTS ($O - C$ ANALYSIS)

11.11.1 Period Determination

One of the methods of analysis is the determination of the moments of the characteristic points t_k only (cf. Tsesevich, 1973; Kreiner et al., 2001).

They may be subdivided into two main types: the maxima or minima (extrema) and the moments of crossings by the approximation of some constant level (e.g., the γ velocity [mean] by the curve of the radial velocity). The main expressions were discussed in Section 11.2. AAVSO prefers to use the abbreviation ToM (times of minimum), but it may be extended to times of maximum and also to times of any other characteristic point.

The simplest mathematical model for $O - C$ analysis is

$$t_k = T_0 + P \cdot E_k + \epsilon_k, \tag{11.85}$$

where T_0 is called "the initial epoch," P is the "period," E_k is the "cycle number," and ϵ_k is the residual, which is often called "$O - C$."

In the simplest case, $P = t_{k+1} - t_k$ is the ToM corresponding to subsequent cycles, e.g., that shown in Fig. 11.6. For better precision, $P = (t_n - t_1)/(n - 1)$.

If the (integer) cycle numbers E_k are correct, then one may determine corrections ΔT_0 and ΔP using the LS, and then new statistically optimal values $T_0 + \Delta T_0)$ and $P + \Delta P$.

If the moments t_k are rare enough, there may be problems with determination of the cycle numbers E_k. In this case, one may apply the greatest common divisor (GCD) algorithm, similar to that proposed by Euclid for integer numbers: $\mathrm{GCD}(a, b) = \mathrm{GCD}(a, b \bmod a)$, swapping the numbers, if needed, for $a < b$, and repeating until $\mathrm{GCD}(a, 0) = a$. For noninteger numbers, this may be replaced by computing the pairs of differences $\Delta_k = t_{k+1} - t_k$, and then sorting in increasing order and trying to find a (generally noninteger) GCD of all these differences, which will be equal to the period (Tsesevich, 1973).

A computer algorithm to carry out periodogram analysis was presented by Dumont et al. (1978). Andronov (1988) studies the properties of the test function.

The periodogram analysis for all data points was proposed by Andronov (1991). The initial epoch is set to $T_0 = t_1$. The test periods were chosen as $P_E = (t_n - t_1)/E$, where trial values of E started from $n - 1$ to some reasonably small values, e.g., $P_E = 0.1^{\mathrm{d}}$. Then T_0 and P_E are corrected using the LS, and the phases are recomputed, so the test function is a (weighted) sum of squares of the residuals.

For fast computations, one even may not make the corrections – in this case, the routine may be realized in the electronic tables even without programming.

This method is effective for stable periodic variations and small statistical errors of t_k. However, it may produce a "saw tooth" periodogram in the case of large phase deviations. This is a typical situation for cataclysmic variables, where the intervals between the outbursts vary by few dozen percent (Andronov and Shakun, 1990), or the minima have large shifts (as, e.g., in the cataclysmic variable TT Ari, Kim et al., 2009).

Jetsu and Pelt (1996) had taken into account accuracies σ_k of the individual time points t_k under a common assumption of uncorrelated statistical errors. Then the accuracy of the time differences is

$$\sigma_{kj} = \sigma[t_k - t_j] = \sigma_k^2 + \sigma_j^2. \tag{11.86}$$

11.11.2 Period Changes

To study changes of the period, the ToM are to be determined during a long-time monitoring. The main mechanisms for the period variations in eclipsing binary stars are the mass transfer (accretion and excretion due to stellar wind); the magnetic stellar wind and gravitational radiation; and the changes in the internal structure in a component. There may be changes of the $O - C$, which are related not to the period variations, but to other mechanisms like the light-time effect due to presence of a third body (a star or a planet) or multiple objects. The ToM may be shifted due to the presence of stellar spots. For binary systems at eccentric orbits, there may be an apsidal motion (cf. Tsesevich, 1973; Kopal, 1959). The discussion on the statistics of the minima compiled in "An Atlas of $O - C$ Diagrams of Eclipsing Binary Stars" (Kreiner et al., 2001) was presented by Kim et al. (2003). A catalogue of 623 systems with eccentric orbits and their classification was presented by Kim et al. (2018). Some system show complicated variations due to few acting mechanisms.

There are few online catalogues of ToM, e.g.,

- http://var2.astro.cz/ocgate/,
- http://www.as.ap.krakow.pl/o-c,
- http://www.bav-astro.eu/index.php/veroeffentlichungen/service-for-scientists/lkdb-engl,
- http://www.aavso.org/bob-nelsons-o-c-files,
- http://www.aavso.org/observed-minima-timings-eclipsing-binaries

The compilations of ToM are often published in journals, e.g., in the "Open European Journal on Variable Stars" (cf. Paschke, 2018). The relation between period and ToM of variable stars was discussed by Kopal and Kurth (1957). Introducing the functions $t(E)$ ($M(E)$ in their notation) and $P(E)$ as the time and period corresponding to the cycle number E, one may distinguish the "discrete" $P(E) = t(E + 1) - t(E)$ and "continuous" $P(E) = dt(E)/dE$ definitions. They discussed some models. In the "continuous" model,

$$t(E) = T_0 + \int_0^E P(E)\,dE. \quad (11.87)$$

Assuming the simplest model for the period variations,

$$t(E) = T_0 + P_0 \cdot E + Q \cdot E^2, \quad (11.88)$$

we get $P(E) = P_0 + 2QE$ and $dP/dE = 2Q$, $\dot{P} = dP/dt = (dP/dE)/(dt/dE) = 2Q/(P_0 + 2QE)$. The characteristic timescale of the period variations is defined usually as $\tau = P/|\dot{P}|$, which is equal for this case to $\tau = P^2/(2|Q|)$, and is thus dependent on time. For $\dot{P} = \text{const}$, the period is $P(E) = P_0 \cdot \exp(\dot{P} \cdot E)$, and

$$t(E) = T_0 + P_0 \cdot (\exp(\dot{P} \cdot E) - 1), \quad (11.89)$$

where the index 0 corresponds to the value at $E = 0$ (cf. Andronov and Chinarova, 2013).

Having a model $t(E)$, one may define $E(t)$ as an inverse function. It is suitable for compute phases (which are decimal parts of $E(t)$) for plotting phase curves. For parabolic $O - C$, the option to compute phases is included in the software MCV (Andronov and Baklanov, 2004) and MAVKA (Andrych and Andronov, 2019). However, the phases may be computed also for more complicated models (e.g., Kim et al., 2005).

11.12 AUTOCORRELATION AND CROSS-CORRELATION ANALYSIS

11.12.1 Continuous and Discrete Regular Signals

Autocorrelation analysis is one of classical methods for data analysis, which was described in hundreds of textbooks (e.g., Box et al., 2015, Blackman and Tukey, 1959). Astronomical applications were reviewed by Deeming (1970).

Classical ACF is defined as $r = R_u/R_0$, where

$$R_u = \frac{1}{N_u} \sum_{k=1}^{n-u} (x_k - \bar{x})(x_{k+u} - \bar{x}). \quad (11.90)$$

This is applicable to evenly distributed time series $t_k = t_1 + (k - 1)\delta$. Obviously, $r_0 = 1$ for any variable signal. For the simplest case of pure noise with a variance σ_0^2, the mathematical expectation is 0 for $u \neq 0$. Sample ACFs differ from this value; the statistical properties were studied by Andronov (1994b).

Here we introduced a new variable N_u. For $N_u = N$, the ACF is called the "biased" one. However, the number of summands in the sum is equal to $n - u$, thus the "unbiased" mathematical expectation is for $N_u = n - u$.

Correlation length is defined as the smallest positive root of the equation $r_u = 0$. It may be expressed in units of time, $\tau_0 = \delta \cdot u$. For a pure sinusoidal signal with a period P, $r_u = \cos(2\pi\delta u/P)$ and $\tau_0 = P/4$. The first nonzero maximum corresponds to $u_{max} = P/\delta$. For quasiperiodic variations, it gives an estimate of some characteristic cycle length.

For the shot noise or autoregressive (AR) process of the first degree, $r_u = \psi^u$ and thus it does not cross zero if the number of observations n is large. However, the removal of the sample mean value $\bar{x}$ from the data leads to a distinct crossing of zero by the ACF (Sutherland et al., 1978), thus leading to wrong interpretation of the character of variability, mainly, the shot noise (AR1), or QPO (AR2).

The bias of the ACF increases with decreasing length of data and more complicated shape of the trend. A correct set of equations for arbitrary length and the number of basic functions were presented by Andronov (1994b).

For the shot noise, the common mathematical model is the Markov process, or the AR-1 model. The theoretical ACF is $r_u = \psi^u = \exp(-u/u_d)$, so the decay time may be determined as $\tau_d = u_d\delta = -\delta = \ln\psi$. So, the ACF decreases by a factor of e for the decay time τ_d. For multicomponent signals, assuming their statistical independence,

$$x_k = \sum_{\beta=1}^{m} x_{\beta k}, \tag{11.91}$$

$$R_u = \sum_{\beta=1}^{m} R_{\beta u}, \tag{11.92}$$

$$r_u = \frac{n}{N_u} \sum_{\beta=1}^{m} \tilde{R}_{\beta 0} \cdot r_{\beta u}, \tag{11.93}$$

where the relative contribution to the ACF $\tilde{R}_{\beta 0} = R_{\beta 0}/R_0$. Using this algorithm, Andronov et al. (2005) proposed a four-component model of the ACF of X-Ray variability of AM Her based on a CHANDRA Observation, and the second component in the shot noise variability of AM Her was discovered.

11.12.2 Bias of ACF due to Trend

For astro- and geosignals, the problem arises with taking into account a sample mean (Sutherland et al., 1978) or a possible trend. Complete study of the influence of trend removal to the resulting ACF for any basic functions was presented by Andronov (1994b). In Fig. 11.13, the ACFs for different lengths are shown for the initial data without trend removal, removal of the sample mean, and removal of the cubic approximation. It is clearly seen that the trend removal significantly affects the ACF and may lead to wrong physical conclusions. For example, numerous studies of AM Her have shown a distinct crossing of zero by the ACF, which is typical for quasiperiodic oscillations, or the AR2 model, rather than for the AR1 model, or the shot noise.

11.12.3 Irregularly Spaced Signals

For irregularly spaced signals, the ACF may be computed in two ways.

(1) The data are interpolated, and then the ACF is computed for such an artificial dataset. This is typically used for single missing points, e.g., deleted because of an outlying value, or an absence of measurement,

If the point x_k is missing, it may be replaced by values using a local approximation – linear or cubic: $\tilde{x}_k = (x_{k-1} + x_{k+1})/2$ and $\tilde{x}_k = (4 \cdot (x_{k-2} + x_{k+2}) - (x_{k-1} + x_{k+1}))/6$. If the accuracy σ is the same for these points, the accuracy of the interpolated value is $\sigma/\sqrt{2}$ and $\sigma\sqrt{17/18}$, respectively. Such an option is included in the software MCV (Andronov and Baklanov, 2004).

Although the first value is more accurate for uncorrelated noise, one may recommend the second one in the case of suggested fast variations of the signal at a timescale of a few δ, where δ is time resolution. Otherwise the maxima and minima should be flattened.

(2) The second approach is to split the data into nonoverlapping intervals and to computed mean values. Then the new time series are equidistant in time, and the analysis may be carried out.

For example, in the series of papers based on the observations from the AAVSO database, there is an approach to make 10^{d} means, i.e., to split the interval into 10^{d} pieces and treat them as one point. The weak points of such method may be in a case of significant variations during these intervals, and the point(s) may correspond to an "effective time," which differs from the mid-interval.

The third method for the general case of distribution of times t_k is not to make time bins for the observations, but to determine individual time differences and round them off to an integer number of shifts. This corresponds to a "rectangular" time window. One may also apply a structure function like in the method for the periodogram analysis.

Another approach is to compute not the sums of deviations from the fit, but the sum of squares of differences between the pairs. So the minimum of Φ is to be found, rather than the maximum of the correlation coefficient r.

The cross-correlation function may be defined in a similar way – typically by using regular time series, or making such series by linear/cubic interpolation or computing a mean in the small intervals.

Contrary to the ACF, the CCF is generally not symmetric, and it may be used for estimates for the time lags between two signals An example may be the time lag between the daily temperature and the solar heating at some place, or the light echo due to reemission of the light from a Supernova explosion by distant clouds in the vicinities.

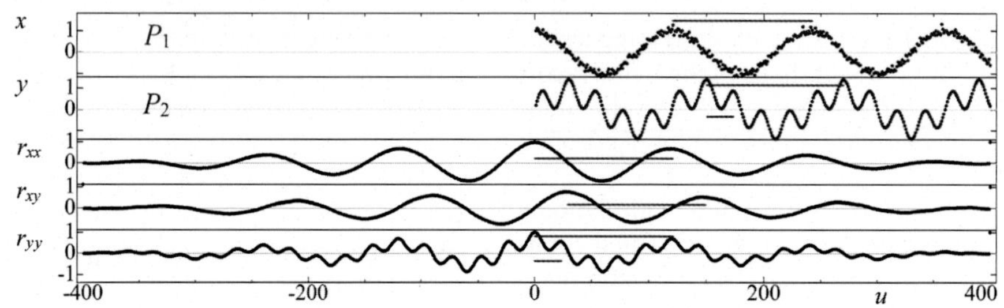

FIG. 11.13 The model signals x_k, y_k: cosine (one period P_1 with some noise) and sine functions (two periods P_1 and P_2) and the corresponding ACF, r_{xx}, r_{yy}, and the CCF r_{xy}. The long and short horizontal lines show periods P_1 and P_2, respectively.

In Fig. 11.13, the CCF for the model harmonic oscillations with some noise is shown.

The vertical lines correspond to maxima. Definitely, for periodic functions, the peaks at the CCF repeat periodically with the same period. However, due to the different number of the summands $n - u$, the highest peak is expected to be closer to zero.

11.13 PRINCIPAL COMPONENT ANALYSIS AND RELATED METHODS

Singular value decomposition (SVD) and principal component analysis (PCA) are described in many textbooks (e.g., Forsythe et al., 1977; Press et al., 2007). Its applications to astronomical time series were reviewed, e.g., by Andronov (2003); Andronov et al. (2003b); Mikulášek (2007b).

The classical PCA is based on the analysis of the covariation matrix for the data obtained in different channels. It is very closely related to the SVD for the matrix $\epsilon_{\alpha k}$ (typically only a mean value is removed, but generally one may propose to subtract a smoothing value):

$$\epsilon_{\alpha k} = \frac{x_{\alpha k} - x_{C\alpha k}}{\sigma_{C\alpha k}}. \tag{11.94}$$

Typically, in simpler models, $x_{C\alpha k} = \bar{x}_\alpha$, just using a sample mean for each channel α.

The main idea is to apply filtering to the time series by incomplete restoration of the sum, decreasing the number of members of the sum from m to L.

11.13.1 Principal Component Analysis: Multichannel Signals

PCA is used for multichannel signals, when, for the same moment of time t_k, there are multiple values of signal $x_{\alpha k}$.

The main idea of the analysis using SVD or PCA is that the accuracy is equal for all measurements in all channels and all data. For data with very different scatter in different channels, there may be some scaling, e.g., by normalizing the data as $x_{ak} = (x_{ak} - x_{0ak})/\sigma_a$. In the two-dimensional case, this leads to a bisector independently on the value of the correlation coefficient.

The slope of the line is generally different either from the regression line, or the line of the orthogonal regression. For the three-dimensional, the situation becomes less undetermined. One may make a partial restoration of the function and use the values $\sigma_\alpha[O - C]$ for scaling. In this case, some iterations are needed to get a self-consistent solution.

In Fig. 11.14, the part of the light curve of SS Cyg, the prototype dwarf nova, is shown. There are large seasonal gaps, when the star is not visible during the night, but the measurements are nearly simultaneous. Four filters (channels) are used. To prepare the data for the PCA, the observations are to be converted to "pseudo-simultaneous" ones. It may be done by neglecting the time difference between the observations in different channels. Or, alternately, one may use interpolation of the signal from one channel to times of another one, using local cubic polynomial. This is implemented in the software MCV (Andronov and Baklanov, 2004).

In Fig. 11.15, the principal components of variability of the signal shown in Fig. 11.14, are presented. One may note that the components 1 and 2 show distinct variability, despite very different amplitudes. The components 3 and 4 have very low amplitudes, and may be interpreted as an observational noise.

In Fig. 11.16, we compare color indices computed using the PCA, and from the approximations using the "running parabola" approximation with a statistically optimal value of the filter half-width $\Delta t = 0.007^d$, which corresponds to SNR = 78.

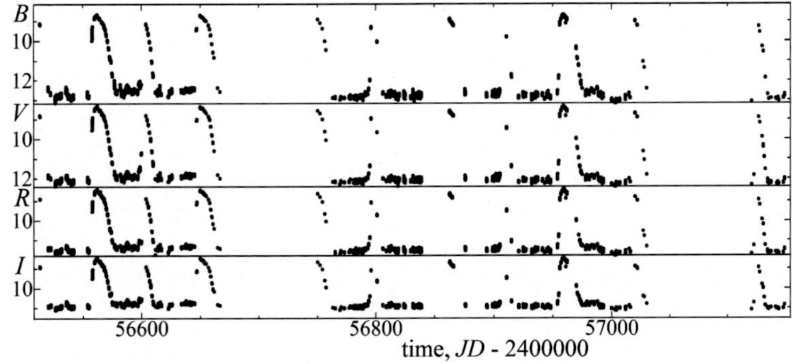

FIG. 11.14 Multicolor BVRI observations of the cataclysmic variable SS Cyg obtained by Robert James (AAVSO code JM).

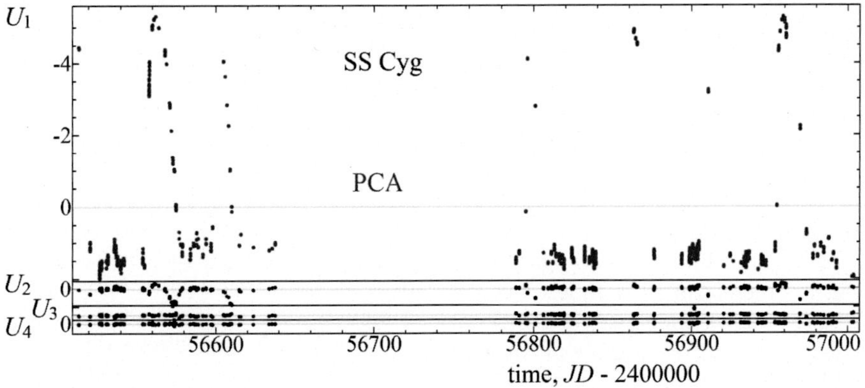

FIG. 11.15 Principal components $U_{\alpha k}$ as a dependence on t_k for the observations of the cataclysmic variable SS Cyg (Fig. 11.14) smoothed using the running parabolae with $\Delta t = 0.07^d$. The right part of the graph was automatically omitted, as there were only single observations in each filter during these nights.

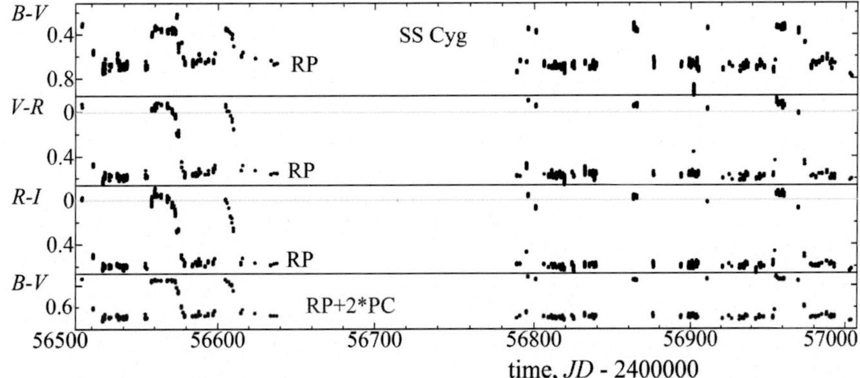

FIG. 11.16 Color indices based on the data smoothed using the running parabolae (up). The bottom graph is the first color index, additionally smoothed using the PCA with the number of principal components $L = 2$. The amplitude of the smoothed curve has decreased by a factor of ≈ 2, removing two outliers.

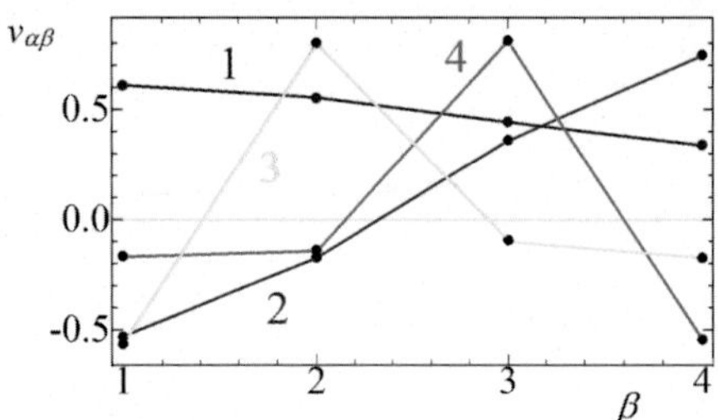

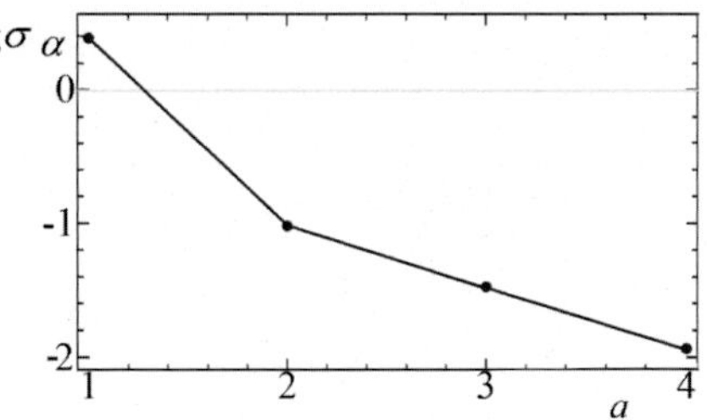

FIG. 11.17 *Left:* The eigenvectors for SS Cyg of the correlation matrix. *Right:* Dependence of the corresponding singular values σ_α on the number of principal components.

In Fig. 11.17, the eigenvectors $v_{\alpha\beta}$ obtained using the PCA, are shown. There is a clear near-linear dependence for the first two components. As the channels are BVRI, the effective wavelength of the signal increases with the number of the component. At the right part of this figure, there is a dependence of the singular value on the number of channel. It shows that the components 3 and 4 have low amplitudes, and may be neglected.

11.13.2 Case of Noisy Channels of Simultaneous Signals

This particular case is applied for measurements of the same signal, which is present in all channels, but the noise is different, as well as the shift, e.g., the changes of the apparent stellar magnitude of a sample of constant stars due to variable extinction and air mass.

Some kind of this approach was introduced by Andronov and Baklanov (2004) in the method of the artificial comparison star, which was implemented in the software MCV. Applications and a description of this method were presented by Kim et al. (2004).

The method of artificial comparison star allows to improve accuracy of the comparison star typically by a few dozen percent. Contrary to the "ensemble photometry" method by Honeycutt (1992), in our method, the weights for the weighted mean are computed after the iterations, and do not depend on the brightness in an approximation.

11.13.3 Singular Spectrum Analysis (SSA)

This method was implemented in the program called Caterpillar (Golyandina et al., 2001), in which L (the number of the principal components for restoration) is defined by the user, so there are no automatic recommendations on what is to be taken into account and what should be neglected as the noise.

The uni-channel (or "one-channel") signal is used to create a pseudomultichannel dataset with m channels. Then the principal components are computed, and the user can choose which of the components will be used for partial restoration of the signal. This is some kind of smoothing. The coefficients may also be used for a short-term forecast. It should be noted that the matrix used for PCA has the components from the autocovariation function.

11.13.4 Effective Amplitudes of Low, Fast, and Noise Variability

A related method was proposed by Tremko et al. (1996). The signal is split into three parts – smooth variations with, e.g., orbital phase (periodic or aperiodic dependent on the method of smoothing), faster correlated variability (e.g., quasiperiodic oscillations [QPOs] or flickering), and an uncorrelated noise.

For the residuals of the observations, the covariation matrix may be computed, the mathematical expectation of which is

$$\mu_{\alpha\beta} = \sigma_{c\alpha} \cdot \sigma_{c\beta} + \sigma_{n\alpha}^2 \cdot \delta_{\alpha\beta}, \tag{11.95}$$

where the Kronecker symbol $\delta_{\alpha\beta} = 1$ if $\alpha = \beta$; otherwise it is 0. Here $\sigma_{c\alpha}$ and $\sigma_{n\alpha}$ are standard deviations for correlated contribution and uncorrelated noise, respectively. There are $2n$ independent values of these parameters and $n \cdot (n+1)/2$ independent values of the matrix, taking into account its symmetry, $\mu_{\beta\alpha} = \mu_{\alpha\beta}$. For $m = 3$, both numbers are equal to 6, thus one may easily estimate all parameters.

For larger m, the parameters may be determined using nonlinear LS (Tremko et al., 1996). This method allows to estimate variances of the correlated signal, so also to estimate the r.m.s. amplitude of the signal $\sigma_{c\alpha}$, as well as of the correlated noise $\sigma_{n\alpha}$.

Such method is an effective tool for cataclysmic variables, where there are aperiodic events like QPO or flickering, and $\sigma_{c\alpha}$ are proportional to the amplitude of the signal in the channel α of any shape. Tremko et al. (1996) used these values to estimate color indices of different contributions to variability, and thus the temperatures.

11.14 CONCLUSIONS

- To obtain new statistically significant astrophysical results, one needs adequate mathematical models.
- The solution is needed for signals with multiple components, which are of finite length, and the observations generally are irregularly spaced.
- The variety of types of variability needs elaboration of special methods, specific for such signals.
- The careful estimate of the FAP prevents publications of many "beautiful discoveries."
- The multicomponent character of signals needs complete models instead of "step-by-step" simplified ones.
- Solution can be achieved by development of an expert system of advanced complementary algorithms and programs, which improve existing methods.
- Depending on the stability of the periodic variations, the following methods are recommended: the global fits (for the best stability), local fits of an intermediate filter half-width Δt and periodic shapes (patterns) for "nearly periodic" oscillations ("running sines," wavelet), and local fits of smaller Δt, which are effective for nonharmonic and chaotic-like variations.
- The observations at different longitudes are very important, as they effectively dump side-bands of the spectral window, i.e., decreasing the peaks with "object-observer" periodic beats.

ACKNOWLEDGMENTS

The data from the AAVSO (http://aavso.org) and AFOEV were used for illustration. "The SAO/NASA Astrophysics Data System" (http://adsabs.harvard.edu) was used. We acknowledge fruitful recommendations by Petr Škoda, Irina B. Vavilova, Yaroslav S. Yatskiv, Lidia L. Chinarova, Gottfried Schwarz, and an anonymous referee. The monitoring of the first star of our sample - AM Her - was initiated by Prof. V. P. Tsessevich (1907–1983) (Andronov, 2017). Since then, more than 387 papers (cited in the ADS) have been published, with a total number of studied variable stars exceeding 2000. Special thanks go to coauthors of 19+ joint papers: L. L. Chinarova, S. V. Kolesnikov, N. M. Shakhovskoy (1931–2011), L. S. Kudashkina, V. I. Marsakova, V. V. Breus, Yonggi Kim, P. Mason, and S. Zoła.

REFERENCES

AAVSO. http://aavso.org.

Anderson, T.W., 1958. An Introduction to Multivariate Statistical Analysis. John Wiley & Sons, New York, p. 374.

Anderson, T.W., 2003. An Introduction to Multivariate Statistical Analysis, 3rd Ed. Wiley, 739 p. ISBN: 0471360910, 9780471360919.

Andronov, I.L., 1987. Smoothing the "smoothing" cubic spline functions. Publications of the Astronomical Institute of the Czechoslovak Academy of Sciences 70, 161. 1987PAICz..70..161A.

Andronov, I.L., 1988. Period search using the DMRT method – the properties of the test function. Astronomische Nachrichten 309, 121.

Andronov, I.L., 1991. Determination of the period of a variable signal from the times of characteristic events. Kinematika I Fizika Nebesnyh Tel 7 (2), 78.

Andronov, I.L., 1994a. (Multi-)frequency variations of stars. Some methods and results. Odessa Astronomical Publications 7, 49.

Andronov, I.L., 1994b. Autocorrelation function bias owed to a limited number of de-trended observations. Applications to autoregressive models with noise. Astronomische Nachrichten 1994 (315), 353–370.

Andronov, I.L., 1997. Method of running parabolae: spectral and statistical properties of the smoothing function. Astronomy & Astrophysics. Supplement Series 125, 207–217.

Andronov, I.L., 1998. Wavelet analysis of time series by the method of least squares with supplementary weights. Kinematika I Fizika Nebesnyh Tel 14, 490.

Andronov, I.L., 1999. Wavelet analysis of the irregularly spaced time series. In: Priezzhev, V.B., Spiridonov, V.P. (Eds.), Self-Similar Systems. Dubna, Russia, 29 July–August, 1998. Joint Inst. Nucl. Res. 57; ADS: 1999sss..conf...57A.

Andronov, I.L., 2003. Multiperiodic versus noise variations: mathematical methods. In: ASP Conference Series, vol. 292, p. 391.

Andronov, I.L., 2005. Advanced methods for determination of arguments of characteristic events. In: ASP Conference Series, vol. 335, p. 37.

Andronov, I.L., 2012. Phenomenological modeling of the light curves of algol-type eclipsing binary stars. Astrophysics 55, 536.

Andronov, I.L., 2017. Odessa Scientific School of researchers of variable stars: from V. P. Tsesevich (1907–1983) to our days. Odessa Astronomical Publications 30, 252.

Andronov, I.L., Andrych, K.D., 2014. Determination of size of the emitting region in eclipsing cataclysmic variable stars. Odessa Astronomical Publications 27, 38.

Andronov, I.L., Baklanov, A.V., 2004. Algorithm of the artificial comparison star for the CCD photometry. Astronomical School's Report 5, 264.

Andronov, I.L., Chinarova, L.L., 1997. On statistical properties of test functions in nonparametric methods for periodogram analysis. Kinematics and Physics of Celestial Bodies 13 (6), 55–65.

Andronov, I.L., Chinarova, L.L., 2001. Determination of characteristic time scales in semi-regular stars: comparison of different methods. Odessa Astronomical Publications 14, 113–115.

Andronov, I.L., Chinarova, L.L., 2003. Statistical study of semi-regular variables: scalegram-based characteristics. In: ASP Conference Series, vol. 292, p. 401.

Andronov, I.L., Chinarova, L.L., 2013. Method of running sines: modeling variability in long-period variables. Częstochowski Kalendarz Astronomiczny 10, 171.

Andronov, I.L., Kulynska, V.P., 2020. Computer modeling of irregularly spaced signals. Statistical properties of the wavelet approximation using a compact weight function. Annales Astronomiae Novae 1, 167–178. arXiv:1912.13096 [astro-ph.IM].

Andronov, I.L., Marsakova, V.I., 2006. Variability of long-period pulsating stars. I. Methods for analyzing observations. Astrophysics 49, 370.

Andronov, I.L., Shakun, L.I., 1990. Astrophysics and Space Science 169, 237.

Andronov, I.L., Tkachenko, M.G., 2013. Comparative analysis of numerical methods of determination of parameters of binary stars. Case of spherical components. Odessa Astronomical Publications 26, 204.

Andronov, I.L., et al., 1997. Scalegram analysis of the variability of the polar AM HER. Odessa Astronomical Publications 10, 15.

Andronov, I.L., et al., 2003a. Inter-Longitude Astronomy project: some results and perspectives. Astronomical and Astrophysical Transactions 22, 793.

Andronov, I.L., et al., 2003b. In: NATO Science Series II, vol. 105, 325. 2003whdw.conf..325A.

Andronov, I.L., et al., 2005. Four-component model of the auto-correlation function of AM Her based on a CHANDRA observation. In: ASP Conference Series, vol. 330, p. 407.

Andronov, I.L., et al., 2008. Astronomy & Astrophysics 486, 855. 2008A&A...486..855A.

Andronov, I.L., et al., 2014. Inter-Longitude Astronomy project: long period variable stars. Advances in Astronomy and Space Physics 4, 3.

Andronov, I.L., et al., 2017a. Instabilities in interacting binary stars. ASP Conference Series 511, 43.

Andronov, I.L., et al., 2017b. Comparative analysis of phenomenological approximations for the light curves of eclipsing binary stars with additional parameters. Astrophysics 60, 57.

Andrych, K.D., et al., 2015. "Asymptotic parabola" fits for smoothing generally asymmetric light curves. Odessa Astronomical Publications 28, 158.

Andrych, K.D., et al., 2017. Statistically optimal modeling of flat eclipses and exoplanet transitions. The "Wall-Supported Polynomial" (WSP) algoritms. Odessa Astronomical Publications 30, 57.

Andrych, K.D., et al., 2020. MAVKA: program of statistically optimal determination of phenomenological parameters of extrema. Parabolic spline algorithm and analysis of variability of the semi-regular star Z UMa. Journal of Physical Studies 24, 1902. arXiv:1912.07677 [astro-ph.SR], pp. 1–18.

Andrych, K.D., Andronov, I.L., 2019. MAVKA: software for statistically optimal determination of extrema. Open European Journal on Variable Stars 197, 65.

Bendat, J.S., Piersol, A.G., 2010. Random Data, Analysis and Measurement Procedures. John Wiley and Sons, New York.

Blackman, R.B., Tukey, J.W., 1959. The Measurement of Power Spectra. Dover Publications, Inc., NY.

Bódi, A., Szatmáry, K., Kiss, L.L., 2016. Periodicities of the RV Tauri-type pulsating star DF Cygni: a combination of Kepler data with ground-based observations. Astronomy & Astrophysics 596, A24, 8 pp.

Bopp, B.W., Evans, D.S., Laing, J.D., Deeming, T.J., 1970. Six spectroscopic binary stars. Monthly Notices of the Royal Astronomical Society 147, 355–366. https://doi.org/10.1093/mnras/147.4.355.

Box, G.E.P., et al., 2015. Time Series Analysis: Forecasting and Control, 5th Edition. Wiley, p. 712.

Breus, V.V., 2007. Programs for data reduction and optimization of the system work. Odessa Astronomical Publications 20, 32.

Butters, O.W., West, R.G., Anderson, D.R., et al., 2010. The first WASP public data release. Astronomy & Astrophysics 520, L10. https://doi.org/10.1051/0004-6361/201015655.

Cherepashchuk, A.M., 1993. Parametric models in inverse problems of astrophysics. Astronomy Reports 37 (6), 585–594.

Chinarova, L.L., 2010. Wavelet analysis of 173 semi-regular variables. Odessa Astronomical Publications 23, 25.

Chinarova, L.L., Andronov, I.L., 2000. Catalogue of main characteristics of pulsations of 173 semi-regular stars. Odessa Astronomical Publications 13, 116.

Clarke, D., 2002. String/Rope length methods using the Lafler-Kinman statistic. Astronomy & Astrophysics 386, 763–774. https://doi.org/10.1051/0004-6361:20020258.

Deeming, T.J., 1970. Stochastic variable stars. The Astronomical Journal 75, 1027. https://doi.org/10.1086/111056.

Deeming, T.J., 1975. Fourier analysis with unequally-spaced data. Astrophysics and Space Science 36, 137.

Devlen, A., 2015. A new variable star in Perseus: GSC 3692-00624. Open European Journal on Variable Stars 171, 1.

Drake, A.J., et al., 2009. First results from the Catalina Real-Time Transient Survey. The Astrophysical Journal 696, 870.

Dumont, T., Morguleff, N., Rutily, B., Terzan, A., 1978. A search for periodicities in the pulsations of Delta Scuti stars. IV. A new method of computing the period of a complex signal. Astronomy & Astrophysics 69, 65.

Dworetsky, M.M., 1983. A period-finding method for sparse randomly spaced observations or "How long is a piece of string?". Monthly Notices of the Royal Astronomical Society 203, 917–924. https://doi.org/10.1093/mnras/203.4.917.

Efron, B., 1979. Bootstrap methods: another look at the jackknife. The Annals of Statistics 7, 1–26.

Efron, B., Tibshirani, R.J., 1993. An Introduction to the Bootstrap. Chapman & Hall. ISBN 978-0412042317, pp. 436.

Fisher, R.A., 1954. Statistical Methods for Research Workers, 12th Ed. Oliver and Boyd, Edinburgh, pp. xv, 356.

Forsythe, G.E., Malcolm, M.A., Moler, C.B., 1977. Computer Methods for Mathematical Computations. Prentice Hall, p. 270.

Foster, G., 1995. The cleanest Fourier spectrum. The Astronomical Journal 109, 1889–1902.

Fourier, J., 1822. Theorie Analytique de la Chaleur. Firmin Didot. (Reissued by Cambridge University Press, 2009; ISBN 978-1-108-00180-9).

GAIA, 2019. http://gea.esac.esa.int/archive/.

Golyandina, N., Nekrutkin, V., Zhigljavsky, A., 2001. Analysis of Time Series Structure. SSA and Related Techniques. Chapman Hall. ISBN 1-58488-194-1, p. 308.

Hamming, R.W., 1997. Digital Filters, Third Edition. Dover Publications, p. 304.

Haykin, S., 1999. Neural Networks and Learning Machines, 3rd Ed. Prentice Hall, NJ. ISBN 978-0-13-147139-9, 906 p.

Høg, E., et al., 2000. The Tycho-2 catalogue of the 2.5 million brightest stars. Astronomy & Astrophysics 355, L27–L30.

Honeycutt, R.K., 1992. CCD ensemble photometry on an inhomogeneous set of exposures. Publications of the Astronomical Society of the Pacific 104, 435.

Isobe, T., et al., 1990. Linear regression in astronomy. I. The Astrophysical Journal 364, 104.

Jetsu, L., Pelt, J., 1996. Searching for periodicity in weighted time point series. Astronomy & Astrophysics. Supplement Series 118, 587.

Jurkevich, I., 1971. A method of computing periods of cyclic phenomena. Astrophysics and Space Science 13, 154.

KEPLER, 2019. www.nasa.gov/mission_pages/kepler/.

Kim, C.-H., Kreiner, J.M., Nha, L.-S., 2003. Statistics of times of minimum light of 1140 eclipsing binary stars. Astrophysics and Space Science Library 298, 127.

Kim, Y., et al., 2004. CCD photometry using multiple comparison stars. Journal of the Astronomy and Space Sciences 21 (3), 191–200.

Kim, Y., et al., 2005. Orbital and spin variability of the intermediate polar BG CMi. Astronomy & Astrophysics 441, 663.

Kim, Y., et al., 2009. Nova-like cataclysmic variable TT Arietis. QPO behaviour coming back from positive superhumps. Astronomy & Astrophysics 496, 765.

Kim, Y., et al., 2018. A comprehensive catalog of galactic eclipsing binary stars with eccentric orbits based on eclipse timing diagrams. The Astrophysical Journal. Supplement Series 235, 41.

Kopal, Z., 1959. Close Binary Systems. Chapman & Hall, London. Bibcode: 1959cbs..book.....K.

Kopal, Z., Kurth, R., 1957. The relation between period and times of the maxima or minima of variable stars. Zeitschrift für Astrophysik 42, 90.

Kreiner, J.M., Kim, Chun-Hwey, Nha, Il-Seong, 2001. An Atlas of $O-C$ Diagrams of Eclipsing Binary Stars. Wydawnictwo Naukowe Akademii Pedagogicznej, Cracow, Poland.

Lafler, J., Kinman, T.D., 1965. An RR Lyrae star survey with Ihe Lick 20-inch astrograph II. The calculation of RR Lyrae periods by electronic computer. The Astrophysical Journal. Supplement Series 11, 216–222. https://doi.org/10.1086/190116.

Lawson, C.L., Hanson, R.J., 1974. Solving Least Squares Problem. Prentice - Hall Inc., Englewood Cliffs, New Jersey.

Levenberg, K., 1944. A method for the solution of certain non-linear problems in least squares. Quarterly of Applied Mathematics 2 (2), 164–168. https://doi.org/10.1090/qam/10666.

Lipunov, V.M., et al., 2010. Master robotic net. Advances in Astronomy, 349171, 6 pp.

Lomb, N.R., 1976. Least-squares frequency analysis of unequally spaced data. Astrophysics and Space Science 39, 447–462. https://doi.org/10.1007/BF00648343.

Mann, S., Haykin, S., 1991. The chirplet transform: a generalization of Gabor's Logon Transform. In: Proc. Vision Interface, pp. 205–212.

Marquardt, D., 1963. An algorithm for least-squares estimation of nonlinear parameters. SIAM Journal on Applied Mathematics 11 (2), 431–441. https://doi.org/10.1137/0111030.

Marsakova, V.I., Andronov, I.L., 1996. Local fits of signals with asymptotic branches. Odessa Astronomical Publications 9, 127.

Mikulášek, Z., 2007a. The benefits of the orthogonal LSM models. Odessa Astronomical Publications 20, 138.

Mikulášek, Z., 2007b. Principal component analysis – an efficient tool for variable stars diagnostics. Astronomical and Astrophysical Transactions 26, 63.

Mikulášek, Z., 2015. Phenomenological modelling of eclipsing system light curves. Astronomy & Astrophysics 584, A8, 13 pp.

Olspert, N., Pelt, J., Käpylä, M.J., Lehtinen, J., 2018. Estimating activity cycles with probabilistic methods. I. Bayesian generalised Lomb-Scargle periodogram with trend. Astronomy & Astrophysics 615, A111. https://doi.org/10.1051/0004-6361/201732524.

Paczynski, B., 1986. Gravitational microlensing by the galactic halo. The Astrophysical Journal 304, 1. ADS:1986ApJ...304....1P.

Paschke, A., 2018. A list of minima and maxima timings. Open European Journal on Variable Stars 191, 1.

Paunzen, E., Vanmunster, T., 2016. Peranso – light curve and period analysis software. Astronomische Nachrichten 337, 239–246.

Pelt, J., 1975. Methods for Search of Variable Star Periods. Tartu Astrofüüs. Obs. Teated, Nr. 52, 24 p., = Preprint No. 5.

Pojmanski, G., 2002. The All Sky Automated Survey. Catalog of variable stars. I. 0h - 6h quarter of the southern hemisphere. Acta Astronomica 52, 397–427. 2002AcA....52..397P.

Prabhu, K.M.M., 2014. Window Functions and Their Applications in Signal Processing. CRC Press, Boca Raton, FL. ISBN 978-1-4665-1583-3.

Press, W.H., Teukolsky, S.A., Vetterling, W.T., Flannery, B.P., 2007. Numerical Recipes. The Art of Scientific Computing. Cambridge University Press, p. 1262.

Renson, P., 1978. Method for finding the periods of variable stars. Astronomy & Astrophysics 63, 125–129.

Roberts, D.H., Lehar, J., Dreher, J.W., 1987. Time series analysis with clean – part one – derivation of a spectrum. The Astronomical Journal 93 (4), 968–989.

Samus, N.N., Kazarovets, E.V., Durlevich, O.V., Kireeva, N.N., Pastukhova, E.N., 2017. General catalogue of variable stars: version GCVS 5.1. Astronomy Reports 61, 80.

Scargle, J.D., 1982. Studies in astronomical time series analysis. II. Statistical aspects of spectral analysis of unevenly spaced data. The Astrophysical Journal 263, 835–853. https://doi.org/10.1086/160554.

Schaffer, J., 2015. What not to multiply without necessity. Australasian Journal of Philosophy 93 (4), 644–664.

Scheffe, H., 1959. The Analysis of Variance. Wiley, p. 477.

Shao, J., Tu, D., 1996. The Jackknife and Bootstrap, 2nd corrected printing. Springer. ISBN 978-1461269038, pp. 517.

Shul'Berg, A.M., 1971. Close Binary Systems with Spherical Components. Nauka, Moskva, p. 246.

Stellingwerf, R.F., 1978. Period determination using phase dispersion minimization. The Astrophysical Journal 224, 953.

Street, R.A., et al., 2003. SuperWASP: wide angle search for planets. In: ASP Conference Series, vol. 294, pp. 405–408.

Sturges, H.A., 1926. The choice of a class interval. Journal of the American Statistical Association 21 (153), 65–66.

Sutherland, P.G., et al., 1978. Short-term time variability of Cygnus X-1. II. The Astrophysical Journal 219, 1029.

Terebizh, V.Yu., 1998. Similarity law in spectral estimation of a time series. V. Astrophysics 41, 198–201.

TESS, 2019. heasarc.gsfc.nasa.gov/docs/tess/.

Tikhonov, A.N., 1963. On the solution of ill-posed problems and the method of regularization. Doklady Akademii Nauk SSSR 151, 501.

Tkachenko, M.G., Andronov, I.L., Chinarova, L.L., 2016. Phenomenological parameters of the prototype eclipsing binaries Algol, β Lyrae and W UMa. Journal of Physical Studies 20, 4902.

Tolstov, G.P., 2012. Fourier Series. Courier Corporation, p. 352.

Tremko, J., et al., 1996. Periodic and aperiodic variations in TT Arietis. Results from an international campaign. Astronomy & Astrophysics 312, 121.

Tsesevich, V.P. (Ed.), 1973. Eclipsing Variable Stars. J. Wiley, New York.

Udalski, A., Kubiak, M., Szymanski, M., 1997. Optical gravitational lensing experiment. OGLE-2 – the second phase of the OGLE project. Acta Astronomica 47, 319–344.

Vavilova, I.B., et al., 2012. Astroinformation resource of the Ukrainian virtual observatory: joint observational data archive, scientific tasks, and software. Kinematics and Physics of Celestial Bodies 28, 85.

Vavilova, I.B., et al., 2017. UkrVO astroinformatics software and web-services. In: Astroinformatics, Proceedings of the International Astronomical Union, IAU Symposium, vol. 325, p. 361.

Wehlau, William, Leung, Kam-Ching, 1964. The multiple periodicity of Delta Delphini. The Astrophysical Journal 139, 843.

Wermuth, N., 2011. Multivariate statistical analysis. In: Lovric, M. (Ed.), International Encyclopedia of Statistical Science, Part 13. Springer, New York, pp. 915–920.

Wozniak, P.R., et al., 2004. Northern sky variability survey: public data release. The Astronomical Journal 127, 2436–2449. ADS:2004AJ....127.2436W.

Wright, E.L., et al., 2010. The Wide-field Infrared Survey Explorer (WISE): mission description and initial on-orbit performance. The Astronomical Journal 140, 1868.

ZTF. https://irsa.ipac.caltech.edu/Missions/ztf.html.

CHAPTER 12

Learning in Big Data: Introduction to Machine Learning

KHADIJA EL BOUCHEFRY, PHD • RAFAEL S. DE SOUZA, PHD

12.1 BRIEF HISTORY OF MACHINE LEARNING

12.1.1 Introduction

Machine learning (ML) is concerned with algorithms and techniques that allow computers to learn. The ML approach covers main domains, such as data mining, difficult to program applications, and software applications. It is a collection of a variety of algorithms that can provide multivariate, nonlinear, nonparametric regression or classification. The remarkable simulation capabilities of the ML-based methods have resulted in their extensive applications in science and engineering. Recently, the ML techniques have found many applications in astronomy and the geosciences and remote sensing. More specifically, these techniques are proved to be practical for cases where the system's deterministic model is computationally expensive or there is no deterministic model to solve the problem.

Part of what makes ML so broadly applicable is the diversity of ML algorithms capable of performing very well under messy, real-world conditions. Despite, and perhaps because of, this versatility, uptake of ML applications has lagged behind traditional statistical techniques in astronomy and geosciences.

12.1.2 The Modern History of Artificial Intelligence and ML

ML is seen as a subset of artificial intelligence that develops dynamic algorithms capable of data-driven decisions, in contrast to models that follow static programming instructions (Nilsson, 2010). It is almost impossible to cover one field without involving the other. The following provides brief details on the development of both fields.

- 1950: Alan Turing created the world-famous Turing Test (Turing, 1950). This test is fairly simple – for a machine or a computer to be intelligent, it has to be able to convince a human that it is a human and not a computer.
- 1951: Minsky (1952) and Dean Edmonds developed the first artificial neural network (ANN), called the Stochastic Neural Analog Reinforcement Computer (SNARC), which consisted of 40 neurons.
- 1952: Arthur Samuel, American pioneer in the field of computer gaming and artificial intelligence, created the first computer program which could learn as it ran. It was a game which played checkers, which quickly learned how to play a better game than Arthur Samuel could himself.
- 1956: Artificial intelligence gained its first name in the 1956 Dartmouth conference. John McCarthy and two senior scientists, Claude Shannon and Nathan Rochester, organized a workshop in 1956 at Dartmouth College for researchers interested in neural nets, automated theory, and the study of intelligence. The workshop's proposal (McCarthy et al., 1955) for the conference included the following declaration: *"every aspect of learning or any other feature of intelligence can be so precisely described that a machine can be made to simulate it."* The 10 participants included (see Fig. 12.1)[1] Arthur Samuel (IBM[2]), Trenchard More (Princeton), and Oliver Selfridge and Ray Solomonoff (MIT[3]). Herbert Simon and Allen Newell, two attendees from Carnegie Tech.[4] already had an ongoing project, the Logic Theorist (Newell and Simon, 1956), which was regarded after the workshop as the first artificial intelligence system.
- 1958: Frank Rosenblatt designed the first ANN for computers, called perceptron. The main goal of this was pattern and shape recognition (Rosenblatt, 1958).
- 1959: Widrow and Hoff (1960) built models called "ADALINE" (recognizes binary pattern) and "MADALINE" (first neural network used to eliminate echoes on phone lines). Both models use what is called Multiple ADAptive LINear Elements. Few years later

[1] https://www.scienceabc.com/innovation/what-is-artificial-intelligence.html.
[2] International Business Machines Corporation.
[3] Massachusetts Institute of Technology.
[4] Now Carnegie Mellon University (CMU).

Knowledge Discovery in Big Data from Astronomy and Earth Observation. https://doi.org/10.1016/B978-0-12-819154-5.00023-0

FIG. 12.1 The founding fathers of artificial intelligence. Courtesy of Scienceabc.com.

Marvin Minsky and Seymour Papert proved the perceptron to be limited. At about the same time period, Arthur Samuel first defined ML as a field of study that gives computers the ability to learn without being explicitly programmed. Despite the success of MADALINE, there was not much progress until late in 1980, for many reasons (e.g., popularity and simplicity of the Von Neumann architecture).

- 1967: Cover et al. (1967) introduced the nearest neighbor (NN) rule. This rule was mentioned earlier as "minimum distance classifier" by Nilsson (1965) and as "proximity algorithm" by Sebestyen (1962).
- 1982: Hopfield (1982) presented a paper to the National Academy of Sciences where he suggested to create more useful machines (using bidirectional lines) instead of neurons. Furthermore, a joint US–Japan conference on cooperative/competitive neural networks took place in 1982 at which Japan announced their fifth-generation computing involving artificial intelligence.
- 1985: The American Institute of Physics started an annual meeting of Neural Networks in Computing. This was followed by the first International Conference on Neural Networks by the Institute of Electrical and Electronic Engineers (IEEE) in 1987.
- 1997: Mitchell (1997) defined ML as any computer program that improves its performance at some task through experience.
- 2012 Kevin Murphy defined ML as methods that can automatically detect patterns in data, and then use them to predict future data or other outcomes of interest. The use of the term has been growing steadily since 1980.

ML and artificial intelligence took separate paths since the late 1970s and early 1980s. A number of large projects were developed, such as GoogleBrain (2012), AlexNet (2012), DeepFace (2014), OpenAI (2015), Amazon Machine Learning Platform (2015), and ResNet, to name a few. Today ML is widely used in several applications all around us in the classification of junk/spam e-mails, fraud detection, language translation, speech recognition, medical diagnosis, social media services, financial services, data analysis, etc. ML is revolutionizing the world and will continue to grow in the coming years.

12.1.3 What Is Learning?

The basic premise of ML is that a machine (i.e., algorithm or model) is able to make new predictions based on data. Some algorithms are supervised, meaning they are shown data a priori and then make predictions about new data based on the previous data. Some are unsupervised, meaning they can make predictions with no a priori data. Some are a combination of the two, (i.e., semisupervised). The basic technique behind all ML methods is an iterative combination of statistics and error minimization or reward maximization, applied and combined to varying degrees. Many ML algorithms iteratively check all or a very high number of possible outcomes to find the best result, defined by the user for the problem at hand. The potentially high number of iterations is prohibitive of manual calculations and is a large part of the reason why these methods are only now widely available to individual researchers. The key notion is that flexible, automatic approaches are used to detect patterns within the data, with a primary focus on making predictions on future data.

12.1.4 Why Use Machine Learning Instead of Traditional Statistics?

The advantage of ML over traditional statistical techniques, especially in geosciences, is the ability to model highly dimensional and nonlinear data with complex interactions and missing values. Astronomy and geoscience data specifically are known to be nonlinear and highly dimensional with intense interaction effects; yet, methods that assume linearity and are unable to cope with interaction effects are still being used. To make these methods work, researchers cope in various ways, including (1) data transformations, which can limit the interpretability of the final results, (2) decompose/recompose methods to break up the system into bits with fewer complicated dynamics, or (3) assuming linearity without any modification of the data. Several comparative studies have already shown that ML techniques can outperform traditional statistical approaches in a wide variety of problems (Levine et al., 1996; Prasad et al., 2006; Zhao et al., 2011); however, comparing techniques can be difficult and requires careful consideration (Fielding, 2006).

The exact division between ML methods and traditional statistical techniques is not always clear and ML methods are not always better than traditional statistics. For example, a system may not be linear, but a linear approximation of that system may still yield the best predictor. The exact method(s) must be chosen based on the problem at hand and a metaapproach that considers the results of multiple algorithms may be best.

12.2 TYPES OF LEARNING

The first step in applying ML is teaching the algorithm using a training dataset. The training dataset is a collection of independent variables with the corresponding dependent variables. The machine uses the training data to learn how the independent variables (input) relate to the dependent variable (output). Later, when the algorithm is applied to new input data, it can apply that relationship and return a prediction. After the algorithm is trained, it needs to be tested to get a measure of how well it can make predictions from new data. This requires another dataset with independent and dependent variables, but the dependent variables are not provided to the learner. The algorithm predictions are compared to the withheld data to determine the quality of the predictions. This process requires a dataset that is large enough to be split in two for training and testing. The type of ML method, the size and nature of the training and test dataset, and the evaluation method should be chosen to optimize the trade-off between bias and accuracy to give a meaningful result for the problem at hand.

ML algorithms can be classified into many different paradigms, based on the desired outcome of the algorithm. In the following, we provide brief descriptions of four learning paradigms.

12.2.1 Supervised Learning

Supervised learning is one of the most widely used ML algorithms. In supervised learning, the training data you use are already labeled. These training data are used to infer a learning algorithm or mapping function from the input variable (X) to the output variable (Y). The correct answers or desired outputs (labels), here, are already known, given a labeled set of input–output pairs, $M = \{\sum(X_i, Y_i)\}_i^N$; N is simply the number of training examples. The training input X_i is a **d**-dimensional vector or numbers also known as features, or attributes. The input X_i can be an image, an email message, a time series, a molecular shape, or a graph. The output Y_i, also known as a response variable, is a categorical or nominal variable for a classification problem or real value for a regression problem. Classification algorithms and regression techniques (see Fig. 12.2) are two types of supervised learning widely used to develop predictive models.

12.2.2 Unsupervised Learning

Unsupervised learning (also known as knowledge discovery) uses unlabeled, unclassified, and categorized training data. The main goal of unsupervised learning is to discover hidden and interesting patterns in unlabeled data. Unlike supervised learning, unsupervised learning methods cannot be directly applied to a regression or a classification problem as one has no idea what the values for the output might be. Clustering is the most com-

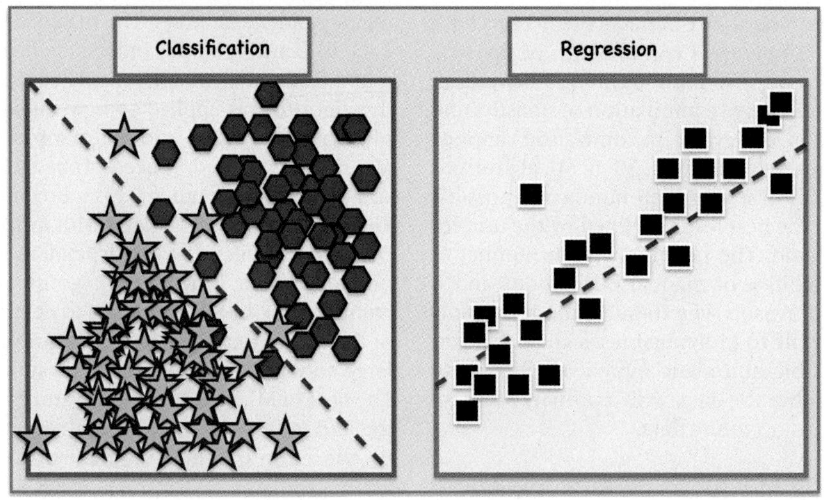

FIG. 12.2 Supervised learning.

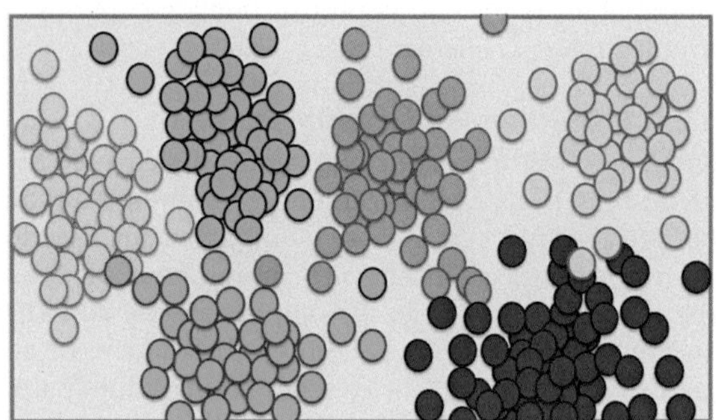

FIG. 12.3 Unsupervised learning.

mon unsupervised learning algorithm used to explore the data analysis to find hidden patterns or groupings in the data (Fig. 12.3). Applications for cluster analysis include gene sequence analysis, market research and object recognition. Common algorithms used in unsupervised learning include clustering, anomaly detection, neural networks, and approaches for learning latent variable models.

12.2.3 Semisupervised Learning

Semisupervised learning is a combination of supervised and unsupervised ML methods. Semisupervised learning algorithms (see Fig. 12.4, *left*) make use of partially labeled training data – typically a small amount of labeled data with a large amount of unlabeled data. Semisupervised algorithms are trained on a combination of labeled and unlabeled data. This is very useful for improving the learning accuracy.

12.2.4 Reinforcement Learning

Reinforcement learning (Sutton et al., 1998) is a type of dynamic programming that trains algorithms using a system of reward and penalty. The learning system, called agent in this context, learns with an interactive environment. The agent selects and performs actions and receives rewards by performing correctly and penalties for performing incorrectly. In reinforcement learning the agent learns by itself, without the intervention from a human, the best strategy to maximize reward in a particular situation using dynamic programming. Unlike unsupervised learning, reinforcement learning is different in terms of goals, while the goal in unsupervised learning is to find a suitable action model that would maximize the total cumulative reward of the agent. Fig. 12.4 (*right*) represents the basic idea and elements involved in reinforcement learning model. Typical practical applications of reinforcement learning include the building of artificial intelligence for playing computer games, robotics and industrial automation, text summarizing engines, dialogue agent (text, speech), etc.

12.2.5 Active Learning

Vanilla supervised learning makes use of available labeled data to fit a learning model. Active learning (AL) adds an extra layer of flexibility by giving the learner a degree of control to select few, but highly informative, instances to be labeled and added to the training set (Settles, 2012; Balcan et al., 2009; Cohn et al., 1996). A typical AL begins with a small labeled set L, selects one

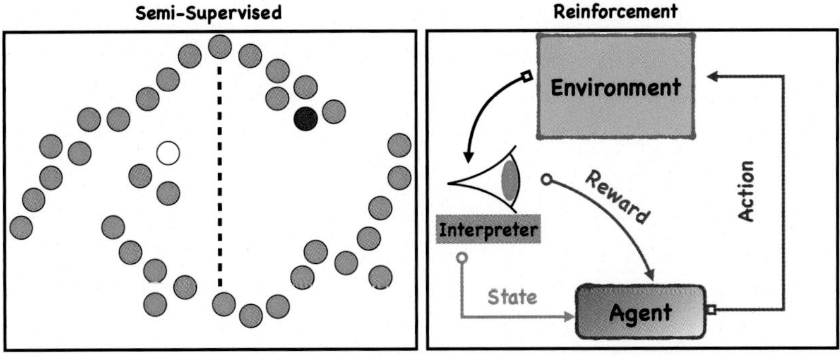

FIG. 12.4 *Left*: Semisupervised learning. *Right*: Reinforcement learning.

or more informative query instances from a large unlabeled pool U, learns from these labeled queries (which are then added to L), and repeats. In this way, the learner aims to achieve high accuracy with as little labeling effort as possible. Thus, AL can be valuable in domains where unlabeled data are readily available, but obtaining training labels is expensive.

There are many variants of AL; examples include *query synthesis*, where the learner can query instances from any region of the input (i.e., feature) space $\mathcal{X}$; or *stream-based selective sampling*, where instances are sampled according to the underlying sample distribution. The most common approach to AL is known as *pool-based learning*, where one assumes the existence of a dataset T_u from which unlabeled instances can be queried. A key concept in AL is that each query is associated with a cost; a trade-off then exists between improving the current model by adding more labeled instances and minimizing the overall cost needed to acquire the corresponding class labels. AL is currently a hot topic in astronomy for classifying cosmic explosions known as Type Ia supernovae (Ishida et al., 2019; Ishida, 2019).

12.3 MACHINE LEARNING ALGORITHMS

There are many ML algorithms (see Fig. 12.5) in the literature; the following provides a brief description of the most widely used algorithms.

12.3.1 Naive Bayes Classifier

Naive Bayes (NB) is a well-known statistical learning algorithm recommended as a base-level classifier for comparison with other algorithms (Guyon, 2009). NB estimates class conditional probabilities by naively assuming that for a given class the inputs are independent of each other. This assumption yields a discrimination function indicated by the products of the joint probabilities that the classes are true given the inputs. NB reduces the problem of discriminating classes to finding class conditional marginal densities, which represent the probability that a given sample is one of the possible target classes (Molina et al., 1994). NB performs well against other alternatives unless the data contain correlated inputs (Hastie et al., 2005; Witten et al., 2016).

12.3.2 k-Nearest Neighbors

The k-nearest neighbors (kNN) algorithm (Cover et al., 1967) is a very simple nonparametric algorithm widely used for classification and regression. kNN is an instance-based learner (also known as lazy learning) that does not train a classification model until provided with samples to classify (Kotsiantis, 2007). The main principle of kNN during classification is that individual test samples are compared locally to k neighboring training samples in variable space, and their category is identified according to the classification of the nearest k neighbors. Neighbors are commonly identified using a Euclidian distance metric (Altman, 1992) between the investigated data point and the k neighbors. Predictions are based on a majority vote cast by neighboring samples (Kotsiantis, 2007; Witten et al., 2016). The k value (number of closest neighbors) is usually small so as not to include a lot of other data points which will obscure the true nature of the data point in question. As high k can lead to overfitting and model instability; appropriate values must be selected for a given application (Hastie et al., 2005).

The NN is the simplest form of kNN when k = 1. Fig. 12.6 shows the kNN decision rule for k = 1 and k = 5 for a set of samples divided into two classes. In Fig. 12.6A, an unknown sample is classified by using only one known sample; in Fig. 12.6B more than one

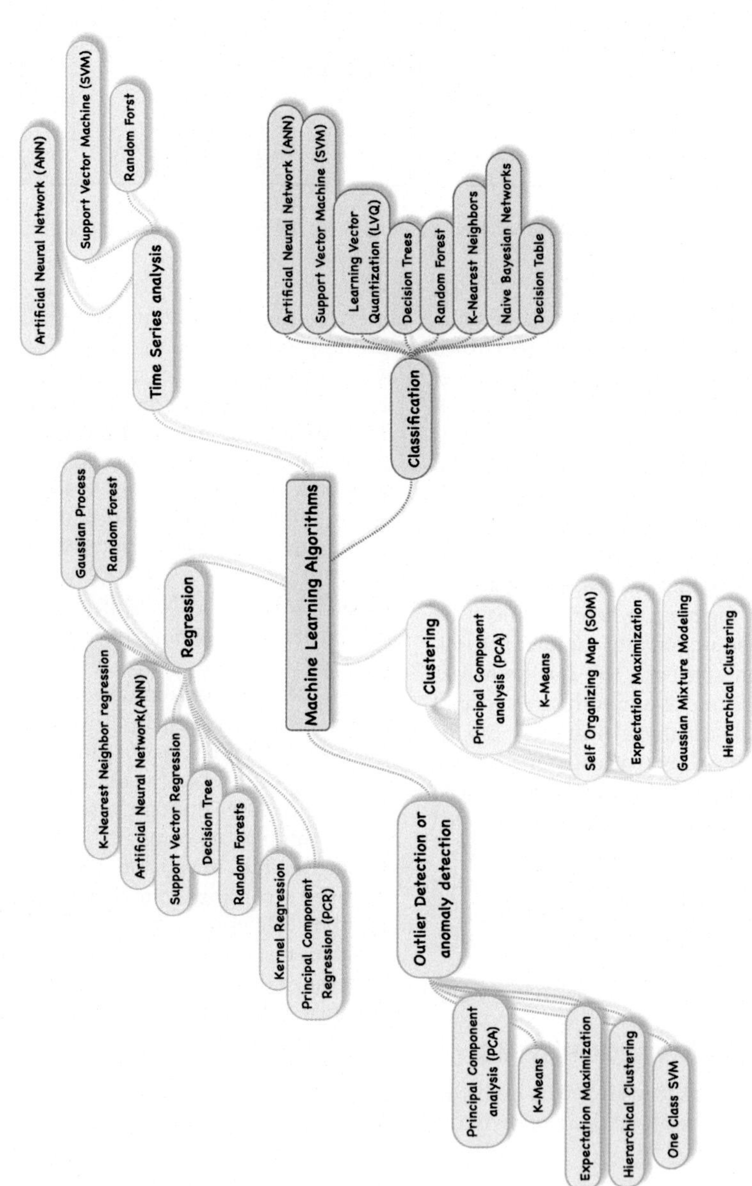

FIG. 12.5 A map of different types of machine learning algorithms.

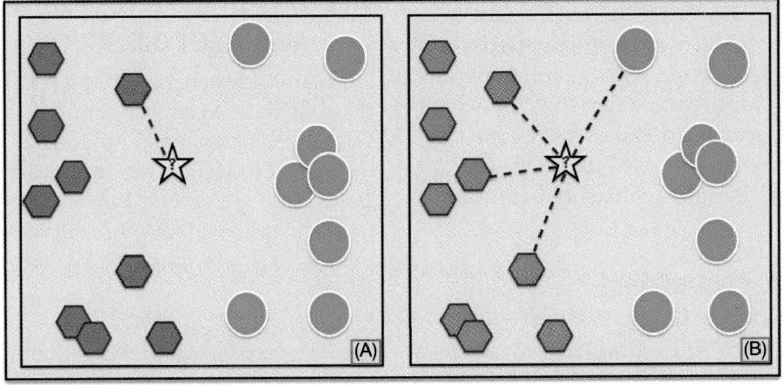

FIG. 12.6 The 1-NN decision rule. (A) The point star is assigned to the class on the left. (B) The kNN decision rule, with k = 4: the point (star) is assigned to the class on the left as well.

known sample is used. In the last case, the parameter k is set to 4, so that the closest four samples are considered for classifying the unknown one. Three of them belong to the same class, whereas only one belongs to the class. In both cases, the unknown sample is classified as belonging to the class on the left.

12.3.3 Support Vector Machine

Support vector machines (SVMs) are among the most popular classification schemes in the field of ML. SVM, formally described by Vapnik (1998), has the ability to define nonlinear decision boundaries in high-dimensional variable space by solving a quadratic optimization problem (Hsu et al., 2016; Karatzoglou et al., 2006). The key idea of this concept is to identify the hyperplane separating different classes such that the induced distance between the hyperplane and the training patterns is maximal. Basic SVM theory states that for a nonlinearly separable dataset containing points from two classes there are an infinite number of hyperplanes that divide classes. The selection of a hyperplane that optimally separates two classes (i.e., the decision boundary) is carried out using only a subset of training samples known as support vectors. The maximal margin M (distance) between the support vectors is taken to represent the optimal decision boundary. In nonseparable linear cases, SVM finds M while incorporating a cost parameter C, which defines a penalty for misclassifying support vectors. High values of C generate complex decision boundaries in order to miss-classify as few support vectors as possible (Karatzoglou et al., 2006). For problems where classes are not linearly separable, SVM uses an implicit transformation of input variables using a kernel function, which allows SVM to separate nonlinearly separable support vectors using a linear hyperplane (Yu et al., 2012). Selection of an appropriate kernel function and kernel width, s, are required to optimize performance for most applications (Hsu et al., 2016). SVM can be extended to multiclass problems by constructing $c(c-1)/2$ binary classification models, the so-called one-against-one method, which generates predictions based on a majority vote (Hsu and Lin, 2002; Melgani and Bruzzone, 2004).

12.3.4 Random Forest

Random forest (RF), developed by Breiman (2001), is an ensemble classification scheme that utilizes a majority vote to predict classes based on the partition of data from multiple decision trees. RF grows multiple trees by randomly subsetting a predefined number of variables to split at each node of the decision trees and by bagging. Bagging generates training data for each tree by sampling with replacement a number of samples equal to the number of samples in the source dataset (Breiman, 1996). RF implements the Gini index to determine a best split threshold of input values for given classes. The Gini index returns a measure of class heterogeneity within child nodes as compared to the parent node (Breiman, 2017; Waske et al., 2009). RF requires the selection of mtry parameter, which sets the number of possible variables that can be randomly selected for splitting at each node of the trees in the forest.

12.3.5 Artificial Neural Network

ANNs have been widely used in science and engineering problems. They attempt to model the ability of biological nervous systems to recognize patterns and objects. The ANN basic architecture consists of networks of primitive functions capable of receiving multiple weighted inputs that are evaluated in terms of their success at discriminating the classes in Ta. Different types of primitive functions and network configurations result in varying models (Hastie et al., 2005). During training, network connection weights are adjusted if the separation of inputs and predefined classes incurs an error. Convergence proceeds until the reduction in error between iterations reaches a decay threshold (Kotsiantis, 2007).

12.3.6 Multilayer Perceptron

The multilayer perceptron (MLP) is an effective and powerful algorithm for solving supervised learning problems. Artificial neurons are a simplified version of biological neurons showing many interesting and useful properties in the research field. The starting point in ANN was set in 1943 by Warren McCulloch and Walter Pitts, who studied networks of binary threshold elements.

12.3.7 Dimensionality Reduction

Dimensionality reduction, or variable reduction techniques, simply refers to the process of reducing the number or dimensions of features in a dataset. It is commonly used during the analysis of high-dimensional data (e.g., multipixel images of a face or texts from an article, astronomical catalogues, etc.). Many statistical and ML methods have been applied to high-dimensional data, such as vector quantization and mixture models, generative topographic mapping (Bishop et al., 1998), and principal component analysis (PCA), to list just a few. PCA is one of the most popular algorithms used for dimensionality reduction (Pearson, 1901; Wold et al., 1987; Dunteman, 1989; Jolliffe and Cadima, 2016).

It is an unsupervised learning technique of dimensionality reduction also known as Karhunen–Loève transform, generally applied for data compression and visualization, feature extraction, dimensionality reduction, and feature extraction (Bishop, 2000). It is defined as a set of data being projected orthogonally onto lower-dimensional linear space (called the principal subspace) to maximize the projected data's variance (Hotelling, 1933). Other common methods of dimensionality reduction worth mentioning are independent component analysis (Comon, 1994), nonnegative matrix factorization (Lee and Seung, 1999), self-organized maps (Kohonen, 1982), isomaps (Tenenbaum et al., 2000), t-distributed stochastic neighbor embedding (van der Maaten and Hinton, 2008), Uniform Manifold Approximation and Projection for Dimension Reduction (McInnes et al., 2018), and autoencoders (Vincent et al., 2008).

12.4 MACHINE LEARNING IN ASTRONOMY AND GEOSCIENCES

For many researchers, ML is a relatively new paradigm that has only recently become accessible with the development of modern computing. While the adoption of ML methods in astronomy and geosciences has been slow, there are several published studies using ML in these disciplines. The following is a brief review of some published applications of ML in astronomy and geosciences.

12.4.1 Case Studies in Astronomy

Astronomy is faced with a data avalanche driven by the enormous technological advances in telescopes and detectors, electronics, and computer technologies (Ball et al., 2008), with an exponentially increasing amount of astronomical data. The complexity and dimension of astronomical data are likewise growing rapidly. Extracting information from such data becomes a critical and challenging problem. The volumes of astronomical data amount to many terabytes, even petabytes, from which catalogues or images of many millions, or even billions of objects are extracted. What is more important is that much of the science will be done purely in the catalogue domain of individual or federated sky surveys. Moreover, with the upcoming next generation radio telescope facilities, such as the Square Kilometre Array (SKA) and its related pathfinders (e.g., SKA Pathfinder), tens of millions of radio sources will be detected. Handling this amount of data is not possible through manual studies: automation of data processing is therefore essential and a new generation of algorithms using ML and other data mining techniques are needed. Astronomical data can be described by the four Vs: volume, variety, velocity, and value.

- **Volume**: refers to the amount of data. Astronomical surveys produce data measurable in terabytes, petabytes, and even exabytes. This high volume poses challenges for capture, cleaning, curation, integration, storage, processing, indexing, searching, sharing, transferring, mining, analysis, and visualization. Such amounts of data cannot be processed in depth in a classical manner.
- **Velocity**: refers to the speed at which the data are generated, transmitted, processed, and analyzed.
- **Variety**: refers to the type and nature of the data. Astronomical data mainly include images, spectra, time series data, and simulation data. In astronomy, the data collected from different telescopes have their own formats. Moreover, data from astronomical surveys have a thousand or more features, causing a large dimensionality problem. In addition, data have

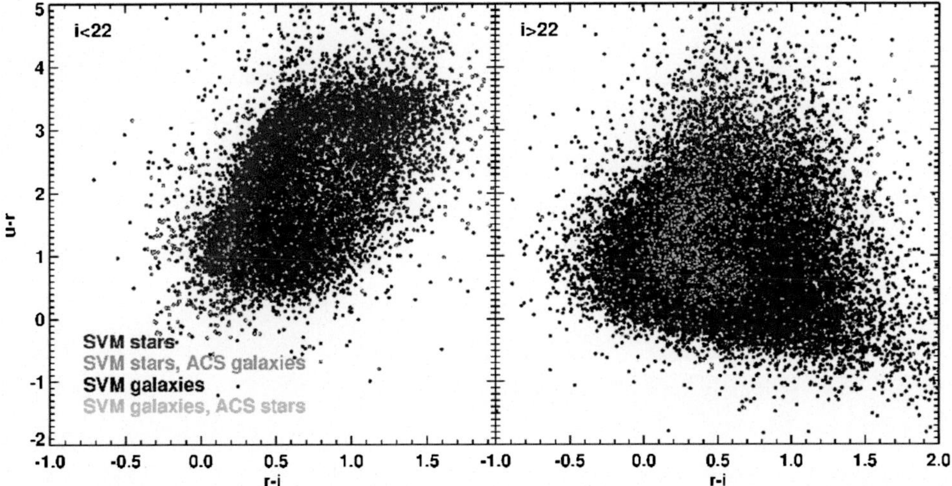

FIG. 12.7 Results of the GA-SVM algorithms used by Heinis et al. (2016) to classify objects into stars and galaxies: red points show objects classified as stars by the GA-SVM, and blue points show objects classified as galaxies. Orange and cyan points show misclassified objects: orange points are galaxies according to the catalogue of Leauthaud et al. (2007) classified as stars, and cyan points are stars classified as galaxies (Heinis et al., 2016).

many data types, i.e., structured, semistructured, unstructured, and mixed.

- **Value**: is the end of the game; after addressing a number of Vs, it is very important to ensure getting value from the data. It is interesting in astronomy to discover unexpected objects and phenomena.

The following provides a review on applications of ML in astronomy and the usage of these algorithms in different areas and for different problems as a whole. Common ML applications cover a wide range, including dimensionality reduction, classification, regression, prediction, clustering, and filtering.

12.4.1.1 Object Classification

12.4.1.1.1 Star Galaxy Classification Accurate classification of astrophysical objects from astronomical images is an essential process that provides fundamental understanding of the astronomical surveys in the field of astronomy. Specifically, separating foreground stars from background galaxies is a crucial and vital step for many astronomical research topics, from galactic science to cosmology. While conventional morphological classification techniques separate point sources (mostly stars) from resolved sources (galaxies) using selections in magnitude-radius space or similar variables (MacGillivray et al., 1976; Kron, 1980; Yee, 1991), for brighter sources, morphology has proven to be a sufficient tool metric for classification. At fainter magnitudes unresolved galaxies and blended sources contaminate the galaxy sample, because distant and/or faint sources start to merge into single detected objects with superiors shapes. Hence classification of stars and galaxies at faint magnitudes can introduce spurious correlations in galaxy surveys (Ross et al., 2011) and will hamper the study of stellar distributions (Drlica-Wagner et al., 2015). ML methods proved to be very efficient at separating stars from galaxies at fainter magnitudes (Weir et al., 1995).

Several ML algorithms and neural networks have been applied to the problem of star galaxy separation. ANNs were first applied to the problem of star galaxy classification in the work of Odewahn et al. (1992) and have become a core part of the astronomical image processing software SExtractor (Bertin and Arnouts, 1996). Bai et al. (2019) applied a random forest to the star/galaxy/quasar classification in the Large Sky Area Multi-Object Fiber Spectroscopic Telescope and SDSS. Plewa (2018) used an RF classification to classify stars in the galactic center. Heinis et al. (2016) utilized a new combination of two ML methods to star galaxy separation (Fig. 12.7), and Vasconcellos et al. (2011) explored the efficiency of 13 different DT algorithms applied to photometric objects in the SDSS Release Seven in order to separate them into stars and galaxies. The 13 DT different algorithms were provided by the Wekato Envi-

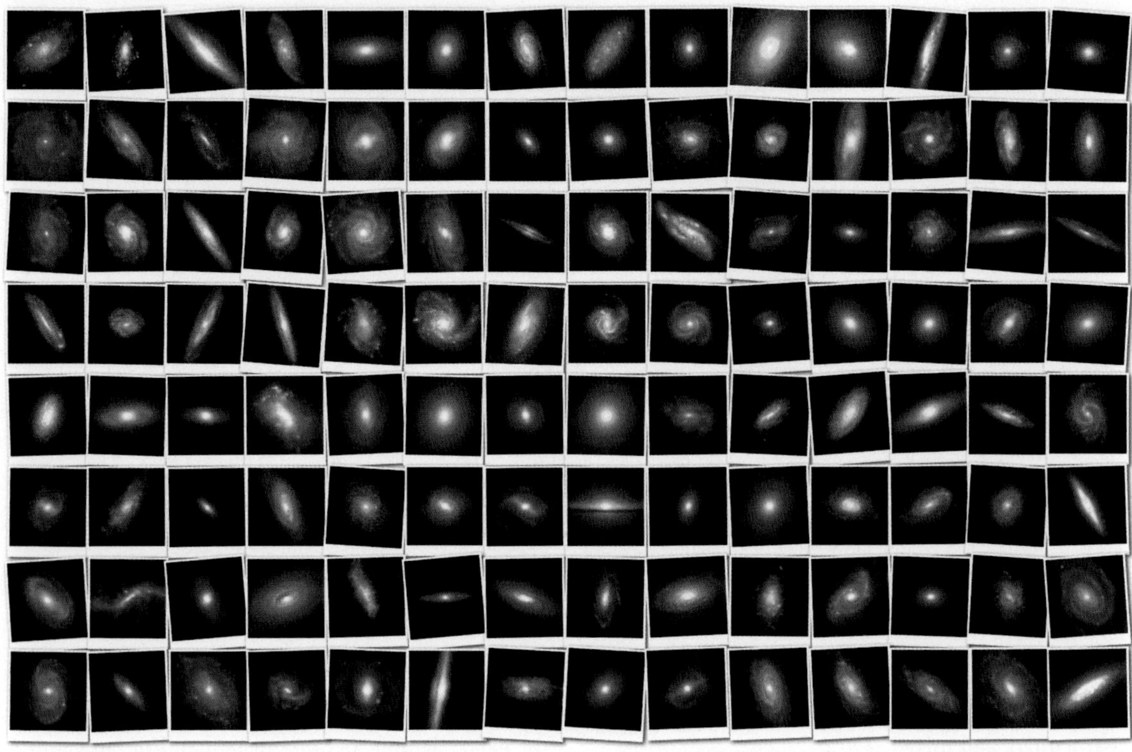

FIG. 12.8 A sampling of galaxies with various morphological types and colors. Credit: Zsolt Frei, Institute of Physics, Eötvös University, Budapest.

ronment for Knowledge Analysis (WEKA) data mining tool.

12.4.1.1.2 Galaxy Morphology Astronomers use the term "morphology" to classify galaxies into different groups based on their visual appearance (see Fig. 12.8). The morphological classification of galaxies divides them into elliptical, spiral, barred spiral, and irregular, along with various subclasses (Hubble, 1926). This classification system is called the Hubble sequence and was invented by Hubble (1926). Galaxy morphology is a complex phenomenon that correlates to the underlying physics. It is correlated with the galaxy color indices, luminosity, and other parameters as de Vaucouleurs radius, inverse concentration index, etc. Understanding the relation between galaxies and their environment is a crucial part of understanding how galaxies form and evolve.

Automatic galaxy classification has been addressed by various authors and different approaches, such as neural networks (Domínguez Sánchez et al., 2018; Kim and Brunner, 2017; Ball et al., 2004; Bazell, 2000; Goderya and Lolling, 2002; Lahav, 1997), Bayesian networks (Dobrycheva et al., 2017; Bazell and Aha, 2001; de la Calleja and Fuentes, 2004), and decision trees (Barchi et al., 2017; Ball et al., 2004; Owens et al., 1996). Recently Dai and Tong (2018) trained three CNNs to classify galaxies, and to also explore the high-dimensional abstract feature representations and how these features help to understand the galaxy image data themselves. Tao et al. (2018) explored the automated galaxy spectral classification using three supervised ML algorithms (logistic regression, RF, and linear SVM). These algorithms were applied on a dataset of 10,000 galaxy spectra of SDSS fourth Data Release. Turner et al. (2019) applied the unsupervised k-means clustering to a sample of 7338 local universe galaxies in order to address the bimodality of galaxies in the local universe. Morphological classifications of galaxies have been tackled with ML algorithms in numerous papers using different approaches, like PCA, SVM, deep neural networks

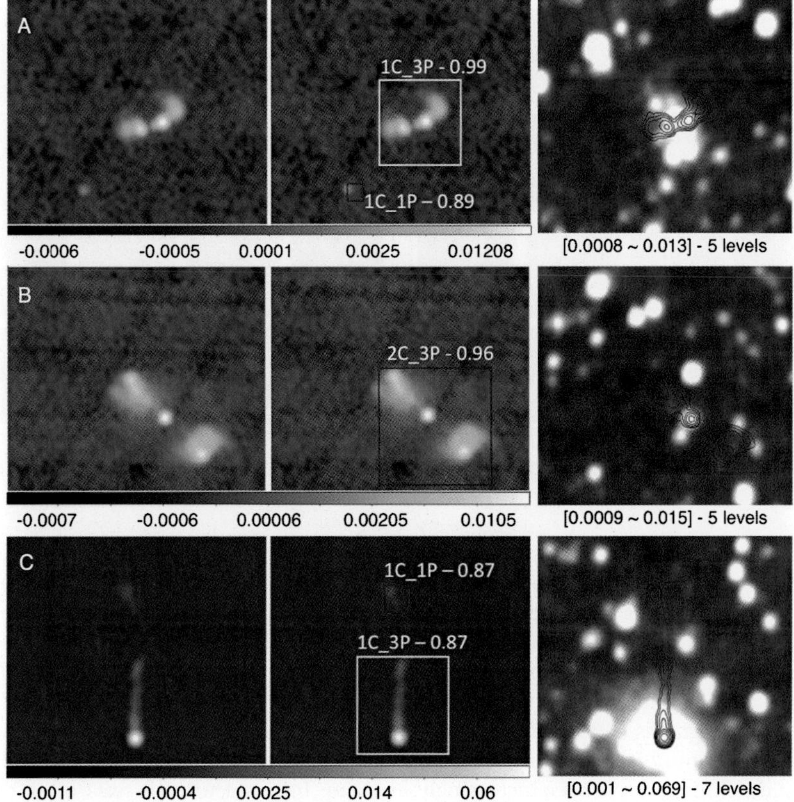

FIG. 12.9 Three classification examples (A, B, and C) on RGZ subjects – each of them $3' \times 3'$ in size, FIRST J081700.6+571626, FIRST J070822.2+414905, and FIRST J083915.7+285125 (Wu et al., 2019).

(Beck et al., 2014; Karampelas, 2013; Yip et al., 2004; Connolly et al., 1995), etc.

In addition, ML algorithms were also used to classify galaxies observed in the radio part of the electromagnetic spectrum. Radio galaxies can be classified into FRI morphologies (edge-darkened) and FRII (edge-brightened) morphologies (Fanaroff and Riley, 1974); several approaches and attempts have been made recently to classify and identify radio sources through automated techniques. Polsterer et al. (2015) used the self-organizing Kohonen map dimensionality reduction technique to classify radio galaxies. Aniyan and Thorat (2017) used CNNs to classify radio galaxies into FRI, FRII, and bent-tailed radio galaxies. Alhassan et al. (2018) extended the later work to compact sources (see Fig. 12.10). Alger et al. (2018) used NNs to classify the host galaxies of radio sources. Recently Wu et al. (2019) applied NNs to classify radio sources into different radio morphologies (Fig. 12.9), and Ma et al. (2019) applied CNN to classify 14,245 radioactive galactic nuclei, achieving 93% precision.

12.4.1.2 Photometric Redshift

The distance to an astronomical object is one of the most important quantities that we want to measure. In extragalactic studies, the best way of estimating source distance is via its redshift. Redshifts are an essential tool for constraining dark matter and dark energy contents of the universe (Serjeant, 2014), for the identification of galaxy clusters and groups (Annunziatella et al., 2016; Capozzi et al., 2009), to reconstitute the universe large-scale structure (Aragon-Calvo et al., 2015), for mapping the galaxy color–redshift relationships (Masters et al., 2015), and for the classification of astronomical sources (Brescia et al., 2012), to quote just a few applications. Spectroscopic methods provide the most accurate measure of redshift. However, with the enormous technological advances in telescopes detectors, electronics,

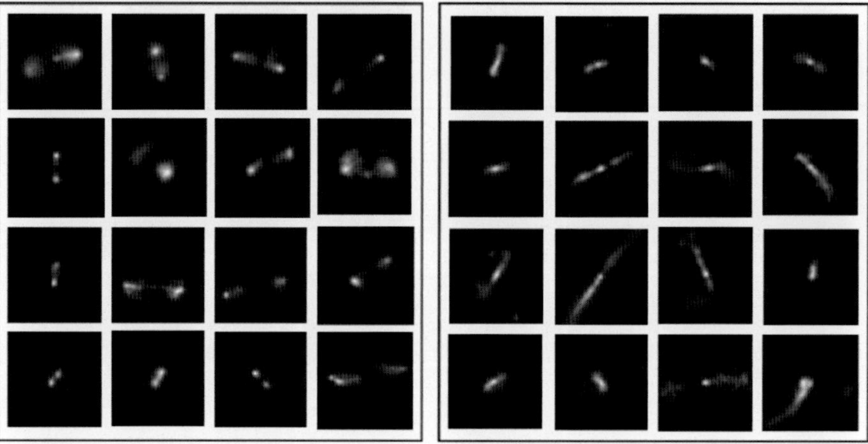

FIG. 12.10 Morphological classification of FIRST radio sources into FRI and FRII morphologies radio sources (Alhassan et al., 2018).

and computer technologies, and the upcoming next-generation radio facilities, such as SKA and its related pathfinder, billions of photometric data will be produced. Obtaining very accurate spectroscopic redshifts for such amounts of data requires very long exposure times on dedicated spectrographs and is time consuming and costly (note on faint objects). For this reason, cheaper methods and techniques have been increasingly implemented to estimate redshifts of astronomical objects (i.e., galaxies and quasars). Without entering into much detail, photometric redshift estimation methods can be divided into template fitting techniques, ML techniques, or some hybrid of the two (Hoyle et al., 2015a).

A wide plethora of ML techniques have been successfully used to estimate photometric redshifts for galaxies, quasars, and X-ray sources (Mountrichas et al., 2017). ANN is one of the popular techniques used for the evaluation of photometric redshifts (Bilicki et al., 2018; Hoyle et al., 2015b; Sadeh et al., 2016; Yèche et al., 2010; Abdalla et al., 2008; Oyaizu et al., 2008; Blake et al., 2007; Collister and Lahav, 2004; Firth et al., 2003). Elliott et al. (2015) employed a semisupervised gamma regression model to handle non-Gaussianities in photo-z scatter. Kind and Brunner (2013) used a combination of prediction trees and RF techniques for estimating photometric redshifts. Brescia et al. (2013) applied the MLP with Quasi Newton Algorithm (MLPQNA) to evaluate photometric redshifts of quasars with the datasets from four different surveys (SDSS, GALEX, UKIDSS, and WISE). Way and Klose (2012) applied self-organizing map (SOM) to estimate photometric redshifts. Gerdes et al. (2010) estimated photometric redshifts for galaxies using a boosted decision tree method.

12.4.1.3 Data Mining Software and Tools

- **DAMEWARE**[5] (DAta Mining and Exploration Web Application REsource) is an innovative, general purpose, web-based distributed platform for data mining infrastructure (Brescia et al., 2014), capable to work on massive datasets and compliant with virtual observatory standards. DAMEWARE offers a number of supervised and unsupervised ML algorithms, including MLP trained by backpropagation learning rule, MLPQNA, SVM, self-organizing features maps, k-means, and probabilistic principal surfaces, to name a few. It has been used in astrophysics for photometric redshift evaluation, globular cluster search and classification, AGN classification, and multiepoch sky survey.
- **AstroML**[6] is a Python module developed for ML (VanderPlas et al., 2012) and data mining, which is built on numpy, scipy, scikit learn, mtplotlib, and astropy. It is distributed under the 3-clause BSD license. AstroML includes a growing library of statistical and ML routines in Python and several uploaded open astronomical datasets. AstroML aims to provide the astronomical community with fast Python algorithms and routines for statistical data analysis in astronomy and astrophysics and an easy to use interface.

[5] http://dame.dsf.unina.it.

[6] https://github.com/astroML/astroML.

- **Weka**[7]/**AstroWeka**[8] stands for Waikato environment for knowledge analysis. Weka is an open source, easy to use, and user-friendly for applied ML algorithms. It has graphical user interface and also a command line interface where all features of the software can be used from the command line. It is a useful tool when working with massive datasets where scripting helps in the automation of the work. AstroWeka is a set of extensions to Weka focusing on astronomical data mining tasks. Both Weka and AstroWeka contain ML algorithms for data preprocessing, classification, regression, clustering, association rules, and visualization.

12.4.2 Case Studies in Geoscience

Massive amounts of data are stored in almost all disciplines. Remote sensing is not an exception, since very large time series or high-resolution satellite and aerial images are sources of valuable information in geosciences. How to extract useful information from these Big Data sources is not, by contrast, an easy task due to the computational and infrastructural costs involved. Very powerful approaches have been developed in the context of advanced ML and Big Data analytics during the last few years. Such approaches deal with large datasets, considering all samples and measurements. With them, advanced ML and Big Data methods for extracting relevant patterns, high-performance computing, or data visualization are being nowadays successfully applied to the field of remote sensing. There have been few but significant contributions to this, in the context of Big Data science and Big Data, by interrogating relevant approaches in data fusion and spatiotemporal modeling in geosciences. These contributions are real-world case studies.

ML algorithms use an automatic inductive approach to recognize patterns in data. Once learned, pattern relationships are applied to other similar data in order to generate predictions for data-driven classification and regression problems. ML algorithms have been shown to perform well in situations involving the prediction of categories from spatially dispersed training data and are especially useful where the process under investigation is complex and/or represented by a high-dimensional input space (Kanevski et al., 2009). The basic premise of supervised classification is that it requires training data containing labeled samples representing what is known about the inference target (Kotsiantis, 2007; Ripley and Hjort, 1996; Witten et al., 2016). The ML algorithm architecture and the statistical distributions of observed data guide the training of classification models, which is usually carried out by minimizing a loss (error) function (Kuncheva, 2014; Marsland, 2014). Trained classification models are then applied to similar input variables to predict classes present within the training data (Hastie et al., 2005; Witten et al., 2016).

The majority of published research focusing on the use of ML algorithms for the supervised classification of remote sensing data has been for the prediction of land cover, environmental protection, agriculture, vegetation classes, and so on (e.g., Foody and Mathur, 2004; Ham et al., 2005; Huang et al., 2002; Song et al., 2012). These studies use multi- or hyperspectral reflectance imagery as inputs and training data are sourced from manually interpreted classes. Hyperspectral image (HSI) classification is an active and very hot topic in remote sensing commonly used to classify objects by their spectral feature. The main purpose of HSI is to obtain the spectrum for each pixel in the image of a scene and then classify the pixels into one or several classes. Recently, Guo et al. (2018, 2019) applied both the SVM and kNN algorithms respectively combined with a guided filter for HSI classification. Among other ML algorithms used in remote sensing, CNNs have been applied to scene classification (Cheng et al., 2018) and HSI classification (Cheng et al., 2018; Zhao and Du, 2016). Persello et al. (2019) combined CNN with a globalization and grouping algorithm to detect field boundaries (see Fig. 12.11) and Mboga et al. (2017) used CNNs to perform automatic detection of informal settlements over Dar-es-Salaam and other areas in Tanzania.

ML algorithms such as RF, SVM, and ANN are commonly compared in terms of their predictive accuracies to more traditional methods of classifying remote sensing data, such as the maximum likelihood classifier (MLC). In general, RF and SVM outperform ANN and MLC, especially when faced with a limited number of training samples and a large number of inputs and/or classes.

Common to all remote sensing image classification studies is the use of geographical referenced input data containing collocated pixels specified by coordinates linked to a spatial reference frame. Nevertheless, inputs used in the majority of studies cited do not include reference to the spatial domain. This is equivalent to carrying out the classification task in geographic space where samples are only compared numerically (Gahegan, 2000). To date few investigations have evalu-

[7] https://www.cs.waikato.ac.nz/ml/weka/.
[8] http://astroweka.sourceforge.net.

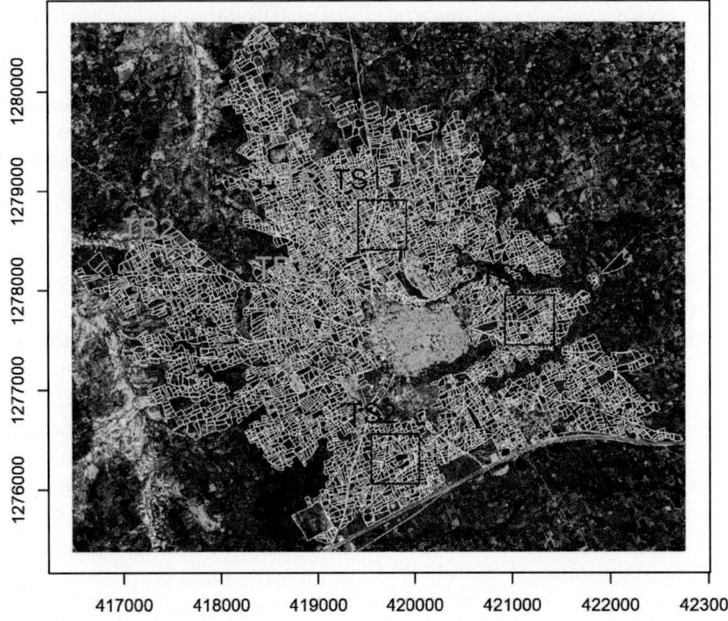

FIG. 12.11 False color composite (bands 7, 5, 3) of the WorldView-3 image acquired over the Kofa study area on 25 September 2015. (C) 2015 DigitalGlobe, Inc., Longmont CO USA 80503. The image is in WGS 84 UTM zone 32 projection, units are meters. Field polygons and lines are shown in yellow. Training and testing tiles are indicated by the green and blue squares, respectively (Persello et al., 2019).

ated the performance of ML algorithms in conjunction with the input of spatial coordinates.

12.4.3 Simple Case Study in Geology: Supervised Classification of Lithology

ML algorithms are programs of data-driven inference tools that offer an automated means of recognizing patterns in high-dimensional data. Hence, there is much scope for the application of ML algorithms to the rapidly increasing volumes of remotely sensed geophysical data for geological mapping and other problems.

Cracknell and Reading (2014) carried out a rigorous comparison of five ML algorithms, representing the five general learning strategies: NB statistical learning algorithms, kNN instance-based learners, RF logic-based learners, SVMs, and ANNs – Perceptrons (Kotsiantis, 2007). This was undertaken in the context of a supervised lithology classification task using widely available and spatially constrained remotely sensed geophysical data. They make a further comparison of ML algorithms based on their sensitivity to variations in the degree of spatial clustering of training data, and their response to the inclusion of explicit spatial information (spatial coordinates).

The work identifies RF as a good first choice algorithm for the supervised classification of lithology using remotely sensed geophysical data. RF is straightforward to train, computationally efficient, highly stable with respect to variations in classification model parameter values, and as accurate as, or substantially more accurate than, the other ML algorithms trialed. The results of the study indicate that as training data become increasingly dispersed across the region under investigation, ML algorithms' predictive accuracy improves dramatically. The use of explicit spatial information generates accurate lithology predictions but should be used in conjunction with geophysical data in order to generate geologically plausible predictions. ML algorithms, such as RF, are valuable tools for generating reliable first-pass predictions for practical geological mapping applications that combine widely available geophysical data.

12.4.4 Common Properties

Astronomy and geosciences are characterized by massive amounts of data that are sources of valuable information. The application of ML methods in astronomy and geosciences has already demonstrated the potential for increasing the quality and accelerating the pace of science. One of the more obvious ways ML does this is by coping with data gaps, especially in geosciences. The Earth is undersampled. Where possible, ML allows a researcher to use data that are plentiful or easy to collect to infer data that are scarce or hard to collect. Another important way ML can fill in data gaps is through downscaling and performing spatial databases. There will never be enough research funding to sample everything all of the time. ML can be a method for addressing the data gaps that prevent scientific progress.

ML can accelerate the pace of science by quickly performing complex classification tasks normally performed by a human. A bottleneck in many astro and geoscience workflows are the manual steps performed by an expert, usually a classification task such as identifying a rock type or an object. Rather than having all of the data classified by an expert, the expert only needs to review enough data to train and test an algorithm. Expert annotation can be even more time consuming when the expert must search through a large volume of data, like a sensor stream, for a desired signal. This bottleneck has been addressed for some types of classifications, finding relevant data in sensor streams, and building a reference knowledge base. In addition to relieving a bottleneck, ML methods can sometimes perform tasks more consistently than experts can, especially when there are many categories and the task continues over a long period of time. In these cases, ML methods can improve the quality of science by providing more quantitative and consistent data.

As discussed above, ML techniques can perform better than traditional statistical methods in systems that are poorly represented by linear models, but a direct comparison of performance between ML techniques and traditional statistical methods is difficult because there is no universal measure of performance and results can be very situation-specific (Fielding, 2006). The true measure of the utility of a tool is how well it can make predictions from new data and how well it can be generalized to new situations. Highly significant P-values, R2 values, and accuracy measurements may not reflect this. If the accuracy is not significantly improved using ML, it may be better to use a traditional method that is more familiar and accepted by peers. Best practice is to test multiple methods (including traditional statistics) while probing the trade-off between bias and accuracy and choose the tool that is most useful. In many natural systems, where nonlinear and interaction effects are common, an ML-based model is more useful and can improve science by building better models. Individual researchers need to select a method based on the specific problem and the data at hand.

12.5 SCALABLE MACHINE LEARNING ALGORITHMS

12.5.1 What Is a Scalable Machine Learning Algorithm?

Scalable ML algorithms (SMLAs) are algorithms which can deal with any large amount of datasets, without consuming tremendous amounts of resources like memory. SMLAs occur when statistics, systems, ML, and data mining are combined into flexible, often nonparametric, and scalable techniques for analyzing massive amounts of data. It is an important and challenging aspect of many ML projects aiming for fast computations and analysis of gigantic sets of data.

12.5.2 Scalable Clustering

Clustering has long been used to organize a dataset of objects into classes or clusters of similar type (observations, features, vectors, or objects) of unlabeled data. It is a method of unsupervised learning useful for exploring datasets that are not yet well understood. There are several algorithms that can be used to cluster a dataset, each with its own advantages and disadvantages. The most popular and commonly used clustering algorithms are k-means (Xu et al., 1993; Jain et al., 1999) and K-medoids. It is fast, easy to understand, and available everywhere (i.e., implemented in many ML toolkits). There are many clustering algorithms (see Fig. 12.12) in the data mining literature. Broadly speaking, clustering methods can be classified into the following categories.

12.5.2.1 Hierarchical Methods

Hierarchical clustering methods are methods of cluster analysis which create a hierarchical decomposition of the given datasets. Hierarchical clustering methods are classified into divisive (top-down) and agglomerative (bottom-up), depending on whether the hierarchical decomposition is formed in a bottom-up or top-down fashion. An agglomerative clustering starts with a singleton (one object) cluster and then successively merges pairs of clusters until all clusters have been merged into one big cluster containing all objects. Divisive clustering is a reverse approach of agglomerative clustering; it starts with one cluster of the data and then partitions the appropriate cluster. Although hierarchical clustering is easy to implement and applicable to any attribute type, they are very sensitive to outliers and do not work with missing data. Moreover, initial seeds have a strong

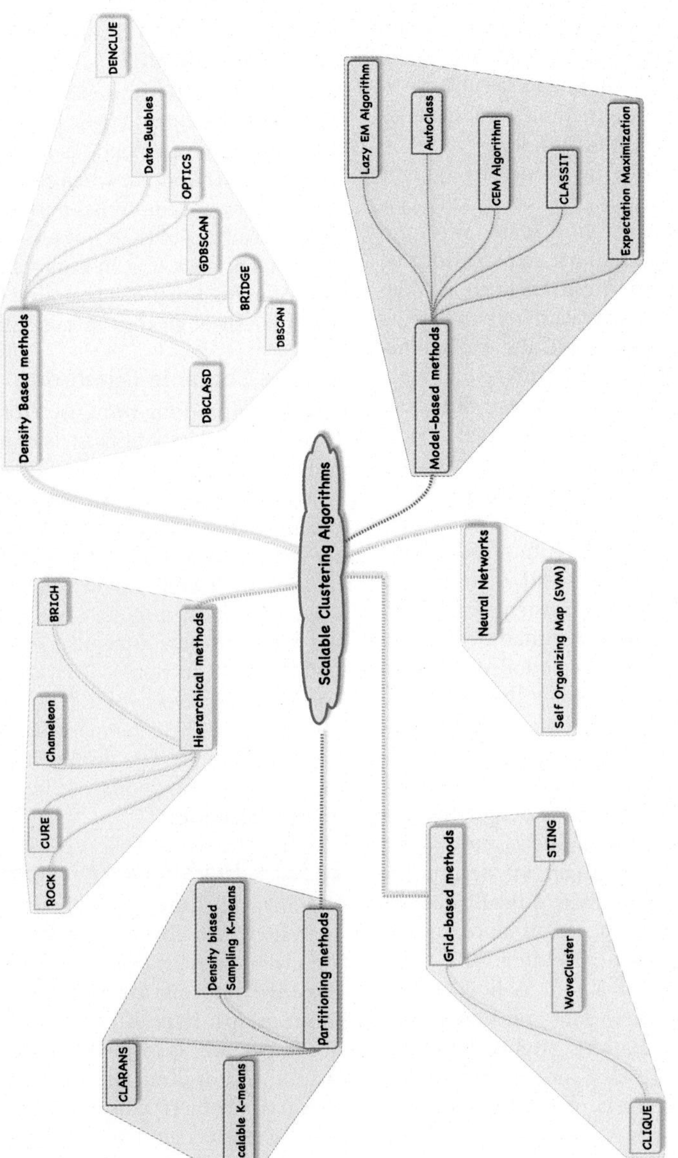

FIG. 12.12 Scalable algorithms.

impact on the final results (involving lots of arbitrary decisions).

BRICH (balanced iterative reducing and clustering using hierarchies) is a scalable clustering method designed for handling very large datasets. It uses statistical methods to compress the data such that an existing clustering algorithm could use the summaries rather than the original data to reduce the necessary number of computations (only one scan of the data is necessary). The BRICH clustering algorithm consists of two main phases of operation. The first phase (building the CF tree) loads the data into memory by building a cluster feature (CF) tree and the second phase (global clustering) applies an existing algorithm on the leaves of the CF tree.

CURE (clustering using representatives) uses random sampling and partitioning to reliably find clusters of arbitrary shape and size. It clusters random samples of the database in an agglomerative fashion. Each random sample is represented by a fixed number c of representative data points and then shrunk gradually towards the center of the cluster by a specific fraction α. CURE produces high-quality clusters of complex shapes and sizes but does not handle categorical attributes.

12.5.2.2 Density-Based Methods

The density-based clustering (Ankerst et al., 1999; Ester et al., 1996) method uses kernel density estimation (Hwang et al., 2003) to identify areas where points are concentrated and where they are separated by areas that are empty or sparse. The density-based methods are effective when analyzing datasets that contains high levels of noise or when datasets are arbitrary-shaped. Among density-based algorithms, **DBSCAN** (Density-Based Spatial Clustering and Application with Noise) is one of the most well-known clustering algorithms, proposed by Ester et al. (1996), and capable of detecting clusters with different shapes and sizes, but it fails to detect clusters with different densities. To improve **DBSCAN**, Sander et al. (1998) introduced the generalized **GDBSCAN** algorithm enabling clustering of point objects using both spatial and nonspatial attributes. Dash et al. (2001) proposed **BRIDGE** to enable **DBSCAN** to handle large datasets efficiently by improving the partitions generated by k-means. Other density-based algorithms include **OPTICS** (Ordering Points To Identify the Clustering Structure), introduced by Ankerst et al. (1999), **Data Bubbles**, introduced by Breunig et al. (2001) to speed up **OPTICS**, and **DENCLUE** (DENsity-based CLUstEring), developed by Hinneburg et al. (1998), to name a few.

12.5.2.3 Grid-Based Methods

Grid-based methods divide the object space into a finite number of cells that form a grid structure and then cluster them. They assign objects to the appropriate grid cell and compute the density of each cell and eliminate cells whose density is below a defined threshold. In this approach, clustering complexity depends on the number of populated grid cells and not on the number of objects in the dataset.

STING (STatistical INformation Grid approach) is another grid-based multiresolution clustering algorithm in which the spatial area is divided into rectangular cells (i.e., using latitude and longitude) and employs a hierarchical structure (Wang and Muntz, 1997). Each cell at a high level is portioned to form a number of cells of the next lower level. It exhibits high performance but the accuracy is relatively low. To improve the accuracy and efficiency of the grid clustering algorithm, Liu et al. (2017) proposed the DSM (Diagonal grid Searching and Merging) algorithm, in which the grid is divided into space and then divided into "valid grid" or "invalid grid" according to the grid containing at least the predetermined threshold parameter. **CLIQUE** (CLustering In QUEst), introduced by Agrawal et al. (1999), is a representative algorithm based on density and grid techniques for clustering in high-dimensional data space. **Wave Cluster**, proposed by Sheikholeslami et al. (2000), is a multiresolution clustering algorithm used to find clusters in very large spatial databases. This approach clusters objects using a wavelet transform method. **MAFIA** (Merging of Adaptive Intervals Approach to Spatial Data Mining) is a parallel subspace clustering algorithm using adaptive computation of the finite intervals in each dimension.

12.5.3 Scalable Prediction: Classification and Regression

The objective of classification is to analyze the input data and to develop an accurate description or model for each class using the features present in the data. The class descriptions are used to classify future test data for which the class labels are unknown. Given the size of present-day data, classifiers that scale well and can handle training data of this magnitude have been developed (Fernández-Delgado et al., 2014; Muja and Lowe, 2014; Shafer et al., 1996). The ability to classify large training datasets can also improve the classification accuracy (Breiman et al., 1984; Bradley et al., 1998).

There has been an increasing interest in developing regression models for large datasets that are both accurate and easy to interpret. Novel regression tree construction algorithms have been proposed that are accurate and can also truly scale to very large datasets (Breiman, 2017; Breiman et al., 1984; Bradley et al., 1998; Gehrke et al., 1998; Li et al., 2000). More recently, apart from classification and regression, extreme learning machine (ELM) has recently been extended for clustering, feature selection, representational learning, and many other learning tasks. Due to its remarkable efficiency, simplicity, and impressive generalization performance, ELM has been applied in a variety of domains, such as biomedical engineering, computer vision, system identification, and control and robotics (Huang et al., 2015).

12.5.4 Scalable Pattern Mining

A current focus of intense research in pattern classification is the combination of several classifier systems, which can be built following either the same or different models and/or datasets building (Woźniak et al., 2014).

Data mining systems aim to discover patterns and extract useful information from facts recorded in databases. A widely adopted approach to this objective is to apply various ML algorithms to compute descriptive models of the available data. One of the main challenges in this area is the development of techniques that scale up to large and possibly physically distributed databases.

With the growing popularity of shared resources, large volumes of complex data of different types are collected automatically. Traditional data mining algorithms generally have problems and challenges including huge memory cost, low processing speed, and inadequate hard disk space. As a fundamental task of data mining, sequential pattern mining (SPM) is used in a wide variety of real-life applications. However, it is more complex and challenging than other pattern mining tasks, i.e., frequent itemset mining and association rule mining, and also suffers from the above challenges when handling large-scale data. To solve these problems, mining sequential patterns in a parallel or distributed computing environment has emerged as an important issue with many applications. Gan et al. (2019) provide an in-depth survey of the current status of parallel SPM, including detailed categorization of traditional serial SPM approaches, and state-of-the-art parallel SPM.

12.6 SCALABLE ML FRAMEWORKS

12.6.1 Apache Spark

Apache Spark is a popular open source platform for Big Data processing (Zaharia et al., 2012). It is best known for its speed in memory primitives. Apache Spark is highly scalable ML, well suited for iterative ML tasks. Moreover, new ML algorithms are constantly being added and the existing ones are enhanced. Spark Ecosystem comprises multiple components, including Spark SQL, Spark Streaming, MLLib (Machine Learning), and GraphX (see Fig. 12.13). These components are built on top of the Spark Core Engine. The following provides a brief introduction to Spark components.

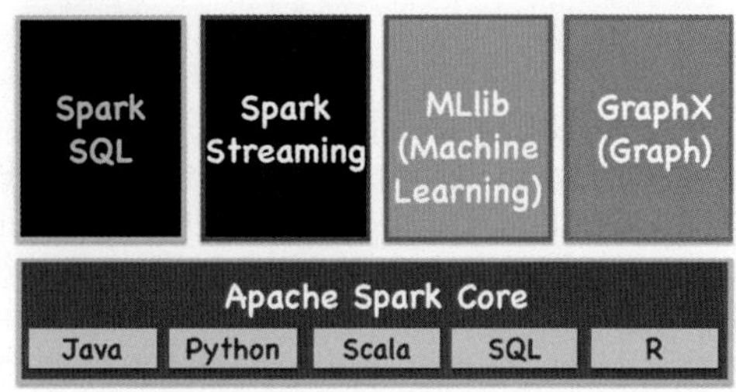

FIG. 12.13 Apache Spark Ecosystem – Spark Core, Spark SQL, Spark Streaming, MLlib, and GraphX.

12.6.1.1 Components of Apache Spark

- **Apache Core** Spark Core is the base framework of Apache Spark. The key features of Apache Spark Core are task dispatching, scheduling, basic I/O functionalities, and fault recovery. It is based on what is called resilient distributed datasets (RDDs, Zaharia et al., 2012). An RDD is an immutable distributed collection of datasets partitioned across a set of nodes of the cluster that can be recovered if a partition is lost, thus providing fault tolerance. It also provides inbuilt memory computing and referencing datasets stored in external storage systems.
- **Spark SQL** is a component on top of Spark Core that introduces DataFrames, which is a new data abstraction method for structured and semistructured data. Spark SQL offers the user the possibility to run SQL queries in the Spark programs, to read data from an existing Hive installation, and to easily write RDDs out to hive tables or parquet files. It also includes SQL language support with command line interface, columnar storage, and code generation to make faster queries.
- **Spark Streaming** is the core of Spark's API extension that allows data engineers and data scientists to process real-time data from different sources (e.g., Kafka, Flume, Amazon Kinesis, etc.) by dividing them into mini batches which are then quickly processed by the Spark engine to generate the final stream of results in batches.
- **Machine learning Library (MLlib)** is Apache Spark's scalable ML primitives. MLlib allows the user to ap-

ply ML to large amounts of datasets with no scalability concerns. This library consists of a number of popular algorithms and utilities such as classification, regression, clustering, collaborative filtering, optimization, dimensionality reduction, and feature extraction. Spark MLLib can be easily integrated with other Spark components such as Spark SQL, Spark Streaming, and DataFrames. The library supports Java, Scala, and Python as part of Spark applications.

- **Spark GraphX** is a component for graphs and graph-parallel computation. Spark GraphX allows the user to view, transform, and join interchangeably both graphs and collections with RDDs efficiently. It also allows the users to write and custom iterative graph algorithms using Pregel abstraction (Malewicz et al., 2010).

12.6.2 Flink ML

Apache Flink,[9] formerly known as Stratosphere (Alexandrov et al., 2014), is essentially a recent open source framework designed to process both streams and batch workloads. Apache Link features several libraries typically embedded in an API and not fully self-contained.

- **Gelly** is a library for graph processing system in Apache Flink. The library features graph analysis on Flink applications using a set of methods and utilities. The user can analyze huge graphs using Apache Flink API in a distributed fashion with Gelly.
- **Flink ML** is the ML library for Apache Flink. This library provides a set of scalable ML algorithms and supports a number of algorithms, including supervised learning, unsupervised learning, data preprocessing, recommendation, outlier selection, pipelines, and other utilities.
- **Table API and SQL** are two simple APIs for accessing streaming data featured by Apache Flink. It supports Java, Scalable Dataset, and Datastream APIs. This API allows the user to create tables from existing Datasets and Datastreams or from external data sources.
- **FlinkCEP** is the complex event processing library. This library provides an API to analyze and detect complex events patterns in continuous streaming data.

12.7 INFERENCE AND LEARNING IN ASTRONOMY AND GEOSCIENCES

At a high level, the goal of artificial intelligence is to replace algorithms that are explicitly programmed by people with algorithms that learn on their own. Hand-coding routines are highly inefficient, error-prone, and not adaptable to new information. Instead, deep learning is used to learn features and patterns that best represent data automatically. The process by which these systems learn is called training.

Training refers to the process of creating an ML algorithm. Training involves the use of a deep learning framework (e.g., TensorFlow) and a training dataset. Astronomical and geoscience data provide a source of training data that scientists can use to train ML models for a variety of use cases.

Inference refers to the process of using a trained ML algorithm to make a prediction. Astronomical and geoscience data can be used as the input to a trained ML model, enabling predictions that can guide decision logic. An important development in ML is the emergence of ML inference servers (also known as inference engines and inference servers). The ML inference server executes the model algorithm and returns the inference output.

12.8 SUMMARY

The geosciences study complex interactions between systems in order to infer understanding and make predictions. ML-based methods have an advantage over traditional statistical methods in studying these systems because the former do not impose unrealistic assumptions (such as linearity), are capable of inferring missing data, and can reduce long-term expert annotation burden. Thus, a wider adoption of ML methods in geosciences has the potential to greatly accelerate the pace and quality of science. Despite these advantages, ML techniques have not enjoyed wide spread adoption in geosciences. This is largely due to (1) a lack of communication and collaboration between the ML research community and natural scientists, (2) a lack of easily accessible tools and services, and (3) the requirement for robust training and test datasets. These impediments can be overcome through financial support for collaborative work and the development of tools and services facilitating ML use. Geoscientists who have not yet used ML methods can be introduced to these techniques through simple and easy to implement ML methods that perform well.

REFERENCES

Abdalla, F.B., Amara, A., Capak, P., Cypriano, E.S., Lahav, O., Rhodes, J., 2008. Photometric redshifts for weak lensing tomography from space: the role of optical and near infrared

[9] https://flink.apache.org.

photometry. Monthly Notices of the Royal Astronomical Society 387 (3), 969–986.

Agrawal, R., Gehrke, J.E., Gunopulos, D., Raghavan, P., 1999. Automatic subspace clustering of high dimensional data for data mining applications. In: Proc. of 1999 ACM SIGMOD International Conference on Management of Data, pp. 61–72.

Alexandrov, A., Bergmann, R., Ewen, S., Freytag, J., Hueske, F., Heise, A., Kao, O., Leich, M., Leser, U., Markl, V., Naumann, V., Peters, M., Rheinländer, A., Sax, M., Schelter, S., Höger, M., Tzoumas, K., Warneke, D., 2014. The stratosphere platform for big data analytics. The VLDB Journal 23.

Alger, M.J., Banfield, J.K., Ong, C.S., Rudnick, L., Wong, O.I., Wolf, C., Andernach, H., Norris, R.P., Shabala, S.S., 2018. Radio Galaxy Zoo: machine learning for radio source host galaxy cross-identification. Monthly Notices of the Royal Astronomical Society 478, 5547–5563.

Alhassan, W., Taylor, A.R., Vaccari, M., 2018. The FIRST Classifier: compact and extended radio galaxy classification using deep Convolutional Neural Networks. Monthly Notices of the Royal Astronomical Society 480, 2085–2093.

Altman, N.S., 1992. An introduction to kernel and nearest-neighbor nonparametric regression. American Statistician 46 (3), 175–185.

Aniyan, A.K., Thorat, K., 2017. Classifying radio galaxies with the convolutional neural network. The Astrophysical Journal. Supplement Series 230, 20.

Ankerst, M., Breunig, M.M., Kriegel, H.-P., Sander, J., 1999. Optics: ordering points to identify the clustering structure. In: ACM Sigmod Record, vol. 28. ACM, pp. 49–60.

Annunziatella, M., Mercurio, A., Biviano, A., Girardi, M., Nonino, M., Balestra, I., Rosati, P., Bartosch Caminha, G., Brescia, M., Gobat, R., Grillo, C., Lombardi, M., Sartoris, B., De Lucia, G., Demarco, R., Frye, B., Fritz, A., Moustakas, J., Scodeggio, M., Kuchner, U., Maier, C., Ziegler, B., 2016. CLASH-VLT: environment-driven evolution of galaxies in the $z = 0.209$ cluster Abell 209. Astronomy & Astrophysics 585, A160.

Aragon-Calvo, M.A., van de Weygaert, R., Jones, B.J.T., Mobasher, B., 2015. Submegaparsec individual photometric redshift estimation from cosmic web constraints. Monthly Notices of the Royal Astronomical Society 454 (1), 463–477.

Bai, Y., Liu, J., Wang, S., Yang, F., 2019. Machine learning applied to star-galaxy-QSO classification and stellar effective temperature regression. The Astronomical Journal 157 (1), 9.

Balcan, M.F., Beygelzimer, A., Langford, J., 2009. Agnostic active learning. Journal of Computer and System Sciences 75 (1), 78–89.

Ball, N.M., Brunner, R.J., Myers, A.D., Strand, N.E., Alberts, S.L., Tcheng, D., 2008. Robust machine learning applied to astronomical data sets. III. Probabilistic photometric redshifts for galaxies and quasars in the SDSS and GALEX. The Astrophysical Journal 683, 12–21.

Ball, N.M., Loveday, J., Fukugita, M., Nakamura, O., Okamura, S., Brinkmann, J., Brunner, R.J., 2004. Galaxy types in the Sloan Digital Sky Survey using supervised artificial neural networks. Monthly Notices of the Royal Astronomical Society 348, 1038–1046.

Barchi, P.H., da Costa, F.G., Sautter, R., Moura, T.C., Stalder, D.H., Rosa, R.R., de Carvalho, R.R., 2017. Improving galaxy morphology with machine learning. ArXiv e-prints, arXiv: 1705.06818.

Bazell, D., 2000. Feature relevance in morphological galaxy classification. Monthly Notices of the Royal Astronomical Society 316, 519–528.

Bazell, D., Aha, D.W., 2001. Ensembles of classifiers for morphological galaxy classification. The Astrophysical Journal 548, 219–223.

Beck, R., Dobos, L., Csabai, I., 2014. Quantifying correlations between galaxy emission lines and stellar continua using a PCA-based technique. In: Heavens, A., Starck, J.-L., Krone-Martins, A. (Eds.), Statistical Challenges in 21st Century Cosmology. In: IAU Symposium, vol. 306, pp. 301–303.

Bertin, E., Arnouts, S., 1996. SExtractor: software for source extraction. Astronomy & Astrophysics. Supplement Series 117, 393–404.

Bilicki, M., Hoekstra, H., Brown, M.J.I., Amaro, V., Blake, C., Cavuoti, S., de Jong, J.T.A., Georgiou, C., Hildebrandt, H., Wolf, C., Amon, A., Brescia, M., Brough, S., Costa-Duarte, M.V., Erben, T., Glazebrook, K., Grado, A., Heymans, C., Jarrett, T., Joudaki, S., Kuijken, K., Longo, G., Napolitano, N., Parkinson, D., Vellucci, C., Kleijn, G.A.V., Wang, L., 2018. Photometric redshifts for the Kilo-Degree Survey. Machine-learning analysis with artificial neural networks. Astronomy & Astrophysics 616, A69.

Bishop, C., 2000. Pattern Recognition and Machine Learning. Springer, New York.

Bishop, C.M., Svensén, M., Williams, C.K., 1998. GTM: the generative topographic mapping. Neural Computation 10 (1), 215–234.

Blake, C., Collister, A., Bridle, S., Lahav, O., 2007. Cosmological baryonic and matter densities from 600000 SDSS luminous red galaxies with photometric redshifts. Monthly Notices of the Royal Astronomical Society 374 (4), 1527–1548.

Bradley, P.S., Fayyad, U.M., Reina, C., et al., 1998. Scaling clustering algorithms to large databases. In: KDD, vol. 98, pp. 9–15.

Breiman, L., 1996. Bagging predictors. Machine Learning 24 (2), 123–140.

Breiman, L., 2001. Random forests. Machine Learning 45 (1), 5–32.

Breiman, L., 2017. Classification and Regression Trees. Routledge.

Breiman, L., Friedman, J., Olshen, R., Stone, C., 1984. Classification and regression trees. Wadsworth Int. Group 37 (15), 237–251.

Brescia, M., Cavuoti, S., D'Abrusco, R., Longo, G., Mercurio, A., 2013. Photometric redshifts for quasars in multi-band surveys. The Astrophysical Journal 772 (2).

Brescia, M., Cavuoti, S., Longo, G., Nocella, A., Garofalo, M., Manna, F., Esposito, F., Albano, G., Guglielmo, M., D'Angelo, G., Di Guido, A., Djorgovski, S.G., Donalek, C., Mahabal, A.A., Graham, M.J., Fiore, M., D'Abrusco, R., 2014.

DAMEWARE: a web cyberinfrastructure for astrophysical data mining. Publications of the Astronomical Society of the Pacific 126, 783.

Brescia, M., Cavuoti, S., Paolillo, M., Longo, G., Puzia, T., 2012. The detection of globular clusters in galaxies as a data mining problem. Monthly Notices of the Royal Astronomical Society 421 (2), 1155–1165.

Breunig, M.M., Kriegel, H.-P., Kröger, P., Sander, J., 2001. Data bubbles: quality preserving performance boosting for hierarchical clustering. In: ACM SIGMOD Record, vol. 30. ACM, pp. 79–90.

Capozzi, D., de Filippis, E., Paolillo, M., D'Abrusco, R., Longo, G., 2009. The properties of the heterogeneous Shakhbazyan groups of galaxies in the SDSS. Monthly Notices of the Royal Astronomical Society 396 (2), 900–917.

Cheng, G., Li, Z., Han, J., Yao, X., Guo, L., 2018. Exploring hierarchical convolutional features for hyperspectral image classification. IEEE Transactions on Geoscience and Remote Sensing 56 (11), 6712–6722.

Cohn, D.A., Ghahramani, Z., Jordan, M.I., 1996. Active learning with statistical models. Journal of Artificial Intelligence Research 4 (1), 129–145.

Collister, A.A., Lahav, O., 2004. ANNz: estimating photometric redshifts using artificial neural networks. Publications of the Astronomical Society of the Pacific 116 (818), 345–351.

Comon, P., 1994. Independent component analysis, a new concept? Signal Processing 36 (3), 287–314.

Connolly, A.J., Szalay, A.S., Bershady, M.A., Kinney, A.L., Calzetti, D., 1995. Spectral classification of galaxies: an orthogonal approach. The Astronomical Journal 110, 1071.

Cover, T.M., Hart, P., et al., 1967. Nearest neighbor pattern classification. IEEE Transactions on Information Theory 13 (1), 21–27.

Cracknell, M.J., Reading, A.M., 2014. Geological mapping using remote sensing data: a comparison of five machine learning algorithms, their response to variations in the spatial distribution of training data and the use of explicit spatial information. Computers & Geosciences 63, 22–33.

Dai, J.-M., Tong, J., 2018. Visualizing the hidden features of galaxy morphology with machine learning. ArXiv e-prints, arXiv:1807.05657.

Dash, M., Liu, H., Xu, X., 2001. '$1 + 1 > 2$': merging distance and density based clustering. In: Proceedings Seventh International Conference on Database Systems for Advanced Applications. DASFAA 2001. IEEE, pp. 32–39.

de la Calleja, J., Fuentes, O., 2004. Machine learning and image analysis for morphological galaxy classification. Monthly Notices of the Royal Astronomical Society 349, 87–93.

Dobrycheva, D.V., Vavilova, I.B., Melnyk, O.V., Elyiv, A.A., 2017. Machine learning technique for morphological classification of galaxies at $z \leq 0.1$ from the SDSS. ArXiv e-prints, arXiv:1712.08955.

Domínguez Sánchez, H., Huertas-Company, M., Bernardi, M., Tuccillo, D., Fischer, J.L., 2018. Improving galaxy morphologies for SDSS with Deep Learning. Monthly Notices of the Royal Astronomical Society 476, 3661–3676.

Drlica-Wagner, A., Bechtol, K., Rykoff, E.S., Luque, E., Queiroz, A., Mao, Y.-Y., Wechsler, R.H., Simon, J.D., Santiago, B., Yanny, B., Balbinot, E., Dodelson, S., Fausti Neto, A., James, D.J., Li, T.S., Maia, M.A.G., Marshall, J.L., Pieres, A., Stringer, K., Walker, A.R., Abbott, T.M.C., Abdalla, F.B., Allam, S., Benoit-Lévy, A., Bernstein, G.M., Bertin, E., Brooks, D., Buckley-Geer, E., Burke, D.L., Carnero Rosell, A., Carrasco Kind, M., Carretero, J., Crocce, M., da Costa, L.N., Desai, S., Diehl, H.T., Dietrich, J.P., Doel, P., Eifler, T.F., Evrard, A.E., Finley, D.A., Flaugher, B., Fosalba, P., Frieman, J., Gaztanaga, E., Gerdes, D.W., Gruen, D., Gruendl, R.A., Gutierrez, G., Honscheid, K., Kuehn, K., Kuropatkin, N., Lahav, O., Martini, P., Miquel, R., Nord, B., Ogando, R., Plazas, A.A., Reil, K., Roodman, A., Sako, M., Sanchez, E., Scarpine, V., Schubnell, M., Sevilla-Noarbe, I., Smith, R.C., Soares-Santos, M., Sobreira, F., Suchyta, E., Swanson, M.E.C., Tarle, G., Tucker, D., Vikram, V., Wester, W., Zhang, Y., Zuntz, J., DES Collaboration, 2015. Eight ultra-faint galaxy candidates discovered in year two of the Dark Energy Survey. The Astrophysical Journal 813, 109.

Dunteman, G.H., 1989. Principal Components Analysis. Sage.

Elliott, J., de Souza, R.S., Krone-Martins, A., Cameron, E., Ishida, E.E.O., Hilbe, J., COIN Collaboration, 2015. The overlooked potential of Generalized Linear Models in astronomy-II: Gamma regression and photometric redshifts. Astronomy and Computing 10, 61–72.

Ester, M., Kriegel, H.-P., Sander, J., Xu, X., 1996. A density based algorithm for discovering clusters in large spatial databases with noise. In: KDD, vol. 96.

Fanaroff, B.L., Riley, J.M., 1974. The morphology of extragalactic radio sources of high and low luminosity. Monthly Notices of the Royal Astronomical Society 167, 31P–36P.

Fernández-Delgado, M., Cernadas, E., Barro, S., Amorim, D., 2014. Do we need hundreds of classifiers to solve real world classification problems? Journal of Machine Learning Research 15 (1), 3133–3181.

Fielding, A.H., 2006. Cluster and Classification Techniques for the Biosciences. Cambridge University Press.

Firth, A.E., Lahav, O., Somerville, R.S., 2003. Estimating photometric redshifts with artificial neural networks. Monthly Notices of the Royal Astronomical Society 339 (4), 1195–1202.

Foody, G.M., Mathur, A., 2004. A relative evaluation of multiclass image classification by support vector machines. IEEE Transactions on Geoscience and Remote Sensing 42 (6), 1335–1343.

Gahegan, M., 2000. On the application of inductive machine learning tools to geographical analysis. Geographical Analysis 32 (2), 113–139.

Gan, W., Lin, J.C.-W., Fournier-Viger, P., Chao, H.-C., Yu, P.S., 2019. A survey of parallel sequential pattern mining. ACM Transactions on Knowledge Discovery from Data 13 (3), 25.

Gehrke, J., Ramakrishnan, R., Ganti, V., 1998. RainForest—a framework for fast decision tree construction of large datasets. In: VLDB, vol. 98, pp. 416–427.

Gerdes, D.W., Sypniewski, A.J., McKay, T.A., Hao, J., Weis, M.R., Wechsler, R.H., Busha, M.T., 2010. ArborZ: photometric redshifts using boosted decision trees. The Astrophysical Journal 715 (2), 823–832.

Goderya, S.N., Lolling, S.M., 2002. Morphological classification of galaxies using computer vision and artificial neural networks: a computational scheme. Astrophysics and Space Science 279, 377–387.

Guo, Y., Han, S., Li, Y., Zhang, C., Bai, Y., 2018. K-nearest neighbor combined with guided filter for hyperspectral image classification. Procedia Computer Science 129, 159–165.

Guo, Y., Yin, X., Zhao, X., Yang, D., Bai, Y., 2019. Hyperspectral image classification with SVM and guided filter. EURASIP Journal on Wireless Communications and Networking 2019 (1), 56.

Guyon, I., 2009. A practical guide to model selection. In: Proc. Mach. Learn. Summer School Springer Text Stat., pp. 1–37.

Ham, J., Chen, Y., Crawford, M.M., Ghosh, J., 2005. Investigation of the random forest framework for classification of hyperspectral data. IEEE Transactions on Geoscience and Remote Sensing 43 (3), 492–501.

Hastie, T., Tibshirani, R., Friedman, J., Franklin, J., 2005. The elements of statistical learning: data mining, inference and prediction. The Mathematical Intelligencer 27 (2), 83–85.

Heinis, S., Kumar, S., Gezari, S., Burgett, W.S., Chambers, K.C., Draper, P.W., Flewelling, H., Kaiser, N., Magnier, E.A., Metcalfe, N., Waters, C., 2016. Of genes and machines: application of a combination of machine learning tools to astronomy data sets. The Astrophysical Journal 821, 86.

Hinneburg, A., Keim, D.A., et al., 1998. An efficient approach to clustering in large multimedia databases with noise. In: KDD, vol. 98, pp. 58–65.

Hopfield, J.J., 1982. Neural networks and physical systems with emergent collective computational abilities. Proceedings of the National Academy of Sciences 79 (8), 2554–2558.

Hotelling, H., 1933. Analysis of a complex of statistical variables into principal components. Journal of Educational Psychology 24 (6), 417.

Hoyle, B., Rau, M.M., Bonnett, C., Seitz, S., Weller, J., 2015a. Data augmentation for machine learning redshifts applied to Sloan Digital Sky Survey galaxies. Monthly Notices of the Royal Astronomical Society 450 (1), 305–316.

Hoyle, B., Rau, M.M., Zitlau, R., Seitz, S., Weller, J., 2015b. Feature importance for machine learning redshifts applied to SDSS galaxies. Monthly Notices of the Royal Astronomical Society 449 (2), 1275–1283.

Hsu, C.-W., Chang, C.-C., Lin, C.-J., 2016. A practical guide to support vector classification. Technical Report. National Taiwan University, Taiwan.

Hsu, C.-W., Lin, C.-J., 2002. A comparison of methods for multiclass support vector machines. IEEE Transactions on Neural Networks 13 (2), 415–425.

Huang, C., Davis, L., Townshend, J., 2002. An assessment of support vector machines for land cover classification. International Journal of Remote Sensing 23 (4), 725–749.

Huang, G., Huang, G.-B., Song, S., You, K., 2015. Trends in extreme learning machines: a review. Neural Networks 61, 32–48.

Hubble, E.P., 1926. Extragalactic nebulae. The Astrophysical Journal 64, 321–369.

Hwang, J.N., Lay, S.R., Lippman, A., 2003. Nonparametric multivariate density estimation: a comparative study. Science 42 (10), 2795–2810.

Ishida, E.E.O., 2019. Machine learning and the future of supernova cosmology. Nature Astronomy 3, 680–682.

Ishida, E.E.O., Beck, R., González-Gaitán, S., de Souza, R.S., Krone-Martins, A., Barrett, J.W., Kennamer, N., Vilalta, R., Burgess, J.M., Quint, B., Vitorelli, A.Z., Mahabal, A., Gangler, E., COIN Collaboration, 2019. Optimizing spectroscopic follow-up strategies for supernova photometric classification with active learning. Monthly Notices of the Royal Astronomical Society 483, 2–18.

Jain, A., Murty, M., Flynn, P., 1999. Data clustering: a review. ACM Computing Surveys 31, 264–323.

Jolliffe, I.T., Cadima, J., 2016. Principal component analysis: a review and recent developments. Philosophical Transactions - Royal Society. Mathematical, Physical and Engineering Sciences 374 (2065), 20150202.

Kanevski, M., Timonin, V., Pozdnukhov, A., 2009. Machine Learning for Spatial Environmental Data: Theory, Applications, and Software. EPFL Press.

Karampelas, A., 2013. Unsupervised spectral classification of synthetic galaxies using Principal Components Analysis. In: 11th Hellenic Astronomical Conference, pp. 24–25.

Karatzoglou, A., Meyer, D., Hornik, K., 2006. Support vector machines in R. Journal of Statistical Software 15 (9), 1–28.

Kim, E.J., Brunner, R.J., 2017. Star-galaxy classification using deep convolutional neural networks. Monthly Notices of the Royal Astronomical Society 464, 4463–4475.

Kind, M.C., Brunner, R.J., 2013. TPZ: photometric redshift PDFs and ancillary information by using prediction trees and random forests. Monthly Notices of the Royal Astronomical Society 432 (2), 1483–1501.

Kohonen, T., 1982. Self-organized formation of topologically correct feature maps. Biological Cybernetics 43 (1), 59–69.

Kotsiantis, S.B., 2007. A review of classification techniques. Informatica 31, 249–268.

Kron, R.G., 1980. Photometry of a complete sample of faint galaxies. The Astrophysical Journal. Supplement Series 43, 305–325.

Kuncheva, L.I., 2014. Combining Pattern Classifiers: Methods and Algorithms. John Wiley & Sons.

Lahav, O., 1997. Artificial neural networks as a tool for galaxy classification. In: Di Gesu, V., Duff, M.J.B., Heck, A., Maccarone, M.C., Scarsi, L., Zimmerman, H.U. (Eds.), Data Analysis in Astronomy, pp. 43–51.

Leauthaud, A., Massey, R., Kneib, J.-P., Rhodes, J., Johnston, D.E., Capak, P., Heymans, C., Ellis, R.S., Koekemoer, A.M., Le Fèvre, O., Mellier, Y., Réfrégier, A., Robin, A.C., Scoville, N., Tasca, L., Taylor, J.E., Van Waerbeke, L., 2007. Weak gravitational lensing with COSMOS: galaxy selection and shape measurements. The Astrophysical Journal. Supplement Series 172, 219–238.

Lee, D.D., Seung, H.S., 1999. Learning the parts of objects by nonnegative matrix factorization. Nature 401, 788–791.

Levine, E., Kimes, D., Sigillito, V., 1996. Classifying soil structure using neural networks. Ecological Modelling 92 (1), 101–108.

Li, K.-C., Lue, H.-H., Chen, C.-H., 2000. Interactive tree-structured regression via principal Hessian directions. Journal of the American Statistical Association 95 (450), 547–560.

Liu, F., Ye, C., Zhu, E., 2017. Accurate grid-based clustering algorithm with diagonal grid searching and merging. IOP Conference Series: Materials Science and Engineering 242, 012123.

Ma, Z., Xu, H., Zhu, J., Hu, D., Li, W., Shan, C., Zhu, Z., Gu, L., Li, J., Liu, C., Wu, X., 2019. A machine learning based morphological classification of 14,245 radio AGNs selected from the Best–Heckman sample. The Astrophysical Journal. Supplement Series 240 (2), 34.

MacGillivray, H.T., Martin, R., Pratt, N.M., Reddish, V.C., Seddon, H., Alexander, L.W.G., Walker, G.S., Williams, P.R., 1976. A method for the automatic separation of the images of galaxies and stars from measurements made with the COSMOS machine. Monthly Notices of the Royal Astronomical Society 176, 265–274.

Malewicz, G., Austern, M.H., Bik, A.J., Dehnert, J.C., Horn, I., Leiser, N., Czajkowski, G., 2010. Pregel: a system for large-scale graph processing. In: Proceedings of the 2010 ACM SIGMOD International Conference on Management of Data. ACM, pp. 135–146.

Marsland, S., 2014. Machine Learning: An Algorithmic Perspective. Chapman and Hall/CRC.

Masters, D., Capak, P., Stern, D., Ilbert, O., Salvato, M., Schmidt, S., Longo, G., Rhodes, J., Paltani, S., Mobasher, B., Hoekstra, H., Hildebrandt, H., Coupon, J., Steinhardt, C., Speagle, J., Faisst, A., Kalinich, A., Brodwin, M., Brescia, M., Cavuoti, S., 2015. Mapping the galaxy color–redshift relation: optimal photometric redshift calibration strategies for cosmology surveys. The Astrophysical Journal 813 (1), 53.

Mboga, N., Persello, C., Bergado, J., Stein, A., 2017. Detection of informal settlements from VHR images using convolutional neural networks. Remote Sensing 9 (11), 1106.

McCarthy, J., Minsky, M.L., Rochester, N., Shannon, C.B., 1955. Proposal for the Dartmouth Summer Research Project for Artificial Intelligence. Technical Report. Dartmouth College, Hanover, NH.

McInnes, L., Healy, J., Melville, J., 2018. UMAP: uniform manifold approximation and projection for dimension reduction. ArXiv e-prints, arXiv:1802.03426.

Melgani, F., Bruzzone, L., 2004. Classification of hyperspectral remote sensing images with support vector machines. IEEE Transactions on Geoscience and Remote Sensing 42 (8), 1778–1790.

Minsky, M., 1952. A Neural-Analogue Calculator Based Upon a Probability Model of Reinforcement. Harvard University Psychological Laboratories, Cambridge, Massachusetts.

Mitchell, T., 1997. Machine Learning, McGraw-Hill International Editions. McGraw-Hill.

Molina, R., De la Blanca, N.P., Taylor, C., 1994. Modern statistical techniques. In: Machine Learning, Neural and Statistical Classification, pp. 29–49.

Mountrichas, G., Corral, A., Masoura, V.A., Georgantopoulos, I., Ruiz, A., Georgakakis, A., Carrera, F.J., Fotopoulou, S., 2017. Estimating photometric redshifts for X-ray sources in the X-ATLAS field using machine-learning techniques. Astronomy & Astrophysics 608, A39.

Muja, M., Lowe, D.G., 2014. Scalable nearest neighbor algorithms for high dimensional data. IEEE Transactions on Pattern Analysis and Machine Intelligence 36 (11), 2227–2240.

Newell, A., Simon, H.A., 1956. A complex information processing system. I.R.E. Transactions on Information Theory 2, 61–79.

Nilsson, N., 1965. Learning Machines: Foundations of Trainable Pattern-Classifying Systems. McGraw-Hill, New York.

Nilsson, N.J., 2010. The Quest for Artificial Intelligence: A History of Ideas and Achievements. Cambridge University Press, New York, NY.

Odewahn, S.C., Stockwell, E.B., Pennington, R.L., Humphreys, R.M., Zumach, W.A., 1992. Automated star/galaxy discrimination with neural networks. The Astronomical Journal 103, 318–331.

Owens, E.A., Griffiths, R.E., Ratnatunga, K.U., 1996. Using oblique decision trees for the morphological classification of galaxies. Monthly Notices of the Royal Astronomical Society 281.

Oyaizu, H., Lima, M., Cunha, C.E., Lin, H., Frieman, J., Sheldon, E.S., 2008. A galaxy photometric redshift catalog for the Sloan Digital Sky Survey Data Release 6. The Astrophysical Journal 674 (2), 768–783.

Pearson, K., 1901. LIII. On lines and planes of closest fit to systems of points in space. The London, Edinburgh, and Dublin Philosophical Magazine and Journal of Science 2 (11), 559–572.

Persello, C., Tolpekin, V., Bergado, J., de By, R., 2019. Delineation of agricultural fields in smallholder farms from satellite images using fully convolutional networks and combinatorial grouping. Remote Sensing of Environment 231, 111253.

Plewa, P.M., 2018. Random forest classification of stars in the Galactic Centre. Monthly Notices of the Royal Astronomical Society 476 (3), 3974–3980.

Polsterer, K.L., Gieseke, F., Igel, C., 2015. Automatic galaxy classification via machine learning techniques: Parallelized rotation/flipping INvariant Kohonen maps (PINK). In: Taylor, A.R., Rosolowsky, E. (Eds.), Astronomical Data Analysis Software and Systems XXIV (ADASS XXIV). In: Astronomical Society of the Pacific Conference Series, vol. 495, p. 81.

Prasad, A.M., Iverson, L.R., Liaw, A., 2006. Newer classification and regression tree techniques: bagging and random forests for ecological prediction. Ecosystems 9 (2), 181–199.

Ripley, B.D., Hjort, N., 1996. Pattern Recognition and Neural Networks. Cambridge University Press.

Rosenblatt, F., 1958. The perceptron: a probabilistic model for information storage and organization in the brain. Psychological Review 65 (6), 386.

Ross, A.J., Ho, S., Cuesta, A.J., Tojeiro, R., Percival, W.J., Wake, D., Masters, K.L., Nichol, R.C., Myers, A.D., de Simoni, F.,

Seo, H.J., Hernández-Monteagudo, C., Crittenden, R., Blanton, M., Brinkmann, J., da Costa, L.A.N., Guo, H., Kazin, E., Maia, M.A.G., Maraston, C., Padmanabhan, N., Prada, F., Ramos, B., Sanchez, A., Schlafly, E.F., Schlegel, D.J., Schneider, D.P., Skibba, R., Thomas, D., Weaver, B.A., White, M., Zehavi, I., 2011. Ameliorating systematic uncertainties in the angular clustering of galaxies: a study using the SDSS-III. Monthly Notices of the Royal Astronomical Society 417, 1350–1373.

Sadeh, I., Abdalla, F.B., Lahav, O., 2016. ANNz2: photometric redshift and probability distribution function estimation using machine learning. Publications of the Astronomical Society of the Pacific 128 (968), 104502.

Sander, J., Ester, M., Kriegel, H., Xu, X., 1998. Density-based clustering in spatial databases: the algorithm GDBSCAN and its application. Data Mining and Knowledge Discovery 2 (2), 169–194.

Sebestyen, G.S., 1962. Decision-Making Processes in Pattern Recognition. Macmillan, New York.

Serjeant, S., 2014. Up to 100,000 reliable strong gravitational lenses in future dark energy experiments. The Astrophysical Journal Letters 793 (1), L10.

Settles, B., 2012. Active Learning. Morgan & Claypool.

Shafer, J., Agrawal, R., Mehta, M., 1996. SPRINT: a scalable parallel classifier for data mining. In: VLDB, vol. 96. Citeseer, pp. 544–555.

Sheikholeslami, G., Chatterjee, S., Zhang, A., 2000. Wavecluster: a wavelet-based clustering approach for spatial data in very large databases. The VLDB Journal 8 (3–4), 289–304.

Song, X., Duan, Z., Jiang, X., 2012. Comparison of artificial neural networks and support vector machine classifiers for land cover classification in Northern China using a SPOT-5 HRG image. International Journal of Remote Sensing 33 (10), 3301–3320.

Sutton, R.S., Barto, A.G., et al., 1998. Introduction to Reinforcement Learning. MIT Press, Cambridge.

Tao, Y., Zhang, Y., Cui, C., Zhang, G., 2018. Automated spectral classification of galaxies using machine learning approach on Alibaba Cloud AI platform (PAI). ArXiv e-prints, arXiv: 1801.04839.

Tenenbaum, J.B., Silva, V.D., Langford, J.C., 2000. A global geometric framework for nonlinear dimensionality reduction. Science 290 (5500), 2319–2323.

Turing, A.M., 1950. I. Computing machinery and intelligence. Mind LIX (236), 433–460.

Turner, S., Kelvin, L.S., Baldry, I.K., Lisboa, P.J., Longmore, S.N., Collins, C.A., Holwerda, B.W., Hopkins, A.M., Liske, J., 2019. Reproducible k-means clustering in galaxy feature data from the GAMA survey. Monthly Notices of the Royal Astronomical Society 482, 126–150.

van der Maaten, L., Hinton, G., 2008. Visualizing data using t-SNE. Journal of Machine Learning Research 9, 2579–2605.

VanderPlas, J., Connolly, A.J., Ivezic, Z., Gray, A., 2012. Introduction to astroML: machine learning for astrophysics. In: Proceedings of Conference on Intelligent Data Understanding (CIDU), pp. 47–54.

Vapnik, V., 1998. Statistical Learning Theory. John Wiley & Sons. Inc., New York, p. 736.

Vasconcellos, E.C., de Carvalho, R.R., Gal, R.R., LaBarbera, F.L., Capelato, H.V., Frago Campos Velho, H., Trevisan, M., Ruiz, R.S.R., 2011. Decision tree classifiers for star/galaxy separation. The Astronomical Journal 141, 189.

Vincent, P., Larochelle, H., Bengio, Y., Manzagol, P.A., 2008. Extracting and composing robust features with denoising autoencoders. In: Proceedings of the International Conference on Machine Learning.

Wang, W., Muntz, R., 1997. Sting: a statistical information grid approach to spatial data mining. In: Proceedings of the Twenty-Third International Conference on Very Large Data Bases, pp. 186–195.

Waske, B., Benediktsson, J.A., Árnason, K., Sveinsson, J.R., 2009. Mapping of hyperspectral AVIRIS data using machine-learning algorithms. Canadian Journal of Remote Sensing 35 (sup1), S106–S116.

Way, M.J., Klose, C.D., 2012. Can self-organizing maps accurately predict photometric redshifts? Publications of the Astronomical Society of the Pacific 124 (913), 274–279.

Weir, N., Fayyad, U.M., Djorgovski, S., 1995. Automated star/galaxy classification for digitized POSS-II. The Astronomical Journal 109, 2401.

Widrow, B., Hoff, M.E., 1960. Adaptive switching circuits. Technical Report. Stanford Univ Ca Stanford Electronics Labs.

Witten, I.H., Frank, E., Hall, M.A., Pal, C.J., 2016. Data Mining: Practical Machine Learning Tools and Techniques. Morgan Kaufmann.

Wold, S., Esbensen, K., Geladi, P., 1987. Principal component analysis – chemometrics and intelligent laboratory systems 2. In: IEEE Conference on Emerging Technologies & Factory Automation (EFTA), vol. 2, pp. 37–52.

Woźniak, M., Graña, M., Corchado, E., 2014. A survey of multiple classifier systems as hybrid systems. Information Fusion 16, 3–17.

Wu, C., Wong, O.I., Rudnick, L., Shabala, S.S., Alger, M.J., Banfield, J.K., Ong, C.S., White, S.V., Garon, A.F., Norris, R.P., Andernach, H., Tate, J., Lukic, V., Tang, H., Schawinski, K., Diakogiannis, F.I., 2019. Radio Galaxy Zoo: CLARAN – a deep learning classifier for radio morphologies. Monthly Notices of the Royal Astronomical Society 482 (1), 1211–1230.

Xu, L., Krzyzak, A., Oja, E., 1993. Rival penalized competitive learning for clustering analysis, RBF net and curve detection. IEEE Transactions on Neural Networks 4, 636–649.

Yèche, C., Petitjean, P., Rich, J., Aubourg, E., Busca, N., Hamilton, J.C., Le Goff, J.M., Paris, I., Peirani, S., Pichon, C., Rollinde, E., Vargas-Magaña, M., 2010. Artificial neural networks for quasar selection and photometric redshift determination. Astronomy & Astrophysics 523, A14.

Yee, H.K.C., 1991. A faint-galaxy photometry and image-analysis system. Publications of the Astronomical Society of the Pacific 103, 396–411.

Yip, C.W., Connolly, A.J., Szalay, A.S., Budavári, T., SubbaRao, M., Frieman, J.A., Nichol, R.C., Hopkins, A.M., York, D.G., Okamura, S., Brinkmann, J., Csabai, I., Thakar, A.R., Fukugita, M., Ivezić, Ž., 2004. Distributions of galaxy spectral types in the Sloan Digital Sky Survey. The Astronomical Journal 128, 585–609.

Yu, L., Porwal, A., Holden, E.-J., Dentith, M.C., 2012. Towards automatic lithological classification from remote sensing data using support vector machines. Computers & Geosciences 45, 229–239.

Zaharia, M., Mosharaf, M., Tathagata, D., Ankur, D., Justin, M., Murphy, M., Michael, J., Scott, S., Ion, S., 2012. Resilient distributed datasets: a fault-tolerant abstraction for in-memory cluster computing. In: NSDI.

Zhao, W., Du, S., 2016. Spectral–spatial feature extraction for hyperspectral image classification: a dimension reduction and deep learning approach. IEEE Transactions on Geoscience and Remote Sensing 54 (8), 4544–4554.

Zhao, K., Popescu, S., Meng, X., Pang, Y., Agca, M., 2011. Characterizing forest canopy structure with lidar composite metrics and machine learning. Remote Sensing of Environment 115 (8), 1978–1996.

CHAPTER 13

Deep Learning – an Opportunity and a Challenge for Geo- and Astrophysics

CHRISTIAN REIMERS, MSC • CHRISTIAN REQUENA-MESA, MSC

13.1 INTRODUCTION

Deep learning has been applied achieving state of the art in fields including computer vision, language translation, speech synthesis and speech recognition, social network filtering, medical drug design and image analysis, board game self-learning, and self-driving vehicles, among others. In many of these disciplines, deep learning methods have achieved results similar to, and in some cases exceeded, those of humans experts.

Originally, artificial neural networks were inspired by the nodal design seen in biological neural brains. However, as of today, this resemblance is quite scarce, and newly developed architectures that work well for processing data tend to not inherit new similarities from biological systems. As an example, most used trained neural networks tend to be static and conceived for a single purpose, while biological brains of most organisms are often dynamic and multipurpose.

In this chapter, we will start by explaining the difference between the classical machine learning methods (shallow learning) and the deep learning approach. We will demonstrate the difference on three example works that predict the power generation of solar power plants in the future. The example showcases a shallow learning approach, a simple deep learning approach, and a more complicated generative deep learning approach.

Most deep learning approaches are realized by deep neural networks. The different functionalities of the deep learning approaches are then determined by the different network architectures, often also called topologies. The rest of the chapter is focused on briefly explaining prototypes of deep learning architectures that have found success in many applications. We will shortly present the main idea behind these architectures and point the reader towards applications of these architectures in different scientific fields, in which these architectures have demonstrated their immense potential, as well as towards works in which researchers have adopted these architectures for applications in geophysics and astrophysics.

The aim of this chapter is to point the interested reader to works that have applied deep learning in astro- and geophysics, and to give the reader a broad idea of where deep learning can be applied to great success. For a reader who is interested in an in-depth introduction to deep learning we recommend Goodfellow et al. (2016). For readers who are looking for a hands-on tutorial to deep learning we recommend Chollet and Allaire (2018).

13.2 THE DIFFERENCE BETWEEN SHALLOW LEARNING AND DEEP LEARNING

In this section we first point out the difference between deep learning and the more classical shallow learning. Then we will continue with an example that demonstrates these differences as well as the difference between the first deep learning approaches and newer deep learning approaches.

The classical machine learning methods, which we will call shallow learning methods throughout this chapter, are three-step methods. The first step is data collection, then a feature selection step is carried out, and the third step is the regression step. There is an assumption underlying this approach. After we have collected data, we perform two steps, the feature selection step and the regression or classification step. (See Fig. 13.1.) To perform the feature selection first, we need to assume that the choice of optimal features can be based on the data alone and is not based on any results or intermediate results of the regression or classification step. More formally, we assume that the quality of the features is independent of the optimization.

This assumption is the main difference between shallow learning and deep learning. In deep learning, it is assumed that the features cannot be chosen independently of the regression or classification. Instead of choosing them by hand using prior knowledge and data exploration, as in the example above, in deep learning the algorithms choose features automatically in paral-

Knowledge Discovery in Big Data from Astronomy and Earth Observation. https://doi.org/10.1016/B978-0-12-819154-5.00024-2

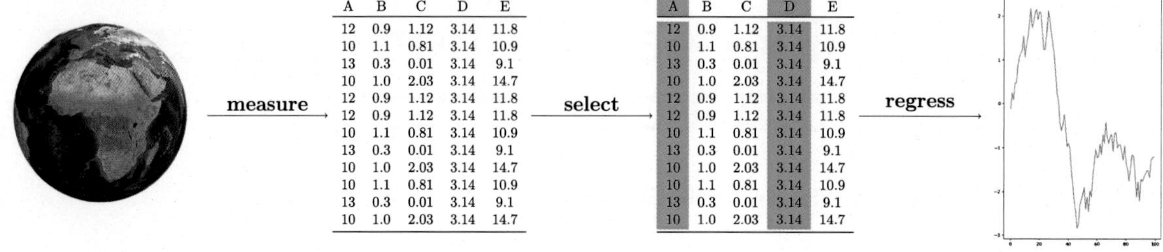

FIG. 13.1 The three-step process of machine learning. First, collect data, then select features, and lastly use the features to regress the variable of interest.

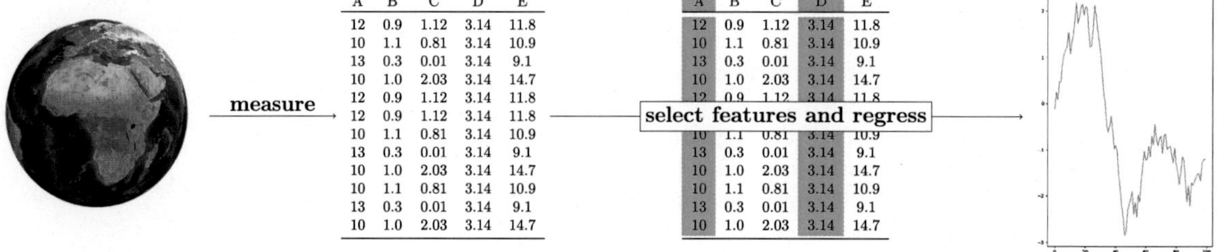

FIG. 13.2 The two-step process of deep learning. First, collect data, and then the feature selection and regression steps are combined into one step.

lel to the regression or classification. The algorithm is provided with all data that are connected to a task and not just the most relevant features. During the regression or classification, not just the function that maps the features onto the variable to predict, gets optimized but also the way the features are calculated from the data is adapted. Hence, deep learning is only a two-step method. (See Fig. 13.2.)

To illustrate the difference, we present some work on predicting solar power generation from weather data. The amount of power generated by renewable energy sources is based on the current weather situation. To guarantee local power quality and net stability, power grid operators need to be able to plan ahead. Hence, predicting the amount of power that will be generated by a solar power plant in the future is an important task. (See Fig. 13.3.)

The problem is hard to solve using a model of the physical process, since the information on the weather is needed with very high spatial resolution, while the phenomena that have to be taken into account to adequately model weather are often on a global scale. This congruence of high spatial resolution and global dependencies makes modeling difficult and gives the edge to statistical methods and methods of machine learning.

We highlight three machine learning approaches for this problem. We start with the work of Sharma et al. (2011), who used linear methods and support vector machines (SVMs). Then we discuss the work of Gensler et al. (2016), who used autoencoders (Section 13.4.3.1) and LSTM (Section 13.4.2) to tackle the problem. Finally, we talk about the work of Khodayar et al. (2018), who applied a variational autoencoder (VAE) (Section 13.4.3.1) to the problem.

We first present an approach using shallow machine learning to solve this problem. This first approach is described in Sharma et al. (2011), where the authors predict the amount of energy produced by a solar panel given weather forecast information. First, they collect data. They use the weather forecast of the National Weather Service[1] and data from an extended weather station on the roof of the Computer Science building at the University of Massachusetts Amherst. The authors then select the features they think are most relevant for the prediction task. They do this by checking correlations and by including their prior knowledge. They choose to only use the weather forecast at the exact location and the day of the year for which they want to predict features. In their paper, they give reasoning on why they expect these features to be relevant. Then they perform a linear regression to find the function that maps these meaningful features onto the solar intensity. In a second approach, they use a kernel SVM. Here,

[1] www.weather.gov.

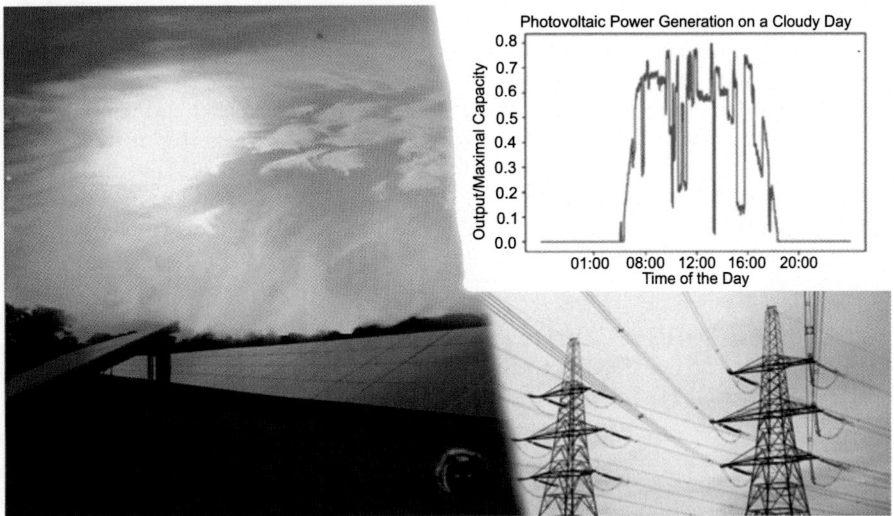

FIG. 13.3 For the stability of a power grid it is indispensable to accurately predict the amount of power that is fed to the grid in advance. Regenerative generators are difficult to predict since they depend on weather phenomena. Predicting these phenomena in high time resolution is a crucial task for effective operation of power grids. Top right: The power yield of a photovoltaic system on a cloudy day.

again, prior knowledge is used to select a kernel that maps the inputs onto the features. Prior knowledge is used to select meaningful features from the data.

We compare this to the deep learning approach on the same problem presented in Gensler et al. (2016). The authors of this work use different deep learning models, namely, autoencoders and LSTM, to solve the problem. They collect data from a numerical weather simulation. Instead of selecting features as described above, they use every variable within their numerical weather prediction model as input. The prediction performance improved compared to physical models and shallow learning approaches, but the interpretation of the results is difficult. While the authors found that their approach underestimates the produced power, they are not able to identify why this is the case.

A more complex approach is selected in Khodayar et al. (2018). The authors introduce two changes compared to the approach above. First, they take into account the randomness that is inherent to predictions on future weather. Therefore, they approximate a distribution instead of one expected value. Second, they predict the power yield in many different locations at the same time. They do not only predict the distribution for the solar irradiance in the individual locations, but they approximate one joint distribution. They use deep learning to identify dependencies between the distributions automatically and train a neural network to approximate the function that maps a multimodal standard normal distribution onto this joint distribution.

Merging the two steps, the features selection step and the regression or classification step, together might seem like a minor change, but the consequences are extensive. On the one hand, deep learning is significantly stronger than shallow learning in many applications, and on the other hand, new challenges arise. The number of layers of the combined feature selection and regression or classification network is higher than the number of layers of the simple regression or classification network. Many parameters have to be optimized in parallel, which leads to technical problems such as vanishing gradients. Therefore, more sophisticated algorithms and more hardware resources are needed to optimize these huge neural networks. Another problem is the interpretability of the automatically extracted features. The hand-picked features in shallow learning are often simple, well understood, and easy to interpret. In deep learning, features are selected purely based on their usefulness for the regression or classification step and not on their interpretability. Hence, the algorithms end up with complicated features that are hard to interpret. To understand the features extracted by a deep learning algorithm has become its own field of research. For an introduction into this field we refer the reader to Montavon et al. (2018).

A lot of terms in machine learning and especially in deep learning are related to neurology. While biological neurons are definitely often used as a motivation, it is important to note that deep learning is not aiming to build or simulate a human brain. Deep learning is a method to perform feature selection and regression at the same time. It is neither trying to create a human-like general artificial intelligence nor trying to replicate biological processes as closely as possible.

13.3 WHY IS DEEP LEARNING A GOOD FIT FOR THE DATA SCIENCE PROBLEMS IN ASTRO- AND GEOPHYSICS

In this section, we will discuss why deep learning is a good method to tackle the tasks of astro- and geophysics.

Several reasons make deep learning a useful approach for several tasks in geo- and astrophysics. The main reason is that using physical methods to tackle the problem is becoming increasingly harder as we try to expand the boundaries of what we know.

The amount of data available and relevant to tasks in geo- and astrophysics is often enormous and ever growing. These data are coming from satellites, observatories, and models of increasing accuracy and resolution. Handling the amounts of data is a challenge for classical methods. But to construct handcrafted features out of the terabytes of data is not just difficult because of the amount of data.

In both astro- and geophysics, relevant effects happen on different scales. Models of coarse resolution might not be able to adequately model local effects. Models of fine resolutions might not be able to cover a wide enough region to correctly model global effects and influences. In the introduction example of predicting the power yield of photovoltaic systems a correct model needs to be able to model clouds that might form multiple hundreds or thousands of kilometers away but the correct prediction for a location might differ significantly from the prediction for a location only few kilometers away. Further, local geographics such as mountains or rivers might change the path of clouds. Therefore, a model must be able to provide high spatial resolution for a huge area.

While this problem of high resolution and wide range exists for the spatial resolution, it is even worse for the temporal resolution of processes. For example, the continental plates drift with a speed below 0.1 meter per year. On the one hand, many relevant effects of tectonic drift happen on a timescale of million of years. On the other hand, one of the crucial consequences of tectonic drift is earthquakes, which we want to predict on a timescale of minutes and hours rather than on a timescale of days.

Not only the spatial and temporal resolution is different across observations; many tasks use not just one sensor but a network of sensors. Data from a multitude of sensors have to be used. Often the distribution of observations available is heavily skewed in time and space. Some regions contain many observations, while others are observed very rarely. These effects have to be considered in the decision on which features to use.

An example of a network of sensors being used in astrophysics is the laser interferometer gravitational wave observatory (LIGO). LIGO detects and studies gravitational waves of astrophysical origin by operating three multikilometer interferometers at two widely separated sites in the United States (Abbott et al., 2009). Arrays of radio telescopes are used to peer deep into the universe. In climate science, networks are used to measure global effects, such as CO_2 fluxes (Baldocchi, 2008). For a second example, the usefulness of sensor networks in climate science is explained in Tsonis et al. (2006).

Not only data from different sensors of the same type have to be considered, but also multimodal data can support research in many tasks. Different kinds of observations might need to be processed in very different ways to be utilized best. A summary on the challenges of multimodal data fusion in, for example, remote sensing can be found in Gómez-Chova et al. (2015). Deep neural networks are a good method to combine different types of features, as demonstrated in, for example, Ngiam et al. (2011) and Srivastava and Salakhutdinov (2012).

Additionally, many classical methods in physics depend on conducting experiments in which some variable is varied while all other variables are held constant. It is, however, very difficult to conduct this kind of experiment on the scale of geo- or astrophysics. A lot of research relies on observational data.

This necessitates complex features that are adapted to the data. It is a very difficult task to select useful handcrafted features. Hence, the automated feature selection and the good scalability of deep learning algorithms is useful and makes them the best choice in many situations. (See Fig. 13.4.)

The most successful methods of deep learning are developed in the field of computer vision. To transfer the methods to the fields of astro- and geophysics is a research task on its own but a lot of progress is made in these directions, as described in Reichstein et al. (2019).

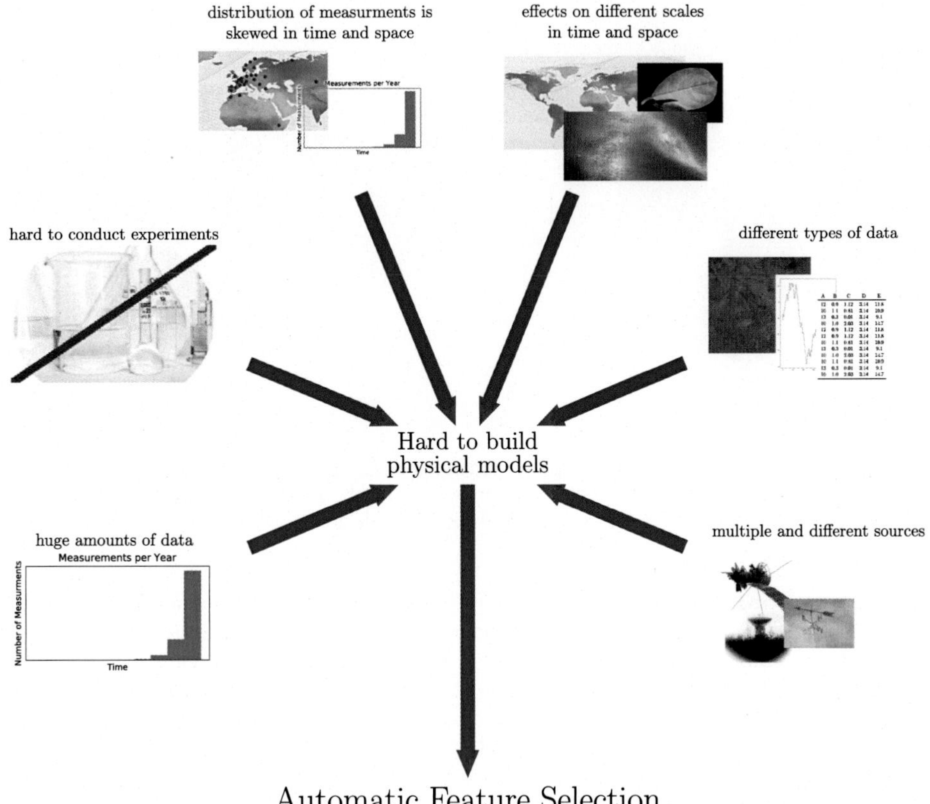

FIG. 13.4 The amount of data, the different scales in time and space, the skewness of the distribution of measurements, the different sources and types of measurements that need to be considered, and the fact that experiments are hard to conduct make many systems in geo- and astrophysics extremely hard to model. Hence, deep learning with its automated feature selection is often a good choice.

13.4 DEEP LEARNING MODELS

13.4.1 Convolutional Neural Networks

Convolutional neural networks are the first deep learning models that received a lot of attention due to their impressive performance in applications of computer vision.

The main idea behind convolutional neural networks is to extract local features from the data. In a convolutional layer, the similarity between small patches of the image and some learned kernels is calculated. Then, in a pooling layer, the values of pixels that are close are grouped and combined into one pixel. This reduces the complexity of the data, leading to less computation and to a feature selection that is robust to small changes. Pooling layers and convolutional layers are alternated. The value of a single pixel after the first pooling layer represents a whole batch of pixels in the original image and after two pooling layers, every value represents a batch of batches of pixels in the original image. The first pair of convolution and pooling layer combines the pixels of the image to local features and the subsequent pairs of convolution and pooling layers combines these local features to more and more global features. These global features are then used for regression. As mentioned above, not every regressor can be used because we need the error signal from the regressor to adapt the kernels used in the convolutional layers. (See Fig. 13.5.)

The first of these applications is image classification. The goal of image classification is to classify images into different classes given their content. The first use of deep learning for this task is described in LeCun et al. (1998). The authors of this paper use convolutional neural networks to classify handwritten digits on images into the classes from zero to nine. This was the first time that an automated system could perform handwritten digit recognition on a level sufficient for a commercial ap-

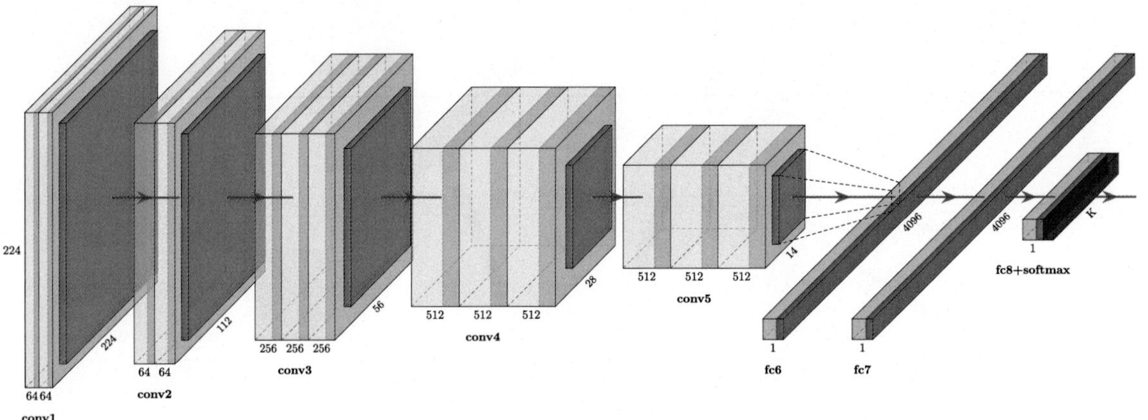

FIG. 13.5 An image of the VGG network developed in Simonyan and Zisserman (2014). For visualization we used the project from Iqbal (2018). The yellow boxes represent convolutional layers, the red boxes represent max-pooling layers, and the violet boxes represent fully connected layers.

plication. They first collected a dataset of handwritten digits and scaled them to 28 × 28 pixel black and white images. They used two convolutional layers and two pooling layers that reduce four pixels to one pixel containing the maximal value of the four pixels.

The breakthrough application for convolutional neural networks, however, is Krizhevsky et al. (2012), which used a convolutional neural network to win the ImageNet Large Scale Visual Recognition Challenge 2012 (ILSVRC2012). Given more than a million example images from different domains and in different resolutions, the contestants of this challenge have to build a system that is able to classify images of an unseen test set into 1000 classes including very broad classes like "broom" and "pretzel," as well as very fine-grained classes like "water jug" versus "whiskey jug" or different dog breeds like "Old English sheepdog" or "Bouvier des Flandres". (See Fig. 13.6.) This mix of coarse-grained and fine-grained classification makes the problem especially hard. The authors presented a deep learning approach that outperformed all other machine learning techniques, establishing convolutional neural networks as the state of the art in image classification and subsequently many other tasks of computer vision, like localization, semantic segmentation, or detection.

The most common measure to document success in this challenge is the Top-5. It states the percentage of examples in the unseen test set, for which the algorithm predicted the correct label within its first five guesses. The method presented in Krizhevsky et al. (2012) reported a Top-5 of 15.3% while the best non-deep learning-based method reported a Top-5 of 26.2%. This challenge demonstrated the power of convolutional neural networks in extracting meaningful features for classification in situations where different classes look very different as well as situations where differences are subtle. This makes deep learning a good choice in applications where classification is based on an image.

The task of classifying objects from images has application in geo- and astrophysics. Any two-dimensional array of data can be understood as an image and, therefore, makes deep convolutional neural networks a reasonable choice. Two-dimensional arrays are common in geo- and astrophysics. They appear in remote sensing in measurements that are on the surface two-dimensional surface of the Earth, or as the output of many measuring techniques for space observation. Often the resulting images must be classified into discrete events or a regression task is carried out. The problems that occur are often similar to the problems described for the ILSVRC2012. The relevant information is often accompanied by irrelevant information, similar to the background of the real-world images, and often the relevant effects are in part coarse-grained and in part fine-grained.

An obvious application of image classification in geophysics is remote sensing. As mentioned above, deep learning is a good answer to many of the challenges in remote sensing. Examples of successful applications of deep learning in remote sensing can be found, for example, in Cao et al. (2019), where the authors use deep learning to retrieve the total absorption coefficient $a(\lambda)$ for inland water bodies. The complexity of inland wa-

FIG. 13.6 Classifying images is a hard problem; often images of the same class look very different because they are taken in front of different backgrounds and the object of interest is in a different pose. On the other hand, images of different classes might look similar. Therefore, a deep learning algorithm has to extract well-suited features that perform in the coarse-grained task as well as the fine-grained task.

ter bodies and the low signal-to-noise ratio of sensors with wider spectral bandwidth make the analytical estimation of $a(\lambda)$ hard. The deep neural network approach outperforms the semianalytic approach QAA-750E (Xue et al., 2019) by a factor of two.

In Liu et al. (2016) convolutional neural networks are used to identify extreme weather events. As input, they use maps of different climatological features. The extreme weather events they identify include tropical cyclones, atmospheric rivers, and weather fronts. This demonstrates that the input data do not need to be images, but any data that have a two-dimensional structure can be used, like maps or complex imaging methods. (See Fig. 13.7.)

In Alhassan et al. (2018), Aniyan and Thorat (2017), Domínguez Sánchez et al. (2018), the authors present the classifiers for automated classification of radio sources. It is based on a trained deep convolutional neural network to automate the morphological classification of compact and extended radio sources.

If data are not in a two-dimensional grid, convolutional neural networks can often be applied after using preprocessing. In Padarian et al. (2019) the output of a diffuse reflectance infrared spectroscopy is represented as a two-dimensional spectrogram, showing the reflectance as a function of wavelength and frequency. Convolutional neural networks are then used to predict six different soil properties.

While the strongest convolutional neural network applications are on two-dimensional data, the idea of convolutions can be applied to any number of dimensions as long as the structure of the data as an n-dimensional Euclidean vector space is meaningful. An example can be found in George and Huerta (2018), where the authors used two different one-dimensional convolutional neural networks on time series data. The input data are pieces of a time series and neighboring values are consecutive measurements. The authors use the first convolutional neural network to detect gravitational waves. If a signal is detected it is passed to the second convolutional neural network to estimate the parameters of the gravitational wave.

Another task that highlights the strength of convolutional neural networks is object detection. While in classification the method only classifies a certain object in the image, in object detection the method additionally outputs a bounding box around the objects in the image. The main approach is to slide a window over the image and classify every region of the image separately. Regions that lead to a high classification score are used as candidates. These candidates are then combined to detections using nonmaximum suppression.

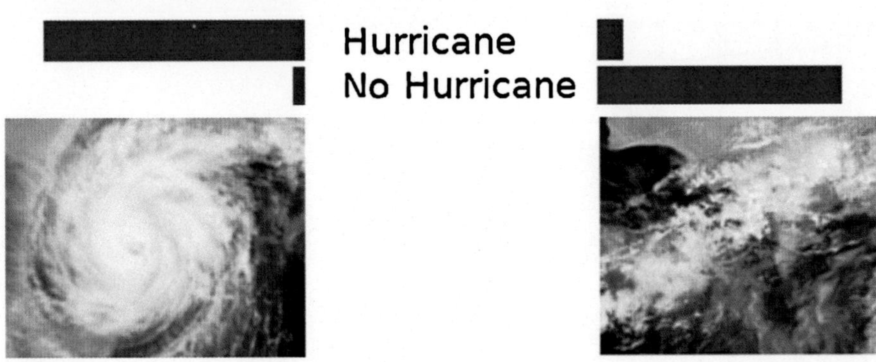

FIG. 13.7 Given an image, a neural network determines the likelihood that this cloud formation is created if a hurricane is present and the likelihood that this cloud formation is created if no hurricane is present. The classification is done by selecting the event with the highest likelihood.

For more details, see, for example, Girshick et al. (2014). While this is the classical approach, other approaches exist (Redmon and Farhadi, 2018). Since object detection is a straightforward extension of classification, it is used in many applications that use convolutional neural networks for classification.

For extreme whether detection it is, for example, demonstrated in Racah et al. (2017), for the localization of the host galaxy for a given radio component it is demonstrated in Alger et al. (2018). For remote sensing, many papers exist describing object detection using convolutional neural networks and proposing solutions for many problems, such as, for example, the different scales of the objects detected (Deng et al., 2018).

For some tasks, even bounding boxes are not enough and we require a pixel-by-pixel classification of the image. This task is called semantic segmentation and convolutional neural networks are the state of the art for it. For details on semantic segmentation of images see, for example, Fulkerson et al. (2009).

Similar to the object detection task the semantic segmentation task can be used in the same application in which we perform classification. For an example in remote sensing, see Kemker et al. (2018).

In conclusion, even though convolutional neural networks were developed for computer vision tasks, they can be used in many fields. In Reichstein et al. (2019) the authors describe how many tasks in Earth system science can be mapped to computer vision problems.

13.4.2 Recurrent Neural Networks

A different method of deep learning is recurrent neural networks. Recurrent neural networks are neural networks that, in addition to its inputs, use an internal state to perform a task. Since the new internal state is calculated from the old internal state and the input, it can be understood as one part of the output of the neural network. If we understand it in this way, recurrent neural networks get part of their output as input for the next time step. This recurrent behavior gave recurrent neural networks their name. The information that is stored in this internal state is automatically selected by the algorithm. Recurrent neural networks are, hence, a deep learning approach. The internal state, often referred to as "memory," makes them useful for sequential data, such as time series, audio, or text, and gives them the ability to handle inputs of different sequence length. (See Fig. 13.8.) This memory allows recurrent neural networks to predict processes of different frequencies that happen at the same time. (See Fig. 13.9.)

There are two main architectures that are used in almost every application of recurrent neural networks: long-short term memory (LSTM) (Hochreiter and Schmidhuber, 1997) and gated recurrent unit (GRU) (Cho et al., 2014). Both of these use every time step to calculate an output and to update the internal state. To perform inference on an input, normally only the network's output after the last step is used. Recurrent neural networks have demonstrated amazing performance in many tasks such as natural language processing and time series prediction. Time series prediction has many applications in astro- and geophysics where many interesting systems have unobserved states that cannot be determined from the measurements at one point in time but need past observations to be determined.

An example of such use of LSTM is rainfall intensity nowcasting. In Xingjian et al. (2015) the authors use a convolutional LSTM to forecast a fixed length of

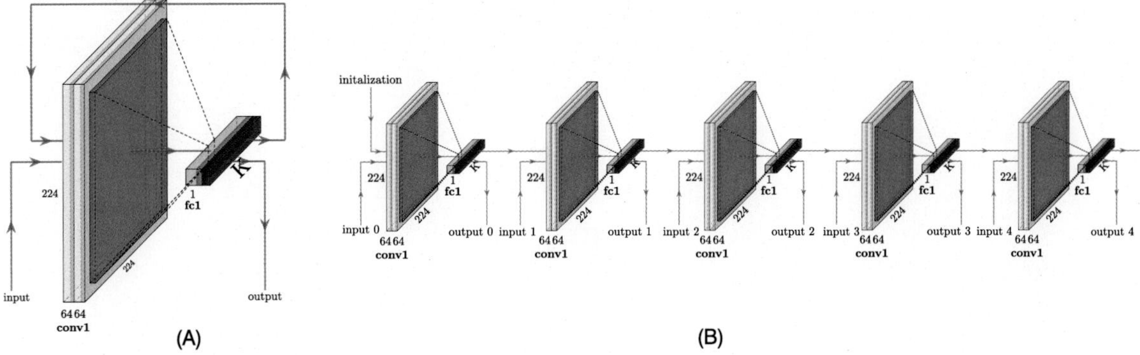

FIG. 13.8 (A) A recurrent neural network is a network that gets as additional input the state of a memory (represented by the top arrow) and generates as an additional output a new state for the memory. (B) A recurrent neural network can be "rolled out." The network can be understood as many copies of the same network connected through the memory.

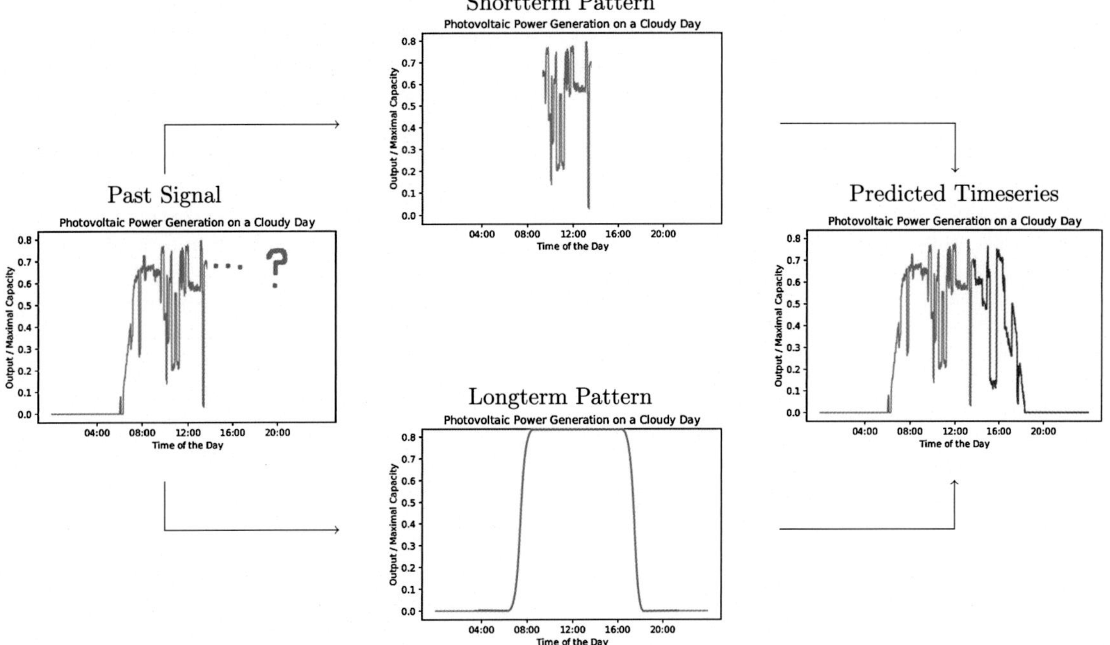

FIG. 13.9 Prediction of time series often include slow, long-term patterns and fast, short-term patterns.

future radar maps from observed radar echo sequences. In Racah et al. (2017) the authors use LSTMs to identify extreme weather events. They argue that some extreme weather events cannot be identified from the state of the system only, but temporal dynamics need to be taken into account. An example of these extreme weather events are extratropical cyclones. (See Fig. 13.7.)

13.4.3 Generative Models

Deep generative models are models capable of mapping a probabilistic latent space into a sample of the distribution of the target data set, i.e., they can generate novel data samples. Currently, generative networks are a very active research area that bloomed with the advent of VAEs in 2013 (Kingma and Welling, 2013) and gener-

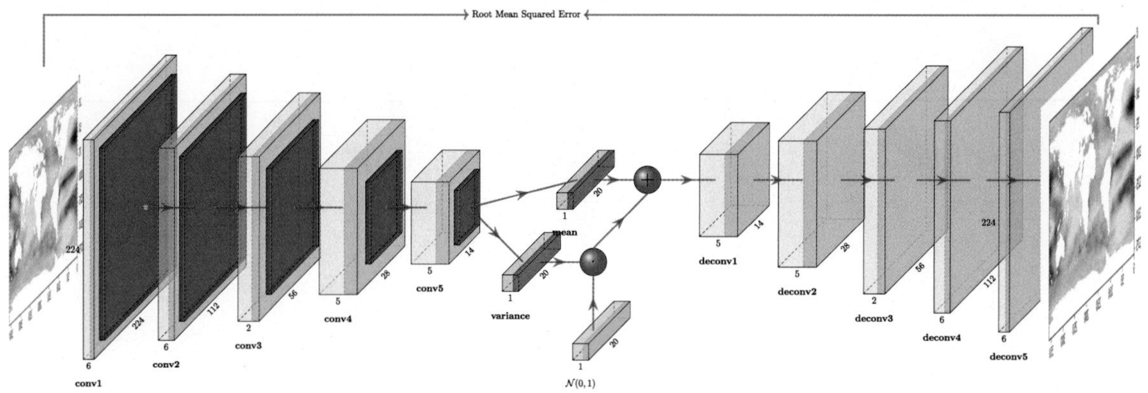

FIG. 13.10 An example architecture of a VAE. First the image is encoded into a vector of noise and a vector of variance, then a random value is drawn from Gaussian distributions with this mean and variance, and, finally, the image is reconstructed from these random values.

ative adversarial networks (GANs) in 2014 (Goodfellow et al., 2014). The common characteristic of generative networks is the mapping from a known arbitrarily selected distribution (typically Gaussian) onto the distribution of the dataset of interest. The algorithm automatically learns to generate features belonging to the dataset using a probabilistic latent space as prior. Generative models come in many shapes, making use of widely different architectures and training procedures. We will focus mainly on VAEs and GANs as these are the most extended, are repeatedly used in the fields of geoscience and astrophysics, and have the most potential.

Generative models have the benefit of not needing labeled data, and therefore they are capable of making use of most of the existing dataset in some way. However, their use is not as straightforward as that of the supervised methods.

13.4.3.1 Variational Autoencoders

An autoencoder consists of an encoder, a decoder, and a loss function. Both the encoder and the decoder are networks. The input of the encoder is a data sample and its output is the latent representation of that sample of lower dimensionality (hence the name "encoder"). The decoder takes the latent representation and projects it back to reconstruct the original sample. In a sense, the latent representation is a bottleneck in the network; hence, the encoder must learn an efficient representation of the original data since the information at the bottleneck is all that is available to reconstruct the sample by the decoder. In a VAE, the lower-dimensional space is stochastic. We can sample from the assumed distribution of the latent space and feed it to the decoder to create novel samples that belong to the original data distribution. In order to train a VAE, two losses are used. On one side the reconstruction loss (how much information is lost?) is used to force the network to learn a reconstruction as lossless as possible. On the other side, the Kullback–Leibler divergence is used to force the latent space's distribution to be similar to the assumed one, often Gaussian. (See Fig. 13.10.)

VAEs are used both for data dimensionality reduction and as generative models for novel data samples. Given the variational constraint, VAEs create efficient latent representations and can be used as powerful unsupervised feature extraction. These features can later be used by other statistical methods that initially could not deal with the high dimensionality of the original data samples.

VAEs have been applied to climate datasets in order to simplify complex temporal dynamics over time into color-coded regions over the planet, making its study more accessible to experts by Tibau et al. (2018). They present a framework to extract a kernel function for a kernel-PCA from data. The kernel is in this case not chosen through prior knowledge but instead learned from the data itself.

The use of an autoencoder to create a more meaningful features representation is also demonstrated in Qian et al. (2018). The authors of this paper use a convolutional autoencoder to reduce the dimensionality of prestack seismic data stratigraphy. The latent representations are afterwards clustered to generate seismic facies maps.

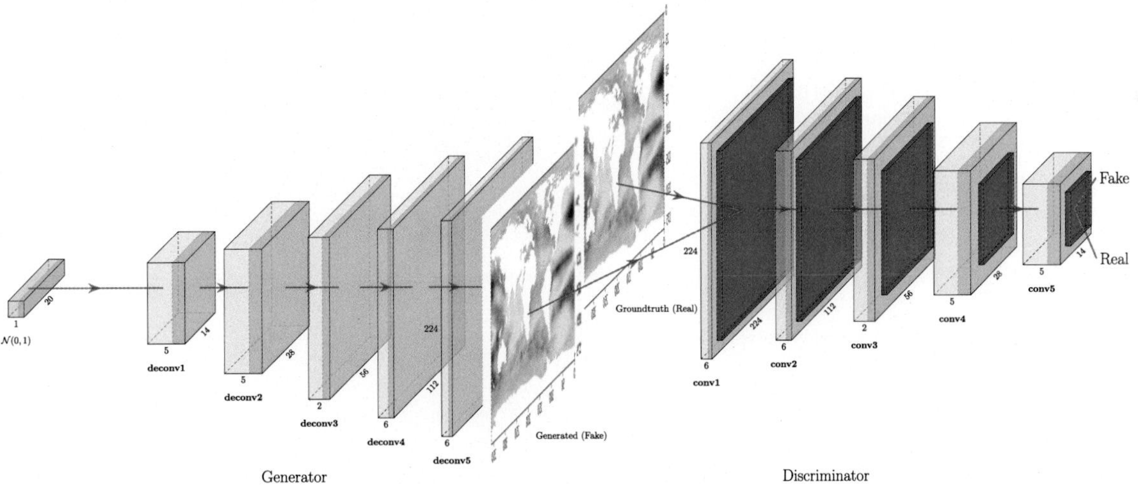

FIG. 13.11 An example architecture of a generative adversarial network. The image is decoded by the generator from a vector of noise, and then the auxiliary discriminator determines whether it fits with the real samples.

An example for the use of a VAE for unsupervised dimensionality reduction is demonstrated in Huijse et al. (2018) for data of transient candidates or in Karmakar et al. (2018) for stellar cluster detection. Here the authors reduce the dimensionality and then build a Gaussian mixture model on the latent space variables.

A geophysical application is demonstrated in Li and Misra (2017). The authors want to predict NMR T2 Distribution without an NMR prediction tool. Therefore, they first use a VAE to produce a compact representation of the NMR T2 Distributions in a feature space and afterwards use a neural network to predict these features from other measurements that are easier to obtain.

The generative part of VAE was for example demonstrated in Regier et al. (2015). The authors build a VAE to generate astronomical, optical telescope images. For an example closer to geophysics, we refer the reader to the paper of Khodayar et al. (2018), which we discussed in the introduction.

13.4.3.2 Generative Adversarial Networks

GANs are composed of a generator and a discriminator network (Fig. 13.11). The generator takes as input a probabilistic latent space and maps it into a sample belonging to the distribution of the target dataset. During training the discriminator takes samples that are real, belonging to the ground truth, and samples that are fake, created by the generator. Its task is to discriminate which ones belong to the ground truth. The loss of the discriminator is used to improve the generator. In a sense, the two networks are in a constant battle, where the generator learns to create more convincing samples to fool the discriminator, while the discriminator is constantly improving in order to correctly discern the real from the generated samples.

In Ravanbakhsh et al. (2017) the authors use a GAN and a VAE to create realistic galaxy images. They claim that the generation using these methods is a reliable alternative to the expensive acquisition of high-quality observations.

In Rodríguez et al. (2018) the authors use a GAN to speed up N-body simulations. The GAN creates realistic two-dimensional image snapshots from the N-body simulation.

There exist hybrid generative models such as the VAEGANs. This architecture consists of an autoencoder and a discriminator network that is trained to tell the original input and the reconstruction apart. Hence, the architecture is effectively an autoencoder, but the training procedure is that of a GAN. This method was applied in astrophysics, for example, in Schawinski et al. (2017) to denoise images of galaxies. The advantage of VAEGANs over regular GANs is the higher ability to capture precisely a meaningful latent space, since classical GAN training is prone to mode collapse. VAEGANs can also be used as a dimensionality reduction method.

In the geoscience field, GANs have been used repeatedly towards the end of the 2010s. Special attention has been given to remote sensing processing. Difficult

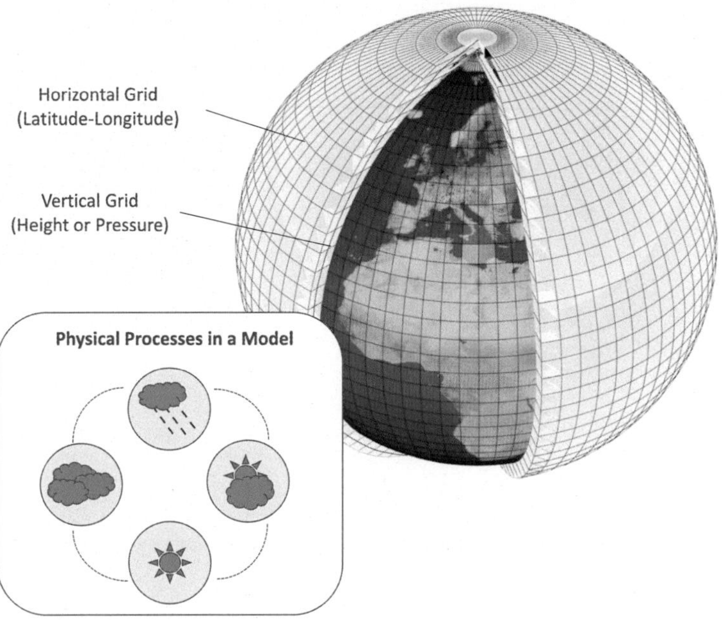

FIG. 13.12 Climate models typically discretize the Earth's surface and its atmosphere into cells. The models make use of this simplification and describe how matter and energy move through the cells. The models that run at a finer resolution can achieve better accuracy at the expense of a higher computational cost. Nowadays, deep learning poses as an alternative to speed up the subgrid processes by emulating the behavior of the finer processes. The biggest challenge is to develop deep learning models that follow the physical constraints of the real system.

tasks such as the unsupervised semantic classification of satellite imagery have been tackled using a GAN over unlabeled satellite imagery and making use of the discriminator as a feature extractor (Lin et al., 2017). Hyperspectral image classification with very few samples has also benefited from the development of semisupervised methods involving a GAN architecture. In He et al. (2017) the discriminator of the GAN is used as a classificator of the labeled class for the real images but it still functions as a discriminator for the generated ones. By making the discriminator multitask, they effectively are augmenting the size of the dataset with unlabeled samples, thus increasing performance especially when there is low availability of data.

Some other more creative uses of GANs simulate the growth of cities generating synthetic urban patterns that reproduce the complex spatial organization of cities, while being able to quantitatively recover certain key high-level urban spatial metrics (Albert et al., 2018). On climate models, deep learning might be able to power research further than the resolution of the model grids (Fig. 13.12). Subgrid processes are one of the main sources of uncertainty in climate and weather models as of today. With adapted deep learning models it is possible to emulate the subgrid processes, reducing greatly the uncertainty while not greatly affecting the run-time. This idea is explored by Rasp et al. (2018), acknowledging the potential of GANs to create stochastic machine learning-based parametrizations that capture the variability of the system from the training data.

One especially useful architecture of GANs are the conditional GANs (Mirza and Osindero, 2014). The variables in the latent distribution of conditional GANs are not only automatically selected features, but some of them are handcrafted or hand-selected predictors for the samples. This allows the user to produce realistic samples that have the desired characteristics. As an example, see Requena-Mesa et al. (2018). The authors demonstrate the ability of GANs to generate landscape images from environmental conditions (see Fig. 13.13). The generative network is able to create photo-interpretable satellite pictures for hypothetical climate scenarios. The images contain visible features that are not directly determined by the environmental variables but are important for the samples to behave as real landscapes.

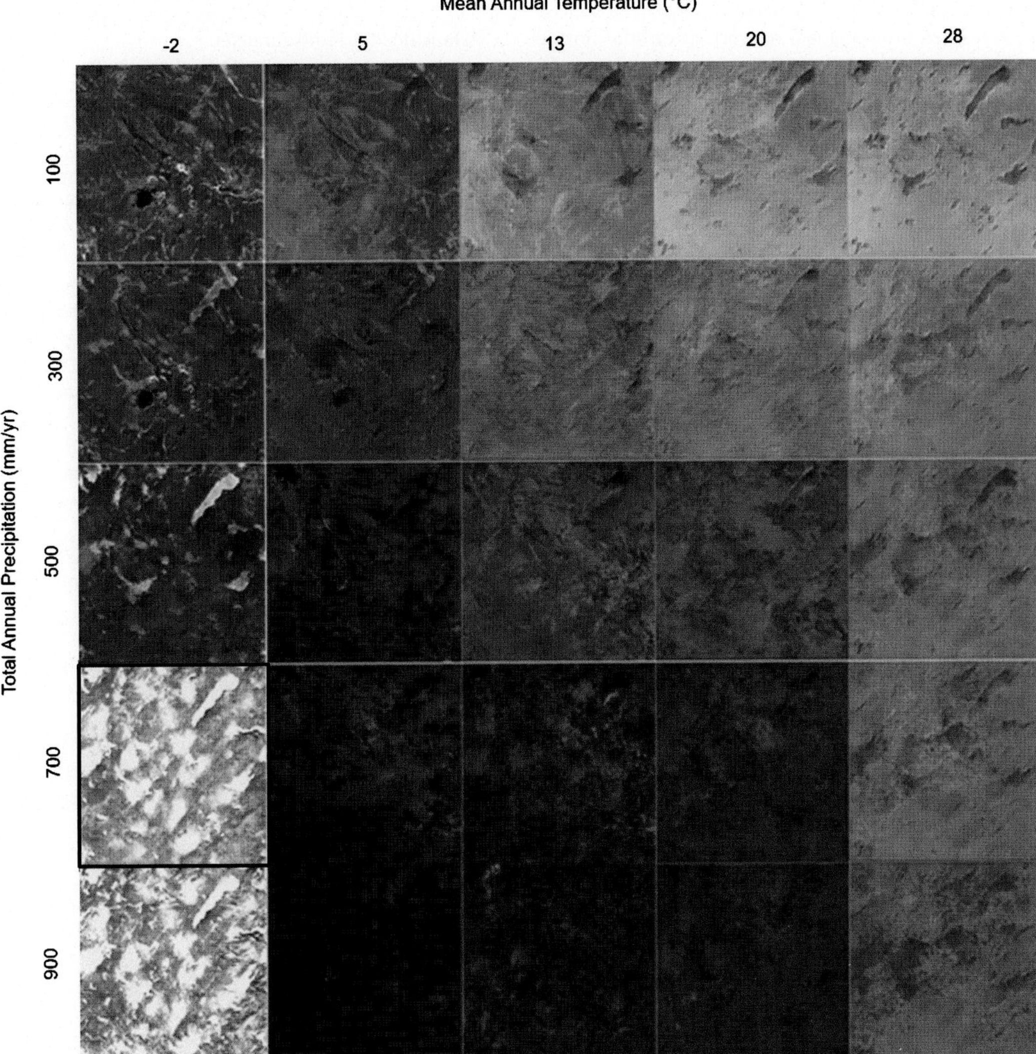

FIG. 13.13 Landscapes generated by a conditional GAN showing how a region in Northern Canada might look under a different climate. The region originally looks like the sample in black. All images displayed are false color infrared, highlighting vegetation cover in red.

REFERENCES

Abbott, B., Abbott, R., Adhikari, R., Ajith, P., Allen, B., Allen, G., Amin, R., Anderson, S., Anderson, W., Arain, M., et al., 2009. LIGO: the laser interferometer gravitational-wave observatory. Reports on Progress in Physics 72 (7), 076901.

Albert, A., Strano, E., Kaur, J., González, M., 2018. Modeling urbanization patterns with generative adversarial networks. In: IGARSS 2018 - 2018 IEEE International Geoscience and Remote Sensing Symposium. IEEE, pp. 2095–2098.

Alger, M., Banfield, J., Ong, C., Rudnick, L., Wong, O., Wolf, C., Andernach, H., Norris, R., Shabala, S., 2018. Radio Galaxy Zoo: machine learning for radio source host galaxy cross-identification. Monthly Notices of the Royal Astronomical Society 478 (4), 5547–5563.

Alhassan, W., Taylor, A., Vaccari, M., 2018. The first classifier: compact and extended radio galaxy classification using deep convolutional neural networks. Monthly Notices of the Royal Astronomical Society 480 (2), 2085–2093.

Aniyan, A., Thorat, K., 2017. Classifying radio galaxies with the convolutional neural network. The Astrophysical Journal. Supplement Series 230 (2), 20.

Baldocchi, D., 2008. 'Breathing' of the terrestrial biosphere: lessons learned from a global network of carbon dioxide flux measurement systems. Australian Journal of Botany 56 (1), 1–26.

Cao, Z., Ma, R., Duan, H., Xue, K., 2019. Effects of broad bandwidth on the remote sensing of inland waters: implications for high spatial resolution satellite data applications. ISPRS Journal of Photogrammetry and Remote Sensing 153, 110–122.

Cho, K., Van Merriënboer, B., Bahdanau, D., Bengio, Y., 2014. On the properties of neural machine translation: encoder-decoder approaches. ArXiv preprint, arXiv:1409.1259.

Chollet, F., Allaire, J., 2018. Deep Learning with R. Manning Publications Company.

Deng, Z., Sun, H., Zhou, S., Zhao, J., Lei, L., Zou, H., 2018. Multi-scale object detection in remote sensing imagery with convolutional neural networks. ISPRS Journal of Photogrammetry and Remote Sensing 145, 3–22.

Domínguez Sánchez, H., Huertas-Company, M., Bernardi, M., Tuccillo, D., Fischer, J., 2018. Improving galaxy morphologies for SDSS with deep learning. Monthly Notices of the Royal Astronomical Society 476 (3), 3661–3676.

Fulkerson, B., Vedaldi, A., Soatto, S., 2009. Class segmentation and object localization with superpixel neighborhoods. In: 2009 IEEE 12th International Conference on Computer Vision. IEEE, pp. 670–677.

Gensler, A., Henze, J., Sick, B., Raabe, N., 2016. Deep learning for solar power forecasting—an approach using AutoEncoder and LSTM neural networks. In: 2016 IEEE International Conference on Systems, Man, and Cybernetics (SMC). IEEE, pp. 002858–002865.

George, D., Huerta, E., 2018. Deep neural networks to enable real-time multimessenger astrophysics. Physical Review D 97 (4), 044039.

Girshick, R., Donahue, J., Darrell, T., Malik, J., 2014. Rich feature hierarchies for accurate object detection and semantic segmentation. In: Proceedings of the IEEE Conference on Computer Vision and Pattern Recognition, pp. 580–587.

Gómez-Chova, L., Tuia, D., Moser, G., Camps-Valls, G., 2015. Multimodal classification of remote sensing images: a review and future directions. Proceedings of the IEEE 103 (9), 1560–1584.

Goodfellow, I., Bengio, Y., Courville, A., 2016. Deep Learning. MIT Press. http://www.deeplearningbook.org.

Goodfellow, I., Pouget-Abadie, J., Mirza, M., Xu, B., Warde-Farley, D., Ozair, S., Courville, A., Bengio, Y., 2014. Generative adversarial nets. In: Ghahramani, Z., Welling, M., Cortes, C., Lawrence, N.D., Weinberger, K.Q. (Eds.), Advances in Neural Information Processing Systems 27. Curran Associates, Inc., pp. 2672–2680.

He, Z., Liu, H., Wang, Y., Hu, J., 2017. Generative adversarial networks-based semi-supervised learning for hyperspectral image classification. Remote Sensing 9 (10), 1042.

Hochreiter, S., Schmidhuber, J., 1997. Long short-term memory. Neural Computation 9 (8), 1735–1780.

Huijse, P., Astorga, N., Estévez, P.A., Pignata, G., 2018. Latent representations of transient candidates from an astronomical image difference pipeline using variational autoencoders. In: ESANN.

Iqbal, H., 2018. PlotNeuralNet https://github.com/HarisIqbal88/PlotNeuralNet.

Karmakar, A., Mishra, D., Tej, A., 2018. Stellar cluster detection using GMM with deep variational autoencoder. In: 2018 IEEE Recent Advances in Intelligent Computational Systems (RAICS). IEEE, pp. 122–126.

Kemker, R., Salvaggio, C., Kanan, C., 2018. Algorithms for semantic segmentation of multispectral remote sensing imagery using deep learning. ISPRS Journal of Photogrammetry and Remote Sensing 145, 60–77.

Khodayar, M., Mohammadi, S., Khodayar, M., Wang, J., Liu, G., 2018. Convolutional graph auto-encoder: a deep generative neural architecture for probabilistic spatio-temporal solar irradiance forecasting. ArXiv preprint, arXiv:1809.03538.

Kingma, D.P., Welling, M., 2013. Auto-encoding variational Bayes. ArXiv preprint, arXiv:1312.6114.

Krizhevsky, A., Sutskever, I., Hinton, G.E., 2012. ImageNet classification with deep convolutional neural networks. In: Advances in Neural Information Processing Systems, pp. 1097–1105.

LeCun, Y., Bottou, L., Bengio, Y., Haffner, P., et al., 1998. Gradient-based learning applied to document recognition. Proceedings of the IEEE 86 (11), 2278–2324.

Li, H., Misra, S., 2017. Prediction of subsurface NMR T2 distributions in a shale petroleum system using variational autoencoder-based neural networks. IEEE Geoscience and Remote Sensing Letters 14 (12), 2395–2397.

Lin, D., Fu, K., Wang, Y., Xu, G., Sun, X., 2017. MARTA GANs: unsupervised representation learning for remote sensing image classification. IEEE Geoscience and Remote Sensing Letters 14 (11), 2092–2096.

Liu, Y., Racah, E., Correa, J., Khosrowshahi, A., Lavers, D., Kunkel, K., Wehner, M., Collins, W., et al., 2016. Application of deep convolutional neural networks for detecting extreme weather in climate datasets. ArXiv preprint, arXiv:1605.01156.

Mirza, M., Osindero, S., 2014. Conditional generative adversarial nets. ArXiv preprint, arXiv:1411.1784.

Montavon, G., Samek, W., Müller, K.-R., 2018. Methods for interpreting and understanding deep neural networks. Digital Signal Processing 73, 1–15.

Ngiam, J., Khosla, A., Kim, M., Nam, J., Lee, H., Ng, A.Y., 2011. Multimodal deep learning. In: Proceedings of the 28th International Conference on Machine Learning (ICML-11), pp. 689–696.

Padarian, J., Minasny, B., McBratney, A., 2019. Using deep learning to predict soil properties from regional spectral data. Geoderma Regional 16, e00198.

Qian, F., Yin, M., Liu, X.-Y., Wang, Y.-J., Lu, C., Hu, G.-M., 2018. Unsupervised seismic facies analysis via deep convolutional autoencoders. Geophysics 83 (3), A39–A43.

Racah, E., Beckham, C., Maharaj, T., Kahou, S.E., Prabhat, M., Pal, C., 2017. ExtremeWeather: a large-scale climate dataset for semi-supervised detection, localization, and understanding of extreme weather events. In: Advances in Neural Information Processing Systems, pp. 3402–3413.

Rasp, S., Pritchard, M.S., Gentine, P., 2018. Deep learning to represent subgrid processes in climate models. Proceedings of the National Academy of Sciences 115 (39), 9684–9689.

Ravanbakhsh, S., Lanusse, F., Mandelbaum, R., Schneider, J., Poczos, B., 2017. Enabling dark energy science with deep generative models of galaxy images. In: Thirty-First AAAI Conference on Artificial Intelligence.

Redmon, J., Farhadi, A., 2018. YOLOv3: an incremental improvement. ArXiv preprint, arXiv:1804.02767.

Regier, J., Miller, A., McAuliffe, J., Adams, R., Hoffman, M., Lang, D., Schlegel, D., Prabhat, M., 2015. Celeste: variational inference for a generative model of astronomical images. In: International Conference on Machine Learning, pp. 2095–2103.

Reichstein, M., Camps-Valls, G., Stevens, B., Jung, M., Denzler, J., Carvalhais, N., et al., 2019. Deep learning and process understanding for data-driven Earth system science. Nature 566 (7743), 195.

Requena-Mesa, C., Reichstein, M., Mahecha, M., Kraft, B., Denzler, J., 2018. Predicting landscapes as seen from space from environmental conditions. In: IGARSS 2018 – 2018 IEEE International Geoscience and Remote Sensing Symposium. IEEE, pp. 1768–1771.

Rodríguez, A.C., Kacprzak, T., Lucchi, A., Amara, A., Sgier, R., Fluri, J., Hofmann, T., Réfrégier, A., 2018. Fast cosmic web simulations with generative adversarial networks. Computational Astrophysics and Cosmology 5 (1), 4.

Schawinski, K., Zhang, C., Zhang, H., Fowler, L., Santhanam, G.K., 2017. Generative adversarial networks recover features in astrophysical images of galaxies beyond the deconvolution limit. Monthly Notices of the Royal Astronomical Society. Letters 467 (1), L110–L114.

Sharma, N., Sharma, P., Irwin, D., Shenoy, P., 2011. Predicting solar generation from weather forecasts using machine learning. In: 2011 IEEE International Conference on Smart Grid Communications (SmartGridComm). IEEE, pp. 528–533.

Simonyan, K., Zisserman, A., 2014. Very deep convolutional networks for large-scale image recognition. ArXiv preprint, arXiv:1409.1556.

Srivastava, N., Salakhutdinov, R.R., 2012. Multimodal learning with deep Boltzmann machines. In: Advances in Neural Information Processing Systems, pp. 2222–2230.

Tibau, X.-A., Requena-Mesa, C., Reimers, C., Denzler, J., Eyring, V., Reichstein, M., Runge, J., 2018. SupernoVAE: VAE based kernel PCA for analysis of spatio-temporal Earth data. In: Climate Informatics Workshop 2018, pp. 73–77.

Tsonis, A.A., Swanson, K.L., Roebber, P.J., 2006. What do networks have to do with climate? Bulletin of the American Meteorological Society 87 (5), 585–596.

Xingjian, S., Chen, Z., Wang, H., Yeung, D.-Y., Wong, W.-K., Woo, W.-c., 2015. Convolutional LSTM network: a machine learning approach for precipitation nowcasting. In: Advances in Neural Information Processing Systems, pp. 802–810.

Xue, K., Ma, R., Duan, H., Shen, M., Boss, E., Cao, Z., 2019. Inversion of inherent optical properties in optically complex waters using Sentinel-3a/OLCI images: a case study using China's three largest freshwater lakes. Remote Sensing of Environment 225, 328–346.

CHAPTER 14

Astro- and Geoinformatics – Visually Guided Classification of Time Series Data

ROMAN KERN, PHD • TAREK AL-UBAIDI, MSC • VEDRAN SABOL, PHD • SARAH KREBS, MSC • MAXIM KHODACHENKO, PHD • MANUEL SCHERF, MSC

14.1 INTRODUCTION

Machine learning and *eScience* are two hot topics of the hour, mainly motivated by the scientific progress and technological advancements; however, an in-depth understanding is often lacking (Marcus, 2018). Nevertheless, looking at predominantly data-driven sciences such as space science the benefits are quite obvious. According to the *IEEE eScience Conference Series*,[1] eScience "...promotes innovation in collaborative, computationally- or data-intensive research across all disciplines, throughout the research lifecycle," thus Big Data and data-driven science in general are crucial aspects. This led Turing award winner Jim Gray to even proclaim eScience as the fourth paradigm of science, adding the data-driven paradigm to the existing (empirical, theoretical, and computational), thus stressing the immense importance of improved approaches for knowledge generation and advanced data analysis tools that in many cases have the *potential to revolutionize the scientific workflow, by speeding up certain processes by several orders of magnitude*.[2] As we will show, content-based search tools enhanced by machine learning can search through and analyze huge amounts of data (in the gigabyte to terabyte range) in minutes and sometimes seconds – tasks that would have taken several months can thus be reduced to days or even hours, while at the same time allowing for much deeper and more thorough investigations of scientific data. This frees up immense scientific resources that were until recently too often bound with rather uninspiring routine tasks.

In principle, the content-based search technologies at hand can be transferred to any time-based signal with manageable effort and thus are readily applicable for a variety of scientific topics. In space science these approaches will only increase in importance, since current missions produce an ever growing amount of observational data. According to NASA's Earth science data systems program executive Kevin Murphy, already in 2016 NASA was generating 12.1 terabyte of data every single day via space- and ground-based sensors. But things will only get better from here (or worse, from the perspective of engineers who have to design systems to store and process all these data, if there are no appropriate tools readily available), once future missions equipped with optical laser systems for data transfer go online. Then, some missions could generate up to a staggering 24 terabyte of data in a *single day*, which is about 38 times the amount of data produced by the entire *Cassini* space mission to Saturn and its moon Titan (approximately 635 gigabyte) during its mission duration of almost 20 years. The increase can be partly attributed to higher resolution of sensors used as well as an overall increase of targets and observation time. In view of the fact that NASA alone is looking at an annual growth of its data archives of between five to 10 petabyte (one *petabyte* is roughly a thousand *terabytes*) per year, it can be expected that the situation is similar for other major space agencies such as ESA, Roscosmos, or JAXA.

It is clear that in view of the several petabytes of data that are added to the archives annually, building new ways to visualize, analyze, and interpret huge quantities of information is of the highest priority and machine learning and artificial intelligence in general will play a pivotal role in the process of extracting scientific knowledge. A survey of work done with regard to content-based search of time-based signals reveals that most applications are centered around audio and video data – with the majority of research in the field of content-based search in general dealing with image data. Akisato Kimura et al., for example, presented a method

[1] https://escience-conference.org/.

[2] As an example we could cite the search for transits of potential exoplanets by manually going through the available light curve data.

Knowledge Discovery in Big Data from Astronomy and Earth Observation. https://doi.org/10.1016/B978-0-12-819154-5.00025-4

for a quick and accurate audio search that uses dimensionality reduction of histogram features, where base features are extracted at every sampled time step, e.g., every 10 ms (Kimura et al., 2008). A general approach for time series classification using multistage deep learning for multivariate time series classification problems was presented by Nijat Mehdiyev et al. in 2017. After extracting the features from the time series data in an unsupervised manner by using stacked *LSTM autoencoders*, a deep feedforward neural network was applied to classify the data and make predictions. The aim was to eliminate the necessity for domain expert knowledge to determine useful features, which is a very time consuming and error-prone task. The system's performance was assessed using data obtained from a steel manufacturer (Mehdiyev et al., 2017). It is obvious that data containing time-based signals, making up a significant part of the available scientific "raw material," will tremendously benefit from these kinds of information retrieval and analysis techniques. In particular when extended with a capability to detect specific patterns of interest based on supervised machine learning, as the prototype described in Section 14.4, highly efficient tools for the analysis and automatic classification of time-based scientific observational or simulation data can be created.

The concept of content-based search or similarity search in general is related to the basic idea of *query by example* (QBE), a term coined by Moshé M. Zloof at *IBM Research* in the mid-1970s (Zloof, 1975). In short, QBE offers a simplified interface (i.e., graphical query language) to the users that allows them to define search conditions by providing example elements and conditions without deeper knowledge of the inner workings of the respective database or information system. Similar to the parser in a QBE interface that typically expresses the high-level search conditions, i.e., visual tables and examples provided by the user in a query language such as SQL that can then be executed, the example signal provided by the user in a graphical interface is resolved, using a comprehensive set of extracted base features that form a kind of "alphabet" for the time-based data that are searched (see Fig. 14.1 and Sections 14.3.2 and 14.6 for further details).

In other words, the data are brought into a representation suitable to be managed within an inverted index, by adapting and extending symbolic representation techniques. Depending on the nature of the data, these representation techniques may range from simple binning methods to methods operating in the frequency domain, e.g., *wavelet analysis*. Contemporary search techniques allow storing additional information alongside the indexed data,[3] e.g., to encode the quantification error and other statistics (e.g., trends) into the inverted index alongside with the indexing structures. The actual search is then executed using off-the-shelf search engines as, e.g., *Apache Lucene*.[4] Additionally, in order to provide automatic classification of time-based signals, the data will be processed using methods from the fields of sequential pattern mining in combination with outlier detection to arrive at temporal patterns of interest and identify a trade-off between frequent and "surprising" patterns. These methods are representative of unsupervised machine learning techniques which do not rely on additionally labeled training examples. However, due to their unsupervised nature, many of the identified patterns will not be of interest for domain-specific, scientific questions, i.e., they will also include glitches and artifacts caused by the specific means of data acquisition along with statistical artifacts, etc. In order to limit the identified patterns to a set of candidates that are of greatest relevance to space scientists looking at specific phenomena, techniques from the field of supervised machine learning are then integrated – a procedure where human annotators provide labeled examples as references. The input for this training process is hence provided in a collaborative effort to the respective community by users of the system, who are using the search interface to also annotate signals or patterns of interest (also see Section 14.4.3). The annotations are in turn discussed among all users of the system and hence validated, to ensure that any automatic classification given has a high likelihood of being correct as well as relevant to specific scientific questions at hand. Once the system is trained using an appropriate amount of examples and assuming that a sufficient level of generalization can be achieved by the network, the expectation is that it will be able to identify these patterns in any measurement or simulation data provided, with a high degree of confidence.

In the following sections we will give an overview of the *MESSENGER* data used to demonstrate the potential of content-based signal search and analysis. Furthermore, the architecture and basic design decisions with regard to preprocessing, search, and reranking will be outlined. A section discussing the basics of time signal classification and forecasting, referencing some of the most promising approaches, will complement the description of the architecture. The visual interface will be presented using screen shots of a prototype implementation and at the end of the chapter a discussion

[3] Known as the *payload feature* in *Apache Lucene*.

[4] https://lucene.apache.org/.

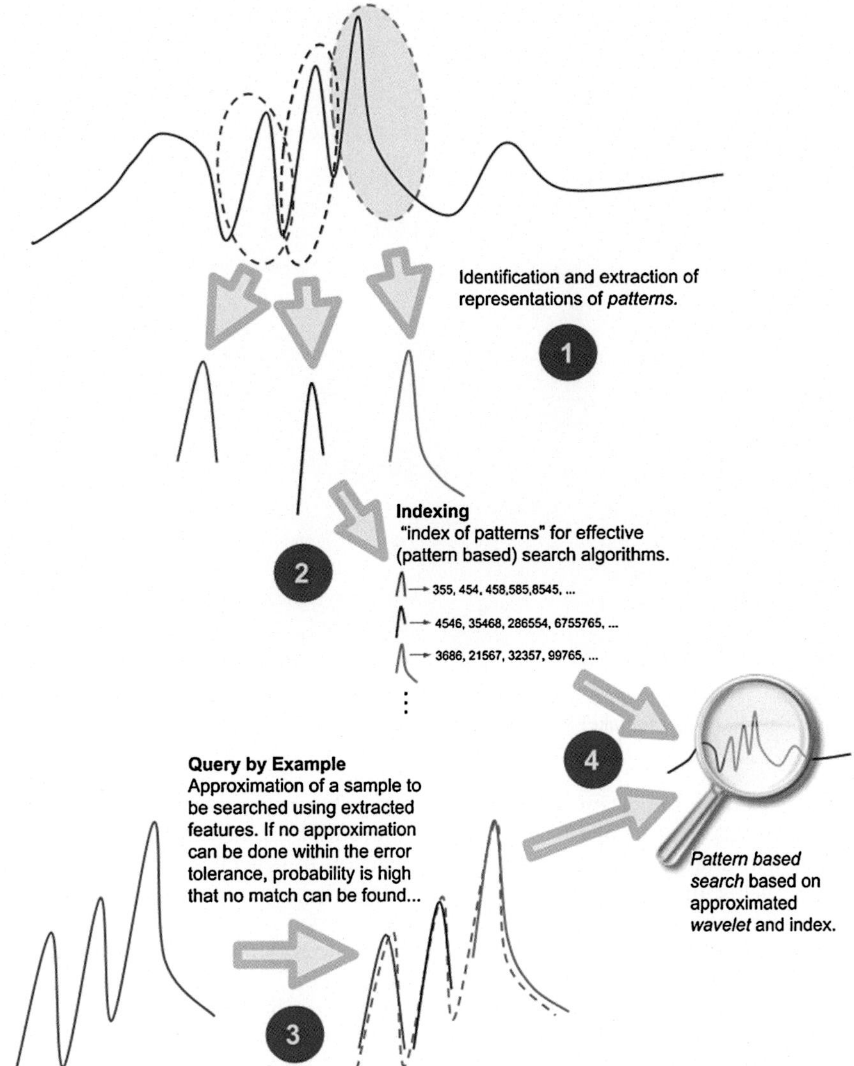

FIG. 14.1 Basic concept of query by example or content-based search in conjunction with a search index where patterns are (1) identified, (2) stored in an *inverted index*, (3) retrieved via a *fuzzy lookup*, and (4) displayed to the user.

of user interaction and scalability will be given, including thoughts on the relevance of these new technologies for space science as well as a tentative outlook on future developments.

14.2 THE MESSENGER DATA

The example uses magnetometer data obtained by the NASA mission MESSENGER (*MErcury Surface, Space ENvironment, GEochemistry, and Ranging*), which was launched on 3 August 2004 and arrived at Mercury in January 2008. MESSENGER went on to become the first spacecraft to enter into an orbit around the planet closest to our Sun on 18 March 2011. With its more than 4000 orbits around Mercury, the mission achieved 100% mapping of the planet surface and a detailed characterization of Mercury's magnetic field – it ended on 30 April 2015 after MESSENGER deorbited as planned and impacted the planet's surface.

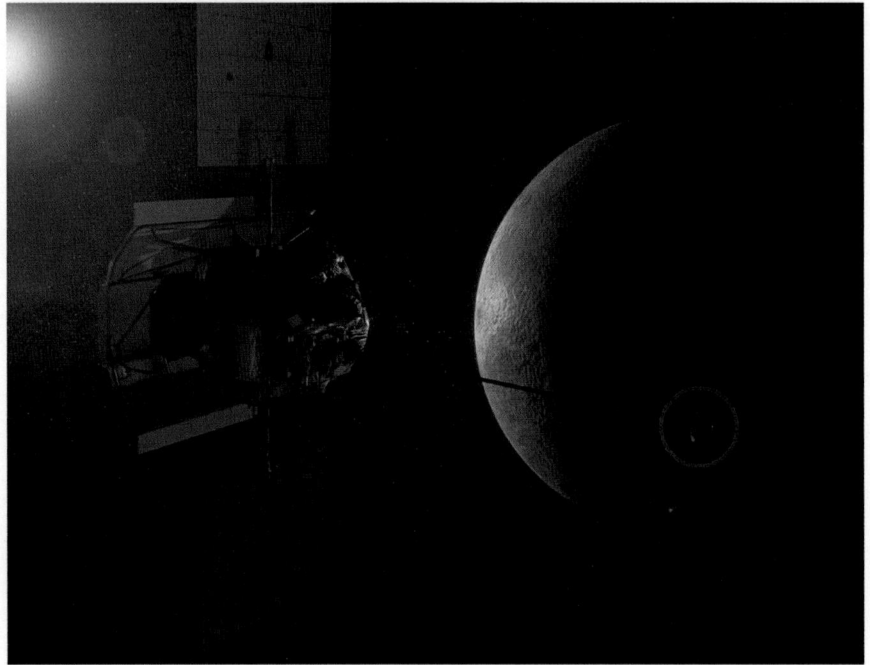

FIG. 14.2 Artist's rendering of MESSENGER orbiting Mercury with its MAG instrument (source: NASA, https://commons.wikimedia.org/wiki/File:MESSENGER.jpg).

One of MESSENGER's most important instruments was its magnetometer (MAG). In general, a magnetometer measures the strength and in most cases also the direction of magnetic fields, including those near the Earth and other astronomical bodies such as planets of the solar system or comets. MESSENGER's magnetometer was a low-noise, tri-axial fluxgate instrument located at the end of a 3.6 meter boom; see Fig. 14.2 (Anderson et al., 2007). The instrument was used to map Mercury's magnetic field and to determine its structure, infer its origin, and search for regions of magnetized rocks in the crust.

Using measurement data of MESSENGER's MAG and simulations of the *Hermean* magnetic field environment with the *paraboloid magnetosphere model*, it has been shown that Mercury possesses a small intrinsic magnetic dipole, which is shifted from the planet's center (Alexeev et al., 2010). MESSENGER has provided a vast amount of magnetic field data around Mercury, which have yet to be studied in more detail. Of high interest in that respect is information regarding the details of the planetary magnetosphere structure, shape, and size, as these characteristics are influenced substantially by the solar wind and further processes related to solar activity. Revealing the mechanisms and physical nature of this influence is crucial for a deeper understanding of complex phenomena related to the solar-planetary interaction in general, and specifically space weathering on Mercury. As a first step, the correct detection of the events when the spacecraft enters and passes through the planetary magnetosphere is crucial. Therefore, the focus lies on MESSENGER's magnetosphere crossings at Mercury to showcase the capabilities of the proposed content-based search tool that is being used, to automatically extract these crossings from the data. Using the search tool, this can be done much faster than a scientist would ever be able to do by going through the whole data record manually.

Initially, we also considered to detect magnetopause crossings, which are crossings of the region that forms the boundary between the planetary magnetic field and the solar wind; see Fig. 14.3. This region is determined by the balance between the pressure of the planetary magnetic field and the dynamic total pressure of the solar wind. However, due to the highly turbulent state of the plasma and the magnetic field in this region, it is not a simple task to detect the magnetopause itself, since the respective patterns are rather stochastic and thus hard to generalize. At the same time, a spacecraft orbiting a planetary object with a magnetic field will periodically

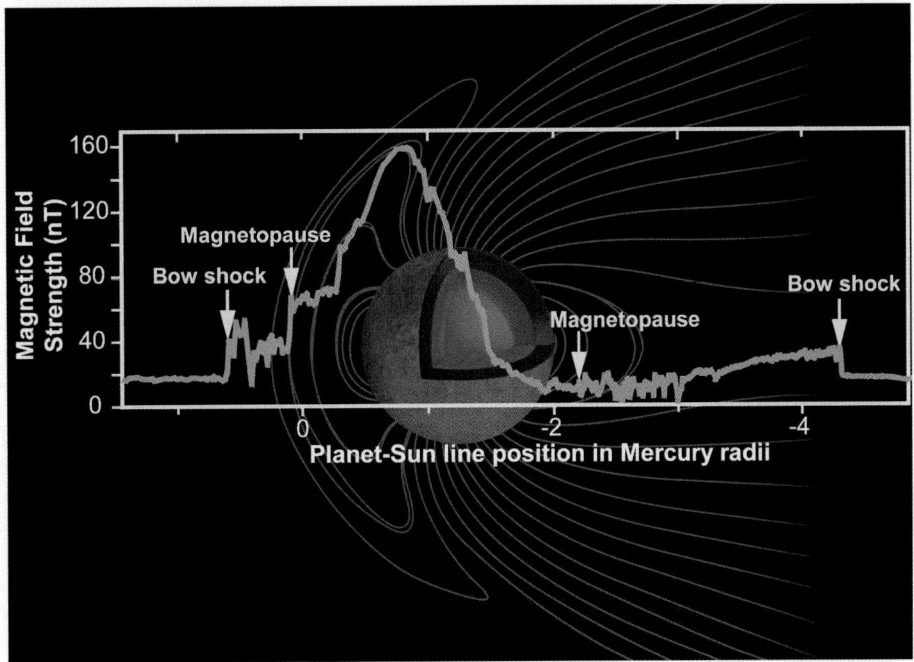

FIG. 14.3 Mercury's magnetic field and magnetopause (source: NASA, https://commons.wikimedia.org/wiki/File:Mercury_Magnetic_Field_NASA.jpg).

pass through the planetary magnetosphere, depending on the exact orbit of the satellite. These crossings will create a specific and consistent pattern in the magnetometer data and can thus be identified visually. The focus was hence shifted to the detection of the complete magnetosphere crossing of the spacecraft, i.e., the period when MESSENGER dips into the magnetic field of Mercury, with a magnetopause crossing located at the beginning and the end of each of these events which turned out to be detectable quite effectively.

To assess the performance of the developed tool, we looked at a relatively small portion of the magnetometer data (10 days, 48 megabyte of data), which include approximately 20 crossings. When comparing to the speed of manual detection, we also assumed that the data is already preprocessed and prepared for visualization in a scientific tool such as CDPP-AMDA.[5]

14.3 SYSTEM ARCHITECTURE

14.3.1 Scalability

Datasets are continuously growing, both in terms of size and in terms of resolution. As such, our system needs to gracefully scale with the amount of data. This quality attribute (Bass et al., 2003) is therefore an important requirement in the selection of used technological components in our system. Historically, one of the main driving forces behind many Big Data solutions was web search. Here the demand is to crawl and index the whole web and make it retrievable by users expecting answers to their search requests in subsecond time spans.

Consequently, building a time series search solution on technologies originally designed to also provide a search infrastructure on web-scale appears to be a promising approach. In order to be able to effectively use these technologies, one has to solve the challenge of transforming the search in time series to a task of full-text retrieval.

14.3.2 Indexing Time Series With Apache Lucene

One of the most important components of our systems is an indexing and retrieval component, which is needed to retrieve for a given segment of consecutive observations (i.e., time series fragment) all time series of a corpus, which contains at least a single similar subsequence. This setting is related to the query-by-example paradigm. Typically the reference time series segment

[5] http://amda.irap.omp.eu/.

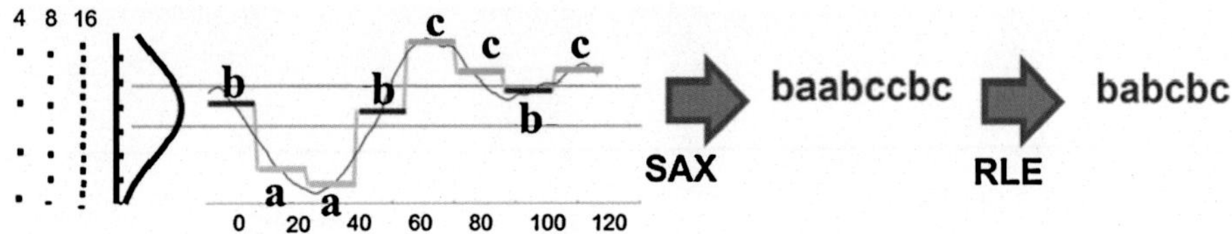

FIG. 14.4 Signal transformation into text domain for indexing.

represents an event the user is interested in, aiming to retrieve all time series that contain such events.

For the information retrieval component we have chosen to use Apache Lucene,[6] a popular Java-based information retrieval library and a central component of many information retrieval solutions, like Apache Solr[7] and ElasticSearch.[8] The Apache Lucene library is open source, provides a rich set of functionality, and offers a high degree of flexibility. There are three specific features implemented by this library, which are crucial to our application: (i) support for multiple fields/facets, (ii) payloads, and (iii) phrase queries. In the following the indexing and search functionality for time series will be described in further detail.

In order to index time series data with a full-text search engine, the data need to be transformed into a representation suitable for efficient retrieval. Applying techniques for symbolic approximation, such as, e.g., SAX, as described in more detail in the following sections, are the most promising starting points to achieve this objective. One of the critical parameters of this method is the size of the vocabulary of symbols. Since an optimal number for this parameter is highly dependent on the phenomenon as encoded into the observations of the time series, using just a single number will not work best in all scenarios. To that end, we run multiple symbolic approximations with varying settings of the symbol vocabulary. These different representations are indexed as separate fields within the Lucene index. Fields can also be seen as separate, independent indices that can be queried at simultaneously. This feature is common among contemporary search engines and is often referred to as *facets*.

The so-called *payload feature*, at the other hand, is not commonly found in information retrieval libraries, but essential for our implementation. This feature allows to store arbitrary data alongside the posting list. The posting list is the main data structure used by mainstream search engines in order to organize the indexed terms. In this data structure each unique term (i.e., words) is stored alongside all its occurrences. This information is then used to retrieve all matching documents, but also to sort the documents according to a relevance score. As such, all information required for ranking needs to be present at this stage for efficient retrieval. The payload allows to integrate time series-specific information, for example the error introduced via the discretization step in the symbolic approximation stage. As such, a highly compressed version of the time series can be used to retrieve a list of candidate time series and a more fine-grained representation is then used for ranking.

The third necessary feature to allow retrieval in time series is the so-called phrase query. This feature allows to restrict the search to indexed documents (i.e., time series) that match a sequence of observations, typical for segments of time series. Here Apache Lucene also provides the option for a so-called *slop value*.[9] This parameter is useful in cases where the indexed time series contains additional observations within the time series, which should be ignored.

The combination of the three features together with their governing parameters allow for a great flexibility to control the "fuzziness" of the retrieval process, starting with retrieval of time series that exactly match the reference to relaxed versions in terms of time and domain. In summary, the search can be configured to be scale-invariant, translation-invariant, and robust with regard to outliers and additional observations.

14.3.3 Related Pattern Search in Signal Data

The following illustrations show how a time-based signal is processed and then searched for specific patterns. First the data are transformed into a representation suitable to be integrated into an off-the-shelf text-based search engine (e.g., Apache Lucene), using *symbolic aggregate approximation* (SAX); see Fig. 14.4.

Then the signal is queried for a specific pattern (also processed via SAX) and the result is visualized accordingly; see Fig. 14.5. Results are ranked according to similarity with the example signal.

[6] http://lucene.apache.org/.
[7] http://lucene.apache.org/solr/.
[8] https://www.elastic.co/.

[9] The *slop* is an edit distance between respective positions of terms as defined in a query and the positions of terms in a document.

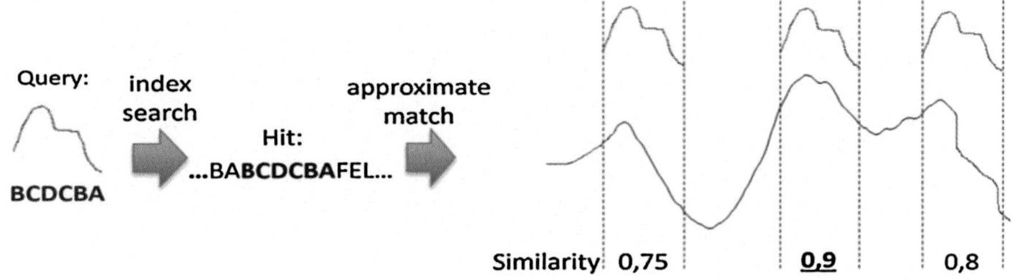

FIG. 14.5 Signal searching in the text domain with subsequent visualization of the found signal patterns.

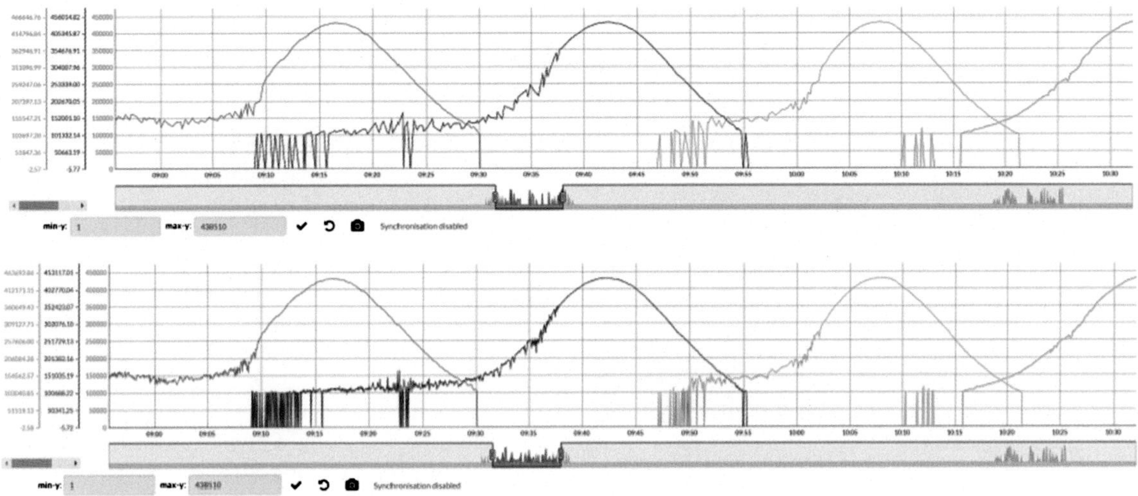

FIG. 14.6 Level of detail rendering for exploration of large signal data.

14.4 VISUAL INTERFACE

In the following an overview of the prototype for signal search is given, showcasing its visual interface using screen shots and briefly discussing its main components and related concepts.

14.4.1 Large-Scale Signal Visualization

Signal data easily grow very large, for example when data are collected over a longer period of time and sampled with high frequency. The resulting amount of data required to represent a single channel may grow into hundreds of megabytes or even gigabytes. Client applications – especially those run in a browser – cannot handle these amounts of data efficiently, making smooth exportation and navigation very slow or even impossible. Aggregation data techniques combined with level of detail (LOD) rendering are usually employed in such cases. A well-known example of a visual system that illustrates this concept is *Google Maps*; the whole dataset is stored on the server in all its detail, where it is aggregated hierarchically to reduce its complexity and size. When the user zooms in, the data are shown in maximum detail (e.g., houses), but as soon as the user starts zooming out, the amount of detail would become too large for the visualization to handle. Instead, the details are replaced with a coarser version of the data, and the further the user zooms out, the lower the amount of shown details becomes. This mechanism ensures that the amount of data that must be transferred to and handled by the client is always sufficiently small to support smooth user interaction. For visualization of signals we employ specialized downsampling techniques; see Fig. 14.6.

In addition to the above case, scaling to a large number of data channels often becomes necessary in modern environments, where myriads of sensors are continuously generating data streams. In such situations, line charts tend to become cluttered very quickly by overlaying lines, as seen in Fig. 14.7.

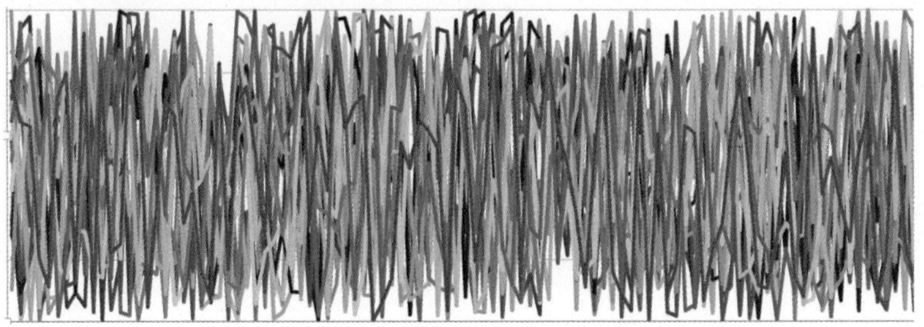

FIG. 14.7 Line chart clutter.

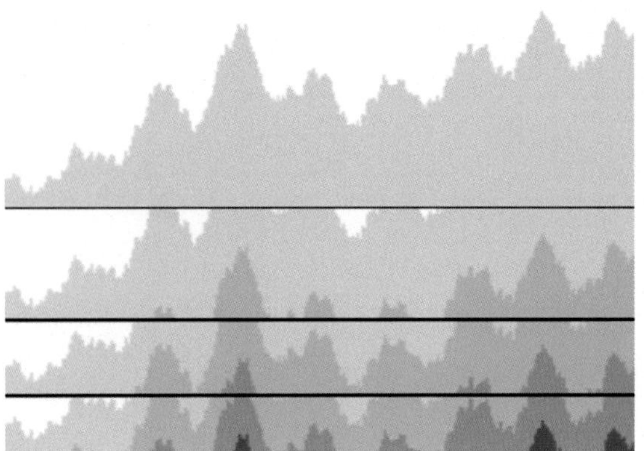

FIG. 14.8 Constructing horizon charts (using Cubism.js library).

Compressed visualization metaphors, such as *horizon charts* (Heer et al., 2009) and pixel-based views, are designed to handle such cases, enabling users to read the visualization and recognize patterns. In Fig. 14.8 the principle of constructing *horizon charts* can be seen: vertical space for a single channel is reduced without any loss in resolution by overlying higher values over the lower ones and displaying the overlays in a darker color tone.

The image shows the uncompressed view of a signal on top. When reducing the vertical space, more additional layers become necessary (three in the bottom version) to display the full amplitude of the same data. Vertical space saved in this way are used to visualize many signal channels through horizon charts, stacked above each other (see Fig. 14.9). Note that different colors – red and blue – are used to discern between the positive and negative values, further reducing the amount of vertical space by a factor of two.

To maximize the number of signals shown on screen, further reduction of the vertical space is possible down to very few or even a single pixel. In the resulting pixel-based view the values of each signal are represented only through the color shade.

14.4.2 Finding Related Signal Patterns

When users explore signal data and stumble upon peculiar signal patterns, they might want to find out if and under which conditions such patterns occur in the dataset. By selecting a pattern of interest in the visualization and sending it to a search system, as described in Section 14.3, users can reduce a large signal space down to a subset containing patterns similar to the selected one.

The interface shown in Fig. 14.10 supports query pattern selection, triggering a similarity search (*query by example*) and exploring the search results. The query pattern shown on top (in blue) is selected with a simple *control + mouse-drag* interaction. The retrieved signals are shown underneath the query pattern. Hits which are also highlighted in blue are ranked by similarity to the search pattern, where various user-defined similarity coefficients (DTW, SWALE, Pearson, etc.) can be selected. Note how particular signal features of the original pattern influence the ranking: on top, the query pattern features a specific shape on its right end, while the found patterns are sorted depending on whether they contain a similar shape or not. By scrolling further down, additional, less similar hits are dynamically loaded from the search engine. As single signals can contain multiple hits, another dedicated interface, shown in Fig. 14.11, is used to navigate between them. The highest-ranked hit is focused on initially (placed in the center), with direct navigation between other hits being possible using the arrows on the right-hand area. Alternatively, the user can zoom and pan along the time axis until the next hit is found – however, this may be time consuming for large data. Clicking on the star will refocus on the top hit in that particular signal. The "i" button will show additional metadata describing the signal and the hits.

FIG. 14.9 Horizon charts showing a large number of channels simultaneously (created with D3.js library).

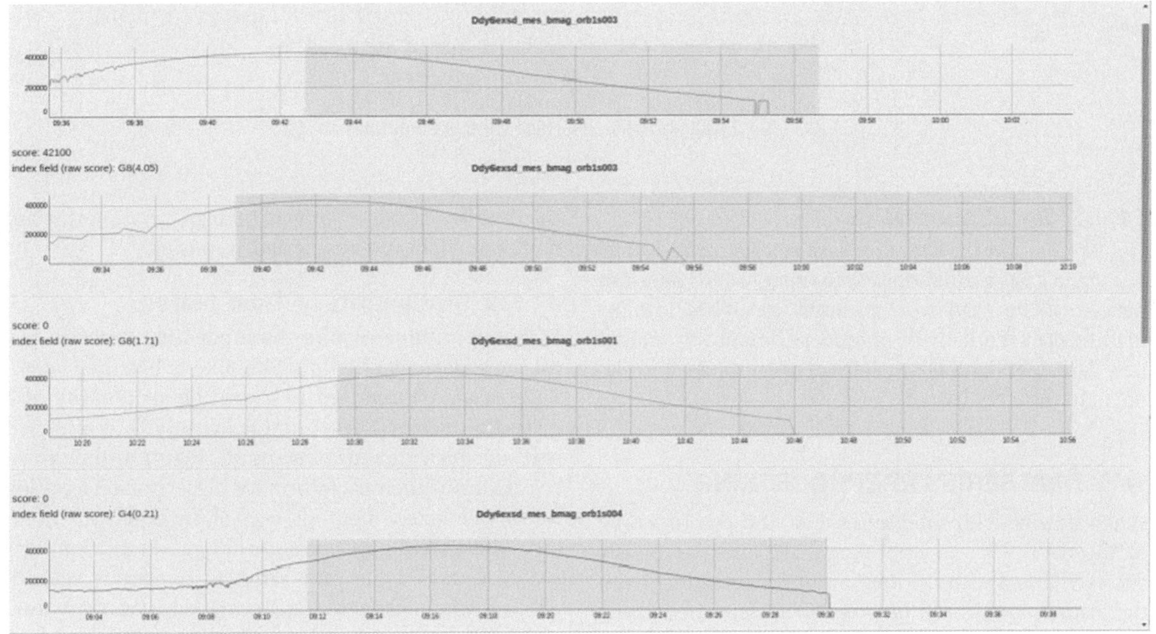

FIG. 14.10 Similar pattern search for signal data showing search results ranked by similarity to the search pattern (on top).

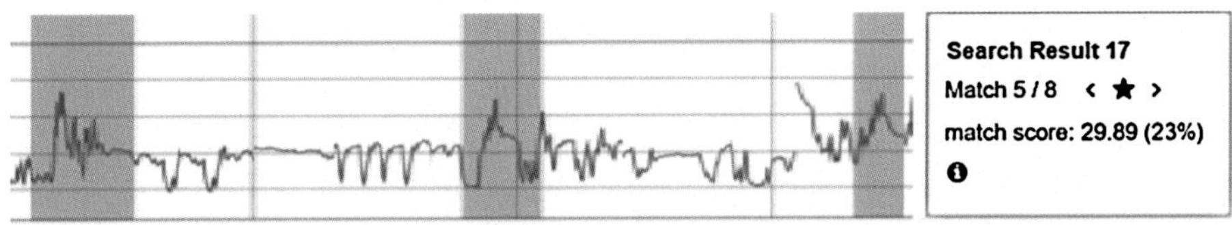

FIG. 14.11 Navigating between multiple hits within a signal.

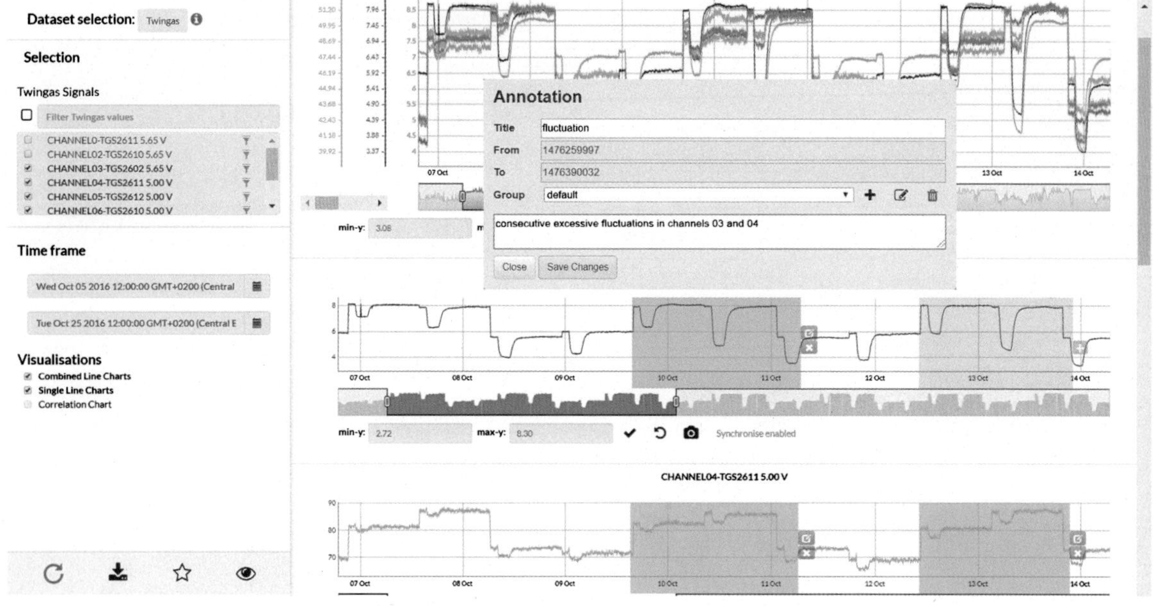

FIG. 14.12 Collaborative interface for signal annotation.

14.4.3 Signal Annotation

Fig. 14.12 shows the interface used to explore and *annotate* signals in a collaborative manner. These data can henceforth be used to, e.g., apply *supervised learning* methods to eventually be able to automatically analyze new data, highlighting any known phenomena and signal patterns included.

14.5 TIME SERIES PREPROCESSING

When dealing with time series data, the preprocessing of the data plays an important role, and is typically strongly linked with the data mining algorithms. Thus, the actually conducted preprocessing steps are heavily influenced by the algorithms used in the consecutive pipeline.

14.5.1 Normalization of Time Series

Normalization and standardization are common techniques not limited to time series. Especially when working with algorithms that are sensitive to the range of input values (e.g., SVMs, neural networks), this preprocessing step plays an important role. The most frequently used techniques are *min-max normalization* and *z-score standardization*, i.e., transformation to have zero mean and unit standard deviation. If the distribution of the time series data is known, then an appropriate standardization should be selected accordingly.

14.5.2 Stationarity of Time Series

Many algorithms require the input time series to fulfill the stationary criterion. Often, tests like the *Dickey–Fuller test* are employed to assess the stationarity of a given time series (Dickey and Fuller, 1979). While such methods provide satisfying results, their run-time complexity limits their usefulness for bigger datasets, calling for more efficient methods, even at the expense of accuracy. Methods like *autoregressive integrated moving average* (ARIMA) expect the time series to be weakly stationary to work. In order to transform a nonstationary time series into a representation suitable for such methods, a number of techniques have been applied. Differencing the time series (multiple times) is a common and simple approach, which is widely used and part of the popular *Box–Jenkins method* (Box et al., 2015).

14.5.3 Trends in Time Series

Trends in time series are another aspect which also influences the stationarity. To this end, a number of different approaches can be applied to identify and remove long-ranging trend components. One of the most common approaches is to compute a *linear regression* over the time series. Another type of popular method are filters, e.g., *high-pass filters*. Here, the estimation of the filters' parameters needs to be made with care. If the observed phenomena represent some natural growth processes, exponential models are best suited to capture the trend.

Based on the insight that the choice of parameters of the trend removal has a big impact on the overall processing pipeline, several means to integrate expert knowledge have been proposed (Taylor and Letham, 2018).

14.5.4 Periodicity and Seasonality in Time Series

Apart from transforming the time series into a stationary representation, the *Box–Jenkins model* also advises to remove seasonal and periodic components (Box et al., 2015). While seasonality usually refers to a single, dominant repeating pattern of fixed frequency, periodicity is more loosely defined based on periodic events, which are not necessarily occurring at a fixed frequency.

Toller and Kern (2017) proposed a robust method for seasonality detection in the presence of various influences, like trends and noise. Their approach is based on detecting the zero distances of the autocorrelation. Identifying periodic events in time series is an active field, with varying methods being researched, ranging from particle filters (Ghosh et al., 2017) to greedy methods based on a scoring heuristic (Yuan et al., 2017).

14.6 TIME SERIES REPRESENTATION

Since raw time series will often be of high resolution, in the time and value domain, approaches are helpful to reduce the amount of data to arrive at a representation, which is sufficiently small for efficient processing, while still capturing the relevant information. Here, the tasks of time series representation is often performed as a way of numerosity reduction, i.e., operations to achieve a more compact representation of the original observations of the time series.

14.6.1 PAA

Numerosity reduction can be achieved via a discretization in the time domain. Multiple consecutive observations of equal length are aggregated and replaced by a representation of lower complexity. Often this representation is just a single scalar, computed via a descriptive statistic, like the sample mean. This procedure is associated with the term *piecewise aggregate approximation* (PAA). The computational complexity of such procedure is linear with respect to the number of observations, which lends itself to be used in practical applications.

14.6.2 SAX

Apart from the time domain, numerosity reduction can also be applied in the value domain of the time series. Here the value of the individual observations are discretized and assigned to a number of finite bins. Each of the bins is then assigned to a symbol, with the number of bins representing the vocabulary size.

The *symbolic aggregate approximation* (SAX) representation can effectively be used for dimensionality reductions, where a lower bound can be computed (Lin et al., 2003).

14.6.3 Piecewise Linear Representation

The piecewise linear representation of a time series is an approximation of the original observations, splitting the time series into varying-length linear segments. In general, this approach can be further split into two different tasks: (i) linear interpolation, where the piecewise representation is continuous, and (ii) linear regression, where there might be discontinuities between the segments.

In order to identify the linear segments, either top-down or bottom-up approaches are applicable (Keogh et al., 2004), similar to the divisive and agglomerative clustering, respectively.

14.6.4 Windowed Approach to Time Series

Generally, time series can be seen as an indefinite stream of observations linked with a time stamp. For the general case, approaches that require the full time series to be available for computation will not work. On the other hand, when processing a time series in an online or streaming manner, where just the most current observation is processed, relevant information might be left out.

To strike a balance between resource consumption and efficiency, often sliding window approaches are employed (Gama, 2010), which can be seen as a special data structure that follows the first-in first-out principle. While there are many different ways to make use of windowed approaches (Babcock et al., 2002), typically windows are chosen to capture a fixed amount of consecutive observations, referred to as the window size. The actual value of this parameter plays an important role in the behavior of the final processing pipeline.

Another parameter of the windowed approach is the overlap between two consecutive windows. Often, the term tumbling window is used for sliding windows without overlap, with the downside of not having any redundancy between the observed windows. If the overlap is too big, unwanted side effects may occur in the processing of the time series; for example, the grouping of similar windows may not work as expected (Keogh and Lin, 2005).

14.7 TIME SERIES SIMILARITY

In the practical applications presented in this chapter, one is often faced with the challenge of assessing the similarity of two given time series. Using the *Euclidean* norm is one of the most straightforward solutions to this task. This method can also be extended from the univariate case to the multivariate case by the use of the *Frobenius* norm. Such approaches work well in cases where the time series are perfectly aligned and there are no temporal or local deviations.

To cater for cases where one of the time series may deviate locally, methods like *dynamic time warping* (DTW) have been proposed (Berndt and Clifford, 1994), which is highly popular due to its ability to also find a high similarity, if there are nonlinear transformations in the time domain of the time series. One downside of this approach is the high computational complexity when applied without constraints. Attempts to achieve similar behavior with higher efficiency have been made (Morse and Patel, 2007). Complementary, the DTW method has also been combined with piecewise linear approximation to achieve improved performance and run-time behavior (Li et al., 2011).

14.8 PATTERN MINING IN TIME SERIES

In our goal to achieve an effective and efficient service for *content-based search* in time series, we also investigate methods to uncover patterns in data. In the general case one cannot assume what the nature of these patters can be, one can start at opposite ends of the spectrum: highly frequent patterns and patterns that only occur sporadically. To distinguish infrequent patterns from random noise or other influences, the removal of outliers is necessary.

14.8.1 Outlier Detection in Time Series

Outlier detection, often also referred to as anomaly detection, is the task to identify observations *that deviate so much from other observations as to arouse suspicion that they were generated by a different mechanism* (Hawkins, 1980). In a frequently cited survey, Aggarwal (2015) categorized the types of outliers into three classes: (i) point outliers, (ii) collective outliers, and (iii) contextual outliers. Specifically for temporal data, the review article by Gupta et al. (2014) highlights the two main types of outliers in time series data: (i) point outliers and (ii) subsequence outliers.

Outlier detection algorithms vary in the way the time series is represented. Laptev et al. (2015) derived a number of features based on descriptive statistics and found that long-term and short-term features work equally well, while the combination of features provided the overall best results. The SAX representation of time series has been used to detect outliers in a consecutive sequence of observations (Keogh et al., 2005).

Generally speaking, the method of outlier detection will also depend on the actual data at hand. Here, a visual inspection of the data using appropriate tools is of help.

14.8.2 Frequent Patterns

The identification of frequent patterns is an important task in data mining. Arguably, the most well-known example is the application for basket analysis, where the objective is to find commonly bought items from the transaction logs of a (grocery) store. Given a suitable representation of the time series, a vast number of varying pattern mining approaches can be applied to detect frequent subsequences. The two main basic approaches for frequent pattern mining are *Apriory* (Agrawal and Srikant, 1994) and *FP-Growth* (Han et al., 2004). Since then, many different algorithms have been proposed based on these general approaches.

14.8.3 Surprising Patterns

The task of identifying "surprising" patterns in time series is closely linked to the tasks of frequent pattern mining and outlier detection. In the absence of domain expertise for what a surprising pattern constitutes, the frequency of a pattern given a reference corpus can be taken as a proxy measure. As such, surprising pattern mining can be seen as a special case of collective outlier detection. Keogh et al. (2002) proposed an approach on this notion of surprise. They applied a symbolic representation and utilized suffix trees for faster retrieval of frequency estimates.

14.9 TIME SERIES MODELING AND CLASSIFICATION

The technological base for many time series analysis algorithms is the successful modeling of the time series. One way to validate a successful model is to measure its performance on forecasting future observations. Thus a good fitting model that is able to forecast the future behavior is also a good starting point for tasks like outlier detection and classification.

14.9.1 Time Series Forecasting

Time series forecasting is the task of predicting one or several future observations in a time series, given a certain number of past observations. Widely used linear approaches for time series forecasting are ARIMA and its different variations, which are based on the *Box–Jenkins* principle (Box et al., 2015). Even though these models are sufficient for many application scenarios, many time series show nonlinear patterns, and thus various nonlinear approaches for time series modeling have been proposed. Deep learning algorithms are inherently nonlinear and hence allow for modeling the relations between features in a highly complex way. *Recurrent neural networks* (RNNs), such as *long short-term memory* (LSTM) networks, are well suited for this task, as they include knowledge from past time steps by construction. In fact, LSTMs were specifically designed to overcome the shortcomings of traditional RNNs, which tended to only make use of the most recent observations and thus failed to model longer-ranging dependencies. LSTMs have been used for time series forecasting in many application domains, e.g., load forecasting (Liu et al., 2017), stock price prediction (Roondiwala et al., 2017), *e-commerce* (Chniti et al., 2017), medicine (Zhang and Nawata, 2018), and weather forecasting (Zaytar and Amrani, 2016).

Recently, more and more approaches leveraging convolutional architectures have been proposed for sequence modeling tasks. *Long- and short-term time series network* (LSTNet) (Lai et al., 2017) uses a combination of a *convolutional neural network* (CNN) and an RNN to perform multivariate time series forecasting. The intention of this approach is to leverage the strengths of both the convolutional layer (to extract short-term local dependency patterns among the multidimensional variables) and the recurrent layer (to discover long-term patterns for time series trends). In the evaluation of real-world data with complex mixtures of repetitive patterns, LSTNet achieved significant performance improvements over that of several state-of-the-art baseline methods, as, e.g., autoregression models and *gated recurrent unit* (GRU) networks.

Yin et al. (2017) performed a systematic comparison of CNNs and RNNs on a wide range of representative natural language processing tasks, such as sentiment classification, answer selection, and part-of-speech tagging. They concluded that RNNs perform well and robustly in a broad range of tasks, except when the task is essentially a key phrase recognition task. For sentiment classification, which they examined in more detail, they found that GRU and CNN are comparable when sentence lengths are small, but then the advantage of GRU over CNN increases for longer sentences. While this work was originally designed for textual data, the question which of these two competing approaches is more suitable also applies to time series.

Bai et al. (2018) presented a systematic evaluation of generic convolutional and recurrent architectures for sequence modeling. The models were evaluated across a broad range of standard tasks that are commonly used to benchmark RNNs. The results indicated that a simple convolutional architecture, called *temporal convolutional network* (TCN), outperforms canonical RNNs such as LSTM networks across a diverse range of tasks and datasets, while demonstrating longer effective memory.

In summary, there is no clear conclusion yet which kind of deep learning models (e.g., recurrent, convolutional, or hybrid) are suited best for time series forecasting. Hence, for a certain task, different models may be applied to compare their performance.

14.9.2 Classification and Clustering of Unsegmented Time Series

If the time series to be classified are subsequences of a larger time series, the subsequences need a clearly defined start and end point, i.e., the larger time series needs to be presegmented before performing the classification. In ideal settings, expert users would annotate the beginning and end of each observed phenomenon. However, this is impractical in many real-world applications, where it is often necessary to classify or cluster subsequences of unsegmented time series. Hence, approaches that perform the segmentation of a time series and assign the segments to one of a given number of predefined classes (classification) or not a priori known clusters (clustering) as a single step have been proposed.

An approach to this utilizing deep learning was introduced by Graves et al. (2006). The problem with applying RNNs directly to sequence labeling is that the standard neural network objective functions are defined separately for each point in the training sequence, i.e., RNNs can only be trained to make a series of independent label classifications. This means that the training data must be presegmented, and that the network outputs must be postprocessed to give the final label sequence. Thus, Graves et al. proposed to interpret the network outputs as a probability distribution over all possible label sequences, conditioned on a given input sequence. Given this distribution, an objective function can be derived that directly maximizes the probabilities of the correct labelings. Since the objective function is differentiable, the network can then be trained with standard backpropagation through time. This approach removes the need for presegmented training data and postprocessed outputs, and models all aspects of the sequence within a single network architecture.

A model-based segmenting and clustering method utilizing *Markov random fields* (MRFs), named *Toeplitz Inverse Covariance-based Clustering* (TICC), was introduced by Hallac et al. (2017). Each cluster in the TICC method is defined by a correlation network, or MRF, characterizing the interdependencies between different observations in a typical subsequence of that cluster. Based on this graphical representation, TICC simultaneously segments and clusters the time series data. For example, in a cluster corresponding to a "turn" in an automobile, this MRF might show how the brake pedal at a generic time t might affect the steering wheel angle at time $t + 1$. The MRF of a different cluster, such as "slowing down," will have a very different dependency structure between these two sensors. The cluster's MRFs are learned by estimating a sparse *Gaussian* inverse covariance matrix. To solve the TICC problem, an expectation-maximization-like approach is used, where the data are iteratively clustered and then the cluster parameters are updated. While the method involves solving a highly nonconvex maximum likelihood problem, it showed to be able to find a (locally) optimal solution very efficiently in practice.

14.9.3 Classification on Segmented Time Series

Time series classification deals with classifying an (already segmented) time series into one of a given number of classes, i.e., map a (possibly multivariate) sequence to a class label. To do so, the state of the art is mainly based on approaches utilizing convolutional networks.

For example, multiscale CNNs (Cui et al., 2016) were specifically designed for classifying (univariate) time series. Their first layer contains multiple branches that perform various transformations of the time series, including those in the frequency and time domains, extracting features of different types and timescales. Zheng et al. introduced multichannel deep CNNs (Zheng et al., 2014) for the classification of multivariate time series. Each channel of the model takes a single dimension of a multivariate time series as input and learns features individually. Finally, the model combines the learned features of each channel and feeds them into a multilayer perceptron to perform classification.

Fully convolutional networks (FCNs) in general are networks that take input of arbitrary size and produce correspondingly sized output with efficient inference and learning (Long et al., 2015). They have commonly been applied to dense prediction problems, such as semantic segmentation. Wang et al. (2017) proposed an FCN model consisting of three convolutional blocks to extract features from time series. The output is passed through a global average pooling layer and the final label is then produced by a *softmax* layer. The model achieved premium performance compared to other state-of-the-art approaches. Karim et al. (2018) extended the FCN approach with LSTM submodules, which significantly enhanced the performance with a nominal increase in model size and minimal preprocessing of the dataset.

14.10 CONCLUSIONS

Looking at our showcase example using MESSENGER data (i.e., time series) the advantages of leveraging machine learning techniques are obvious. Searching the data using the prototype tool, providing *content-based search* based on a search index and additional (machine-learning-based) layers for fine-ranking the results is *faster by several orders of magnitude* than the current "state of the art," i.e., manual approach, while at the same time providing high accuracy and generality with regard to the complexity of the investigated signal. Looking at an index with the approximate size of the complete dataset of the MESSENGER dataset, a single search takes about 0.5 seconds to be completed (including coarse ranking). Typically, four or five queries over different quantization spaces may be necessary to obtain an adequate amount of hits, which are then fine-ranked, yielding a typical search time of 2 to 5 seconds. Longer queries or unfavorable search patterns in general can lead to significantly longer search times; however, in practical use we have *rarely experienced search times exceeding 30 seconds*. In addition to search applications, machine learning approaches can also provide efficient methods to filter out unwanted data (i.e., signal noise or technical artifacts) or preselect data based on certain scientific criteria and hence significantly reduce efforts involved in handling large volumes of scientific data.

Applications for machine learning and artificial intelligence in space exploration are manifold and not constrained to pure data analytics tasks whatsoever. As shown in Amedeo et al. (2007), artificial intelligence-based systems like *MEXAR2*[10] can significantly improve

[10] *Mars Express AI Tool* – Institute for Cognitive Science and Technology (ISTC-CNR), Italy.

the data downlink for interplanetary space probes, by synthesizing the information for optimized, or "smart," data transmission. Another strain of applications is focused on predictive systems as, e.g., the *Deep Flare Net* (DeFN) developed by two teams under the umbrella of *NASA's Frontier Development Lab*[11] that even outperformed *NOAA*'s existing system for predicting flares (Nishizuka et al., 2018). In actual data analysis tasks, machine learning technologies can not only be applied to time-based data very successfully, as shown in this example, but of course also to the analysis of images – another major data source for planetary sciences as well as a "classical" application of machine learning. Besides the aforementioned DeFN, which was trained on observation images taken during 2010–2015 by the *Solar Dynamic Observatory*, there are countless further topics in space science that have already been shown to benefit from machine learning, ranging from meteor detection and the identification of lunar craters to the hunt for extrasolar planets. In a recent study, astronomers at UC Santa Cruz leveraged deep learning to analyze images of galaxies, in order to understand the birth and evolution of galaxies (Huertas-Company et al., 2018). These examples give an impression of how widespread possible applications of machine learning techniques are in space science, while at the same time making it virtually impossible to predict where this journey will lead the (space) science community just a couple of years from now. However, the use of machine learning and (further down the road) artificial intelligence-based science tools might well yield a plethora of new discoveries that are lurking in existing and recently acquired data, which will in turn lead the way to new scientific topics and investigations. New developments with regard to computing hardware that support machine learning processing with ample memory bandwidth and highly optimized matrix operations, etc., as, e.g., *Intel's Nervana*,[12] will speed up training and execution of neural network systems and provide massive processing power also for smaller projects and teams.

REFERENCES

Aggarwal, C.C., 2015. Outlier analysis. In: Data Mining. Springer, pp. 237–263.

Agrawal, R., Srikant, R., 1994. Fast algorithms for mining association rules in large databases. In: Proceedings of the 20th International Conference on Very Large Data Bases. VLDB'94, pp. 487–499.

[11] FDL is part of an effort by NASA to assess machine learning technologies, https://frontierdevelopmentlab.org/.

[12] https://ai.intel.com/.

Alexeev, I., Belenkaya, E., Slavin, J., Korth, H., Anderson, B., Baker, D., Boardsen, S., Johnson, C., Purucker, M., Sarantos, M., Solomon, S., 2010. Mercury's magnetospheric magnetic field after the first two messenger flybys. Icarus 209 (1), 23–39.

Amedeo, C., Cortellessa, G., Denis, M., 2007. Mexar2: AI solves mission planner problems. IEEE Intelligent Systems 22, 12–19.

Anderson, B., Acuna, M., Lohr, D., et al., 2007. The magnetometer instrument on MESSENGER. Space Science Reviews 131, 417–450.

Babcock, B., Babu, S., Datar, M., Motwani, R., Widom, J., 2002. Models and issues in data stream systems. In: Proceedings of the Twenty-First ACM SIGMOD-SIGACT-SIGART Symposium on Principles of Database Systems. ACM, pp. 1–16.

Bai, S., Kolter, J.Z., Koltun, V., 2018. An empirical evaluation of generic convolutional and recurrent networks for sequence modeling. CoRR arXiv:1803.01271.

Bass, L., Clements, P., Kazman, R., 2003. Software Architecture in Practice. Addison–Wesley Professional.

Berndt, D.J., Clifford, J., 1994. Using dynamic time warping to find patterns in time series. In: KDD Workshop. Seattle, WA, vol. 10, pp. 359–370.

Box, G.E.P., Jenkins, G.M., Reinsel, G.C., Ljung, G.M., 2015. Time Series Analysis: Forecasting and Control. John Wiley & Sons.

Chniti, G., Bakir, H., Zaher, H., 2017. E-commerce time series forecasting using LSTM neural network and support vector regression. In: Proceedings of the International Conference on Big Data and Internet of Things. BDIOT2017. ACM, New York, NY, USA, pp. 80–84. http://doi.acm.org/10.1145/3175684.3175695.

Cui, Z., Chen, W., Chen, Y., 2016. Multi-scale convolutional neural networks for time series classification. CoRR arXiv:1603.06995.

Dickey, D.A., Fuller, W.A., 1979. Distribution of the estimators for autoregressive time series with a unit root. Journal of the American Statistical Association 74 (366a), 427–431.

Gama, J., 2010. Knowledge Discovery From Data Streams. Chapman and Hall/CRC.

Ghosh, A., Lucas, C., Sarkar, R., 2017. Finding periodic discrete events in noisy streams. In: Proceedings of the 2017 ACM on Conference on Information and Knowledge Management. ACM, pp. 627–636.

Graves, A., Fernández, S., Gomez, F., Schmidhuber, J., 2006. Connectionist temporal classification: labelling unsegmented sequence data with recurrent neural networks. In: Proceedings of the 23rd International Conference on Machine Learning. ICML'06. ACM, New York, NY, USA, pp. 369–376. http://doi.acm.org/10.1145/1143844.1143891.

Gupta, M., Gao, J., Aggarwal, C.C., Han, J., 2014. Outlier detection for temporal data: a survey. IEEE Transactions on Knowledge and Data Engineering 26 (9), 2250–2267.

Hallac, D., Vare, S., Boyd, S., Leskovec, J., 2017. Toeplitz inverse covariance-based clustering of multivariate time series data. In: Proceedings of the 23rd ACM SIGKDD International Conference on Knowledge Discovery and Data

Mining. KDD'17. ACM, New York, NY, USA, pp. 215–223. http://doi.acm.org/10.1145/3097983.3098060.

Han, J., Pei, J., Yin, Y., Mao, R., 2004. Mining frequent patterns without candidate generation: a frequent-pattern tree approach. Data Mining and Knowledge Discovery 8 (1), 53–87.

Hawkins, D.M., 1980. Identification of Outliers, vol. 11. Springer.

Heer, J., Kong, N., Agrawala, M., 2009. Sizing the horizon: the effects of chart size and layering on the graphical perception of time series visualizations. In: CHI'09 Proceedings of the SIGCHI Conference on Human Factors in Computing Systems, pp. 1303–1312.

Huertas-Company, M., Primacj, J., Dekel, A., Koo, D., Lapiner, S., Ceverino, D., Simons, R., Synder, M.G., Bernadi, M., Chen, Z., Domínguez-Sánchez, H., Lee, C., Margalef-Bentabol, B., Tuccillo, D., 2018. Deep learning identifies high-z galaxies in a central blue nugget phase in a characteristic mass range. The Astrophysical Journal 858 (2).

Karim, F., Majumdar, S., Darabi, H., Chen, S., 2018. LSTM fully convolutional networks for time series classification. IEEE Access 6, 1662–1669.

Keogh, E., Chu, S., Hart, D., Pazzani, M., 2004. Segmenting time series: a survey and novel approach. In: Data Mining in Time Series Databases. World Scientific, pp. 1–21.

Keogh, E., Lin, J., 2005. Clustering of time-series subsequences is meaningless: implications for previous and future research. Knowledge and Information Systems 8 (2), 154–177.

Keogh, E., Lin, J., Fu, A., 2005. HOT SAX: Efficiently finding the most unusual time series subsequence. IEEE, pp. 226–233.

Keogh, E., Lonardi, S., Chiu, B.Y.-c., 2002. Finding surprising patterns in a time series database in linear time and space. In: Proceedings of the Eighth ACM SIGKDD International Conference on Knowledge Discovery and Data Mining. ACM, pp. 550–556.

Kimura, A., Kashino, K., Kurozumi, T., 2008. A quick search method for audio signals based on a piecewise linear representation of feature trajectories. IEEE Transactions on Audio, Speech, and Language Processing 16 (2), 396–407.

Lai, G., Chang, W., Yang, Y., Liu, H., 2017. Modeling long- and short-term temporal patterns with deep neural networks. CoRR arXiv:1703.07015.

Laptev, N., Amizadeh, S., Flint, I., 2015. Generic and scalable framework for automated time-series anomaly detection. In: Proceedings of the 21st ACM SIGKDD International Conference on Knowledge Discovery and Data Mining. ACM, pp. 1939–1947.

Li, H., Guo, C., Qiu, W., 2011. Similarity measure based on piecewise linear approximation and derivative dynamic time warping for time series mining. Expert Systems with Applications 38 (12), 14732–14743.

Lin, J., Keogh, E., Lonardi, S., Chiu, B., 2003. A symbolic representation of time series, with implications for streaming algorithms. In: Proceedings of the 8th ACM SIGMOD Workshop on Research Issues in Data Mining and Knowledge Discovery. ACM, pp. 2–11.

Liu, C., Jin, Z., Gu, J., Qiu, C., 2017. Short-term load forecasting using a long short-term memory network. In: 2017 IEEE PES Innovative Smart Grid Technologies Conference Europe. ISGT-Europe, pp. 1–6.

Long, J., Shelhamer, E., Darrell, T., 2015. Fully convolutional networks for semantic segmentation. In: 2015 IEEE Conference on Computer Vision and Pattern Recognition. CVPR, pp. 3431–3440.

Marcus, G., 2018. Deep learning: a critical appraisal. CoRR arXiv:1801.00631.

Mehdiyev, N., Lahann, J., Emrich, A., Enke, D., Fettke, P., Loos, P., 2017. Time series classification using deep learning for process planning: a case from the process industry. Procedia Computer Science 114 (C), 242–249.

Morse, M.D., Patel, J.M., 2007. An efficient and accurate method for evaluating time series similarity. In: Proceedings of the 2007 ACM SIGMOD International Conference on Management of Data. ACM, pp. 569–580.

Nishizuka, N., Sugiura, K., Kubo, Y., Den, M., Ishii, M., 2018. Deep Flare Net (DeFN) model for solar flare prediction. The Astrophysical Journal 858 (2).

Roondiwala, M., Patel, H., Varma, S., 2017. Predicting stock prices using LSTM. International Journal of Science and Research (IJSR).

Taylor, S.J., Letham, B., 2018. Forecasting at scale. The American Statistician 72 (1), 37–45.

Toller, M., Kern, R., 2017. Robust parameter-free season length detection in time series.

Wang, Z., Yan, W., Oates, T., 2017. Time series classification from scratch with deep neural networks: a strong baseline. In: 2017 International Joint Conference on Neural Networks. IJCNN, pp. 1578–1585.

Yin, W., Kann, K., Yu, M., Schütze, H., 2017. Comparative study of CNN and RNN for natural language processing. CoRR arXiv:1702.01923.

Yuan, Q., Shang, J., Cao, X., Zhang, C., Geng, X., Han, J., 2017. Detecting multiple periods and periodic patterns in event time sequences. In: Proceedings of the 2017 ACM on Conference on Information and Knowledge Management. ACM, pp. 617–626.

Zaytar, M.A., Amrani, C.E., 2016. Sequence to sequence weather forecasting with long short-term memory recurrent neural networks. International Journal of Computer Applications 143 (11), 7–11. http://www.ijcaonline.org/archives/volume143/number11/25119-2016910497.

Zhang, J., Nawata, K., 2018. Multi-step prediction for influenza outbreak by an adjusted long short-term memory 146, 1–8.

Zheng, Y., Liu, Q., Chen, E., Ge, Y., Zhao, J.L., 2014. Time series classification using multi-channels deep convolutional neural networks. In: WAIM.

Zloof, M., 1975. Query by example. In: AFIPS'75 Proceedings of the May 19–22, 1975, National Computer Conference and Exposition. ACM, pp. 431–438.

CHAPTER 15

When Evolutionary Computing Meets Astro- and Geoinformatics

ZAINEB CHELLY DAGDIA, DR • MIROSLAV MIRCHEV, DR

15.1 INTRODUCTION

On the origins of primitive life, biological life has been evolving every year. From the beginning, unicellular life organisms gradually mutated to complex life forms (multicellular organisms). This progressive change is a result of the process of genetic evolution. In the process of evolution, a series of natural changes cause organisms (or species) to arise, adapt to and familiarize with their environment, and finally go extinct. In this concern, Darwin's theory of evolution (Darwin, 2004) asserts that species survive through a process named "natural selection." Those species that successfully adapt, or evolve, to meet the changing requirements of their natural habitat thrive, while those that fail to evolve and reproduce die off (Goodwin, 1982).

Evolutionary computation includes a set of approaches that seek to emulate the mechanism of natural selection described in Darwin's theory, with the aim of solving complex optimization problems. In real life, there are countless applications that require optimization such as in business, in economics, in astronomy, and in geoscience. In these kinds of applications, there are many processes that can be potentially optimized. Optimization could occur in the minimization of time, cost, and risk, or in the maximization of profit, quality, and efficiency. In astronomy and geoscience, we often need to minimize the difference from a model output to some observed data, or maximize the distance between different classes of pixels representing objects found in an image. In this chapter, we present a palette of evolutionary computation techniques that tend to solve the complex and difficult optimization problems encountered in astro- and geoinformatics as these will be the main application areas studied in this chapter. More precisely, we focus on the challenges that arise in astro- and geoinformatics specifically when it comes to dealing with the large amount of acquired data that became easily accessible given the emergent technologies. In this concern, we introduce parallel evolutionary computation techniques that can solve astro- and geoinformatics optimization problems as they can speed up computation, solve large problems, and find better solutions.

To model and solve astro- and geoinformatics optimization problems, we have to first understand the basic concepts of optimization models and related solution methods. This chapter introduces related concepts, models, and solution methods of optimization, including genetic algorithms, evolutionary strategy, evolutionary programming, genetic programming, ant colony optimization, particle swarm optimization, and artificial immune system optimization approaches.

This chapter is organized as follows. Section 15.2 introduces the basic concepts of an optimization problem including its standard formulation, the types of an optimization problem, and the multiobjective optimization model. Section 15.3 presents the structure of an evolutionary computation algorithm and a range of evolution operators. Section 15.4 addresses evolutionary computation algorithms and some bio-inspired optimization approaches by their definition, classification, models, and theories. Section 15.5 introduces the context of Big Data and discusses a set of parallel frameworks dedicated for evolutionary computing algorithms. Section 15.6 discusses the applications of evolutionary computing techniques in astro- and geoinformatics, and Section 15.7 presents a summary of the main points highlighted in this chapter.

15.2 THE OPTIMIZATION PROBLEM

Optimization is a key requirement for making decisions and in analyzing astro- and geoinformatics systems. Formally, an optimization problem is defined in terms of a set of parameters and restrictions. The parameters chosen to describe the design of a specific system include one or many decision makers, the system's variables, single or several objectives to be achieved, and a set of structural restrictions known as constraint conditions. Several optimization studies are, indeed, formulated as

Knowledge Discovery in Big Data from Astronomy and Earth Observation. https://doi.org/10.1016/B978-0-12-819154-5.00026-6

problems aiming at finding the best solution(s) from among the set of all feasible solutions, i.e., solutions that satisfy all the constraints of the optimization problem.

The variables, also called decision variables, reflect the system's components for which we want to find the sought values. In model parameter estimation or data fitting, for instance, the variables are the parameters of the given model, whether the model is based on some known physical laws or it uses some generic functions.

The notation of an optimization problem also implies that there is some objective function or functions that can be improved either by performing a minimization or a maximization action, and that can also be used as a quantitative measure of effectiveness of the system. The objective function is also called fitness function, merit function, or cost function. For instance, in fitting experimental data to a model, we may want to minimize the total deviation of the observed data from the data predicted with the model.

Finally, the constraints are the functions that describe the relationships between the system's variables. They define the allowable values to be taken by the variables. For example, in parameter estimation the values could be constrained within an expected range that is known in advance.

15.2.1 Standard Formulation

Any astro- or geoinformatics problem can be defined as an optimization problem. The domain of the function to be optimized is called the search space ($\mathbf{S}$). The goodness of any solution in $\mathbf{S}$ is measured with a fitness function ($\mathbf{O} : \mathbf{S} \rightarrow \mathbb{R}$). A fitness landscape that maps from a configuration space into the real numbers $\mathbb{R}$ may be considered as a triple ($\mathbf{S}$, $\mathbf{O}$, $\mathbf{D}$), where $\mathbf{D}$ is a metric defined on $\mathbf{S}$.

In general, optimization problems comprise setting a vector $\mathbf{X}$ of decision variables (x_i) of a system in order to optimize (either minimize or maximize) some fitness function $\mathbf{O}(\mathbf{X})$. In some cases, this is achieved subject to the satisfaction of a set of constraint conditions, namely, the equality constraints $h_k(\mathbf{X})$, the inequality constraints $g_j(\mathbf{X})$, and with $(x_i)^L$ and $(x_i)^U$ corresponding to the lower and upper bounds of the variable x_i, respectively. A solution x_i satisfying the $(P + Q)$ constraints is called "feasible" and the set of all feasible solutions defines the feasible search space denoted by Ω. Assuming the nonlinearity of an optimization problem and with respect to a set of constraints, its mathematical modeling that deals with the search for a minimum of a nonlinear function $\mathbf{O}(\mathbf{X})$ of m variables can be outlined as follows:

$$\min \mathbf{O}(\mathbf{X}), \ \mathbf{X} = (x_1, x_2, \ldots, x_m)^T \in \mathbf{S}, \tag{15.1}$$

$$\text{subject to} \begin{cases} h_k(\mathbf{X}) = 0, \ k = \{1, 2, \ldots, Q\}, \\ g_j(\mathbf{X}) \le 0, \ j = \{1, 2, \ldots, P\}, \\ (x_i)^L \le (x_i) \le (x_i)^U, \ i = \{1, 2, \ldots, m\}. \end{cases} \tag{15.2}$$

Without any loss of generality, minimization is assumed in Eq. (15.1) as any minimization problem can be equivalently transformed into a maximization problem by a simple modification of the fitness function. This can be achieved, for example, by $-\mathbf{O}(\mathbf{X})$, $\frac{1}{\text{constant} + \mathbf{O}(\mathbf{X})}$, or some other means. In Eqs. (15.1) and (15.2), there are no specific conditions tied to the system's variables type. However, the formalization of an optimization problem is attached to the optimization problem features and variants. These will be detailed in the following section.

15.2.2 Types of Optimization Problems

Categorizing the optimization problem is an essential step in the optimization process. This is because algorithms for solving optimization problems are tailored to a specific type of problems. These can be classified in terms of the number of decision makers, the nature or type of the decision variables, the number of constraints, the number of objective functions, the linearity of the problem, and the uncertainty tied to the optimization model. These factors require careful thoughts along with mathematical details while designing the optimization models. It is also important to mention that for each specific type of any optimization problem, there is a set of dedicated optimization algorithms that explicitly deal and handle the particular nature and variants of the problems' components. In what follows, we provide the key variants, features, and types classifying the optimization models into distinctive various optimization problem types.

15.2.2.1 The Number of Decision Makers

As previously defined, an optimization problem aims at finding the best solution(s) among the set of all feasible solutions. In such a case, when decision making is emphasized, the problem supports a human Decision Maker (DM) to find the most preferred optimal solution according to his/her subjective preferences (Branke et al., 2008). Based on the assumption that a single solution to the problem must be identified to be implemented in practice, the DM, who is expected to be an expert in the problem domain, has to make his/her decision. In a more formal way, we deal with an optimization problem in the case when either one decision maker is involved or when no preference methods are required, i.e., no DM is expected to be available. Otherwise, the problem that we deal with is concerned with a game that can be either cooperative or noncooperative, depending on the decision makers' perspectives (Myerson, 2013).

15.2.2.2 The Type of the Decision Variables

The formalization of an optimization model is usually tied to the nature or type of the system's decision variables. In some cases, such formalization only makes sense when the system's variables take values from a discrete set; usually a subset of integers. Conversely, some other optimization models can be only formalized via variables that take any real value. Models with discrete variables are known as "discrete optimization" problems while models with continuous variables are called "continuous optimization" problems.

In discrete optimization, some or all of the variables may be binary, i.e., restricted to the values 0 and 1, integers for which only integer values may be taken, or more abstract objects drawn from sets with several (finite) elements. Within this category, we may consider two divisions of optimization problems. The first branch is called "integer programming," where the discrete set of the feasible solutions is a subset of integers. This class of models is mostly common as many real-life applications are modeled with discrete variables as their handled resources are indivisible, e.g., image pixels belonging to a certain object (Bertsimas and Weismantel, 2005). The second branch is called "combinatorial optimization," where the discrete set is a set of objects or combinatorial structures such as assignments, combinations, routes, schedules, or sequences (Papadimitriou and Steiglitz, 1998; Schrijver, 2003).

In continuous optimization, the system's variables in the optimization model take real values. In general, continuous optimization problems tend to be simpler and easier to solve in comparison to discrete optimization problems. This can be explained by the fact that smoothing the fitness function and the constraint conditions' values at a point x can be used to infer information about some other points in the neighborhood of x. It is essential to mention that continuous optimization has an important connection to discrete optimization because many discrete optimization algorithms often require continuous optimization problems to be solved as subproblems or relaxations.

In some other cases, it is possible that an optimization model is formalized using different variables' types. In this case, we refer to a "mixed-integer optimization" problem where the decision variables are both discrete and continuous. This class of optimization problems presents a generalization of both the discrete optimization problems and the continuous optimization problems (Pochet and Wolsey, 2006).

15.2.2.3 The Number of Constraints

One further distinction in optimization problems is between problems which are defined using a set of constraints on the system's variables, called "constrained optimization," and problems in which there are no constraints on the decision variables, called "unconstrained optimization." We refer to a constrained optimization problem when there are explicit constraints on the decision variables. These constraint conditions may vary from simple bounds to systems of equalities and inequalities that model complex relationships among the variables. The formalization of a constrained optimization problem was given in Section 15.2.1. In unconstrained optimization problems, the model may be based on a reformulation of constrained optimization problems in which the constraints are replaced by a penalty term in the fitness function.

15.2.2.4 The Number of Objective Functions

In real-world applications, an optimization problem may have either a single objective function, multiple objective functions, or even no objective function. "Feasibility problems" are those problems aiming at finding values for the variables that satisfy the model constraint conditions with no specific objective to optimize. We may, also, refer to "multiobjective optimization" problems when involving more than one objective function to be optimized simultaneously. These objectives are often conflicting and incommensurable. Usually, there is no single solution that is optimal with respect to all the used objective functions at the same time, but rather many different designs exist which are incomparable. Consequently, contrary to "single-objective optimization" problems, where we look for the solution presenting the best performance, the resolution of a multiobjective optimization problem gives rise to a set of compromise solutions presenting the optimal trade-offs between two or more conflicting objectives. In practice, problems with multiple objectives are often reformulated as single objective problems by either forming a weighted combination of the different objectives or by replacing some of the objectives by constraints (Coello et al., 2007; Deb, 2001). Defining multiobjective optimization problems requires further mathematical definitions; these will be given in Section 15.2.3.

15.2.2.5 The Linearity

An optimization problem may be categorized, indeed, as a linear problem or as a nonlinear problem. A linear optimization problem can be defined as solving an optimization problem in which the objective function(s) and all associated constraint conditions are linear. As all linear functions are convex, linear optimization problems are intrinsically simpler and easier to solve than general nonlinear problems, in which the resolution becomes more complex and the decision space is nonconvex. There are several types of nonlinear optimization problems where for many of them the objective function may have many locally optimal solutions, i.e., solutions that are optimal (either maximal or minimal) within a neighboring set of candidate solutions. Finding the best of all such minima, i.e., the global solution which is the optimal solution among all possible solutions, not just those in a particular neighborhood of values, is often difficult.

15.2.2.6 The Uncertainty Tied to the Optimization Model

One further possible classification of optimization problems is in terms of the randomness or uncertainty tied to the data dealt with. In this concern, two main branches can be emphasized, namely, the "deterministic optimization" problems and the "stochastic optimization" problems. In deterministic optimization problems, it is supposed that the problem dealt with presents data which are known accurately. Nevertheless, in practice and in many real-world problems, the given data cannot be known with such certainty for several reasons, e.g., due to measurement and model errors, or if the data are for some predicted quantities for some period in the future. In general, deterministic optimization algorithms are unidirectional, i.e., there exists at most one way to proceed (otherwise, the algorithm gets terminated), and do not use random numbers in any step of execution. On the other hand, in stochastic optimization problems or optimization under uncertainty, the uncertainty or a concept of probability is incorporated into the model that employs at least one instruction or at least one operation that makes use of random numbers. Efficient optimization methods can be applied when the system's parameters are known with certain bounds. In such a case, the aim is to find a solution that is feasible for all data and optimal in some sense. To deal with such context, stochastic optimization models replies on the fact that probability distributions governing the data either are known or can be estimated. Hence, the aim is to find some policy that is feasible for all (or almost all) possible data instances and optimizes the expected performance of the model.

15.2.3 The Multiobjective Optimization Problem

Nowadays, most astro- and geoinformatics real-world problems that are encountered in practice often involve multiple objectives to be minimized or maximized simultaneously with respect to a set of constraints. Hence, multiobjective optimization in such fields that offer highly complex search spaces has become a standard practice. Calling for this specific class of optimization problems and its related optimization algorithms has become essential, as they require little domain information to operate, they are easy to use, and most importantly, they are known to be flexible. The general form of a multiobjective optimization problem is based on the same definitions of the set of constraints presented in Section 15.2.1, with the following objective function (Deb, 2001):

$$\text{minimize } \mathbf{O(x)} = [O_1(x), O_2(x), \ldots, O_M(x)]^T. \quad (15.3)$$

The resolution of a multiobjective optimization problem yields a set of trade-off solutions, called "Pareto optimal" solutions or "nondominated" solutions, and the image of this set in the objective space is called the "Pareto front." Hence, the resolution of a multiobjective optimization problem consists of approximating the whole Pareto front. In the following, we give the key background definitions related to multiobjective optimization.

Definition 15.1 (Pareto optimality). A solution $x^* \in \Omega$ is Pareto optimal if $\forall x \in \Omega$ and $I = \{1, 2, \ldots, M\}$ either $\forall m \in I$ we have $O_m(x) = O_m(x^*)$ or there is at least one $m \in I$ such that $O_m(x) > O_m(x^*)$. The definition of Pareto optimality states that x^* is Pareto optimal if no feasible vector x exists which would improve some objectives without causing a simultaneous worsening in at least one other one.

Definition 15.2 (Pareto dominance). A solution $u = (u_1, u_2, \ldots, u_n)$ is said to dominate another solution $v = (v_1, v_2, \ldots, v_n)$ (denoted by $O(u) \preceq O(v)$) if and only if $O(u)$ is partially less than $O(v)$. In other words, $\forall m \in \{1, 2, \ldots, M\}$ we have $O_m(u) \leq O_m(v)$ and $\exists m \in \{1, 2, \ldots, M\}$, where $O_m(u) < O_m(v)$.

Definition 15.3 (Pareto optimal set). For a given multiobjective optimization problem $O(x)$, the Pareto optimal set is $P^* = \{x \in \Omega | \neg \exists x' \in \Omega, O(x') \preceq O(x)\}$.

Definition 15.4 (Pareto optimal front). For a given multiobjective optimization problem $O(x)$ and its Pareto optimal set P^*, the Pareto front is $PF^* = \{O(x), x \in P^*\}$.

When solving a multiobjective optimization problem, the aim is to find not one, but the set of solutions representing the best possible trade-offs among the objectives, i.e., the so-called Pareto optimal set.

15.3 EVOLUTIONARY COMPUTATION

In evolutionary computation, the derived evolutionary optimization algorithms use the main principles and mechanisms inferred from the Darwinian ideas of natural selection and population, which were presented in Section 15.1. These evolutionary mathematical models operate on a population composed of a set of individuals or chromosomes. Each individual represents a potential solution to the problem being solved and is codified according to the problem's requirements, as highlighted in Section 15.2.2. The goodness of an individual is represented by the objective function, and the restrictions to the problem reflect how apt that individual is to survive in that environment (Eberbach, 2005). Those individuals having lower fitness values are gradually eliminated by the dominant competitors. Within a population and for each individual, probabilistic operators, typically chromosomal cross-over and mutation, are applied over these individuals (parents) to produce new features in the chromosome. These new features represent new individuals (offspring) that maintain some properties of their ancestors which are conserved or are eliminated via selection. This evolution process repeats itself during a certain number of cycles or generations, where species continuously strive to reach a specific genetic structure of the chromosomes that maximize their probability of survival in a given environment. This process continues until an acceptable result is achieved, i.e., the maximal fitness solution is found. These key mechanisms of evolutionary optimization algorithms are defined in what follows.

15.3.1 Basic Structure of an Evolutionary Algorithm

When defining any evolutionary algorithm, the first steps are usually the most critical ones. These comprise the encoding of the candidate solution and the definition of the objective function(s). The encoding and the fitness function(s) are tied to a specific problem, and hence, adequate choices must be taken to guarantee the success of the algorithm. To make these choices, knowledge about the problem dealt with and about the expected solution should be used. Once the problem is defined, the next step is the application of the evolutionary algorithm itself. A basic representation of the different steps of a general evolutionary algorithm is shown in Fig. 15.1.

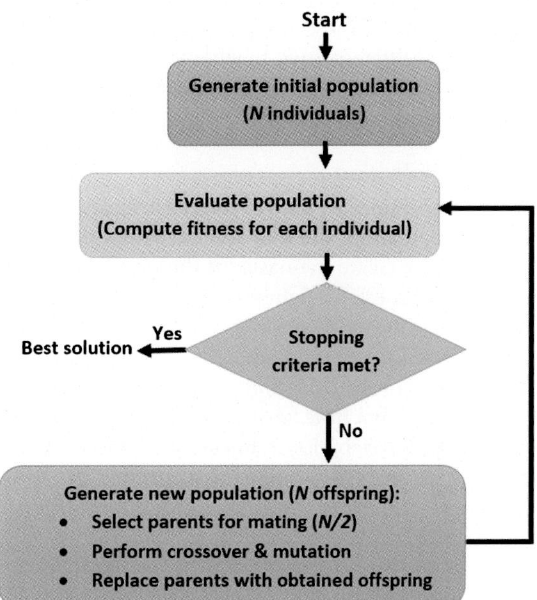

FIG. 15.1 Flowchart of a basic evolutionary algorithm.

The algorithm starts with creating randomly an initial population of chromosomes of size N. The objective function is then evaluated for each chromosome to determine the chromosome's fitness. Based on this fitness value, a termination criterion is evaluated specifying if a solution of the desired quality was found or a specific number of iterations was run. In case where the termination criterion is not satisfied, some chromosomes will be selected (some parents of size $N/2$) and then reproduced via the cross-over and the mutation mechanisms, resulting in offspring. The new chromosomes will replace the old ones producing a new generation, i.e., the parents. This process continues until the termination criterion is satisfied. Finally, the fittest chromosome will be decoded, producing the best solution of the problem.

15.3.2 Evolution Operators

As presented in Fig. 15.1, the key components of any evolutionary algorithm are selection, cross-over, and mutation. Each of these components can be realized in a number of different ways. In what follows, we will give a general overview on how to perform these basic genetic operators.

15.3.2.1 Selection

Selection operators are used to select a proportion of the existing population, the individuals, with a predefined probability to create the basis for the next new generation. It is a process designed to ensure that promising solutions of the population get a greater probability to be selected for mating, and hence forcing the population to improve over time. There are several selection operators that have been proposed in the literature (Goldberg and Deb, 1991; Blickle and Thiele, 1996), and it is worth mentioning that a number of studies have been conducted to compare them. In Blickle and Thiele (1996), it has been concluded that the best selection operator is problem-dependent. In the following, we will highlight the most common selection operators used in practice, mainly in astro- and geoinformatics real-world problems.

Roulette Wheel Selection In roulette selection, each member of the population is selected according to its fitness value. The higher the fitness, the higher the probability of being selected. In such a process, and within a given population P and for each individual x_i, the related selection probability $P_s(x_i)$ is defined as

$$P_s(x_i) = \frac{O(x_i)}{\sum_{j=1}^{N} O(x_j)}. \tag{15.4}$$

The population is then mapped onto a roulette wheel, where each chromosome x_i has its $P_s(x_i)$ proportional slice of area. To select an individual, a random number is generated and the individual whose slot spans the random number is then selected. The wheel is then spun N times, where N is the cardinality of the population. This is to create the next generation of parents that will undergo the simulated reproduction process of cross-over and mutation.

Tournament Selection In a tournament selection, a user-defined number of chromosomes or tournament pool is chosen at random from the current population. The chromosomes in this tournament pool compete with each other and the one with the best fitness value is selected to be a parent for production of the next generation.

Ranking Selection The ranking selection approach is based on the idea that individuals are sorted according to their fitness values. Once they are sorted, rank-based weights are assigned to each chromosome from one generation to another in a way that the rank of each chromosome defines how likely it is to be selected to be a parent for the next new generation. In the literature, there are several linear and exponential rank-based weighting schemes that were proposed (Blickle and Thiele, 1996) and that can be applied with respect to the targeted context. After the application of a specific selected rank-based weighting function to the sorted population, simple roulette or any other selection operator can be used.

Truncation Selection Similar to the ranking selection approach, truncation selection also uses a sorting technique to gain knowledge of the rank order of a given population. More precisely, the truncation selection operator truncates the population by only looking at a fixed number of the highest performing chromosomes. From this specific subset, a random selection technique can be applied where each highest performing chromosome has an equal chance of being selected (Mühlenbein and Schlierkamp-Voosen, 1993). The truncation selection operator has the drawback of additional computational complexity due to the requirement of ranking the population based on the fitness value during each generation.

Supplement to Selection: Elitism In some cases, due to the application of genetic operators, valuable genetic information may be destroyed during the search. When such case occurs, there is no guarantee whether the evolutionary algorithm will be able to rediscover this lost information or not. To prevent such loss, the concept of elitism is introduced. Elitism retains the best members, i.e., a small proportion of the fittest individuals, from the current population and ensures that they pass onto the new generation so that the valuable information remains intact within the population, and hence reduces the genetic drift. It is important to mention that the degree of elitism should be carefully adjusted. This is because a higher proportion, for instance, may cause a rise in selection pressure and hence lead to a premature convergence. On the other hand, elitism is a very useful practice that can significantly increase the performance of any evolutionary algorithm that uses any selection technique.

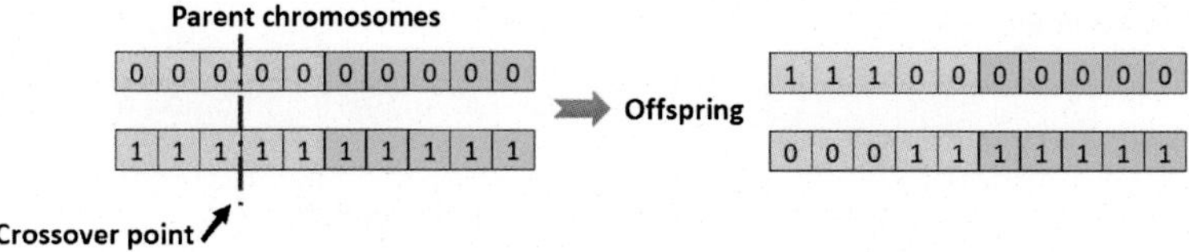

FIG. 15.2 A single-point cross-over operator.

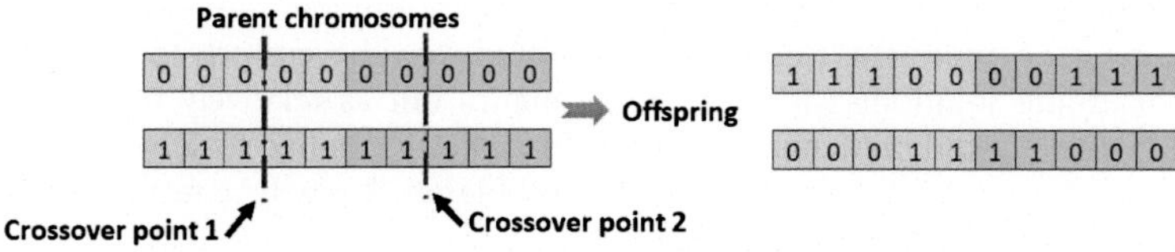

FIG. 15.3 A multipoint cross-over operator.

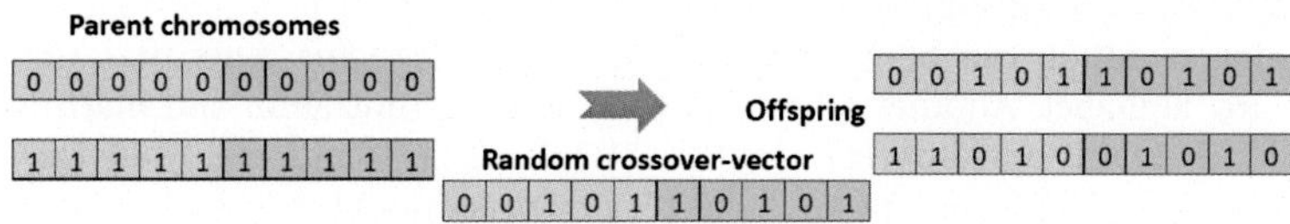

FIG. 15.4 A uniform cross-over operator.

15.3.2.2 Cross-Over

Based on the evolutionary algorithm selection process, the most fit chromosomes should be always selected. However, it is possible that these chromosomes may be represented several times in the upcoming generations, hence leading to a population that is entirely composed of a number of copies of the same candidate solution. If the initial population is not large enough, then this might be a problem as there is no guarantee that the initial population contains a global optimal solution, or even a solution that is considered good enough for the problem being solved. In this case, the evolutionary algorithm will converge to a population filled with duplicates of the best solution that is originally attained in the initial population. To overcome this limitation, cross-over operators as reproduction techniques were introduced and are considered as key components of any evolutionary algorithm to efficiently evolve populations toward optimal points. A detailed study of these can be found in De Jong and Spears (1990). In what follows, we will elucidate the most commonly used cross-over operators for binary encoding.

Single-Point Cross-Over Single-point cross-over is a technique where the selected parent population, i.e., the two mating chromosomes, is cut at a randomly selected location, called the pivot point or cross-over point. At this cut, the genetic information to the left (or right) of the point is swapped between the two parent chromosomes to produce two offspring chromosomes (children). The single-point cross-over operator is illustrated in Fig. 15.2.

Multipoint Cross-Over Contrary to the single-point cross-over, multipoint cross-over works with more than one pivot point. This increases the extent of disruption of the original parent chromosomes. As described in Fig. 15.3, the genetic information between two selected cross-over points that is to the right of an even (or odd) number of pivot points is swapped to produce two unique offspring individuals.

Uniform Cross-Over The uniform cross-over operator looks at one specific gene at a time. It produces a random cross-over vector filled with binary values where 1 indicates that a specific gene location is swapped between parents, and 0 indicates that each parent retains that specific gene. This process can be seen in Fig. 15.4.

15.3.2.3 Mutation

The mutation operator typically occurs after one of the cross-over techniques has been applied to the chosen parent chromosomes. This genetic process perturbs one or more components (genes) of a selected chromosome

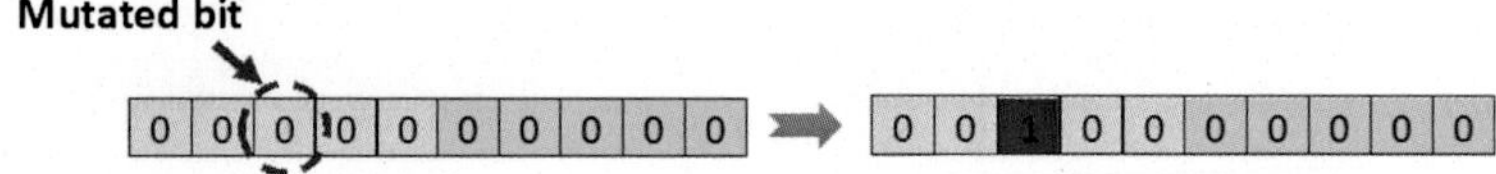

FIG. 15.5 A bitwise mutation operator.

and is regulated by a predefined mutation probability. Mutation is mainly applied to restore lost information or import unexplored genetic components into the population in order to distribute solutions widely across the search space and therefore avoid premature convergence.

Different mutation schemes were proposed in the literature, such as bit-flipping, where the operator simply flips a bit from 0 to 1 or from 1 to 0 with certain probability. Fig. 15.5 illustrates a simple bit-flipping mutation process. The bit-flipping mutation can be generalized to mutate strings of any alphabet. Another scheme is the uniform mutation, where this operator replaces the value of the chosen gene with a uniform random value selected between the user-specified upper and lower bounds for that gene. The shrink mutation operator, for instance, is another scheme which adds a random number taken from a Gaussian distribution with mean equal to the original value of each decision variable characterizing the entry parent vector. Some other mutation operators can be found in Voigt and Anheyer (1994).

15.4 EVOLUTIONARY COMPUTING METAHEURISTICS

Based on the discussed concepts of evolutionary computation, numerous possibilities arise to design and implement advanced evolutionary algorithms, i.e., metaheuristics which are able to efficiently explore complex search spaces in order to solve complex optimization problems. In the literature, a wide range of different evolutionary optimization algorithms have been proposed, and a full review of these can be found in Deb (2001), Talbi (2009), which is beyond the scope of this chapter. All of these algorithms are similar in their basic approach and in making use of the evolutionary concepts; they mainly differ in the way they represent the information. In what follows, we will give an overview of the main evolutionary optimization metaheuristics that are the most commonly used in astro and geo real-world problems. We will give general descriptions of the algorithms and with respect to the specific problem requirements, the metaheuristic can be designed for single-objective problems, multiobjective problems, or for any other type of optimization algorithm as discussed in Section 15.2.2. We will also emphasize some other evolutionary approaches and bio-inspired algorithms that have much to offer to astro- and geoinformatics but as relatively new paradigms, they are not commonly applied to solve astro and geo optimization problems. This will be further discussed in Section 15.7.

15.4.1 Genetic Algorithms

Genetic algorithms (GAs) have been firstly proposed in Holland (1992) to understand the adaptive processes of natural systems. After that, they have been applied to solve optimization and machine learning problems (Holland and Goldberg, 1989; De Jong, 2005). The classical versions of GAs use a binary representation, i.e., the chromosome representation is based on a binary string of fixed length. However, nowadays, GAs make use of several other types of representations, e.g., nominal-valued discrete variables where each nominal value is encoded as a bit string, integer or real-valued representations, order-based representations or chromosomes of variables length, and many more (Chambers, 2000).

For the selection process, the algorithm uses a probabilistic selection that is originally the roulette wheel operator. A replacement selection is also performed, i.e., the selection of survivors of both the parent and the offspring populations. Specifically, a generational replacement is used where the parents are replaced systematically by the offspring. Concerning the reproduction process, it is traditionally made via the cross-over and the mutation operators with a fixed probability for each of them. However, the algorithm emphasizes more the importance of the cross-over operator than that of the mutation operator. The cross-over operator is based on the single-/multipoint or uniform cross-over, while the mutation is generally bit-flipping.

15.4.2 Evolutionary Strategy

Evolution strategies (ESs) were originally proposed in Rechenberg (1981). Unlike GAs, which are mostly applied to discrete optimization problems, ESs are mostly applied to continuous optimization, where representations are based on real-valued vectors. The first ES applications (Klockgether and Schwefel, 1970) include real-valued parameter shape optimization. They usually apply the elitism concept and a Gaussian distributed mutation. Cross-over is rarely applied. In ES, the representation of an individual is made by its genetic material and by a so-called strategy parameter which defines the behavior of the individual in its related environment. The genetic material is represented by floating-point variables while the strategy parameter is, generally, defined by the standard deviation of a Gaussian distribution associated with each (variable of the) individual.

In many ESs, the selection operator is mainly deterministic and based on the fitness ranking. Two types of mutation operators are commonly used, namely, the discrete mutation, in which the gene value of the offspring is the gene value from the parent, and intermediate mutation, in which the mid-point between the gene value of the parents gives the gene value of the offspring. The mutation operator has a special implementation in ES as it mutates both the strategy parameter and the genetic material. Hence, the evolution process evolves the genetic material and the strategy parameter at the same time; accordingly, ES is considered to be a "self-adaptive" mechanism (Meyer-Nieberg and Beyer, 2007). The main advantage of ESs is their efficiency in terms of time complexity.

15.4.3 Evolutionary Programming

Evolutionary programming (EP) was first proposed in Fogel (1998). This paradigm emphasizes on mutation and does not use cross-over. Classical EPs have been developed to evolve finite-state automata in such a way that they were capable of solving time series prediction problems and more generally evolving learning machines (Fogel, 1998). Modern EPs have been applied to solve continuous optimization problems using real-valued representations. As ESs, EPs use Gaussian distributed mutations and the self-adaptation paradigm. In EP, the parent selection mechanism is deterministic, and the survivor selection process (replacement) is probabilistic and based on a stochastic tournament selection (Eiben et al., 2003).

15.4.4 Genetic Programming

Genetic programming (GP) (Koza, 1992) is seen as an extension of the generic model, GA, in which the structures in the population are not fixed-length strings that encode candidate solutions to a problem (linear representation) but programs expressed as syntax trees, i.e., nonlinear representation based on trees. In Koza (1992), computer programs (solutions) were encoded using LISP and their representations are S-expressions where the leaves are terminals and the internal nodes are operators (functions). The definition of the leaves and the operators are strictly tied to the targeted application being solved.

In GP, generally, the parent selection is a fitness proportional and the survivor selection is a generational replacement. The cross-over operator exchanges parts of two parent trees, resulting in two new trees, and the mutation randomly changes a function of the tree into another function, or a terminal into another terminal. One of the main problems in GP is the uncontrolled growth of trees, which is a phenomenon called "bloat." Indeed, GPs need a huge population and then they are very computationally intensive.

15.4.5 Other Evolutionary Algorithms and Bio-Inspired Approaches

Several other astro and geo real-world problems are based on some other evolutionary approaches. Among these, we can mention differential evolution, coevolutionary algorithms, and some other nature-inspired metaheuristics, such as swarm intelligence and artificial immune systems, which may also be used to solve complex optimization problems.

15.4.5.1 Differential Evolution

Differential evolution (DE) (Storn and Price, 1997; Price et al., 2006) is one of the most successful metaheuristics for continuous optimization. The algorithm uses a population of parameter vectors that encode the problem and are initially chosen uniformly random from the search space. During each generation an update is attempted for each vector in the following way, where the vector that we try to update is called a target vector. First, a mutation vector is created by randomly choosing three parameter vectors from the population, different from the target vector, and adding a scaled difference from the first two to the third vector. The difference is scaled using a mutation scaling factor. A cross-over operation is then performed between the obtained mutation vector and the target vector to create a trial vector. In the selection operation, if the trial vector has a better fitness than the target vector, a replacement is made for the next generation. This idea has been integrated in a novel cross-over operator of two or more solutions and a self-referential mutation operator to direct the search towards good solutions (Talbi, 2009). Several DE variants are possible, where these algorithms mainly differ in the way parents are selected and in the form in which cross-over and mutation take place. A detailed survey of various DE algorithms can be found in Das and Suganthan (2011), Das et al. (2016).

15.4.5.2 Coevolutionary Algorithms

A coevolutionary algorithm is based on the concept of natural complementary evolution of closely allied species (Durham, 1991). In nature, various species represented by a collection of similar individuals coevolve based on their phenotype. A coevolutionary algorithm, unlike traditional evolutionary algorithms where a population is composed of a single species, may be seen as a competitive-cooperative paradigm that involves different interacting populations, where each represents a given species, together optimizing coupled objectives. The competitive coevolutionary approaches rely on the idea that different populations compete in solving the global problem. In such schemes, the individual fitness defines a competition. Each population aims at minimizing a local cost specified by a local fitness function. The competition between different populations leads to an equilibrium in which the local objectives cannot be improved and hopefully the global objective is achieved (Talbi, 2009). On the other hand, the cooperative models reflect the cooperative behavior of the various populations to solve the problem. In such schema, a population evolves a subcomponent of the solution. Based on the cooperative interaction between the populations, a global solution arises from the assembled species' subcomponents. The fitness of an individual of a given species will depend on its ability to cooperate with individuals from other populations (Talbi, 2009).

15.4.5.3 Swarm Intelligence

Swarm intelligence is an innovative intelligent paradigm which was successfully applied to solve optimization problems. It took its inspiration from the collective behavior of social swarms in nature such as flocks of birds, honey bees, schools of fish, and ant colonies (Bonabeau et al., 1999). Specifically, swarm intelligence is based on the common behavior of these species that compete for food. The main features of swarm intelligence-based algorithms are their simplicity and their particle aspect. They are based on agents, i.e., insects or swarm individuals, which are relatively unsophisticated and which cooperate together by making movements in the decision space to achieve tasks necessary for their survival. Among the most effective swarm intelligence-based algorithms used in astro and geo real-world optimization problems are ant colony and particle swarm optimization. These will be detailed in what follows.

Ant Colony Optimization Algorithms Ant colony optimization (ACO) algorithms are based on the idea of imitating the foraging behavior of real ants to solve complex optimization tasks such as transportation of food and finding shortest paths to the food sources (Dorigo and Di Caro, 1999). In nature, ants communicate by means of chemical trails, called "pheromone." This substance assists ants in finding the shortest paths between their nest and food. In a natural observation, ants usually wander randomly. When they find food, they return to their nest while laying down pheromone trails on the ground. This chemical, if found by other ants, will not keep them wander at random, but will help them to follow the trail and to quickly return to their nest, i.e., this trail will guide the other ants toward the target point. Meanwhile, these ants will reinforce the path if they find food. However, as ants have to travel the path back and forth and as the pheromone has to evaporate, the path becomes less prominent. In such situations, ants will look for the path having a higher density of pheromone. This means that this particular path was visited by more ants and is definitely the shortest path to take. Based on this inspiration, ACO algorithms can be seen as multiagent systems in which each single agent is inspired by the behavior of a real ant. In the literature, there are numerous successful implementations of the ACO metaheuristic. A review of their applications to a wide range of different optimization problems can be found in Dorigo and Stützle (2003).

Particle Swarm Optimization Another successful swarm intelligence model is particle swarm optimization (Kennedy and Eberhart, 1995; Clerc, 2010). It draws inspiration from the sociological behavior of natural organisms such as bird flocking and fish schooling to find a place with sufficient food. Within these swarms' populations, a synchronized behavior using local movements emerges without any dominant control. Each individual within its community (population) is moved to a good area based on its fitness for the environment, i.e., its flexible velocity (position change) in the search space. Indeed, based on the particle's memory, the best position the individual has ever visited in the search space is remembered. Following this natural observation, the movement of swarms is seen as an aggregated acceleration towards their best previously visited position and towards the best particle of a topological neighborhood, i.e., the social influence between the particles. This phenomenon led to several efficient particle swarm optimization algorithms which are mainly applied to solve optimization problems (Clerc, 2010).

We describe one possible PSO implementation similar to the original one proposed in Kennedy and Eberhart (1995). The particles are randomly scattered at the beginning in an n-dimensional space. Each particle is characterized by its position C_i and its velocity v_i, which is initially zero. Each particle remembers its fittest value and position P_i and also the fittest value and position G_i from the entire swarm and at each iteration every particle updates its velocity as $v_i(k+1) = \alpha v_i(k) + \beta \text{rand}()(P_i - C_i) + \gamma \text{rand}()(G_i - C_i)$, which is then added to its current position. The function rand() generates a random number from the range $(0, 1)$, while the parameters α, β, and γ can be used to balance between the inertial, cognitive, and social influences.

15.4.5.4 Artificial Immune Systems

The study and design of artificial immune systems (AISs) represent a relatively new area of research that tries to build computational systems that are inspired by the natural immune system (De Castro and Timmis, 2002). This growing field has been mainly applied to data mining problems (DasGupta, 1993), but lately its application to optimization problems has been rapidly increasing (Chandrasekaran et al., 2006; Cutello and Nicosia, 2002; Coello and Cortés, 2005).

AISs are based on the human immunological concepts. The natural immune system is a network of cells, tissues, and organs that work together to defend the body against attacks by "foreign" invaders that are trying to do it harm. This main task is achieved thanks to its capability to recognize the presence of infectious foreign cells and substances, known as "nonself" elements, and to respond to them by eliminating them or neutralizing them. This distinction between the "nonself" and the body's "self" cells is based on a process called "self–nonself discrimination" (Janeway, 1992). The nonself elements, also called "antigens," are mainly microbes; tiny organisms such as bacteria, parasites, and fungi. All of these can, under the right conditions, cause damage and destruction to parts of the body and if these were left unchecked, the human body would not be able to function appropriately. Thus, it is the purpose of the immune system to act as the body's own army. More precisely, the immune system does not rely on one single mechanism to deter invaders, but instead uses many strategies. The main division between the strategies is that between innate immunity and adaptive immunity. The innate immunity is the first line of defense against invading antigens. It is those parts of the immune system that work no matter what the damage is caused by. They are always at work and do not need to have seen the offending invader before to be able to start attacking it. The innate immune system includes anatomical barriers, secretory molecules, and cellular components. In addition, the innate immune system employs a different group of cells, e.g., antigen presenting cells (APCs), to eliminate threats or to interact with the rest of the immune system. The second line of defense is the adaptive immune system, which affords protection against reexposure to the same pathogen. The adaptive immune system is called into action against pathogens that are able to evade or overcome innate immune defenses. The cells of the adaptive immune system are mainly B cells and T cells, but there are also other important parts of the adaptive immune system, such as the complement cascade and the production of antibodies. These mentioned elements of the immune system do not work separately, but all work together in a cooperative fashion via specific immune proteins called "cytokines." An inspiration from these remarkable immune properties led to the conception and the design of artificial immune systems exhibiting similar functionalities. These systems are discussed in what follows.

Clonal Selection Theory Clonal selection theory (Burnet et al., 1959) is used to clarify the basic response of the adaptive immune system to antigenic stimuli. Clonal selection involves two main concepts i.e., are cloning and affinity maturation. More precisely, it establishes the idea that only those cells capable of recognizing an antigen will proliferate, while other cells are selected against. Clonal selection calls both B and T cells. When B cell antibodies bind with an antigen, cells become activated and differentiated either to be plasma cells or memory cells. The closer the matching between an antibody and a specific antigen is, the stronger is the bond. This property is called affinity. Plasma cells make large amounts of a specific antibody that work against a specific antigen to destroy it. Memory cells remain with the host and promote a rapid secondary response. However, before this process, clones of B cells are produced and undergo somatic hypermutation. Consequently, diversity is introduced into the B cell population. Moreover, a selection pressure is performed, which implies a survival of the cells with higher affinity. Let us note that clonal selection is a kind of an evolutionary process, where an antibody represents a solution, the affinity defines the fitness function, and the antigen represents the value of the objective function to optimize. The cloning process is seen as the reproduction of solutions, the somatic hypermutation represents the mutation of a solution, and the affinity maturation represents the mutation and the selection of best solutions. Based on this theory, various clonal selection algorithms have been proposed in the literature, and most of them are devoted to optimization problems. A detailed description and comparison of AIS clonal selection algorithms can be found in Ulutas and Kulturel-Konak (2011).

The Self–Nonself Theory The self–nonself theory is able to tell the difference between what is foreign and potentially harmful, and what is actually a part of its own system. The representative self–nonself theories are negative selection and positive selection. The purpose of the negative selection theory is to provide tolerance for self-cells. During the generation of T cells, receptors are made through a pseudorandom genetic rearrangement process. Then, they undergo a censoring process in the thymus, called negative selection. There, T cells that react against self-proteins are destroyed; thus, only those that do not bind to self-proteins are allowed to leave the thymus (worse solutions are removed to get optimal solutions). These matured T cells then circulate throughout the body to perform immunological functions and protect the body against foreign antigens. As for the positive selection theory, it works as the opposite of the negative selection process. An inspiration from the negative selection and positive selection theories gave rise to numerous AIS algorithms which are mainly used for classification to solve optimization problems (Cao et al., 2007; Gao et al., 2008).

Immune Network Theory The immune system is a network of cells and antibodies that have a profound sense of self and the ability to remember and learn. The immune network theory states that the recognizers of the immune system, the B cells and antibodies, not only recognize foreign particles, but also recognize and interact with each other. This created network is based on interconnected B cells for antigen recognition. The strength of the B cells connections is directly proportional to the affinity that they share. Indeed, B cells can both stimulate and suppress each other in order to stabilize the network. These characteristics of the immune network not only maintain the diversity of the antibody population effectively but also facilitate self-organization and regulation in the biological immune system. Basic concepts of the immune network theory are implemented, leading to several immune algorithms dedicated to solving optimization problems (Hajela and Yoo, 1999).

Danger Theory Danger theory (Matzinger, 2001) is a new theory which has become popular amongst immunologists. It was proposed to explain current anomalies in the understanding of how the immune system recognizes foreign invaders. The central idea in danger theory is that the immune system does not respond to nonself but to danger. Thus, just like the self–nonself theory, it fundamentally supports the need for discrimination. However, it differs in the answer to what should be responded to. Instead of responding to foreignness, the immune system reacts to danger based on environmental context (signals) rather than the simple self–nonself principle. Specifically, the dangerous antigens stimulate the production of danger signals by stimulating cellular stress or cell death. Those signals are recognized by APCs that recognize these signals and based on these phenomena the immune system detects the danger zone and then evaluates the danger. By defining the danger zone to calculate the danger signals for each antibody, the algorithm adjusts antibodies' concentrations through its own danger signals and then triggers immune responses of self-regulation. Consequently, the population diversity can be maintained.

15.5 PARALLEL EVOLUTIONARY COMPUTING METAHEURISTICS FOR BIG DATA

In real-world applications, optimization problems are usually NP-hard and are CPU time and/or memory consuming. Hence, the application of metaheuristics is essential as the algorithms help in finding the appropriate solutions to the problem being solved while reducing the computational complexity of the search process considerably. Although the use of metaheuristics permits such gain, the used objective functions and the constraint resource requirements (e.g., CPU, memory) remain intensive, specifically when the size of the search space becomes huge. To deal with these limitations, in recent years, several parallel and distributed computer architectures have emerged for the design of metaheuristics (Alba and Tomassini, 2002).

The design of parallel metaheuristics can be realized in different ways. However, it can be categorized into three main levels, namely, the algorithmic level, the iteration level, and the solution level. At the algorithmic level, metaheuristics can be either independent or cooperative. If the metaheuristics are independent, i.e., the different metaheuristics are executed without any cooperation, the search will be equivalent to the sequential execution of the metaheuristics in terms of the quality of solutions. Nevertheless, in the cooperative model, i.e., the different algorithms are exchanging information related to the search with the intent to compute better and to have solutions that are more robust, the behavior of the metaheuristics is altered to enable the improvement of the quality of solutions. On the other hand, at the iteration level, the main idea is the parallelization of each iteration of metaheuristics. This design is based on the distribution of the handled solutions, where the behavior of the metaheuristic is not altered.

More precisely, it is the generation and the evaluation of the neighborhood (or candidate solutions), which is done in a parallel way as this step presents the most computation-intensive part of the metaheuristic. The same design can be also based on the distribution of the population. In this case, the operations commonly applied (cross-over and mutation) to each of the population elements are performed in parallel. The main objective of the iteration level is to speed up the algorithm by reducing the search time. The third design is the solution level, where the main focus is the parallelization of the evaluation of a single solution (objective and/or constraints) of the search space. Similar to the iteration level, in this design, the behavior of the metaheuristic is not altered.

Several parallel computer architectures have been proposed in recent years to represent an effective strategy for the design and implementation of parallel metaheuristics. Among the most popular architectures, we mention the master–slave model, which is mainly used at the iterative level and the solution level, the island model, and the cellular model. The two latter architectures are the most widely known parallel algorithmic-level models and the most commonly used for evolutionary algorithms. In the master–slave model, a single machine represents the master and it distributes the workload for executing operations to several other machines, named "slaves." As the selection and the replacement are usually sequential procedures, as they need a global management of the population, it is the master who performs these tasks. The operations, like mutation, cross-over, and the evaluation of the fitness function, are performed by the associated slaves. This is because these operations can be performed independently and they are often among the most expensive operations. In this model, the master sends the partitions, i.e., the subpopulation, to the workers, who in turn return the newly evaluated solutions to the master.

In the island model, which is also called "distributed evolutionary algorithms," the population of each run is considered as an island. Some research papers define island models based on subpopulations that together form the population of the whole island model. In island models, the evolution of the island happens in an independent manner. However, sometimes, solutions are exchanged between the different islands in a process called migration. The main idea is to create this migration topology which is seen as a directed graph with islands as its nodes and directed edges connecting two islands. Selected individuals from each island are sent off to neighboring islands, i.e., islands that can be reached by a directed edge in the topology. These individuals, named migrants, are involved in the target island after an additional selection process. In this way, the different islands can communicate, exchange, and compete with one another. When some islands get stuck in low-fitness regions of the search space, they will be taken over by individuals from other more successful islands. This helps to coordinate search, to focus on the most promising regions of the search space, and to use the available resources efficiently.

The other well-known parallel model for evolutionary algorithms is the cellular model. Cellular models can be seen as a special case of the island models with a more fine-grained form of parallelization. In such model, the island is composed of a single individual, and is called a cell, which explains the term cellular evolutionary algorithms. Similar to the island models, the cells in cellular models are connected by a fixed topology, e.g., rings, torus graphs, etc. Each individual in a cell is only allowed to mate with its neighbors in the topology. This communication happens in every generation. The main difference to island models is that there is no evolution that happens in the cell itself. The improvements can only be obtained by cells interacting with each other. However, it is possible that an island can interact with itself. In this design, the overlapped small neighborhood helps in exploring the search space because a slow diffusion of solutions through the (sub)population provides a kind of exploration, while exploitation takes place within each neighborhood. Fig. 15.6 shows a general scheme of a basic island model with six islands (on the left), and two cellular models reflecting a torus graph (on the right) and a complete graph (in the middle).

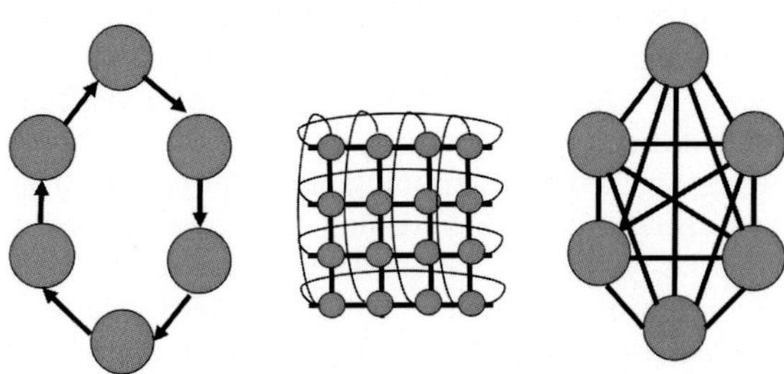

FIG. 15.6 Left: The traditional parallel island evolutionary algorithm. Right: Cellular models for evolutionary algorithms.

The discussed parallel and distributed computing architectures as well as many others have the ability to reduce the search time and hence help designing real-time and interactive optimization methods, improve the quality of the obtained solutions, improve the robustness in terms of solving in an effective manner different optimization problems and different instances of a given problem, and allow to solve large-scale instances of complex optimization problems. A review of the distributed computer architectures, of parallel evolutionary algorithms, and of their characteristics can be found in Alba (2005), Alba et al. (1999), Gong et al. (2015).

15.6 PRACTICAL APPLICATIONS OF EVOLUTIONARY COMPUTING METAHEURISTICS IN THE CONTEXT OF ASTRO- AND GEOINFORMATICS

Evolutionary computing has a wide range of practical applications across many different research areas, and in this section we show some examples from astro- and geoinformatics as well as some other related fields. For example, in Arias-Montano et al. (2012) the reader can find many applications of evolutionary computing in aeronautical and aerospace engineering. Our goal is to motivate the reader to indulge in application of the algorithms and to provide an insight into their characteristics when applied to real-world problems. We have grouped the various applications by the algorithm that is used, although some of the works examine or combine several different algorithms. For instance, one interesting work (Civicioglu, 2012) addresses the problem of mapping geocentric Cartesian into geodetic coordinates with many different metaheuristics and proposes a new one that outperforms all the other for that particular application.

Genetic Algorithm From all metaheuristics, GAs are probably the most applied. One popular article demonstrating the application of GAs to astronomy and astrophysics is Charbonneau (1995). The article describes three different applications of GAs: fitting rotation curves in galaxies, determining pulsation periods from stars' spectral lines, and creating wind models for certain solar-type stars. All examples use single-point crossover, uniform mutations, and ranking selection, and all variables are encoded using integers. The first problem is representing rotation curves in galaxies from available data. A rotation curve is the variation of velocity around the center of the galaxy, which can be modeled as a function of the distance from the galaxy center. The velocity usually has four components: bulge, disk, interstellar gas, and halo components. Brightness profiles can be determined from observations for the first two components, which can be converted into mass using mass-to-light ratios that are left as two adjustable parameters. The halo component requires two parameters, i.e., a velocity dispersion and a characteristic length scale. Two constraint strategies have been examined and resulted in different solutions, one with only positivity requirements, and another with the parameters constrained in particular positive ranges. The parameters are then learned by applying a GA with a population of 100 individuals over 100 generations, performing crossover with a rate of 0.65 and a mutation rate of 0.003. The second problem falls within the wider class of fitting time series data, which is often done in astronomy. The specific presented problem is fitting time series data to a multiperiodic signal that is a sum of sinusoidal functions. Each of these functions are characterized by an amplitude, phase, and frequency, which all together constitute a set of parameters that need to be determined. The authors have used a "cleaning algorithm" in which they first identify the dominant periodical function using power spectra analysis and fit it to the data. Then, the previously fitted function is subtracted from the original data and the procedure is repeated until the remaining data can be considered as noise. The method has been applied to pulsation periods in δ Scuti stars using seven sinusoidal components and a GA with a population of 200 individuals over 1000 generations with a variable mutation rate in conjunction with elitism. The third problem consists of finding solutions to a magnetohydrodynamic wind model, where an initial root finding problem is transformed into a minimization problem with six parameters, which can then be addressed appropriately using a population of 100 individuals over 500 generations with variable mutation rate and elitism. All three problems have been solved using a similar approach with very little code changes from one to another. Later, this approach has also been successfully applied in a parallel manner to study white dwarfs and extract their physical and structural characteristics by observing their pulsation frequencies (Metcalfe and Charbonneau, 2003). The parallelization is done using the master–slave model, where the master process performs all the calculations related to the GA, while the slave processes evaluate the fitness function.

In Orfila et al. (2002), the authors have demonstrated how GA can be applied in developing a dynamical model of the solar cycles to monthly and yearly observations, which in turn can be used to predict its long-term future behavior. The authors try to extract the structure of the model from the observed data. Therefore, they generate a population of 120 equations with random combinations of state variables, parameters, and the four basic mathematical operations, and try to find the best possible description using a GA over 10,000 generations. A regular roulette wheel selection is employed, with a small percentage of uniform mutations, and the problem is encoded using character strings.

Another work (Wahde, 1998) has explored how these algorithms can be used for estimation of a large number of parameters in a pair of interacting galaxies. The interactions can be modeled using 11 parameters, or in a simplified version with seven. A particular example for NGC 5194 and NGC 5195 was given in Wahde and Donner (2001), which uses a population of 50 individuals over 70 generations (higher numbers do not bring improvements), with single-point cross-over, uniform mutations with a constant rate, rank-based selection, and elitism, in order to solve the problem encoded using integers.

In Vachier et al. (2012), a GA was successfully applied for deriving the orbital parameters for several binary asteroid systems, such as 22 Kalliope, 3749 Balam and 50000 Quaoar. First, a simple Keplerian model is applied to do a wide-space search to find an initial solution. Then, the whole physical dynamical system behavior is represented as a more detailed N-body problem, which is fitted to observations with a search starting from the previous solution. The algorithm uses a population of 60 individuals and a rank-based selection that takes the fittest 40, and performs a cross-over to create a new offspring population in which a mutation is induced in two individuals. The authors also introduce search space resizing that resembles a cataclysmic extermination in which after about 20 generations all individuals are exterminated and the search space is shrunk by discarding the parts without any individuals. The algorithm terminates if a predefined acceptable error range is reached, if the fitness difference between the best and worst individual is below 10^{-2}, or if a predefined maximum number of generations have passed (between 500 and 1000).

GAs have been used for various problems in geoscience. In Maulik and Bandyopadhyay (2003), a GA is used for land cover fuzzy clustering with a dynamic number of clusters from remote sensing imagery, and the algorithm has been applied for Calcutta and Mumbai, India. The cluster membership is represented using a matrix with elements denoting if a pixel belongs to a particular cluster in a fuzzy manner, i.e., a pixel can be a member of more clusters at the same time. The centers of each cluster are then the optimization variables whose coordinates are encoded using a string of real values with variable length as the number of clusters can vary. The objective function is to minimize the Xie–Beni index, i.e., the ratio between the total within-cluster variation and the minimal intercluster center separation. The selection is done using the roulette wheel with additional elitism. The cross-over operator uses a single-point cut, but it is slightly different from the classical version as the string length is variable. The cluster centers are considered to be indivisible, the cut position can be different in the parents, and it is required that each offspring has at least two clusters. The mutations are uniform. The GA is run using a population of 20 individuals, cross-over and mutation rates of 0.8 and 0.01, over a fixed number of 100 generations, with an upper bound of 20 for the number of clusters.

GAs can be used for efficiently training or constructing artificial neural networks (ANNs), which could serve various tasks. For training an ANN with a GA the network weights are the optimized variables, while the objective function is the difference between the ANN outputs and some desired outputs. Unlike the classical backpropagation algorithm, GAs can search the space of weights more exhaustively and avoid being trapped in local minima. However, we should keep in mind that GAs need more computational time. In Jain and Srinivasulu (2004), a GA was used for training an ANN representing rainfall-runoff. Beside the classical black box approach, the authors also consider an incorporation of some physical knowledge into the ANN through additional inputs, thus making it a kind of a gray box model. The mapping of rainfall into runoff is a very complex nonlinear phenomenon that has two principal flow components, from the surface and from the subsurface. The gray box model incorporates a base flow component, an infiltration component, a soil moisture accounting component, and an ANN component. Both black box and gray box approaches are examined with data from the Kentucky River basin from 26 years. The GA is coded with real values. A tournament selection is used for choosing parents in the next generation combined with an elitism that preserves the best solution from each generation. Then, a single-point cross-over that allows to control how near the children are from their parents allowing a better control over the range of weight values. The mutation is bounded to some predefined ranges and its distribution is controllable. A population of 290 individuals is used with cross-over and mutation rates of 0.9 and 0.01. The algorithm terminates if the error falls within some acceptable range or the maximal number of generations is reached. The authors have compared several models, both black box and gray box, using backpropagation and GA training, and have concluded that a GA-trained gray box ANN outperforms the rest, particularly in estimating low-magnitude flows.

In Yan and Minsker (2006) another approach combining GA and ANN is used for groundwater remediation design. Two case studies have been considered, one simpler hypothetical case for testing and a real-world case at the Umatilla Chemical Depot in Hermiston, Oregon. The goal is to find an appropriate setup of the pump and treatment system in order to minimize its costs under its current capacity until the chemical contaminants are cleaned up. The authors develop a hybrid approach called adaptive neural network genetic algorithm (ANGA). This approach uses an initial set of trial designs that are first evaluated using simulation models of flow and transport, and the results are stored in a cache. The cache stores the evaluations of the design fitness, to avoid recalculation of the designs if they appear again. The existing designs are then used for the creation of a new generation of designs by a GA. After a large number of simulations, the evaluation of new designs can be carried out using ANNs instead of a simulation model. Unlike in the previous example, here the ANNs are trained using the backpropagation algorithm, and the GA is used to combine completely different model designs. The ANNs are initially trained and then retrained completely when required. We shortly describe the GA setup only for the real-world example. A combination of tournament and roulette wheel selection is used. The population consists of 160 individuals with cross-over and mutation rates of 0.5 and 0.00625. The algorithm is considered to have converged if the difference between individuals of the same generation is within some range, and in this case typically that has varied between 60 and 80. Most of the constraints are included as linear penalty terms in the cost function, while one of them is enforced directly.

In D'Ambrosio et al. (2006) parallel GAs have been used for estimating the parameters in a cellular automata model of debris flows, which have been then used for analyzing the May 1998 Curti-Sarno case in Italy. The fitness function takes values from [0, 1] and it shows how similar the simulated events are to the actual areal observations. The problem was encoded into the GA directly by representing the parameters using real values, which were constrained by previous experience from manual experimentation. The population consists of 200 individuals over 100 generations, and from each generation to the next, only the worst 15 are replaced, which is a steady-state and elitist evolution. The new offspring is obtained by a binary tournament selection without replacement, where in each comparison the fitter individual has a fixed probability of 0.6 for winning. The cross-over is single-point with a rate of 0.8, while the mutation is uniform with a rate of 0.125. The authors have also parallelized the GA for execution over several CPUs using a master–slave model, where the master processor performs all the operations except for the calculation of the fitness values, which is done by the slave processors. Typically the fitness calculation is the most time consuming, so it is the operation that benefits the most from parallelization, although the other operations could be parallelized as well. The rest of the technical details about the parallelization can be found in D'Ambrosio et al. (2006).

Many geoscience-related engineering problems have benefited from the application of GAs. One example are adaptation and mitigation strategies in water supply systems to the world's climate change, which take into account greenhouse gas (GHG) emissions (Paton et al., 2014). As a case study, the Southern Adelaide System in South Australia has been analyzed, which consists of three local reservoirs and water brought from a distant river, with additional potential of including three desalination plants, reusing storm water, and introducing home rainwater containers. The studied problem involves minimizing the system's vulnerability, cost, and GHG emissions, under some reliability and availability constraints. In order to evaluate many different design alternatives, the system's behavior was simulated using a water resources model named WaterCress, which incorporates supply and demand. The authors have generated 1000 time series of 30 year rainfall data, from which they have chosen the 10 most representative for the optimization and the rest for evaluation of solutions. The optimization was performed using a so-called water system multiobjective GA, proposed in Wu et al. (2009). The authors have experimented with a range of parameters of the GA and as most appropriate they have found having a population of 150 over 150 generations, with cross-over and mutation rates of 0.9 and 0.1, respectively. The problem was encoded using a string of integer values, which are suitable for discrete decisions, and real values were required. The next generation was produced using a special type of constrained tournament selection with elitism. As the problem is multiobjective, a set of Pareto optimal solutions was obtained, and a detailed comparative analysis of all solutions from the Pareto front was provided.

In Wang et al. (2018b) the authors have presented how GAs can be used for optimizing large-scale computations of astronomical data, particularly for the Sloan Digital Sky Survey,[1] which has more than 125 terabyte of data. The data consist of a huge amount of rich telescope images covering more than a third of the sky,

[1] http://www.sdss.org.

including spectra for millions of objects. The problem is finding an optimal task scheduling strategy for the many computational procedures that are required for the processing of the astronomical data in a fog computation setup. It is a min-max combinatorial problem in which the authors minimize the maximal time for execution of each of the individual computation steps. An encoding with real values and a population of 100 individuals are used. A roulette wheel selection is performed with supplemental elitism with the five best-fitted individuals from each generation. Two-point cross-over with a rate of 0.8 and two-point mutation with a rate of 0.02 are used. For convergence speedup a local search is introduced, which tries to balance the workload among the fog computational nodes. The algorithm finishes after a fixed number of 2500 generations have passed. The authors performed the optimization using data of one single day of about 200 gigabyte.

Another interesting application of GA is in the design of wind farms (Wang et al., 2018a) where the algorithm optimizes the turbine positions and the turbine wind hub heights. In Gonzalez et al. (2018) the authors have employed a GA to optimize the operation of a hybrid renewable energy system combining wind, photovoltaic, and forest biomass energy sources.

Evolutionary Strategy From the various types of evolutionary strategies (ESs), covariance matrix adaptation ES (CMA-ES) is probably the most applied one because of its ease of use and efficiency. The CMA-ES algorithm was proposed in Hansen and Ostermeier (2001), while in Hansen and Kern (2004) it was shown that it can find the global optima for many types of multimodal standard test functions. The algorithm starts with an initial parental set of individuals, also called parameter vectors. At each iteration a part of the parental vectors are selected and randomly intermediately recombined to create a new set of candidate parameter vectors. The candidate vectors are then mutated by sampling an n-dimensional normal distribution of the form $N(0, \mathbf{C})$, where $\mathbf{C}$ is an adaptive covariance matrix. The parameter vectors are then ranked and the fittest ones are selected for the next generation. The algorithm uses the concept of elitism, because at each generation it could leave some of the parameter vectors nonrecombined and nonmutated, if they remain fitter than the other candidate vectors. Most of the internal strategy parameters of the algorithm are self-adaptive, and do not need special initialization efforts. The mutations tend to favor the search directions from the previous steps, thus creating an evolution path, as it is known that a random search that tends to move away from its initial location is generally faster.

In Quast et al. (2004), the authors applied the CMA-ES method in their procedure of estimating the variability of the fine-structure constant in the cosmos. They used observations of the spectra of the quasar HE 0515-4414 taken by the ultraviolet and visual Echelle spectrograph from the ESO very large telescope, and then applied the many-multiplet technique with some parameter values obtained by CMA-ES. A similar approach was also used in Quast et al. (2005) for decomposition of quasar spectra into individual line profiles. A CMA-ES with 100 parents and 200 offspring was used and most of the parameters of the CMA-ES were set to their default initial values. The optimization runs are terminated after 100,000 evaluations of the objective function, which is a normalized residual sum of squares. The algorithm was examined with several test cases of synthesized data with noise and its performance was compared with other classical deterministic optimization methods. The results showed that unlike the other methods, CMA-ES can find the global optimum without any special initialization.

Similarly, in Mirchev et al. (2012) it was also shown how CMA-ES can outperform the classical optimization methods in finding a better optimum of the objective function. The problem there is in fitting the coupling coefficients of an interactive ensemble of imperfect models to a given set of "truth" data. As a case study the chaotic Lorenz 63 attractor was used, which has a behavior that is suggestive to that of the atmosphere. As a reference "truth" a Lorenz attractor with its typical values was used, while a set of three Lorenz attractors with perturbed parameter values was used to mimic imperfect models. The idea of this approach is to be able to combine multiple climate models developed at different institutions, which provide predictions with variable accuracy both in space and time. The CMA-ES was applied with its default values, and the self-adaptiveness of its internal strategy proved to work very well.

In Chwatal and Raidl (2007), an algorithm named Exoplanet Orbit Determination by Evolutionary Strategies (ExOD-ES) has been specially developed for finding extrasolar planets based on spectral observations of the central star's radial velocity. The algorithm applies the classical operations of recombination, mutation, and selection, but with several problem-specific modifications. The exploration of the search space is restricted to long-lasting systems. An intermediate recombination of the parameters of the internal strategy is done for all individuals, while the optimized parameters are recombined only in 10% of the individuals. The mutation operator takes into account the Hill stability criterion. The selection operation should choose the fittest individuals from each generation, but it is modified to slightly favor newly created solutions with larger mutations, which are at the beginning of the evolution path, in order to allow for new planets to be added more easily, by reserving special places for them in the next generation. The algorithm was successfully applied to the υ-Andromedae and the 55-Cancri systems using a population of 50 individuals and 5000 offspring candidates in each generation, and it converged over about 100 generations.

Genetic and Evolutionary Programming A combination of genetic and evolutionary programming was applied in Li et al. (2004) for characterizing radial brightness of elliptical galaxies using a dataset of 18 brightness profiles of elliptical galaxies from the Coma cluster. The authors first apply GP to find a functional form of the profiles and then use EP to fit the parameters of the obtained function to the observed data. The functional form is restricted to consist of the operations $\{+, -, *, /, \sin, \cos, \exp, \log\}$, some real constants from $[-10, 10]$, and a variable radius r along the major elliptical axis. The fitness function is a combination of the individual's fit to the observations and the length of the expression. The fit to the data is represented by the amount of hits, which is the number of points generated by the function that are within some small predefined tolerance range of 0.005, called "hits criterion." Obviously, shorter expressions should be preferred as typically they also generalize better. The GP is run with a population size of 6000 and cross-over and mutation rates of 0.9 and 0.01, respectively. A tournament selection is applied of size 6. The initial depth of the expressions is limited to 6, while later it is allowed to grow up to 17. The GP is terminated after a predefined maximal time of 6 hours has passed or a maximal number of 100 generations is reached. After finding an acceptable functional form an EP is applied for fitting the parameter values using a Cauchy mutation with a rate of 0.9, a population of 10,000, tournament selection of size 6, and a fitness function based solely on the amount of hits with a hits criterion of 0.005. The EP terminates after a fixed number of 150 generations.

Differential Evolution In Maulik and Saha (2009), the authors explore the applicability of DE to the problem of fuzzy clustering in remote sensing imagery, which was previously addressed using a GA in Maulik and Bandyopadhyay (2003). The authors develop a modified differential evolution (MoDE) algorithm based on the classical DE approach, and they particularly apply it to fuzzy clustering. Similarly as in Maulik and Bandyopadhyay (2003), the problem is encoded by a vector containing the n-dimensional coordinates of all cluster centers and a special fitness function is used for evaluating how good the clustering is. The modification is introduced by allowing two types of mutation process: one where the three parent vectors are randomly chosen (as in the classical DE), and a second one in which the difference is calculated between globally and locally best vectors and the result is added to a randomly chosen third vector. The modification should bring a faster convergence toward the global optimum. At each mutation a random decision is made about which mutation process is applied, but the probability is controlled by a parameter that favors the second type less and less as the generations pass. A classical cross-over is used between the mutation vector and the target vector to produce a trial vector that would replace the target vector if it has a better fitness. The proposed MoDE is shown to perform better than the classical DE, GA (Maulik and Bandyopadhyay, 2003), and several other methods for fuzzy clustering, using several test functions and generated and real data. For both DE and MoDE, a population of 20 individuals over 100 generations are used, with a cross-over rate of 0.8 and a mutation scaling factor of 0.8. The GA is run with 20 individuals over 100 generations with cross-over and mutation rates of 0.8 and 0.3, respectively. The MoDEFC is then successfully applied to three large datasets, two of the cities of Calcutta and Mumbai obtained with the Indian Remote Sensing Satellites, and another one of Calcutta obtained with the SPOT system.

The problem of scheduling the James Webb Space Telescope, which should be launched in 2020 after several delays, is addressed in Giuliano and Johnston (2008) using a multiobjective DE-based algorithm named generalized differential evolution 3 (GDE3) (Kukkonen and Lampinen, 2005). The scheduling is divided in long-term and short-term phases and the authors address the latter one. The problem objectives are minimizing the schedule gaps, the momentum accumulation, and the missed observation opportunities. The GDE3 algorithm uses the same mutation and cross-over steps as in the classical DE, which was developed for single-objective optimization, while the selection step is modified to be suitable for multiobjective optimization. The trial and target vectors are compared and if any of them dominates the other, it is selected for the next generation, otherwise both vectors are kept and the population is reduced using solutions sorting and crowding distance. Solutions sorting is done by placing the solutions in ranks, such that all solutions in the higher ranks are dominated by the solutions from the lower ranks, while the solutions at the same rank are nondominated among themselves. The crowding distance is used to differentiate between the mutually nondominated solutions by favoring solutions from noncrowded regions. At the end, a set of solutions are obtained forming the Pareto front. Several parameter settings and strategies are explored, such as minimizing all objectives at once, one by one, or in groups.

The classical DE approach was applied in Bazi et al. (2014) in the process of classification of hyperspectral images using a method called extreme machine learning (EML). The DE algorithm was used during the model selection step that is required by the EML method, and it was run with a population of 10 vectors over 100 function evaluations and a crossover rate and a mutation scaling factor of 0.9. The whole approach was successfully applied for land cover classification in several image datasets from Indiana, the Kennedy Space Center in Florida, Washington DC, and the University of Pavia in Italy.

In another interesting study (Olds et al., 2007), a classical DE was applied for designing interplanetary missions including complicated aspects such as parking orbit determination and multiple gravity assists. A thorough analysis was conducted for examining and tuning the parameters of the algorithm using a large number of trials, and it was found that using a population of 28 individuals with a cross-over rate of 0.8 and a random mutation scaling factor drawn from $[-1, 1]$ results in the best outcome for this problem. Several example missions were studied, like Cassini, Galileo, crewed Mars mission, and a theoretical sample-return mission to the Tempel 1 comet.

Ant Colony Optimization In Zhang et al. (2011), two types of ACO algorithms were examined for extracting end-members from hyperspectral images obtained by remote sensing. The problem consists of detecting pixels (end-members) capturing a single ground object, and then estimating the presence proportion of the end-members into the mixed pixels. The images were represented by a weighted directed graph that is used for selecting the end-member pixels by finding a path that goes through all of their corresponding vertices. The ants move through the graph and keep a Tabu table of vertices visited in the past. Two types of objective functions were considered, resulting in two different ACOs: one where the numbers of end-members and image bands are equal, and another one that allows a mismatch where several bands can be combined into a single end-member through screening. A population of 30 ants was used with a pheromone dissipation factor of 0.99. The algorithm terminates if a maximum number of 10,000 iterations have passed or if the same optimal path is reached after three consecutive iterations. The algorithm was successfully applied to some simulated data as well as to the AVIRIS dataset obtained from Cuprite, Nevada, USA.

Particle Swarm Optimization One interesting application of PSO was given in Ruiz et al. (2015), where the algorithm was used for estimating the free parameters of the SAG semianalytic model of galaxy formation. The model uses merger trees crated by simulation of Lambda cold dark matter for generating galaxy populations. Seven model parameters were left for estimation while the others were set to some given values. Observational statistics of the stellar mass and the masses of supermassive black holes in the galaxy center were used to form two data constraint relations. The PSO parameters were set to $\alpha \approx 0.72$ and $\beta = \gamma \approx 1.193$. Instead of positioning the particles randomly a maximin Latin hypercube technique was used, which places the particles more evenly distributed. The particle positions were constrained within a certain range, and the velocity was inverted when the boundary was reached. The velocities were constrained to some maximal values. A population of 30 particles was used over 150 iterations. A stopping convergence criterion was also defined based on the distance of the particles from the globally fittest one. The same problem was also solved using Monte Carlo Markov chains and the solutions are comparable, but PSO needs one order of magnitude less time to reach it.

In Shaw and Srivastava (2007), the general problem of inverting geophysical data was addressed, which basically is fitting model parameters given some observed set of data. Both GA and PSO were considered and compared to a ridge regression (RR). Both GA and PSO were run with a population of 300 over 100 generations/iterations. The GA was run with a single-point cross-over at a rate of 0.6 and a mutation rate of 0.02. The PSO was set with parameters $\beta = 2.8$ and $\gamma = 3.1$, while α was varied from 0.1 to 0.05 by a drop of 99% at each iteration. The algorithms were examined using synthetic data as well as real data obtained with vertical electric sounding and magnetotellurics. The quality of the solutions as well as the required computational time of GA and PSO are comparable. On the other hand, RR can find a similarly good solution very fast, but only if it is initialized from a position near the global optimum. In another work (Ali Ahmadi et al., 2013), a hybrid approach combining GA and PSO was used for training an ANN that predicts a reservoir permeability.

Artificial Immune System An interesting algorithm was developed in Zhong et al. (2006), which uses unsupervised learning based on an artificial immune system for classification in hyperspectral images obtained by remote sensing. Each image was represented as a feature vector consisting of all pixels and in the AIS approach it was considered as an antigen. The goal was to classify the image pixels into a given number of classes. The developed unsupervised artificial immune classifier (UAIC) starts with a random greedy initialization of antibody cells and corresponding memory cells. By an iterative process of selection, cloning, and mutation of the antibodies a better affinity is developed over time. At each iteration a number of antibodies are replaced at a given displace rate. In the mean time, the population of memory cells also evolves in order to represent the classes of antibodies appropriately. As a stopping criterion a threshold is used of the percentage of pixels changing class between two iterations, or if a maximal number of iterations is reached. The algorithm was tested using images from Wuhan and Xiaqiao, with four and seven classes, respectively. A set of 10 antibodies was used with a clonal rate of 15, a displace rate of 0.1, a termination threshold of 3%, and a maximal number of 20 iterations. The performance of the algorithm was shown to be better than that of other algorithms, like k-means, fuzzy k-means, ISODATA, and self-organized map.

15.7 DISCUSSION AND CONCLUSIONS

In this chapter, we gave an overview of various methods from evolutionary computation and provided some interesting examples of their applications to real-world problems in astronomy and geoscience, hoping to motivate the reader in employing them to the problems they are facing. Comparisons of the different algorithms can be made from many aspects, but typically the appropriateness of the method depends on the particular application and how the problem is formulated. Therefore, it is better to first explore the previous similar experiences for the specific problem, before deciding which algorithm is worth trying. All these methods are inspired from various evolutionary and other natural processes happening on Earth. On the other hand, in Rashedi et al. (2009), the authors found inspiration in collection of masses and their mutual gravitational interactions to formulate an interesting metaheuristic optimization method called gravitational search algorithm (GSA). There are many different versions and modifications of GSA, such as inclusion of black hole operators in Doraghinejad and Nezamabadi-pour (2014). Maybe this chapter could also motivate some of the readers to indulge in developing new metaheuristic algorithms by drawing inspiration from their fields of expertise.

Evolutionary computing brings plenty of benefits in facing optimization challenges. As was outlined in Fogel (1997), generally they are conceptually simple, have a wide range of applications, allow for external knowledge to be incorporated, can be combined with other methods like machine learning and classical optimization, can be self-adaptive to the particular problem and to dynamical changes, provide better results than the classical optimization methods in many real applications, and can be applied to complex unsolved problems. Moreover, they can be easily parallelized, which is of great importance due to the rapid developments of computer architectures, distributed computing, and GPUs in the last two decades. Among the different applications of parallel evolutionary algorithms, within the different distributed technologies, we can mention the work proposed in Nebro et al. (2008), where authors proposed a parallel GA for solving the DNA fragment assembly problem in a computational grid. The authors proved that their proposed distributed genetic approach is very promising in taking advantage of a grid system to solve large DNA fragment assembly problem instances. Also, in Alba and Luque (2006), the same real-world problem was considered but from a different perspective. In this work, authors tend to analyze the behavior of a parallel GA over different local area network (LAN) technologies. The aim of this study is to show the potential impact in the search mechanics when shifting between LANs and to show the actual power and utility of the proposed distributed GA to solve the DNA fragment assembly problem. Another work dealing with high-dimensional data is the work proposed in Chu and Zomaya (2006), where authors proposed a parallel ACO for three-dimensional protein structure prediction using the HP lattice model, and results show the performance of the distributed ACO approach in terms of accuracy and network scalability. Another optimization problem dealing with Big Data is the work proposed in Luque and Alba (2011b). In this study, authors focused on decision making associated with workforce planning. More precisely, a parallel GA for the workforce planning problem is proposed where two sets of decisions are considered, namely, selection and assignment. The first step of decisions consists in selecting a small number of employees from a large number of available workers, while the second decision consists in assigning this staff to the tasks to be performed. The main objective is to minimize the costs associated to the human resources needed to fulfill the work requirements. An effective workforce plan is an essential tool to identify appropriate workload staffing levels and justify budget allocations so that organizations can meet their objectives (Luque and Alba, 2011b). Another interesting application of parallel evolutionary algorithms is the well-known natural language processing problem of part-of-speech tagging. In Luque and Alba (2011a), different parallel metaheuristic approaches such as the parallel GA and the parallel simulated annealing were considered. The study highlights the high performances achieved by the parallel algorithms for complex tagging scenarios, and states the singular advantages for every technique. For further applications of various metaheuristics for global optimization of large-scale problems can be found in Mahdavi et al. (2015); Alba (2005); Alba et al. (1999); Gong et al. (2015). A detailed survey of the various recent distributed evolutionary algorithms and models can be found in Gong et al. (2015), including cloud-, Map/Reduce-, and GPU-based implementations.

Recently, artificial immune systems as alternative metaheuristics to evolutionary algorithms have been investigated to solve astro- and geoinformatics optimization problems. However, we noticed that their applications to these research areas is still limited. This might be explained by the recent emergence of this area of research. It is worth mentioning that artificial immune systems, as discussed in Section 15.4, are a very diverse area of research, ranging from the modeling of immune systems to the development of algorithms for specific applications, and hence they have much to offer for astro- and geoinformatics problem optimization. However, when it comes to Big Data, there are still some theoretical issues that need to be further explored, and among these we mention the development of unified frameworks, convergence, and scalability. Therefore, more works are emerging such as in Dagdia (2018a, 2018b) to better fit artificial immune system techniques to real-world applications, particularly when it comes to handling Big Data. This might also attract the attention of researchers to either adapt or develop new artificial immune system techniques, and fit them to their astro- and geoinformatics Big Data optimization problems.

ACKNOWLEDGMENT

This work is part of a project that has received funding from the European Union's Horizon 2020 research and innovation program under the Marie Skłodowska-Curie grant agreement No. 702527.

REFERENCES

Alba, E., 2005. Parallel Metaheuristics: A New Class of Algorithms, vol. 47. John Wiley & Sons.

Alba, E., Luque, G., 2006. Performance of distributed gas on DNA fragment assembly. In: Parallel Evolutionary Computations. Springer, pp. 97–115.

Alba, E., Tomassini, M., 2002. Parallelism and evolutionary algorithms. IEEE Transactions on Evolutionary Computation 6 (5), 443–462.

Alba, E., Troya, J.M., et al., 1999. A survey of parallel distributed genetic algorithms. Complexity 4 (4), 31–52.

Ali Ahmadi, M., Zendehboudi, S., Lohi, A., Elkamel, A., Chatzis, I., 2013. Reservoir permeability prediction by neural networks combined with hybrid genetic algorithm and particle swarm optimization. Geophysical Prospecting 61 (3), 582–598.

Arias-Montano, A., Coello, C.A.C., Mezura-Montes, E., 2012. Multiobjective evolutionary algorithms in aeronautical and aerospace engineering. IEEE Transactions on Evolutionary Computation 16 (5), 662–694.

Bazi, Y., Alajlan, N., Melgani, F., AlHichri, H., Malek, S., Yager, R.R., 2014. Differential evolution extreme learning machine for the classification of hyperspectral images. IEEE Geoscience and Remote Sensing Letters 11 (6), 1066–1070.

Bertsimas, D., Weismantel, R., 2005. Optimization Over Integers, vol. 13. Dynamic Ideas, Belmont.

Blickle, T., Thiele, L., 1996. A comparison of selection schemes used in evolutionary algorithms. Evolutionary Computation 4 (4), 361–394.

Bonabeau, E., Dorigo, M., Theraulaz, G., 1999. Swarm Intelligence: From Natural to Artificial Systems, 1st Ed. Oxford University Press.

Branke, J., Deb, K., Miettinen, K., 2008. Multiobjective Optimization: Interactive and Evolutionary Approaches, vol. 5252. Springer Science & Business Media.

Burnet, Sir Frank Macfarlane, et al., 1959. The Clonal Selection Theory of Acquired Immunity. Vanderbilt University Press, Nashville.

Cao, X., Qiao, H., Xu, Y., 2007. Negative selection based immune optimization. Advances in Engineering Software 38 (10), 649–656.

Chambers, L.D., 2000. The Practical Handbook of Genetic Algorithms: Applications. Chapman and Hall/CRC.

Chandrasekaran, M., Asokan, P., Kumanan, S., Balamurugan, T., Nickolas, S., 2006. Solving job shop scheduling problems using artificial immune system. The International Journal of Advanced Manufacturing Technology 31 (5–6), 580–593.

Charbonneau, P., 1995. Genetic algorithms in astronomy and astrophysics. The Astrophysical Journal. Supplement Series 101, 309.

Chu, D., Zomaya, A., 2006. Parallel ant colony optimization for 3D protein structure prediction using the HP lattice model. In: Parallel Evolutionary Computations. Springer, pp. 177–198.

Chwatal, A.M., Raidl, G.R., 2007. Determining orbital elements of extrasolar planets by evolution strategies. In: International Conference on Computer Aided Systems Theory. Springer, pp. 870–877.

Civicioglu, P., 2012. Transforming geocentric Cartesian coordinates to geodetic coordinates by using differential search algorithm. Computers & Geosciences 46, 229–247.

Clerc, M., 2010. Particle Swarm Optimization, vol. 93. John Wiley & Sons.

Coello, C.A.C., Cortés, N.C., 2005. Solving multiobjective optimization problems using an artificial immune system. Genetic Programming and Evolvable Machines 6 (2), 163–190.

Coello, C.A.C., Lamont, G.B., Van Veldhuizen, D.A., et al., 2007. Evolutionary Algorithms for Solving Multi-Objective Problems, vol. 5. Springer.

Cutello, V., Nicosia, G., 2002. An immunological approach to combinatorial optimization problems. In: Ibero-American Conference on Artificial Intelligence. Springer, pp. 361–370.

Dagdia, Z.C., 2018a. A distributed dendritic cell algorithm for big data. In: Proceedings of the Genetic and Evolutionary Computation Conference Companion, GECCO '18. ACM, New York, NY, USA, pp. 103–104. URL: http://doi.acm.org/10.1145/3205651.3205701.

Dagdia, Z.C., 2018b. A scalable and distributed dendritic cell algorithm for big data classification. Swarm and Evolutionary Computation 50, 100432. https://doi.org/10.1016/j.swevo.2018.08.009.

Darwin, C., 2004. On the Origin of Species, 1859. Routledge.

Das, S., Mullick, S.S., Suganthan, P.N., 2016. Recent advances in differential evolution—an updated survey. Swarm and Evolutionary Computation 27, 1–30.

Das, S., Suganthan, P.N., 2011. Differential evolution: a survey of the state-of-the-art. IEEE Transactions on Evolutionary Computation 15 (1), 4–31. https://doi.org/10.1109/TEVC.2010.2059031.

DasGupta, D., 1993. An overview of artificial immune systems and their applications. In: Artificial Immune Systems and Their Applications. Springer, pp. 3–21.

De Castro, L.N., Timmis, J., 2002. Artificial Immune Systems: A New Computational Intelligence Approach. Springer Science & Business Media.

De Jong, K., 2005. Genetic algorithms: a 30 year perspective. In: Perspectives on Adaptation in Natural and Artificial Systems, p. 11.

De Jong, K.A., Spears, W.M., 1990. An analysis of the interacting roles of population size and crossover in genetic algorithms. In: International Conference on Parallel Problem Solving From Nature. Springer, pp. 38–47.

Deb, K., 2001. Multi-Objective Optimization Using Evolutionary Algorithms, vol. 16. John Wiley & Sons.

Doraghinejad, M., Nezamabadi-pour, H., 2014. Black hole: a new operator for gravitational search algorithm. International Journal of Computational Intelligence Systems 7 (5), 809–826.

Dorigo, M., Di Caro, G., 1999. Ant colony optimization: a new meta-heuristic. In: Evolutionary Computation, 1999. CEC 99. Proceedings of the 1999 Congress on, vol. 2, pp. 1470–1477.

Dorigo, M., Stützle, T., 2003. The ant colony optimization metaheuristic: algorithms, applications, and advances. In: Handbook of Metaheuristics. Springer, pp. 250–285.

Durham, W.H., 1991. Coevolution: Genes, Culture, and Human Diversity. Stanford University Press.

D'Ambrosio, D., Spataro, W., Iovine, G., 2006. Parallel genetic algorithms for optimising cellular automata models of natural complex phenomena: an application to debris flows. Computers & Geosciences 32 (7), 861–875.

Eberbach, E., 2005. Toward a theory of evolutionary computation. BioSystems 82 (1), 1–19.

Eiben, A.E., Smith, J.E., et al., 2003. Introduction to Evolutionary Computing, vol. 53. Springer.

Fogel, D.B., 1997. The advantages of evolutionary computation. In: BCEC, pp. 1–11.

Fogel, D.B., 1998. Artificial Intelligence through Simulated Evolution. Wiley-IEEE Press.

Gao, X.-Z., Ovaska, S.J., Wang, X., 2008. A GA-based negative selection algorithm. International Journal of Innovative Computing, Information & Control 4 (4), 971–979.

Giuliano, M.E., Johnston, M.D., 2008. Multi-objective evolutionary algorithms for scheduling the James Webb space telescope. In: ICAPS, pp. 107–115.

Goldberg, D.E., Deb, K., 1991. A Comparative Analysis of Selection Schemes Used in Genetic Algorithms. Foundations of Genetic Algorithms, vol. 1. Elsevier, pp. 69–93.

Gong, Y.-J., Chen, W.-N., Zhan, Z.-H., Zhang, J., Li, Y., Zhang, Q., Li, J.-J., 2015. Distributed evolutionary algorithms and their models: a survey of the state-of-the-art. Applied Soft Computing 34, 286–300.

Gonzalez, A., Riba, J.-R., Esteban, B., Rius, A., 2018. Environmental and cost optimal design of a biomass–Wind–PV electricity generation system. Renewable Energy 126, 420–430.

Goodwin, B.C., 1982. Development and evolution. Journal of Theoretical Biology 97 (1), 43–55.

Hajela, P., Yoo, J.S., 1999. Immune network modelling in design optimization. In: New Ideas in Optimization. McGraw-Hill Ltd., UK, pp. 203–216.

Hansen, N., Kern, S., 2004. Evaluating the CMA evolution strategy on multimodal test functions. In: International Conference on Parallel Problem Solving From Nature. Springer, pp. 282–291.

Hansen, N., Ostermeier, A., 2001. Completely derandomized self-adaptation in evolution strategies. Evolutionary Computation 9 (2), 159–195.

Holland, J.H., 1992. Adaptation in Natural and Artificial Systems: An Introductory Analysis With Applications to Biology, Control, and Artificial Intelligence. MIT Press.

Holland, J., Goldberg, D., 1989. Genetic Algorithms in Search, Optimization and Machine Learning. Addison-Wesley, Massachusetts.

Jain, A., Srinivasulu, S., 2004. Development of effective and efficient rainfall-runoff models using integration of deterministic, real-coded genetic algorithms and artificial neural network techniques. Water Resources Research 40 (4).

Janeway Jr., C.A., 1992. The immune system evolved to discriminate infectious nonself from noninfectious self. Immunology Today 13 (1), 11–16.

Kennedy, J., Eberhart, R., 1995. Particle swarm optimization. In: Neural Networks, 1995. Proceedings, IEEE International Conference on, vol. 4, pp. 1942–1948.

Klockgether, J., Schwefel, H.-P., 1970. Two-phase nozzle and hollow core jet experiments. In: Proc. 11th Symp. Engineering Aspects of Magnetohydrodynamics. California Institute of Technology, Pasadena, CA, pp. 141–148.

Koza, J.R., 1992. Genetic Programming II, Automatic Discovery of Reusable Subprograms. MIT Press, Cambridge, MA.

Kukkonen, S., Lampinen, J., 2005. GDE3: the third evolution step of generalized differential evolution. In: Evolutionary Computation, 2005. The 2005 IEEE Congress on, vol. 1. IEEE, pp. 443–450.

Li, J., Yao, X., Frayn, C., Khosroshahi, H.G., Raychaudhury, S., 2004. An evolutionary approach to modeling radial brightness distributions in elliptical galaxies. In: International Conference on Parallel Problem Solving From Nature. Springer, pp. 591–601.

Luque, G., Alba, E., 2011a. Natural language tagging with parallel genetic algorithms. In: Parallel Genetic Algorithms. Springer, pp. 75–89.

Luque, G., Alba, E., 2011b. Parallel genetic algorithm for the workforce planning problem. In: Parallel Genetic Algorithms. Springer, pp. 115–134.

Mahdavi, S., Shiri, M.E., Rahnamayan, S., 2015. Metaheuristics in large-scale global continues optimization: a survey. Information Sciences 295, 407–428.

Matzinger, P., 2001. Essay 1: the danger model in its historical context. Scandinavian Journal of Immunology 54 (1–2), 4–9.

Maulik, U., Bandyopadhyay, S., 2003. Fuzzy partitioning using a real-coded variable-length genetic algorithm for pixel classification. IEEE Transactions on Geoscience and Remote Sensing 41 (5), 1075–1081.

Maulik, U., Saha, I., 2009. Modified differential evolution based fuzzy clustering for pixel classification in remote sensing imagery. Pattern Recognition 42 (9), 2135–2149.

Metcalfe, T.S., Charbonneau, P., 2003. Stellar structure modeling using a parallel genetic algorithm for objective global optimization. Journal of Computational Physics 185 (1), 176–193.

Meyer-Nieberg, S., Beyer, H.-G., 2007. Self-adaptation in evolutionary algorithms. In: Parameter Setting in Evolutionary Algorithms. Springer, pp. 47–75.

Mirchev, M., Duane, G.S., Tang, W.K., Kocarev, L., 2012. Improved modeling by coupling imperfect models. Communications in Nonlinear Science and Numerical Simulation 17 (7), 2741–2751.

Mühlenbein, H., Schlierkamp-Voosen, D., 1993. Predictive models for the breeder genetic algorithm I. Continuous parameter optimization. Evolutionary Computation 1 (1), 25–49.

Myerson, R.B., 2013. Game Theory. Harvard University Press.

Nebro, A.J., Luque, G., Luna, F., Alba, E., 2008. DNA fragment assembly using a grid-based genetic algorithm. Computers & Operations Research 35 (9), 2776–2790.

Olds, A.D., Kluever, C.A., Cupples, M.L., 2007. Interplanetary mission design using differential evolution. Journal of Spacecraft and Rockets 44 (5), 1060–1070.

Orfila, A., Ballester, J., Oliver, R., Alvarez, A., Tintoré, J., 2002. Forecasting the solar cycle with genetic algorithms. Astronomy & Astrophysics 386 (1), 313–318.

Papadimitriou, C.H., Steiglitz, K., 1998. Combinatorial Optimization: Algorithms and Complexity. Courier Corporation.

Paton, F., Maier, H., Dandy, G., 2014. Including adaptation and mitigation responses to climate change in a multiobjective evolutionary algorithm framework for urban water supply systems incorporating GHG emissions. Water Resources Research 50 (8), 6285–6304.

Pochet, Y., Wolsey, L.A., 2006. Production Planning by Mixed Integer Programming. Springer Science & Business Media.

Price, K., Storn, R.M., Lampinen, J.A., 2006. Differential Evolution: A Practical Approach to Global Optimization. Springer Science & Business Media.

Quast, R., Baade, R., Reimers, D., 2005. Evolution strategies applied to the problem of line profile decomposition in QSO spectra. Astronomy & Astrophysics 431 (3), 1167–1175.

Quast, R., Reimers, D., Levshakov, S.A., 2004. Probing the variability of the fine-structure constant with the VLT/UVES. Astronomy & Astrophysics 415 (2), L7–L11.

Rashedi, E., Nezamabadi-Pour, H., Saryazdi, S., 2009. GSA: a gravitational search algorithm. Information Sciences 179 (13), 2232–2248.

Rechenberg, I., 1981. Evolutionsstrategie: Optimierung technischer Systeme nach Prinzipien der biologischen Evolution. John Wiley, New York.

Ruiz, A.N., Cora, S.A., Padilla, N.D., Domínguez, M.J., Vega-Martínez, C.A., Tecce, T.E., Orsi, Á., Yaryura, Y., Lambas, D.G., Gargiulo, I.D., et al., 2015. Calibration of semi-analytic models of galaxy formation using particle swarm optimization. The Astrophysical Journal 801 (2), 139.

Schrijver, A., 2003. Combinatorial Optimization: Polyhedra and Efficiency, vol. 24. Springer Science & Business Media.

Shaw, R., Srivastava, S., 2007. Particle swarm optimization: a new tool to invert geophysical data. Geophysics 72 (2), F75–F83.

Storn, R., Price, K., 1997. Differential evolution—a simple and efficient heuristic for global optimization over continuous spaces. Journal of Global Optimization 11 (4), 341–359.

Talbi, E.-G., 2009. Metaheuristics: From Design to Implementation, vol. 74. John Wiley & Sons.

Ulutas, B.H., Kulturel-Konak, S., 2011. A review of clonal selection algorithm and its applications. Artificial Intelligence Review 36 (2), 117–138.

Vachier, F., Berthier, J., Marchis, F., 2012. Determination of binary asteroid orbits with a genetic-based algorithm. Astronomy & Astrophysics 543, A68.

Voigt, H.-M., Anheyer, T., 1994. Modal mutations in evolutionary algorithms. In: Evolutionary Computation, 1994. IEEE World Congress on Computational Intelligence, Proceedings of the First IEEE Conference on. IEEE, pp. 88–92.

Wahde, M., 1998. Determination of orbital parameters of interacting galaxies using a genetic algorithm-description of the method and application to artificial data. Astronomy & Astrophysics. Supplement Series 132 (3), 417–429.

Wahde, M., Donner, K., 2001. Determination of the orbital parameters of the M 51 system using a genetic algorithm. Astronomy & Astrophysics 379 (1), 115–124.

Wang, L., Cholette, M.E., Zhou, Y., Yuan, J., Tan, A.C., Gu, Y., 2018a. Effectiveness of optimized control strategy and different hub height turbines on a real wind farm optimization. Renewable Energy 126, 819–829.

Wang, X., Veeravalli, B., Rana, O.F., 2018b. An optimal task-scheduling strategy for large-scale astronomical workloads using in-transit computation model. International Journal of Computational Intelligence Systems 11 (1), 600–607.

Wu, W., Simpson, A.R., Maier, H.R., 2009. Accounting for greenhouse gas emissions in multiobjective genetic algorithm optimization of water distribution systems. Journal of Water Resources Planning and Management 136 (2), 146–155.

Yan, S., Minsker, B., 2006. Optimal groundwater remediation design using an adaptive neural network genetic algorithm. Water Resources Research 42 (5).

Zhang, B., Sun, X., Gao, L., Yang, L., 2011. Endmember extraction of hyperspectral remote sensing images based on the ant colony optimization (ACO) algorithm. IEEE Transactions on Geoscience and Remote Sensing 49 (7), 2635–2646.

Zhong, Y., Zhang, L., Huang, B., Li, P., 2006. An unsupervised artificial immune classifier for multi/hyperspectral remote sensing imagery. IEEE Transactions on Geoscience and Remote Sensing 44 (2), 420–431.

CHAPTER 16

Multiwavelength Extragalactic Surveys: Examples of Data Mining

IRINA VAVILOVA, DR • DARIA DOBRYCHEVA, DR • MAKSYM VASYLENKO, MSC • ANDRII ELYIV, DR • OLGA MELNYK, DR

16.1 INTRODUCTION

Multiwavelength astronomy provides a huge amount of data from different sources over the whole electromagnetic spectrum. Recently it has been supplemented with data from neutrino astronomy and gravitational wave astronomy (multimessenger astronomy). Since the end of the 20th century the all-sky and large-area observational surveys as well as their catalogued databases enriched and continue to enrich our knowledge of the Universe. Astronomy has entered the Big Data era, when these data are combined/compiled into numerous archives. Each archive contains the observational data that were obtained in a specific spectral range, for which ground-based or space-born telescopes were developed (see, for example, Chapter 5 in this book and commentaries by Brunner et al. (2002) on the nature of astronomical data). However, modern astrophysics forces to study astrophysical objects across the whole electromagnetic spectrum, because different physical processes make themselves felt at different wavelengths (different substructures of celestial bodies radiate at different wavelengths). An identification of such substructures and processes requires a very wide knowledge from high-energy astrophysics to the radio decameter astronomy, wherein X-ray, gamma ray, and radio sources need to be identified with their optical counterparts.

16.2 THE AUTOMATED MORPHOLOGICAL CLASSIFICATION FOR THE SDSS GALAXIES

Since 2000, the Sloan Digital Sky Survey (SDSS) (York et al., 2000) has collected the most data that have been amassed in the entire history of astronomy.[1]

[1] See Chapter 5 in this book.

Now, its archive contains about 170 terabytes of information, with most of these data about galaxies. Astronomers who are directly involved in the SDSS identified the problem of morphological galaxy classification "as one of the most cumbersome areas in celestial classification, and the one that has proven the most difficult to automate" (Kasivajhula et al., 2007).

Sense of galaxy morphological classification. Such substructures of galaxies as the central region with active nucleus and supermassive black hole, bulge and bar, spiral arms, halo with a dark matter component, star formation regions, intergalaxy medium, jets, disk, rings, and others features compose the important building blocks, which are related to galaxy morphology, dynamics (mass distribution) and kinematics, and have a decisive role in our understanding of galaxy formation and evolution. Galaxy morphological classification on large-scale datasets allows us to reduce classification errors and to improve statistics of the known morphological types of galaxies at different redshifts. A good introduction to the classification algorithms for astronomical tasks, including the machine learning methods for galaxy morphological classification, one can find in works by Buta and McCall (1999), de la Calleja and Fuentes (2004), Feigelson and Babu (2006), Ball and Brunner (2010), Ivezić et al. (2014) as well as in Chapters 12 and 13 of this book.

To imagine better this interesting scientific problem, we remind in Table 16.1 the numerical morphological Hubble stage, which was introduced by G. de Vaucouleurs in 1959 and improved by him later in "Global Physical Parameters of Galaxies" in 1994. Various morphological types of galaxies are illustrated in Fig. 16.1, where images in optical range were taken from the SDSS. It is important to underline that morphological type (T) of a galaxy correlates with stellar population

Knowledge Discovery in Big Data from Astronomy and Earth Observation. https://doi.org/10.1016/B978-0-12-819154-5.00028-X

TABLE 16.1
Numerical Hubble stage.

Hubble stage T	−6	−5	−4	−3	−2	−1	0	1	2
de Vaucouleurs class	cE	E	E^+	$S0^-$	$S0^0$	$S0^+$	$S0/a$	Sa	Sab
Hubble class		E			$S0$		$S0/a$	Sa	$Sa-b$

Hubble stage T	3	4	5	6	7	8	9	10
de Vaucouleurs class	Sb	Sbc	Sc	Scd	Sd	Sdm	Sm	Im
Hubble class	Sb	$Sb-c$		Sc		$IrrI$	$IrrI$	

and star formation history, bulge/disk luminosity ratio, mass concentration, interstellar media (chemical abundance), and nuclear activity properties.

A very good pedagogical review with discussion of the major methods in which galaxies are studied morphologically and structurally is given by Conselice (2014) that "includes the well-established visual method for morphology; Sercic fitting to measure galaxy sizes and surface brightness profile shapes; non-parametric structural methods including the concentration (C), asymmetry (A), clumpiness (S) (CAS method), the Gini/M_{20} parameters, as well as newer structural indices." We remind that concentration C index (the logarithmic ratio of the radii containing 90% and 50% of the light, R_{90}/R_{50}) is tightly related to morphology, A (asymmetry) to merging, and S to star formation rate (CAS parameters).

Visual and automated classification. Notwithstanding, below we would like to underline briefly several works, where different approaches were developed and great efforts were made to identify the morphological types of galaxies, first of all, from the SDSS, in the visual and/or the automated modes. We note that many machine learning methods were actively involved to disentangle this problem.

During the 1990s, the artificial neural network (ANN) algorithm was widely used for automatic morphological classification of galaxies since the very large extragalactic datasets have been constructed. The classification accuracy (or the success rate) of the ANN ranged from 65% to 90%, depending on the mathematical subtleties of the applied methods and the quality of galaxy samples. One of the first such works was made by Storrie-Lombardi et al. (1992) with a feedforward neural network and dealt with classification of 5217 galaxies onto five classes (E, SO, Sa-Sb, Sc-Sd, and Irr) with a 64% accuracy. A detailed comparison of human and neural classifiers was presented by Naim et al. (1995), who used principal component analysis to test various architectures to classify 831 galaxies: the best result was obtained with an r.m.s. deviation of 1.8 T-types. Summarizing the first attempts, Lahav et al. (1996) concluded that "the ANNs can replicate the classification by a human expert almost to the same degree of agreement as that between two human experts, to within 2 T-type units" (see Table 16.1).

Later de la Calleja and Fuentes (2004) developed a method which combines two machine learning algorithms: Locally Weighted Regression and ANN. They tested it with 310 images of galaxies from the NGC catalogue and obtained an accuracy of 95.11% and 90.36%, respectively. Ball et al. (2004) using the supervised ANN derived that it may be applied without human intervention for the SDSS galaxies (correlations between predicted and actual properties were around 0.9 with r.m.s. errors of order 10%). Andrae et al. (2010) classified the SDSS bright galaxies with a probabilistic classification algorithm and obtained that it produces reasonable morphological classes and object-to-class assignments without any prior assumptions.

As for the visual morphological classification conducted during the last years, we note a very powerful study by Banerji et al. (2010), where galaxies classified by the Galaxy Zoo Project into three classes (early types, spirals, "spam" objects) have formed a training sample for morphological classifications of galaxies in the SDSS DR6 (http://data.galaxyzoo.org). These authors convincingly showed that using a set of certain galaxy parameters, the neural network is able to reproduce the human classifications to better than 90% for all these classes and that the Galaxy Zoo catalogue (GZ1) can serve as a training sample.

Totally, hundreds of thousands of volunteers were involved into the Galaxy Zoo project to achieve visual classification of a million galaxies in the SDSS. Most of their results have found good scientific application. For

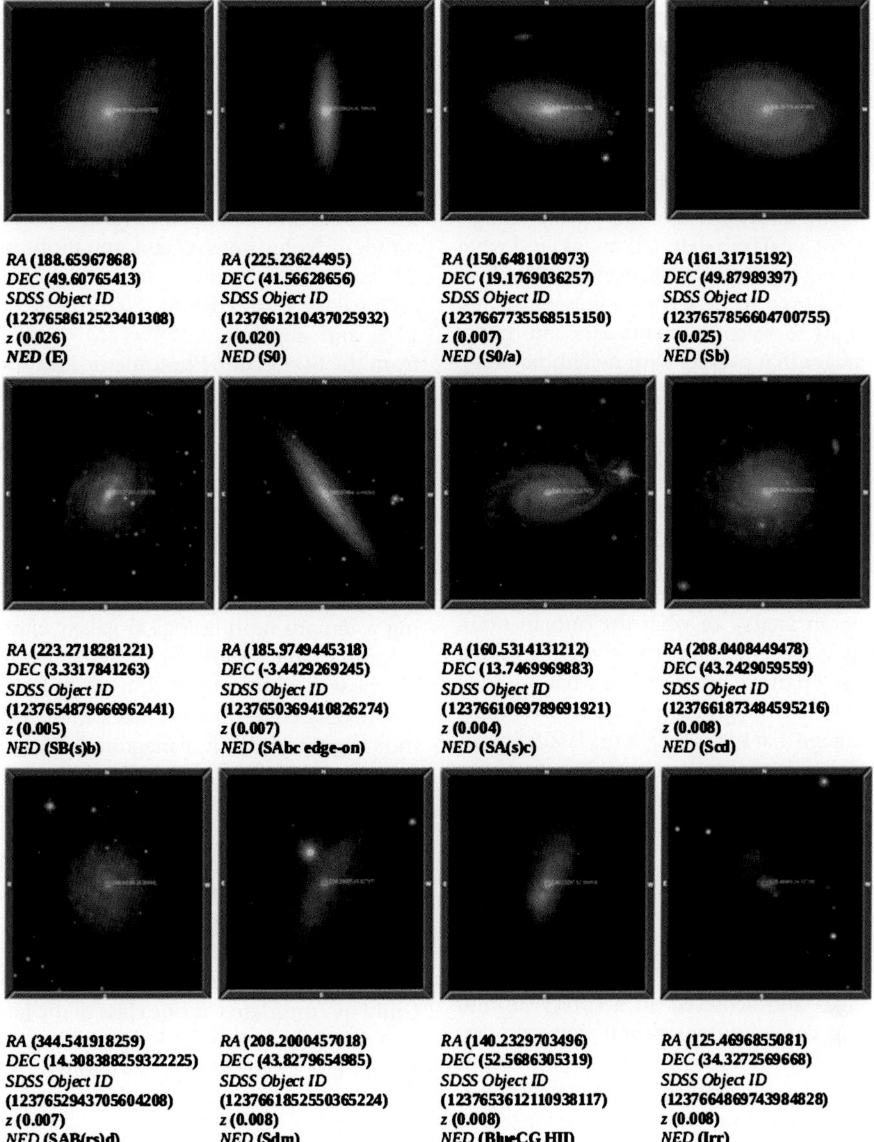

FIG. 16.1 Galaxies of different morphological types from the SDSS sample.

example, using the raw imaging data from the SDSS that was available in the GZ1, and the hand-picked features from the SDSS, Kates-Harbeck (2012) applied a logistic regression classifier and attained 95.21% classification accuracy. Willett et al. (2013) issued a new catalogue of morphological types from the Galaxy Zoo Project (GZ2) in synergy with the SDSS DR7, which contains more than 16 million morphological classifications of 304,122 galaxies and their finer morphological features (bars, bulges, and the shapes of edge-on disks, as well as parameters of the relative strengths of galactic bulges and spiral arms). Another approach was developed by Nair and Abraham (2010), who prepared the detailed visual classifications for 14,034 galaxies in the SDSS DR4 at $z < 0.1$, which can be used as a good training sample for calibrating the automated galaxy classification algorithms. A morphology catalogue of the SDSS galaxies was generated with the Wndchrm image analysis utility and nearest neighbor classifier by Kuminski and Shamir (2016). These authors pointed out that about 900,000 of the instances classified as spiral galaxies and about 600,000 of those classified as elliptical galaxies have a statistical agreement rate of about 98% with the Galaxy Zoo classification.

"One approach to automated classification is to ask what set of analytic or empirical components (bulge, disk) best represent a galaxy's detected image, and what the expected errors (say in the χ^2 sense) are. The limitation here is that even in perfectly ordinary galaxies, the fitted forms for these components vary, and many galaxies have images that overlap with neighbors or are dotted with brilliant star-forming regions. A quite different approach is taken by neural-network schemes. Here, one defines a set of input values based on the galaxy image, and trains the code using a large set of galaxies classified by eyeball (usually by several sets of eyeballs for a consistency check). The code then finds the set of hidden connections needed to give these outputs, and can apply this mapping to any further data desired. This is thought to be an analog of what the human brain does in learning to recognize patterns, though working backwards, it is not particularly clear just what the code is responding to in the image, except that it looks most like the typical image that it was taught to classify in this way. Neural net classifiers seem to be statistically about as good as human ones, which is especially impressive if one considers that people may fold in all sorts of outside knowledge as to redshifts and pass bands in their estimates" (Keel, 2007).

Recently Murrugarra and Hirata (2017) evaluated the convolutional neural network to classify galaxies from the SDSS onto two classes as ellipticals/spirals using their images and achieved an accuracy around 90%–91%. Using convolutional neural networks, especially the inception method, Wahaono and Azhari (2018) conducted classification into three general categories: ellipticals, spirals, and irregulars. They used 710 images (206 E, 320 Sp, 184 Irr) and obtained that images which underwent image processing showed a rather poor testing accuracy compared to not using any form of image processing. Their best testing accuracy was 78.3%. Both supervised and unsupervised methods were applied by Jain et al. (2016) to study the Galaxy Zoo dataset of 61,578 preclassified galaxies (spiral, elliptical, round, disk). They found that the variation of galaxy images is correlated with brightness and eccentricity, the random forest method gives a best accuracy (67%), meanwhile its combination with regression to predict the probabilities of galaxies associated with each class allows to reach a 94% accuracy.

Examples on binary and ternary morphological classification of the SDSS galaxies. Let us use the well-known fact that galaxy morphological type is correlated with the color indices, luminosity, de Vaucouleurs radius, inverse concentration index (R_{50}/R_{90}), etc. For example, let us combine visual classification and the two-dimensional diagrams of color indices g–i and one of the aforementioned parameters as "color-absolute magnitude," "color-inverse concentration index," "color-de Vaucouleurs radius," and "color-scale radius" for each galaxy with redshifts $0.02 < z < 0.06$, visual $m_r < 17.7$, and absolute $-24^m < M_r < -17^m$ magnitudes from the SDSS DR9. Photometric and spectral parameters of each object as well as their images are available through the SDSS web site. As a result, we can discover possible criteria for separating the galaxies into three classes (Melnyk et al., 2012), i.e., early types (E) – elliptical and lenticular; spiral (S) – Sa – Scd types; late spiral (LS) – Sd – Sdm types and irregular Im/BCG galaxies. One can see in Fig. 16.2 that the "color indices vs inverse concentration indexes" diagrams allow making a ternary morphological galaxy classification with a good accuracy (98% for E, 88% for S, and 57% for LS classes). The combinations of (1) color indices g–i and inverse concentration index $R50/R90$ and (2) color indices g–i and absolute magnitude M_r gives the best result: 143,263 E class, 112,578 S class, 61,177 LS class for the sample of the SDSS galaxies at $z < 0.1$ (Dobrycheva et al., 2017).

We can apply different machine learning methods[2] for providing a binary automated morphological classification for the same sample of the SDSS DR9 galaxies as in the above case of photometric diagrams. Why is it binary one? Because (1) S and LS classes of galaxies could be considered as one class of the late type galaxies L at the Hubble stage and (2) an accuracy for classification of late spirals LS was low enough (57%).

The first step is to prepare a training galaxy sample based on the SDSS DR9 and selected randomly with different redshift and luminosity from the total sample for the following visual classification. The second step is a training of the classifier. With this aim, we can use the absolute magnitudes M_u, M_g, M_r, M_i, and M_z, all the kinds of color indices M_u–M_r, M_g–M_i, M_u–M_g, M_r–M_z, and inverse concentration indexes $R50/R90$ to the center in each photometric band.

Using our own code in Scikit Learn Python[3] to predict correctly the galaxy morphology (late and early types) we verified several machine learning methods

[2]See, for example, Chapter 12 in this book.
[3]https://scikit-learn.org/.

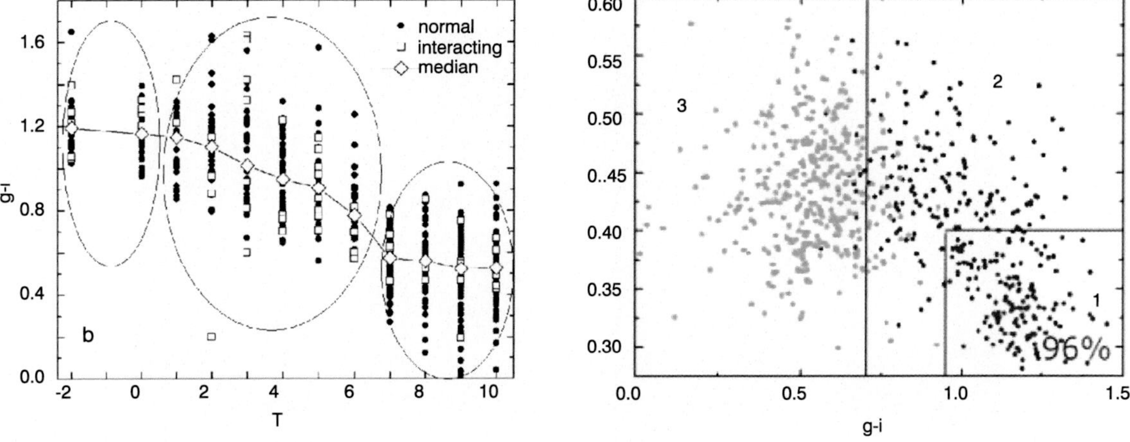

FIG. 16.2 (Left) The dependence of the morphological types T on the color indices *g–i* for 730 galaxies from the SDSS DR5. (Right) The inverse concentration index R50/R90 as functions of color indices for these galaxies; the red circles correspond to early types (−2 to 0), the blue circles to spirals (1–6), and the green circles to late type spiral and irregular galaxies (see, also, Table 16.1). The lines define regions into which a maximum (more than 90%) number of galaxies of morphological types (−2 to 0), (1 to 6), and (7 to 10), respectively, fall (or with a minimum number of the missclassified morphological types.

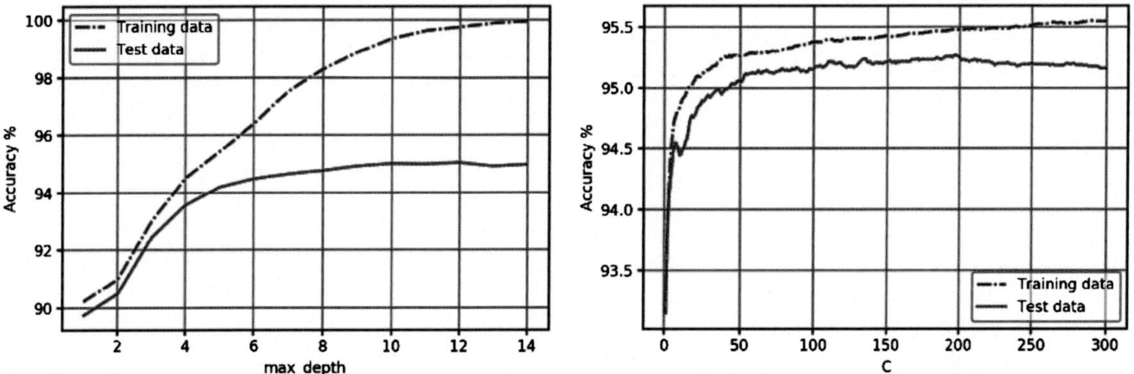

FIG. 16.3 Dependence of prediction accuracy for different machine learning methods of the automated morphological classification with training SDSS galaxy sample: (left) for the random forest classifier on the parameter "max depth"; (right) for the support vector machine classifier on the "C" parameter.

for binary morphological classification of the SDSS galaxies. With this aim we used the sample of 60,561 galaxies from the SDSS DR9 survey with a redshift of $0.02 < z < 0.06$ and absolute magnitudes of $-24^m < M_r < -19.4^m$. Among the machine learning methods were the following: naive Bayes, random forest, support vector machines, logistic regression, and the *k*-nearest neighbor algorithm. Prediction accuracy was evaluated for each of these methods for the training galaxy sample and reached the following values (all the above-mentioned classifiers include the k-fold cross-validation method):

naive Bayes classifier: 0.89 (E – 0.92, L – 0.82) $\pm$ 0.01;

k-nearest neighbors classifier: 0.945 (E – 0.9389, L – 0.958) $\pm$ 0.006;

logistic regression classifier: 0.949 (E – 0.968, L – 0.911) $\pm$ 0.006;

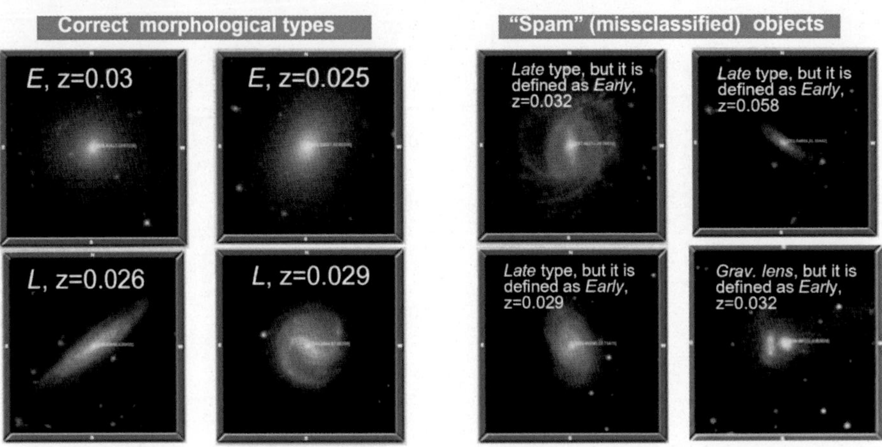

FIG. 16.4 Images of SDSS galaxies. Left: With correctly classified morphology. Top: Early type. Bottom: Late type. Right: with the misclassified morphology. Top and left bottom: Late types, which are classified as early types. Right bottom: gravitational lens classified as early type galaxy.

random forest classifier: 0.955 (E – 0.967, L – 0.928) $\pm$ 0.003;

support vector machine classifier: 0.964 (E – 0.961, L – 0.969) $\pm$ 0.006.

It turned out (Fig. 16.3) that the methods of random forest and support vector machine provide the highest accuracy (Vasylenko et al., 2019). Examples of images of galaxies with a correct classification on the early and late types are given in Fig. 16.4 (left panel).

The problem points arise when we have cases of the face-on and edge-on galaxies (Fig. 16.4, right panel). Most of these galaxies are misclassified as elliptical galaxies (early type). The good thing is that this approach allow us to recover gravitational lenses (point-like sources, arcs) and most of such misclassifications are also among elliptical galaxies. So, we have overestimated the number of elliptical and underestimated the number of spiral galaxies (about 10%).

But this problem can be solved when we form training samples through several steps (pretraining, fine-tuning, and classification). The step of fine-tuning should include the limitations on the axes-ratio for elliptical galaxies and additional photometry parameters for the face-on galaxies, as well as trainings with images and spectral features of galaxies which requires a specific algorithm with deep learning methods.

The distribution of the SDSS galaxies at $0.02 < z < 0.1$ with the automated morphological binary classification (early and late types) is given in Fig. 16.5.

Last remarks. The machine learning methods are indispensable assistants in solving morphological classification since their first application to tackle this problem with the ANN algorithm (Storrie-Lombardi et al., 1992). They are also effective for reconstruction of the Zone of Avoidance, distance modulus for local galaxies, gravitational lenses search, and other important tasks. The race in accuracy of machine learning methods leads to the search for the most effective among them and to the selection of the most reliable galaxy parameters (photometry, spectra, images), which can be used to determine galaxy morphology. Note that the diversity of the morphological types (Hubble stage, optical range), which we discussed in this subsection, is observed at redshifts $z < 3$; at larger redshifts the other approaches and algorithms should be applied. Also, another frontier in classification problems is approaching when we consider galaxies in the ultraviolet, infrared, or radio ranges (see, for example, Buta et al., 2010; Bell and Salim, 2011; Banfield et al., 2015; Smith and Donohoe, 2019), when their parameters and images should be complemented or cross-matched with optical counterparts.

16.3 ZONE OF AVOIDANCE OF THE MILKY WAY

The data incompleteness in dependence on the wavelength at which galaxies are sampled says that there are important problems in the sky area obscured by our galaxy. This sky area is the so-called Zone of Avoidance (ZoA) of the Milky Way (see, for example, Fig. 16.5, where several large-scale structures as the SDSS Great Wall, SDSS voids (Mao et al., 2016), CfA2 Great Wall, Great Attractor, and the Zone of Avoidance are pointed

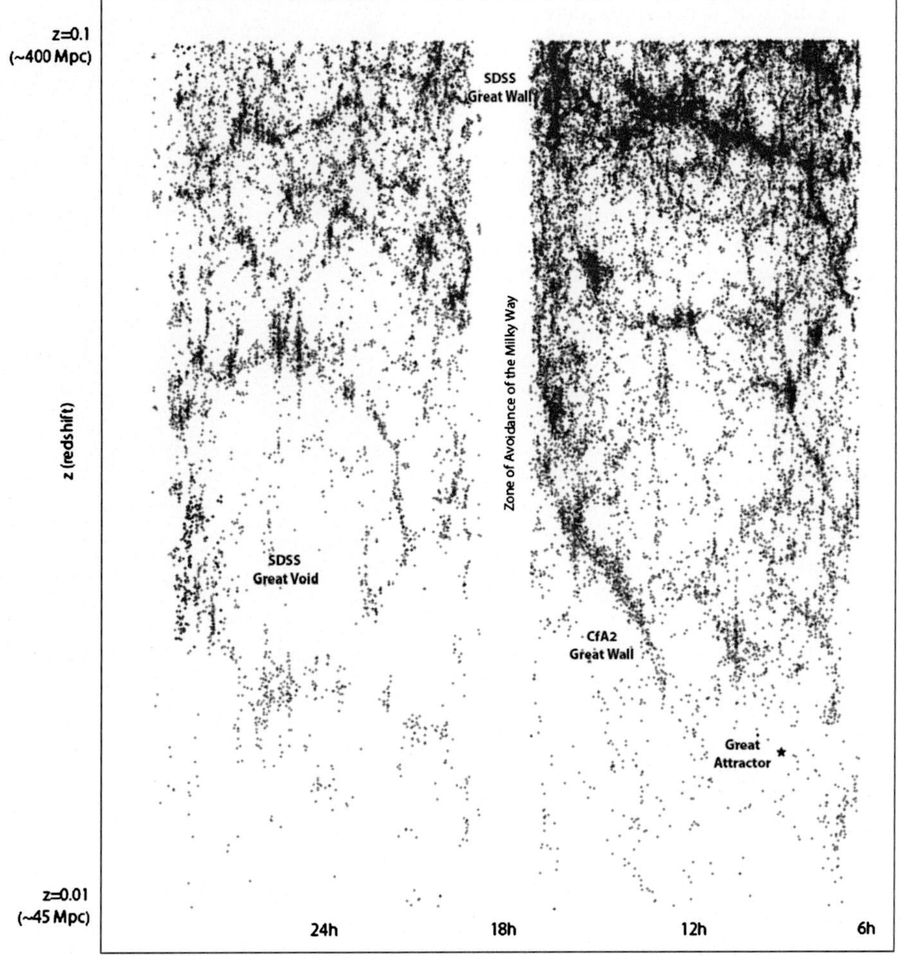

FIG. 16.5 Distribution of 48,651 galaxies from the SDSS DR9 at $0.02 < z < 0.1$ ($\delta = \pm 5°$) with determined morphological classification. Red dots, elliptical galaxies and lenticulars ($E - S0a$, N = 24916); blue dots, spirals, late spirals, and irregulars ($Sa - Irr$, N = 23735). Several large-scale structures (SDSS Great Wall, SDSS voids, CfA2 Great Wall, Great Attractor) and the Zone of Avoidance of the Milky Way are pointed out. At distances of more than 200 Mpc, the Universe becomes "gradually" homogeneous and isotropic.

out). "Why is it of interest to know the galaxy distribution behind the Milky Way, and why is it not sufficient to study galaxies and their large-scale distribution away from the foreground "pollution" of the Milky Way? To understand the dynamics in the nearby Universe and answer the question whether the dipole in the Cosmic Microwave Background (CMB) and other velocity flow fields (e.g. towards the Great Attractor) can be fully explained by the clumpy galaxy/mass distribution, whole-sky coverage is essential" (cited by Kraan-Korteweg and Lahav, 2000).

Brief history. The English astronomer Proctor (1878) firstly noted the Zone of Avoidance of the Milky Way as a Zone of Few Nebulae. Later, in papers by Stratonoff (1900), Easton (1904), Sanford (1917), Charlier (1922) and other authors, who used mostly isopleths as a cosmographic method (contour maps "number of galaxies per the sky area"), the presence of this zone in the distribution of galaxies became obvious. A first definition of the Zone of Avoidance was proposed by Shapley (1961) as the region delimited by "the isopleth of five galaxies per square degree from the Lick and Harvard surveys" (compared to a mean of 54 galaxies/square degree found in unobscured regions by Shane and Wirtanen, 1967).

Due to the incomplete sampling in the area of absorption, on the basis of which the velocity field is constructed, we cannot determine its homogeneity, which gives an error in the definite direction of motion of our Galaxy by this method. We can assume that there are a significant number of galaxies in this zone (Fig. 16.5) based on discrepancy between the vectors of movement of galaxies of the Local Group relative to the coordinate system associated with the cosmic microwave background (CMB) radiation. The Zone of Avoidance is also heterogeneous because the Solar System is not located in the center of our Galaxy.

Due to the small number of known objects, decreasing the brightness of the extragalactic objects when we approach the galactic equator, increasing the concentration of stars on the line of sight, which results in increasing the overlap of the extragalactic object with the star, the extragalactic astronomers usually avoid this area (Kraan-Korteweg and Lahav, 2000).

The problem can be solved by either direct or indirect techniques. Under direct methods we understand the observations of whole-sky surveys in different spectral ranges near the galactic equator ($b \in [-20°, +20°]$). Indirect methods consist in applying the mathematical simulation and data mining methods to fill the Zone of Avoidance as well as to determine the gravitational potentials of the nearest galaxies in order to predict the positions of galaxies and galaxy systems in the area of Milky Way absorption. Great attention is also focused on the machine learning technique.

Multiwavelength observations. Direct methods. Since the 1970s the Zone of Avoidance has decreased significantly due to studies in the infrared and radio spectral ranges (due to the decrease in the amount of light absorption with increasing wavelength, the Zone of Avoidance becomes more transparent in these spectral ranges).

First of all, on 29 September 1967, Italian astronomer P. Maffei discovered the elliptical galaxy Maffei 1 together with the spiral galaxy Maffei 2 in the Zone of Avoidance. He used a hypersensitized I-N photographic plate for the infrared range and exposed it with the Schmidt telescope at Asiago Observatory (see the paper by Maffei, 2003 for a review of his own works). Maffei 1 is located 0.55° from the galactic plane in the middle of the Zone of Avoidance ($\alpha = 02^h\ 36^m\ 35.4^s$, $\delta = +59°\ 39'\ 19''$, $m = 11.14 \pm 0.06$ in the V-band). Maffei 1 would be one of the largest and brightest elliptical galaxies in the sky (about 3/4 the size of the full moon) if there were no 4.7^m of extinction (a factor of about 1/70) in the visible range (Fig. 16.6).

FIG. 16.6 Galaxies Maffei 1 (down right) and Maffei 2 (top left), discovered by P. Maffei in the Zone of Avoidance.

Maffei's discovery promoted a lively discussion in those times about possible membership of these galaxies to the Local Group. In 1970 Spinrad suggested that Maffei 1 is a nearby heavily obscured giant elliptical galaxy and estimated the distance to Maffei 1 as 1 Mpc (Local Group member?). In 1983 this estimate was revised up to $2.1^{+1.3}_{-0.8}$ Mpc by Buta and McCall (Maffei 1 is outside the Local Group!). In 2001, Davidge and van den Bergh used adaptive optics to observe the brightest AGB stars in Maffei 1 and concluded that the distance is $4.4^{+0.6}_{-0.5}$ Mpc. A latest determination of the distance to Maffei 1 is 2.85 ± 0.36 Mpc, which is based on the recalibrated luminosity/velocity dispersion relation for E-galaxies and the updated extinction. It proves that Maffei 1 is a key member of a nearby galaxy group named Maffei Group, where among other members are the giant spiral galaxies IC342 and Maffei 2. Maffei 1 has also a small satellite spiral galaxy Dwingeloo 1 as well as a number of dwarf satellites like MB1. The IC 342/Maffei Group is one of the closest galaxy groups to the Milky Way (Huchtmeier et al., 1995; Karachentsev et al., 2003). The larger (≥ 3 Mpc) distances reported in

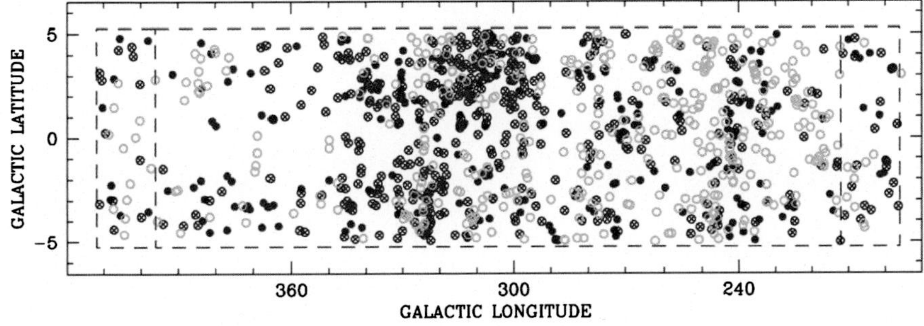

FIG. 16.7 Distribution in galactic coordinates of the 1036 galaxies detected in the deep HI ZOA survey. Open circles, $V_{hel} < 3500$; circled crosses, $3500 < V_{hel} < 6500$; filled circles, $V_{hel} > 9500$ km/s (Kraan-Korteweg et al., 2003, Open Astronomy).

the past 20 years would imply that Maffei 1 has never been close enough to the Local Group to significantly influence its dynamics.

The current notion of the Zone of Avoidance has changed in the 1990s and was connected with exploration of the infrared satellite IRAS and the releases of 2MASS survey as well as with several projects in the radio range. If it was previously believed that this area closes an observer about 20% of the spatial distribution of galaxies in the optical range, which leads to an incomplete catalogue of galaxies near the Galactic Plane, then this value is now about 10%. Completeness of Zone of Avoidance galaxy catalogues as a function of the foreground extinction is as follows: optical Zone of Avoidance surveys are complete to an apparent diameter of $D = 14''$, where the diameters correspond to an isophote of 24.5 mag/arcsec^2 for extinction levels less than $A_B = 3.0^m$.

Because of the transparency of the galaxy to the 21 cm radiation of neutral hydrogen, systematic HI-surveys are particularly powerful in mapping large-scale structures in this part of the sky. The redshifted 21 cm emission of HI-rich galaxies are readily detectable at lowest latitudes and highest extinction levels and the signal will furthermore provide immediate redshift and rotational velocity information. Observations of the neutral hydrogen (21 cm) in the frame of the DOGS project revealed the Dwingeloo 1 (Kraan-Korteweg et al., 1994) and Dwingeloo 2 (Burton et al., 1996) galaxies in this zone (see, for example, Huchtmeier et al., 1995; Buta and McCall, 1999; Karachentsev, 2005 on the estimates of their kinematic and dynamic parameters). Supplementary to these surveys, the Parkes Multibeam HI ZOA Survey as a systematic deep blind HI survey of the Southern Milky Way was begun in 1997 with the Multibeam receiver at the 64 m Parkes telescope. Surveys were centered on the Southern Galactic Plane: $196° \leq l \leq 52°$, $|b| \leq 5°$. The coverage in redshift space was $-1200 < V_{hel} < 12700$ km/s (see, for example, Saurer et al., 1997). Distribution of the 1036 galaxies in galactic coordinates detected in the deep HI ZOA survey is shown in Fig. 16.7.

It should be noted that the absence of a signal does not always indicate the absence of a galaxy, but may be associated with a low HI content (Lahav et al., 1998). This method is slow and requires a lot of time, but the conjunction of HI surveys and 2MASS will greatly increase the current census of galaxies hidden behind the Milky Way. In 2000, Jarrett et al. (2000) reported on the detection of newly discovered sources from 2MASS. There were also identification results of the HI spectra of galaxies which were observed by the IRAS (Lu et al., 1990).

The Milky Way is transparent to the hard X-ray emission above a few keV, and because the rich clusters are strong X-ray emitters. Since the X-ray luminosity is roughly proportional to the cluster mass as $L_X \propto M^{3/2}$ or M^2, depending on the still uncertain scaling law between the X-ray luminosity and temperature (see, for example, Babyk and Vavilova, 2012; Babyk and Vavilova, 2013; Babyk and Vavilova, 2014 and references therein), the massive clusters hidden by the Milky Way should be gravitationally stable through their X-ray emission (Kocevski et al., 2004; Ebeling et al., 2001). The clusters are primarily composed of early-type galaxies, which are not recovered by infrared galaxy surveys or by systematic HI surveys; that is why the method is particularly interesting (see Fig. 16.8).

The inhomogeneous distributed mass of matter in the Zone of Avoidance surrounding the Local Group may cause the unbalanced gravity toward the Local Group in one direction. Despite the fact that the result-

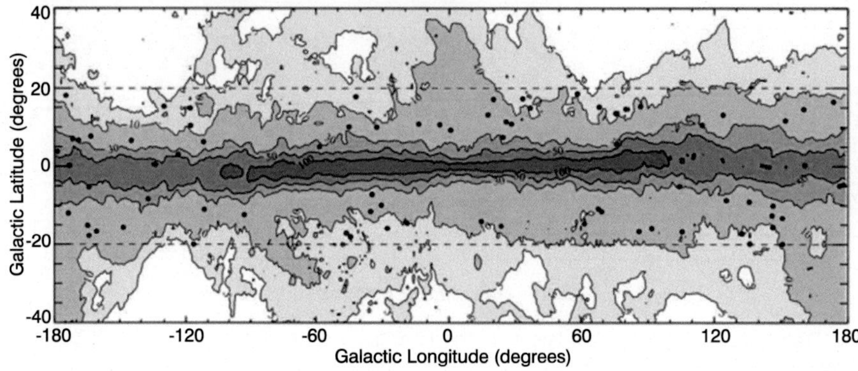

FIG. 16.8 Distribution in galactic coordinates of the 76 by Ebeling et al. (2002) so far spectroscopically confirmed X-ray clusters (solid dots) of which 80% were previously unknown. Superimposed are galactic HI column densities in units of 10^{20} cm^{-2} (Dickey and Lockman, 1990). Note that the region of relatively high absorption ($N_{HI} > 5 \times 10^{21}$ cm^{-2}) actually is very narrow and that clusters could be identified to very low latitudes (Kraan-Korteweg and Lahav, 2000, Open Astronomy).

ing vector of velocity of the Local Group lies within 20° of the observed cosmic background dipole, the calculations remain ambiguous (Karachentsev et al., 2013; Kashibadze et al., 2018), partly because galaxies in the Zone of Avoidance are not taken into account (Vavilova, 2000; Erdoğdu and Lahav, 2009).

A dipole known by CBM studies is the asymmetry of the radiation temperature. It is the heating of 0.1% of CMB radiation in comparison with the average in one direction and in the same cooling in the opposite direction. The COBE (1989–1990) studies indicated that the Milky Way and the Local Group are moving at a velocity $\sim V_p = 627$ km/s to ($l = 276°$, $b = 30°$), towards the Hydra constellation (Kogut et al., 1993). This motion determines the distribution of matter M_i in the Local Group and the cosmological parameter Ω_0 (Giovanelli and Haynes, 1989): $\vec{V}_p \propto \frac{\Omega_0^{0.6}}{b} \sum_i \frac{M_i}{r_i^2} r_i$. Filling the zone $|b| \leq 20°$ by galaxies changes the direction of movement measured in the volume of 6000 km/s by 31° (Kolatt and Dekel, 1997; Vasylenko and Kudrya, 2017). Nearby unknown galaxies in the Zone of Avoidance can make a larger contribution to the definition of a vector of collective velocity than whole clusters over long distances: $\vec{V}_p \propto \sum_i 10^{-0.4m} r_i$.

This discrepancy between the direction on the dipole and the expected velocity vector made it necessary to introduce the concept of "attractors" (the Great Attractor at a distance of about 60 Mpc; see also Fig. 16.5). Perseus-Pisces and the Great Attractor overdensity lie at similar distances on opposite sides of the Local Group and are partially obscured by the Zone of Avoidance. The Zone of Avoidance is fully incomplete at low galactic latitudes in the larger Galactic Bulge area ($l \approx 0° \pm 90°$). Even if the obscured galaxies can be identified, it is very difficult to determine their redshifts because of the higher extinction levels. Since the method involves uniform filling of the sky by the galaxies of the field, and chaotic filling them with nonreal objects leads to the formation of nonexistent fields, attempts to solve the problem of the incompatibility of the vector apex motion of the Local Group determined by the CMB and the velocity field did not give a positive result.

So, the multiwavelength surveys of the Zone of Avoidance in the last decades were aimed at addressing such key problems as the cosmological questions about the dynamics of the Local Group, the possible existence of nearby hidden massive galaxies, the dipole determinations based on luminous galaxies, the continuity and size of nearby superclusters, and the mapping of cosmic flow fields (a very comprehensive review is given by Kraan-Korteweg and Lahav, 2000).

Machine learning. Indirect methods. The solution of these problems is possible also by indirect methods, which include the methods of signal processing applied to obscured and incomplete data; indirect estimates of averaged variables; the mask inversion using Wiener filtering in spherical harmonic analysis; reconstruction of the projected galaxy distribution in infrared, radio, and X-ray spectral ranges; two-dimensional Wiener reconstruction to three dimensions; methods of Voronoi mosaic, cluster, and fractal analysis; and machine learning techniques.

The last successful results of analysis of the spatial distribution of galaxies and their systems in the areas surrounding the Milky Way Zone of Avoidance based on the 2MASS Tully–Fisher Survey and the HI observational surveys are presented in works by Said et al. (2014, 2016b, 2016a), where the optimized Tully–Fisher relation for measured distances and peculiar velocities is developed for dust-obscured galaxies. But it remains a complex and unresolved problem, as well as the estimation of the "invisible" content of the spatial galaxy distribution

The problem of Zone of Avoidance reconstruction is related to dealing with gaps in the spectroscopic observations to restore homogeneous sky coverage. Classical three-dimensional reconstruction of the extragalactic objects behind the Milky Way to preserve the coherence of the large scale structure was triggered by the search of the Great Attractor in the 1990s (Kraan-Korteweg, 2005). And reconstruction of missing information could be oriented towards the observations of galaxies and their systems that surround the Zone of Avoidance (Courtois et al., 2012; Sorce et al., 2017).

The existence of unobserved zones in scale comparable to the size of investigated zones can have a serious impact on the study of galaxy properties and local environments. In this case, the local and deterministic recovery of the missing data is needed (Cucciati et al., 2006). For small-scale reconstruction techniques such as the following are common: direct cloning (Elyiv, 2006), wavelet analysis (Vavilova, 1997), cluster analysis (Gregul et al., 1991; Vavilova and Melnyk, 2005), randomized cloning of objects into unobserved areas or application of Wiener filtering (Lahav et al., 1994; Branchini et al., 1999), Voronoi tessellation (Melnyk et al., 2006; Elyiv et al., 2009; Dobrycheva et al., 2014). Cucciati et al. (2014) proposed two algorithms that use photometric redshift of target objects and assign redshifts based on the spectroscopic redshifts of the nearest galaxies. A Wiener filter applied in this work was very efficient also to reconstruct the continuous density field instead of individual galaxy positions. These methods can clearly separate underdense from overdense regions on scales of 5 h^{-1} Mpc at moderate redshifts $0.5 < z < 1.1$, which is important for studies of cosmic variance and rare population galaxy systems.

There are limits of optical observations of extended objects due to random and systematic noise from detector, the telescope system, and the sky background. Schawinski et al. (2017) estimated a possibility to recover artificially degraded images with a high noise using state-of-the-art methods of machine learning, namely, deep learning – generative adversarial networks (GANs). It works better than simple deconvolution.

Generative adversarial neural networks as the type of unsupervised machine learning algorithms were first invented by Goodfellow et al. (2014). The main idea of these classes of algorithms are two neural networks contesting with each other. First, a neural network called "generative" (typically a deconvolutional one) generates candidate images and a second neural network (a convolutional discriminative one) evaluates them. The generative network trains to transfer from a space of features to a particular data distribution. At the same time the discriminative network discriminates between the produced candidates and real examples. Schawinski et al. (2017) applied the GAN to 4550 galaxies from the SDSS DR12. The authors have proved that this method can reliably recover features in images of galaxies and can go well beyond the limitation of deconvolutions. As the training sample they used image pairs: one original image of a galaxy and the same image artificially degraded (convolved with PSF). In general, the GAN learns how to recover the degraded image by minimizing the difference between the recovered and true images. With this purpose, the authors used a second neural network, whose aim was to distinguish the synthetic recovered image from the true image. These two neural networks are trained simultaneously. Therefore, by training on higher-quality images, the GAN method can learn how to recover information from the lower-quality data by building priors. Such approach has a potential for recovering partially damaged images with gaps and dead CCD chips. The algorithm of reconstruction of three-dimensional structures behind the Zone of Avoidance with the modified GAN method is presented in Fig. 16.9 and described by Vavilova et al. (2018).

We have just one unique sample of galaxies, i.e., just one set for training, which is a principal problem. In the approach described above we cannot use a set of many images for training. One solution could be to prepare the mock catalogues from numerical simulations, which reproduce a target sample. In this case we may generate as many pairs as possible – a real survey and a survey with Zone of Avoidance. Additionally, the position of the Zone of Avoidance could be randomized over the survey field. A goal of generative ANN will be to generate galaxy distributions and their properties in the Zone of Avoidance from a latent space of features. At the same time, a discriminative network will compare the obtained survey with the real one and evaluate how realistic it is. The generative network produces better surveys with iteration, while the discriminative

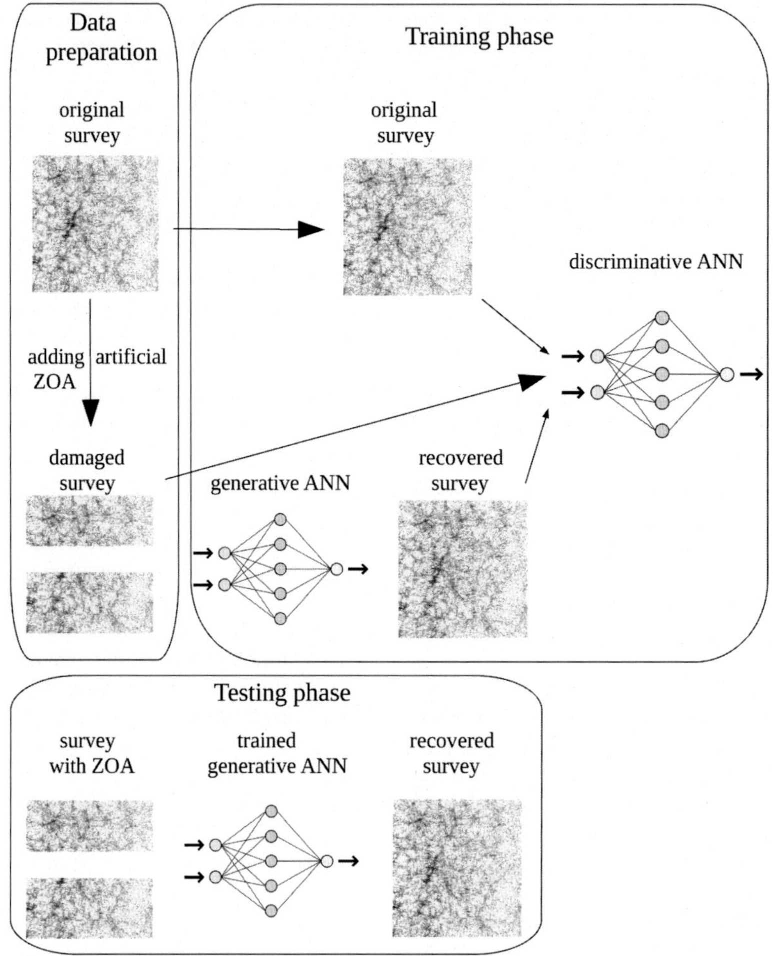

FIG. 16.9 Scheme of the data preparation, the training and testing phases for the Zone of Avoidance recovering by the GAN method. The input is a set of mock surveys from which the artificial Zone of Avoidance was generated to train the GAN. A generative ANN is used to recover surveys in the at the testing phase (Vavilova et al., 2018).

one becomes more experienced at labeling the synthetic ones. In such a way the system learns the sophisticated loss functions automatically without its predefinition.

To apply the algorithm, we should prepare a sample of galaxies surrounding the Zone of Avoidance, which is complete by stellar magnitudes. To get a three-dimensional spatial distribution of galaxies in this sample, we must obtain their photometric redshifts and to divide this sample on the slices by coordinates, taking into account the cosmological parameters. Each of these slices should contain a real distribution and the damaged image (part of the Zone of Avoidance region), which will require darning. The preliminary step how the algorithm works and restores a galaxy distribution should be conducted and tested with subsamples of real galaxies selected from the nondamaged regions.

16.4 FLUX VARIABILITY OF THE BLAZAR 3C 454.3

Another good example for illustration of the data mining from multiwavelength astronomical databases is related to the active galactic nuclei (AGNs). We explain it with the blazar 3C 454.3 (see, all-sky view taken with Fermi/LAT in Fig. 5.16), which is one of the brightest AGNs at all frequencies.[4]

This blazar is located in direction of Alpha Pegasi (Markab) at the distance of 7.7 Gly (redshift $z = 0.859001 \pm 0.000170$); right ascension is $\alpha = 22^h\ 53^m\ 57.7^s$, declination $\delta = +16°\ 08'\ 53.6''$. It has a strong flux variability at all wavelengths from gamma ray to radio. The spectral energy distribution of 3C 454.3 displays the two peaks typical for AGNs, one in the infrared and optical, and the other in the X-ray and gamma ray. The spectral characteristics of these peaks are determined by two radiation mechanisms – synchrotron radiation by relativistic electrons and inverse-Compton scattering of "soft" photons on relativistic electrons.

Observations have also established that a single radiation mechanism operates from the radio to the optical spectrum. This was first confirmed directly when correlations were found between flux variations at different frequencies during the development of a major flare in 2005–2006 (Volvach et al., 2007). Variations in the flux of 3C 454.3 were observed on scales from days to a year, which were repeated in the optical and radio spectra. It was shown that both the duration of the flare (about a year) and individual features of the flare were the same in these two frequency ranges. This is possible only if a single mechanism is generating the radiation in these different ranges. Thus, it was established that both radio and optical emission is produced by the jet. The delay between the flares in the optical and millimeter ranges was about ten months, with about the same delay observed for centimeter wavelengths. The frequency dependence of the delays and the intervals between flares can be used to predict future flares in this object in various frequency ranges.

For example, the three flares in the blazar 3C 454.3 were observed during 2005–2010 and allowed to determine their locations in the jet from gamma ray to radio range and to estimate a size of the Stromgren zone for

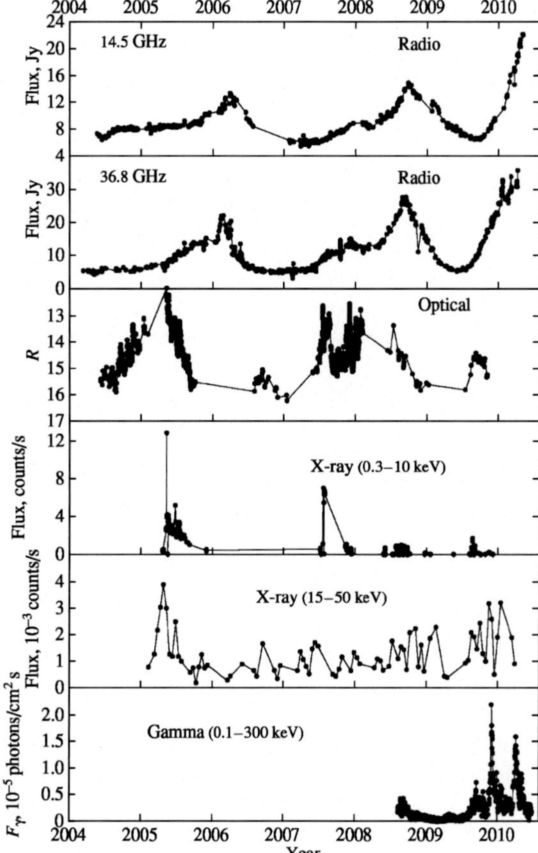

FIG. 16.10 Light curves for the blazar 3C 454.3 at various frequencies from radio to gamma ray ranges, obtained for the observational period of 2004–2010 (Volvach et al., 2011).

sources of ionization associated with a binary supermassive black hole in the central region (Volvach et al., 2011).

Light curves for the blazar 3C 454.3 in spectral ranges from radio to gamma ray obtained for the observational period of 2004–2010 years are presented in Fig. 16.10.

- Radio observations were conducted at 22.2 and 36.8 GHz with the 22 m radio telescope of the Crimean Astrophysical Observatory using modulation radiometers (Efanov et al., 1979). Observations at 4.8, 8, and 14.5 GHz were carried out on the 26 m telescope of the University of Michigan Radio Astronomy Observatory (Aller et al., 2001). Observations at 37 GHz were obtained using the 14 m telescope of the Metsahovi Radio Observatory of Aalto University. The data for radio range obtained with different telescopes are in good agreement and sup-

[4]Blazar is an active galaxy nucleus, which is located at the center of the elliptical galaxy and has a relativistic jet oriented close to the line of sight with the observer. The name *blazar* was coined in 1978 by E. Spiegel to denote the combination of BL Lac objects and of optically violently variable quasars. Being one of the powerful sources of emission, blazars are characterized by high polarization and very rapid fluctuations in brightness. Among well-known blazars are BL Lacertae, 3C 454.3, 3C 273, PKS 2155-304, S5 0014+81 (TeV Blazar with the most supermassive black hole, $10^9\ \odot$), and others. The unique blazar TXS 0506+056, which is a source of high-energy neutrinos, was discovered in the frame of the IceCube project (icecube.wisc.edu) in July, 2018 (see, for example, Overbye Dennis, "It Came From a Black Hole, and Landed in Antarctica – For the first time, astronomers followed cosmic neutrinos into the fire-spitting heart of a supermassive blazar," 12 July 2018, The New York Times).

plement each other during the long-term monitoring period.

- Optical data were obtained from the WEBT archive at the Osservatorio Astronomico di Torino of the Istituto Nazionale Di Astrofisica as a part of the WEBT program (Whole Earth Blazar Telescope) and were supplemented with observational data from the 70 cm telescope of the Crimean Astrophysical Observatory (Sergeev et al., 2005).
- The gamma ray observational data of 3C 454.3 from 24 April 2005 through 18 September 2010 were provided from the Swift spacecraft Burst Alert Telescope (BAT), which operates at 15–195 keV, and from the X-ray Telescope (XRT), which operates at 0.3–10 keV. These data are available through the HEASARC web site.[5] The energy range for the 3C 454.3 light curve obtained during the Swift/BAT transient monitoring program is 15–50 keV (Fig. 16.10). X-ray observations at 2–10 keV range were obtained with the Rossi X-ray Timing Explorer (RXTE). To exclude short-period variability and reduce uncertainty in the measured fluxes these light curves could be averaged over time intervals of one day, which corresponds to 15 orbital periods of the RXTE satellite (Chesnok et al., 2009). The data from the Fermi gamma ray telescope (the main instrument is the Large Area Telescope) were used to calculate light curves from 100 MeV to 300 GeV (lower panel of Fig. 16.10) of 3C 454.3, which is identified with the gamma ray source 1FGL 2253.9+1608 (Abdo et al., 2010).

Using the Shuster method one can conduct a harmonic analysis of the flux variations of 3C 454.3 from radio to gamma ray ranges and derive a unified law for the frequency-dependent delays of the flares. The double character of the flares in the period 2005–2010 may indicate the passage of a companion of the central supermassive black hole through the accretion disk at the pericenter, with the disk oriented at some angle to the orbit of the companion (Vol'Vach et al., 2011).

The American Association of Variable Star Observers installed a "Light Curve Generator for 3C 454.3," which is available through http://www.aavso.org/, where everybody can find periods of its outbursts, brightening to a peak apparent magnitude of 13.4 in June 2014. Using SIMBAD entry for "3C 454.3" one can access to all the available multiwavelength data on this blazar.

Instead of conclusion to the chapter. Summarizing what is written above, we emphasize that the current multiwavelength databases help us not only to study deeply the known phenomena in the universe, but also to discover new features that have not been seen before. Current data in various archives save a lot of time needed for downloading, analyzing, collecting, and describing soft observational data. Astronomical data are very heterogeneous, allowing everyone to find in this variety a solution to the new puzzles of the Universe.

[5] http://heasarc.gsfc.nasa.gov/docs/swift/results/transients/weak/.

REFERENCES

Abdo, A.A., et al., 2010. Fermi Large Area Telescope first source catalog. The Astrophysical Journal. Supplement Series 188 (2), 405–436. https://doi.org/10.1088/0067-0049/188/2/405. arXiv:1002.2280 [astro-ph.HE].

Aller, M.F., Aller, H.D., Hughes, P.A., 2001. The longterm centimeter-band total flux and linear polarization properties of the Pearson-Readhead Survey Sources. In: American Astronomical Society Meeting Abstracts, vol. 199. American Astronomical Society Meeting Abstracts, p. 138.15.

Andrae, R., Melchior, P., Bartelmann, M., 2010. Soft clustering analysis of galaxy morphologies: a worked example with SDSS. Astronomy & Astrophysics 522, A21. https://doi.org/10.1051/0004-6361/201014169. arXiv:1002.0676 [astro-ph.CO], 2010.

Babyk, Iu.V., Vavilova, I.B., 2012. The distribution of baryon matter in the nearby X-ray galaxy clusters. Odessa Astronomical Publications 25, 119.

Babyk, Iu.V., Vavilova, I.B., 2013. Comparison of optical and X-ray mass estimates of the Chandra galaxy clusters at $z < 0.1$. Odessa Astronomical Publications 26, 175.

Babyk, I., Vavilova, I., 2014. The Chandra X-ray galaxy clusters at $z < 1.4$: constraints on the evolution of L_X–T–M_g relations. Astrophysics and Space Science 349 (1), 415–421. https://doi.org/10.1007/s10509-013-1630-z.

Ball, N.M., Brunner, R.J., 2010. Data mining and machine learning in astronomy. International Journal of Modern Physics D 19 (7), 1049–1106. https://doi.org/10.1142/S0218271810017160. arXiv:0906.2173 [astro-ph.IM].

Ball, N.M., Loveday, J., et al., 2004. Galaxy types in the Sloan Digital Sky Survey using supervised artificial neural networks. Monthly Notices of the Royal Astronomical Society 348 (3), 1038–1046. https://doi.org/10.1111/j.1365-2966.2004.07429.x. arXiv:astro-ph/0306390 [astro-ph].

Banerji, Manda, et al., 2010. Galaxy Zoo: reproducing galaxy morphologies via machine learning. Monthly Notices of the Royal Astronomical Society 406 (1), 342–353. https://doi.org/10.1111/j.1365-2966.2010.16713.x. arXiv:0908.2033 [astro-ph.CO].

Banfield, J.K., et al., 2015. Radio Galaxy Zoo: host galaxies and radio morphologies derived from visual inspection. Monthly Notices of the Royal Astronomical Society 453 (3), 2326–2340. https://doi.org/10.1093/mnras/stv1688. arXiv:1507.07272 [astro-ph.GA].

Bell, Keaton, Salim, S., 2011. Automatic morphological classification of galaxies in SDSS/GALEX based on catalog photometry. In: American Astronomical Society Meeting Abstracts, vol. 217. American Astronomical Society Meeting Abstracts, p. 335.33.

Branchini, E., et al., 1999. A non-parametric model for the cosmic velocity field. Monthly Notices of the Royal Astronomical Society 308 (1), 1–28. https://doi.org/10.1046/j.1365-8711.1999.02514.x. arXiv:astro-ph/9901366 [astro-ph].

Brunner, R.J., Djorgovski, S.G., Prince, T.A., Szalay, A.S., 2002. Massive datasets in astronomy. In: Abello, James, Pardalos , Panos M., Resende, Mauricio G.C. (Eds.), Handbook of Massive Data Sets. In: Massive Computing, vol. 4. Springer, Boston, MA. ISBN 978-1-4613-4882-5.

Burton, W.B., et al., 1996. Neutral hydrogen in the nearby galaxies Dwingeloo 1 and Dwingeloo 2. Astronomy & Astrophysics 309, 687–701. arXiv:astro-ph/9511020 [astro-ph].

Buta, Ronald J., McCall, Marshall L., 1999. The IC 342/Maffei Group revealed. The Astrophysical Journal. Supplement Series 124 (1), 33–93. https://doi.org/10.1086/313255.

Buta, Ronald J., Sheth, Kartik, et al., 2010. Mid-infrared galaxy morphology from the Spitzer Survey of Stellar Structure in Galaxies (S^4G): the imprint of the De Vaucouleurs revised Hubble–Sandage classification system at 3.6 μm. The Astrophysical Journal. Supplement Series 190 (1), 147–165. https://doi.org/10.1088/0067-0049/190/1/147. arXiv:1008.0805 [astro-ph.CO].

Charlier, C., 1922. How an infinite world may be built up. Arkiv för Matematik, Astronomi och Fysik 16 (22), 1–34.

Chesnok, N.G., Sergeev, S.G., Vavilova, I.B., 2009. Optical and X-ray variability of Seyfert galaxies NGC 5548, NGC 7469, NGC 3227, NGC 4051, NGC 4151, Mrk 509, Mrk 79, and Akn 564 and quasar 1E 0754. Kinematics and Physics of Celestial Bodies 25 (2), 107–113. https://doi.org/10.3103/S0884591309020068.

Conselice, Christopher J., 2014. The evolution of galaxy structure over cosmic time. Annual Review of Astronomy and Astrophysics 52, 291–337. https://doi.org/10.1146/annurev-astro-081913-040037. arXiv:1403.2783 [astro-ph.GA].

Courtois, Hélène M., et al., 2012. Three-dimensional velocity and density reconstructions of the local universe with Cosmicflows-1. The Astrophysical Journal 744 (1), 43. https://doi.org/10.1088/0004-637X/744/1/43. arXiv:1109.3856 [astro-ph.CO].

Cucciati, O., Granett, B.R., et al., 2014. The VIMOS Public Extragalactic Redshift Survey (VIPERS). Never mind the gaps: comparing techniques to restore homogeneous sky coverage. Astronomy & Astrophysics 565, A67. https://doi.org/10.1051/0004-6361/201423409. arXiv:1401.3745 [astro-ph.CO].

Cucciati, O., Iovino, A., et al., 2006. The VIMOS VLT Deep Survey: the buildup of the colour-density relation. Astronomy & Astrophysics 458 (1), 39–52. https://doi.org/10.1051/0004-6361:20065161. arXiv:astro-ph/0603202 [astro-ph].

de la Calleja, J., Fuentes, O., 2004. Machine learning and image analysis for morphological galaxy classification. Monthly Notices of the Royal Astronomical Society 349 (1), 87–93. https://doi.org/10.1111/j.1365-2966.2004.07442.x.

Dickey, John M., Lockman, Felix J., 1990. H I in the galaxy. Annual Review of Astronomy and Astrophysics 28, 215–261. https://doi.org/10.1146/annurev.aa.28.090190.001243.

Dobrycheva, D.V., Melnyk, O.V., et al., 2014. Environmental properties of galaxies at $z < 0.1$ from the SDSS via the Voronoi tessellation. Odessa Astronomical Publications 27, 26.

Dobrycheva, D.V., Vavilova, I.B., et al., 2017. Machine learning technique for morphological classification of galaxies at $z < 0.1$ from the SDSS. arXiv:1712.08955 [astro-ph.GA].

Easton, C., 1904. La distribution des nebeleuses et leurs relation avec le systeme galactique. Astronomische Nachrichten 166, 131.

Ebeling, H., et al., 2001. Discovery of a very X-ray luminous galaxy cluster at $z = 0.89$ in the Wide Angle ROSAT Pointed survey. The Astrophysical Journal Letters 548 (1), L23–L27. https://doi.org/10.1086/318915. arXiv:astro-ph/0012175.

Ebeling, Harald, Mullis, Christopher R., Brent Tully, R., 2002. A systematic X-ray search for clusters of galaxies behind the Milky Way. The Astrophysical Journal 580 (1), 774–788. https://doi.org/10.1086/343790.

Efanov, V.A., Moiseev, I.G., Nesterov, N.S., 1979. A 1.35 CM wavelength survey of extragalactic radio sources. Izvestiya Ordena Trudovogo Krasnogo Znameni Krymskoj Astrofizicheskoj Observatorii 60, 3–13.

Elyiv, Andrii, 2006. UHECRs deflections in the IRAS PSCz catalogue based models of extragalactic magnetic field. ArXiv e-prints, arXiv:astro-ph/0611696 [astro-ph].

Elyiv, A., Melnyk, O., Vavilova, I., 2009. High-order 3D Voronoi tessellation for identifying isolated galaxies, pairs and triplets. Monthly Notices of the Royal Astronomical Society 394 (1), 1409–1418. https://doi.org/10.1111/j.1365-2966.2008.14150.x. arXiv:0810.5100 [astro-ph].

Erdoğdu, Pirin, Lahav, Ofer, 2009. Is the misalignment of the Local Group velocity and the dipole generated by the 2MASS Redshift Survey typical in Λ cold dark matter and the halo model of galaxies? Physical Review D 80 (4), 043005. https://doi.org/10.1103/PhysRevD.80.043005. arXiv:0906.3111 [astro-ph.CO].

Feigelson, E.D., Babu, G.J., 2006. Statistical Challenges in Astronomy. Springer. ISBN 978-0-387-21529-7.

Giovanelli, Riccardo, Haynes, Martha P., 1989. A 21-cm survey of the Pisces-Perseus supercluster. IV. Addenda to the declination zone 21.5 degrees to 33.5 degrees. The Astronomical Journal 97, 633. https://doi.org/10.1086/115010.

Goodfellow, Ian J., et al., 2014. Generative adversarial networks. ArXiv e-prints, arXiv:1406.2661 [stat.ML].

Gregul, A.Ia., Mandzhos, A.V., Vavilova, I.B., 1991. The existence of the structural anisotropy of the Jagiellonian field of the galaxies. Astrophysics and Space Science 185 (2), 223–235. https://doi.org/10.1007/BF00643190.

Huchtmeier, W.K., et al., 1995. Two new possible members of the IC342-Maffei1/2 group of galaxies. Astronomy & Astrophysics 293, L33–L36.

Ivezić, Željko, Connolly, Andrew J., VanderPlas, Jacob T., Gray, Alex, 2014. Statistics, Data Mining, and Machine Learning in Astronomy: A Practical Python Guide for the Analysis of Survey Data. Prinseton University Press. ISBN 9780691151687.

Jain, Archa, Gauthier, Alexandre, Noordeh, Emil, 2016. Galaxy morphology classification. In: CS 229 Machine Learning Final Projects. Stanford University, 6 p.

Jarrett, T.H., et al., 2000. 2MASS extended source catalog: overview and algorithms. The Astronomical Journal 119 (5), 2498–2531. https://doi.org/10.1086/301330. arXiv:astro-ph/0004318.

Karachentsev, I.D., 2005. The Local Group and other neighboring galaxy groups. The Astronomical Journal 129 (1), 178–188. https://doi.org/10.1086/426368. arXiv:astro-ph/0410065.

Karachentsev, I.D., et al., 2003. Distances to nearby galaxies around IC 342. Astronomy & Astrophysics 408, 111–118. https://doi.org/10.1051/0004-6361:20030912.

Karachentsev, Igor D., Makarov, Dmitry I., Kaisina, Elena I., 2013. Updated nearby galaxy catalog. The Astronomical Journal 145 (4), 101. https://doi.org/10.1088/0004-6256/145/4/101. arXiv:1303.5328 [astro-ph.CO].

Kashibadze, O.G., Karachentsev, I.D., Karachentseva, V.E., 2018. Surveying the local supercluster plane. Astrophysical Bulletin 73 (2), 124–141. https://doi.org/10.1134/S1990341318020025.

Kasivajhula, Siddhartha, Raghavan, Naren, Shah, Hemal, 2007. Morphological galaxy classification using machine learning. In: CS 229 Machine Learning Final Projects. Stanford University, 5 p.

Kates-Harbeck, Julian, 2012. Galaxy image processing and morphological classification using machine learning. In: APS April Meeting Abstracts, vol. 2012, E1.075.

Keel, W.C., 2007. The Road to Galaxy Formation. Springer-Verlag Berlin Heidelberg. ISBN 978-3-540-72534-3.

Kocevski, D.D., Ebeling, H., Mullis, C.R., 2004. Clusters in the Zone of Avoidance. In: Mulchaey, J.S., Dressler, A., Oemler, A. (Eds.), Clusters of Galaxies: Probes of Cosmological Structure and Galaxy Evolution, p. 26. arXiv:astro-ph/0304453 [astro-ph].

Kogut, A., et al., 1993. Dipole anisotropy in the COBE differential microwave radiometers first-year sky maps. The Astrophysical Journal 419, 1. https://doi.org/10.1086/173453. arXiv:astro-ph/9312056.

Kolatt, Tsafrir, Dekel, Avishai, 1997. Large-scale power spectrum from peculiar velocities. The Astrophysical Journal 479 (2), 592–605. https://doi.org/10.1086/303894.

Kraan-Korteweg, Renée C., 2005. Cosmological structures behind the Milky Way. Reviews in Modern Astronomy 18, 48–75. https://doi.org/10.1002/3527608966.ch3. arXiv:astro-ph/0502217 [astro-ph].

Kraan-Korteweg, R.C., Loan, A.J., et al., 1994. Discovery of a nearby spiral galaxy behind the Milky Way. Nature 372 (6501), 77–79. https://doi.org/10.1038/372077a0.

Kraan-Korteweg, R.C., Staveley-Smith, L., et al., 2003. The Universe behind the Southern Milky Way. ArXiv e-prints, arXiv:astro-ph/0311129 [astro-ph].

Kraan-Korteweg, Renée C., Lahav, Ofer, 2000. The Universe behind the Milky Way. The Astronomy and Astrophysics Review 10 (3), 211–261. https://doi.org/10.1007/s001590000011. arXiv:astro-ph/0005501 [astro-ph].

Kuminski, Evan, Shamir, Lior, 2016. A computer-generated visual morphology catalog of ~3,000,000 SDSS galaxies. The Astrophysical Journal. Supplement Series 223 (2), 20. https://doi.org/10.3847/0067-0049/223/2/20. arXiv:1602.06854 [astro-ph.GA].

Lahav, Ofer, et al., 1998. Galaxy candidates in the Zone of Avoidance. Monthly Notices of the Royal Astronomical Society 299 (1), 24–30. https://doi.org/10.1046/j.1365-8711.1998.01686.x. arXiv:astro-ph/9707345 [astro-ph].

Lahav, O., Fisher, K.B., et al., 1994. Wiener reconstruction of all-sky galaxy surveys in spherical harmonics. The Astrophysical Journal Letters 423, L93. https://doi.org/10.1086/187244. arXiv:astro-ph/9311059 [astro-ph].

Lahav, O., Naim, A., et al., 1996. Neural computation as a tool for galaxy classification: methods and examples. Monthly Notices of the Royal Astronomical Society 283, 207. https://doi.org/10.1093/mnras/283.1.207. arXiv:astro-ph/9508012.

Lu, N.Y., et al., 1990. Identifying galaxies in the Zone of Avoidance. The Astrophysical Journal 357, 388. https://doi.org/10.1086/168929.

Maffei, Paolo, 2003. My researches at the infrared doors. Memorie Della Società Astronomica Italiana 74, 19.

Mao, Q., et al., 2016. A cosmic void catalog of SDSS DR12 BOSS galaxies. The Astrophysical Journal 835. https://doi.org/10.3847/1538-4357/835/2/161.

Melnyk, O.V., Dobrycheva, D.V., Vavilova, I.B., 2012. Morphology and color indices of galaxies in pairs: criteria for the classification of galaxies. Astrophysics 55 (3), 293–305. https://doi.org/10.1007/s10511-012-9236-7.

Melnyk, O.V., Elyiv, A.A., Vavilova, I.B., 2006. The structure of the Local Supercluster of galaxies detected by three-dimensional Voronoi's tessellation method. Kinematika i Fizika Nebesnykh Tel 22 (4), 283–296. arXiv:0712.1297 [astro-ph].

Murrugarra, J.H., Hirata, N.S.T., 2017. Galaxy image classification, pp. 01–04. URL: http://sibgrapi2017.ic.uff.br/e-proceedings/assets/papers/WIP/WIP8.pdf.

Naim, A., et al., 1995. A comparative study of morphological classifications of APM galaxies. Monthly Notices of the Royal Astronomical Society 274 (4), 1107–1125. https://doi.org/10.1093/mnras/274.4.1107. arXiv:astro-ph/9502078.

Nair, Preethi B., Abraham, Roberto G., 2010. A catalog of detailed visual morphological classifications for 14,034 galaxies in the Sloan Digital Sky Survey. The Astrophysical Journal. Supplement Series 186 (2), 427–456. https://doi.org/10.1088/0067-0049/186/2/427. arXiv:1001.2401 [astro-ph.CO].

Proctor, R., 1878. The Universe of Stars. Longmans, Green and Co., London, 41.

Said, K., Kraan-Korteweg, R.C., Jarrett, T.H., 2014. Galaxy peculiar velocities in the Zone of Avoidance. ArXiv e-prints, arXiv:1410.2992 [astro-ph.GA].

Said, Khaled, Kraan-Korteweg, Renée C., Jarrett, T.H., et al., 2016a. NIR Tully–Fisher in the Zone of Avoidance - III. Deep NIR catalogue of the HIZOA galaxies.

Monthly Notices of the Royal Astronomical Society 462 (3), 3386–3400. https://doi.org/10.1093/mnras/stw1887. arXiv:1607.08596 [astro-ph.GA].

Said, Khaled, Kraan-Korteweg, Renée C., Staveley-Smith, Lister, et al., 2016b. NIR Tully–Fisher in the Zone of Avoidance - II. 21 cm H I-line spectra of southern ZOA galaxies. Monthly Notices of the Royal Astronomical Society 457 (3), 2366–2376. https://doi.org/10.1093/mnras/stw105. arXiv: 1601.07162 [astro-ph.GA].

Sanford, R.F., 1917. On some relation of the spiral nebulae to the Milky Way. Lick Observatory Bulletin 297, Part IX, 80.

Saurer, W., Seeberger, R., Weinberger, R., 1997. Penetrating the "zone of avoidance": IV. An optical survey for hidden galaxies in the region $130° \leq l \leq 180°$, $-5° \leq b \leq +5°$. Astronomy & Astrophysics. Supplement Series 126, 247–250. https://doi.org/10.1051/aas:1997385.

Schawinski, Kevin, et al., 2017. Generative adversarial networks recover features in astrophysical images of galaxies beyond the deconvolution limit. Monthly Notices of the Royal Astronomical Society 467 (1), L110–L114. https://doi.org/10.1093/mnrasl/slx008. arXiv:1702.00403 [astro-ph.IM].

Sergeev, S.G., et al., 2005. Lag-luminosity relationship for interband lags between variations in B, V, R, and I bands in active galactic nuclei. The Astrophysical Journal 622 (1), 129–135. https://doi.org/10.1086/427820.

Shane, C.D., Wirtanen, C.A., 1967. The distribution of galaxies. Publications of the Lick Observatory XXII, Pt. I.

Shapley, H., 1961. Galaxies. Harvard Univ. Press, Cambridge, 159.

Smith, Michael D., Donohoe, Justin, 2019. The morphological classification of distant radio galaxies explored with three-dimensional simulations. Monthly Notices of the Royal Astronomical Society 490 (1), 1363–1382. https://doi.org/10.1093/mnras/stz2525. arXiv:1909.03905 [astro-ph.GA].

Sorce, Jenny G., et al., 2017. Predicting structures in the Zone of Avoidance. Monthly Notices of the Royal Astronomical Society 471 (3), 3087–3097. https://doi.org/10.1093/mnras/stx1800. arXiv:1707.04267 [astro-ph.CO].

Stratonoff, W., 1900. Etudes sur la structure de l'univers. Publications de l'Observatoire astronomique et physique de Tachkent. 2, 136P.

Storrie-Lombardi, M.C., et al., 1992. Morphological classification of galaxies by artificial neural networks. Monthly Notices of the Royal Astronomical Society 259, 8P. https://doi.org/10.1093/mnras/259.1.8P.

Vasylenko, M.Yu., et al., 2019. Verification of machine learning methods for binary morphological classification of galaxies from SDSS. Odessa Astronomical Publications 32, 46–51. https://doi.org/10.18524/1810-4215.2019.32.182538.

Vasylenko, M.Yu., Kudrya, Yu.N., 2017. Dipole bulk velocity based on new data sample of galaxies from the catalogue 2MFGC. Advances in Astronomy and Space Physics 7, 6–11. https://doi.org/10.17721/2227-1481.7.6-11.

Vavilova, I.B., 1997. Cluster and wavelet analysis for detachment of the structure of galaxy cluster: comparison. Data Analysis in Astronomy, 297–302.

Vavilova, I.B., 2000. Wavelet analysis as approach to recognize abundance zone in galaxy distribution. Kinematika i Fizika Nebesnykh Tel Supplement 3, 155.

Vavilova, I.B., Elyiv, A.A., Vasylenko, M.Yu., 2018. Behind the Zone of Avoidance of the Milky Way: what can we restore by direct and indirect methods? Russian Radio Physics and Radio Astronomy 23 (4), 244–257. https://doi.org/10.15407/rpra23.04.244.

Vavilova, Iryna, Melnyk, Olga, 2005. Voronoi tessellation for galaxy distribution. In: Mathematics and Its Applications. Proceedings of the Institute of Mathematics of the NAS of Ukraine, vol. 55, pp. 203–212.

Volvach, A.E., Larionov, M.G., et al., 2011. Flare activity of the blazar 3C 454.3 from gamma to radio wavelengths in 2004–2010. Kosmichna Nauka i Tekhnologiya 17 (2), 68–76. https://doi.org/10.15407/knit2011.02.068.

Volvach, A.E., Volvach, L.N., et al., 2007. The variability of a 3C 454.3 blazar over a 40-year period. Astronomy Reports 51 (6), 450–459. https://doi.org/10.1134/S1063772907060030.

Vol'Vach, A.E., et al., 2011. Multi-frequency studies of the non-stationary radiation of the blazar 3C 454.3. Astronomy Reports 55 (7), 608–615. https://doi.org/10.1134/S1063772911070092.

Wahaono, M.A.R., Azhari, S.N., 2018. Classification of galaxy morphological image based on convolutional neural network. International Journal of Advanced Research in Science, Engineering and Technology 5 (6), 6066–6073.

Willett, Kyle W., et al., 2013. Galaxy Zoo 2: detailed morphological classifications for 304 122 galaxies from the Sloan Digital Sky Survey. Monthly Notices of the Royal Astronomical Society 435 (1), 2835–2860. https://doi.org/10.1093/mnras/stt1458. arXiv:1308.3496 [astro-ph.CO].

York, Donald G., et al., 2000. The Sloan Digital Sky Survey: technical summary. The Astronomical Journal 120 (3), 1579–1587. https://doi.org/10.1086/301513. arXiv:astro-ph/0006396.

CHAPTER 17

Applications of Big Data in Astronomy and Geosciences: Algorithms for Photographic Images Processing and Error Elimination

LUDMILA PAKULIAK, PHD • VITALY ANDRUK

17.1 FLATBED SCANNERS AS DIGITIZERS FOR ASTRONOMIC PHOTOGRAPHIC MATERIAL

Not each astronomical institution which has in its possession glass collections of photographic plates can afford the creation of the special digitizer for the digitization of their collections. The attempts to apply more available commercial scanners for those purposes were made since these scanners had appeared. But the results could not match the requirements both on precision in positions and on photometric determinations. Besides, the process with these facilities was highly time consuming, making the digitization of entire glass collections very difficult in the foreseeable time interval. Fortunately, progress in digitizing technology in the 2000s has given new model lines of powerful scanners, suitable under some conditions for the aims of astronomy. Special methods and algorithms of digitization and processing of digitized images allow publishers of glass collections to achieve acceptable results with flatbed scanners in observatories of Germany, Bulgaria, Latvia, Ukraine, Uzbekistan, and the Russian Federation.

Scanning photographic material allows you to get all 100% of the data at the disposal of the researcher, whose main task is to extract the data with maximum completeness and accuracy, determined by the specifics of the material itself. At the time of the beginning of the digitization of archives, there was practically no software available that could handle large images with linear dimensions up to several tens of thousands of pixels containing hundreds of thousand of objects. Additional problems were created both by commercial scanners with errors caused by the peculiarities of their mechanics and by the properties of the plates themselves, on which, under the conditions of observational programs, several exposures of different duration with an offset between them could be superimposed on each other. For the basis of the programs for processing archive images, the modules of the MIDAS/ROMAFOT software package, designed to obtain photometric data from CCD frames and distributed under a free license, were taken. These modules have been reworked for tasks that include, in addition to photometric, positional definitions, taking into account the characteristics of large stellar fields.

By now, the processing of large arrays of scanned plates has already been successfully implemented and catalogues of positions and stellar magnitudes of objects have been obtained for various observational programs: the Photographic Survey of the Northern Sky (FON program) (Andruk et al., 2017; Kazantseva et al., 2015; Mullo-Abdolov et al., 2017); Saturn's satellites (Yizhakevych et al., 2017); observations of Uranus and Neptune (Protsyuk et al., 2014); Pluto (Eglitis and Andruk, 2017); and the observations of the first epoch for obtaining the proper motions of stars in the vicinity of open clusters (Pakuliak et al., 2016).

17.2 ALGORITHM OF CORRECTION FOR SCANNER ERRORS

The software package realizes the next scheme for processing and reduction of astronomical plates with star fields to extract the useful information from digitized images, which consists of the following main steps:

- digitizing of astronegatives on commercial scanners such as Microtek and Epson with scanning mode 1200 dpi, a gray scale of 16-bit color,
- preprocessing procedures to form the input data for the software package,

Knowledge Discovery in Big Data from Astronomy and Earth Observation. https://doi.org/10.1016/B978-0-12-819154-5.00029-1

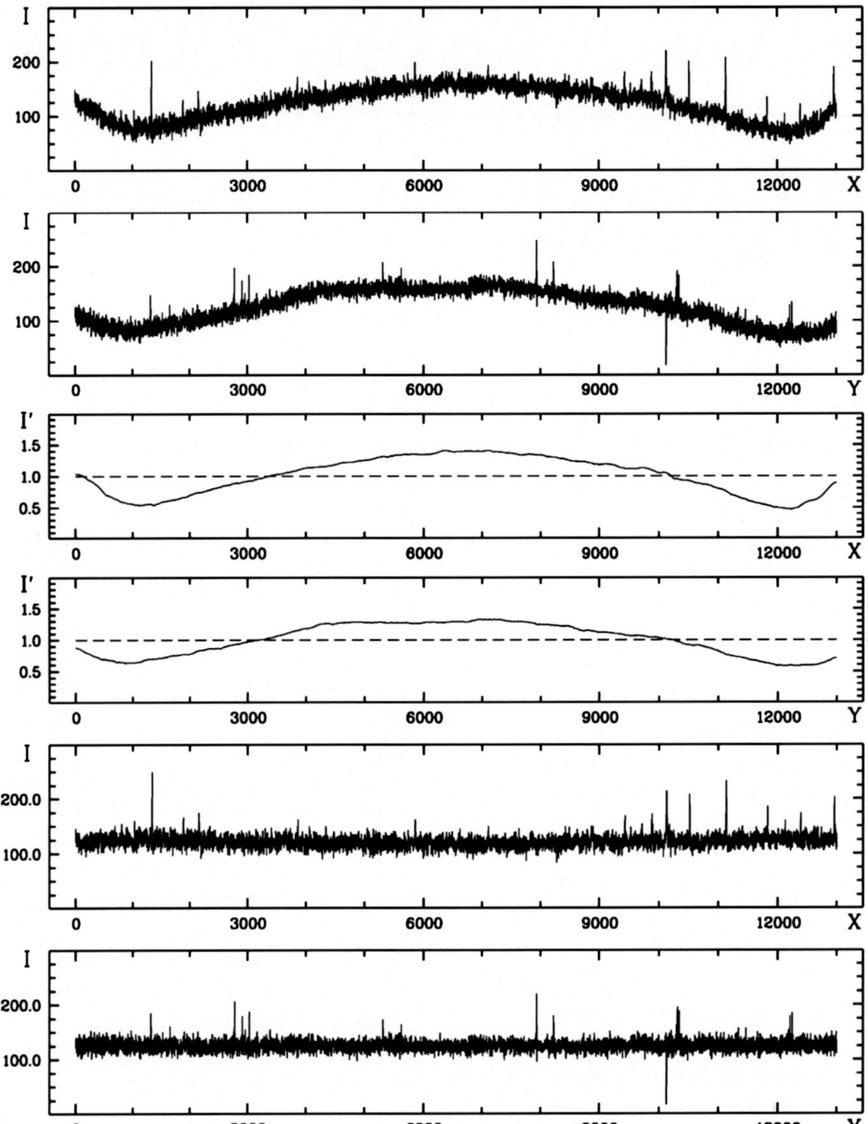

FIG. 17.1 An example of central photometric sections along the X, Y coordinates of a digital image of a wide-angle astrograph.

- calculation of the rectangular coordinates and the photometric instrumental values in the MIDAS/ROMAFOT environment for all objects registered on the astronegative,
- separation of registered objects (if necessary) by exposure,
- identification of stars of the reference catalogue by their rectangular and equatorial coordinates,
- astrometric reduction of all objects into the reference frame at the epoch of the plate exposure with the assessment of the accuracy,
- photometric reduction of instrumental stellar magnitudes to a system of photoelectric standards with the assessment of the accuracy,
- Final analysis of the list of objects with the rejection of fictitious and erroneous images.

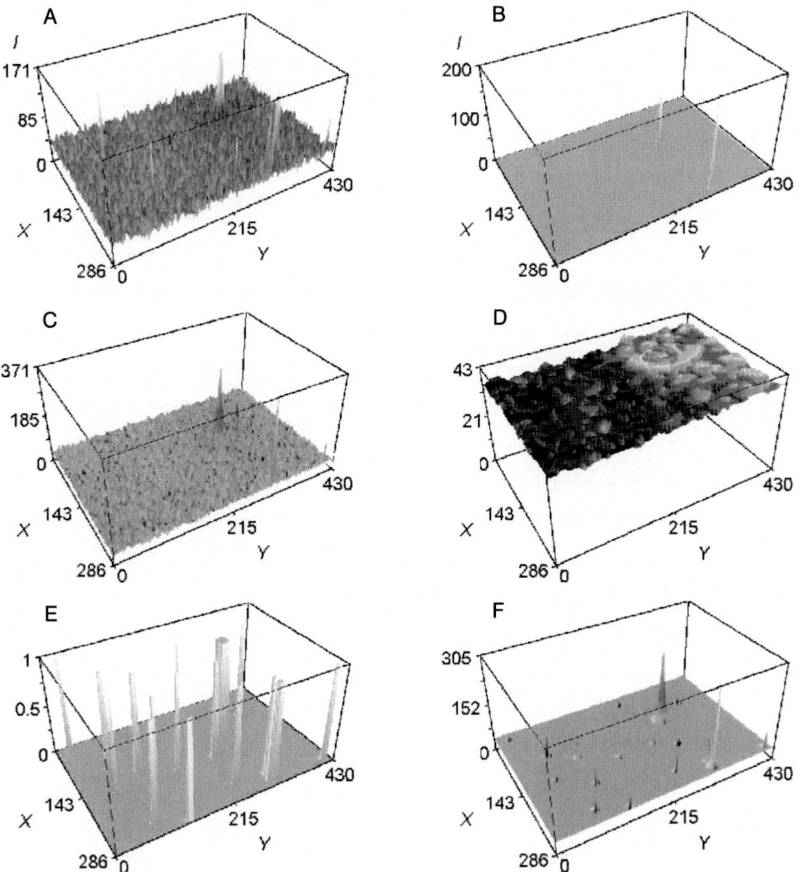

FIG. 17.2 The preprocessing of the digitized plate of a 1.2 m Schmidt telescope.

Techniques and methods of error elimination used in this type of processing are demonstrated for plates of two telescopes with different optical scheme generated the diverse sets of aberrations.

Preprocessing procedures of a digitized image include the photometric equalization of a digitized star field, which is done using MIDAS software modules to prepare the input data for ROMAFOT. The equalization is made over the plate data only without involving the auxiliary flat-field images.

Fig. 17.1 shows an example of central photometric sections along the X, Y coordinates of a digital image of a wide-angle astrograph. Two top panels show the data for the primary scan. The middle panels show a normalized cross-section for the envelopes of a flat field of the plate. The two bottom panels give the resulting cross-sections after correcting the primary scan for the envelope of the flat field. In more detail, Fig. 17.2 depicts this process for the area of the plate of a 1.2 m Schmidt telescope. (A) The three-dimensional projection of the scanned area. (B) The plotted complemented vertices for the overexposed stars. (C) The sum of the two previous projections. (D) The flat field for this area of the image (stars are removed). (E) The apertures for the objects. (F) The final input image for ROMAFOT processing.

The objective "surprise" encountered when having used commercial scanners for photographic plate digitization was the "wobble" effect that appeared in pixel coordinates due to mechanics of scanner. Six consecutive scans of one wide-angle astrograph plate were processed to evaluate the impact of scanner errors on astrometric and photometric determinations. For each scan, trends of differences in the coordinates of objects between consecutive scans and average scan were constructed. Fig. 17.3 demonstrates the manifestation of

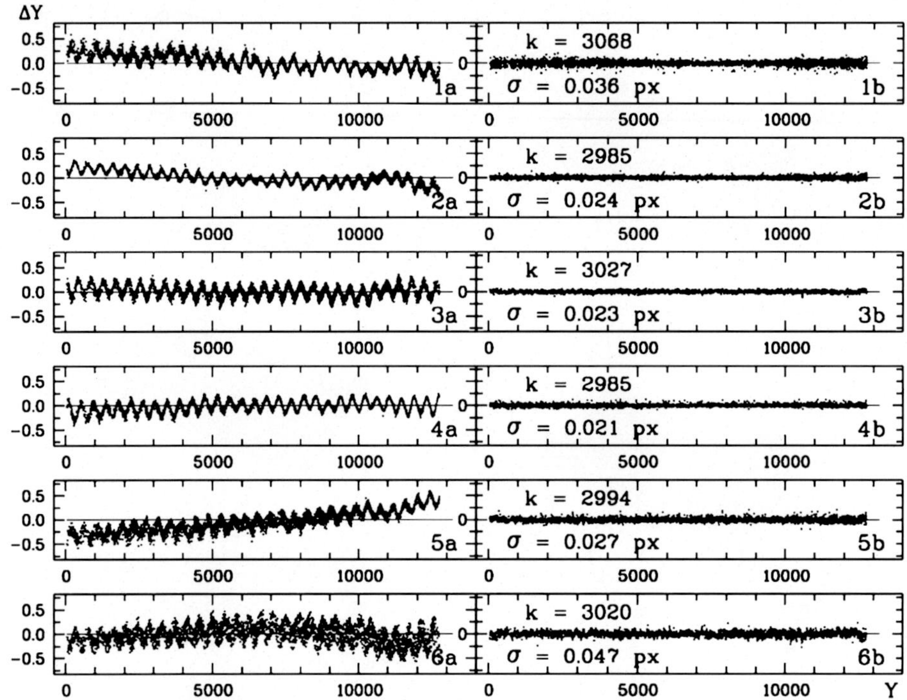

FIG. 17.3 "Wobble" effect on pixel coordinates and results of its correction.

the specific saw-edged pattern as a result of scanner mechanics operation which causes internal errors in fixing the coordinates of the objects. Here, on the right side of the figure the dispersion of the differences is shown after the correction for the scanner systematics for each scan.

As seen from Fig. 17.3, the flatbed scanners have systematic mechanical errors which are especially large along the Y (DEC) coordinate, i.e., along the direction of the moving of linear CCD. The amplitude of differences between the calculated equatorial coordinates and their catalogued values can reach several arcsec for some models of scanners. The algorithm of their correction used in series of projects for different models of commercial scanners is as follows:

- The length of a plate along the scan direction L_Y in pixels is divided by the number of reference stars. The resulted value is taken as the first step of approximation $S = L_Y/K$. The latter means that at last one reference star should be on each step of the length S.
- If the step S contains two or more reference stars, then the reference point is calculated for the middle point of the step as an average deviation $\Delta X = \Delta_{RA}/M$ and $\Delta Y = \Delta_{DEC}/M$ of stars from their true position on the plate image. Here, M is the size of a pixel in the scale of the plate.
- If there are no any reference stars in the step, the differences of the reference points ΔX, ΔY are calculated by interpolation between two neighboring steps.

17.3 BIG PHOTOGRAPHIC DATA AND THEIR ERRORS

It may seem that too much attention paid to such thorough exclusion of errors is superfluous. Big Data obtained in semiautomatic or automatic surveys or in space missions and transformed into catalogues of stellar characteristics using different computational algorithms always contain both systematic and random errors. Often this factor is crucial as it determines the time interval or sample characteristics when or where the data are reliable. For example, a dense network of reference stars with accurate positions is utterly required for newly discovered near-Earth objects. Their observations are conducted with CCD frames which have small field dimensions and a low number of reference stars. If

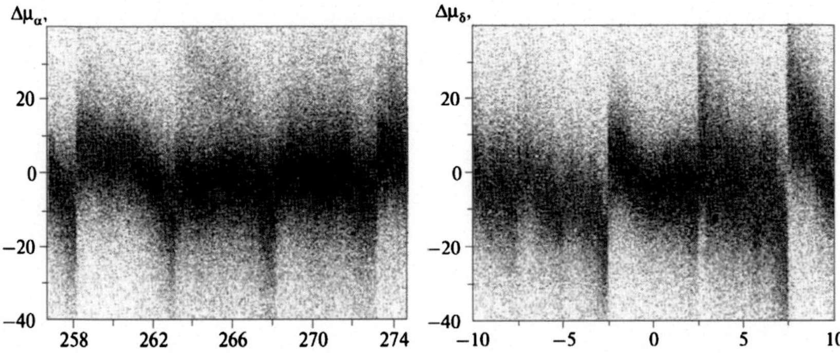

FIG. 17.4 The individual differences of proper motions of stars (XPM-UCAC3.0) in selected fields as a function of RA and Dec (Fedorov et al., 2010b).

there is any danger of a collision with the Earth, then a quickly determined accurate orbit is essential.

Independent sets of Big Data help to analyze errors and propose methods of their elimination.

In 2008–2010 at the Institute of Astronomy of the V. N. Karazin National University of Kharkiv (Ukraine) the *XPM catalogue* was created on the basis of combined data from the Two-Micron All Sky Survey (2MASS) and USNO-A2.0 catalogues. This catalogue was intended to derive the absolute proper motions of about 280 million stars distributed all over the sky, excluding a small region near the Galactic Center, in the magnitude range $12 < B < 19^m$. Proper motions were derived from the 2MASS Point Sources and USNO-A2.0 catalogue positions with a mean epoch difference of about 45 years for the Northern hemisphere and about 17 years for the Southern one. The zero point of the absolute proper motion frame (the "absolute calibration") was specified with the use of about 1.45 million galaxies from 2MASS. The generated catalogue contains the International Celestial Reference System (ICRS) positions of stars for the J2000 epoch, original absolute proper motions, as well as B, R, J, H, and K magnitudes (Fedorov et al., 2009, 2010a). The XPM catalogue is one of the independent realizations of ICRS in the optical and near-infrared ranges. The comparison of the XPM with another ICRS realization, catalogue UCAC-3 (The Third US Naval Observatory CCD Astrograph Catalogue), showed that in UCAC-3 in the fields of 5o × 5o there have been significant systematic distortions in proper motions (Fig. 17.4) in the form of sharp discontinuities.

This factor is crucial for astrometric tasks. For example, positional observations of artificial satellites, potentially dangerous objects, and other solar system bodies are carried out in consecutive series on apparent arcs 5 to 20 degrees long in CCD frames of small dimensions. The presence of systematic distortions of proper motions' field in the widely used reference catalogues (USNO, UCAC) leads to significant errors in positions of studied objects.

ACKNOWLEDGMENTS

The work of authors of this chapter was partially supported in frame of the budgetary program of the NAS of Ukraine "Support for the development of priority fields of scientific research" (CPCEL 6541230).

REFERENCES

Andruk, V., et al., 2017. On the concept of the enhanced FON catalog compilation. Odessa Astronomical Publications 30, 159. https://doi.org/10.18524/1810-4215.2017.30.114411.

Eglitis, Ilgmars, Andruk, Vitaly, 2017. Processing of digital Plates1.2m of Baldone Observatory Schmidt telescope. Open Astronomy 26 (1), 7–17. https://doi.org/10.1515/astro-2017-0006.

Fedorov, P.N., Akhmetov, V.S., Bobylev, V.V., et al., 2010a. An investigation of the absolute proper motions of the XPM catalogue. Monthly Notices of the Royal Astronomical Society 406 (3), 1734–1744. https://doi.org/10.1111/j.1365-2966.2010.16830.x. arXiv:1006.5195 [astro-ph.GA].

Fedorov, P.N., Akhmetov, V.S., Shulga, A.V., 2010b. The reference coordinate systems in the modern astrometry. Kosmìčna Nauka ì Tehnologìâ 16 (6), 68–74. https://doi.org/10.15407/knit2010.06.068.

Fedorov, P.N., Myznikov, A.A., Akhmetov, V.S., 2009. The XPM catalogue: absolute proper motions of 280 million stars. Monthly Notices of the Royal Astronomical Society 393 (1), 133–138. https://doi.org/10.1111/j.1365-2966.2008.14168.x.

Kazantseva, L.V., et al., 2015. Processing results of digitized photographic observations of Pluto from the collections of the Ukrainian Virtual Observatory. Kinematics and Physics of Celestial Bodies 31 (1), 37–54. https://doi.org/10.3103/S0884591315010031.

Mullo-Abdolov, A., et al., 2017. Investigation of the Microtek Scanmaker 1000XL Plus scanner of the Institute of Astrophysics of the Academy of Sciences of the Republic of Tajikistan. Odessa Astronomical Publications 30, 186. https://doi.org/10.18524/1810-4215.2017.30.114407.

Pakuliak, L.K., et al., 2016. FON: from start to finish. Odessa Astronomical Publications 29, 132. https://doi.org/10.18524/1810-4215.2016.29.85140.

Protsyuk, Yu.I., et al., 2014. Compiling catalogs of stellar coordinates and proper motions via coprocessing of archival photographic and modern CCD observations. Kinematics and Physics of Celestial Bodies 30 (6), 296–303. https://doi.org/10.3103/S0884591314060051.

Yizhakevych, O.M., Andruk, V.M., Pakuliak, L.K., 2017. Photographic observations of Saturn's moons at the MAO NAS of Ukraine in 1961–1990. Kinematics and Physics of Celestial Bodies 33 (3), 142–148. https://doi.org/10.3103/S0884591317030035.

CHAPTER 18

Big Astronomical Datasets and Discovery of New Celestial Bodies in the Solar System in Automated Mode by the CoLiTec Software

SERGII KHLAMOV, PHD • VADYM SAVANEVYCH, DSC

18.1 INTRODUCTION

The fast technological progress provokes the creation of big amounts of information. There are a lot of different fields of science that use high-dimensional datasets for analysis. One of them is astronomy.

So, in what way can the big datasets be fed? These data can be fed in different forms, for example, files stream, video stream, physical data saved on different servers, virtual observatories, or even astroplates. All these data can be received from the networks of automated ground- and space-based observation systems or even from old astronomical archives.

For example, the Pan-STARRS in Hawaii contains two telescopes with 1.8 m aperture. Both of them are equipped with the largest CCD camera, which records about 1.4 billions of pixels per image. Each image requires about 2 gigabytes of storage and exposure times will be up to one minute. So, more than 10 terabytes of data are obtained every night (Denneau et al., 2013).

Also the Large Synoptic Survey Telescope (LSST) in Chile currently is under construction. It will take images of the full sky every few nights. There will be about 200,000 of uncompressed images per year, which equals 1.28 petabytes (Tuell et al., 2010).

Such new survey projects and networks of automated ground- and space-based observation systems lead to the fast growing of astronomical datasets that can be provided in the different form and volume.

When we have a lot of astronomical data, what can we do with it? There are a lot of directions of research in astronomy, including the solar system, variable stars, asteroids, comets, near-Earth objects, satellites, and others.

A huge amount of the raw datasets are needed to be processed, but the trend of data growth is ahead of computing abilities of the existing machines. So, using the data mining approach is necessary for all goals that are connected to the processing of big amounts of data. The data mining approach is very useful for the optimization of data stream processing. It allows using only necessary input data to improve the computing abilities of machines.

In one frame there are many thousands of objects, and series of frames can contain hundreds of frames. So, big amounts of data must be processed for the analysis and research.

We can also encounter data mining problems with these big astronomical datasets. How can we solve them?

We need the data preprocessing methods and data reduction models to simplify input datasets by reducing unnecessary information.

Software for the processing of big datasets in automated mode is necessary for the most effective astronomical observations. These datasets can be fed as the surveys given as series of frames.

18.2 BIG ASTRONOMICAL DATA PROCESSING

Modern astronomical systems and telescopes, e.g., Panoramic Survey Telescope and Rapid Response System (Pan-STARRS) (Denneau et al., 2013) and Large Synoptic Survey Telescope (LSST) (Tuell et al., 2010), allow taking a lot of frames of considerable sky area in one night.

This is a big amount of raw datasets that should be processed. The main goal of this can be achieved using the data mining approach.

This approach is provided by the CoLiTec software (Khlamov et al., 2016b), which allows processing of input datasets or streams in real-time. This is a complex system for astronomical data processing, which was cre-

Knowledge Discovery in Big Data from Astronomy and Earth Observation. https://doi.org/10.1016/B978-0-12-819154-5.00030-8

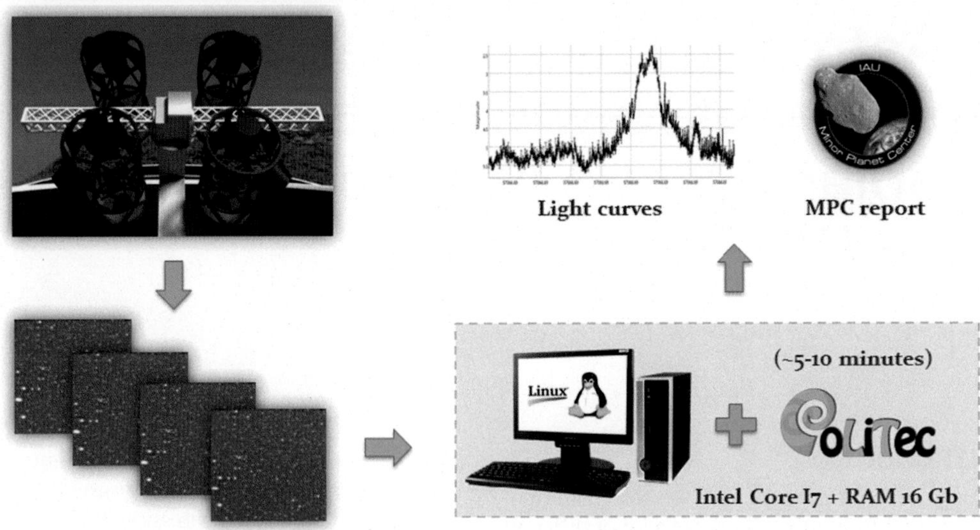

FIG. 18.1 The brief sequence of online processing by CoLiTec software.

ated by the CoLiTec project[1] (Khlamov et al., 2018). This system includes features, user-friendly tools for processing control, results reviewing, integration with online catalogues, and a lot of different computational modules that are based on the developed methods. The visual confirmation of results after processing is also available.

More features of CoLiTec software are described in a detailed manner in the following list:

- working with a very wide field of view (up to 10 square degrees);
- automated calibration and cosmetic correction;
- FrameSmooth software for brightness equalization;
- automated rejection of the worst observations;
- fully automatic robust algorithm of astrometric and photometric reduction;
- automated objects rejection with bad or unclear measurements;
- automated detection of faint moving objects (SNR > 2.5);
- automated detection of very slow and very fast objects (from 0.7 to 40.0 pixels/frame);
- LookSky – processing results viewer with user-friendly GUI;
- multithreaded processing support;
- multicore systems support with the ability to manage individual treatment processes;
- processing pipeline managed by On-Line Data Analysis System (OLDAS);
- decision system for the processing of results that allows adapting the user settings and inform the user about correct results at each stage of processing;
- data control during processing with using the subject mediator.

These features allow effective using of CoLiTec software at the different observatories in the world. All possibilities of the CoLiTec software are provided by the different computational methods that have been developed by authors and implemented in it (Khlamov et al., 2016a; Savanevych et al., 2018).

With the help of the CoLiTec software you can process the observational data that are continuously formed during observation (online stream). The processing pipeline includes corrupted data rejection, brightness equalization of frames, astrometry, photometry, detection of the moving objects, and others.

The main directions of the CoLiTec project are the following:

- CoLiTecVS software for automated reduction of photometric observations (Kudzej et al., 2019);
- CoLiTec virtual observatory (VO) (Pohorelov et al., 2016);
- CoLiTec software for automated detection of the solar system objects (SSOs) (Khlamov et al., 2018);
- FrameSmooth software for brightness equalization and background alignment (Dubovský et al., 2017).

The brief sequence of the online processing is presented in Fig. 18.1.

[1] http://neoastrosoft.com.

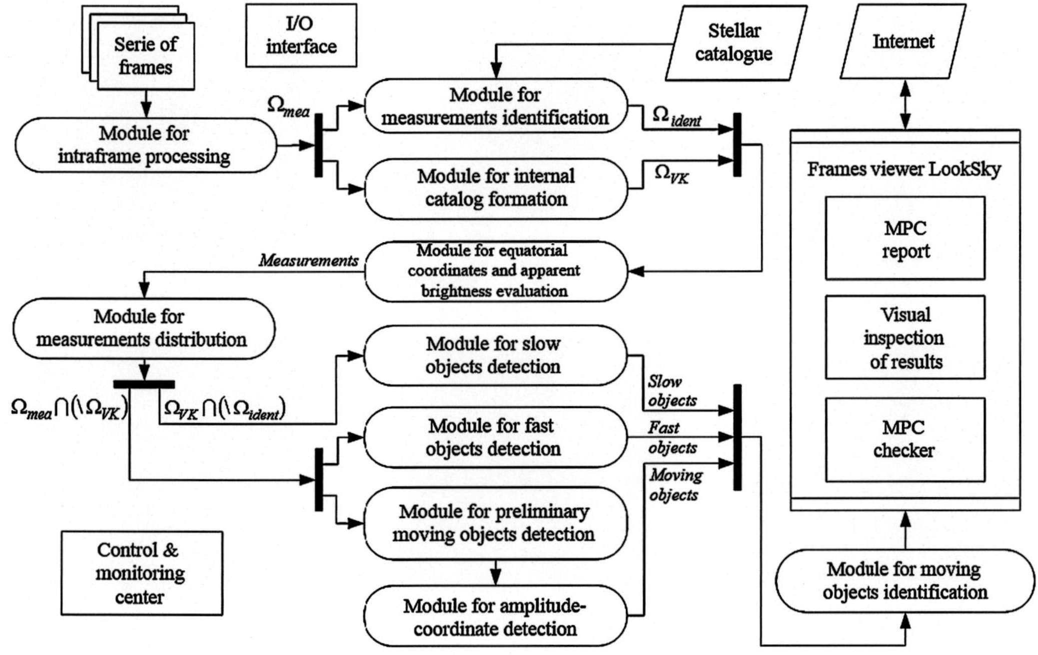

FIG. 18.2 Detailed functional scheme of the CoLiTec software workflow.

Images are saved from the telescope, then they are processed by CoLiTec software in different modes, and we can get different necessary results such as light curves and appropriate reports (MPC, AAVSO, etc.).

The more detailed functional scheme of the CoLiTec software workflow is presented in Fig. 18.2.

There are a lot of different types of telescope aberration that can cause corrupted astronomical data, for example, diffraction rays, motion blur, vignetting, flare light, and coma (Fig. 18.3). Data with aberrations are unnecessary information. So, removing these data during the preprocessing stage allows increasing the quality of processing and reducing the execution time.

OLDAS is an especially significant part of the CoLiTec software. Using OLDAS one can process the datasets and streams as soon as they are successfully saved on the storage or uploaded to the server. So, this is a near-real-time data processing and confirmation assigning of the most interesting objects at the night of their preliminary discovery. This approach allows speeding up of the processing with prevention of data collisions. Also, it provides the immediate notification about emerging issues for the user.

Also, OLDAS provides the ability to process Big Data in real-time. For example, the dataset that includes frames can be used for real-time photometry. The result of processing can be represented as light curves of the investigated variable stars. These light curves will be created and visualized on the server.

According to the data mining approach, CoLiTec software performs the following.

Preprocessing During the preprocessing step CoLiTec software in OLDAS mode starts with the input dataset or stream as soon as they are successfully received. These raw data will be moderated before use in the computational process. At this stage unsupported and corrupted frames will be rejected. Only useful information from datasets will be used in the computing process.

Clustering The remaining useful information in the dataset will be categorized into clusters with the help of specified attributes. CoLiTec software uses the following attributes: equatorial coordinates, telescope, filter type, object under investigation, and others. According to these attributes the appropriate input data from the set will be separated into subsets with similar data.

Classification After clustering subsets, data will be classified by applying known astronomical structures of the raw data as specified in the Flexible Image Transport

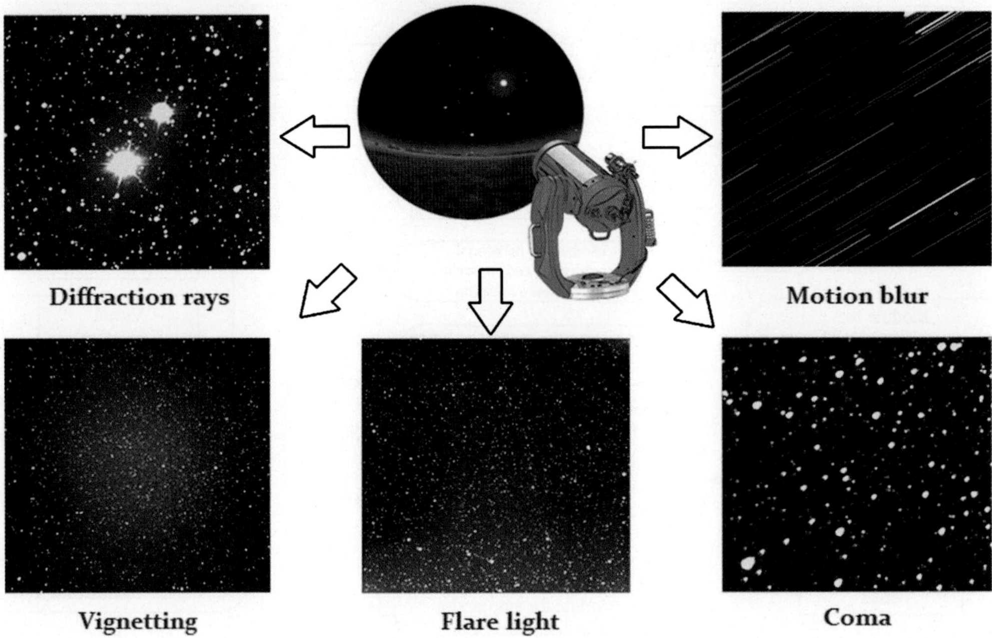

FIG. 18.3 Different types of telescope aberration (diffraction rays, motion blur, vignetting, flare light, coma).

System (FITS) standard by NASA (Wells and Greisen, 1979). FITS is the most commonly used digital file format in astronomy. It is designed specifically for scientific data and includes various astrometric, photometric, or calibration information as the image metadata. After input dataset classification, the FITS files are sent to the processing pipeline.

Identification While the processing pipeline starts receiving the classified FITS files, it identifies types of them. For example, the FITS file could be a raw light frame, or it could be a service master frame that is used in the frame's calibration (e.g., bias, dark, dark-flat, flat). If this is a raw light frame the processing pipeline starts the computing process.

Processing The computing process consists of two stages: intraframe and interframe processing.

Intraframe processing is designed to estimate the position of all objects (stars, galaxies, asteroids, comets) in the frame at current moments. Also calibration, background alignment, and brightness equalization are performed at this stage.

Brightness equalization of images is based on the inverse median filtration and the alignment of interference substrate on the large CCD frames without the use of calibration (master) frames (bias, dark, dark-flat, flat) (Fig. 18.4).

In OLDAS mode the calibration process has the following workflow:

- searching online for the frames in specified directories;
- searching for the required additional frames (bias, dark, dark-flat, flat) if they are not specified;
- creating the appropriate master frames;
- calibration using appropriate master frames;
- applying of the inverse median filter.

The processing results after calibration and brightness equalization are presented in Fig. 18.5.

The next stage of intraframe processing is the identification of all objects (stars, galaxies, asteroids, comets) in the frame at the current moment and the estimation of their positions (Fig. 18.6). For astrometric reduction, first we need to recognize pixels that are related only to the real object's signal. Then software removes all unnecessary pixels from the input dataset to reduce the number of measurements for processing.

The estimation of the object's positions is based on the new iteration method for accurate estimation, which is implemented in the CoLiTec software. This method is based on the subpixel Gaussian model of a discrete object image and operates by continuous

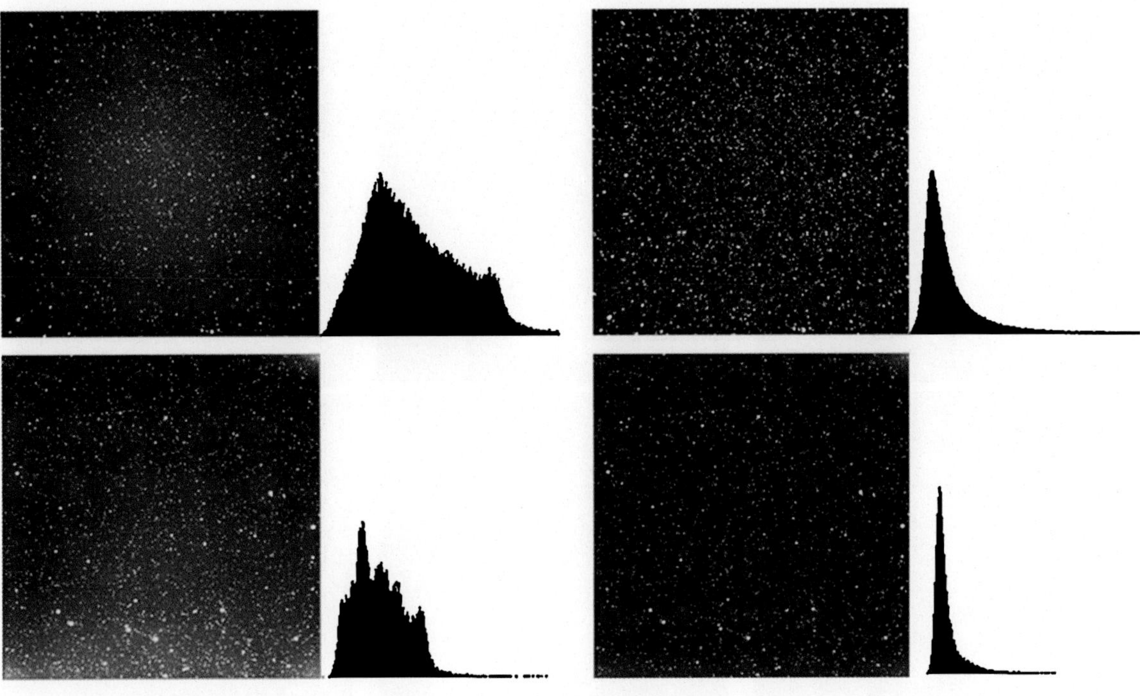

FIG. 18.4 Brightness equalization with histograms.

parameters (asteroid coordinates) in a discrete observational space (the set of pixels potential) in a CCD frame (Savanevych et al., 2015b). In the model of this method the coordinate distribution of the photons hitting a pixel of the CCD frame is known a priori, while the associated parameters are determined from a real digital image of the object. The developed method is more flexible in adapting to any form of the object's image. It has high measurement accuracy along with a low calculating complexity due to a maximum likelihood procedure, which is implemented to obtain the best fit instead of a least-squares method and Levenberg–Marquardt algorithm for the minimization of the quadratic form Savanevych et al. (2015b).

The next stage is a photometric reduction, which includes the estimation of the object's apparent brightness from its signal amplitude (Fig. 18.7). This stage is performed for all real objects in series of frames.

After estimation of the object's position and apparent brightness, the software starts frames identification with the appropriate known stellar catalogues (Ochsenbein et al., 2000).

This is a very difficult procedure because these catalogues contain billions of objects with appropriate information, such as:

- position in the spherical coordinate system;
- errors in determining of the coordinates;
- stellar magnitude in the different photometric bands (brightness of the object);
- standard errors of stellar magnitude;
- proper motions and other useful information.

The main goal of this stage is to understand to what part of the sky these frames are related. The coordinate identification of objects in large stellar catalogues is very complicated. In general, this complication is due to the large random and systematic errors in determining the object's coordinates. Also the complication can be caused by the significant difference between the epochs of observations and the stellar proper motions (Akhmetov et al., 2017). That is why the stellar catalogues with the highest accuracy are required for object identification with high quality (Fedorov et al., 2018; Akhmetov et al., 2018).

Interframe processing is used to detect and estimate objects' trajectories. The core of the CoLiTec software consists of preliminary object detection based on ac-

Image with flash

Image without flash

Raw image

Processing result (without CoLiTec)

Processing result (with CoLiTec)

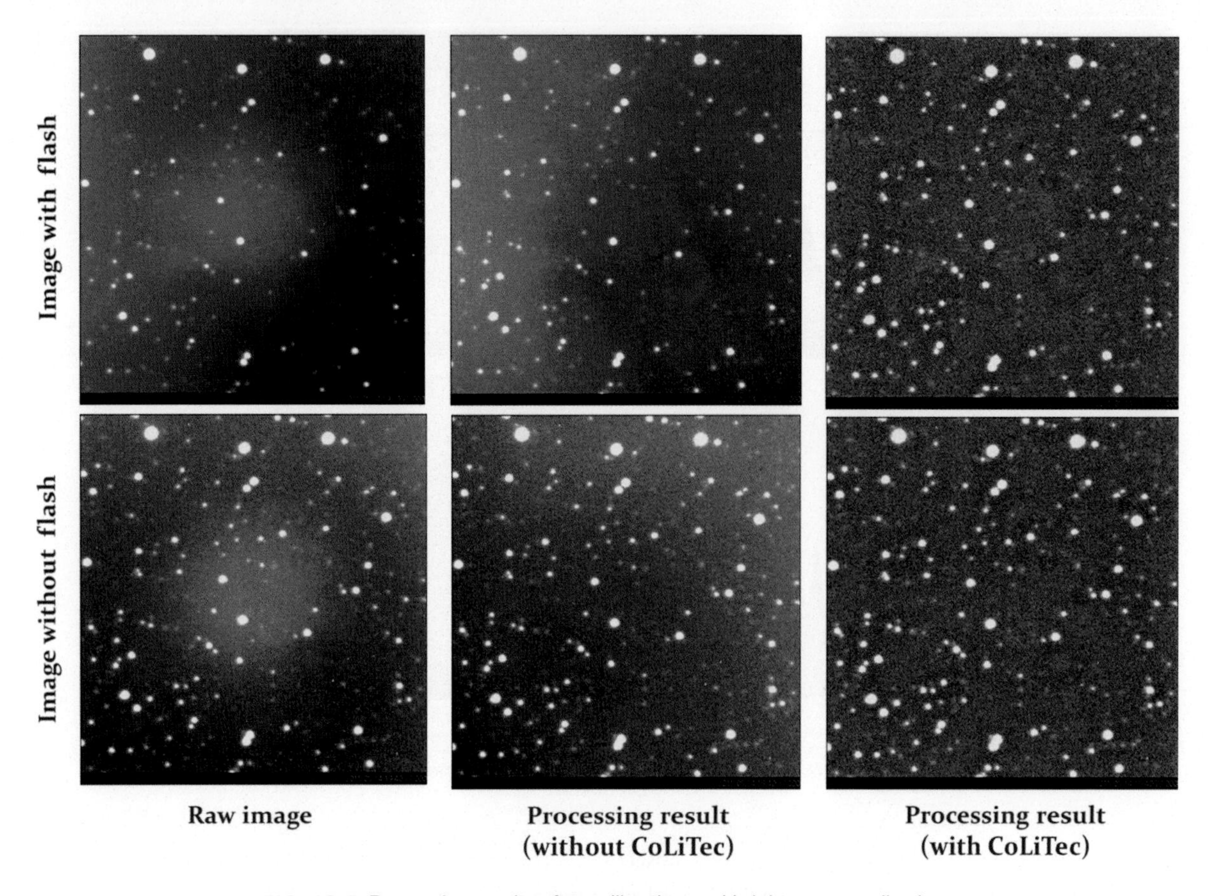

FIG. 18.5 Processing results after calibration and brightness equalization.

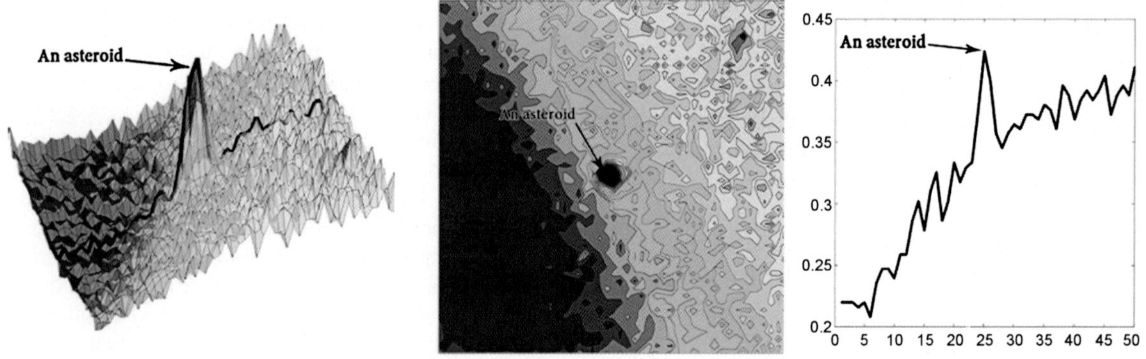

FIG. 18.6 Object identification and estimation of its position.

cumulation of statistics that are proportional to the signal's energy along possible object motion paths (Fig. 18.8). Such accumulation is performed by multi-valued transformation of the object's coordinates that is equivalent to the Hough space–time transformation.

After estimation of the object's trajectories, the software starts identification of object trajectories detected

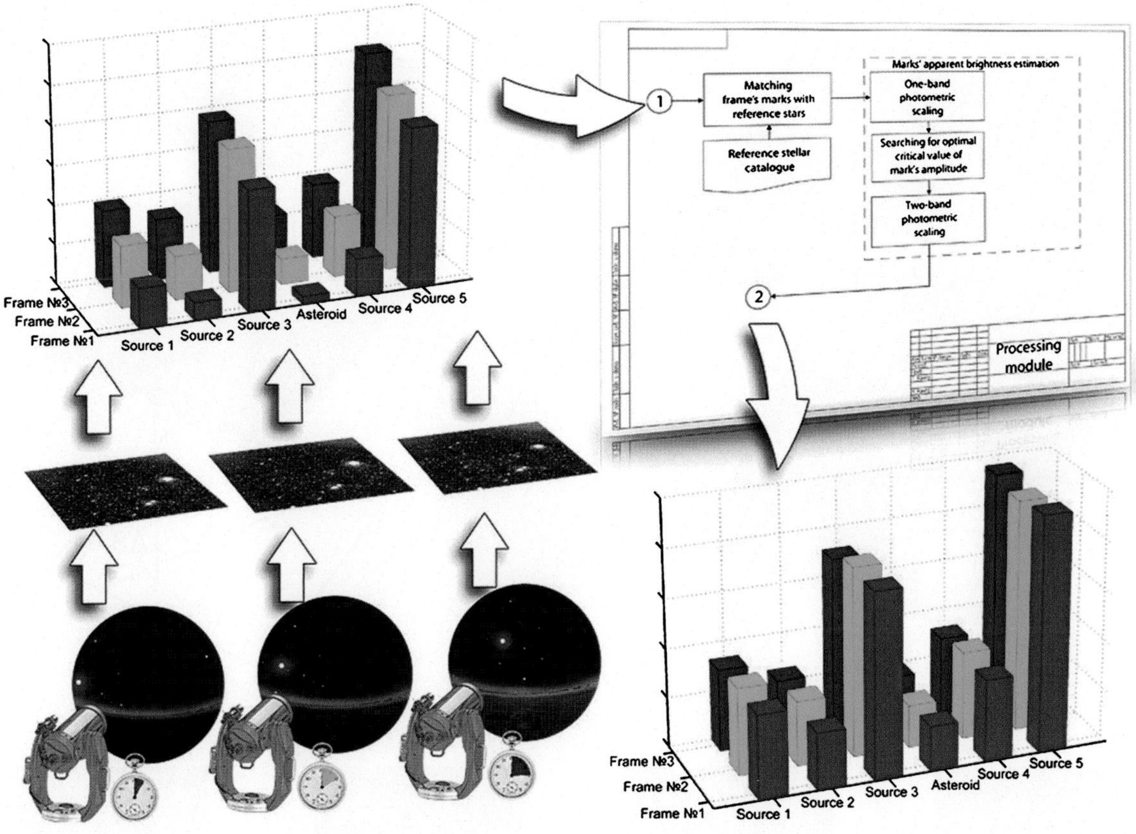

FIG. 18.7 Estimation of an object's apparent brightness from its signal amplitude.

in all frames of series with large international databases that contain information about the billions of trajectories.

CoLiTec software has the ability to detect the very slow, very fast objects and objects with near-zero apparent motion (Khlamov et al., 2016b, 2018). The last one is objects whose interframe shifts in series of CCD frames during the observation are commensurate with the errors in measuring of their positions. These objects have velocities of apparent motion between CCD frames not exceeding three r.m.s. errors of measurements of their positions.

The detection of objects with a near-zero apparent motion in series of CCD frames in the CoLiTec software is performed by a new computational method, which is based on the Fisher f-criterion instead of using the traditional decision rules that are based on the maximum likelihood criterion (Savanevych et al., 2018).

The range of visible velocities of moving objects for detection is from 0.7 to 40.0 pixels per frame. For example, the faster NEO is K12C29D (40 pixels per frame) or the object with near-zero apparent motion is ISON C/2012 S1 comet (0.8 pixels per frame) (MPC, 2012) (Fig. 18.9).

At the moment of discovery of sungrazing comet C/2012 S1 (ISON), its magnitude was equal to 18.8 m, and its coma had 10 arc seconds in diameter, which corresponds to 50,000 km at a heliocentric distance of 6.75 a.u. The size of the comet's image in CCD frame was about 5 pixels (Fig. 18.10, left panel).

Within 26 min of the observation, the comet's image was moved by 3 pixels in the series of four CCD frames (Fig. 18.10, right panel). So, the apparent motion velocity of the ISON comet at the moment of discovery was equal to 0.8 pixels per frame (Savanevych et al., 2018).

All methods that are implemented in the CoLiTec software lead to increases in astrometry accuracy indicators and star photometry quality as well as the quality indicators of asteroid and comet detection.

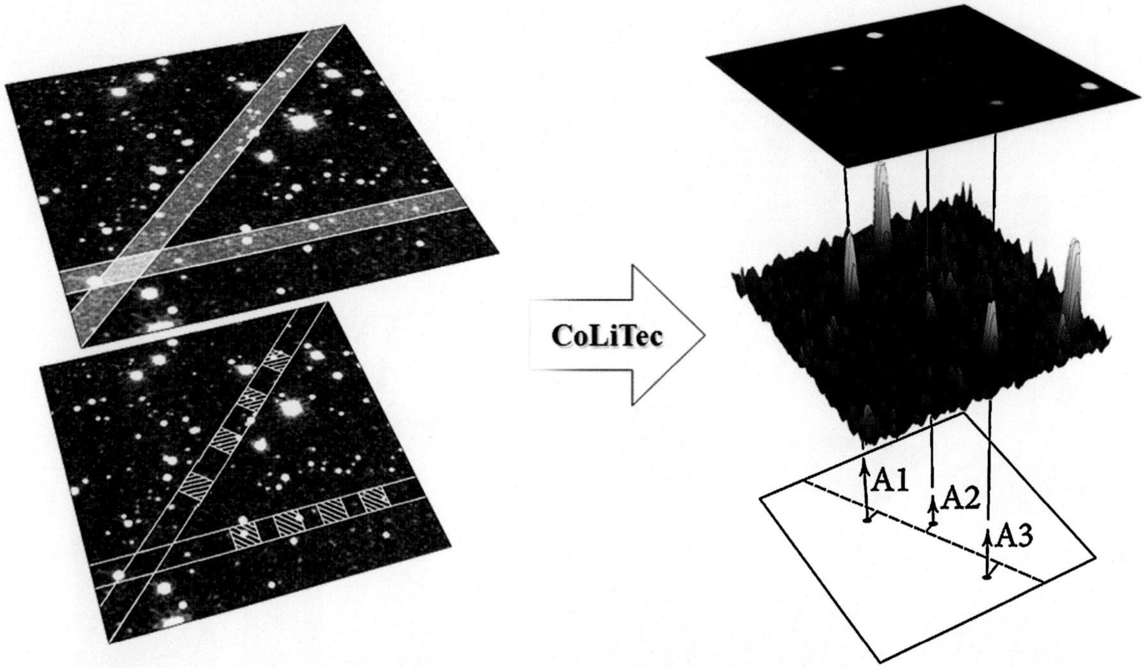

FIG. 18.8 Detection of the moving object.

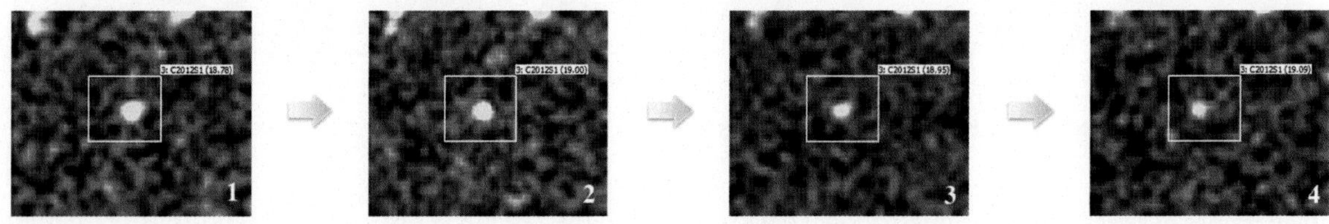

FIG. 18.9 Sungrazing comet C/2012 S1 (ISON) in a series of four CCD frames.

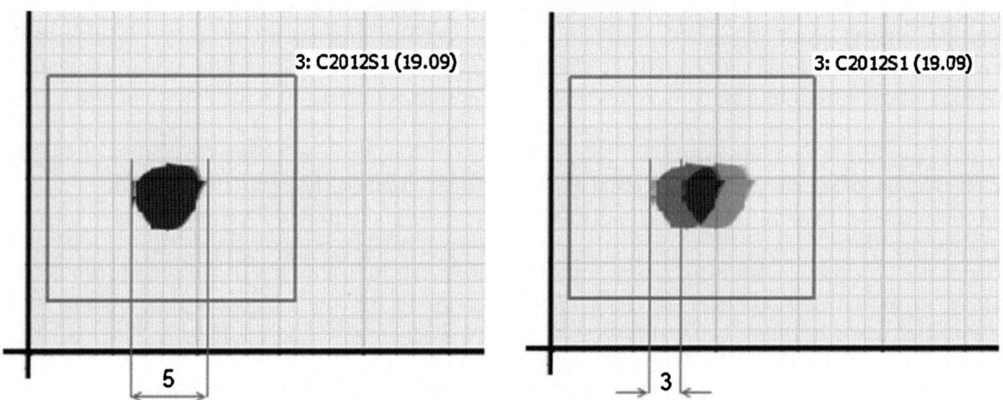

FIG. 18.10 Images of C/2012 S1 (ISON) comet in CCD frames. The image size is 5 pixels. The shift of the comet's image between the first and the fourth CCD frame is 3 pixels.

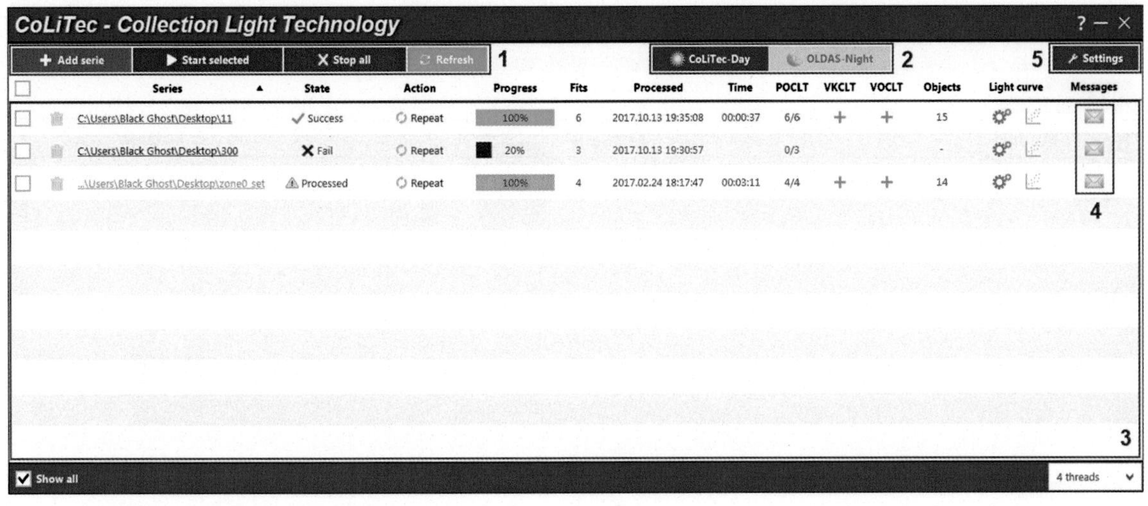

FIG. 18.11 CoLiTec Control Center main window: 1 – control buttons; 2 – processing modes; 3 – working area; 4 – messages during the processing; 5 – program settings.

The CoLiTec software guarantees not only a high efficiency of dataset processing but also a high astrometric and photometric accuracy of the data measurements (Savanevych et al., 2015b, 2018). Full reliability of the detection of moving objects is retained up to the lower limit of SNR equal to three units in the case of a minimum series consisting of four frames, with no star covering of the asteroid (Savanevych et al., 2015b).

The comparison of statistical characteristics of the object's positional CCD measurements with CoLiTec (Khlamov et al., 2016b) and Astrometrica (Raab, 2012) software in the same set of test CCD frames has demonstrated that the limits for reliable positional CCD measurements with CoLiTec software are wider than those with Astrometrica, in particular for the area of extremely low SNR (Savanevych et al., 2015a).

CoLiTec software equipped with the CoLiTec Control Center (3C) is a system for monitoring the processing messages with a detailed logging of handling process and tracking system of all running modules (Figs. 18.11 and 18.12). 3C allows the correct managing and terminating processes at any stage without data losses.

Also, CoLiTec software includes a pipeline for digital video processing. It is presented in the form of a flexible platform for receiving and processing video in any resolution. The pipeline allows for easy integration of the different modules for improving the image quality and detection of the moving objects.

Summarization After pipeline processing, CoLiTec software provides the different forms of dataset representation, including results visualization and generation of report to different services.

CoLiTec software is equipped with a modern viewer of obtained results with a user-friendly GUI. LookSky viewer can be run without the main program (Fig. 18.13). It can be used for independent review of the processing results by CoLiTec during data processing of the main program.

With help of the blinking method researchers and observers cannot carefully analyze all interesting objects in the frame. The particularly serious difficulty for this is the analysis of frames from wide-field telescopes with a huge aperture. Because about several tens of asteroids with slight shine can be present simultaneously in a telescope's field of view, automated frame series processing is necessary for the modern astronomer.

Also, the CoLiTec project developed the CoLiTecVS software – the universal system for automated data reduction that allows the astronomer to create and store the light curves of investigated variable stars without manual data handling between processing steps (Parimucha et al., 2019).

At first, the astronomer with a telescope and specified software takes images of the interesting sky areas where the investigated variable stars are located. These raw images will be moderated before processing. At this stage faulty and unsupported frames will be rejected by OLDAS as soon as they are formed. This technique allows greatly speeding up the processing and provides

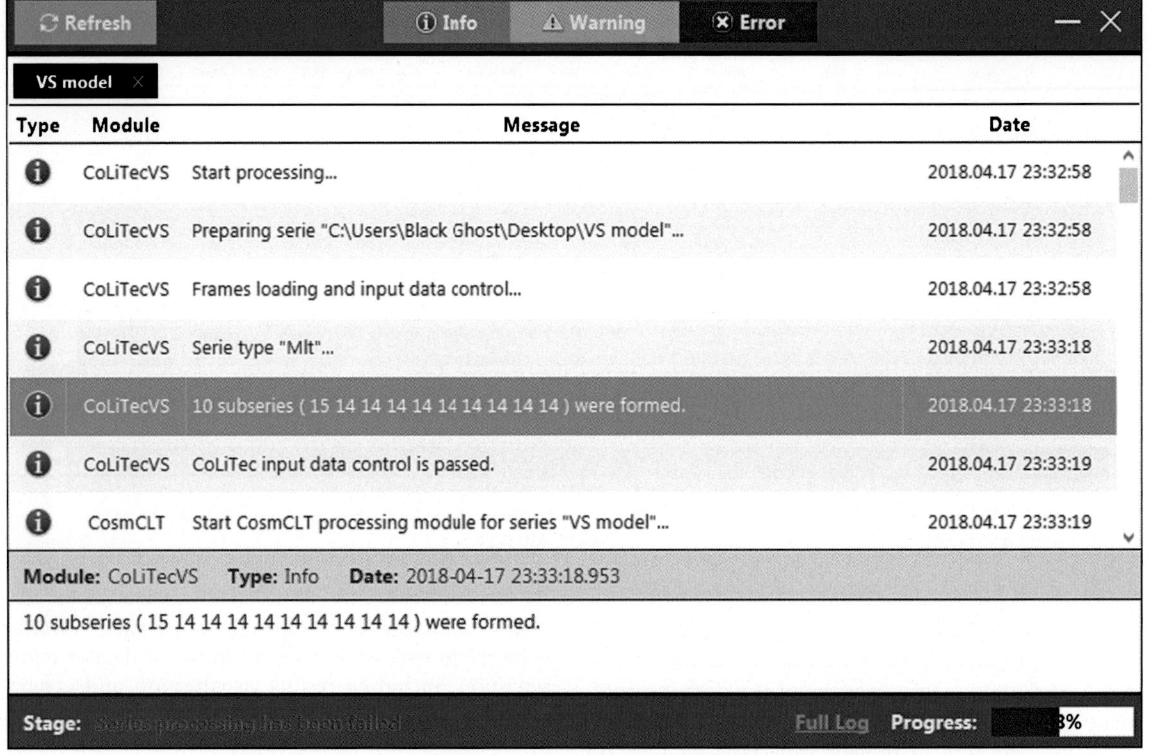

FIG. 18.12 CoLiTec Control Center window with processing messages.

the user with the immediate notification of emerging issues. Also, OLDAS determines the frames' affiliation to the series and telescopes.

CoLiTecVS has the following workflow (Fig. 18.14):

- forming the series of frames with the investigated variable star;
- brightness equalization of frames using calibration master frames and an inverse median filter;
- assessment of the brightness of the investigated star using the ensemble photometry method;
- preparing the task file with selected comparison stars;
- processing of the photometric observations;
- light curve creation of the investigated variable star.

Created light curves can be viewed and analyzed using the specific module of CoLiTecVS: Plots Viewer.

The online dynamics of light curve creation for the specific investigated variable star in Plots Viewer is presented in Fig. 18.15.

The International Virtual Observatory Alliance (IVOA) integrates enormous amounts of data that contain various spectral ranges carrying unique information about celestial objects, which gives possibilities of their use in specific scientific applications to perform astrometric/photometric reduction and moving object detection, such as implemented in the CoLiTec software (Khlamov et al., 2016b).

According to IVOA recommendations for the photometric data storing and VO technologies, the CoLiTec project has created the VO software as a part of the international astronomical community.

The first step was creating the software for storing and publication of CCD frames, which allows to archive and search frames by specified parameters (coordinates). External access to the storage is provided by its own web interface and can be accessed through Aladin by the whole international astronomical community (Fig. 18.16).

The next step was creation of the automatic frames loader, which is accessible via the web interface and allows moderating and then processing frames using OLDAS. Processed frames are stored in the database, from which they are published through an SIAP protocol.

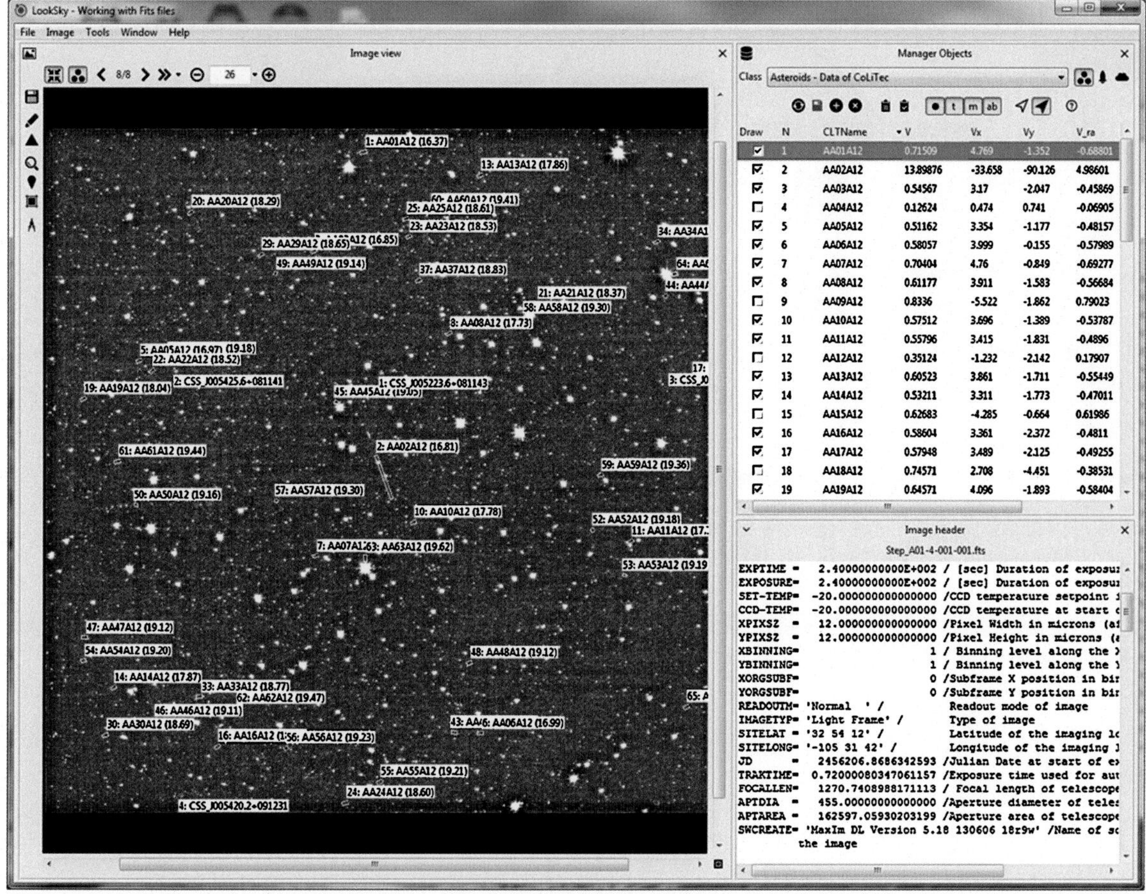

FIG. 18.13 LookSky: visual inspection of processed results.

Then software for the frame storage of CoLiTecVO was extended with the software for light curve storage. Additionally, cross-references between light curves and appropriate frames were implemented (Pohorelov et al., 2016). This allows receiving all relevant information such as required frames of specified sky areas, light curves of the investigated variable stars, or photometric measurements (apparent brightness) from one place.

The main purpose of CoLiTecVO is collecting the astronomical observation results of ground- and space-based instruments and providing astronomers with powerful and easy-to-use instruments for access to the collected data (Vavilova et al., 2012; Pohorelov et al., 2016). The web interface represents light curves as a graph of variation in brightness measurements. It is plotted by measurements from the archive storage or data streams from the different catalogues and/or telescopes (Fig. 18.17).

Processing workflow and pipeline sequence of the VO\ server contains a full stack of processing modules and features that are included in the CoLiTec software (Dubovský et al., 2017; Khlamov et al., 2018).

Also, CoLiTecVO software provides the ability to flexibly convert the values from existing IVOA formats of photometric data. The developed system involves the ability for automatically publishing the light curves uploaded to VO in the international catalogues of variable stars.

18.3 SUMMARY

The CoLiTec software is widely used in a number of observatories in the world. Using this software in the

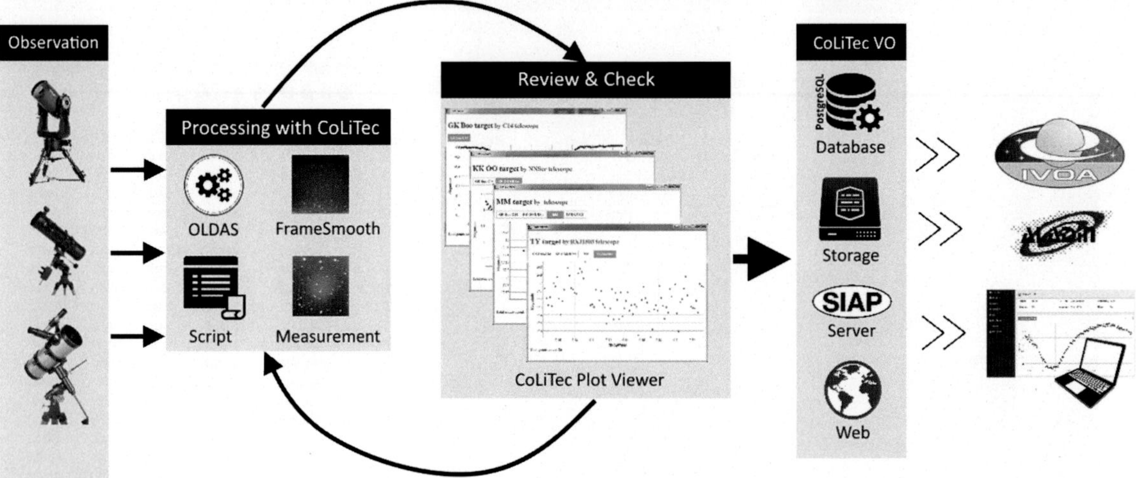

FIG. 18.14 CoLiTecVS workflow for the automated photometric reduction and light curve creation of the investigated variable stars.

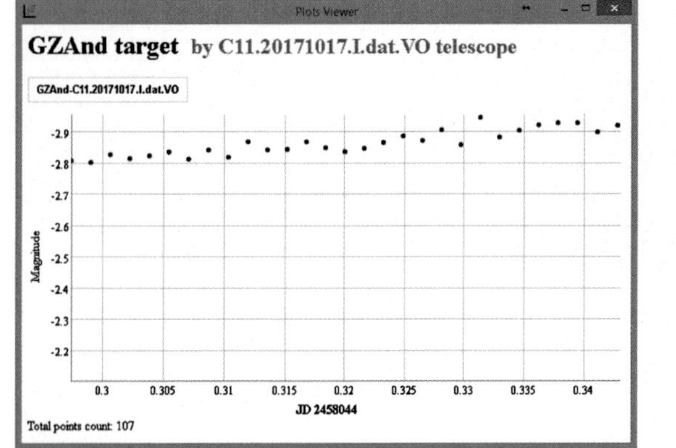

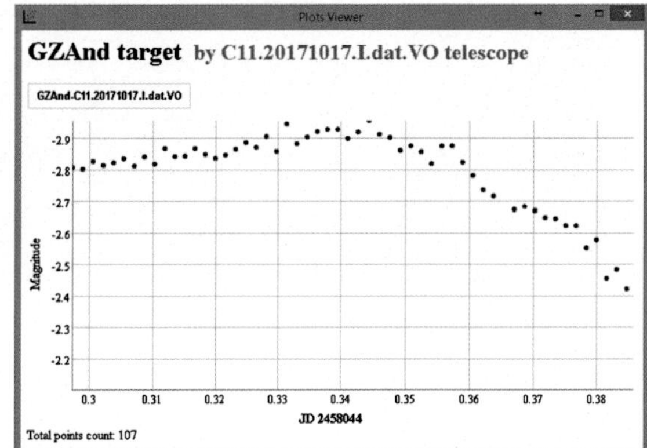

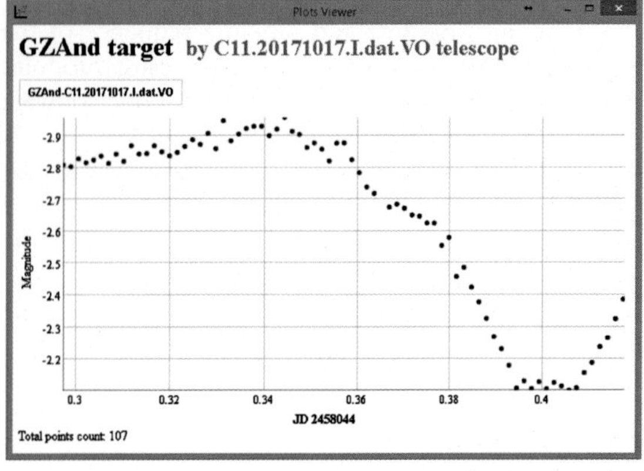

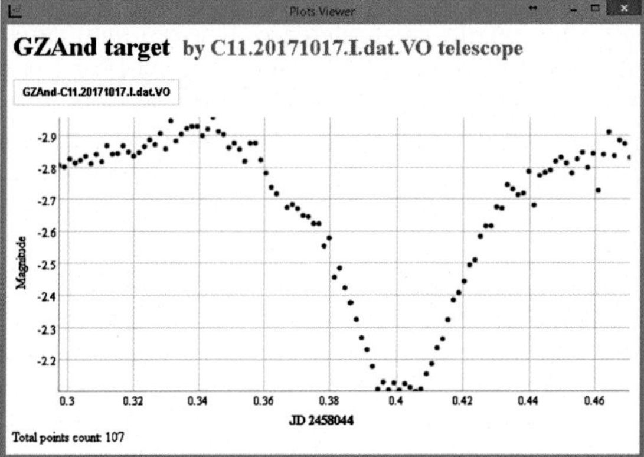

FIG. 18.15 Online dynamics of light curve creation in Plots Viewer.

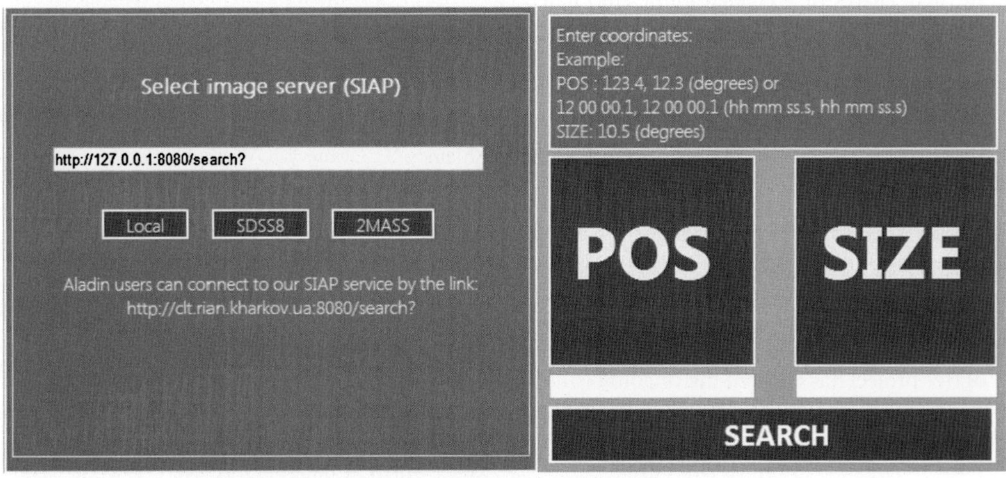

FIG. 18.16 Software for storage and publication of CCD frames.

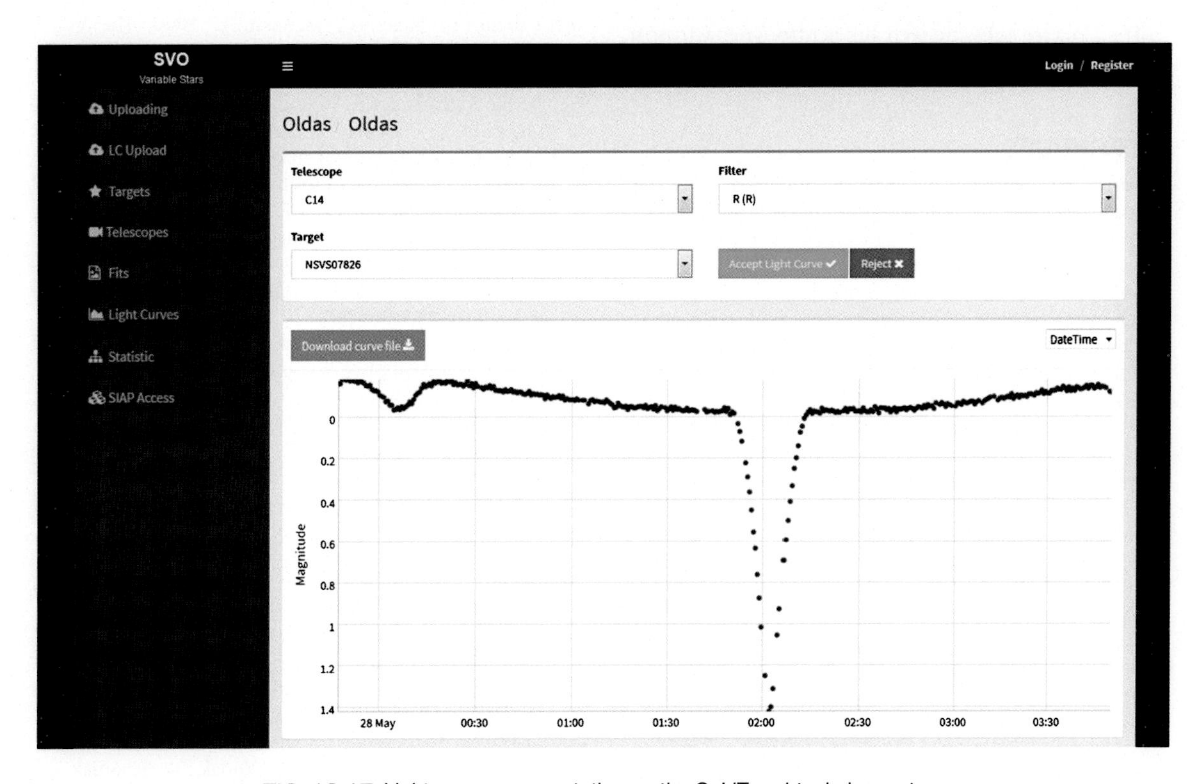

FIG. 18.17 Light curve representation on the CoLiTec virtual observatory.

CIS for the first time, asteroids and comets were discovered and the light curves were created in the automated mode.

It has been used in about 700,000 observations, during which four comets (C/2011 X1, Elenin, MPC, 2011a; P/2011 NO1, Elenin, MPC, 2011b; C/2012 S1, ISON, MPC, 2012; and P/2013 V3, Nevski, MPC, 2013) were discovered. The comet C/2010 X1 (Elenin), discovered on 10 December 2010, was the first comet discovered by a CIS astronomer in automated mode using the CoLiTec software.

In total, the CoLiTec software has been used for the discovery of more than 1600 asteroids, including five NEOs, 21 Trojan asteroids of Jupiter, and one Centaur.

The CoLiTec project has enjoyed the ongoing collaboration for several years with the following organizations: MAO NASU, RI NASU, UkrVO, ISON, Slovakia Virtual Observatory, GOTO project, and NARIT.

In 2014, the CoLiTec software was recommended to all members of the Gaia-FUN-SSO network[2] for analyzing observations as a well-adapted tool for the processing and detection of frames for faint moving objects on CCD frames (Spoto et al., 2018).

The CoLiTec software web site is http://neoastrosoft.com.

REFERENCES

Akhmetov, V., Fedorov, P., Velichko, A., Shulga, V., 2017. The PMA Catalogue: 420 million positions and absolute proper motions. Monthly Notices of the Royal Astronomical Society 469 (1), 763–773.

Akhmetov, V., Khlamov, S., Dmytrenko, A., 2018. Fast coordinate cross-match tool for large astronomical catalogue. In: Conference on Computer Science and Information Technologies. Springer, pp. 3–16.

Denneau, L., Jedicke, R., Grav, T., Granvik, M., Kubica, J., Milani, A., Vereš, P., Wainscoat, R., Chang, D., Pierfederici, F., et al., 2013. The Pan-STARRS moving object processing system. Publications of the Astronomical Society of the Pacific 125 (926), 357.

Dubovský, P., Briukhovetskyi, O., Khlamov, S., Kudzej, I., Parimucha, Š., Pohorelov, A., Savanevych, V., Vlasenko, V., 2017. FrameSmooth software – new tool for the calibration of astronomical images. In: 48th Conference on Variable Stars Research, p. 16.

Fedorov, P., Akhmetov, V., Velichko, A., 2018. Testing stellar proper motions of TGAS stars using data from the HSOY, UCAC5 and PMA catalogues. Monthly Notices of the Royal Astronomical Society 476 (2), 2743–2750.

Khlamov, S., Savanevych, V., Briukhovetskyi, O., Oryshych, S., 2016a. Development of computational method for detection of the object's near-zero apparent motion on the series of CCD-frames. Eastern-European Journal of Enterprise Technologies 2 (9 (80)), 41–48.

Khlamov, S.V., Savanevych, V.E., Briukhovetskyi, O.B., Pohorelov, A.V., 2016b. CoLiTec software – detection of the near-zero apparent motion. Proceedings of the International Astronomical Union 12 (S325), 349–352.

Khlamov, S., Savanevych, V., Briukhovetskyi, O., Pohorelov, A., Vlasenko, V., Dikov, E., 2018. CoLiTec software for the astronomical data sets processing. In: 2018 IEEE Second International Conference on Data Stream Mining & Processing (DSMP). IEEE, pp. 227–230.

Kudzej, I., Savanevych, V., Briukhovetskyi, O., Khlamov, S., Pohorelov, A., Vlasenko, V., Dubovský, P., Parimucha, Š., 2019. CoLiTecVS—a new tool for the automated reduction of photometric observations. Astronomische Nachrichten 340 (1–3), 68–70.

MPC, 2011a. Comet C/2011 X1 (Elenin). URL: http://minorplanetcenter.org/mpec/K10/K10XA1.html.

MPC, 2011b. Comet P/2011 NO1 (Elenin). URL: http://minorplanetcenter.org/mpec/K11/K11O10.html.

MPC, 2012. Comet C/2012 S1 (ISON). URL: http://minorplanetcenter.org/mpec/K12/K12S63.html.

MPC, 2013. Comet P/2013 V3 (Nevski). URL: http://minorplanetcenter.org/mpec/K13/K13V45.html.

Ochsenbein, F., Bauer, P., Marcout, J., 2000. The VizieR database of astronomical catalogues. Astronomy & Astrophysics. Supplement Series 143 (1), 23–32.

Parimucha, Š., Savanevych, V., Briukhovetskyi, O., Khlamov, S., Pohorelov, A., Vlasenko, V., Dubovský, P., Kudzej, I., 2019. CoLiTecVS—a new tool for an automated reduction of photometric observations. Contributions of the Astronomical Observatory Skalnaté Pleso 49, 151–153.

Pohorelov, A., Khlamov, S., Savanevych, V., Briukhovetskyi, A., Vlasenko, V., 2016. Virtual observatory and CoLiTec software: modules, features, methods. Odessa Astronomical Publications 29, 136–140.

Raab, H., 2012. Astrometrica: astrometric data reduction of CCD images. Astrophysics Source Code Library.

Savanevych, V., Briukhovetskyi, A., Ivashchenko, Y.N., Vavilova, I., Bezkrovniy, M., Dikov, E., Vlasenko, V., Sokovikova, N., Movsesian, I.S., Dikhtyar, N.Y., et al., 2015a. Comparative analysis of the positional accuracy of CCD measurements of small bodies in the solar system software CoLiTec and Astrometrica. Kinematics and Physics of Celestial Bodies 31 (6), 302–313.

Savanevych, V., Briukhovetskyi, O., Sokovikova, N., Bezkrovny, M., Vavilova, I., Ivashchenko, Y.M., Elenin, L., Khlamov, S., Movsesian, I.S., Dashkova, A., et al., 2015b. A new method based on the subpixel Gaussian model for accurate estimation of asteroid coordinates. Monthly Notices of the Royal Astronomical Society 451 (3), 3287–3298.

Savanevych, V., Khlamov, S., Vavilova, I., Briukhovetskyi, A., Pohorelov, A., Mkrtichian, D., Kudak, V., Pakuliak, L., Dikov, E., Melnik, R., et al., 2018. A method of immediate detection of objects with a near-zero apparent motion in series of CCD-frames. Astronomy & Astrophysics 609, A54.

[2] https://gaiafunsso.imcce.fr.

Spoto, F., Tanga, P., Mignard, F., Berthier, J., et al., 2018. Gaia Data Release 2—observations of solar system objects. Astronomy & Astrophysics 616, A13.

Tuell, M.T., Martin, H.M., Burge, J.H., Gressler, W.J., Zhao, C., 2010. Optical testing of the LSST combined primary/tertiary mirror. In: Modern Technologies in Space- and Ground-based Telescopes and Instrumentation, vol. 7739. International Society for Optics and Photonics, p. 77392V.

Vavilova, I., Pakulyak, L., Shlyapnikov, A., Protsyuk, Y.I., Savanevich, V., Andronov, I., Andruk, V., Kondrashova, N., Baklanov, A., Golovin, A., et al., 2012. Astroinformation resource of the Ukrainian virtual observatory: joint observational data archive, scientific tasks, and software. Kinematics and Physics of Celestial Bodies 28 (2), 85–102.

Wells, D.C., Greisen, E.W., 1979. FITS—a flexible image transport system. In: Image Processing in Astronomy, p. 445.

CHAPTER 19

Big Data for the Magnetic Field Variations in Solar-Terrestrial Physics and Their Wavelet Analysis

BOZHIDAR SREBROV, ASSOC PROF, DR • OGNYAN KOUNCHEV, PROF, DR • GEORGI SIMEONOV, ASSISTANT

19.1 INTRODUCTION TO BIG MAGNETIC DATA IN SOLAR-TERRESTRIAL PHYSICS

According to the popular understanding of Big Data, in solar-terrestrial physics this would refer to large amounts of measured data satisfying the following properties:

1. their source is not homogeneous,
2. they have large dimension,
3. the size and the format of the data excess the capacity of the conventional tools to effectively capture, store, manage, analyze, and exploit them, and finally,
4. the data have a complex and dynamic relationship.

In the present, public and private institutions are increasingly facing Big Data challenges, and a wide variety of techniques have been developed and adapted to aggregate, manipulate, organize, analyze, and visualize them. The techniques currently applied in big astronomical and Earth observation data usually draw from several fields, including statistics, applied mathematics, and computer science, and institutions that intend to derive value from the data should adopt a flexible, reliable, and multidisciplinary approach. In particular, utilizing Big Data in solar-terrestrial physics and related analytics is expected to improve the performance of prediction mechanisms for extreme geomagnetic events as geomagnetic storms.

Big magnetic data in solar-terrestrial physics are collected at a tick-by-tick level, i.e., at higher frequency. Let us remark that the standard registrations of the geomagnetic field variations have an order of about 0.1–10 mHz (i.e., of periods 1.66 min until 2.77 hours). The analysis of geomagnetic data at higher frequency (here, high frequency is understood from the point of view of the standards in geomagnetism) uncovers the complex structure of irregularities and roughness (i.e., multifractal phenomena) due to huge amounts of microstructure noise. The nonhomogeneity characterized by multifractal phenomena is caused by a large number of instantaneous changes in the geomagnetic field due to geomagnetic storms and various sources of noises as, for example, the low-frequency plasma instability modes. Therefore, mining big geomagnetic data needs to intelligently extract information conveyed at different frequencies. At the present moment the registration of geomagnetic signals has the maximal frequency of seconds; however, the majority of the geomagnetic or ionosound data are still collected at maximal frequency 4 or 5 min, as we will see below.

With the classic assumption of data mining and statistical modeling, data are generated by certain unknown functions representing signals plus random noise (see, e.g., one of the bibles of modern pattern recognition (Bishop, 2006) and the references therein). Analyzing big magnetic data in solar-terrestrial physics is equivalent to extracting the systematic patterns (i.e., approximate the unknown function) conveyed in the data from noise, which is the standard approach of the classic signal processing theory brilliantly presented in the famous handbook Oppenheim and Schafer (2010).

The situation in the modeling of geomagnetic fields and in particular geomagnetic storms falls in the framework of analyzing jump events in Big Data, which has been recently thoroughly studied by various researchers. In particular, it has been studied in the context of financial time series, in the nice research of Sun and Meinl (2012); see also Sun and Yu (2015).

In geomagnetism, jumps are often caused by some unexpected large geomagnetic storms or by predictable changes in the sectorial structure of the interplanetary magnetic field (IMF). Traditional linear denoising methods (e.g., moving average) usually fail to capture these jumps accurately as these linear methods tend to blur them. On the other hand, nonlinear filters are

Knowledge Discovery in Big Data from Astronomy and Earth Observation. https://doi.org/10.1016/B978-0-12-819154-5.00031-X

not appropriate to smooth out these jumps sufficiently, because the patterns extracted by nonlinear filters are not stationary to present long-run dynamic information. The situation is rather similar to that in the case of financial time series, as noted in Sun and Meinl (2012) and in the references therein.

The present chapter deals with research in the global structure of the magnetic phenomena in solar-terrestrial physics, by systematically applying *wavelet analysis* to the available Big Data. We use the *continuous wavelet transform* (CWT) to analyze the structure and the dynamics of geomagnetic storms. The wavelet method is one of a number of multifractal spectrum computing methods and has proven to be a reliable in signal processing, as established in the classical monograph Mallat (2009). It has proven to be very suitable for analyzing time series analysis – for example, smoothing, denoising, and jump detection in diverse areas of science, finance, and economics; see, e.g., Percival and Walden (2000), Gençay et al. (2002), Hubbard (1998), Meinl and Sun (2012), Kounchev (2001).

An important advantage of the wavelet method in analyzing magnetic phenomena in solar-terrestrial physics, where factors of different scales interfere, is that it performs a multiresolution analysis (MRA), in other words, it allows us to analyze the data at different scales (each one associated with a particular frequency passband) at the same time. In this way, wavelets enable us to identify single events truncated in one frequency range as well as coherent structures across different scales. Many recent studies have applied wavelet methods in mining geophysical and geomagnetic data; some recent references are Xu et al. (2008), Xu (2011), Klausner et al. (2016a, 2016b, 2017), Schnepf et al. (2016).

Let us recall that there are in fact two versions of wavelet analysis: CWT and the *discrete wavelet transform* (DWT). In both of them, there is a large variety of wavelet functions by which one may perform the signal decomposition. A common approach in choosing the wavelet function is to use the shortest wavelet filter that can provide reasonable results; see, e.g., Percival and Walden (2000). The main challenge in performing wavelet analysis is to determine the combination of wavelet function, level of decomposition, and threshold rule to reach an optimal smoothness that generally improves the performance of classic models after denoising the data. We have provided a short appendix at the end of the chapter which explains briefly the technical details of wavelet analysis and our choice of wavelet function.

Another major advantage of wavelet analysis (either discrete or continuous) which makes it well adapted to the purposes of Big Data is that, similar to the classical Fourier analysis, there exist very fast algorithms for processing large amounts of data (Mallat, 2009).

Due to space restrictions, in the present chapter we have limited ourselves to just preliminary research of the correlations which exist among the data in the frequency domain (the coefficients of the CWT of the time series). Although the results obtained are very interesting and promising, we have not provided a more detailed statistical analysis (as, e.g., in Sun and Meinl, 2012, Sun and Yu, 2015) which would uncover much deeper and interesting connections between the different factors playing a role in the geomagnetic phenomena. We leave such analysis for follow-up research, which is in progress.

The structure of the chapter is as follows: in Section 19.2 we recall the big picture of the solar-terrestrial mechanism – in quiet and in stormy periods. This has to give an idea to the unexperienced reader about the complexity of the manifestation of a geomagnetic storm and the necessity to apply modern methods of machine/deep learning to Big Data, for a deeper understanding of the phenomena. In Section 19.2.3 we provide a short description of the different components of the ground geomagnetic field provided by Chapman's analysis – the global index D_{St} and local disturbance index DS, which show how complicated the structure of the geomagnetic field on Earth is. In Section 19.2.6 we provide basic information about two famous geomagnetic storms. In Section 19.2.7 we provide basic information about the different types of data used for analyzing the big picture of the magnetic phenomena in solar-terrestrial physics. In Section 19.3 we provide the results of the application of CWT to the three main types of data in the form of simultaneous time series (IMF data, ionospheric TEC data, geomagnetic data), in some quiet days and in days of geomagnetic storms. For every experiment carried out, we provide some empirical observations on the correlations between the CWT coefficients (the frequency domain) for simultaneous time series. In a forthcoming study we will apply the methods of statistical analysis for a more rigorous analysis of these observed correlations between the different types of data.

The large variety of data used, from both solar astronomy and Earth observations, makes our research a contribution to the newly developing area of *Astro-GeoInformatics*.

19.2 MECHANISM OF GENERATING STRONG GEOMAGNETIC STORMS (LONG-PERIOD GEOMAGNETIC FIELD VARIATIONS)

The main purpose of the present study is to analyze different sources of data in solar-terrestrial physics, and to discover correlations between the (relatively) high-frequency geomagnetic variations (wave packages with short period about 0.1 mHz until 10 mHz) which happen not only during geomagnetic storms but also in more quiet periods.

Before carrying out such an analysis we will provide a short outline of the global context of the geomagnetic picture in the solar-terrestrial interactions.

19.2.1 The Big Picture of Solar-Terrestrial Physics - Quiet and Disturbed Geomagnetic Phenomena

In the present section we will outline the big picture of the influence of the solar activity on the IMF, the ionospheric parameters, and the (ground) geomagnetic field.

First of all, we speak about events happening in the magnetosphere, i.e., in the region of space surrounding Earth where the dominant magnetic field is the magnetic field of Earth, rather than the magnetic field of interplanetary space. The magnetosphere is formed by the interaction of solar wind with the Earth's magnetic field. The Earth's magnetic field is continually changing as it is buffeted by the solar wind.

1. In quiet periods the Sun emits a flow of particles (solar wind) having a relatively constant intensity and speed, which start with appr. 50 km/sec and is accelerated to about approximately 360 km/sec close to Earth (which is obtained by the Parker model). The speed is accelerated by a mechanism which is still not known but most probably due to energy transfer in the solar wind.
2. In a disturbed state, the hyperactivity of the Sun causes coronal mass ejection (CME), which increases the amount of charged particles; they have a macrospeed of about 1500–2500 km/sec when they leave the Sun; this speed decreases to about 500–700 km/sec when they approach the Earth.
3. In the interplanetary medium there is a collisionless plasma where the particles are electrons and ions (protons); the *Advanced Composition Explorer (ACE) satellite* collects the data of the *interplanetary magnetic field* (IMF) at a distance of 1.5 million km from the Earth in this plasma (see Fig. 19.1); the most important is the Z component of the IMF called B_Z (which is perpendicular to the ecliptic) which influences the formation of the storm; during the strong storms the component B_Z is negative.
4. After that, the flow of particles approaches the magnetosphere (about 12,000–25,000 km), which is the belt of Van Allen (mathematical figure called torus), where they are caught by the Earth's magnetic field (the Earth's dipole); this looks like a cavern but they are mainly concentrated in the equatorial domain. The so-called *ring current* is formed in the Van Allen belts. In the Van Allen belts one does not have plasma, but there is a kinetic movement of the different particles.
5. Then at about 1000 km from the Earth they enter the plasmasphere and the ionosphere, which is a plasma (partially ionized gas), and the laws describing the plasma state start to work. There are collisions among the particles. Hence, more wave effects may arise compared to the collisionless plasma. The ionosound stations acquire the values of the ionospheric parameters at these heights. For more details, see the excellent exposition in the classical monograph (Mitchner and Kruger, 1973).
6. On quiet days a regular source of disturbances in the Earth's ionosphere is also the solar terminator (at sunrise and sunset).

In Fig. 19.1 we provide the overall picture of different sources of measurements which we have used in our study.

In Fig. 19.2 we provide the magnetic field structure.

19.2.2 Geomagnetic Storms

The geomagnetic storms are by their nature *long-period geomagnetic field variations*.

Geomagnetic storms are phenomena directly related to solar activity. They result from the interaction of the magnetosphere and the ionosphere with changes in the interplanetary conditions caused by closed interplanetary magnetic structures. These structures are formed in the active centers of the Sun's chromosphere as a result of a sudden impulse ejection of the substance in the quiet solar wind called CME.

The storm is mainly characterized by a decrease in the horizontal component H of the geomagnetic field, which encompasses the entire planet during a geomagnetic storm. At low latitudes, for a long time, a current system called the *ring current* is formed around the Earth at a distance of approximately 2 to 4 Earth radii. The nowadays idea of the size and strength of the ring current system gives us reason to view it as a toroid inside the magnetosphere (in the area of Van Allen belts) and formed by particles of the solar wind; see Fig. 19.2. The geomagnetic field in the magnetosphere captures parti-

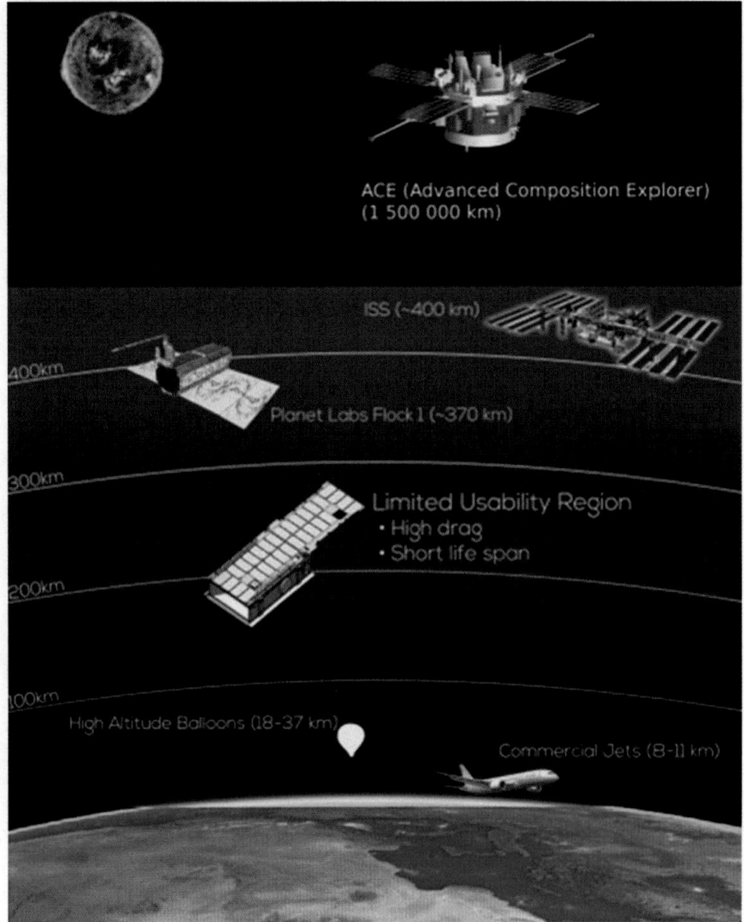

FIG. 19.1 Scheme of the interplanetary space, ionosphere, and Earth, from which we acquire data.

cles from the interplanetary plasma by creating an axisymmetric distribution of these particles in space. This leads to an amplification of the Van Allen radiation belt. The perturbed geomagnetic field associated with the so-called main storm phase in many cases is not axisymmetric. The analysis of the perturbed field shows that it can be represented as a sum of an axisymmetric part and an asymmetric component. This indicates that the proton belt is substantially asymmetric, especially in the early part of the main phase of the storm. Thus, asymmetry appears as an essential feature of ring current formation.

The intensive *proton belt* during a storm significantly deforms the magnetic field of the magnetosphere and changes its structure. In particular, we see that the domain of particle trapping approaches the Earth, in the interplanetary environment, and the polar ray oval shifts in the direction of the equator.

The intensity of the storm strongly depends on the geographic latitude. Thus, the decreasing of the horizontal component of the field during the storm in the different ground magnetic observatories is similar in shape but different in amplitude (see, e.g., Fig. 19.12 with the H component of the geomagnetic field from observatory PAG in Bulgaria, and Fig. 19.13 with the DS index of the geomagnetic field data from observatory SUA in Romania). At low latitudes, this amplitude is larger and decreases as the latitude increases. There is a difference in amplitude of the H component of the magnetic field for the different longitudes. This asymmetry of the field is related to the asymmetry of the ring current considered above.

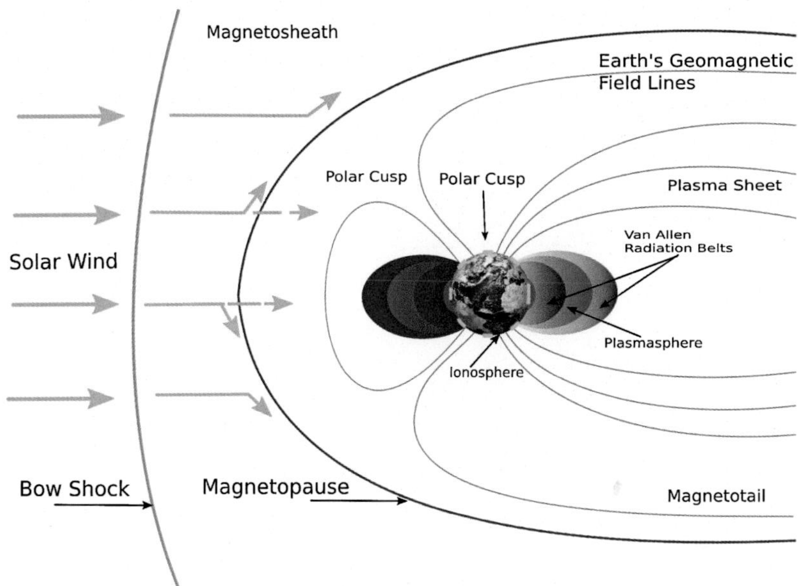

FIG. 19.2 Magnetic field structure.

Geomagnetic storms are very diverse, but they are subdivided into two main types, "standard" and "with sudden commencement." Geomagnetic storms of the second type are characterized by the absence of a pronounced sudden start. However in practice the main features of these storms during the main phase are like those of the standard type. Therefore, the initial contraction of the magnetosphere is not a prerequisite for the occurrence of the main phase of the storm.

For our purposes it is enough to mention that the type of Sun activity (solar chromospheric disturbances near or far from the solar equator) may cause different types of storms. The formation and spread of the interplanetary disturbed structure of plasma and the interplanetary magnetic field is a complex magnetohydrodynamic process that is essentially three-dimensional. The statistical analysis shows that more than 80% of the strong geomagnetic storms are associated with intensive processes in the active centers of the Sun. For average storms, this percentage is smaller and is in the range from 60% to 80%. Briefly, solar astronomy may provide important information which has to be taken into account if predictions are needed.

It happens that important Sun processes *do not have any impact* on the geomagnetic field. This can occur in a relatively small number of cases when the disturbed interplanetary structure does not affect the point of the Earth's orbit in which it is at that moment. The precise understanding of these phenomena requires the joint efforts of astronomers and geophysicists. All this indicates that knowing and predicting geomagnetic storms depends on knowing the propagation of interplanetary disturbances as a hydrodynamic process in three dimensions (Srebrov, 2003). For example, it has been found that the direction of the IMF vector B is essential to unlock the geomagnetic storm mechanism. In particular, if the Z component (denoted by B_z in the coordinate system where the axis Z is perpendicular to the ecliptic) is negative, then this indicates a possibility for a very strong geomagnetic storm.

It would be interesting for the unexperienced reader to hear about the parameters of a simulated *coronal mass ejection* (CME); see details in Srebrov (2003), where a magnetohydrodynamic model is numerically implemented: the release energy is $E = 6.0 \times 10^{22}$ J; the release mass is $M = 2.5 \times 10^{10}$ kg; the initial velocities of the ejected flow are the radial $u_d = 1500$ km/sec and the tangential is $v_d = 0$; the duration of CME is 180 seconds; and the angle of the small conic area associated with the CME is 27°; see a model with these realistic parameters simulated in Srebrov (2003). To understand how the Southern direction of the IMF is formed, we will look at the results of the computer modeling of the disturbance propagation in the interplanetary environment caused by the CME.

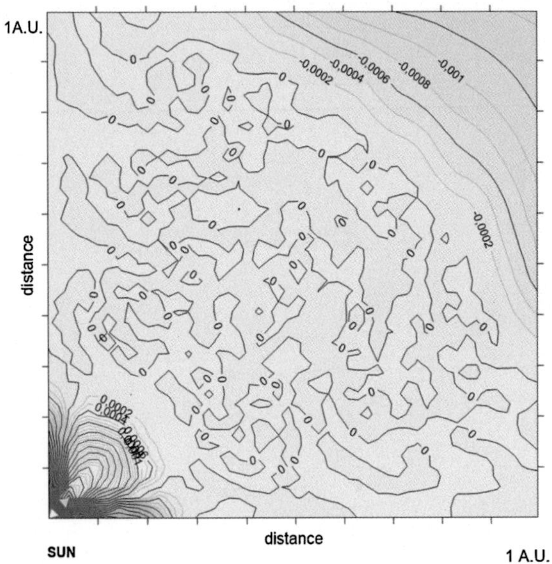

FIG. 19.3 Magnetic field components after CME – contour plot.

As more details are provided in Srebrov (2003), let us shortly describe the dynamics of the simulated disturbance propagation in the interplanetary medium. After the CME happens at $t = 0$ and has duration 3 min, about 40 hours later the disturbance reaches Earth's orbit. In order to provide a deeper perspective on the disturbance caused by the CME, we show the following figures. The magnetic field components are visualized in Fig. 19.3 (see also other similar figures in Srebrov, 2003).

Let us note that Fig. 19.3 visualizes the magnetic field components on the meridional plane. The meridional plane is defined by Sun–Earth as the x axis, and Z is vertical to the ecliptic; in it the tangential velocity v is the component of the velocity along the Z axis.

Further, in Figs. 19.4, 19.5, and 19.6 we provide the radial solar wind velocity U at different times: 10, 30, 50 hours after the CME; note that the value of U in the Figures is dimensionless.

Finally, in Fig. 19.7 we provide the contour plot of the *radial* velocity at 40 hours after the CME.

In Fig. 19.8 we provide the tangential velocity V of the disturbed solar wind at 40 hours after the CME; the values of V are dimensionless.

In Fig. 19.9 we provide the contour plot of the tangential velocity 40 hours after the CME.

As concerns the visualization of the magnetic field, we refer to the paper Srebrov (2003), where detailed figures of the disturbed tangential IMF B_Z are provided. In

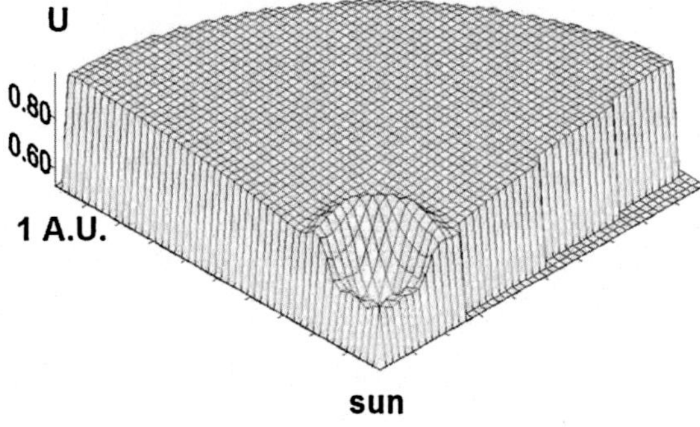

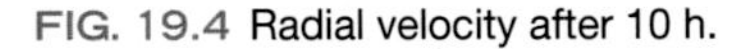

FIG. 19.4 Radial velocity after 10 h.

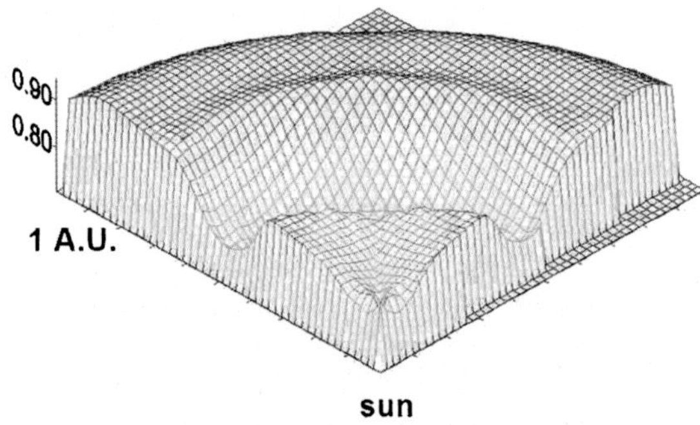

FIG. 19.5 Radial velocity after 30 h.

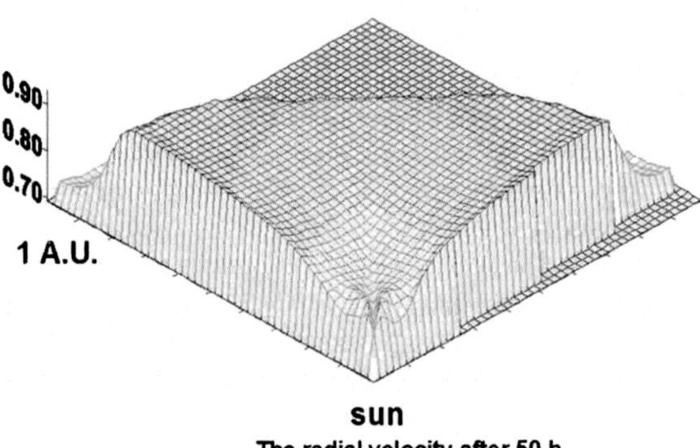

FIG. 19.6 Radial velocity after 50 h.

Srebrov (2003) it is seen that the tangential component of the magnetic field vector B_t (which in fact coincides with the B_Z component in the geocentric Cartesian coordinate system) also has positive and negative values. Thus a structure is formed in which the direction of

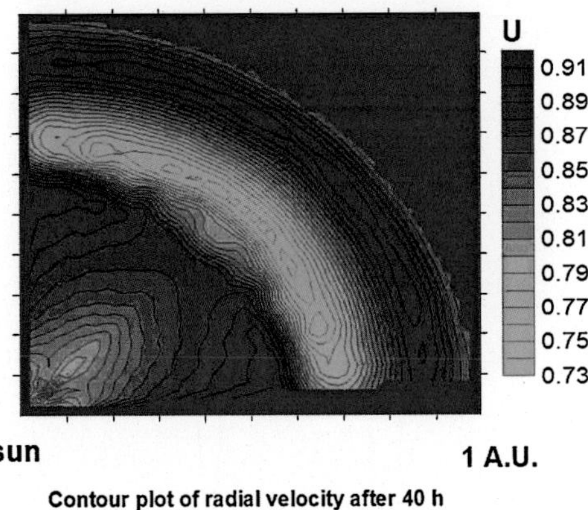

FIG. 19.7 Radial velocity after 40 h – contour plot.

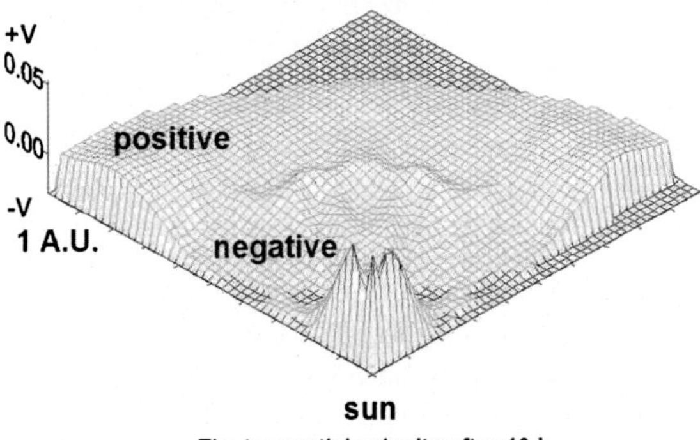

FIG. 19.8 Tangential velocity after 40 h.

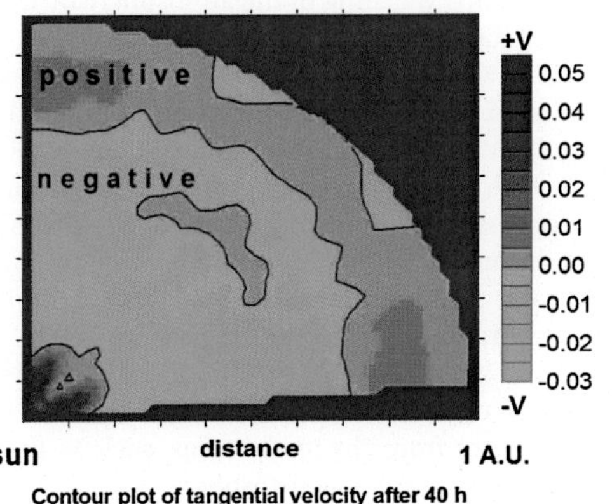

FIG. 19.9 Tangential velocity after 40 h – contour plot.

the disturbed magnetic field vector is changing to the North or to the South (in the same coordinate system mentioned above). As we know, these changes are the ones found to be the main cause of the geomagnetic storms at coupling of the disturbed IMF with the Earth's magnetosphere magnetic field. From the results of the simulations it is visible that the initial relatively closely disturbed conical area, caused by the CME, also expands in the tangential direction during the propagation, and has values which are comparable with the data gathered for example from the observations near the Earth orbit.

The above explanation justifies why the component B_z of the IMF is so important for the understanding of geomagnetic storms.

19.2.3 Ground Geomagnetic Field, and Geomagnetic Activity Index During a Storm

One of the classical models of geomagnetic storm is the Chapman model (Akasofu and Chapman, 1972). It describes the disturbance of the (ground) geomagnetic field variation during a geomagnetic storm.

According to Chapman's analysis (Akasofu and Chapman, 1972), if the time $t = 0$ denotes the sudden start of the storm, then at a certain time point t the disturbed magnetic field vector D measured at a point on Earth's surface (with components denoted by $D(H)$, $D(D)$, and $D(Z)$) can be expressed in a Fourier series expansion as follows:

$$D(\theta, \varphi, t) = c_0(\theta, t) + \sum_{n=-\infty}^{\infty} c_0(\theta, t) \sin(n\varphi + \alpha_n(\theta, t)),$$

where θ is the complement of the geomagnetic latitude to 90°, φ is the geomagnetic longitude, and α_n is the phase angle. The first and second terms in this expression represent the axially symmetrical component of the dipole axis and the asymmetric part of the disturbed field that varies with the longitude φ. These components are referred to as *storm-time variation* (D_{st}) and *local time-dependent disturbance* (DS), respectively, which contain daily regular variation S_r of type S_q and the variance from the asymmetric part of the ring current. So we can write

$$D = D_{st} + DS \qquad (19.1)$$

where DS is a *geomagnetic index* that characterizes local storms. The variation of the horizontal component $D_{st}(H)$ of the D_{st} field is a function of the variable θ. It is larger at low latitudes. The declination $D_{st}(D)$ has little change during the geomagnetic storm. The vertical component $D_{st}(Z)$ also changes slightly compared to the horizontal component. Thus, the D_{st} variation is practically parallel to the Earth's surface except in the areas of the polar cap, where the vertical component has larger positive changes. For that reason the main interest represents the horizontal component $D_{st}(H)$. Hence, we have the following formula for the *horizontal components*:

$$DS(H) = D(H) - D_{st}(H), \tag{19.2}$$

where $D(H)$ is often denoted by $H(t)$ and is called *the horizontal component* of the field.

By neglecting $D_{st}(D)$ and $D_{st}(Z)$ as *small quantities*, it is apparent from the above analysis that the magnitude of the geomagnetic storm is determined mainly by the horizontal component $D_{st}(H)$, and this is the geomagnetic activity index during the storm. It describes the intensity of the symmetrical part of the circular current that occurs in the equatorial area of the magnetosphere during a magnetospheric storm. This is the global part of the geomagnetic storm index. It represents the mean value of the disturbed horizontal component of the geomagnetic field determined by data from several low-level observatories, distributed by geographic lengths. On quiet days, this variation may be around ±20 nT, but during a geomagnetic storm it reaches large negative values of the order of hundreds of nT.

Remark 19.1. The determination of the D_{st} *index* is provided every hour, and is practically determined by taking an average of the data through a consortium of geomagnetic observatories. The process of practical determination of this geomagnetic index is described in detail in the IAGA bulletin.

Remark 19.2. Geomagnetic field variation data in the past decades contained mean *hour* values; however, the present data contain mean *minute* values, and in all INTERMAGNET observatories, the registration is done with *mean second* values. This shows an increase in the information about the geomagnetic field and one may speak about a "Big Data" shift of the measurement paradigm.

19.2.3.1 The D_{st} Index During the 2003 Storm

In order to give an idea about the general form of the D_{st} index during the storm we provide the D_{st} data during the storm on 29 October 2003, downloaded from the World Data Center (WDC) for geomagnetism in Kyoto; the data available cover 3 days, every hour; see Fig. 19.10. In Fig. 19.10 we see that the D_{st} variation has two big decreases due to two different CMEs.

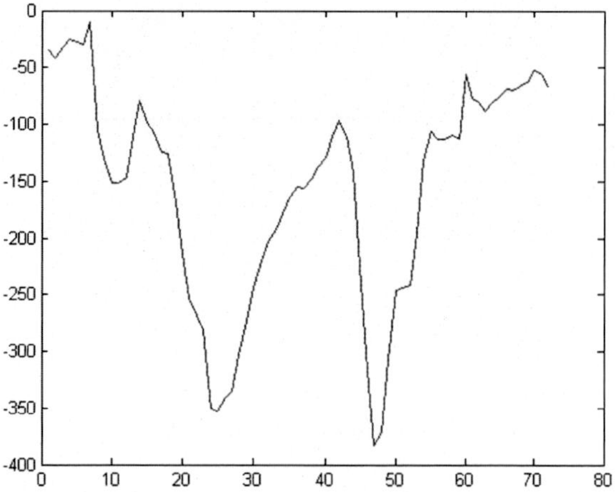

FIG. 19.10 The D_{st} index from 29 October until 1 November 2003.

19.2.4 Ionospheric Parameters From Ionospheric Sounding Stations

Let us recall that the ionosphere of the Earth can be seen as a *conductive layer*. Its motion induces an electromotive force $v \times B$. Here v is the vector of the *drift velocity* of the charged particles (not to be mixed with the tangential macrovelocity of the solar wind!), and B is the geomagnetic field vector in the *ionosphere*.

Let us mention that the ionosphere is influenced by different factors which cause some *long-period (low-frequency)* ionospheric parameter changes:

1. *Solar tidal* movements and movements, caused by the periodic warming of the atmosphere by the Sun, cause changes in the geomagnetic field, called quiet-solar variations. The latter, as noted above, are referred to as S_q. In turbulent conditions, we have a regular variation S_r associated with the local time, which differs from S_q due to the influence on the magnetospheric storm ionosphere. As with S_q, the conditions in the high atmospheric layers are particularly important and essential. For example, the concentration of the charged particles in the ionosphere, which depends on UV ionization, as well as the degree of invasion of charged particles in the ionosphere from the above areas, such as the plasmasphere and the magnetosphere.
2. The distribution in the ionosphere of the electromotive forces, determined by the *lunar tidal* movements of the atmosphere, is approximately fixed in relation to the moon. Thus, the induced current and the resulting magnetic field are also fixed for an observer on the moon. Each magnetic observatory performs one turn per day about this distribution, turning in

a circle defined by the latitude. Therefore, the stations register a variation of the geomagnetic field over time. This variation is called lunar-day and is denoted by L.

19.2.4.1 Data About the Parameters of the Ionospheric Plasma

In the present work we have used data for different parameters of *ionospheric plasma* in a specific local area, as TEC, F2, etc. In the experiments presented we have limited ourselves to the TEC data since it is considered as the most important of all parameters. For example, we use the ionosphere sounding data from ground ionospheric stations, which are chosen to be located near the (ground) geomagnetic field registration points. This allows by comparison of the two "signals" to seek for the presence or absence of a possible correlation between them. Thus, we have the possibility of identifying the origin of individual modes (wave packages) and groups of modes.

The *ionospheric* and geomagnetic data are synchronized in universal time so that we can monitor the presence or absence of simultaneity of the two signals. We have used data on ionospheric parameters with a sampling frequency comparable to the geomagnetic data, namely, every 5 minutes. This is very interesting in terms of decomposition of geomagnetic variations. Of course, the *modes* associated with extraionospheric origin can also be identified and this has already been commented (as for example the geomagnetic pulsations).

Let us note that in recent years, the *ionospheric* data registration also tends to increase the sampling frequency and the sampling has now reached in some stations a period of 5 minutes between two samples. Given that a large number of ionospheric plasma parameters are measured at these stations, we can assume that they may also be referred to as "Big Data" paradigm in this area.

For example, in Fig. 19.11 we provide a graph of the TEC data from the ionosound station in Athens, on a quiet day, 4 January 2018. The variation of the TEC data is shown. We see that the main trend is given by the daily variation of the TEC. The fast oscillations with periods less than 3 hours are observed during the whole day. At midday time we observe modes (wave packages) with a larger amplitude which are apparently of soliton type. This phenomenon has been studied in Belashov and Belashova (2015).

Remark 19.3. An interesting research of the short-period modes (wave packages) in the ionosphere was carried out by V. Belashov, who created the soliton

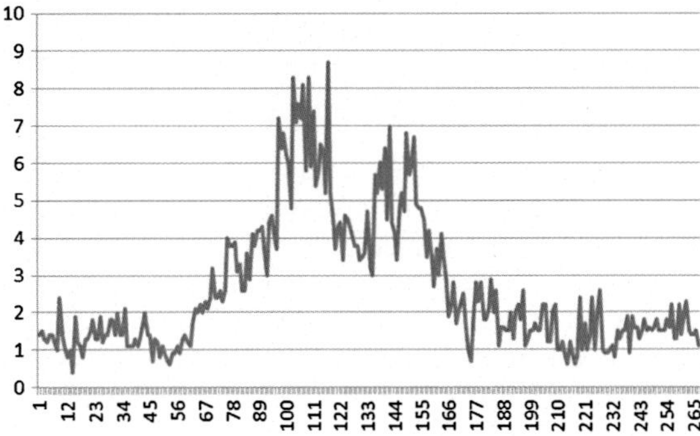

FIG. 19.11 Graph of the TEC data from the ionosound station in Athens, on 4 January 2018.

model for their explanation; see Belashov and Belashova (2015) and references therein. Matching of the model to the TEC values is presented on p. 338. See also Srebrov et al. (2018) for confirming observations.

19.2.5 Emergence of Higher-Frequency Modes in the Ionospheric Parameters and in the IMF Which Are Related to the Ground Geomagnetic Field Variations

By now we have considered the long-periodic geomagnetic variation phenomena. Below we describe shortly the generation of short-period (high-frequency) geomagnetic variations caused by *low-frequency plasma instabilities*.

As already mentioned above, the plasma in the ionosphere, plasmasphere and the interplanetary collisionless plasma medium have considerable instability. Instability also occurs in the radiation belts of the Earth, i.e., in the magnetosphere, although most of the latter do not realize the conditions characterizing the medium as plasma. Some examples of cosmic plasma show the presence of instabilities that produce nonthermal waves and various distribution functions of kinetic particles parameters, especially in the ionosphere and the plasmasphere.

Instabilities can also cause a nonlinear effect of wave propagation in the ionosphere, plasmasphere, and interplanetary medium, and they also cause collisions and acceleration of fast particles in astrophysical plasma.

For the current work, it is important to note the low-frequency instabilities in terms of natural plasma frequencies (magnetohydrodynamic instabilities, fluid instabilities, and drift instabilities). They create the low-frequency modes in the geomagnetic field and in the IMF. The properties of the wave modes are strongly dependent on the frequency. For low-frequency plasma modes, the circular frequency of the waves is much smaller than the natural frequencies as the plasma frequency and the cyclotron frequency of the plasma.

Practically, very often these *low-frequency* plasma modes coincide with the *higher-frequency* geomagnetic field variations considered by us. In this work, we use geomagnetic field registration data on the ground with periods more than 1 minute that are associated with *low-frequency modes* in the Earth's plasma cosmic environment, and also with ring current fluctuations in the magnetosphere.

In *plasma*, the macroinstabilities occur in the low-frequency mode and usually involve the magnetic field (not just the ground geomagnetic field!). Therefore, in the *present research* we use data about the variations of the magnetic field in different areas of the near-Earth space and those generated in the interplanetary space.

As concerns the *short-period (higher-frequency)* ionospheric parameter variations, let us recall that the variation can be decomposed, as a superposition of sources located in the magnetosphere and the ionosphere and also associated with the various tidal movements of the high atmosphere. The high-frequency (ground) geomagnetic field variations may be a result of short-period ionospheric parameters variations.

The above arguments justify the *necessity* to use data on the *ionosphere* status as well as data of the *interplanetary magnetic field* (IMF) and the ground geomagnetic field during the storm.

In the present work, in each case, we choose to study the behavior of the various magnetic field components that are associated with certain processes in the respective area of observation.

As we have mentioned above, the disturbances of the IMF and of the solar wind propagate in the interplanetary medium. Beyond that, in the interplanetary medium different *low-frequency* plasma modes are generated, as a result of the plasma instabilities. Data on IMF variations induced by macroinstability of the interplanetary plasma medium are recorded by satellites located outside the Earth's magnetosphere, as for example the *satellite ACE*, whose data we use. These data contain the values of the components of the magnetic vector as well as the values of its magnitude, measured by different instruments. In order to study these instabilities and related waves in the interplanetary medium, we use data for the IMF with data sampled every 4 minutes. Magnetic modes in these media and in this frequency range are associated with the so-called macroinstability.

Finally, *one of the main objectives of our study* is to identify modes with *short periods* (with frequencies much larger than the natural plasma frequencies), which are generated by microinstabilities, and are caught by the ground geomagnetic field registration and in the plasma ionospheric parameters.

19.2.6 The Strong Geomagnetic Storms in 2003 and 2017 to Be Analyzed

In the present research we apply wavelet analysis to analyze short-period geomagnetic field variations, ionospheric parameter variations, and IMF variations, during the manifestation of the two famous strong geomagnetic storms during the 23rd solar cycle in 2003, and during the 24th solar cycle in 2017.

19.2.6.1 The Storm in 2003

The data obtained from solar astronomy observations have provided the following report.

During the 23rd solar cycle in October and November 2003 there were two very strong storms. One started on October 29 and the other began on November 21. In the last 10 days of October 2003, the lean activity has gone to an extremely high level. On October 18, a large active region (AR), turning North of the solar equator, was designated by NOAA as AR 484. On October 28, AR 484 was located near the sub-Earth point of the solar disk 8° on the East of the central meridian and 16° North latitude. At 11:10 UTC AR 484 produced one of the largest solar flares for the current solar cycle. This flare was classified as X17 (peak X-ray flux 1.7×10^{-3} W/m^2).

An extreme CME with a radial plasma velocity of 2500 km/s was observed. The mass ejected from this CME was in the range of 1.4–2.1 × 10^{13} kg, and the kinetic energy released was 4.2–6.4 × 10^{25} J. The following day, 29 October, AR 484 again produced a large eruption. This peak was named X10 (X-ray flux 10^{-3} W/m^2) at 20:49 UTC. I was targeting the Earth halo at a speed of 2000 km/s and with a kinetic energy of 5.7×10^{25} J. The IMF reached about −50 nT; its normal value, in calm conditions, is 10 times lower. The shock wave of the event on 28 October was determined by the ACE satellite at 05:59 UTC. At 06:13 UTC an SSC pulse was registered, marking the beginning of the sixth storm by the registration stamp (since 1932). On 29 and 30 October the planetary index Kp reached a value of 9. The geomagnetic storm continued until 1 November and had a horizontal component down to around −400 nT. The highest value of the D_{st} index was registered on 30 October at 23:00 UT.

19.2.6.2 The Storm on 7 and 8 September 2017

The other storm considered in the present work is the one on 7 and 8 September 2017. It was one of the most flare-productive periods of the now-waning solar cycle 24. Solar AR2673 and AR2674 both matured to complex magnetic configurations as they transited the disk. AR2673 transformed from a simple sunspot on 2 September to a complex region with order-of-magnitude growth on 4 September, rapidly reaching beta-gamma-delta configuration. On subsequent days the region issued three X-class flares and multiple partial halo ejecta. Combined, the two active regions produced more than a dozen M-class flares. As a parting shot AR2673 produced (1) an X-9 level flare; (2) an associated moderate solar energetic particle event; and (3) a ground-level event, as it arrived at the solar West limb on 10 September. From 4–16 September the radiation environment at geosynchronous orbit was at minor storm level and 100 MeV protons were episodically present in geostationary orbit during that time frame. The early arrival of the CME associated with the 6 September X-9 flare produced severe geomagnetic storming on 7 and 8 September. The full set of events was bracketed by high-speed streams that produced their own minor-to-moderate geomagnetic storming.

19.2.7 Acquired Data for Short-Period Variations of the Geomagnetic Field, the Ionospheric Parameters, and the IMF

As we have explained in Section 19.2, the global picture of the geomagnetic phenomena is very complicated and dynamic. For that reason, for the explanation as well as for the prediction of its dynamics one needs to attract as much as possible observable data, which form the basis of our Big Data analysis.

We analyze high-frequency time series data from different sources. Let us recall that "high-frequency" registrations in geomagnetism are of the order of 0.1–10 mHz (i.e., of periods 1.66 min until 2.77 hours). The following three high-frequency time series were acquired:

1. T_G: Time series for the ground geomagnetic data (from ground geomagnetic observatories, 1 min sampling and 1 sec sampling), in nT. Our main objective is to seek for correlations in the wavelet coefficients of the CWT of the above time series which explain the dynamics of different geomagnetic phenomena.
2. T_{IP}: Time series for the ionospheric parameters – TEC (from ionospheric sounding stations, 5 min sampling), in TEC units.
3. T_{IMF}: Time series for the IMF (from the ACE satellite, 4 s sampling), in nT.

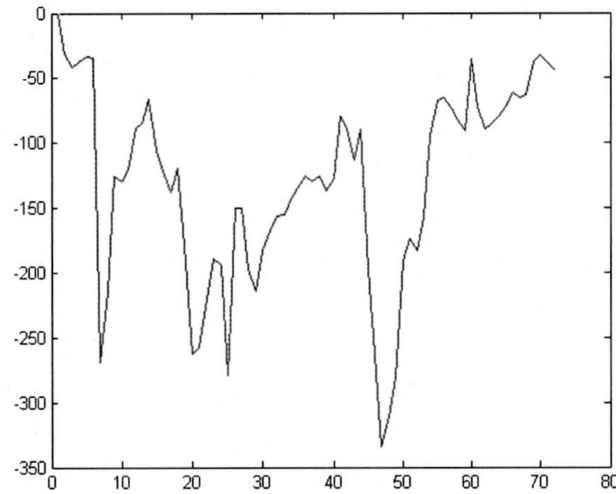

FIG. 19.12 The mean hour values of the H component registered at PAG, 29–31 October 2003.

19.2.8 Data About the Strongly Disturbed Geomagnetic Field in October and November 2003 and September 2017

19.2.8.1 The H Component of the Geomagnetic Data From the Panagyurishte (PAG) Observatory

In Fig. 19.12 we visualize the variation of the H component (see formula (19.2)) registered at the geomagnetic observatory PAG during the geomagnetic storm on 29 October 2003. The data are mean hour values, registered from 0:00 on 29 October until 23:59 on 31 October.

19.2.8.2 The DS Index From the Surlary (SUA) Geomagnetic Data

In Fig. 19.13 we provide the data for the index DS variation (defined in formula (19.1)) during the storm on 29–31 October 2003, registered at the Surlary (SUA) geomagnetic observatory in Romania. The sampling of the data is 1 minute.

We see all details due to the fact that the data are provided every single minute.

Remark 19.4. It is questionable whether it is worth applying the CWT to the DS index given by $H(t) - D_{st}$ (see formulas (19.1) and (19.2)), or directly to the raw data of the H component, since we are in principle seeking for variations of H, but the D_{st} index is provided on an hourly basis. This tends to create artificial jump every hour.

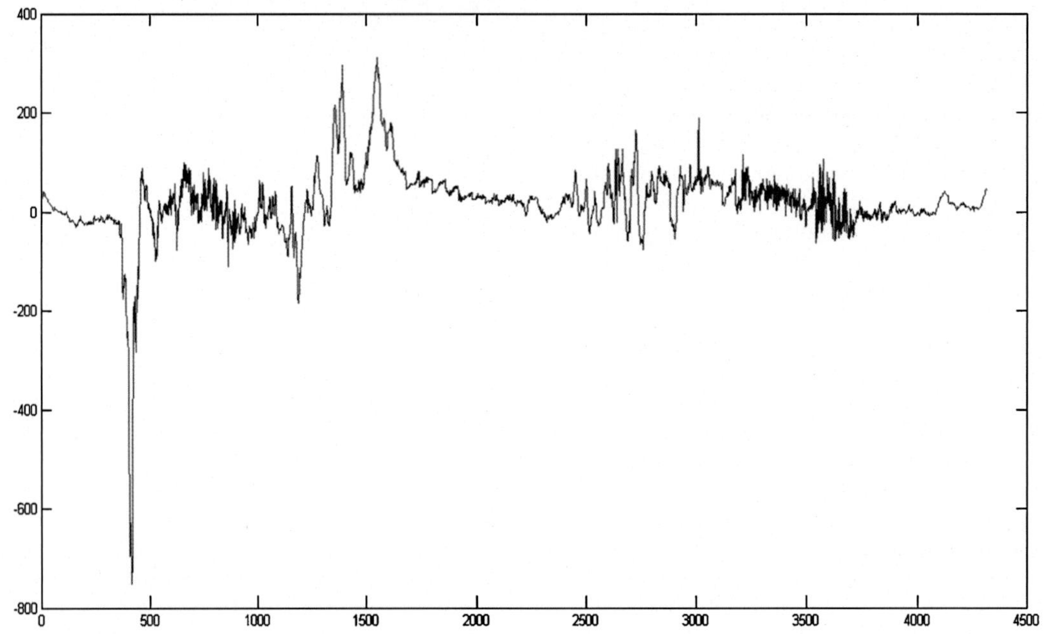

FIG. 19.13 The index DS variation during the period 29–31 October 2003, registered at the Surlary (SUA). The sampling is 1 minute.

19.3 EXPERIMENTS WITH WAVELET ANALYSIS AND CONCLUSIONS

In the experiments to follow, we are motivated by the interest to discover wave packages of short periods and their correlations in the three different sources of data which we have already discussed:

1. (ground) geomagnetic data, from geomagnetic observatories,
2. ionospheric data (TEC values, from ionosound stations),
3. IMF data (from the ACE satellite).

We would like to identify any kind of correlation and causality among them by applying the CWT method.

19.3.1 References on Applications of Wavelet Analysis to Geomagnetism

Before presenting the results of our experiments, we provide some references which might be useful to the reader.

Let us note that in a number of works (Mandrikova et al., 2014; Wei et al., 2004; Zossi de Artigas et al., 2008; Boudouridis and Zesta, 2007; Jach et al., 2006; Katsavrias et al., 2016; Xu et al., 2008), the variation of geomagnetic data (in particular of D_{st}) is analyzed by means of wavelet analysis. Recently, wavelet analysis of geomagnetic field perturbations was widely used in the study of tsunami waves (Klausner et al., 2011, 2014, 2016a, 2016b, 2017; Schnepf et al., 2016). In Jach et al. (2006) and Xu (2011), wavelet analysis of the geomagnetic field is used to define a new index, alternative to D_{st}, but on a minute basis.

19.3.2 Experiments With Data on a Quiet Day, 28 July 2018

In order to have control over the statistical behavior of the geomagnetic data during geomagnetic storms, we have taken data from quiet days from two geomagnetic observatories – the in situ repeat station at Balchik (Bulgaria) and SUA in Romania; the distance between them is about 190 km. This means that they have almost identical conditions, and it is well known that there are no strong magnetic variations from natural or artificial character in the regions.

19.3.2.1 Visualization of the Wavelet Analysis

We have provided a brief summary on the CWT in Appendix 19.A.

In the experiments below we perform CWT to time series $f(j)$, where $f(j) = F\left(\delta t_j\right)$ for some "continuous time series" $F(t)$, and $t_j\delta$ are the sampling times on a uniform mesh, $t_1, t_2, ..., t_N$. Here δ denotes the sampling interval, for example, we have $\delta = 1$ second, 4 sec-

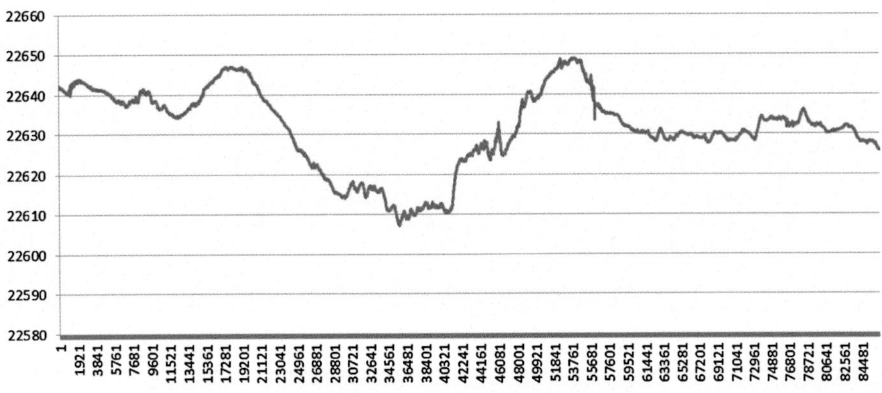

FIG. 19.14 The H component with sampling 1 second at a repeat station in Balchik on 28 July 2018.

onds, 1 minute, 1 hour, etc. We visualize the absolute values $|W_\psi f(a,b)|$ of the CWT coefficients $W_\psi f(a,b)$ defined in formula (19.A.1). In all our experiments the shifts b (visualized on the X axis) run through the full set of indices $1, 2, ..., N$. On the other hand, the nonnegative parameter a in formula (19.A.1), which denotes the scaling parameter (and has the meaning of *periodicity*), is interesting for us mainly for shorter intervals. Hence, in some experiments we consider only periods $a \leq N_1$ for some maximal period $N_1 < N$. The parameter a is visualized on the Y axis. In order to get a better idea of the behavior of the CWT coefficients $W_\psi f(a,b)$, we find it instructive to have the visualization of $|W_\psi f(a,b)|$ both as a *heatmap* and as a *contour* map; see, for example, Fig. 19.15.

19.3.2.2 Experiments With Balchik Geomagnetic Data, 28 July 2018, 1 Second Data

The H component geomagnetic data were collected every second for 24 hours at a repeat station in Balchik (Bulgaria) on 28 July 2018.

We provide the graph of the time series of the data in Fig. 19.14. The daily variation of the field is clearly visible as the main trend, and also some rather permanent short-period variations.

Below, in Fig. 19.15, we provide the experiments, namely, the *heatmap* (top) and the *contour map* (bottom) of the absolute values of the CWT coefficients $|W_\psi f(a,b)|$ of the H component.

In Fig. 19.15 we see that in the CWT of the time series of the Balchik geomagnetic data, some very interesting details are identified. Around midday, some wave packages with periods about 100 min are clearly visible. They show the possibility for soliton-like oscillations, related to the solar terminator, theoretically studied in Belashov and Belashova (2015), and which has been recently observed in the wavelet analysis experiments with geomagnetic data in the paper Srebrov et al. (2018).

19.3.2.3 Experiment With SUA Geomagnetic Data on 28 July 2018

We have performed experiments with the time series formed by the H component of the geomagnetic data from SUA (Surlary, Romania), on a whole day, 28 July 2018; the sampling is every minute. The *heatmap* (top) and the *contour map* (bottom) of the CWT are provided in Fig. 19.16.

Remark 19.5. In Fig. 19.16 we see wave packages of 100 minutes during the whole day, and wave packages with periods below 180 minutes about midday. At midday one observes an intensive process of generation with a period between 100 and 150 minutes. Right at the same time interval (between 600 and 700 minutes) there are wave packages with period 30–50 minutes. One may suggest that the latter phenomenon may be generated by the solar terminator, and have a soliton structure (Belashov and Belashova, 2015).

Remark 19.6. There is a lot of similarity at the scale of 20–60 minutes between the CWT of Balchik data and the CWT of SUA data.

We compare the CWT of the Balchik data in Fig. 19.15 with the CWT of the SUA data in Fig. 19.16. We see that the wave packages with periods about 80 minutes are much more expressed in the Balchik second data than on the SUA minute data. This shows that the oscillation phenomena carry a persistent character. In particular, as was concluded in the paper Srebrov et al. (2018), we may suggest the existence of soliton-like patterns at the periods of 40 to 60 min.

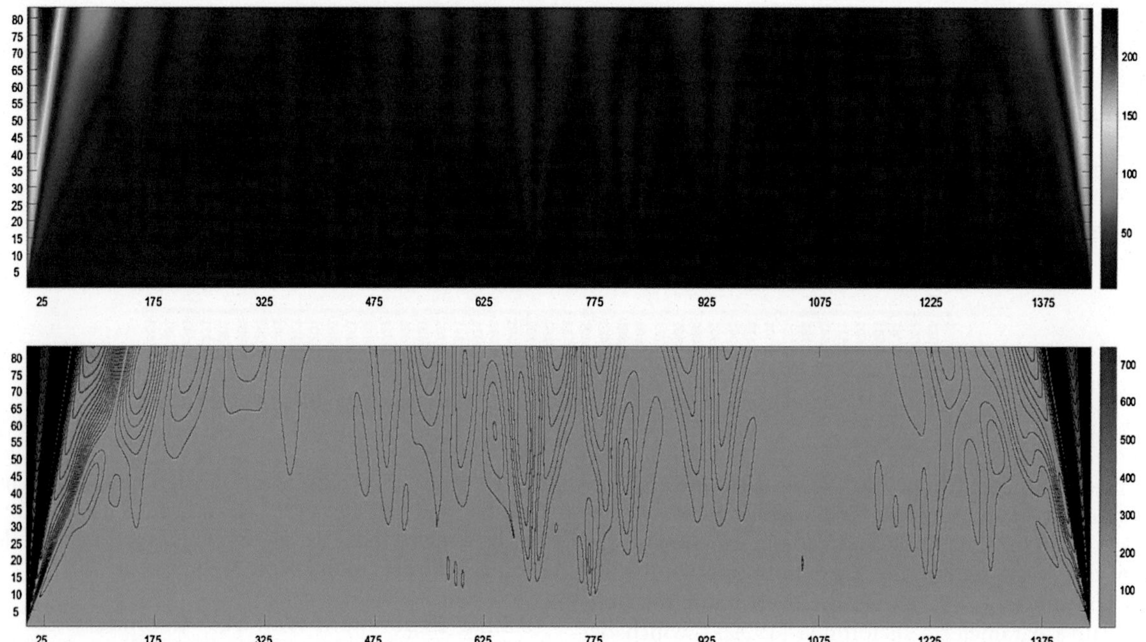

FIG. 19.15 (top) *Heatmap* and (bottom) *contour map* of the CWT of the H component of the Balchik data, on 28 July 2018.

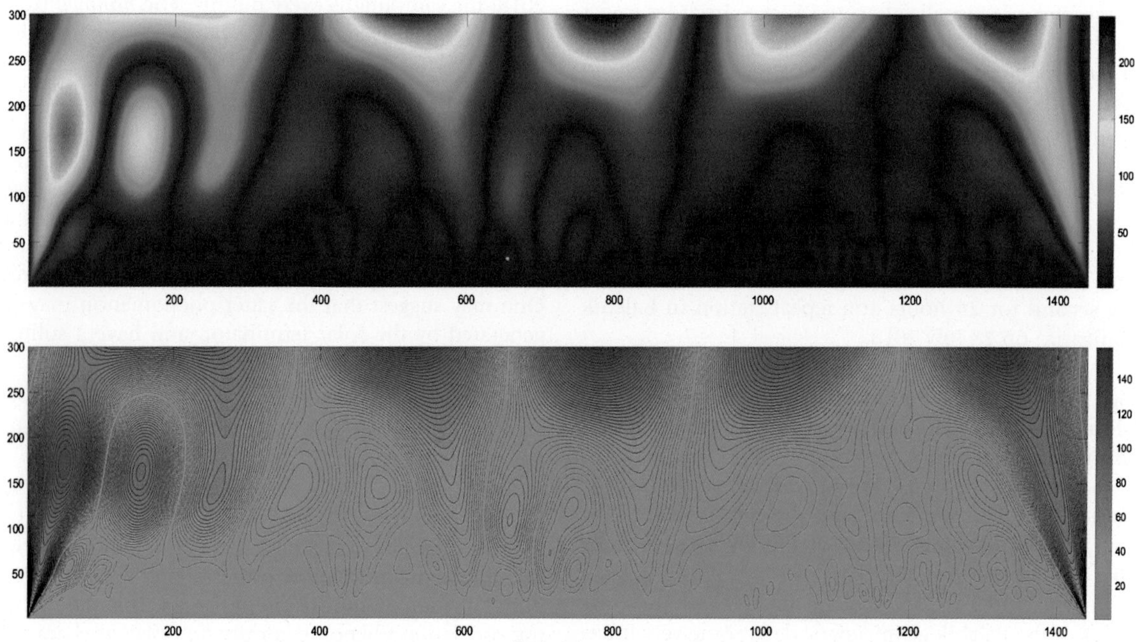

FIG. 19.16 (top) *Heatmap* and (bottom) *contour map* of the CWT of SUA data, 28 July 2018.

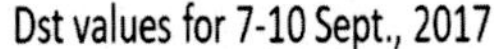

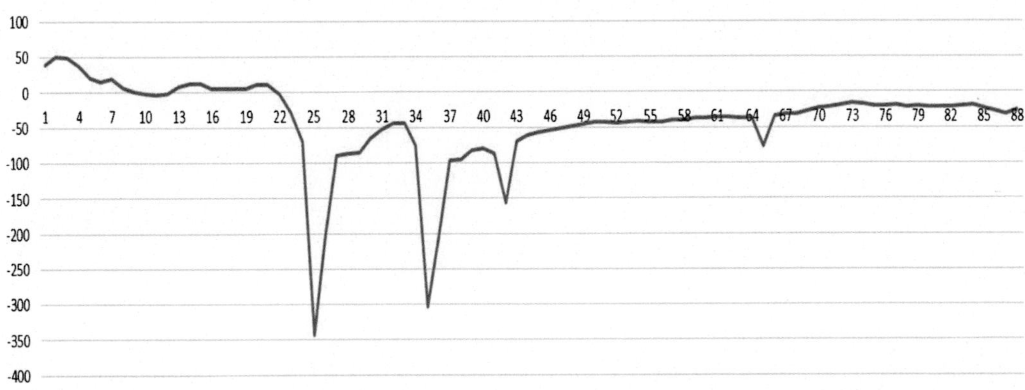

FIG. 19.17 The D_{st} graph for 7–10 September 2017.

Remark 19.7. Conclusions: Wave packages with periods from 10 to 100 min exist during quiet geomagnetic days, which are however predominant at midday.

19.3.3 Experiments With Data for the Geomagnetic Storm on 7 and 8 September 2017

In the following we provide a wavelet analysis of the data from ground geomagnetic field, ionospheric parameters, and IMF, collected during this strong geomagnetic storm.

First we provide the picture of the main trend which is determined by the D_{st} index.

19.3.3.1 The D_{st} for the Period 7–10 September 2017

Fig. 19.17 shows the D_{st} graph for the period.

As we have described the D_{st} index in Section 19.2.3, it shows a very unusual behavior after the storm of 7–10 September 2017.

Remark 19.8. In Fig. 19.17 we see the variation of the D_{st} index during the manifestation of the geomagnetic storm in September, 2017. In the figure we see the two decreases of the magnetic field, caused by two events on the Sun surface, and also a very long recovery phase of the storm during the period 9 and 10 September. This long recovery phase is most probably related to the lack of short-period variations in the geomagnetic records in the observatory on 9 and 10 September. The decay of the ring current cannot create significant geomagnetic variations on the ground in the region of the PAG observatory. However, as we have remarked above, the ionospheric macroinstabilities during these two days cannot create a variation in the ionospheric current system, which itself would create variations to be registered by this ground observatory.

19.3.3.2 Experiments With Geomagnetic Data From PAG, 7–10 September 2017, 1 min Data

We have acquired the H component of the geomagnetic data from the Panagyurishte (PAG) geomagnetic observatory. These data are every 1 minute sampling period. We provide the *heatmap* (top) and the *contour map* (bottom) of the CWT; see Fig. 19.18.

We see that during 7 and 8 September of the geomagnetic storm we may identify wave packages with periodicity 20–100 minutes. However, on 9 and 10 September, one cannot identify wave packages with periods in the interval 20–100 minutes.

19.3.3.3 Experiments With Ionospheric Data From Athens, 7–10 September 2017, 5 min Data

We have taken the ionosound TEC data from the ionosound station in Athens (ATN). The data are measured for four full days, 7–10 September 2017, every 5 minutes (this frequency is the modern standard for sampling of ionosounding data).

In Fig. 19.19 we provide the *heatmap* (top) and the *contour map* (bottom) of the CWT of the TEC time series.

We see that the short-period scales which are interesting for us really show regular patterns. For that reason we have restricted the scales *only to 50*, and we show the results in Fig. 19.20.

We see that during the geomagnetic storm we have a lot of wave packages and regular patterns. However

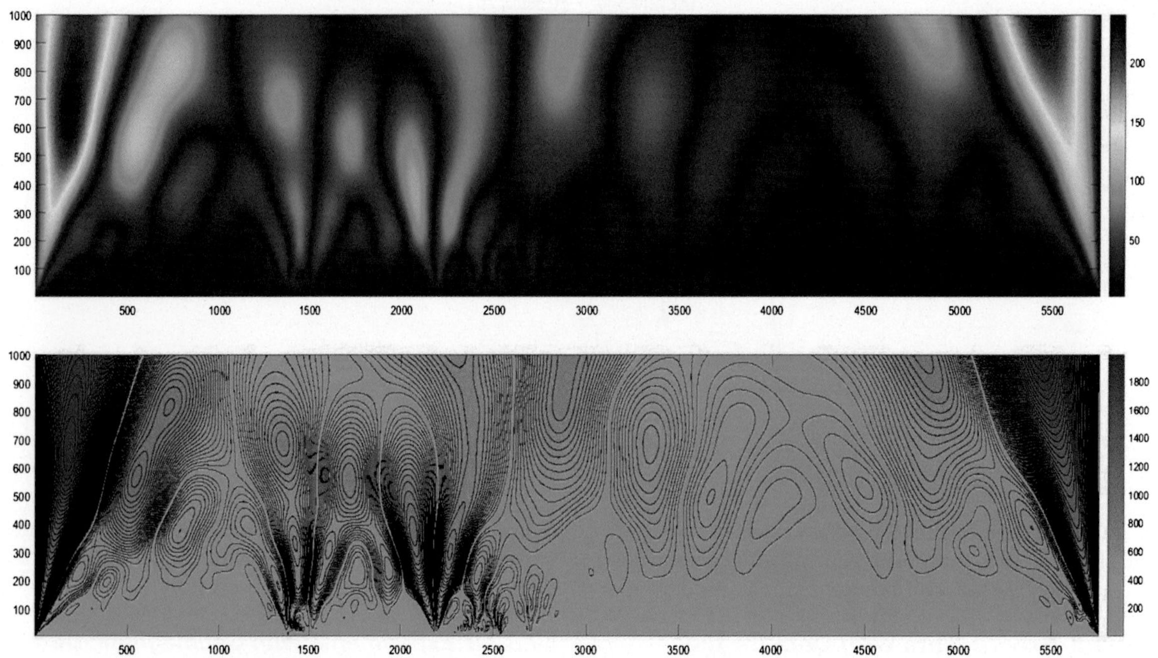

FIG. 19.18 (top) *Heatmap* and (bottom) *contour map* of the CWT for H component of PAG, 7–10 September 2017.

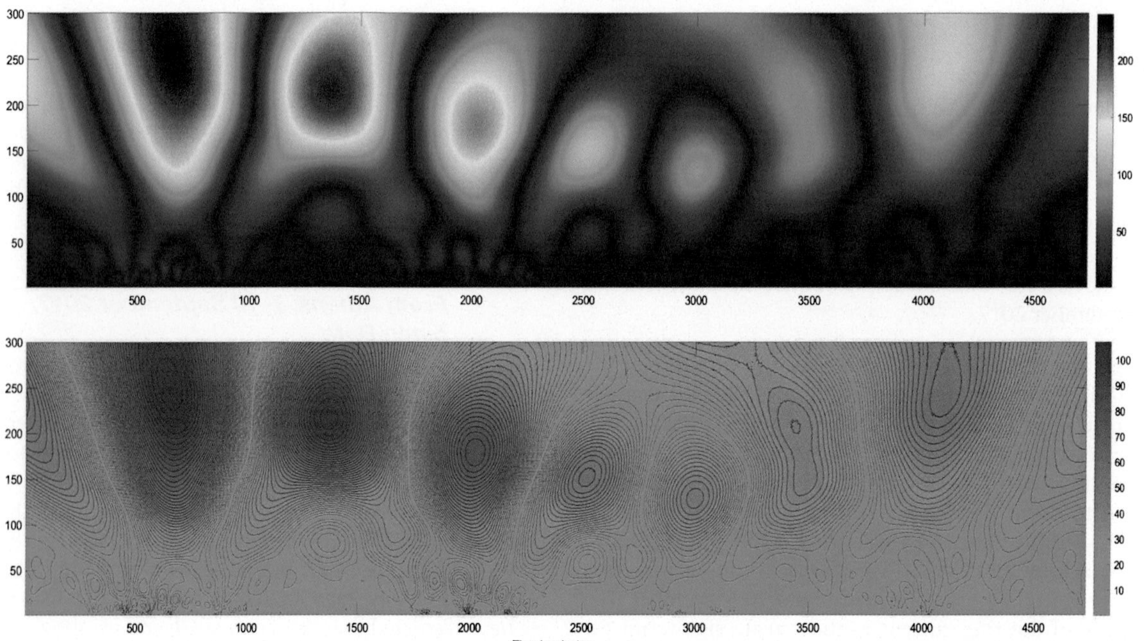

FIG. 19.19 (top) *Heatmap* and (bottom) *contour map* of the CWT of the TEC data, Athens, 7–10 September 2017.

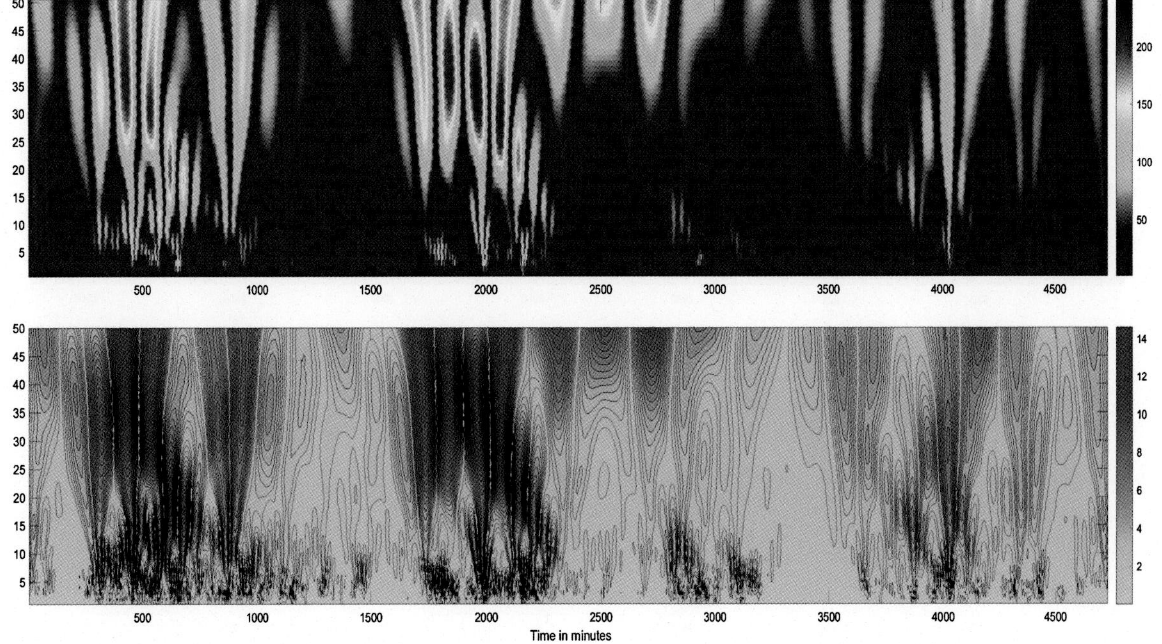

FIG. 19.20 (top) *Heatmap* and (bottom) *contour map* of the CWT of the TEC data, Athens, 7–10 September 2017, limited to short periods ≤ 50 minutes.

what is not less interesting, similar pattern appear during the two days 9 and 10 September (during the recovery phase of the storm), at midday time, having periods 20–50 minutes. This seems to be due to the solar terminator influence, as was suggested in Belashov and Belashova (2015), Schnepf et al. (2016).

19.3.3.4 Experiments With IMF Data From ACE Satellite, 7–10 September 2017, 4 min Data

We retrieved the B_Z component of the IMF from the ACE satellite on 7–10 September 2017, every 4 min data.

In Fig. 19.21 we provide the CWT of the time series containing the values of B_Z; again the *heatmap* is on the top and the *contour map* is on the bottom of the figure.

An interesting observation is that in the ACE data and in the ground PAG data one *cannot identify* any wave packages of periods 20–100 min after the storm, i.e., on 9 and 10 September 2017. This makes us believe that there is a correlation between the two observable values. This is in strong contrast to the ionospheric observations provided in Fig. 19.19 and Fig. 19.20, where such wave packages are available. This shows that in some cases the ionospheric plasma generates short-period modes with a relatively small amplitude. This would explain the lack of similar modes in the ground geomagnetic data of PAG. On the other hand, it is clear that on 9 and 10 September, these short-periodic modes are the result of eigenoscillations of the ionospheric plasma, which are not caused by the influence of the IMF.

19.3.4 Experiments With Data for the 2003 Strong Geomagnetic Storm

Since the geomagnetic storm in 2003 was unusually strong, it has become a handbook example for testing analysis tools. However, in 2003 there were not so many data available. In particular, the *ionosound* data are not available with the present sampling, but only hour data.

We provide below the results for wavelet analysis for the following data:

1. In 2003, for different (ground) geomagnetic ground observatories (SUA) we have 1 minute data for the H component.
2. ACE satellite data for IMF are available at frequency 4 min.

On the other hand, during this storm the ionospheric data are only *mean hour* which is *insufficient* for the present analysis.

First of all, we have the main trend of the magnetic fields provided by the D_{St} data in Fig. 19.10 above.

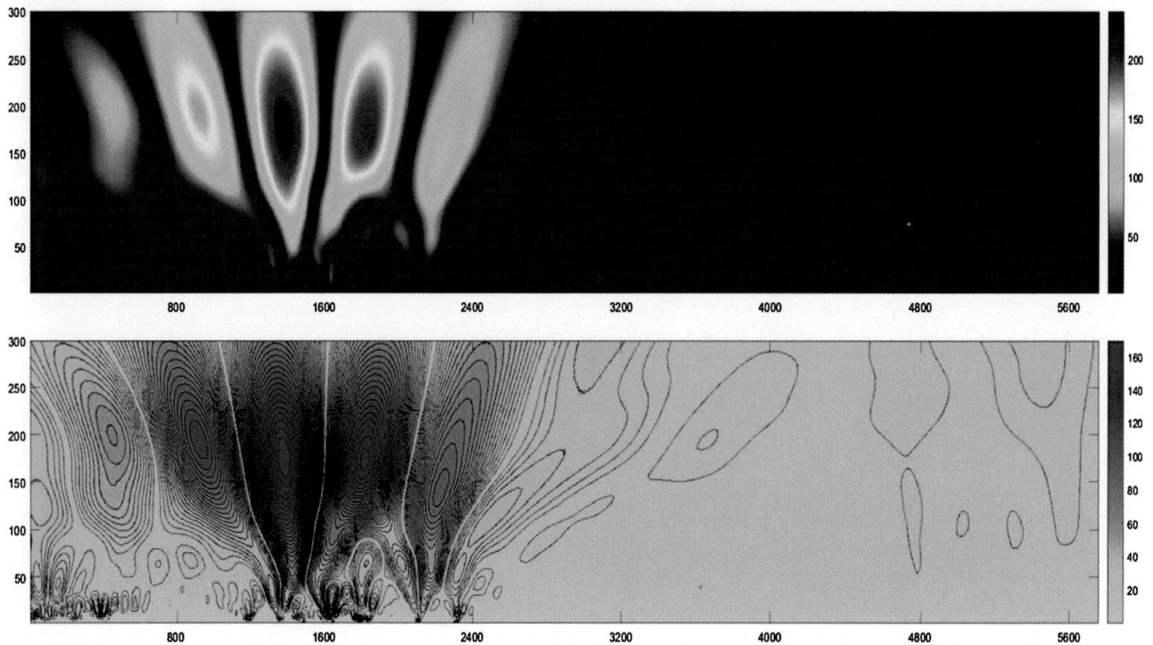

FIG. 19.21 (top) *Heatmap* and (bottom) *contour map*, for CWT of the B_z component of IMF, 7–10 September 2017.

19.3.4.1 Experiments With Geomagnetic Data From SUA, on 28 and 29 October 2003

One may analyze the original source data H component, or DS obtained after subtracting the D_{st} from the data; see formula (19.1). As we said above, since the D_{st} data are given every hour this may create artifacts every whole hour. We provide the contour plot of the CWT for the H component of the geomagnetic field from SUA, 28 and 29 October 2003, sampling 1 minute, in Fig. 19.22.

In Fig. 19.22, we may clearly identify short-period wave packages with periods below 3–4 hours, as well as with periods of 4–12 hours, but also with periods about 24 hours. The last are related with the main and recovery phase of the storm, i.e., with the ring current. The wave packages with periods below 3 hours may be related to the fluctuations of the ring current and the ionospheric instabilities generating such modes. Modes with periods below 100 minutes are generated mainly at midday, and may be related to the solar terminator and are eventually of soliton type, as was mentioned already above; see also Belashov and Belashova (2015), Srebrov et al. (2018).

In Fig. 19.23 we provide the CWT of the DS data.

19.3.4.2 Experiments With Spline Smoothing of D_{st}

We have provided some interesting experiments which show the effect of subtraction of D_{st} after smoothing the D_{st} with splines; see Fig. 19.24.

On the bottom of Fig. 19.24 we have the CWT of the H component and on the top we have the CWT of the $DS = H - \widetilde{D_{st}}$, where $\widetilde{D_{st}}$ is the smoothed D_{st} (which as mentioned is provided by the WDC of geomagnetism in Kyoto on an hourly basis).

This shows that one has to be careful when subtracting the D_{st} index (which is a step function) from the H component since this creates nonsmooth signal and the Fourier or wavelet analyses generate artificial frequencies.

19.3.4.3 Experiments With IMF Data, on 28 and 29 October 2003, 4 min Data

We have provided the CWT for the B_z component of the IMF in Fig. 19.25. Again the *heatmap* of the CWT is shown on top, while the *contour map* is shown on bottom.

In Fig. 19.25 we see various families of wave packages, which may be separated into two types: those with periods less than 100 minutes, and those with periods between 100–450 minutes. Their explanation is related

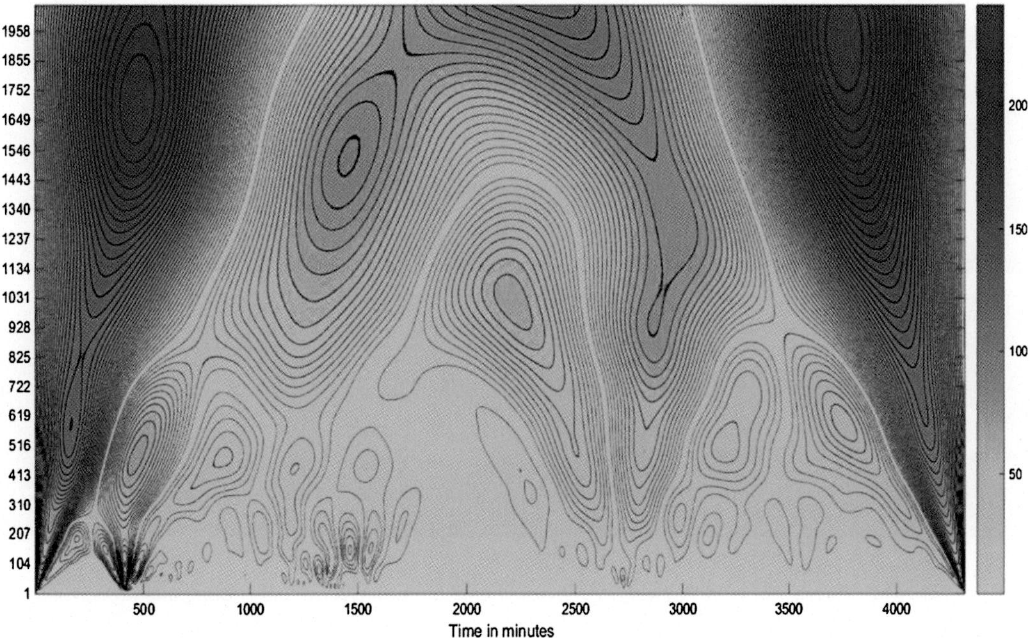

FIG. 19.22 Contour plot of CWT for the H component from SUA, 28 and 29 October 2003.

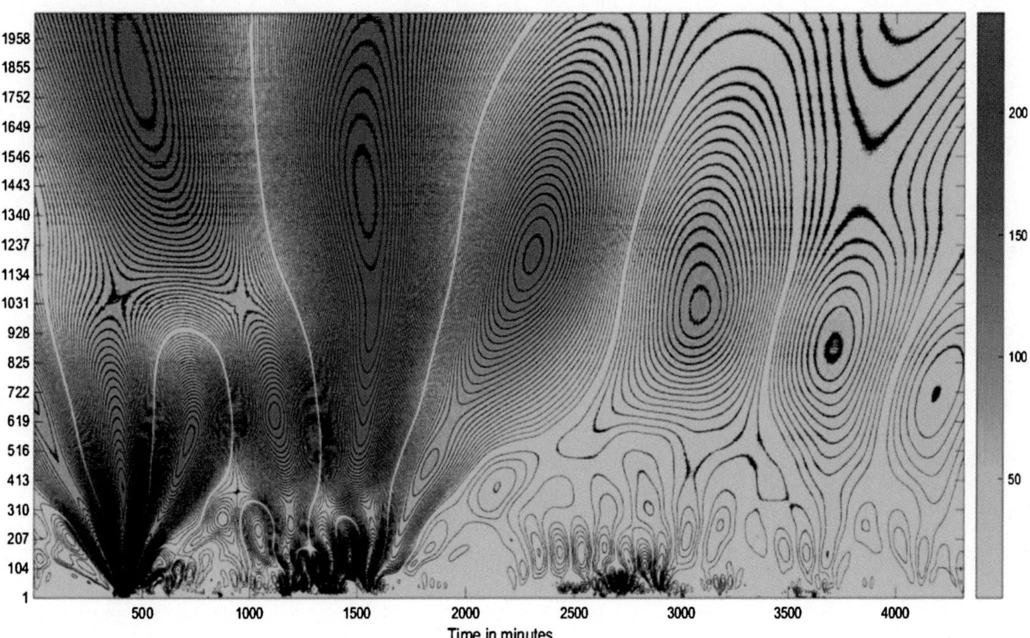

FIG. 19.23 Contour plot of the CWT for the DS index of the geomagnetic field from SUA, 28 and 29 October 2003.

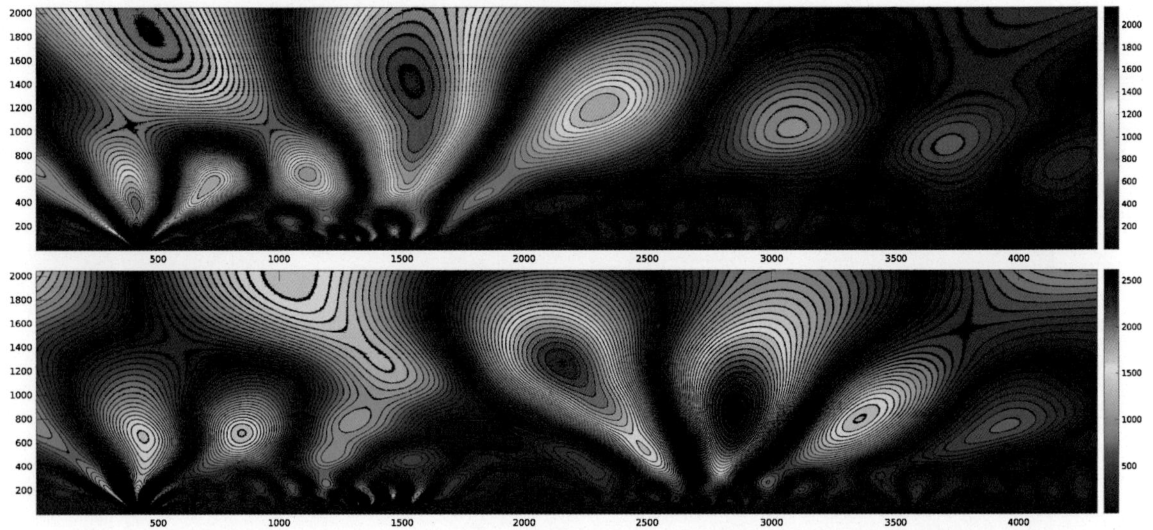

FIG. 19.24 (top) CWT of the $DS = H - \widetilde{D_{st}}$ and (bottom) CWT of the H component.

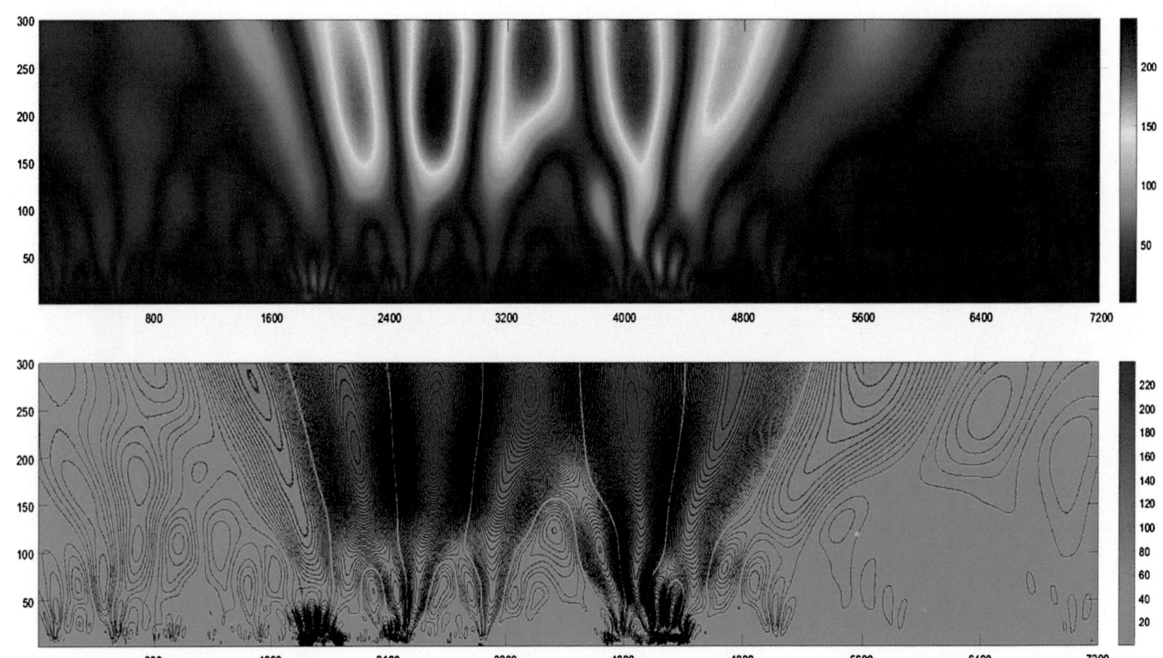

FIG. 19.25 (top) *Heatmap* of the CWT and (bottom) *contour map* for the B_z component of IMF, 28 and 29 October 2003.

to the complex structure of the disturbance of the B_Z component during the strong geomagnetic storm.

19.4 CONCLUSIONS

1. The main objective of the present research is to apply wavelet analysis to Big Data in solar-terrestrial physics, for the investigation of short-period variations of the (ground) *geomagnetic field/ionospheric parameters* in a region with mean geographic latitude. Thus, by applying CWT to large amounts of heterogeneous data (geomagnetic field, ionospheric parameters, and IMF), we have identified modes (wave packages) with different periods, of the order of 20 to a few hundred minutes with a significant amplitude, which is enough to be registered by the equipment in the geomagnetic observatories.
   As is known, in the same range there exist the so-called geomagnetic pulsations, but they have a very low amplitude and exist for a short time only during the night hours for this geographic latitude. Unlike the geomagnetic pulsations, the short-period variations of our interest, have significant amplitude, and are identified in the present research; they are discovered during the whole day and may be divided into modes with periods less than 3 hours and modes which have a period greater than 3 hours.
2. Our analysis of the variations of the geomagnetic field, the ionospheric plasma parameters, and the IMF, has shown persistent short-term periodic events, as wave packages. The short-period modes (wave packages) of the variations which we have identified have a clear explanation (e.g., from plasma physics) and are caused by macroinstabilities in different domains of the near-Earth space environment.
3. We have identified the presence of modes with periods shorter than 3 hours, generated predominantly by the ionospheric plasma, but also similar modes which exist in the IMF. The ability of wavelet analysis to uncover multiresolution structure of the data gave us the possibility to identify short-periodic wave packages in the geomagnetic field variations, in IMF, and in the ionospheric parameters.
4. The present research represents a contribution to the newly developing area of AstroGeoInformatics due to the large spectrum of the analyzed phenomena which belong to solar-terrestrial physics.

ACKNOWLEDGMENTS

The authors thank the editors of this volume, Petr Škoda and Adam Fathalrahman, for their patience and assistance. Thanks extend to Aleksandra Nina (Belgrade) for discussions on the ionosphere. All authors thank the Project on Modern mathematical methods for Big Data, DH 02-13 and Project on Statistical methods for machine learning of data with complex structure, KP-06-N32-8 with Bulgarian NSF, and also the Project SatWebMare with ESA (in the PECS framework). Last but not least, all owe thanks to the COST action BigSkyEarth with EU. OK thanks the Alexander von Humboldt Foundation.

The services of the World Data Center for Geomagnetism, at the Kyoto University, Japan, the INTERMAGNET network, the ACE Science Center at Caltech, and the GIRO center, University of Massachusetts at Lowell, are gratefully acknowledged. MATLAB®/Octave code was written for the experiments.

APPENDIX 19.A WAVELET ANALYSIS AND ITS APPLICATIONS TO GEOMAGNETIC DATA

In the present research we have decided for continuous wavelet transform (CWT). We provide the essentials of the CWT and some useful references for the applications of wavelet analysis.

19.A.1 Technical Stuff

We will say that the integrable function ψ is a *wavelet function* if it satisfies the following properties:

1. the *admissibility* condition holds,

$$0 < C_\psi := \int_{-\infty}^{\infty} \frac{\left|\widehat{\psi}(\omega)\right|^2}{|\omega|} d\omega < \infty,$$

2. the *zero integral* condition holds,

$$\widehat{\psi}(0) = 0,$$

where $\widehat{\psi}(\omega)$ is the Fourier transform of ψ. This condition is equivalent to

$$\int \psi(\omega)\, d\omega = 0.$$

We consider only real-valued functions ψ.

Once the wavelet function ψ is fixed, then for every integrable function f (which is considered to represent the signal) which has a sufficient decay at ∞, and for every two real numbers $a, b \in \mathbb{R}$ with $a > 0$, we may define the CWT $W_\psi f(a, b)$ by putting

$$W_\psi f(a, b) := \frac{1}{\sqrt{a}} \int_{-\infty}^{\infty} f(t)\psi\left(\frac{t-b}{a}\right) dt. \quad (19.A.1)$$

The number a is called *scale*, and b is called *translation* (shift). Recall that the usual definition of the *frequency* k is then given by putting

$$k = \frac{1}{a}.$$

Unlike the usual Fourier transform, where the dimension of the variable of the signal $f(t)$ is transformed into the same-dimensional frequency domain, here we see that the CWT $W_\psi f(a, b)$ depends on two variables. We are able to reconstruct the original signal f from this representation, by means of the *Calderon* inversion formula (Jaffard et al., 2001; Mallat, 2009)

$$f(t) = \frac{1}{C_\psi} \int_{-\infty}^{\infty} \int_{-\infty}^{\infty} \frac{1}{\sqrt{a}} W_\psi f(a, b) \psi \left(\frac{t-b}{a} \right) \frac{da}{a^2} db. \tag{19.A.2}$$

As in the DWT it is always the question to find some reasonable approximation in the Calderon formula which takes into account only the larger values of $|W_\psi f(a, b)|$, which will result in an approximation of the double integral in the equality in formula (19.A.2). Thus, the question is whether it is possible to use just a part of the integration domain. This may be achieved in different ways; one approach is to apply a threshold on the absolute value of the CWT $|W_\psi f(a, b)|$, say, $\varepsilon > 0$, and define the domain

$$D_\varepsilon = \{(a, b) : |W_\psi f(a, b)| < \varepsilon\}$$

and then consider the approximation integral

$$I_\varepsilon := \frac{1}{C_\psi} \iint_{D_\varepsilon} \frac{1}{\sqrt{a}} W_\psi f(a, b) \psi \left(\frac{t-b}{a} \right) \frac{da}{a^2} db,$$

so that the remainder would satisfy

$$|f(t) - I_\varepsilon(t)| \le \delta \qquad \text{for all } t \in \mathbb{R}.$$

Unlike the discrete wavelet transform (DWT) here we do not have a clearly defined multiresolution analysis (MRA), and also there are no father wavelets (scaling functions). However, in general, one may use the wavelets in the discrete wavelet theory and apply them in CWT if they are smooth enough. The CWT is a very convenient tool to detect and characterize singularities in functions, in order to distinguish between noise and signal (see Jaffard et al., 2001; Mallat, 2009). In particular, one may use CWT to study fractal behavior of the signals.

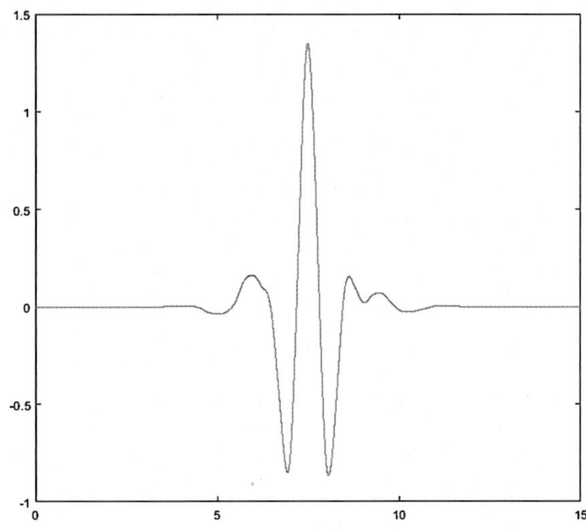

FIG. 19.26 The graph of the sym8 mother wavelet.

In a wide area of applications, people use CWT with a wavelet function ψ equal to the Mexican hat and the Morlet wavelet, although these two functions do not enjoy the usual scheme of MRA as introduced in Mallat (2009), Jaffard et al. (2001). In the present research, after numerous experimentations, we have decided for the *symlet* family of functions, which are a modified version of the Daubechies wavelet family *db*, since they enjoy increased symmetry (Addison, 2011). We have applied the *sym8* wavelet function, provided in Fig. 19.26.[1]

However, it is important to remark that the experiments with many other wavelets ψ have shown that the singularities which we detect by the *symlets* may be analyzed with the same success by applying the other wavelets; completely subjectively, we have found that *sym8* gives on average one of the best possible visual pictures. This fact shows that our observation is due to persistent physical events and may not be an artifact which is due to the particular wavelet which we choose.

19.A.2 CWT of Some Simple Functions

An important control of the wavelet analysis method is to consider the CWT of the simple jump functions, as for example *impulse trains*, which are sums of Dirac delta functions. It gives us idea about the behavior of the CWT of more complicated signals. A main reason to consider these impulse trains is the result of Belashov (Belashov and Belashova, 2015), who has very successfully modeled (better than the traditional IRI model)

[1] http://wavelets.pybytes.com/wavelet/sym8/.

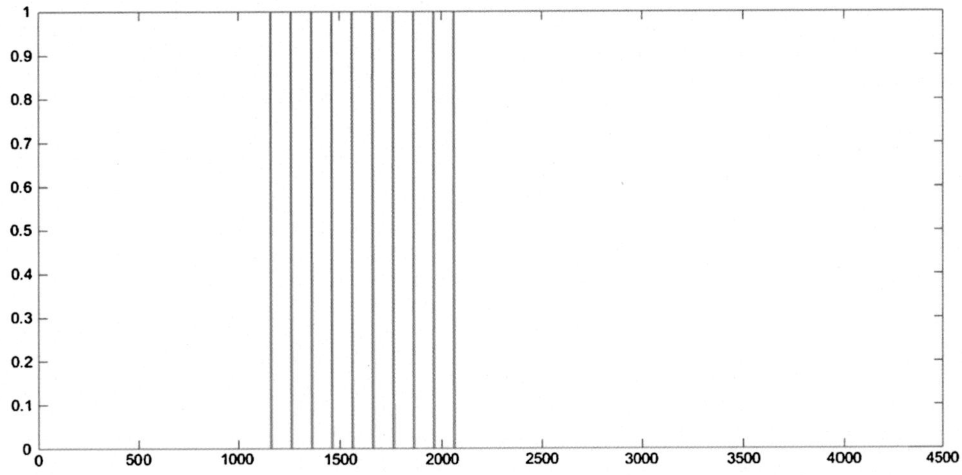

FIG. 19.27 Impulse train.

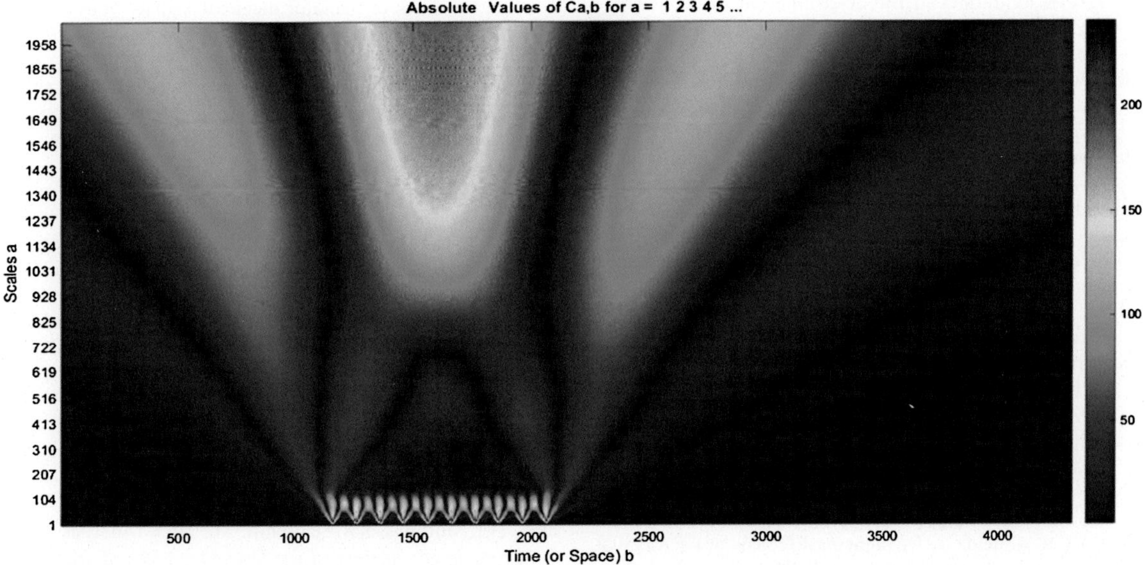

FIG. 19.28 CWT of the impulse train.

the ionospheric data by assuming that the disturbances of the ionosphere are related (even on quiet days!) to wave packages having a soliton character; we have already announced this resemblance of our analysis in Srebrov et al. (2018).

Fig. 19.27 shows the graph of a simple impulse train.

It has the CWT shown in Fig. 19.28. The y axis shows the length of the period (the scale a in the CWT $W_\psi(a, b)$ and the x axis shows the number b).

Remark 19.A.1. From Fig. 19.28 we see that a package of periodic pulses has considerable CWT $|W_\psi(a, b)|$ for lower periods $0 \leq a \leq 105$, but it also shows an "integral effect" and shows a considerable CWT $|W_\psi(a, b)|$ for longer periods $a \geq 1030$. The moral of this observation is that one has to be really careful when solving the inverse problem, i.e., when drawing conclusions about the singularities of the original signal $f(t)$ judging by the large-period behavior of the CWT $W_\psi(a, b)$.

Remark 19.A.2. Another interesting example is provided also on Wikipedia,[2] where CWT is provided of a frequency breakdown signal by using the *symlet* as a wavelet function with five vanishing moments.

REFERENCES

Addison, Paul S., 2011. The Illustrated Wavelet Transform Handbook: Introductory Theory and Applications in Science, Engineering, Medicine and Finance, Second Edition. CRC Press.

Akasofu, S.-I., Chapman, Sydney, 1972. Solar-Terrestrial Physics. Clarendon Press, Oxford.

Belashov, Vasily Yu, Belashova, Elena S., 2015. Dynamics of IGW and traveling ionospheric disturbances in regions with sharp gradients of the ionospheric parameters. Advances in Space Research 56, 333–340.

Bishop, C., 2006. Pattern Recognition and Machine Learning. Springer.

Boudouridis, A., Zesta, E., 2007. Comparison of Fourier and wavelet techniques in the determination of geomagnetic field line resonances. Journal of Geophysical Research 112 (A08205).

Gençay, R., Selçuk, F., Whitcher, B., 2002. An Introduction to Wavelets and Other Filtering Methods in Finance and Economics. Academic Press.

Hubbard, Barbara Burke, 1998. The World According to Wavelets: The Story of a Mathematical Technique in the Making. AK Peters Ltd.

Jach, A., Kokoszka, P., Sojka, J., Zhu, L., 2006. Wavelet-based index of magnetic storm activity. Journal of Geophysical Research 111, A09215.

Jaffard, S., Meyer, Y., Ryan, R.D., 2001. Wavelets. Tools for Science and Technology. SIAM.

Katsavrias, Ch., Hillaris, A., Preka-Papadema, P., 2016. A wavelet based approach to solar–terrestrial coupling. Advances in Space Research 57, 2234–2244.

Klausner, V., Almeida, T., de Meneses, F.C., Kherani, E.A., Pillat, V.G., Muella, M.T.A.H., 2016a. Chile2015: induced magnetic fields on the Z component by tsunami wave propagation. Pure and Applied Geophysics 173, 1463–1478.

Klausner, V., Almeida, T., De Meneses, F.C., Kherani, E.A., Pillat, V.G., Muella, M.T.A.H., 2017. Chile2015: induced magnetic fields on the Z component by tsunami wave propagation. In: The Chile-2015 (Illapel) Earthquake and Tsunami. Pageoph Topical Volumes. Birkhäuser, Cham, pp. 193–208.

Klausner, V., Domingues, M.O., Mendes, O., Papa, A.R.R., 2011. Tsunami effects on the z component of the geomagnetic field. Online version at https://arxiv.org/abs/1108.4893.

Klausner, Virginia, Kherani, Esfhan A., Muella, Marcio T.A.H., 2016b. Near- and far-field tsunamigenic effects on the Z component of the geomagnetic field during the Japanese event. Journal of Geophysical Research.

Klausner, V., Mendes, O., Domingues, M.O., Papa, A.R.R., Tyler, R.H., Frick, P., Kherani, Esfhan A., 2014. Advantage of wavelet technique to highlight the observed geomagnetic perturbations linked to the Chilean tsunami (2010). Journal of Geophysical Research: Space Physics 119, 3077–3093.

Kounchev, O., 2001. Multivariate Polysplines. Applications to Numerical and Wavelet Analysis. Academic Press/Elsevier.

Mallat, S., 2009. A Wavelet Tour of Signal Processing. Academic Press.

Mandrikova, O.V., Solovev, I.S., Zalyaev, T.L., 2014. Methods of analysis of geomagnetic field variations and cosmic ray data. Earth, Planets and Space, 66–148.

Meinl, Thomas, Sun, Edward, 2012. A nonlinear filtering algorithm based on wavelet transforms for high-frequency financial data analysis. Studies in Nonlinear Dynamics and Econometrics 16 (3), 1–24.

Mitchner, M., Kruger, Charles H., 1973. Partially Ionized Gases, 1st Edition. Wiley.

Oppenheim, A., Schafer, R., 2010. Discrete-Time Signal Processing. Pearson.

Percival, Donald B., Walden, Andrew T., 2000. Wavelet Methods for Time Series Analysis. Cambridge University Press.

Schnepf, N.R., Manoj, C., An, C., et al., 2016. Time-frequency characteristics of tsunami magnetic signals from four Pacific Ocean events. Pure and Applied Geophysics 173, 3935–3953.

Srebrov, B.A., 2003. MHD modeling of supersonic, superalfvenic distrubances propagating in the interplanetary plasma and their relationship to the geospace environment. Advances in Space Research 31, 1413–1418.

Srebrov, B., Pashova, L., Kounchev, O., 2018. Study of local manifestations of G5 – extreme geomagnetic storms (29–31 October, 2003) in midlatitudes using geomagnetic data by continuous wavelet transforms. Comptes Rendus de L'Academie Bulgare Des Sciences 71.

Sun, Edward W., Meinl, Thomas, 2012. A new wavelet-based denoising algorithm for high-frequency financial data mining. European Journal of Operational Research 217, 589–599.

Sun, Edward W., Yu, Min-The, 2015. Generalized optimal wavelet decomposing algorithm for big financial data. International Journal of Production Economics 165, 194–214.

Wei, H.L., Billings, S.A., Balikhin, M., 2004. Analysis of the geomagnetic activity of the Dst index and self-affine fractals using wavelet transforms. Nonlinear Processes in Geophysics 11, 303–312.

Xu, Zhonghua, 2011. Study of geomagnetic disturbances and ring current variability during storm and quiet times using wavelet analysis and ground-based magnetic data from multiple stations. Online PhD thesis at https://digitalcommons.usu.edu/etd/984/.

Xu, Z., Zhu, L., Sojka, J.J., Kokoszka, P., Jach, A., 2008. An assessment study of the wavelet-based index of magnetic storm activity (WISA) and its comparison to the Dst index. Journal of Geophysical Research 70, 1579–1588.

Zossi de Artigas, M., Fernandez de Campra, P., Zotto, E.M., 2008. Geomagnetic disturbances analysis using discrete wavelets. Geofísica Internacional 47 (3), 257–263.

[2] https://en.wikipedia.org/wiki/Continuous_wavelet_transform.

CHAPTER 20

International Database of Neutron Monitor Measurements: Development and Applications

D. SAPUNDJIEV, PHD • T. VERHULST, PHD • S. STANKOV, PHD

20.1 INTRODUCTION

Cosmic rays are the particles and the radiation present in the interstellar space originating from natural cosmological processes related to the lifecycle of stellar objects, including the Sun. The basic characteristics of cosmic rays are their composition and energies. Cosmic rays arriving in the interplanetary space are called primary or galactic cosmic rays (GCRs). They include also particles originating from the Sun – solar wind – which have greater intensity and lower energies. The latter are named solar cosmic rays (SCRs). The SCR intensity and energy distribution are related to the primary solar activity variations (solar cycles) and the random activity known as space weather. Because of their greater intensity, the SCRs alter (modulate) the intensity and energy spectrum of the cosmic rays that enter the solar system. When the Sun is active, fewer cosmic rays reach the Earth than during times when the Sun is quiet. As a result, the cosmic rays follows an 11-year cycle like the Sun's activity level but in the opposite direction: higher solar activity corresponds to lower intensity of cosmic rays, and vice versa (Fig. 20.1). Monitoring of this modulation is used for analysis of the interplanetary medium and the solar activity and its effects on the Earth.

On entering the atmosphere, cosmic rays interact with its constituents by a series of cascade reactions, resulting in a large number of secondary particles. Measurements showed that neutrons and protons constitute the largest fraction of the secondary particles at sea level. Thus, to be able to constantly monitor these secondary particles on the ground, a reliable instrument was developed in the 1950s – the so-called neutron monitor (Simpson, 2000). The design of the neutron monitor consists of four main parts (Fig. 20.2): a large proportional counter (tube) filled with boron trifluoride ($^{10}BF_3$), a cylindrical moderator from hydrogen-rich material, a high-atomic number neutron producer made of lead (Pb), and an outer box of hydrogen-rich material – reflector (Hatton and Carimichael, 1964). The protons and the neutrons produced in the atmospheric interactions produce an additional number of neutrons in the lead producer. The neutrons are slowed down (moderated) in the hydrogen-rich cylinder inside the producer in order to increase the detection probability in the detector. The surrounding reflector is used to moderate and reflect neutrons coming from other sources than the cosmic rays (Stoker, 2009).

Neutron monitors have been in continuous operation since the late 1950s. It was soon realized that the scientific value of the data will be much greater if there are multiple stations positioned at different geomagnetic locations. The results from several latitude surveys showed the dependence of neutron monitor measurements on geomagnetic latitude. A practical measure to compare measurements made at different locations on Earth is the geomagnetic cutoff rigidity. Several cosmic rays stations were operating neutron monitors with comparable design, which made possible the use of the geomagnetic field to analyze the primary cosmic ray properties (Simpson et al., 1953). Initially, the data from the available stations were collected in various World Data Centers (WDCs). The first network of operating neutron monitors was initiated during the international geophysical year (1957). During these periods data accessibility and dissemination was tedious and slow, which hindered analysis and research.

Why does monitoring the cosmic rays matter? Cosmic rays have a substantial impact on the Earth's atmosphere by the secondary particles they produce when colliding with atmospheric atoms and by the ionization of atmospheric atoms. Fast charged particles are a source of irradiation, as are X-rays, and therefore can be hazardous. While there seems to be little effect on the ground, aircraft and spacecraft crews are less protected by the atmosphere. As a result, long-time exposure of pilots and cabin crew, especially on transpolar routes, increases the possibility of adverse effects of the cosmic

Knowledge Discovery in Big Data from Astronomy and Earth Observation. https://doi.org/10.1016/B978-0-12-819154-5.00032-1

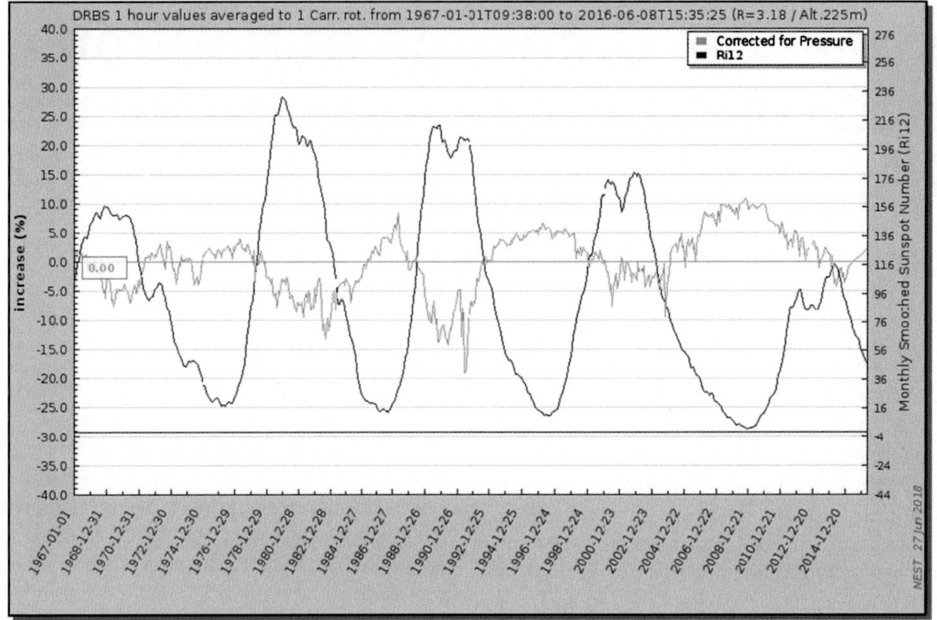

FIG. 20.1 Solar modulation of the cosmic ray intensity based on long-time measurements from the neutron monitor in Dourbes, Belgium (image courtesy of NMDB; smoothed sunspot numbers and monthly sunspot numbers are provided by the Solar Influences Data Analysis Center [SIDC], Royal Observatory of Belgium).

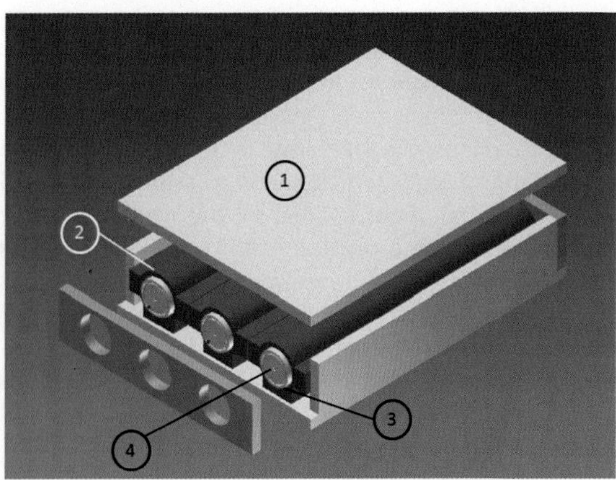

FIG. 20.2 Schematic of a three-tube neutron monitor design (3-NM64): moderator tubes (3) housing the proportional detectors (4), producer (2), and reflector (1).

radiation. Also, short-time increased radiation during strong solar energetic particle (SEP) events can have adverse effects on passengers and avionics. High-frequency (HF) radio communications in the polar regions are also affected by polar cap absorption events due to SEP events. In addition to the adverse effects of the cosmic radiation on astronauts, there is also a potential risk for damage of spacecraft equipment. As the modern society relies more and more on aircraft/space travel, cosmic radiation effects need to be monitored and investigated.

20.2 THE NEUTRON MONITOR DATABASE (NMDB)

20.2.1 The Need for NMDB

The increasing space weather awareness posed more rigorous demands on the data quality, accessibility, and real-time availability. The strong dependence of neutron monitor count rate on the solar cycle and solar activity together with its simple design and low maintenance costs rendered the neutron monitor an important tool in space weather research and monitoring. In 2007 several European and non-European institutions decided to systematize, optimize, and provide reliable data in real-time from the operating stations worldwide and also to provide historical data of ground-based cosmic ray stations continuously measuring the neutron intensity. The idea evolved into a project that was funded for 2 years (2008–2009) by the European Union's Seventh Framework Programme (FP7) for Research and Techno-

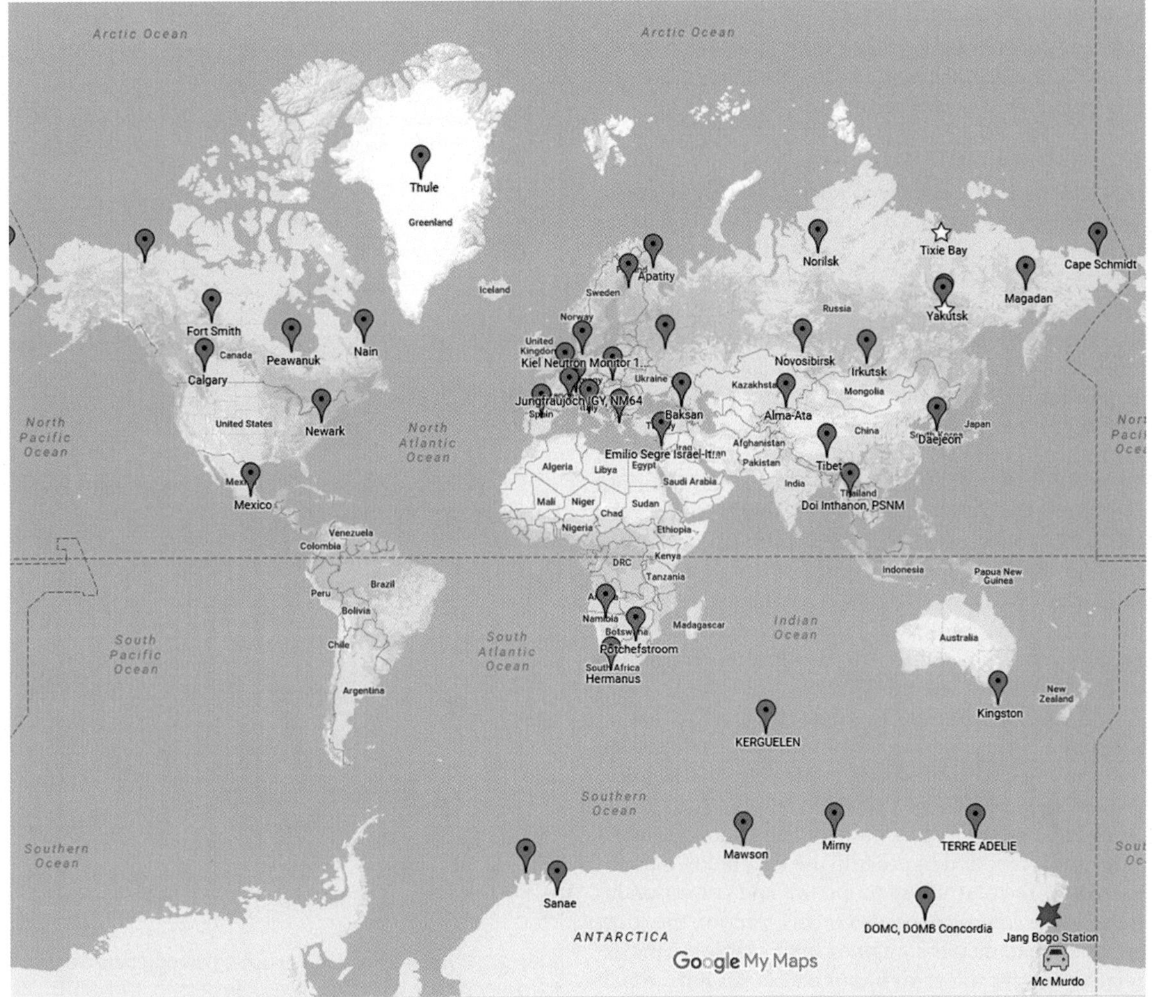

FIG. 20.3 A global map showing the locations of the neutron monitor stations (image courtesy of NMDB).

logical Development (Capacities) (Steigies et al., 2009). The neutron monitor database (NMDB) was thus created[1] and by the end of 2009 there were already 14 stations providing data in real-time.

The main objectives of the project were to create an interactive database that is capable to:

- provide data in standard format,
- provide high-resolution data,
- offer real-time data with a targeted delay of less than 5 min,
- offer easy access to the data,
- provide design for modern registration systems,
- develop online applications using the neutron monitor data.

The project started by first collecting hourly data in the WDC format (now World Data System under the International Council for Science).[2] At that time there were no real-time or high-resolution data, which are required for space weather applications. A year later, in 2008, the NMDB commenced operation by collecting data in near-real-time from several stations. Later on, more and more institutions worldwide have joined the NMDB consortium. As of 2019, the NMDB collects data from more than 60 cosmic ray stations (Fig. 20.3).

[1] www.nmdb.eu.

[2] www.icsu-wds.org.

About half of them (Table 20.1) are providing data in real-time in the rigidity range from 0 at the poles to 15 GV at the geomagnetic equator utilizing the entire GMF for analysis and observations.

20.2.2 The NMDB Database

The NMDB is a centralized state-of-the-art database with distributed mirrors. At the time of writing, it has two master file servers – an active (db01) and a passive database server (db02). Each member of the NMDB consortium provides its data to the active master server. The data are then synchronized with the passive master server. The master server is also responsible for disseminating the data to a mirror server (db04), which has the objective to distribute the data to the three external mirrors of the NMDB, located in Moscow (db10), Athens (db20), and Oulu (db30). To provide sufficient data redundancy each mirror has a complete copy of the NMDB. The fan-out server (db04) to the slave servers: db03 for external users, db05 (a slave server at the ET), and a third slave server, db06, at the neutron monitor. The external users can have data access via the slave db03 in read-only mode. The structure of the NMDB hardware is illustrated in Fig. 20.4.

The data are arranged in several tables. The first table contains detailed information about each station contributing data to the NMDB (Fig. 20.5). This information, coordinates, geomagnetic cutoff, altitude, detector type, building materials, and surrounding materials, etc., is required in order to use the station data for analysis of the cosmic ray properties and correct evaluation of the intensity measurements. Among the most important parameters contained in this table are the average count per solar minimum and maximum, which are used by the NMDB software to calculate the relative intensity change of the station.

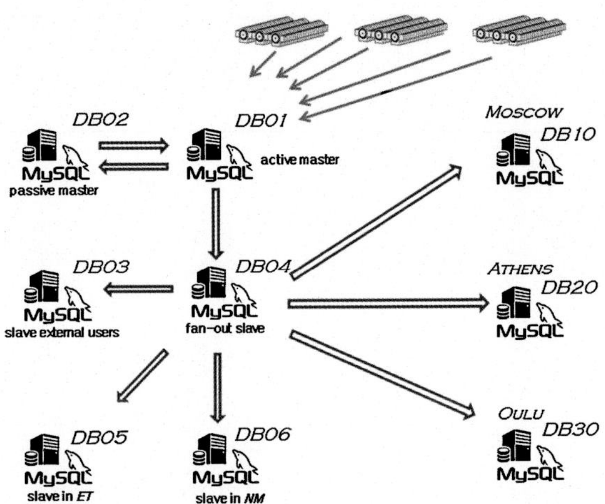

FIG. 20.4 A schematic view of the NMDB structure and data exchange (Steigies, 2016).

```
## -----------------------------------------------------------------------
##describe station_information;
##+-------------------------------+---------------+------+-----+---------+----------------+
##| Field                         | Type          | Null | Key | Default | Extra          |
##+-------------------------------+---------------+------+-----+---------+----------------+
##| ID                            | bigint(20)    | NO   | PRI | NULL    | auto_increment |
##| station_full_name             | varchar(100)  | NO   |     |         |                |
##| station_short_name            | varchar(20)   | NO   |     |         |                |
##| start_date_time               | date          | NO   |     | NULL    |                |
##| end_date_time                 | date          | YES  |     | NULL    |                |
##| head_organization             | varchar(1000) | NO   |     |         |                |
##| principal_investigator        | varchar(200)  | NO   |     |         |                |
##| contact_person                | varchar(200)  | NO   |     |         |                |
##| latitude_deg                  | float         | NO   |     | 0       |                |
##| longitude_deg                 | float         | NO   |     | 0       |                |
##| altitude_m_asl                | float         | NO   |     | 0       |                |
##| geomag_cutoff_GV              | float         | NO   |     | 0       |                |
##| avg_count_per_sec_solar_min   | float         | NO   |     | 0       |                |
##| avg_count_per_sec_solar_max   | float         | NO   |     | 0       |                |
##| ref_pressure_mbar             | float         | NO   |     | 0       |                |
##| barometric_coefficient        | float         | NO   |     | 0       |                |
##| detector_housing              | varchar(1000) | YES  |     |         |                |
##| counter_tube_type             | varchar(100)  | NO   |     |         |                |
##| counter_tube_num              | int(11)       | NO   |     | 1       |                |
##| detector_info                 | varchar(1000) | NO   |     |         |                |
##| pressure_sensor               | varchar(1000) | NO   |     |         |                |
##| clock_info                    | varchar(200)  | NO   |     |         |                |
##| additional_station_info       | varchar(2000) | NO   |     |         |                |
##+-------------------------------+---------------+------+-----+---------+----------------+
```

FIG. 20.5 Table structure containing comprehensive data for every station providing measurements to the NMDB. This information is a prerequisite for correct use of the data for analysis and calculations (image courtesy of NMDB).

20.2.3 Data Contribution and Dissemination

For every station there are three basic tables containing the measured data: the original data table, the revised data table, and the hourly data table. The tables' structure is outlined in Table 20.2.

An additional table per station contains metadata like the timestamp of the first and the last record, the number of records, and the time elapsed from the last recorded measurement. Table 20.3 is used for diagnostics of the health of the individual contributing station and the database as a whole.

Most of the NMDB data providers are performing measurements at 1 min resolution. The raw data consist of the count rate of every detector tube (nine for the DRBS neutron monitor) and the atmospheric pressure measured at the location of the neutron monitor. The raw data are often subjected to an automatic quality check to filter out (remove) noisy tube measurements, and prepare the data to be sent to the NMDB (Sapundjiev et al., 2014, 2016a, 2016b). The data format at this stage consists of the date and time, the pressure (mbar), total counts (i.e., the sum of the individual detector tubes), the total counts corrected for pressure, and the relative intensity change. A portion of the 1 minute data file from the Dourbes neutron monitor is given in Table 20.3.

Every minute the new data are read from this file and sent to the NMDB active master server. Before the NMDB database is populated, a standard quality check (to detect and remove spikes, check pressure values,

TABLE 20.1
NM64-type neutron monitor stations providing data to the NMDB (data courtesy of NMDB).

Station	Rigidity (GV)	Altitude (m)	Number of detectors	Latitude (°)	Longitude (°)
TERA	0.01	32	9	−66.65	140
INVK	0.30	21	18	−68.35	−133.72
THUL	0.30	26	9	76.5	68.7
JBGO	0.30	30	5	−74.6	164.2
TXBY	0.48	0	18	71.36	128.54
NRLK	0.63	0	18	69.26	88.05
APTY	0.65	181	18	67.57	33.4
SNAE	0.73	856	6	−71.667	−2.85
OULU	0.81	15	9	65.05	25.47
KERG	1.14	33	19	−49.35	70.25
YKTK	1.65	105	18	62.01	129.43
MGDN	2.09	220	18	60.04	151.05
KIEL	2.36	54	18	54.3399	10.1199
MOSC	2.43	200	24	55.47	37.32
MCRL	2.46	2000	6	55.47	37.32
NVBK	2.91	163	24	54.48	83
DRBS	3.18	225	9	50.1	4.6
IRKT	3.64	475	18	52.47	104.03
IRK2	3.64	2000	12	52.37	100.55
IRK3	3.64	3000	6	51.29	100.55
LMKS	3.84	263	8	49.2	20.22
JUNG1	4.49	3475	3	46.55	7.98
HRMS	4.58	26	12	−34.43	19.23
BKSN	5.70	1700	6	43.28	42.69
BURE	5.00	2555	3	44.633889	5.907222
BKSN	5.70	1700	6	43.28	42.69
ROME	6.27	0	20	41.86	12.47
PTFM	6.94	1351	12	−26.68	27.09
CALM	6.95	708	15	40.559167	−3.1625
NANM	7.10	2000	18	40.3667	44.25
MXCO	8.28	2274	6	19.33	−260.82
ATHN	8.53	260	6	37.97	23.78
TSMB	9.15	1210	18	−19.2	17.58
DJON	11.22	200	18	36.24	127.22
PSNM	16.80	2565	18	18.59	98.49

TABLE 20.2
Data structure of a single measurement: original, revised, and hourly data.

	Original data	Revised data	Hourly data
start_date_time	yes	yes	yes
length_time_interval_s	yes	yes	
uncorrected	yes	yes	yes
corr_for_efficiency	yes	yes	yes
corr_for_pressure	yes	yes	yes
pressure_mbar	yes	yes	yes
version		yes	

TABLE 20.3
An example of the original minute data measurements from the Dourbes Neutron Monitor.

Date	Time, UTC	p (mbar)	Raw counts	Corrected counts	Relative
...	...	...	...	...	...
30/06/2016	00:00:00	986.65	6214.65	6217	−10.1446
30/06/2016	01:00:00	986.67	6373.02	6377	−7.84114
30/06/2016	02:00:00	986.71	6309.16	6315	−8.7377
...	...	...	...	...	...

etc.) is executed. Following this the data are synchronized with the passive master server and then sent to the slaves and the mirrors (Fig. 20.4).

A second set of data is sent in a similar way to the NMDB. This is the hourly average measurements. An hour average is calculated only if there are at least 75% of the measurements during this hour (e.g., 45 min). The files have the same data format with the difference that the measurement data are averaged over an hour. This allows faster data access and display for data requests over large time intervals. The hourly data are typically transferred at 15 min past an hour. For the stations that have performed analysis of their detector and have calculated the detector efficiency, their data are corrected for efficiency and populated in the corresponding table.

The minute and hourly data just described are corrected and sent to the NMDB automatically. Despite data quality control, it is not 100% immune to errors coming from combinations of noisy tubes that could not be filtered out by the correction algorithms (this is the case when more than half of the tubes are producing noisy and erroneous data). For this purpose every station PI is encouraged to manually revise the minute and hourly data and send it to the corresponding data tables. The easiest way to access the data from the NMDB is to use the NMDB Event Search Tool (NEST), developed and maintained by the Paris Observatory. It is implemented as a web interface on the NMDB web site offering a large number of possibilities for data retrieval and display (Fig. 20.6). The color code shows the stations which are currently online. NEST provides a quick access to the entire set of recognized Ground Level Enhancement (GLE) and Forbush decrease (FD) events. It also offers the possibility to request historical data for all of the participating cosmic rays stations. NEST is optimized for fast data delivery depending on the requested time interval. It offers the possibilities to retrieve the data in an ASCII format and/or as a plot. Useful additional information is available, like the Kp index, the proton flux from the GOES satellite, the sunspot number, and numerous options concerning the data that are requested (raw counts, pressure, and/or efficiency corrected). For advanced users and developers of NMDB-tools, e.g. real-time applications using the data, a read-only account to the database can be set. The

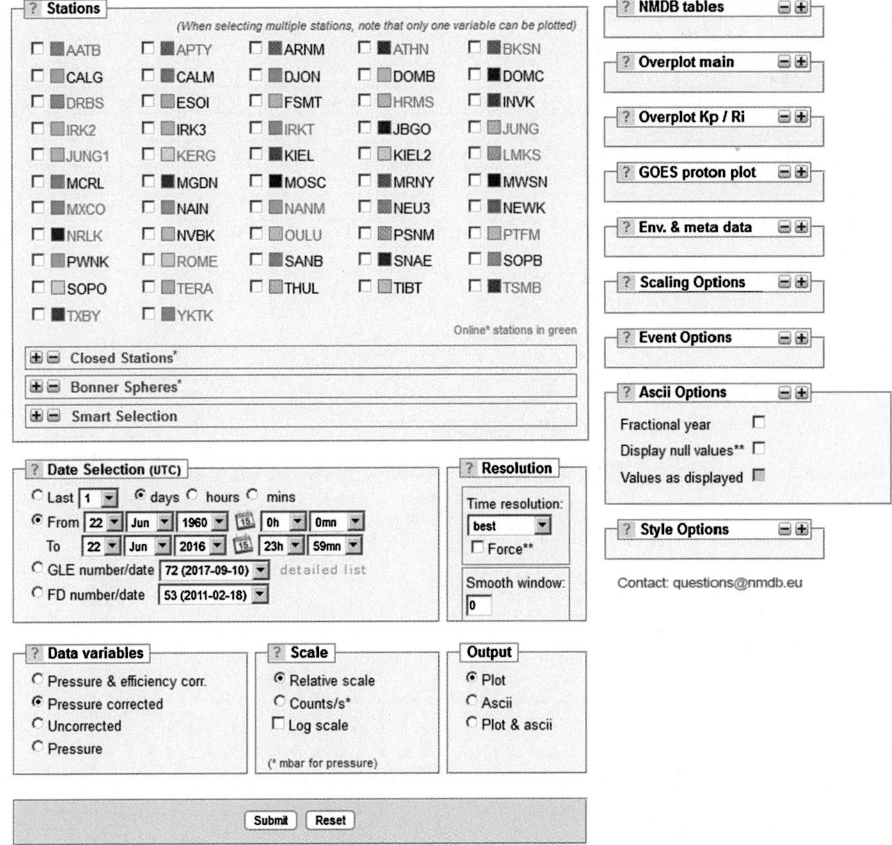

FIG. 20.6 A screen capture of the NMDB Event Search Tool (NEST) – the online interface for data plotting and retrieval developed by the Paris Observatory (image courtesy of NMDB).

web site interface allows a quick check of the cosmic ray intensity worldwide via the Cosmic Rays Now plot provided by NEST. The health of the monitors and the data quality can be checked via the NMDB status tool available from the web site interface. This tool provides useful information for the stations currently online and the data they are providing. The tool has several options to choose from (cf. Fig. 20.7).

20.3 APPLICATIONS

The two principal user groups of the NMDB data are the astrophysicists and space weather forecasters. The former are using the long datasets of historic data in the database to search for periodicities and quasiperiodicities in the intensity of the GCRs (Kudela et al., 2009). The latter use high-resolution real-time NMDB data to develop several research applications for prediction of space weather events (Mavromichalaki et al., 2011). Due to the high energy and stability of the GCR component and the nature of their interaction with solar wind, observations of cosmic ray anisotropies by the global network of neutron monitors can serve as precursors for energetic solar particles directed towards the Earth (Belov et al., 2001, 2017; Dorman et al., 2004). At present the NMDB offers several applications (some of which are provided by external research institutions), such as daily cosmic ray variations, ionization rates and atmospheric dose rates, GLE spectrum, and GLE alert.

20.3.1 Ground Level Enhancements (GLEs) – Detection and Characterization

Thanks to the reliable measurements and stable performance of the neutron monitors, the strong correlation between the GCR flux and the solar activity could be observed (Fig. 20.1). The great number of stations in

Station: DRBS · Table: All

Station: DRBS

Table	Table name	Status	First data	Last data	#Records	Note
Original data (1min)	DRBS_ori	Online	1967-01-01 00:00:00	2018-06-27 20:44:00	4346629	
1 hour data	DRBS_1h	Online	1967-01-01 00:00:00	2018-06-27 19:00:00	439853	
Enviromental data	DRBS_env	Offline			0	Table empty
Revised data	DRBS_rev				0	Table empty
META data	DRBS_meta		1965-10-01 00:00:00	2016-05-18 00:00:00	108	

Current UTC time: 2018-06-27 20:47:10
Used NMDB mirror: **db04.nmdb.eu**

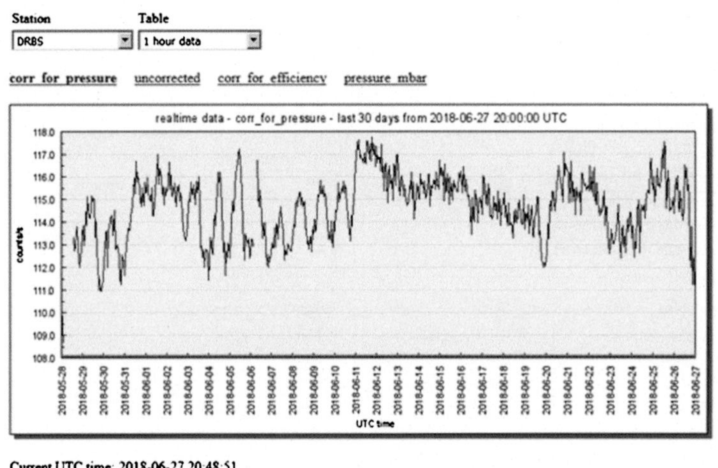

FIG. 20.7 Quick check tool using the metadata. (Top) Quick diagnostics of the data and availability of the Dourbes neutron monitor. (Bottom) Hourly data (as selected from the table above) (image courtesy of NMDB).

the NMDB allows for mapping of the random events within the geomagnetic field. During high solar activity, the Sun is capable of accelerating particles to very high energies, as noticed by ground level observations. These events are registered by the neutron monitors as a GLE or a substantial decrease in the count rate, known as FD (Forbush, 1946). The observed variations in the count rate of the neutron monitors are still a subject of scientific research. The NEST has a quick access option to select and display any of the recognized GLE events as well as quick access shortcuts to the last two or three GLE events (Figs. 20.8 and 20.9). From Fig. 20.8 we can note that the relative increase in the neutron monitor count rate depends strongly on the geomagnetic position of the station for comparable neutron monitor designs (both Dourbes and Oulu operate a nine-tube neutron monitor, 9-NM64, while Rome is running a 20-tube monitor, 20-NM64).

The relative increase in the count rate of a neutron monitor is a function of the geomagnetic rigidity and the atmospheric depth. At a given time instant the relative count rate increase is related to the primary cosmic ray spectrum via the specific yield of the neutron monitor for a given intensity of the secondary particles. Mathematically this is accounted for by the neutron monitor yield function. Three methods for determination of the yield function are utilized (Clem and Dorman, 2000): theoretical calculation, Monte Carlo simulations of particles transports (Wainio et al., 1968; Mishev et al., 2013), and parametrization of experimental latitude surveys (Treiman, 1952; Lockwood and Webber, 1967; Caballero-Lopez and Moraal, 2012), or sometimes combinations of those (Flückiger et al., 2008). The yield function depends on the primary particle type, its direction, and the design of the neutron monitor. The

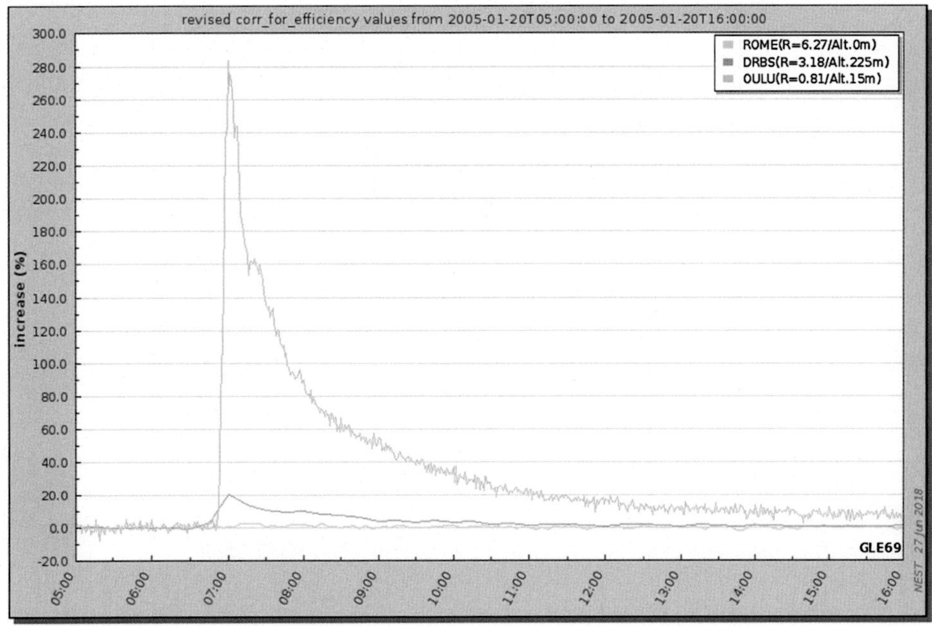

FIG. 20.8 A record of a ground level enhancement, the GLE #69 event that occurred on 20 January 2005 (data from Rome, Oulu, and Dourbes neutron monitors; image courtesy of NMDB).

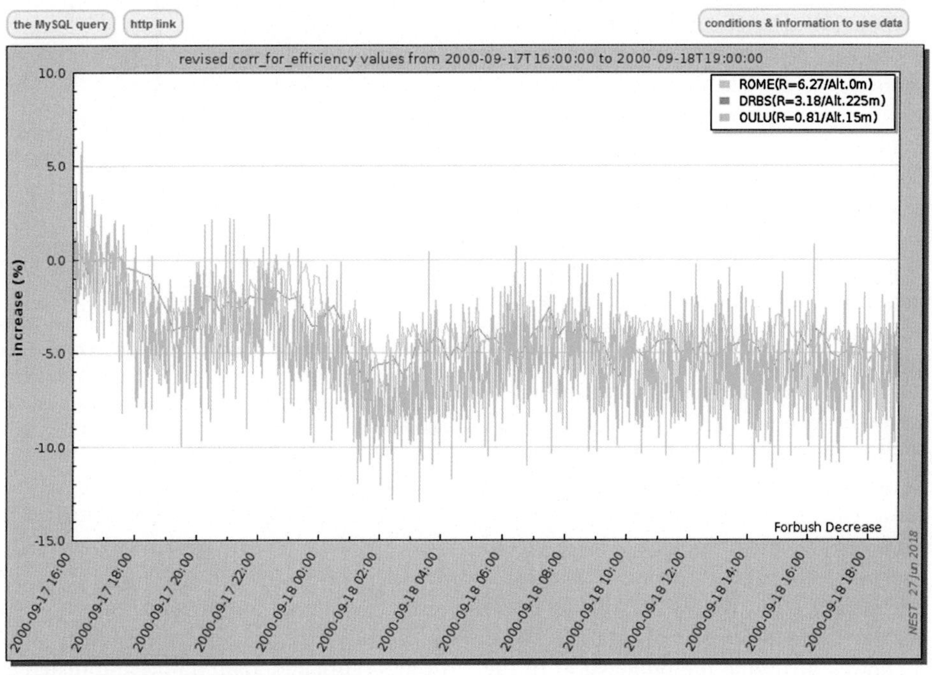

FIG. 20.9 A record of a Forbush decrease, the FD #39 event that occurred on 18 September 2000 (data from Rome, Oulu, and Dourbes neutron monitors; image courtesy of NMDB).

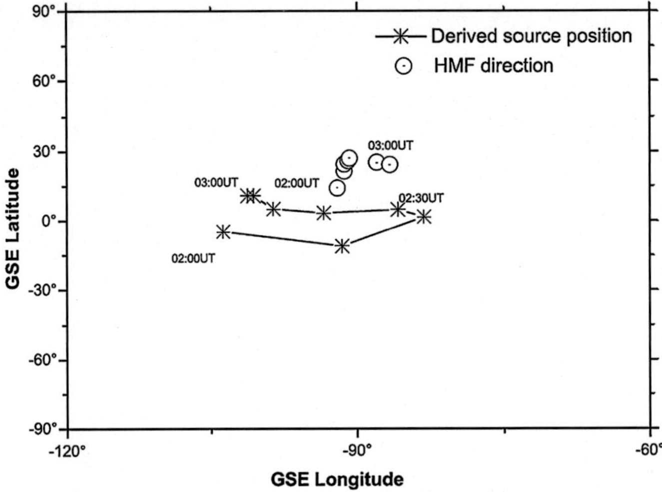

FIG. 20.10 Determination of the apparent source position and its evolution during GLE #71 (Geocentric Solar Ecliptic System) (Mishev et al., 2014) (copyright license No.: 4647560881398).

best results are obtained by calculating the yield function for every neutron monitor used for the analysis using the station-specific data parameters from the NMDB using any transport code (Ferrari et al., 2005; Allison et al., 2016). If data about the design parameters (dimensions and materials) of the station are not available, a more general yield function can be used (e.g., the yield function parametrized for atmospheric depth and rigidity proposed by Flückiger et al., 2008). Many yield functions are determined for specific conditions (sea level, specific solar activity, etc.). In this case the count rate of the different NMDB stations used for the analysis can be normalized to sea level using the 2-attenuation data correction (McCracken, 1962). With the yield function and a model of the solar particle spectrum, we can determine the apparent position of the solar particles and the pitch angle distribution (Debrunner and Lockwood, 1980). The data analysis is complex and requires additional measurements like the properties of the heliospheric magnetic field, the primary GCR spectra (from the GOES satellite), modeling of the atmospheric interactions, etc. The results of a detailed analysis of the GLE #71 from 17 May 2012 performed by Mishev et al. (2014) are illustrated in Figs. 20.10 and 20.11. The availability of high-resolution NMDB data allowed analyzing the progression of the GLE parameters and spectrum analysis like the evolution of the pitch angle during the different phases of the event (Fig. 20.12).

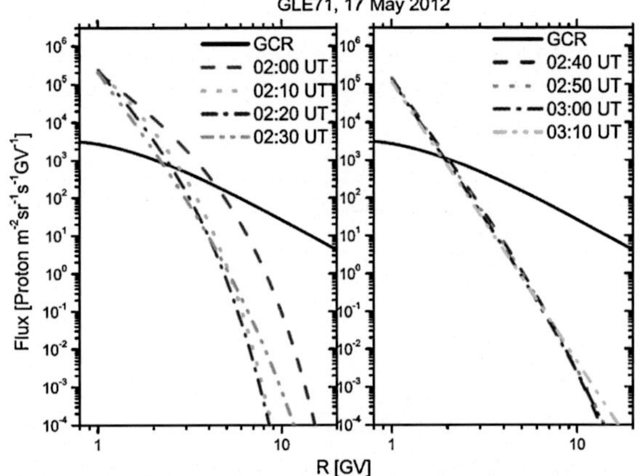

FIG. 20.11 GLE #71 analysis: high-energetic solar particle rigidity spectrum progression during the course of the GLE. The spectra are taken along the symmetry axis of the event (Mishev et al., 2014) (copyright license No.: 4647560881398).

20.3.2 Evaluation of the Radiation Effects on Electronics and Health

The neutron fluence at a given location can be determined. Neutrons in the energy range of 0.01–10 GeV may pose problems to the components of control electronics (airplanes, train traffic, automated cars, etc.) by single event effects (state flips between 0 or 1) (Gordon et al., 2004). A typical problem is the determination of the total neutron fluence during a ground-level en-

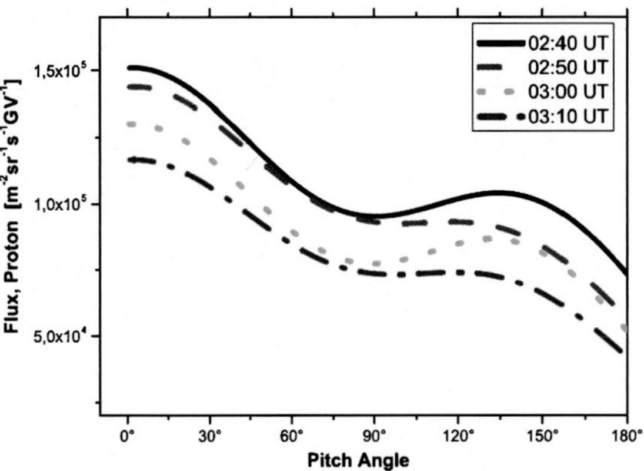

FIG. 20.12 GLE #71: Evolution of the primary particle pitch-angle distribution (Mishev et al., 2014) (copyright license No.: 4647560881398).

hancement event in the vicinity of a neutron monitor station. In this case, the count rate of the neutron monitor has to be corrected for multiplicity (i.e., the number of secondary neutrons produced per primary neutron within the neutron monitor) and for efficiency. The final fluence is then obtained by integration over the continuation of the GLE (work in progress). The results from the GLE analysis can be used to calculate the radiation doses at flight altitudes during a GLE event. The contribution of the GCRs to the effective dose rates was studied during the first space flights and the following missions. The modeling of the GCR spectrum is still under study and affects the results from the calculations of the radiation burden in space (Mrigakshi et al., 2013). The dose rate due to GCRs depends on the solar modulation and the geomagnetic latitude. A typical dose rate due to GCR during a flight between Europe and North America amounts to about 5 μSv/h (Matthiä et al., 2014). The contribution to this dose can be significant during a strong solar particle event (Reitz, 1993; Vainio et al., 2009). An extreme event may result in dose rates of about 3 mSv as calculated for the GLE #05 (Lantos and Fuller, 2004), which is about 3 times the annual occupational dose of 1 mSv. The high resolution data from NMDB allows determining the dose rates and their evolution due to solar energetic particles for a given location (i.e., flight path). The advantage of the NMDB data is that the anisotropies in the dose rates can be determined. For example, the analysis of Mishev et al. (2015) reported dose rates of 150 μSv/h over the North polar region and 1000 μSv/h over the South polar region during the main phase of GLE #69. The calculated dose rates depend on the methods used during the characterization of the GLE. This can introduce differences in the obtained results, as pointed out by Bütikofer and Flückiger (2013).

20.3.3 Space Weather Nowcast and Forecast

After collecting the available historical data from the neutron monitors world-wide and creating the infrastructure of registering and receiving data from high- and low-latitude stations in near-real-time, the efforts were focused on utilizing these data for characterization of the interplanetary space at the present moment (nowcast) and for prediction of possible GLE or SEP events (forecast). Space weather nowcast is possible because the GCRs occupy the interplanetary medium. Changes in the solar activity influence the interplanetary space and therefore the omnipresent GCRs (Kudela et al., 2000; Kudela and Storini, 2006). Hence by monitoring the cosmic rays intensity worldwide we can obtain information about the state of the solar activity. The first systematic and continuous space weather monitoring was the mininetwork of cosmic ray observatories Spaceship Earth, operated by the Bartol University (Bieber et al., 2004). The NMDB has many more stations, providing the opportunity for better evaluation of the current interplanetary conditions. A powerful tool for nowcasting is the evaluation of the cosmic ray anisotropies which are directly related to fluctuations in the interplanetary state (Asipenka et al., 2009; Eroshenko et al., 2009). Geomagnetic storms are also detected by the NMDB. The changes in the local geomagnetic field affect the geomagnetic cutoff rigidity and hence the count rate of the neutron monitor located at this point (Kudela and Bucik, 2005; Kudela and Usoskin, 2004). Changes in the relative intensity of a neutron monitor can be due to energetic solar particles, or a depreciation of the GCR flux due to propagating disturbances which sweep the GCRs, resulting in a decrease in the measured intensity. Solar particles contribute very little to the total flux and cannot be detected by satellites (Dorman et al., 2004). Analysis of such disturbances have pointed out that they result in specific changes of the count rate in the network of neutron monitors called precursors. They can manifest themselves as increase or decrease in the count rate before the arrival of the main phase of the disturbance (Belov et al., 2001, 2003, 2016). These observations have been quantified by Dorman et al. (2004) for utilizing the NMDB real-time measurements for forecasting of GLE occurrences. The GLE alert systems utilizing the NMDB data often run their own mirror of the database to allow a quick access to the available measurements

(IZMIRAN,[3] ANEMOS[4]) (Mavromichalaki et al., 2005; Kuwabara et al., 2006).

20.4 SUMMARY AND OUTLOOK

The main outlook for the future of NMDB is to improve the data availability for real-time applications. The time interval between the measurement and the arrival of the data to the database varies for the different remote stations with slow network infrastructure. The second important target is the improvement of the data quality. Many stations are still not checking basic data (e.g., out of range pressure measurements, noisy spikes in the intensity data, etc.). Cosmic rays are a formidable source of information about the universe and are still an important topic of international research, with many open questions waiting for answer, e.g., under what circumstances and how the charged particles are accelerated to such high energies or speeds. Modern applications frequently employ extensive arrays of monitors. In effect, the observing instrument is not an isolated instrument, but rather an array of instruments. Networking the neutron monitors yields new information in several areas, such as anisotropy, energy spectrum, relativistic solar neutrons, etc.

ACKNOWLEDGMENTS

The NMDB and the research results from the data provided by it received funding from the European Union's Seventh Framework Program, FP7 (2007–2013) under grant agreement No. 213007. Data retrieved via NMDB are the property of the individual data providers. We acknowledge the Rome, Oulu and Dourbes neutron monitors for providing data. The authors thank Dr. Christian Steigies for the helpful discussions.

REFERENCES

Allison, J., Amako, K., Apostolakis, J., et al., 2016. Recent developments in Geant4. Nuclear Instruments & Methods in Physics Research. Section A, Accelerators, Spectrometers, Detectors and Associated Equipment 835, 186–225.

Asipenka, A., Belov, A., Eroshenko, E., et al., 2009. Asymptotic longitudinal distribution of cosmic ray variations in real time as the method of interplanetary space diagnostic. In: Proc. 31st International Cosmic Rays Conference, Lòdž 2009 (ICRC 2009), p. 2.

Belov, A.V., Abunina, M.A., Abunin, A.A., et al., 2017. Vector anisotropy of cosmic rays in the beginning of Forbush effects. Geomagnetism and Aeronomy 57, 541–548.

[3] http://cr0.izmiran.ru/GLE-AlertAndProfilesPrognosing/.
[4] http://cosray.phys.uoa.gr/index.php/glealertplus.

Belov, A.V., Bieber, J.W., Eroshenko, E.A., et al., 2001. Pitch-Angle features in cosmic rays in advance of severe magnetic storms: neutron monitor observations. In: International Cosmic Ray Conference, vol. 9, p. 3507.

Belov, A.V., Bieber, J.W., Eroshenko, E.A., et al., 2003. Cosmic ray anisotropy before and during the passage of major solar wind disturbances. Advances in Space Research 31, 919–924.

Belov, A.V., Eroshenko, E.A., Abunina, M.A., et al., 2016. Behavior of the cosmic ray density during the initial phase of the Forbush effect. Geomagnetism and Aeronomy 56, 645–651.

Bieber, J.W., Evenson, P., Dröge, W., et al., 2004. Spaceship Earth observations of the Easter 2001 solar particle event. The Astrophysical Journal Letters 601 (1), L103–L106.

Bütikofer, R., Flückiger, E.O., 2013. Differences in published characteristics of GLE60 and their consequences on computed radiation dose rates along selected flight paths. Journal of Physics. Conference Series 409, 012166.

Caballero-Lopez, R.A., Moraal, H., 2012. Cosmic-ray yield and response functions in the atmosphere. Journal of Geophysical Research: Space Physics 117 (A12), 103.

Clem, John M., Dorman, Lev I., 2000. Neutron monitor response functions. Space Science Reviews 93 (1), 335–359.

Debrunner, H., Lockwood, J.A., 1980. The spatial anisotropy, rigidity spectrum, and propagation characteristics of the relativistic solar particles during the event on May 7, 1978. Journal of Geophysical Research 85, 6853–6860.

Dorman, L.I., Pustilnik, L.A., Sternlieb, A., et al., 2004. Monitoring and forecasting of great solar proton events using the neutron monitor network in real time. IEEE Transactions on Plasma Science 32 (4), 1478–1488.

Eroshenko, E., Asipenka, A., Belov, A., et al., 2009. Definition of cosmic ray density and anisotropy vector beyond the magnetosphere in real time mode. In: Proc. 31st International Cosmic Rays Conference, Lòdž 2009 (ICRC 2009), p. 2.

Ferrari, A., Sala, P.R., Fassò, A., Ranft, J., 2005. FLUKA: a multi-particle transport code. Technical Report INFN/TC_05/11, SLAC-R-773. CERN.

Flückiger, E.O., Moser, M.R., Pirard, B., Bütikofer, R., Desorgher, L., 2008. A parameterized neutron monitor yield function for space weather applications. In: International Cosmic Ray Conference, vol. 1, pp. 289–292.

Forbush, S.E., 1946. Three unusual cosmic-ray increases possibly due to charged particles from the Sun. Physical Review 70, 771–772.

Gordon, M.S., Goldhagen, P., Rodbell, K.P., et al., 2004. Measurement of the flux and energy spectrum of cosmic-ray induced neutrons on the ground. IEEE Transactions on Nuclear Science 51 (6), 3427–3434.

Hatton, C.J., Carimichael, H., 1964. Experimental investigation of the NM-64 neutron monitor. Canadian Journal of Physics 42, 2443–2472.

Kudela, K., Bucik, R., 2005. Low energy cosmic rays and the disturbed magnetosphere. In: Solar Extreme Events: Fundamental Science and Applied Aspects (SEE-2005), International Symposium. Nor Amberd, Armenia, 26–30 September, 2005.

Kudela, K., Mavromichalaki, H., Papaioannou, A., Gerontidou, M., 2009. On mid-term periodicities in cosmic rays: utilizing the NMDB archive. In: Proc. 31st International Cosmic Rays Conference, Lòdž 2009 (ICRC 2009), p. 2.

Kudela, K., Storini, M., 2006. Possible tools for space weather issues from cosmic ray continuous records. Advances in Space Research 37, 1443–1449.

Kudela, K., Storini, M., Hofer, M.Y., Belov, A., 2000. Cosmic rays in relation to space weather. Space Science Reviews 93, 153–174.

Kudela, K., Usoskin, I.G., 2004. On magnetospheric transmissivity of cosmic rays. Czechoslovak Journal of Physics 54, 239–254.

Kuwabara, T., Bieber, J.W., Clem, J., Evenson, P., Pyle, R., 2006. Development of a ground level enhancement alarm system based upon neutron monitors. Space Weather 4 (10).

Lantos, P., Fuller, N., 2004. Semi-empirical model to calculate potential radiation exposure on board airplane during solar particle events. IEEE Transactions on Plasma Science 32, 1468–1477.

Lockwood, J.A., Webber, W.R., 1967. Differential response and specific yield functions of cosmic-ray neutron monitors. Journal of Geophysical Research 72, 3395–3402.

Matthiä, D., Meier, M.M., Reitz, G., 2014. Numerical calculation of the radiation exposure from galactic cosmic rays at aviation altitudes with the PANDOCA core model. Space Weather 12, 161–171.

Mavromichalaki, H., Gerontidou, M., Mariatos, G., et al., 2005. Space weather forecasting at the new Athens center: the recent extreme events of January 2005. IEEE Transactions on Nuclear Science 52, 2307–2312.

Mavromichalaki, H., Papaioannou, A., Plainaki, C., et al., 2011. Applications and usage of the real-time Neutron Monitor Database. Advances in Space Research 47, 2210–2222.

McCracken, K.G., 1962. The cosmic-ray flare effect: 1. Some new methods of analysis. Journal of Geophysical Research 67 (2), 423–434.

Mishev, A.L., Adibpour, F., Usoskin, I.G., Felsberger, E., 2015. Computation of dose rate at flight altitudes during ground level enhancements no. 69, 70 and 71. Advances in Space Research 55 (1), 354–362.

Mishev, A.L., Kocharov, L.G., Usoskin, I.G., 2014. Analysis of the ground level enhancement on 17 May 2012 using data from the global neutron monitor network. Journal of Geophysical Research: Space Physics 119, 670–679.

Mishev, A.L., Usoskin, I.G., Kovaltsov, G.A., 2013. Neutron monitor yield function: new improved computations. Journal of Geophysical Research: Space Physics 118, 2783–2788.

Mrigakshi, A.I., Matthiä, D., Berger, T., Reitz, G., Wimmer-Schweingruber, R.F., 2013. How Galactic Cosmic Ray models affect the estimation of radiation exposure in space. Advances in Space Research 51, 825–834.

Reitz, G., 1993. Radiation environment in the stratosphere. Radiation Protection Dosimetry 48 (1), 5–20.

Sapundjiev, D., Nemry, M., Stankov, S., Jodogne, J.C., 2014. Data reduction and correction algorithm for digital real-time processing of cosmic ray measurements: NM64 monitoring at Dourbes. Advances in Space Research 53 (1), 71–76.

Sapundjiev, D., Stankov, S., Jodogne, J.C., 2016a. Present status and modernisation of the Dourbes cosmic ray observatory for improved space weather research and forecasting. In: Proc. European Space Weather Week (ESWW). 14–18 November 2016, Ostende, Belgium.

Sapundjiev, D., Steigies, C., Verhulst, T., Jodogne, J.C., Stankov, S., 2016b. Upgrading the Dourbes cosmic ray observatory for research and development of improved space weather monitoring services. In: Proc. European Space Weather Week (ESWW). 23–27 November 2015, Ostende, Belgium.

Simpson, J.A., 2000. The cosmic ray nucleonic component: the invention and scientific uses of the neutron monitor. Space Science Reviews 93 (1), 11–32. IGY neutron monitor development.

Simpson, J.A., Fonger, W., Treiman, S.B., 1953. Cosmic radiation intensity-time variations and their origin. I. Neutron intensity variation method and meteorological factors. Physical Review 90, 934–950.

Steigies, C.T., for the NMDB Consortium, 2016. The Neutron Monitor Database (NMDB) and its applications to space weather. Workshop on cosmic rays and space weather: research activities, service developments, and future strategy. Dourbes, Belgium, https://events.spacepole.be/event/10/sessions/4/attachments/11/60/03-nmdb.pdf.

Steigies, C.T., Klein, K.L., Fuller, N., NMDB Consortium, 2009. www.nmdb.eu: the real-time neutron monitor database. In: Proc. 31st International Cosmic Rays Conference, Lòdž 2009 (ICRC 2009), p. 2.

Stoker, P.H., 2009. The IGY and beyond: a brief history of ground-based cosmic-ray detectors. Advances in Space Research 44, 1081–1095.

Treiman, S.B., 1952. Analysis of the nucleonic component based on neutron latitude variations. Physical Review 86, 917–923.

Vainio, R., Desorgher, L., Heynderickx, D., et al., 2009. Dynamics of the Earth's particle radiation environment. Space Science Reviews 147 (3–4), 187–231.

Wainio, K.M., Colvin, T.H., More, K.A., Tiffany, O.L., 1968. Calculated specific yield functions for neutron monitors. Canadian Journal of Physics 46 (10), S1048–S1051.

CHAPTER 21

Monitoring the Earth Ionosphere by Listening to GPS Satellites

LIUBOV YANKIV-VITKOVSKA, ASSOC PROF, DR • STEPAN SAVCHUK, DSC, PROF

21.1 INTRODUCTION

Research of the composition and structure of the ionosphere is important both for understanding the physics of the processes occurring in it and for solving various radiophysical tasks related to the propagation of radio waves. The processes occurring in the ionosphere are inextricably connected with the processes of interaction in the Earth–Sun system. Information about the state of the near-Earth plasma sheath might serve to identify and investigate cause–effect relations of solar and geomagnetic activity, the processes of the ionosphere interaction with atmospheric processes and earthquakes, etc. Due to the increasing load on global telecommunication systems, as well as widespread use of satellite navigation systems, a practical interest in the ionosphere studies has also substantially increased.

Most of our knowledge about the structure and dynamics of the ionosphere has been obtained by classical radiophysical sounding methods (ionosondes, incoherent scattering radars, space radio sources signal recording, and so on). The present methods are successfully applied nowadays. However, resolving the issues related to detailed diagnostics and prediction of ionospheric disturbances imposes new, more strict requirements to technical specifications of the ionospheric sounding systems, as well as to the quality of the information received. Thus, according to currently known spatial and temporal parameters of ionospheric disturbances, in order to optimally reflect the given disturbances, the means of diagnostics (monitoring) must have a temporal resolution of not less than 10–100 s and a spatial resolution not worse than 10–100 km. No less important is the necessity of simultaneous and uniform measurements. In fact, this means that the ionosphere monitoring system should be involved in the given process, which would operate in continuous mode. Examples of organizing the systems of similar type are meteorological networks, seismic services, and nuclear test detection services.

Currently, there are many methods and tools to study the ionosphere. But not all of them are suitable for global monitoring of its current state. Thus, in particular, physically grounded theoretical numerical models are implemented separately for cases of perturbed and calm ionosphere, which alone cannot claim to adequately reflect the current situation. The empirical models that represent the statistical averaging of a large amount of experimental information are capable of producing the average value of the basic parameters of the ionosphere that are the most typical for the given time and geographical location.

A prerequisite for monitoring the ionosphere state is the use of operational experimental information. One of the perspective opportunities to obtain such experimental information on the ionosphere state is connected with the use of radio signals from global navigation satellite systems (GNSSs): GPS, GLONASS, Galileo, Beidou, and auxiliary subsystems. A navigation signal contains information about the integral number of electrons in the path of its propagation. Today, there are at least 30–40 GNSS satellites at each moment of time in the "visibility" zone of the satellite receiver of the observation station and, in fact, thousands of "satellite-receiver" rays simultaneously shine through the ionosphere (Klobuchar, 1996; Komjathy and Langley, 1996; Tsugawa et al., 2004; Stanislawska et al., 2012). Therefore, to study the ionosphere using GPS satellites, it is necessary to process arrays of big databases.

Considering their promptness and accuracy, GNSS data have been used for several decades to monitor and study ionospheric processes. Active development of remote sensing technologies of the ionosphere with the use of GPS signals began in the late 1980s and was conducted in several directions: development of the technology for constructing global total electron content (TEC) maps (GIM); developing the methods of GPS radiotomography of the ionosphere; developing assimilation models for operative prediction of ionosphere parameters; and development of methods for detection and mathematical modeling of ionospheric disturbances. Some contributions to solving the latter issue were made by the research group of Lviv Poly-

Knowledge Discovery in Big Data from Astronomy and Earth Observation. https://doi.org/10.1016/B978-0-12-819154-5.00033-3

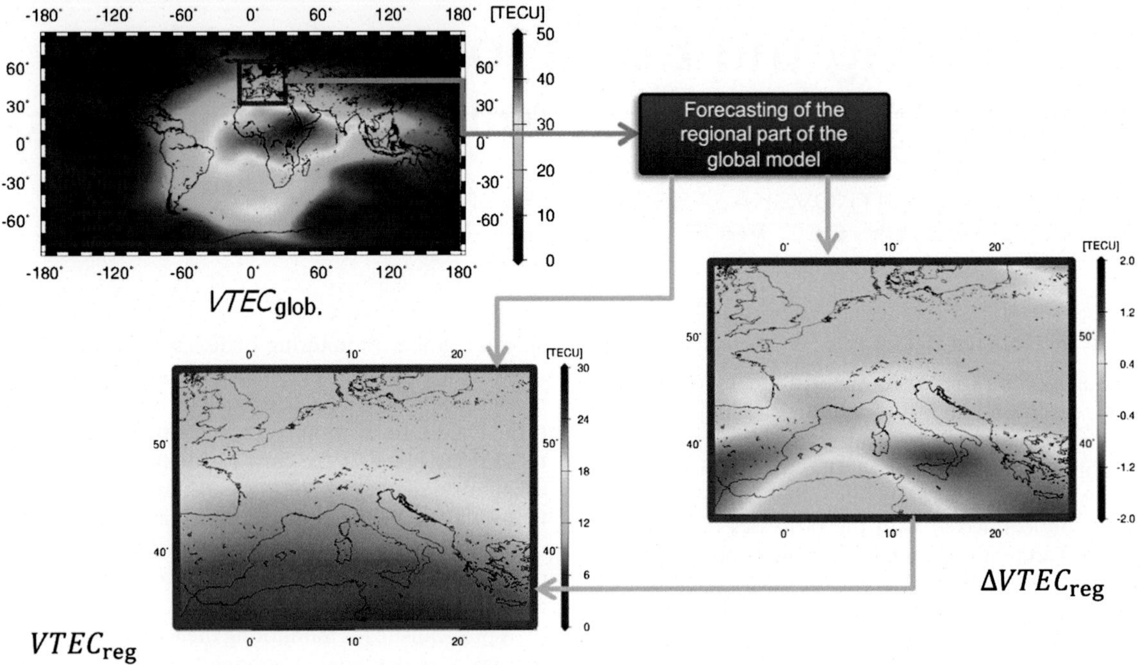

FIG. 21.1 Application: regional real-time VTEC map.

technic with the active participation of the authors (Yankiv-Vitkovska et al., 2016).

Therefore, navigation systems are one of the powerful tools to study global and regional ionospheric processes, as well as variations of the ionosphere characteristics under different conditions. Remote measurements of integral electronic concentration TEC allows not only to monitor the processes of the ionosphere and analyze the ionospheric irregularities movement, but also to evaluate the propagation of radio waves when passing through the Earth's ionosphere. The prototype of the global ionosphere monitoring system currently serves a worldwide GNSS station network from the IGS service and regional/national networks of active GNSS reference stations, and its use in the practice of ionospheric research firstly provides the opportunity to organize global, continuous, totally computerized monitoring of ionospheric disturbances with high temporal and spatial resolution. (See Figs. 21.1 and 21.2.)

The range of works in preparation and processing of GNSS data conducted at the Department of Higher Geodesy and Astronomy of National University Lviv Polytechnic is performed within the frameworks of the WGA LPI project. The organizational chart includes the following structures:

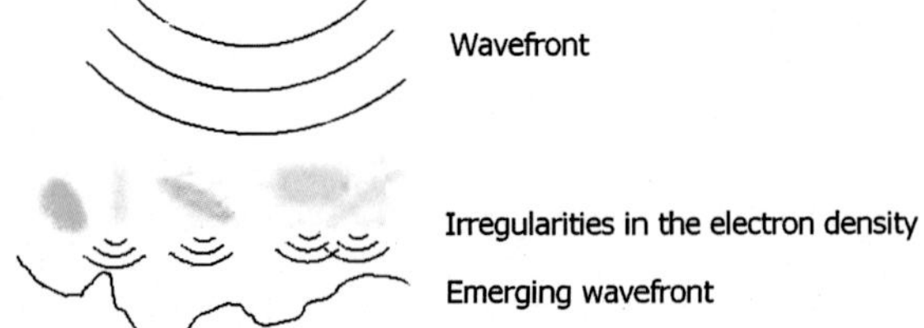

FIG. 21.2 Illustration of the scintillation phenomena.

- Observation stations: GNSS receivers are permanently installed and antennas are attached to geodetic centers, which continuously monitor the satellites and form the database of input data (raw observations).
- Operational centers: The data obtained from the observation stations are checked here, the output data from the receiver format are converted into an independent RINEX format, and the data are archived and uploaded to the data center of the corresponding network via the internet. In most cases, the role of operational centers is performed by the stations themselves through an appropriate set of specialized software or computing network centers.

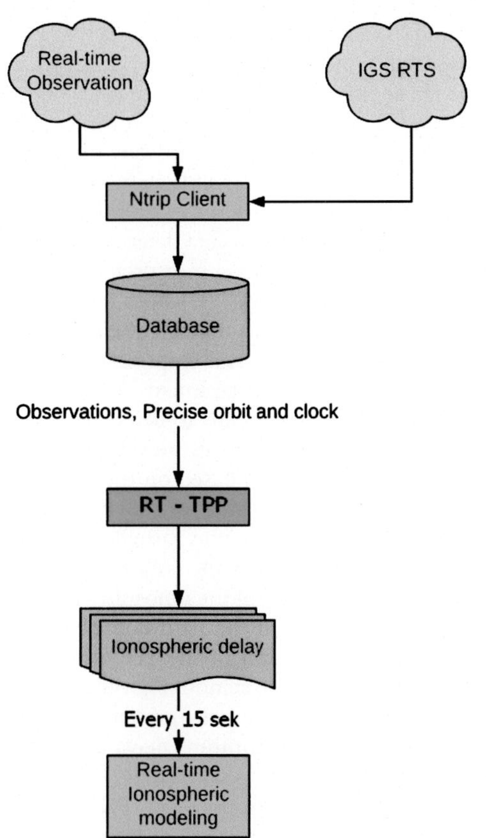

FIG. 21.3 Flowchart of real-time ionospheric modeling. IGS RTS, the International GNSS Service Real-Time Service; TPP, Trimble Pivot Platform.

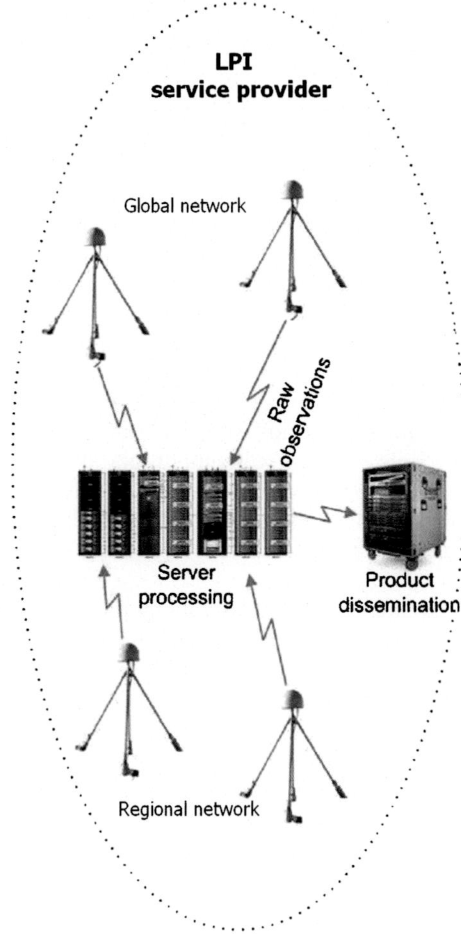

FIG. 21.4 Data analysis center Lviv Polytechnic National University.

- Global network data center – BKG (Bundesamt für Kartographie und Geodäsie) GNSS Data center – collects data from all observation stations included in the IGS/EPN network and places them on its FTP server.[1]
- A regional/national network data center collects data from all observation stations by appropriate sorting (hourly and daily files) included in a specific network of national operators of active GNSS reference stations and operational centers and places it on its FTP server. (See Fig. 21.3.)
- Data analysis center LPI – collects data from all global and regional networks data centers for the purpose of further processing and use (Fig. 21.4).

Quantitative composition of IGS/EPN network observation stations and national networks ZAKPOS, TNT, SKPOS, TPI, ROMPOS continuously changes and fluctuates within 100–130 stations. Daily observation data from different sources come to the server of National University Lviv Polytechnic in RINEX format, where blocks are formed in automatic mode, depending on the day of the current GPS week. Data of IGS/EPN observations and products come in there: GNSS observation files, accurate and navigational satellite ephemeris, Earth orientation parameters, and the parameters of the following: amendments to satellites clocks, satellite and receiver antenna phase centers, etc. The total volume of daily input data is more than 100 megabytes, and the output products are more than 300 megabytes. For example, a daily ionospheric file with TEC values for all observation stations reaches, on average, 115 megabytes.

Solutions to the problems of coordinate-time provision based on continuous GNSS observations is based on the processing of large datasets of code and phase

[1] https://igs.bkg.bund.de/.

measurements. One of the possible additional options for conducting this study is computation of the numerical characteristics of the ionospheric impact on the signals distribution from the satellites – the TEC values. These characteristics reflect the dynamics of the atmosphere ionization that is important in terms of monitoring the circumterrestrial space.

However, the TEC values have one major disadvantage. They are determined with accuracy to arbitrary constant – calibration coefficients (DCB data) that are related with hardware delays of signals in electronic channels of satellite antennas and ground receivers. This fact makes it impossible to interpret measurement results unambiguously. Development of a method for determining the absolute TEC value using regular GNSS observations will simplify the data processing by extracting the results of phase ambiguities and improve the reliability of the received result by applying appropriate mathematical smoothing procedures (Matviichuk, 2000).

The accurate calculation of calibration coefficients is a tedious task but these factors are crucial when determining the TEC value, which has physical importance (incorrect calibration usually results either in a negative TEC value or in a value that exceeds the physically possible maximum). Calculation of calibration coefficients requires the analysis of the long time series of data (continuous observations for a few days) and appropriate computational procedures. For a network of IGS stations and airborne transmitters of navigation satellites, calibration coefficients are regularly published on the internet. Calibration is necessary when using a receiver that is not a part of the IGS network. Improving the accuracy of the calibration algorithm will allow facilitating the data interpretation and reducing the systematic error of the method.

21.2 THE DETERMINATION AND PROCEDURE TRANSFORMATION OF THE IONOSPHERE PARAMETERS WITH GNSS OBSERVATIONS

TEC indicators can be computed due to the automated processing of files with GNSS observation results from each satellite for an individual station. Processing algorithms are based on the use of computed code and phase pseudodistances in the receiver and calibration coefficients (DCB data). Such an algorithm allows one to get the TEC values in two ways: (a) according to the phase measurements only and having used the results of phase ambiguities in the network as a whole beforehand and (b) according to the code measurements only that were smoothed beforehand.

We chose method (a) as the main one. Taking into consideration the procedure for determining the phase ambiguities using the network software, the actual obtaining of TEC values is possible with some delay. In our case, the delay was 15 seconds. This is the minimum time required for absolute determination of integer components of the phase measurements. With this interval, we obtain series of TEC values for each satellite that are observed at a given time on a separate station.

Method (b) was used as a control. This method is not associated with the assessment of ambiguity and uses phase measurements only to detect phase changes over time. Code pseudoranges are smoothed by the functions that are selected based on the approximated phase changes over time. Code pseudoranges smoothed in such a way are used in the formation of the TEC values. One of the problematic issues that can significantly affect the accuracy of the ionospheric parameters is GNSS signal delays in the hardware of the receiver (DCB). For determination of the TEC values, two algorithms are used: a one-station algorithm (a separate permanent station) based on the presented variant, and (b) a multiple-station algorithm (network of active reference stations) based on variant (a).

In the one-station algorithm, the TEC value is determined for an individual station according to the measurements of all satellites during a period of 24 hours. For converting STEC (along the satellite-receiver beam) in vertical VTEC, a vertical single-layer model of the ionosphere is used. This model presupposes that all electrons are concentrated in a thin layer that is located at a certain height above the Earth's surface. The height of the layer is considered to be fixed and equal to 450 km. Geometric factor is used for recalculation.

The basic algorithm of the TEC computation using GNSS observations consists of the following main steps:

1. Obtaining files with observation results for a separate station. As a result, we receive code and phase pseudoranges at two frequencies, L1 and L2.
2. Introduction of additional parameters of the station: its coordinates, satellites cutoff angle, and value for signal differential delay of the receiver station.
3. Obtaining differential code delays for all available satellites at the time of the calculation from the FTP server code.
4. Analysis of the observational data (code and phase measurements, detection and elimination of cyclic phase slips, smoothing code measurements with the phase ones).

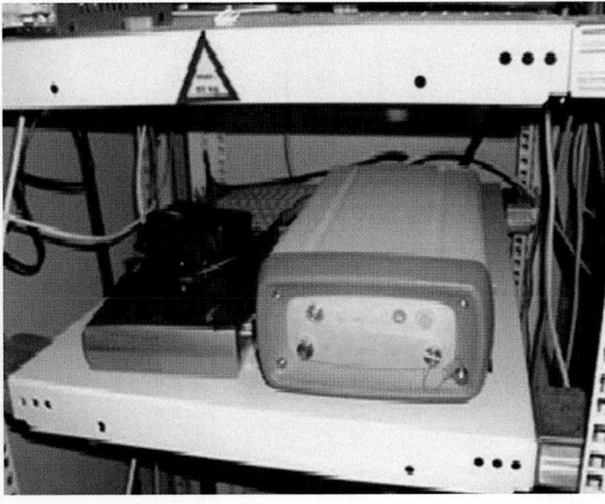

FIG. 21.5 GNSS equipment of the SULP station.

5. Calculation of the slant value of total electron content (STEC) at fixed moments of observation for each GNSS satellite.
6. Determination of the vertical values of total electron content (VTEC) on the fixed points of observation in view of STEC data from all available GNSS satellites.
7. Creation of files with the computation results.

The program for calculation of the STEC and VTEC values was written in the C++ programming language. This program uses:

- already known subprograms (classes) that are used for reading RINEX files, detection, estimation, and elimination of cyclical phase jumps ("cycle slip"), which arise in the process of measurements,
- subprograms that we developed for smoothing code measurements, receiving differential corrections at the time of calculation from the FTP server code, calculation of the horizontal coordinates of the satellite on the observational station, direct calculation of TEC, and subsequent storage of the received data in the new file on the server of Lviv Polytechnic National University.

The entire program was compiled for the Linux operating system and automated for use with observational data of the permanent IGS station SULP (Fig. 21.5). The station software generates hourly observation files with the recording interval of 1 second. Then, these files are automatically directed and executed by the program that we developed. Every hour we receive text files (with record intervals of 1 second) that contain the results of STEC and VTEC values (Yankiv-Vitkovska et al., 2012).

An example of the file with slant TEC values for the SULP station is shown in Fig. 21.6 and with vertical TEC values in Fig. 21.7.

For determination of the spatial TEC distribution, an algorithm of processing GPS measurements for multiple stations was implemented using a network of active reference stations in the Western Ukraine.

The network consists of 17 stations that work under control of specialized software in real-time to provide the RTK services to the wide range of users interested in geodetic areas. Fig. 21.8 shows the location scheme of stations in this network.[2]

Arrays of the STEC and VTEC values are so significant that there is an actual problem of preparing ionosphere parameters for their further analysis and use. To solve this problem, we proposed a technique based on a set of programs that convert VTEC data measurements to a format suitable for the analysis.

A developed program reads data from the text files and "automatically generates" two MATLAB® programs. On the one hand, this is a simple data format change in a text file. In terms of computing, this is a record of data in the form of explicitly declared large numerical arrays. In the first program, discrete functional dependencies of the VTEC change over time (Fig. 21.9) are described using the assignment operator. They are two-dimensional arrays of data, time series, clearly declared in the program text. Then, using these time series, the spline interpolation for all their nodes is constructed (Fig. 21.10).

[2] http://zakpos.zakgeo.com.ua.

h:m:s	PRN01	Z01	PRN02	Z02	PRN03	Z03	•••
0:1:0	12.27	38.5	99999	99999	31.70	76.4	
0:1:1	12.29	38.5	99999	99999	31.91	76.4	
0:1:2	12.30	38.5	99999	99999	31.80	76.4	
0:1:3	12.32	38.5	99999	99999	31.76	76.4	
0:1:4	12.27	38.5	99999	99999	31.79	76.4	
0:1:5	12.28	38.5	99999	99999	31.84	76.5	
0:1:6	12.30	38.5	99999	99999	31.81	76.5	
0:1:7	12.22	38.5	99999	99999	31.84	76.5	
0:1:8	12.25	38.4	99999	99999	31.67	76.5	
0:1:9	12.26	38.4	99999	99999	31.75	76.5	
0:1:10	12.24	38.4	99999	99999	31.84	76.5	
0:1:11	12.26	38.4	99999	99999	31.89	76.5	
0:1:12	12.23	38.4	99999	99999	31.91	76.5	
0:1:13	12.25	38.4	99999	99999	31.89	76.5	
0:1:14	12.27	38.4	99999	99999	31.98	76.5	
0:1:15	12.22	38.4	99999	99999	31.92	76.5	
0:1:16	12.21	38.4	99999	99999	31.83	76.5	
0:1:17	12.26	38.4	99999	99999	31.78	76.5	
0:1:18	12.33	38.4	99999	99999	31.87	76.5	
0:1:19	12.30	38.4	99999	99999	31.77	76.6	

FIG. 21.6 File fragment with STEC values.

h:m:s	VTEC	Number of satellites
0:1:0	11.98	9
0:1:1	11.96	9
0:1:2	11.91	9
0:1:3	11.90	9
0:1:4	11.86	9
0:1:5	11.81	9
0:1:6	11.81	9
0:1:7	11.76	9
0:1:8	11.73	9
0:1:9	11.71	9
0:1:10	11.67	9
0:1:11	11.63	9
0:1:12	11.61	9
0:1:13	11.57	9
0:1:14	11.54	9
0:1:15	11.50	9
0:1:16	11.48	9
0:1:17	11.47	9
0:1:18	11.45	9
0:1:19	11.38	9
0:1:20	11.37	9
0:1:21	11.33	9
0:1:22	11.29	9

FIG. 21.7 File fragment with VTEC values.

The second program implements the approximation of the above-described splines with a wider level in determining the functional dependencies. Weak smoothing is used for this approximation as well. This fact made up for specific errors in the measured data that were caused by the peculiarities of the receiver-antenna electronic channel and close environment to the station. Simultaneously, another approximation spline was calculated with significant smoothing of the data (Fig. 21.11). It is necessary to highlight the daily change of the VTEC rate.

Because of the consistent usage of these two programs, two spline approximation objects are created. One of these programs accurately approximates the measurement data. Another program includes approximation of the smoothed values for the measurement data. Due to this approach, for further VTEC calculations, it is sufficient to read the named objects of approximations from their files and identify VTEC at any moment of time using spline computing functions according to their argument as the defined splines reflect VTEC changes as analytically set functions (Yankiv-Vitkovska, 2013a).

In this research, we used spline approximation to determine the smoothed VTEC values. This approximation is constructed using the approximation function fit of the software tool for curvilinear approximation **"Curve Fitting Toolbox"** in MATLAB. We chose this method of smoothing based on computational experiments that

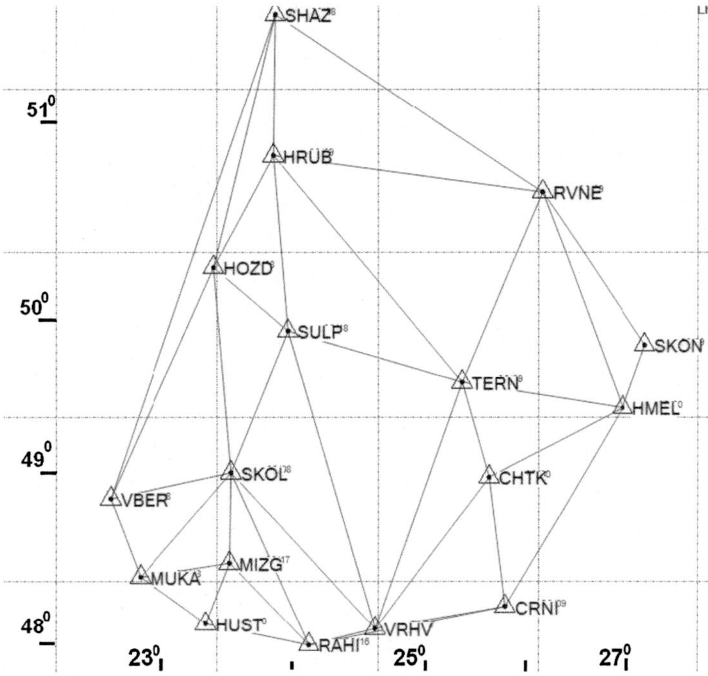

FIG. 21.8 Location scheme of active reference stations.

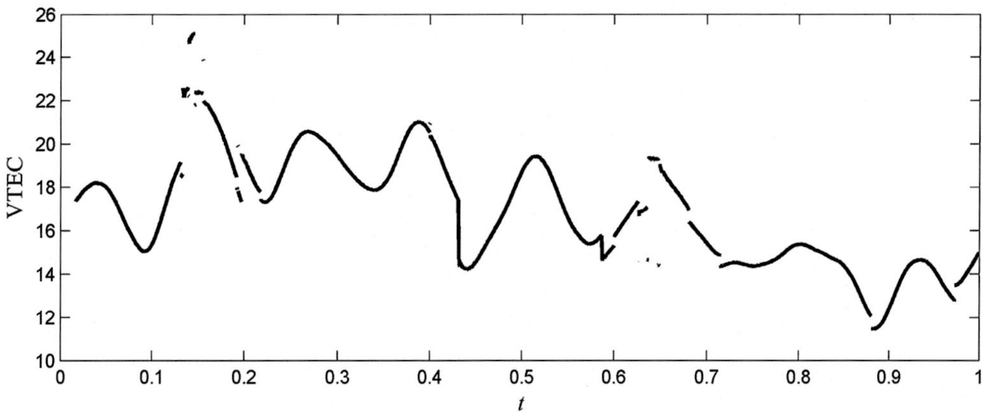

FIG. 21.9 Functional dependencies of VTEC over time.

were aimed at comparing different smoothing methods. In particular, during these experiments, spline approximation methods were used using the function spaps from the "**Spline Toolbox**" and a smoothing function from the "**Curve Fitting Toolbox**."

We would like to mention an important aspect concerning the smoothing function *smooth*. Sometimes, in a short time interval, the STEC value from individual satellites undergoes significant changes. It may seem like a temporary increase in errors on the graph or "amplification" of original "measurement obstacles." Therefore, there is no reason to assume that the measurement error is less (or greater), if the data are defined using a larger number of satellites. Thus, we cannot know a priori in which nodes of the VTEC time interval the error rate is higher and in which it is lower, because smooth-

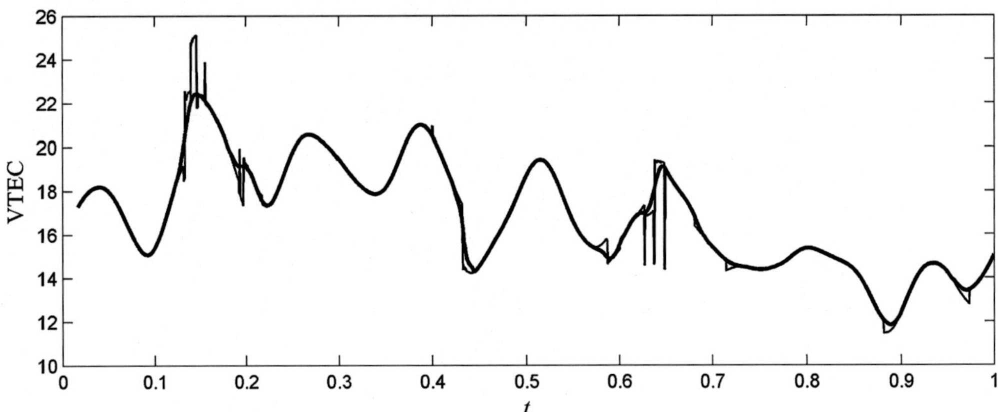

FIG. 21.10 Spline interpolation of the VTEC time series.

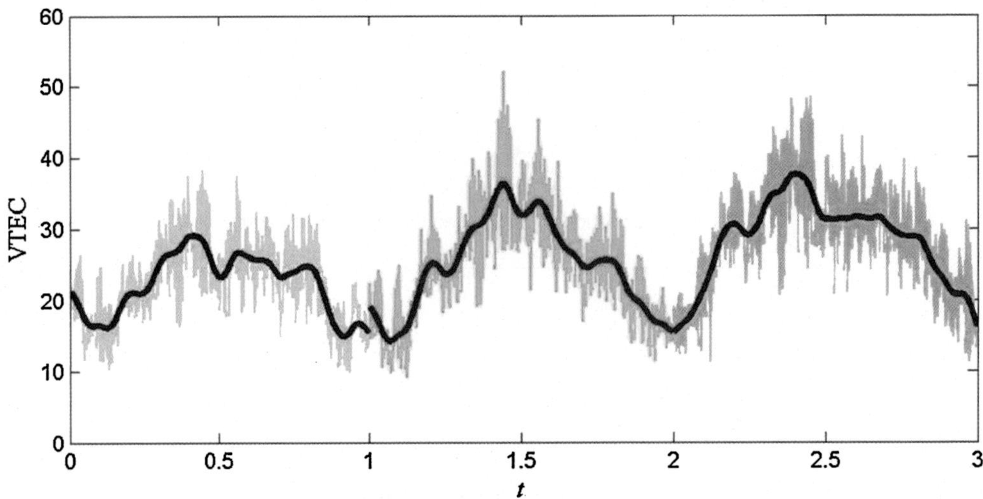

FIG. 21.11 Smoothing VTEC time series by approximating spline.

ing with the smooth function leads to a more severe impact of short-term deviations on the smoothing results than on any other data. This disadvantage can be avoided with applying smoothing approximation using the spline function *fit*. We chose the value of the parameter that describes the intensity of smoothing in an experimental way using qualitative analysis of the VTEC time series and their smoothed values.

21.3 RECOVERY OF THE SPATIAL STATE OF THE IONOSPHERE

For continuously operating reference GNSS stations, the results of the determined ionization identifier TEC that describes the number of ions in the atmosphere on the line between the ground station and the moving satellite accumulate. On the one hand, these data reflect the state of the ionosphere during the observation; on the other hand, it is a substantial tool for accuracy improvement and reliable determination of coordinates of the observation place.

Thus, it was decided to solve a problem of restoring the spatial position of the ionospheric state or its ionization field according to the regular definitions of the TEC identifier, i.e., STEC. The description below shows one of the possible solutions that is based on the application of the regularized approximation of functions with numerous variables.

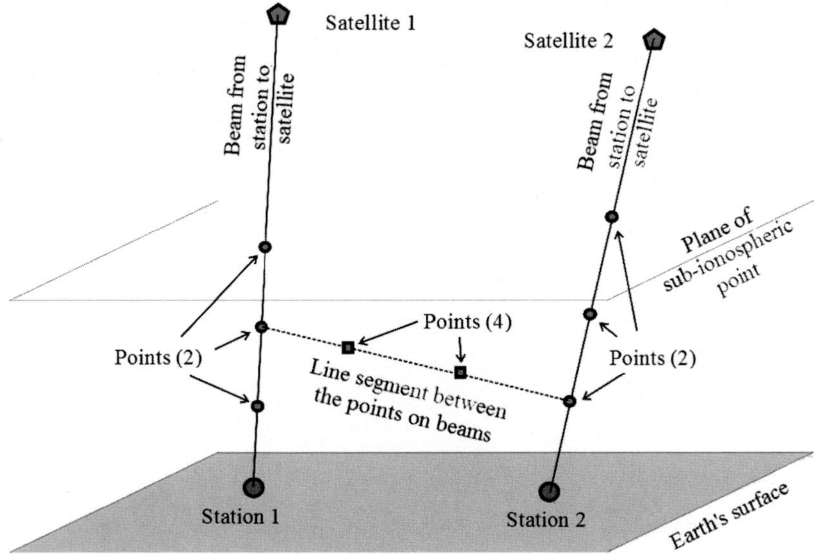

FIG. 21.12 Spatial location of the nodes.

Initial data to restore the ionization field:

- coordinates of reference stations:

$$X_i^{st}(t_k), Y_i^{st}(t_k), Z_i^{st}(t_k)\ (i=\overline{1,n_k},\ k=\overline{1,K}), \tag{21.1}$$

- coordinates of GNSS satellites:

$$X_j^{sp}(t_k), Y_j^{sp}(t_k), Z_j^{sp}(t_k)\ (j=\overline{1,m_k},\ k=\overline{1,K}), \tag{21.2}$$

- STEC values between the station i and satellite j:

$$s_i^j(t_k)\ (i=\overline{1,n_k},\ j=\overline{1,m_k},\ k=\overline{1,K}), \tag{21.3}$$

where t_k is the time of the STEC measurement, K is the number of measurements, i is the station number, j is the satellite number, and n_k, m_k are the numbers of stations and satellites during the measurement k, respectively.

Further, we used data from 19 reference stations in the Western Ukraine. The solution to this problem is to define the ionization field

$$\nu=\nu(x,y,z,t),$$

for the area where the stations are located at (x, y, z) during the time $t \in [t_1, t_k]$.

21.3.1 Restrictions and Assumptions for Use of the GNSS Measurements to Restore the Ionization Field

The coordinates of an individual station i (21.1) and available satellite j (21.2) define the line segment that connects the point on the Earth's surface with the satellite. This line segment comes through the Earth's ionosphere as well. One of the assumptions in this case is that the ionosphere layer has an effective thickness that is defined by the subionospheric point H. According to this assumption, all ionized atoms are located on the surface of some sphere with the radius defined by the subionospheric point.

Let us divide the part of the specified line segment into $N-1$ equal segments, thus getting N equally located nodes that lie on a beam from the station to the satellite below the subionospheric point (Fig. 21.12), i.e.,

$$\bar{x}_{ijl}^k=\bar{x}_{ijl}(t_k),\ \bar{y}_{ijl}^k=\bar{y}_{ijl}(t_k),\ \bar{z}_{ijl}^k=\bar{z}_{ijl}(t_k), \tag{21.4}$$

where $\bar{x}_{ijl}^k, \bar{y}_{ijl}^k, \bar{z}_{ijl}^k$ are spatial coordinates of a point l on a beam between station i and satellite j $(i=\overline{1,n_k}, j=\overline{1,m_k}, l=\overline{1,N}, k=\overline{1,K})$.

Supposing that the state of the ionosphere changes evenly along the beam between the station and the satellite, the ionization field indicator in each node can

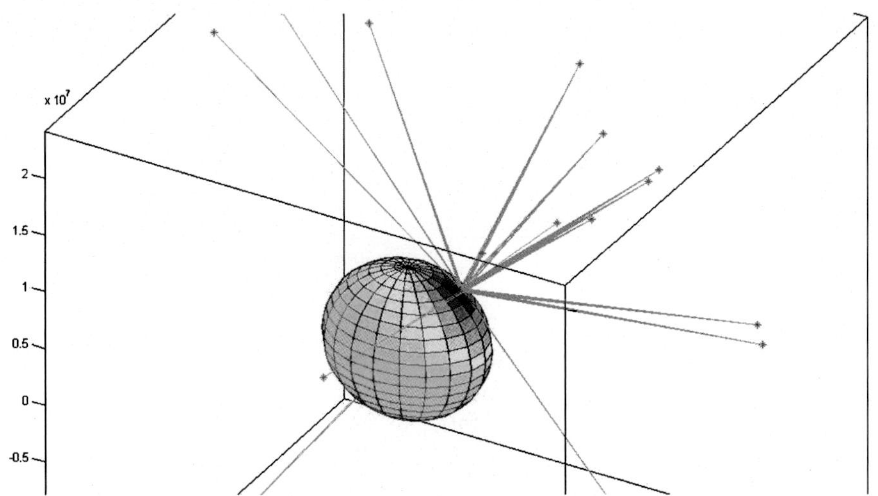

FIG. 21.13 Common view of beams from the station to the satellites according to (21.1) and (21.2).

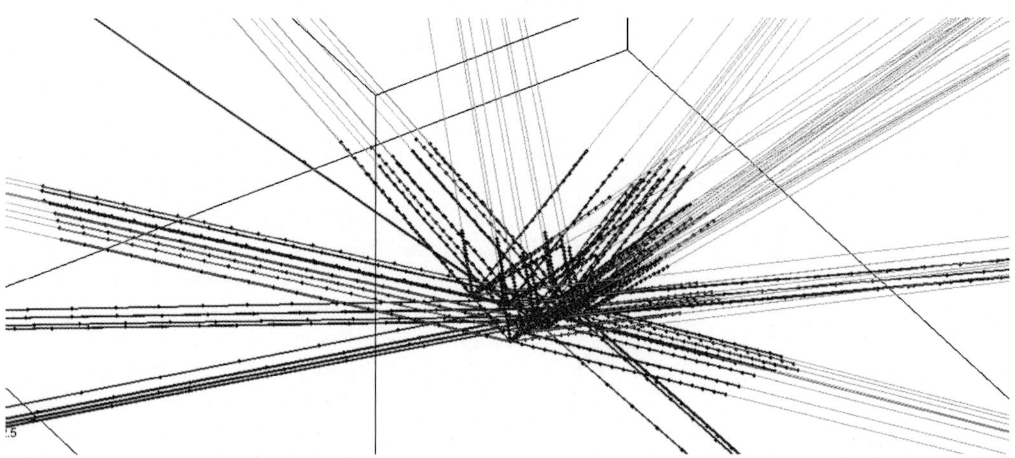

FIG. 21.14 Beams from stations to satellites (in gray), their subionospheric line segments (in red), and points (21.4) on them (in blue).

be described as

$$\bar{v}^k_{ijl} = s_{ij}(t_k)/N,$$
$$(i = \overline{1, n_k},\ j = \overline{1, m_k},\ l = \overline{1, N},\ k = \overline{1, K}), \quad (21.5)$$

where $\bar{v}^k_{ijl}$ is the ionization parameter in point l on a beam between station i and satellite j.

Expression (21.5) describes the ionization values determined experimentally. GNSS observations show that the beams from satellites to stations at one point in time lie mostly in certain directions only, while there are few of them or none at all that lie in other directions (Fig. 21.13).

This fact makes the conditions of the interpolation problem worse. For their improvement, we connected all the points (21.4) on the beams between the stations and satellites and on the formed line segments we defined internal equally located nodes that do not lie on the boundaries of the line segments (see Fig. 21.14), i.e.,

$$\breve{x}^{kr}_{ijpq},\ \breve{y}^{kr}_{ijpq},\ \breve{z}^{kr}_{ijpq}, \quad (21.6)$$

where $\breve{x}^{kr}_{ijpq}, \breve{y}^{kr}_{ijpq}, \breve{z}^{kr}_{ijpq}$ are the spatial coordinates of the point r $(r = \overline{1, M})$ on the line segment between the nodes $(\bar{x}^k_{ijp}, \bar{y}^k_{ijp}, \bar{z}^k_{ijp})$, $(\bar{x}^k_{ijq}, \bar{y}^k_{ijq}, \bar{z}^k_{ijq})$, $p, q \in [1, N]$, $i = \overline{1, n_k}$, $j = \overline{1, m_k}$, $k = \overline{1, K}$, and M is the number of internal nodes on a line segment between the points on the beam.

Ionization parameters in the nodes (21.6) are defined using linear interpolation values of this indicator on a line segment between the points $(\bar{x}^k_{ijp}, \bar{y}^k_{ijp}, \bar{z}^k_{ijp})$, $(\bar{x}^k_{ijq}, \bar{y}^k_{ijq}, \bar{z}^k_{ijq})$, $p, q \in [1, N]$, $i = \overline{1, n_k}$, $j = \overline{1, m_k}$. It is described as

$$\breve{v}^{kr}_{ijpq}, \ (i = \overline{1, n_k},\ j = \overline{1, m_k},\ p, q \in [1, N],\ r = \overline{1, M},\ k = \overline{1, K}). \tag{21.7}$$

Expression (21.5) describes ionization in the nodes (21.4) located along the beams between the stations and satellites. Expression (21.7) describes ionization in the nodes (21.6) that lie between such different beams. It was found that data from expressions (21.4)–(21.7) are not enough to restore the ionization field as approximating functions deviate greatly from the observational ionization values beyond the nodes (21.4), (21.7).

21.3.2 Description of the Method for Determining Ionization Using STEC

For practical purposes, we need to define ionization in the spatial area

$$x \in [x^k_{min}, x^k_{max}],\ y \in [y^k_{min}, y^k_{max}],\ z \in [z^k_{min}, z^k_{max}], \tag{21.8}$$

where the area boundaries are defined by the extreme points of the set of nodes (21.4) that are located on the beams between the stations and satellites, i.e.,

$$x^k_{min} = \min_{ijl} \bar{x}^k_{ijl},\ y^k_{min} = \min_{ijl} \bar{y}^k_{ijl},\ z^k_{min} = \min_{ijl} \bar{z}^k_{ijl},$$
$$x^k_{max} = \max_{ijl} \bar{x}^k_{ijl},\ y^k_{max} = \max_{ijl} \bar{y}^k_{ijl},\ z^k_{max} = \max_{ijl} \bar{z}^k_{ijl}.$$

Let us divide line segments that describe the rectangular area (21.8) into $L - 1$ smaller segments and determine the coordinates of the equally located nodes, i.e.,

$$\hat{x}^k_i,\ \hat{y}^k_i,\ \hat{z}^k_i,\ (i, j, l = \overline{1, L}). \tag{21.9}$$

To restore the ionization field in the area (21.8) according to the data from the expressions (21.4)–(21.7), a new condition needs to be imposed: ionization derivatives with respect to the coordinates must be minimal in the points (21.9). Such condition reduces strong deviations of the approximating function beyond the nodes (21.4), (21.6). It should be noted that the solution to the problem of the ionization field restoration lies in finding the ionization values in the nodes (21.9).

Set of the points (21.4), (21.6) coordinates is denoted by

$$x^k_a = \left\{\bar{x}^k_{ijl}, \breve{x}^{kr}_{ijpq}\right\},\ y^k_a = \left\{\bar{y}^k_{ijl}, \breve{y}^{kr}_{ijpq}\right\},$$
$$z^k_a = \left\{\bar{z}^k_{ijl}, \breve{z}^{kr}_{ijpq}\right\}, \tag{21.10}$$

where $a = \overline{1, A_k}$, and A is the number of points in the expressions (21.4), (21.6) $(i = \overline{1, n_k},\ j = \overline{1, m_k},\ p, q \in [1, N],\ r = \overline{1, M},\ k = \overline{1, K})$.

The set of the ionization identifier values (21.5) and (21.7) is denoted by

$$v^k_a = \left\{\bar{v}^k_{ijl}, \breve{v}^{kr}_{ijpq}\right\}, \tag{21.11}$$

where $a = \overline{1, A_k}$, $i = \overline{1, n_k}$, $j = \overline{1, m_k}$, $p, q \in [1, N]$, $r = \overline{1, M}$, $k = \overline{1, K}$.

Sets (21.10), (21.11) define the discrete dependency of the ionization v^k_a from the values of three spatial coordinates (x^k_a, y^k_a, z^k_a) $(a = \overline{1, A_k})$. This dependency is approximated by the exponential polynomial from numerous arguments, i.e.,

$$v(x, y, x) = P_k(x, y, x).$$

During the calculations, the following polynomials were selected:

$$P_k(x, y, x) = \sum_{i+j+l<R} c^k_{ijl} x^i y^j z^l,$$
$$P_k(x, y, x) = \sum_{|i+j+l|<R} c^k_{ijl} x^i y^j z^l,$$
$$P_k(x, y, z) = \sum_{i+j+l<r} c^k_{ijl} x^{\lambda i} y^{\lambda j} z^{\lambda l},$$
$$P_k(x, y, z) = \sum_{|i+j+l|<R} c^k_{ijl} x^{\lambda i} y^{\lambda j} z^{\lambda l},$$

where R is the exponent of the polynomial $(R = 1, \ldots, 4)$, c^k_I are the coefficients of this polynomial, I are multiindices of these coefficients, and λ is a number close to R $(\lambda \in [R - 0.3; R + 0.3])$. A polynomial with random exponents was also selected:

$$P_k(x, y, z) = \sum c^k_{\xi_x \xi_y \xi_z} x^{\xi_x} y^{\xi_z} z^{\xi_z},$$

where ξ_i $(i = x, y, z)$ equally distributed random numbers. In particular, exponents $\xi_i \in [-0.5, +0.5]$, $\xi_i \in [-1, +1]$ were selected as an initial approximating basis from 50, 100, and 200 polynomial items. The structure of the approximating basis was selected in such a way that the argument exponents are close to 1. It is empirically known that this improves the extrapolation of the simulated values in the nodes (21.9).

Polynomials, rational functions, and generalized polynomials are used for the approximation of functions as well (Vladimirov, 1981):

$$f(x) = \sum_{j=1}^{L} a_j \varphi_j(x),$$

where $f(x)$ is the approximating function, a_j are approximation coefficients, $\varphi_j(x)$ are functions with special properties, in particular, trigonometric and exponential functions, and n is the number of functions. If $\varphi_j(x)$ is a Legendre polynomial, then during its orthogonalization fractional-rational functions that are identical to polynomials with fractional exponents appear (Vasyliev and Symak, 2008).

To find the approximation coefficient c_I^k using the data from expressions (21.10) and (21.11), we used identification problems regularized by minimizing the stabilizing Tikhonov functional (Tikhonov and Arsenin, 1979) and reduction of the approximating basis (Matviichuk et al., 2000). However, such approximation has an acceptable error of approximation only in the identification nodes (21.4), (21.6) and beyond them deviates greatly from the approximated value in the points (21.9).

Thus, to restore the ionization field, additional measures were taken. An artificial argument that depends nonlinearly on x, y, z was added to the arguments of the polynomial, i.e.,

$$P_k(x, y, z, r) = \sum_{|i+j+l+p|<R} c_{ijlp} x^{\lambda i} y^{\lambda j} z^{\lambda l} r^{\lambda p}, \tag{21.12}$$

in particular,

$$r = \sqrt{(x_c^k - x)^2 + (y_c^k - y)^2 + (z_c^k - z)^2},$$

where r is the radius vector and x_c^k, y_c^k, z_c^k are the center coordinates of the parallelepiped (21.9).

The identification problem intended to determine the polynomial coordinates for a separate calculation $k \in [1, K]$ has two conditions:

- approximation of values (21.11) in the points (21.10):

$$\min_{c_I^k} \sum_{a=1}^{A} \left[v_a^k - P_k(x_a, y_a, z_a, r_a) \right]^2 + \alpha \sum_I \left(c_I^k \right)^2, \tag{21.13}$$

- minimization of the polynomial derivative P_k with respect to its argument in the points (21.9):

$$\min_{c_I^k} \sum_{i,j,l,p=1}^{L} \left(0 - \sum_{q=x,y,z,r} P_k^q(\hat{x}_i, \hat{y}_j, \hat{z}_l, \hat{r}_p) \right)^2 + \alpha \sum_I \left(c_I^k \right)^2, \tag{21.14}$$

where c_I^k are polynomial coefficients P_k, I are their multiindices, and P_k^q are polynomial derivatives $P_k (q = x, y, z, r)$,

$$P_k^x = \frac{\partial}{\partial x} P_k(x, y, z, r), \quad P_k^y = \frac{\partial}{\partial y} P_k(x, y, z, r),$$
$$P_k^z = \frac{\partial}{\partial z} P_k(x, y, z, r), \quad P_k^r = \frac{\partial}{\partial r} P_k(x, y, z, r).$$

To solve (21.13), (21.14), the reduction of the approximating basis described in Matviichuk (2000), Yankiv-Vitkovska (2012) was used. To reduce the deviations of the approximating polynomial from the measured ionization values, the first-degree ($R = 1$) polynomial was chosen and minor deviations $\lambda \in [0.7, 1.3]$ were applied (Malachivskyy, 2009; Verlan and Fedorchuk, 2013).

Multiple solving of the problems (21.13) and (21.14) for all measured data $k = \overline{1, K}$ lead to interim conclusions such as the following:

- if polynomial exponents of numerous arguments are close to 1, then an approximation basis found using the reduction of the polynomial exponent while solving the problem (21.13), (21.14) for the values of an individual measurement k (1) $k \in [1, K]$ provides an acceptable approximation for all measurements $k = \overline{1, K}$;
- if the exponent of the approximation polynomial differs greatly from 1 ($\lambda R > 1.5$ or $\lambda R < 0.5$), the reduction of the approximation polynomial exponent for each measurement leads to obtaining different approximation bases. This does not lead to substantial improvement of the approximation accuracy (21.12) and mostly makes the accuracy of expression (21.14) worse.

From the results of these computational experiments, it can be concluded that to restore the ionization field, it is advisable to use the polynomial (21.12) with the exponent $R = 1$ and multiplier λ that is slightly less than 1. For other conditions, we need substantial costs for computational resources to determine the approximation basis and coefficient c_I^k for each of the measurements ($k = \overline{1, K}$) separately.

TABLE 21.1
Exponents of the polynomial arguments determined for most measurements and the values of the approximation coefficients c_l^1 for a measurement $k = 1$.

No.	Exponent x	Exponent y	Exponent z	Exponent r	Coefficient c_l^1
1	0	0	0	0	7.623044
2	0	0	0	0.769231	0.004773
3	0	0	0.769231	0	−0.000469
4	0	0.769231	0	0	−0.003179
5	0.769231	0	0	0	−0.010377
6	0	0	−0.769231	0	−6.092696

21.3.3 Results of the Experimental Restoration of the Changes in the Atmosphere Ionization

To experimentally determine the changes in the ionization field in time, we took $k = 46$ measurements from 272 days in 2013, namely, STEC values with a time interval of 15 seconds during the first 12 minutes from the beginning of the day that were determined during the GNSS observations at 17 continuously operating stations of the ZAKPOS network (Yankiv-Vitkovska, 2013b).

For most measurements ($k = \overline{1, K}$), in the case of the reduction of the approximation polynomial exponent, the same approximation basis was found. The exponents of the polynomial arguments are given in Table 21.1. The full set of coefficients of this approximation basis is determined using the parameters $R = 1$, $\lambda = 1.3^{-1}$, $|i + j + l + p| \leq 1$. Values of the approximation coefficients c_l^k determined for a measurement at time $t_1 = 0$ (the time is defined in the seconds of the day) are provided in Table 21.1 as well.

The obtained solutions to the problems (21.13) and (21.14) are the coefficient values $c_i(t_k)$ ($i = \overline{1, n}$, $k = \overline{1, K}$) at the time t_k where $n = 6$ is the number of these coefficients.

Using cubic Hermite spline for interpolation (function *pchip* in MATLAB), we obtained the approximation coefficients values for each second. Using the interpolation by the fifth-degree spline, their continuous values $c_i(t)$ ($t \in [t_1, t_k]$) were determined. Fig. 21.15 shows the common dependencies of these approximation parameters on time.

Fig. 21.15 shows that parameters gradually change over time. This change depends on angle altitude of certain satellites and their ascent and descent, namely, changes in the number of satellites. Only one parameter $c_5(t)$ changes relatively very quickly (Fig. 21.15D). This indicates that the problem of restoration of $c_i(t)$ during $t \in [t_1, t_k]$ using data (21.1) to determine the reduced (regularized) approximation basis common to all observations $k = \overline{1, K}$ is incorrect. However, the above solution shows that the change speed of $c_5(t)$ is limited. This fact confirms a good choice in the approximation basis for all observations $k = \overline{1, K}$ (during all time $t \in [t_1, t_k]$).

The quality analysis of the graphs $c_i(t)$ ($t \in [t_1, t_k]$) described above is applied to control the adequacy of the solution to the problem (21.13), (21.14) for all $k = \overline{1, K}$. In addition, to evaluate this solution, we applied quality analysis of the determined ionization distribution in the points (21.9). Acceptable solutions to (21.13) and (21.14) are distribution (probability density) of the needed ionization in the points (21.9) that has a central maximum or is close to even or linear distribution. Fig. 21.16 shows common distribution graphs and functions of the ionization distribution $v(x, y, z, t_{12})$ determined using the polynomial (21.12) in the points (21.9).

It can be seen that all the restored ionization values are approximately equal to experimentally measured ionization values. Fig. 21.16 also illustrates a common result of using the approximation basis (see Table 21.1) determined in one point ($k = 2$) to restore ionization in another point ($k = 12$). Such quality analysis of distribution laws for ionization approximation in the points (21.9) is applied to all measurements $k = \overline{1, K}$.

To restore the change in the ionization field, we need to determine continuous dependencies of the area center coordinates (21.9) $x_c(t)$, $y_c(t)$, $z_c(t)$ ($t \in [t_1, t_k]$) from time using approximation by spline. Graphs of the coordinate changes are shown in Fig. 21.17.

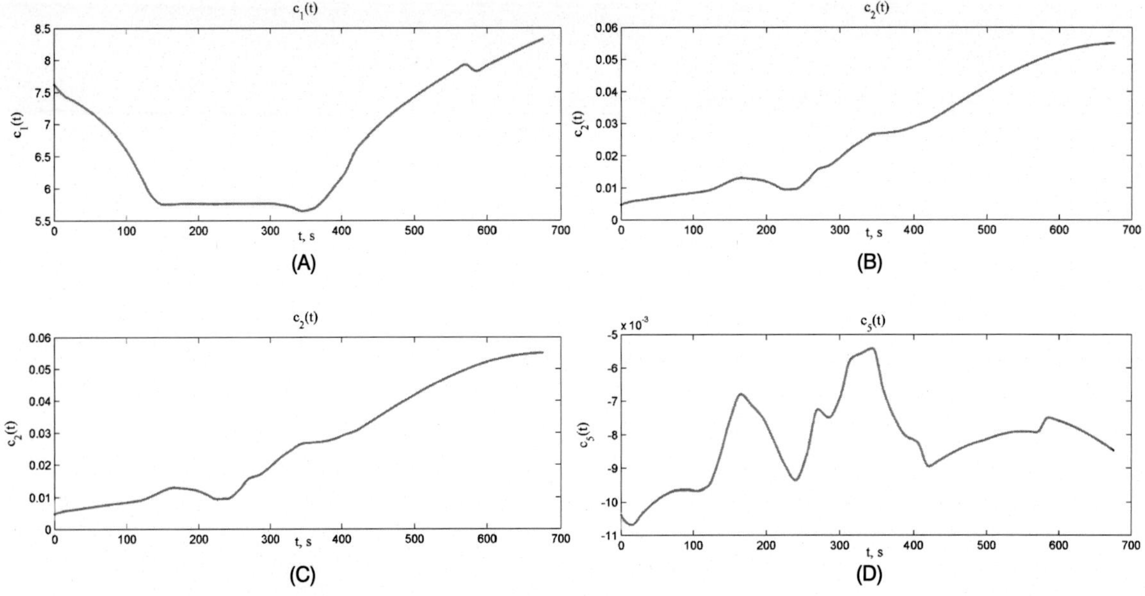

FIG. 21.15 Time dependencies of approximation coefficients: (A) $c_1(t)$, (B) $c_2(t)$, (C) $c_3(t)$, (D) $c_5(t)$.

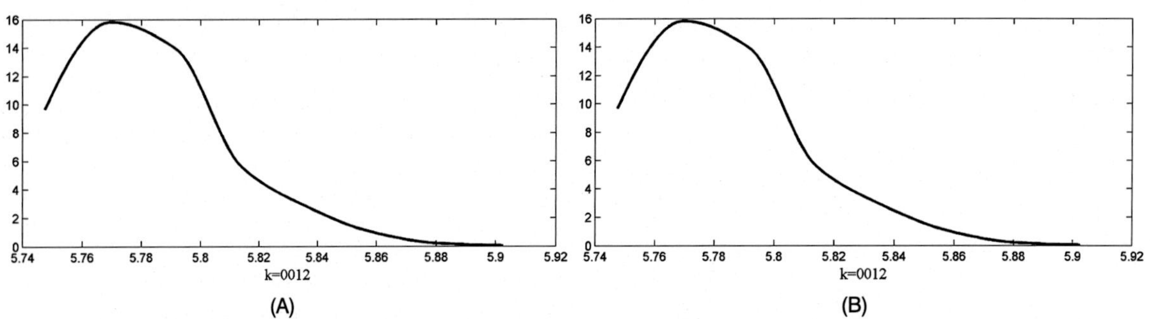

FIG. 21.16 Common view of graphs for (A) probability density and (B) distribution function of the ionization value $v(x, y, z, t_{12})$ restored in the rectangular area (21.16).

Fig. 21.17 shows the shifts of the center of the rectangular area with irregular fluctuations. This can be explained by the movement of satellites and discrete division of subionospheric line segments of the beam from the station to the satellite. This indicates that the ionization field depends on the algorithm parameters.

It should be noted that the area boundaries (21.9) change gradually depending on the satellite movements. Common graphs of change $x_{min}(t_k) = x^k_{min}$, $y_{min}(t_k) = y^k_{min}$, $z_{min}(t_k) = z^k_{min}$, $x_{max}(t_k) = x^k_{max}$, $y_{max}(t_k) = y^k_{max}$, $z_{max}(t_k) = z^k_{max}$ $(k = \overline{1, K})$ are shown in Fig. 21.18.

Extreme values of these boundaries were defined for the ionization field restoration:

$$\bar{x}_{min} = \max_{k\in[1,K]} x^k_{min},$$

$$\bar{y}_{min} = \max_{k\in[1,K]} y^k_{min},$$

$$\bar{z}_{min} = \max_{k\in[1,K]} z^k_{min},$$

$$\bar{x}_{max} = \min_{k\in[1,K]} x^k_{max},$$

$$\bar{y}_{max} = \min_{k\in[1,K]} y^k_{max},$$

$$\bar{z}_{max} = \min_{k\in[1,K]} z^k_{max}.$$

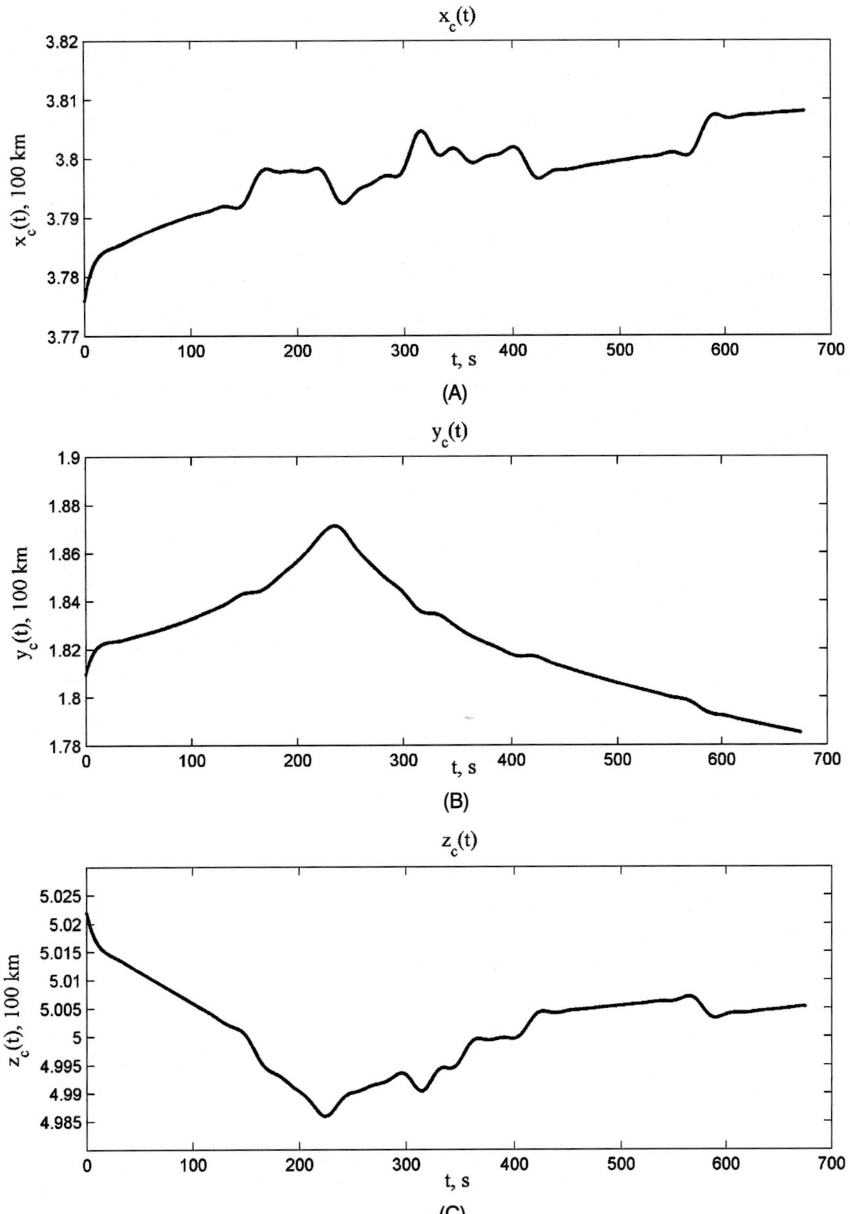

FIG. 21.17 Time dependency graphs of the center of the ionization restoration area (A) $x_c(t)$, (B) $y_c(t)$, (C) $z_c(t)$.

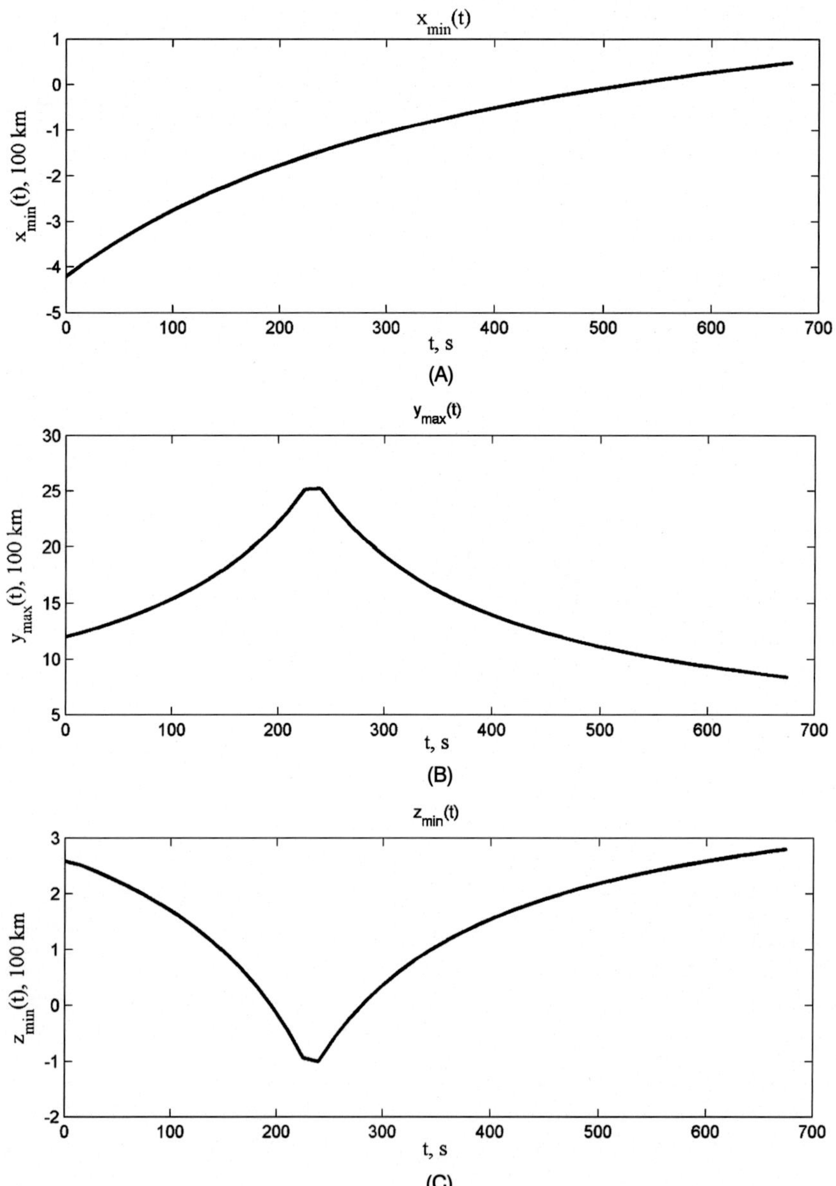

FIG. 21.18 Common view of time dependency graphs of the changes in boundaries of the ionization restoration area: (A) $x_{max}(t)$, (B) $y_{max}(t)$, (C) $z_{max}(t)$.

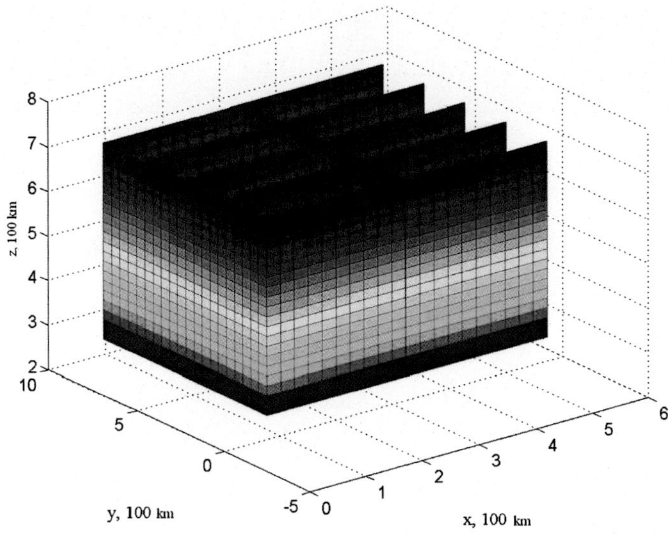

FIG. 21.19 Graphs of the ionization dependency from the spatial coordinates (in 100 km) in the moment $t_{38} = 555$ s.

They describe a rectangular area

$$x \in [\bar{x}_{min}, \bar{x}_{max}],\ y \in [\bar{y}_{min}, \bar{y}_{max}],\ z \in [\bar{z}_{min}, \bar{z}_{max}], \tag{21.15}$$

for which the ionization values during the whole period of observations $t \in [t_1, t_k]$ are restored.

Let us divide the line segments (21.15) into $L-1$ segments and determine the coordinate values for equally located nodes, i.e.,

$$\hat{x}_i,\ \hat{y}_j,\ \hat{z}_l\ (i, j, l = \overline{1, L}). \tag{21.16}$$

Ionization values in the nodes (21.16) are computed using the polynomial (21.12) with parameters $c_i(t)$ $(i = \overline{1, n})$, $x_C(t)$, $y_C(t)$, $z_C(t)$ that depend on time, i.e.,

$$v(x, y, z, r) = \sum_{ijlp} c_{ijlp}(t) x^{\lambda i} y^{\lambda j} z^{\lambda l} r^{\lambda p}, \tag{21.17}$$

where i, j, l, p are the indices of the polynomial coefficients. The exponents of its arguments are determined beforehand based on computational experiments conducted when solving the problems (21.13) and (21.14) (see Table 21.1). The common view of graph that shows the instantaneous value of the ionization field restored in the nodes (21.16) is illustrated in Fig. 21.19. In formula (21.17), continuous-time functions $c_i(t)$ $(i = \overline{1, n})$, $x_C(t)$, $y_C(t)$, $z_C(t)$ are defined using splines. According to formula (21.17), we determine the ionization at an arbitrary time $t \in [t_1, t_k]$ in the arbitrary point (21.15). It was found that most often, the ionization increases with an altitude and there is a shift of spherical areas with reduced or increased ionization. Sometimes such areas stop shifting and start moving in the opposite direction and mix. Restoring the spatial dynamics of ionization (21.17) models complex processes of electric charge movements in the ionized air.

The described method is based on the interpolation of the coefficients of the polynomial from numerous arguments (21.12). It can be used for data in (21.10) and (21.11) that lead to the same reduced approximation basis. This method is described briefly in the algorithm at the end of the chapter.

Using this algorithm, the results of the ionization field restoration were obtained. In particular, this algorithm was applied with the following parameters: number of points on the beam from the station to satellites $N = 3$; number of points between the points on different beams $M = 1$. With such parameters in expression (21.13), the number of approximation nodes exceeds 55,000. The number of nodes along the area boundary of the ionization restoration is $L = 20$. Using this value, the number of minimization nodes of the derivative by the polynomial from numerous arguments in (21.14) is 9261. An increase in the abovementioned parameters causes severe computation complications in problems (21.13) and (21.14) and does not improve the accuracy of its solution.

TABLE 21.2
Statistical characteristics for 46 results of the solution to the problem (21.13) obtained using a common approximation basis.

Name of identifier	Min value	Max value	Average value	Median	Distrib. mode	SD
SD of the absolute approximation error in (21.13)	1.3892	1.4652	1.4198	1.4226	1.3892	0.016779
SD of the relative approximation error in (21.13)	0.15017	0.15737	0.15247	0.15230	0.15017	0.001354
Average ionization in the approximation nodes (21.10) computed from polynomial (21.12)	5.7050	5.8206	5.7377	5.7354	5.7050	0.021957
Average ionization in the approximation nodes (21.10), experimental values (21.11)	5.7049	5.8206	5.7377	5.7352	5.7049	0.021971
Average relative approximation error (21.13) (by module)	0.21556	0.22512	0.21967	0.22016	0.21556	0.002148

Statistical characteristics, which describe 46 results of the solution to the problem (21.13) and (21.14) ($k = \overline{1, K}$, $K = 46$) are provided in Table 21.2.

This table shows the following characteristics:

- standard deviation (SD) of the absolute approximation error in 46 results of the problem (21.13);
- SD of the relative approximation error in 46 results of the problem (21.13);
- average ionization in the approximation nodes (21.10) computed for 46 moments in observation from polynomial (21.12);
- average ionization in the approximation nodes (experimental values from 46 observational moments) (21.11);
- average relative approximation error by module for 46 solutions to the problem (21.13).

For these characteristics, the following parameters were computed: the smallest, largest, and average value, median, distribution mode, and SD of the accuracy identifier as a set of its 46 values. The first two rows of the last column show the SD of 46 approximation problems (21.13). These parameters describe the accuracy of application of common approximation basis (see Table 21.1) in order to restore the changes in the ionization field in time.

In particular, Table 21.2 shows that the average values of the experimental and model (obtained from approximation) values are close for all measurements. The relative accuracy of the problem (21.13) solution for an individual measurement is approximately 21% (with dispersion 0.21% for all measurements). The standard deviation of this error is 15% (with dispersion 0.13% for all measurements). This means that using the common approximation basis, we obtained ionization approximation (21.13) with relatively low accuracy (21%) but the error of such approximation varies a little for each of the measurements. This proves the efficiency of applying the common approximation basis for regularized approximation of the atmosphere ionization when using the polynomial from numerous arguments with coefficients dependent on time. It should be noted that before we added a new argument to the polynomial (21.12), the approximation accuracy was worse. Other ways to expand or change the approximation basis (described above) do not influence the accuracy parameters for ionization field restoration.

21.3.4 Algorithm of the Ionization Field Change Restoration Using the Approximation of the Change in Time of the Coefficients of the Polynomial From Numerous Arguments

1. Obtain the coordinates of the stations (21.1), satellites (21.2), and STEC values (21.3).
2. Determine the altitude of the subionospheric point.
3. Select a number of points N on the beams from the stations to satellites located lower than the subionospheric point.
4. Compute (21.4) the coordinates $\bar{x}^k_{ijl}, \bar{y}^k_{ijl}, \bar{z}^k_{ijl}$ of the points located on the beams between the stations

and satellites lower the subionospheric point ($i = \overline{1, n_k}$, $j = \overline{1, m_k}$, $l = \overline{1, N}$, $k = \overline{1, K}$).

5. Compute (21.5) the value $\bar{v}^k_{ijl}$ in the points (21.4) (located on the beams between the stations and satellites lower than the subionospheric point) that are defined in step 4 ($i = \overline{1, n_k}$, $j = \overline{1, m_k}$, $l = \overline{1, N}$, $k = \overline{1, K}$).
6. Select a number of internal nodes located on the line segments between two points on the beams from stations to satellites.
7. Compute coordinates of internal nodes $\check{x}^{kr}_{ijpq}$, $\check{y}^{kr}_{ijpq}$, $\check{z}^{kr}_{ijpq}$ (21.6) that lie on the segments between two points on the beams from stations to satellites ($r = \overline{1, M}$, $p, q \in [1, N]$, $i = \overline{1, n_k}$, $j = \overline{1, m_k}$, $k = \overline{1, K}$).
8. Using interpolation determine $\check{v}^{kr}_{ijpq}$ (21.7) in the points defined in steps 6 and 7 ($i = \overline{1, n_k}$, $j = \overline{1, m_k}$, $r = \overline{1, M}$, $p, q \in [1, N]$, $k = \overline{1, K}$).
9. Determine boundaries x^k_{min}, y^k_{min}, z^k_{min}, x^k_{max}, y^k_{max}, z^k_{max} (21.8) of the rectangular spatial area with the points $\bar{x}^k_{ijl}$, $\bar{y}^k_{ijl}$, $\bar{z}^k_{ijl}$ (21.4) and defined values $\bar{v}^k_{ijl}$ (21.5) ($i = \overline{1, n_k}$, $j = \overline{1, m_k}$, $l = \overline{1, N}$) for each measurement ($k = \overline{1, K}$).
10. Select a number of points L where the spatial area is divided and limited by the boundaries (21.8) set in step 9.
11. In the rectangular area defined in step 9 determine the coordinates of the equally located points $\hat{x}^k_i$, $\hat{y}^k_j$, $\hat{z}^k_l$ (21.9) ($i, j, l = \overline{1, L}$).
12. Join the sets of the nodes coordinates on the beams from stations to satellites (21.4) and on these beams (21.6) and sets of the correspondent known values (21.5), (21.7) into a combined set x^k_a, y^k_a, z^k_a (21.10), (21.11) ($a = \overline{1, A_k}$) that is experimentally defined discrete functional dependency of the ionization from spatial coordinates for a measurement k ($k = \overline{1, K}$).
13. Determine the best approximation basis common for all measurements $k = \overline{1, K}$ from the results of the problems (21.13), (21.14) obtained using the exponent reduction of the polynomial from numerous arguments (21.12) (Vladimirov, 1981; Vasyliev and Symak, 2008).
14. Solve the problem (21.13), (21.14) for all measurements $k = \overline{1, K}$ using an approximation basis determined in step 13.
15. Based on the results from step 14, using the interpolation by spline, define coefficient dependencies $c_i(t)$ ($i = \overline{1, n}$) of the polynomial (21.12) from time and dependency of the center coordinates $x_C(t)$, $y_C(t)$, $z_C(t)$ of the rectangular area (21.9) from time ($t \in [t_1, t_k]$).
16. Make a quality analysis of changes in the polynomial (21.12) coefficient $c_i(t)$ ($i = \overline{1, n}$) in time ($t \in [t_1, t_k]$).
17. Define approximation errors in the problem (21.13) for all measurements.
18. Make a quality analysis of the probability density (and distribution function) of the restored ionization value $v(x, y, z, t)$ in points (21.9) in the rectangular area (21.8). If these distributions have central maximum or are approximately even or linear, and if approximation errors in the problem (21.13) for all measurements are within acceptable limits, then the determined approximation basis (step 13) can be regarded as acceptable. Otherwise, go to step 3.
19. Define the boundaries $[\bar{x}_{min}, \bar{x}_{max}]$, $[\bar{y}_{min}, \bar{y}_{max}]$, $[\bar{z}_{min}, \bar{z}_{max}]$ (21.15) of the rectangular area (21.16) where the approximation value of the ionization for all measurements $k = \overline{1, K}$ is computed.
20. Determine the coordinates of the equally located points $\hat{x}_i$, $\hat{y}_j$, $\hat{z}_l$ in the rectangular area defined in step 18 ($i, j, l = \overline{1, L}$).
21. Using the coefficients $c_i(t)$ ($i = \overline{1, n}$) in the polynomial (21.12) and the center coordinates $x_C(t)$, $y_C(t)$, $z_C(t)$ of the rectangular area (21.15) that depend on time, determine the ionization value $v(x, y, z, t)$ at an arbitrary time $t \in [t_1, t_k]$ in the arbitrary point of the area (21.15).
22. To represent the results of the ionization restoration graphically, compute the ionization values in the points (21.16), i.e., compute $v(\hat{x}_i, \hat{y}_j, \hat{z}_l, t)$ ($i, j, l = \overline{1, L}$, $t \in [t_1, t_k]$).

21.4 CONCLUSION

The processes occurring in the ionosphere are inextricably connected with the processes of interaction in the Earth–Sun system. Information about the state of the near-Earth plasma sheath might serve to identify and investigate cause–effect relations of solar and geomagnetic activity, the processes of the ionosphere interaction with atmospheric processes and earthquakes, etc. One of the perspective opportunities to obtain such experimental information on the ionosphere state is connected with the use of radio signals from GPS satellites. For the study of the ionosphere using GPS satellites, it is necessary to process arrays of big databases.

For determination of the parameters of ionosphere, an algorithm of processing GPS measurements for multiple stations was implemented using a network of active reference stations in the Western Ukraine. In our research, we used spline approximation to determine the smoothed parameters of ionosphere. We chose this method of smoothing based on computational experiments that were aimed at comparing different smoothing methods. For continuously operating reference GNSS stations, the results of the determined parameters ionosphere TEC on the one hand, these data reflect the state of the ionosphere during the observation; on the other hand, it is a substantial tool for accuracy improvement and reliable determination of coordinates of the observation place. Thus, we decided to solve the problem of restoring the spatial position of the ionospheric state or its ionization field according to the regular definitions of the TEC identifier.

We proposed an algorithm for ionization field change restoration using the approximation of the change in time of the coefficients of the polynomial from numerous arguments. The resulting error indicators show that the developed algorithm gives consistent results for ionization field restoration that do not depend on the ionosphere state, satellites positions, and changes in number of stations in the network used for computations. Instant accuracy of the ionization field restoration is acceptable for our problem. To improve the described method, we need to conduct research to explain the structure of the approximating polynomial and search for additional computation tools to increase the approximation accuracy in the observational nodes and prevent rapid change of the approximating polynomial beyond these nodes.

REFERENCES

Klobuchar, J.A., 1996. Ionospheric effects on GPS. In: Parkinson, B.W., Spilker, J.J. (Eds.), Global Positioning System: Theory and Applications, vol. 1. American Institute of Aeronautics and Astronautics, 370 L'Enfant Promenade, SW. Washington DC, 20024.

Komjathy, A., Langley, R.B., 1996. An assessment of predicted and measured ionospheric total electron content using a regional GPS network. http://gauss.gge.unb.ca/grads/attila/papers/papers.htm.

Malachivskyy, P.S., 2009. Mathematical Modeling of Functional Relationships between Physical Quantities Using Continuous and Smooth Minimax Spline Approximations. The thesis is presented for Dr. Tech. Sci. of the 01.05, 2009. Lviv Polytechnic National University, Lviv.

Matviichuk, Y.M., 2000. Mathematical Macro-Modeling of Dynamic Systems: Theory and Practice. Publishing House of Ivan Franko National University of Lviv, Lviv, p. 214.

Matviichuk, Y.M., Kurhanevych, A., Olyva, O., Pauchok, V., 2000. Prognostic modeling of dynamic systems (macromodel approach). In: Automatics – 2000: International Conference. Lviv, September 11–15, 2000. 7 volumes. – V.7. – Lviv, – 232 pp. 82–87.

Stanislawska, I., Jakowski, N., Béniguel, Y., De Franceschi, G., Hernandez Pajares, M., Jacobsen, Knut Stanley, Tomasik, L., Warnant, R., Wautelet, G., 2012. Monitoring, tracking and forecasting ionospheric perturbations using GNSS techniques. Journal of Space Weather and Space Climate. https://doi.org/10.1051/swsc/2012022.

Tikhonov, A.N., Arsenin, V.Y., 1979. Solutions of Ill-Posed Problems. Nauka, Moscow, p. 288.

Tsugawa, T., Saito, A., Otsuka, Y., 2004. A statistical study of large-scale traveling ionospheric disturbances using the GPS network in Japan. Journal of Geophysical Research 109. https://doi.org/10.1029/2003JA010302.

Vasyliev, V.V., Symak, L.A., 2008. Fractional Calculus and Approximation Methods in the Modeling of Dynamic Systems. Scientific publication, The National Academy of Sciences of Ukraine, Kyiv, p. 256.

Verlan, A.F., Fedorchuk, V.A., 2013. Models of the Dynamics of Electromechanical Systems. The National Academy of Sciences of Ukraine, Pukhov Institute for Modeling in Energy Engineering, Naukova Dumka, Kyiv, p. 222.

Vladimirov, V.S., 1981. Equations of Mathematical Physics. Nauka, Moscow, p. 512.

Yankiv-Vitkovska, L.M., 2012. On the computation of the ionosphere parameters using a special algorithm: first results. Space Science and Technology 18 (6), 73–75.

Yankiv-Vitkovska, L.M., 2013a. On the study of the ionosphere parameters for GNSS-stations SULP, RVNE, and SHAZ. Geodesy, cartography, and aerial photography, Ukrainian interdepartmental scientific-technical collection 78, 172–179.

Yankiv-Vitkovska, L.M., 2013b. Methods of determining the ionosphere parameters in the network of satellite stations in the Western Ukraine. Space Science and Technology 19 (6), 47–52.

Yankiv-Vitkovska, L.M., Matviichuk, Y.M., Savchuk, S.H., Pauchok, V.K., 2012. The research of changes of GNNS stations coordinates by the method of macromodeling. Geodesy and Cartography Bulletin, Kyiv 1 (78), 9–17.

Yankiv-Vitkovska, L.M., Savchuk, S.H., Pauchok, V.K., Matviichuk, Ya.M., Bodnar, D.I., 2016. Recovery of the spatial state of the ionosphere using regular definitions of the TEC identifier at the network of continuously operating GNSS stations of Ukraine. Journal of Geodesy and Geomatics Engineering 1 (9), 37–48.

CHAPTER 22

Exploitation of Big Real-Time GNSS Databases for Weather Prediction

NATALIYA KABLAK, DSC, PROF • STEPAN SAVCHUK, DSC, PROF

22.1 INTRODUCTION

Nowadays, multifunctional systems of high precision positioning and networks of active reference stations allowing to get the coordinates of the site with centimeter-level accuracy in real-time (Real-Time Kinematics, RTK) are widespread. Additionally, in order to find coordinates in real-time to within 1–2 cm, such systems make it possible to solve indirect problems being essential for the development of all Earth sciences: navigation, geodesy, cadastre, meteorology, atmospheric physics, etc.

Among errors of Global Navigation Satellite System (GNSS) measurements of distances to satellites, the component caused by the insufficient precision in the determination of tropospheric delay of an electromagnetic signal in the Earth's neutral atmosphere makes the greatest contribution.

Therefore, the relevance of this chapter is in peculiarities of influence of the neutral atmosphere on measurement results using advanced satellite technology for:

- the coordinate support of the process of establishment of an integrated coordinate database in real-time using modern GNSS technology in differential mode for reliable modeling of atmospheric errors dependent on distance both within minimally possible time intervals and with maximal precision;
- solving problems of meteorology and atmospheric physics such as determination of spatiotemporal changes of weather parameters in the large territory of Ukraine and of the dynamics of the precipitable water vapor (PWV) change.

In accordance with the technical development of GNSS, requirements for accurate consideration of the atmosphere influence are changing. However, there are still no effective (accurate and operational) methods for taking into account the atmosphere influence in RTK technologies that does not allow the full realization of the technical capabilities of GNSS networks. Knowledge on the zenith tropospheric delay makes it possible to obtain valuable information for meteorology, because water vapor in the troposphere is an important component of the equilibrium energy of the atmosphere, which is a significant component (more than 60%) of the natural greenhouse effect. It is from this position that GNSS technologies can help improve the quality of weather forecasting and climate modeling.

Meteorology from the Global Positioning System (GPS), which became an important approach for the remote sensing of atmospheric water vapor, was first proposed in the early 1990s (Bevis et al., 1992; Herring et al., 1992). However, only in the last 10–15 years, significant progress has been achieved (Guerova et al., 2016; Baldysz et al., 2018). Compared to established atmospheric sounding methods, GPS meteorology has great potential for water vapor sensing because it has unique advantages: accessibility in any weather conditions, high accuracy, long-term stability, and high temporal resolution.

The zenith tropospheric delay, which is a major product of GPS meteorology, is commonly used to quantify PWV. The main efforts of the scientists were to assimilate zenith tropospheric delay determined from GPS observations and the precipitable water vapor into numerical weather prediction (NWP) models (Nilsson and Elgered, 2008; Ning and Elgered, 2012; Kablak, 2012). The main purpose of such assimilation is to improve the current state of forecasting heavy rainfall. Experimental studies of tropospheric GPS delay in various physical and geographical areas have shown that GPS meteorology has significant potential in monitoring the amount of water vapor in the Earth's troposphere.

In order to meet the various requirements of meteorological applications, GNSS water vapor data, which are obtained from both post-processing and near-real-time modes, is assimilated into regional and global NWP models. However, several important and time-critical meteorological applications, such as the detection of short-term weather changes, may be useful in obtaining atmospheric state information at a higher update rate, which may result from real-time processing of GNSS data. For example, tropospheric delays, which are determined at intervals of several minutes and with an accuracy of 5–30 mm, are already being applied to models of short-term weather prediction.

Knowledge Discovery in Big Data from Astronomy and Earth Observation. https://doi.org/10.1016/B978-0-12-819154-5.00034-5

NWP, especially short-term prediction, is a problem of large amounts of data that are needed at certain stages of the integration of different data – tropospheric delays and existing prediction models, complicated calculations needed for real-time data processing. At assimilation, large amounts of observational data are taken into account, and nonlinear equations at the grid nodes are solved. Therefore, the biggest problem in detecting short-term weather changes with the use of GNSS observations and NWP models is the availability of large databases. However, the era of information technologies and advances in high-performance computing opens up opportunities to really model the evolution of the Earth's system at a scope that was not previously thought possible.

Significant contributions to the study of the influence of the neutral atmosphere (tropospheric delay) on the propagation of electromagnetic waves (EMWs) were made by Ukrainian scientists A. Ostrovsky, F. Zablotsky, S. Savchuk, Ya. Kostetska, O. Prokopov, M. Myronov, I. Motrunych, I. Shvalagin, N. Kablak, and others.

In Ukraine, theoretical and practical studies of the influence of the atmosphere on propagation of signals from space objects are conducted by the scientists of Lviv, Uzhhorod, Kyiv, and Kharkiv, where whole scientific schools of research in this field have been formed.

22.2 INFLUENCE OF NEUTRAL ATMOSPHERE ON RESULTS OF RANGE FINDING OBSERVATIONS OF ARTIFICIAL SATELLITES

When an EMW propagates from the satellite to the receiver, the atmosphere creates two effects to be taken into account in range finding: a decrease in the speed of wave propagation in the atmosphere (signal delay) and curvature of the signal path (refraction). These errors in range finding observations are taken into account as atmospheric delays in the neutral and ionized parts of the atmosphere. The neutral atmosphere consists of the troposphere and stratosphere. In the neutral atmosphere, a delay is called the tropospheric delay.

For determination of the tropospheric delay, various approaches are applied: the analytical models of tropospheric delay without the use of surface weather parameters (error 4.0–6.0 cm), analytical models of tropospheric delay with the use of surface weather parameters (error 2.5–4.0 cm), aerological sounding of the atmosphere (error 4.0–6.0 cm), and determination of the tropospheric delay from GNSS data (error 0.5–2.0 cm).

The most accurate method is to determine the tropospheric delay from analysis of GNSS data.

In December 2003, in the Main Astronomical Observatory of the National Academy of Sciences of Ukraine, the Analysis Center of GPS data for processing observations of GPS satellites from Ukrainian permanent stations was established. The Transcarpathian Positioning Service ZAKPOS is a local initiative and project of installation of uniform basic infrastructure of differential GNSS (DGNSS) in the region with the computing center in Mukachevo. Regular GNSS observations in reference stations of the ZAKPOS network were started on 4 February 2009 (Kablak and Savchuk, 2013).

In the geodetic sense, the network of active reference stations is an extensive network of permanent stations. These networks differ in their objectives, accuracy, infrastructure, etc. If the network of permanent stations is actually a fundamental network solving scientific and technical problems of the highest accuracy, the network of active reference stations based on the most up-to-date RTK-technology is centralized and maximally automated, allowing actually to get objective data on the location of objects with centimeter accuracy in a single coordinate system and to solve complex problems first of all by qualitative geodetic support of cadastral works (Savchuk et al., 2018).

In the middle of 2010, the ZAKPOS network consisted of 17 stations, in the end of 2010 of 28 stations and, eventually, it became a nationwide network with a new name: UA-EUPOS/ZAKPOS. In 2019, the active reference stations network UA-EUPOS/ZAKPOS processes data from almost 90 GNSS stations located in Ukraine, Poland, Slovak Republic, Hungary, Romania, and Moldova. The use of unknown tropospheric parameters when processing satellite observations allows to find tropospheric delay for each network reference station.

For study in this chapter, we use information on tropospheric delay of a signal from the UA-EUPOS/ZAKPOS network (2010–2012, 2018), data of aerological sounding of the atmosphere (1999–2004, 2012, 2013), and archives of meteorological data in locations of active reference stations (2009, 2012, 2018).

In order to determine the tropospheric delay, one can use the following methods:

- analytical representation – $\Delta\rho_S$,
- aerological sounding of the atmosphere – $\Delta\rho_a$,
- GNSS observations – $\Delta\rho$.

The "contribution" of different layers of the troposphere to the tropospheric delay value is estimated by means of aerological sounding (Kablak et al., 2002, 2004, 2005; Kablak, 2007a, 2007b): a 10 kilometer layer contributes 73.5% of the delay and a 14 kilometer layer gives 85.5%. At the zenith, the total contribution to the atmospheric delay of layers from 50 to 100 km is only 1.2 mm.

In the real conditions of observation, the tropospheric delays determined by means of both analytical models and aerological sounding of the atmosphere are not representative. In our opinion, the main reasons for this unrepresentativity are the following:

1. Analytical models of determining the tropospheric delay and its constituents are developed using the model distribution of meteorological parameters in the troposphere (spherically symmetric atmosphere, gas laws, the standard model of the atmosphere, polytropic model, etc.). Meteorological parameters are measured in observation points with some errors affecting also the accuracy of tropospheric delay determination. Additionally, there are layers of powerful inversion of temperature, chaotic distribution of partial pressure of water vapor with altitude, and so on. None of them can be taken into account by any analytical model of tropospheric delay determination.

   Considering these arguments, we can draw the following conclusions:
   - Any model represents some average state of the atmosphere only.
   - All models have empirical parameters whose actual values may differ significantly from accepted ones.
   - None of the models allows taking into account the presence of surface temperature inversion.
   - The surface layer of air, namely, the lower layer, causes the main distortion.
2. Sounding of the atmosphere occurs within two hours and its result is a single value of tropospheric delay, but during this time, the actual state of the atmosphere can be changed. The strength and direction of wind in the atmosphere may shift the radiosonde from the vertical direction by an angle. The process of measuring meteorological parameters during radiosonde flight is accompanied by some errors. The refractive index at each altitude is calculated by means of empirical formulas, in which the coefficients are determined with an error.

Experimental determination of the zenith tropospheric delays according to data of GNSS observations in stations in Lviv (SULP) and Uzhhorod (UZHL) is performed. Irregular diurnal changes of tropospheric delay are investigated. During short time periods (10 min), changes of tropospheric delays were less than 1 cm, i.e., within the accuracy (Kablak, 2009). In stations, significant changes in tropospheric delay were observed during the day in 1 hour intervals. Any a priori model is not practically able to explain these changes.

Estimation of errors in determining the tropospheric delay and the reasons for decrease of its accuracy are performed. It is shown that errors of zenith tropospheric delay determination by means of the analytical model are within a few centimeters. The accuracy of tropospheric delay from GNSS observations in reference stations is about 1 cm.

Since the time of propagation of electromagnetic signal through the troposphere at GNSS observations is too short, the changes of meteorological parameters in a certain point of the troposphere above the station during this time cannot be significant. Due to these changes, the tropospheric delay in a certain time point is practically determined without error connected with the atmospheric state. Errors can exist when the satellites are at certain zenith distances, and tropospheric delays are reduced to zenith direction. To this the existence of horizontal gradients of the refraction index, i.e., the additional value of tropospheric delay, should be added.

Accuracy of reduction of tropospheric delay to the zenith one depends on the mapping function. According to some authors, error from the mapping function does not exceed 3 cm at $Z = 85°$. Observations of satellites are continuously performed within 1 minute intervals. Measuring pseudoranges between reference station and satellites one can obtain $4, 20, 40, \ldots, 240$ values of tropospheric delay that increases the determination accuracy of tropospheric delay $\Delta\rho$.

22.3 TAKING ATMOSPHERIC DELAY INTO ACCOUNT USING MODERN SATELLITE TECHNOLOGY FOR COORDINATE SUPPORT IN REAL-TIME

In GNSS observations in RTK mode, two options exist for obtaining the differential corrections by the rover station:

- from the network;
- from a single base station.

In order to use RTK technology in topography and surveying practice in determining coordinates of rover stations, it is necessary to take into account the residual, uncompensated impact of the atmosphere, which is caused by the fact that when processing the results of GNSS measurements, tropospheric delay is considered to be either identical at the "base" and "rover" or changing linearly. However, a simple linear interpolation cannot always take into account the influence of the troposphere on the accuracy of the determination

FIG. 22.1 GNSS stations of the UA-EUPOS/ZAKPOS network selected for the processing.

of rover receiver location. Therefore, the concept of uncompensated tropospheric delay appears.

In order to identify unrecompensed tropospheric delays in observations, experimental studies were made in RTK mode. At a selected station of the UA-EUPOS/ZAKPOS network, lengthy coordinate determinations of a site close to it were made using RTK technology in two modes: with the inclusion of the data of the selected station in network solution and with exclusion of them. Taking into account the configuration of the network, we have selected for research the station FRAN (Ivano-Frankivsk). The distance from it to the nearest station of the UA-EUPOS/ZAKPOS network is about 100 km (see Fig. 22.1).

Experimental studies were carried out during three days in 2010 (one day in April, two days in June) from 0:00 to 18:00. Differences between coordinates of reference station FRAN calculated from lengthy static observations (during over 6 months) and from minute RTK observation were calculated. Totally, about 200 such coordinate differences for all four sessions of observation were received (Kablak, 2010).

Experimental results have shown that when considering all the possible influences on the three-dimensional coordinates, the tropospheric delay not taken into account at the distance of 100 km can reach 10 cm during GNSS observations in real-time. The standard error of this method being 2–5 cm can be inadmissible for many applications and requires introduction of respective corrections.

Thus, the problem of the tropospheric delay determination in any point of the network coverage, where rover receiver can be located, has arisen. For this purpose, the authors of this chapter have developed the technology for determination of tropospheric delay at any point of coverage of the active reference stations network UA-EUPOS/ZAKPOS in real-time. The technology is based on the use of instantaneous values of tropospheric delays in reference GNSS stations.

The Transcarpathian Region is divided by grid squares of size 250 m × 250 m. The coordinates of more than 200,000 points distanced by 250 m (here, the range of latitude is 47.9°–49.1°, and the range of longitude is 22.1°–24.6°) were obtained.

Based on 845 time measurements of tropospheric delay in 20 stations of the UA-EUPOS/ZAKPOS network (Fig. 22.1), it was observed that in order to reduce the number of data, minute values of tropospheric delay should be averaged within intervals of 15 minutes. In this case, r.m.s. error is ±2 mm.

The tropospheric delay interpolation on a point location of rover receiver (200,000 nodal points of the regular altitude grid) was made by three methods: (1) the method using quadratic dependence of tropospheric delays on station altitude, (2) the method developed using the exponential dependence of the height of tropospheric delay stations, and (3) the method using exponential dependence and Shepard–Alexandrov method. Within the changing altitudes of reference stations, one can find tropospheric delay for any point location (in plane and height) by means of the proposed interpolation methods. At altitudes exceeding an altitude of reference stations, it is necessary to calculate the tropospheric delay by the developed method based on the following formula:

$$\frac{\Delta\rho_i}{\Delta\rho_b} = \left[C_0 + C_1(B_i - B_b) + C_2(L_i - L_b)\right] \times \exp\left(-\frac{H_i - H_b}{C_3}\right), \quad (22.1)$$

where $\Delta\rho_i$, B_i, L_i, H_i are zenith tropospheric delay and coordinates of ith reference station, and $\Delta\rho_b$, B_b, L_b, H_b are zenith tropospheric delay and coordinates of the base reference station. The coefficients C_0, C_1, C_2, C_3 can be found by the least-squares method (Kablak, 2013a, 2013b).

The results of the research showed that the mean square error of spatial interpolation is about ±2 cm, with a standard deviation of 0.2 cm in the network coverage area.

Fig. 22.2 demonstrates the surface of the tropospheric delay change based on the data of 12 reference stations and interpolation methods for determining the tropospheric delay at 200,000 nodal points of a regular plane-altitude grid on the example of the Transcarpathian Region, on 13 July 2012, 12:55 UTC. Fig. 22.3 shows isolines of the tropospheric delay change.

It should be noted that the Transcarpathian Region is complicated in terms of the tropospheric delay determination (because there are large differences in elevation). The technique developed for making a map of the tropospheric delay gives nice results (with an error of the order of 1.5 cm). On flat terrain, the tropospheric delay calculation accuracy is higher. This method of construction of tropospheric delay change surfaces can be applied to any area covered by the network of reference GNSS stations.

Using the method proposed in this chapter, one can determine the dynamics of tropospheric delay change, and thus the state of the troposphere over the coverage area of the reference stations network. The advantages of this technique are not only speed, clarity and, continuity (by 1 min), but also the fact that it does not require information on weather parameters at locations of reference stations or conducting upper-air sounding of the atmosphere. It is enough just to find the tropospheric delay at reference stations of the network. At the same time, one can quickly find the tropospheric delay in any point with known geodetic coordinates.

The method of constructing surfaces of tropospheric delay change was used to develop the technology of taking into account residual uncompensated tropospheric delays obtained at rover stations in RTK observations. When calculating the coordinates of a rover station with regard to unrecompensed tropospheric delays $\delta(\Delta\rho)$, the residual tropospheric delays in zenith direction approximated with the appropriate sign should be entered to the measured coordinates (Kablak, 2013c):

$$\begin{aligned} X &= (N + H' + \delta(\Delta\rho))\cos B \cos L; \\ Y &= (N + H' + \delta(\Delta\rho))\cos B \sin L; \\ Z &= \left(\frac{b^2}{a^2}N + H' + \delta(\Delta\rho)\right), \end{aligned} \quad (22.2)$$

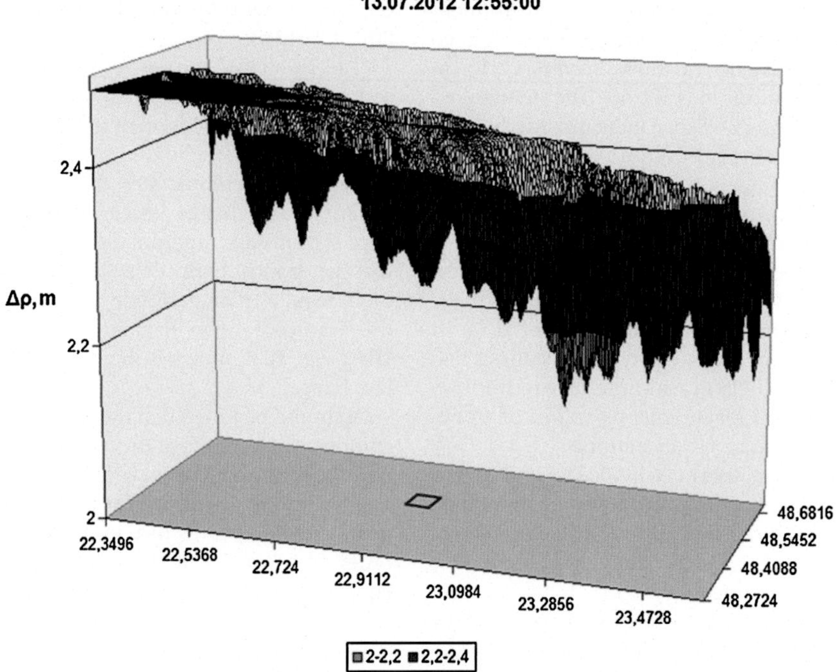

FIG. 22.2 The surface (in longitude and latitude coordinates, in degrees) of the tropospheric delay change in 12 reference stations and nodal grid points (200,000 points) distanced by 250 m.

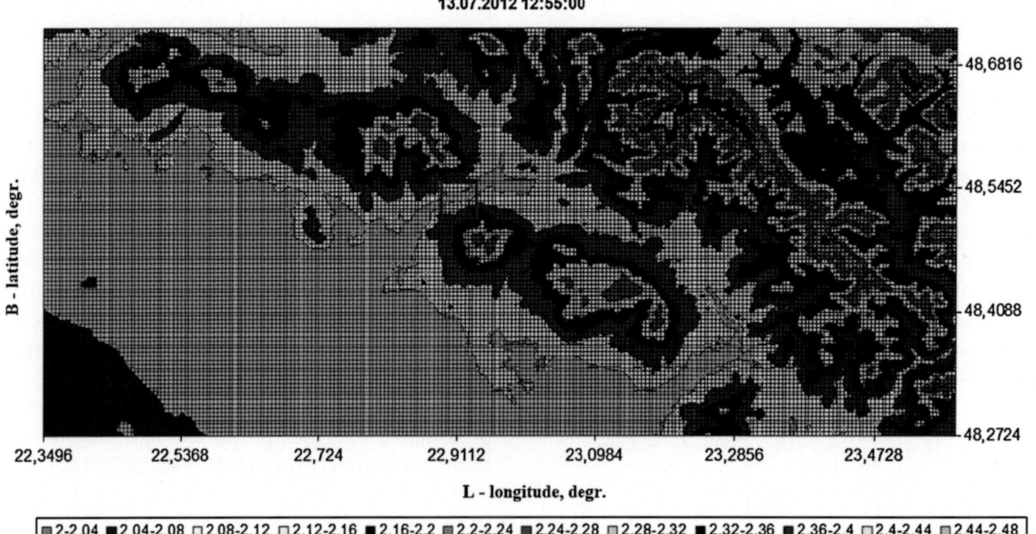

FIG. 22.3 The spatial (in longitude and latitude coordinates, in degrees) distribution of tropospheric delay change (in m) in the form of isolines (13 July 2012, 12:55 UTC).

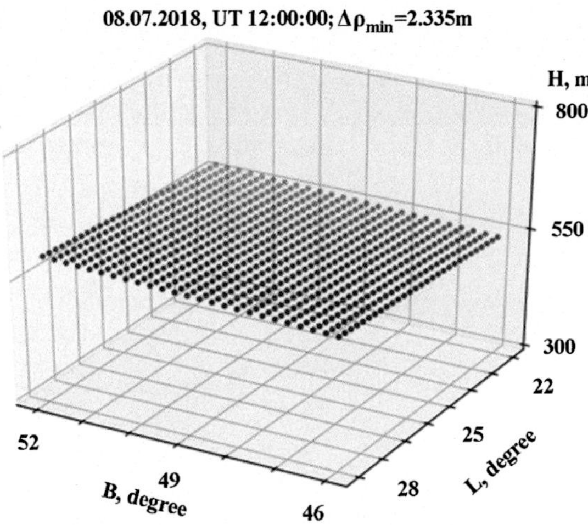

FIG. 22.4 The isosurface of tropospheric delay value ($\Delta\rho_{\min} = 2335$ m) on 8 July 2018, 12:00 UTC.

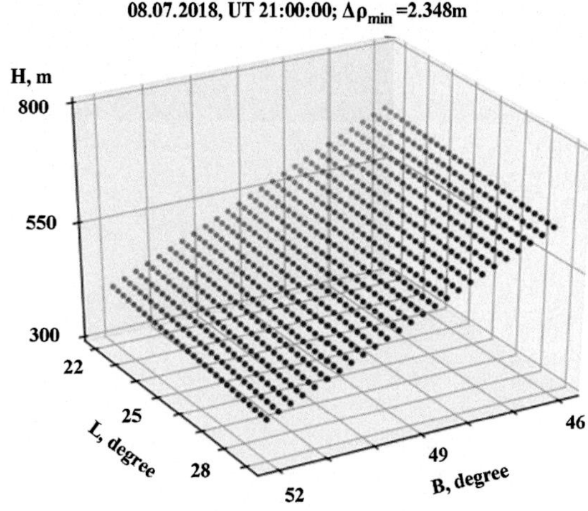

FIG. 22.5 The isosurface of tropospheric delay value ($\Delta\rho_{\min} = 2348$ m) on 8 July 2018, 21:00 UTC.

where $N = \frac{a^2}{\sqrt{a^2 \cos^2 B + b^2 \sin^2 B}}$ is the curvature radius in the first vertical, a, b are semimajor and semiminor axes of the reference ellipsoid (in meters), and H', B, L are ellipsoidal coordinates determined in a rover station.

In order to evaluate the impact of unrecompensed tropospheric delay on the accuracy of the coordinate determination in RTK mode, tropospheric delays at 12 time points for the seven stations of the UA-EUPOS/ZAKPOS network were used. Differences of tropospheric delay obtained from network processing and using approximation (22.1) to conditionally accepted rover station were calculated.

The results of the research show that the magnitude of unrecompensed tropospheric delay is 0.5–1.0 cm, giving an error of the order of 1 cm when determining plane coordinates by expression (22.2).

Efficiency and practical continuity of GNSS observations with determination of tropospheric delay in reference stations in RTK mode and the possibility of calculating by the developed method of determining tropospheric delays in other points of the network coverage area create preconditions for determining the stability dynamics of the atmosphere.

For the first time, the technology of plotting the isosurface of tropospheric delay is developed. The idea of this technology is to find the height above the reference stations with the same values of tropospheric delay (Kablak et al., 2017). For example, in Figs. 22.4 and 22.5, the isosurfaces are shown at two dates.

In order to analyze the situation in this period at specific points of time, we used data of aerological sounding of the atmosphere in the airports of Lviv, Uzhhorod, Satu Mare, Baia Mare, Kosice, Rzeszow, and Budapest at 00:00 UTC and data on weather conditions at the meteorological stations.

Table 22.1 shows the maximal and minimal heights of different points of UA-EUPOS/ZAKPOS coverage area with a fixed value of tropospheric delay and elevation change ΔH on three dates of observations.

Analysis of all data gives reason to say that on 26 May 2018 at 00:00 UTC and 8 July 2018 at 12:00 UTC, there was a stable situation in the troposphere, i.e., above different points of the UA-EUPOS/ZAKPOS network coverage area, the isosurface is characterized by height difference (ΔH) from 55 to 78 m, but on 26 May 2018, 12 June 2018, and 8 July 2018 after 15:00, the troposphere was unstable. Change in elevation (ΔH) is extremely high: at 00:00 UTC it is of the order of 300–500 m. During this period, there were thunderstorms and torrential rains.

A dynamic map of isosurfaces of tropospheric delays over the territory covering the reference stations makes it possible to predict the occurrence and movement of tropospheric disturbances.

22.4 USE OF MODERN SATELLITE TECHNOLOGY IN METEOROLOGY

The main characteristic of the atmosphere as a radio signal propagation environment is an index refraction N. In view of (22.1), we represent the new method for determining the refraction index at any point and at any height above the coverage area of the UA-EUPOS/ZAKPOS network of reference stations. The method of definition of altitudinal distribution of the refraction in-

TABLE 22.1
The heights of the layers of the troposphere above the reference stations of the UA-EUPOS/ZAKPOS network with fixed value $\Delta\rho_{GNSS}$.

Date	UTC	$\Delta\rho_{GNSS}$ (m)	$H_{\min}$ (m)	$H_{\max}$ (m)	ΔH (m)
26.05.2018	00^h00^m	2.325	518.66	597.17	78.51
	03^h00^m	2.319	530.52	627.51	96.99
	06^h00^m	2.323	477.21	664.91	187.7
	09^h00^m	2.318	449.01	762.73	313.72
	12^h00^m	2.322	482.84	792.73	309.89
	15^h00^m	2.333	417.18	681.69	264.51
	18^h00^m	2.322	521.57	660.48	138.91
	21^h00^m	2.318	280.7	747.04	466.34
12.06.2018	00^h00^m	2.335	299.27	695.22	395.95
	03^h00^m	2.332	422.05	674.78	252.73
	06^h00^m	2.343	398.58	622.77	224.19
	09^h00^m	2.337	282.77	672.52	389.75
	12^h00^m	2.326	536.59	700.96	164.37
	15^h00^m	2.344	509.58	716.31	206.73
	18^h00^m	2.352	434.96	794.82	359.86
	21^h00^m	2.336	499.42	699.9	200.48
08.07.2018	00^h00^m	2.328	460.07	686.91	226.84
	03^h00^m	2.332	514.0	581.51	67.51
	06^h00^m	2.336	434.8	591.06	156.26
	09^h00^m	2.334	401.15	624.54	223.39
	12^h00^m	2.335	539.31	595.03	55.72
	15^h00^m	2.337	448.76	654.1	205.34
	18^h00^m	2.348	488.59	600.92	112.33
	21^h00^m	2.348	378.5	679.38	300.88

dex is based on the GNSS observation in RTK mode, namely, the values of tropospheric delays at reference stations (Kablak, 2013a). $N_i(H)$ can be determined by the following formula:

$$N_i(H) = -10^6 \frac{\partial \Delta\rho}{\partial H} = \frac{10^6}{c_3} \Delta\rho_b \left[1 + c_1(B_i - B_b) + c_2(L_i - L_b)\right] \times \exp\left(-\frac{H_i - H_b}{c_3}\right). \quad (22.3)$$

In Fig. 22.6, the dependencies of the refraction index N on height H at station SULP (Lviv) are shown.

At the same time, comparison of altitude profiles of the refraction index found by the method (22.4) (red line) was conducted with the altitude profiles of N, built according to the aerological sounding of the atmosphere (blue line), and using the model representations of weather parameters given in the literature (green line).

The results of the research indicate that the authors' method makes it possible to determine the refraction index at any point within the coverage of the active network of reference stations in real-time observations. Evaluation of the method's accuracy for determining altitudinal distribution of refraction index was conducted using data of meteorological sounding. The r.m.s. error ranged from 5 to 6 N-units.

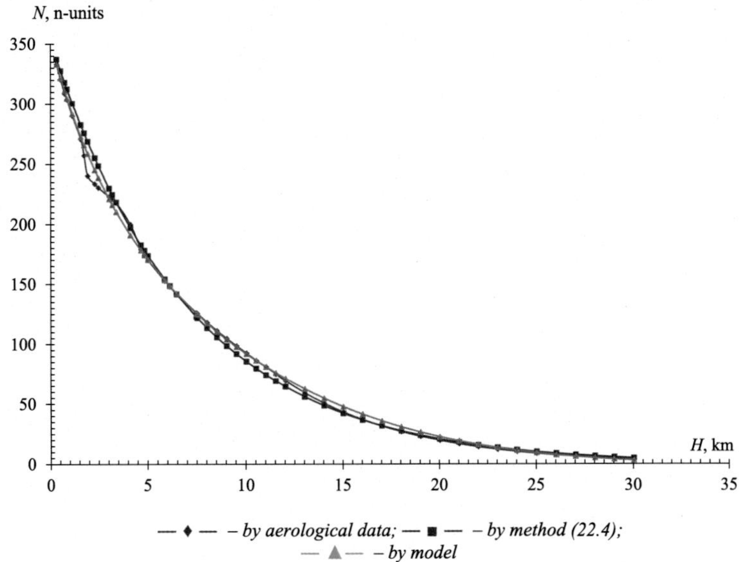

FIG. 22.6 The refraction index dependencies on the height in the reference station SULP (Lviv) on 2 September 2012, 00:00 UTC, as defined by different methods.

The altitude dependence of the refraction index makes it possible to determine the change in partial pressure with altitude. Using the expression of the refraction index for short radio waves proposed by Essen and Frum (Caballero, 2014), the partial pressure at a certain height H of the reference stations can be calculated from the following expression:

$$e(H) = \frac{T(H)N(H) - K_1 p(H)}{K_2 - K_1 + \frac{K_3}{T(H)}}, \quad (22.4)$$

where

P_i is atmospheric pressure (mbar),

e_i is partial pressure of water vapor (mbar),

T_i is air temperature (K),

K_1 is a coefficient characterizing the polarization of molecules of dry air (K·mbar^{-1}),

K_2 is a coefficient taking into account the polarization of water molecules (K·mbar^{-1}), and

K_3 is a coefficient reflecting the effect of changing the electric orientation of polar molecules of water (10^5 K^2·mbar^{-1}).

One can find $N(H)$ using expression (22.3).

Fig. 22.7 shows dependencies of the partial pressure on height above stations SULP and CRNI, defined by (22.4) and (22.3) and by our model representation of temperature and pressure presented in the literature. In the calculations, the results of upper-air sounding in station 33393 Lviv (see http://weather.uwyo.edu/upperair/sounding.html) on 12 July 2012 as well as the tropospheric delays at reference stations of the UA-EUPOS/ZAKPOS network on the same date are used. At different altitudes, nonregular changes in partial pressure are clearly seen.

Actual values of partial pressure on some days differ significantly from the model ones. On these days, strong inversions of temperature and humidity were present.

The combination of data processing of GNSS measurements and additional meteorological information is able to make a model of the troposphere in near-real-time and monitor the state of the troposphere.

The wet component of the tropospheric delay of electromagnetic signal ($\Delta\rho_w$) derived from the data analysis of GNSS observations is used to calculate the integrated water vapor (IWV) content in the atmosphere. As is known, this parameter is crucial for meteorologists, as the content of water vapor in the atmosphere is a key parameter for developing NWP models.

Operational weather forecast is usually based on observations of relative humidity along with pressure and temperature, determined using radiosonde and surface meteorological instruments. One of the main disadvantages of radiosondes is a relatively low accuracy of sensors due to their contamination during flight. That is why the amount of water vapor is also determined by the radiometers. This tool usually provides very accurate data, but the measurements are unreliable in the rain. Additionally, this device is expensive. Radiosonde

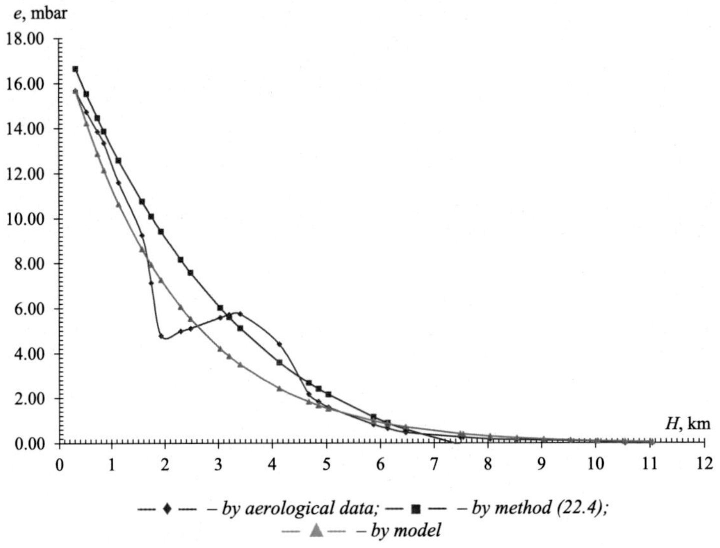

FIG. 22.7 The partial pressure dependencies on height in reference station SULP (Lviv) on 2 September 2012, 00:00 UTC, as defined by different methods.

and ground-based or space radiometers – meters of water vapor – are located at considerable distances from each other, and their discreteness of measurement is low. With such methods of measurement, vertical resolution of determination of the water vapor content in the atmosphere is sufficient, but the spatial and temporal distribution of measured data are very sparse, depending on weather conditions.

A method for evaluating and determining the IWV in the atmosphere (IWV method) by means of GNSS observations is based on the assessment of tropospheric delay ($\Delta\rho$) of GNSS satellite signals. Advantages of this method include the possibility of its implementation using the existing GNSS infrastructure (network of active reference stations with a single center of control) and the fact that water vapor data derived from GNSS measurements do not depend on rain fall and the presence of clouds (Kablak, 2007c).

This chapter represents a new method for determining the IWV, based on the equation of state of wet air. It is proposed to determine the mass of the IWV using altitude dependencies of partial pressure of water vapor (h) and temperature $T(h)$ as follows (Kablak, 2011):

$$\mathrm{IWV}e = \frac{\mu}{R}\int_{h_0}^{h} \frac{e_i(h)}{T_i(h)}\,dh, \qquad (22.5)$$

where $R_w = 461{,}525\ \mathrm{J \cdot kg^{-1} \cdot K^{-1}}$ and μ is the molecular weight of air.

In expression (22.5), the integration is carried out from height of point h_0 to the height of wet air h.

The distribution of IWV differences calculated on the basis of the proposed method and according to GNSS observations for three years in Kyiv is found. We indicate that the magnitude of the difference is less than 1.5 mm, and the error of IWV determination by two methods is $\pm$ 2 mm.

According to the calculated data IWV, the software is developed and the map of the distribution of precipitable water vapor in reference and rover stations in the UA-EUPOS/ZAKPOS network is made. On the basis of maps of tropospheric delay distribution on the Earth's surface (Figs. 22.2 and 22.3), we constructed the map of the spatial distribution of IWV (Fig. 22.8).

On 13 July 2012, values of IWV above locations of active reference stations of the UA-EUPOS/ZAKPOS network were high (24–40 $\mathrm{kg/m^2}$). A summer season is selected because at this time, the wet component of the tropospheric delay is much higher than in winter. As a fragment, we chose of the Transcarpathian Region of Ukraine because this region is of complicated relief (mountains, valleys, rivers). IWV determination error is about $\pm$2 mm.

If one builds the map of IWV change using GNSS observation with steps of 1 minute or 15 minutes, one gets the dynamic map of IWV changes (Kablak and Savchuk, 2012; Kablak et al., 2014; Kablak and Reity, 2017).

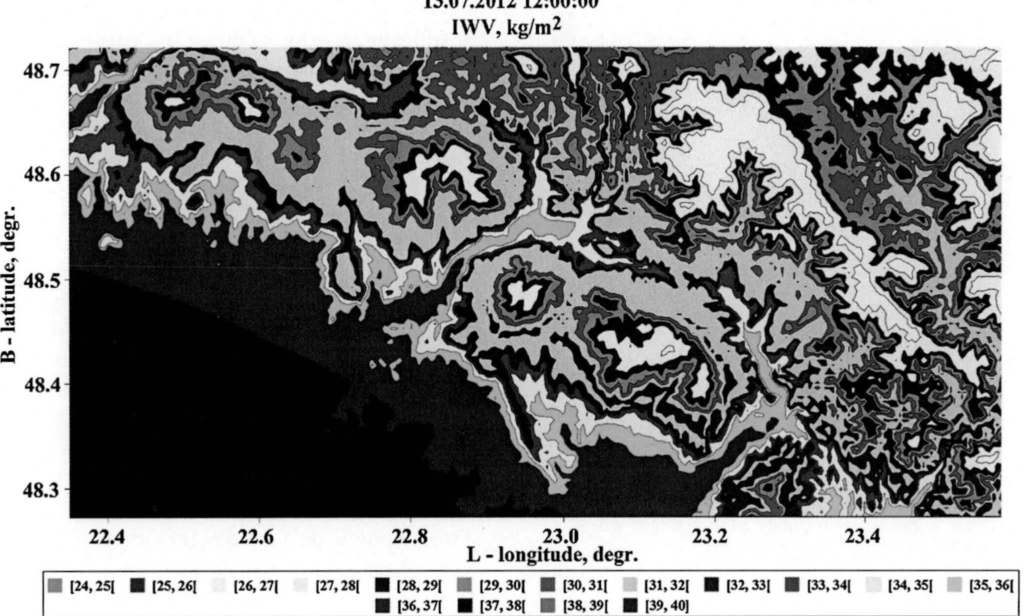

FIG. 22.8 The spatial (in longitude and latitude coordinates, in degrees) distribution of integrated water vapor (IWV) in the form of isolines (13 July 2012, 12:00 UTC).

Continuity and efficiency of GNSS observations in determination of the precipitable water vapor over large areas allows us to determine and predict the dynamics of this indicator. The Transcarpathian Region borders with Hungary, Slovakia, Romania, and Poland. In these countries, the following networks of active reference stations exist: SKPOS (Slovakia), GNSSNET.hu (Hungary), ROMPOS (Romania), and ASG-EUPOS (Poland). Due to the geographical location of the Transcarpathian Region and, therefore, the presence of the UA-EUPOS/ZAKPOS network, as well as cross-border cooperation with European countries, we can make precise and systematic sets of IWV values in large areas that allow us to determine and predict the dynamics of changes in water vapor in real-time (Kablak et al., 2016).

Thus, we have developed the technology for building a dynamic map of precipitable water vapor changes from three-dimensional and four-dimensional fields of tropospheric delay as an example of the Transcarpathian Region, which can be applied in other regions covered by a network of reference stations.

The developed method for graphical and numerical models of three-dimensional and four-dimensional fields of water vapor provides control of speed and direction of movement of water vapor. This process is schematically shown in Figs. 22.9 and 22.10.

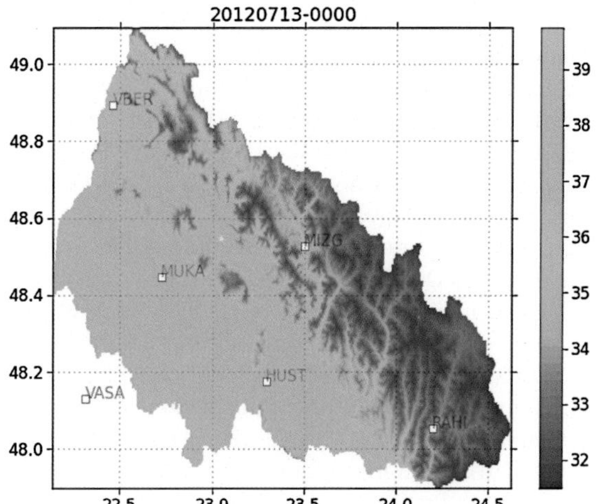

FIG. 22.9 The spatial (in longitude and latitude coordinates in degrees) distribution of integrated water vapor (in mm) on 13 July 2012, 00:00 UTC.

The works complex (processing GNSS observations in the UA-EUPOS/ZAKPOS network, reference stations of national GNSS networks of border countries, and data of meteorological stations) and construction of an operational four-dimensional field of precipitable water vapor will help improve the quality of weather

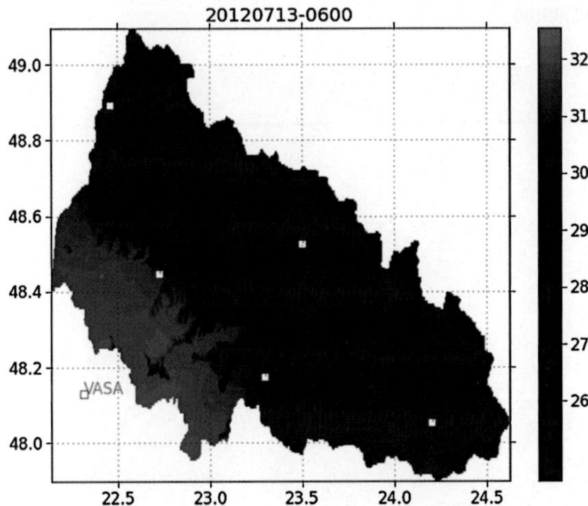

FIG. 22.10 The spatial (in longitude and latitude coordinates in degrees) distribution of integrated water vapor (in mm) on 13 July 2012, 6:00 UTC.

forecast and prevent the natural emergencies of various kinds.

Based on the research, we have developed the system of remote sensing of the atmosphere that will ensure operational and qualitative weather forecasting and prevention of severe weather events and possible environmental threats.

A result of the joint work of Uzhhorod National University (Applicant) and partners – University of Miskolc (Miskolc, Hungary), Vihorlat Observatory (Humenne, Slovakia), the Center for Association Research, Innovation and Technology Transfer «NORDTech» (Baia Mare Romania), and International Association of institutions for Regional Development MAIRR (Uzhhorod) – became the international project HUSKROUA/1101/252 "Space Emergency System – cross-border system for prediction of natural disasters incidents on basis of exploitation of satellite technologies in Hungary, Slovakia, Romania and Ukraine," which has been implemented during 2014–2015 in the framework of the Cross Border Cooperation Programme HU-SK-RO-UA 2007–2013.

22.5 CONCLUSIONS

In order to construct models of atmospheric delay for application and wide implementation of RTK methods into topographic and geodetic practices, as well as the use of GNSS technologies for solving problems of meteorology and atmospheric physics, the technology of the determination of the tropospheric delay at any point in the coverage area of the UA-EUPOS/ZAKPOS network of active reference stations, based on both the use of instantaneous values of tropospheric delays from reference GNSS stations and the proposed method of exponential spatial interpolation, is developed. The results using this technology have shown that the mean square error of spatial interpolation is about 2 cm, with a standard deviation of 0.2 cm, in the coverage area. The surfaces of tropospheric delay change are built based on the data from 12 reference stations and interpolation methods for determining the tropospheric delay at 200,000 nodal points of a regular plane-altitude grid on the example of the Transcarpathian Region of Ukraine. This method of building surfaces of tropospheric delay change can be applied to any area covered by the network of reference GNSS stations. For the first time, the technology of plotting isosurfaces of tropospheric delay is developed. The idea of this technology is to find the height above the reference stations with the same values of tropospheric delay. A dynamic map of isosurfaces of tropospheric delays over the territory covering the reference stations makes it possible to predict the occurrence and movement of tropospheric disturbances, and four-dimensional fields of tropospheric delays can be used to predict the occurrence of natural disasters. Possibilities of exploitation of tropospheric delays determined from GNSS observations to obtain the distribution of weather parameters and precipitable water vapor in the troposphere of the Earth are investigated. The technology for making a dynamic map of precipitable water vapor changes with the help of the Transcarpathian Region's example is developed and can be applied to other regions covered by a network of reference GNSS stations.

The system for remote monitoring of the atmosphere in real-time in the Transcarpathian Region and the basic principles and mathematical models of the application of GNSS technology and ground-based meteorological data to creation of the system of remote monitoring of the atmosphere in real-time in order to protect the environment are developed.

REFERENCES

Baldysz, Z., Nykiel, G., Figurski, M., Araszkiewicz, A., 2018. Assessing the impact of GNSS processing strategies on long-term parameters of 20 years IWV time series. Remote Sensing 10 (4), 496. https://doi.org/10.3390/rs10040496.

Bevis, M., Businger, S., Hering, T., Rocken, C., Anthes, R., Ware, R., 1992. GPS meteorology: remote sensing of atmospheric water vapor using the global positioning system. Journal of Geophysical Research. Atmospheres 97, 15787–15801. https://doi.org/10.1029/92JD01517.

Caballero, R., 2014. Physics of the Atmosphere. IOP Publishing Ltd., Bristol.

Guerova, G., Jones, J., Dousa, J., Dick, G., de Haan, S., Pottiaux, E., Bock, O., Pacione, R., Elgered, G., Vedel, H., Bender, M., 2016. A review of state of the art and future prospects for ground-based GNSS meteorology in Europe. Atmospheric Measurement Techniques 9 (11). https://doi.org/10.5194/amt-9-5385-2016.

Herring, T., Bengtsson, L., Gendt, G., 1992. On the determination of atmospheric water vapor from GPS measurements. Radio Science 108, 157–164. https://doi.org/10.1029/2002JD003235.

Kablak, N.I., 2007a. Modeling the changes of physical parameters of the atmosphere with height. Kinematics and Physics of Celestial Bodies 23 (1), 11–15.

Kablak, N.I., 2007b. Refractive index and atmospheric correction to the distance to the Earth's artificial satellites. Kinematics and Physics of Celestial Bodies 23 (2), 84–88.

Kablak, N.I., 2007c. Study of precipitable water vapour in the atmosphere. Recent Advances in Geodetic Science and Industry 2, 89–94.

Kablak, N.I., 2009. Modern approaches to the determination and use of tropospheric delays of GNSS signals. Geodesy, Cartography and Air Photography 72, 22–27.

Kablak, N.I., 2010. Estimations of the influence of atmosphere in network of active reference GNSS stations. Geodesy, Cartography and Air Photography 73, 17–21.

Kablak, N.I., 2011. Monitoring of precipitable water vapour on base of GNSS data processing. Space Science and Technology 17 (4), 65–73.

Kablak, N.I., 2012. Development of technology determining the influence of tropospheric delay on GNSS observations in real time. Geodesy, Cartography and Air Photography 76, 3–7.

Kablak, N.I., 2013a. Method of determination of the troposphere influence on results of GNSS measurements in network of active reference stations. Recent Advances in Geodetic Science and Industry 1 (25), 62–66.

Kablak, N.I., 2013b. Procedure for determining tropospheric delays in the ZAKPOS/UA-EUPOS network of active reference stations. Kinematics and Physics of Celestial Bodies 29 (4), 202–206.

Kablak, N.I., 2013c. Study of the troposphere state using the results of GNSS measurements in UA-EUPOS/ZAKPOS network of active reference stations. Bulletin of Geodesy and Cartography 4, 3–7.

Kablak, N.I., Kablak, U.I., Shvalagin, I.V., et al., 2002. Modelling troposphere delays radio ranging observations of satellites. Journal of Physical Studies 6 (4), 401–403.

Kablak, N., Kalyuzhny, M., Shulga, A., Vovk, V., 2017. Practical realization of detection of space-time instability of the atmosphere in the UA-EUPOS/ZAKPOS network of active reference stations. Space Science and Technology 23 (1), 54–62. https://doi.org/10.15407/knit2017.01.054.

Kablak, N.I., Klimik, V.U., Shvalagin, I.V., et al., 2004. Monitoring of precipitable water vapour using GPS for weather forecasting. Space Science and Technology 10 (5/6), 163–166.

Kablak, N., Klimyk, V., Shvalagin, I., Kablak, U., 2005. Atmospherical effects on measurements of distance to Earth artificial satellites. Kinematics and Physics of Celestial Bodies 21 (5), 361–364.

Kablak, N., Reity, O., 2017. Application of GNSS Technology to solving meteorology problems. Baltic Surveying 1, 10–16.

Kablak, N., Reity, O., Ştefan, O., Rădulescu, A.T.G.M., Rădulescu, C., 2016. The remote monitoring of Earth's atmosphere based on operative processing GNSS data in the UA-EUPOS/ZAKPOS network of active reference stations. Sustainability 8 (4), 391. https://doi.org/10.3390/su8040391.

Kablak, N.I., Savchuk, S.G., 2012. Remote monitoring of the atmosphere. Space Science and Technology 18 (2), 20–26.

Kablak, N.I., Savchuk, S.G., 2013. Study of spatio-temporal stability of the atmosphere and its influence on the accuracy of the coordinates in UA-EUPOS/ZAKPOS network of active reference stations. Bulletin of Geodesy and Cartography 3, 19–24.

Kablak, N., Savchuk, S., Kalynych, I., Reity, O., 2014. Meteorology monitoring of the precipitable water vapor distribution in the atmosphere based on operational GNSS data processing at reference station network ZAKPOS. Baltic Surveying 1, 67–75.

Nilsson, T., Elgered, G., 2008. Long-term trends in atmospheric water vapor content estimated from ground-based GPS data. Journal of Geophysical Research. Atmospheres 113 (D19101). https://doi.org/10.1029/2008JD010110.

Ning, T., Elgered, G., 2012. Trends in atmospheric water vapor content from ground-based GPS: the impact of elevation cut off angle. IEEE Journal of Selected Topics in Applied Earth Observations and Remote Sensing 5, 744–751. https://doi.org/10.1109/JSTARS.2012.2191392.

Savchuk, S., Kablak, N., Hoptar, A., 2018. Comparison of approaches to the determination of the antiaircraft troposphere delay based on atmospheric radio data and GNSS observation. Geodesy, Cartography and Aerial Photography 88, 24–32. https://doi.org/10.23939/istcgcap2018.02.024.

CHAPTER 23

Application of Databases Collected in Ionospheric Observations by VLF/LF Radio Signals

ALEKSANDRA NINA, PHD

23.1 INTRODUCTION

The ionosphere as a part of the terrestrial atmosphere is very sensitive to numerous external factors. Variable influences coming from both outer space and different domains of the Earth make the ionospheric plasma characteristics time-dependent. The most significant perturber from outer space originates in solar activity whose consequences can be either periodical due to the solar cycle, and seasonal and diurnal owing to the terrestrial motion (Lei et al., 2005; Pham Thi Thu et al., 2011), or transient like those arising from solar flares (McRae and Thomson, 2004; Nina et al., 2011, 2012a; Kolarski et al., 2011; Raulin et al., 2010) and coronal mass ejections (Balan et al., 2008; Bochev and Dimitrova, 2003) among others. Significant transient perturbations can also be induced by phenomena in deep outer space like gamma ray bursts (Fishman and Inan, 1988; Nina et al., 2015) and gamma ray flares (Inan et al., 2007b). In addition, there are some processes in the Earth lithosphere (like volcanic eruptions and earthquakes (Utada et al., 2011; Heki and Ping, 2005)) as well as in the atmosphere (like lightnings (Voss et al., 1998; Collier et al., 2011)) that cause nonperiodic, sporadic disturbances in the ionosphere. All of these events cause spatial and temporal changes in ionospheric plasma whose research requires data that can be obtained from databases collected by different types of observation.

Generally, the technique of atmospheric monitoring depends on the altitude of the considered medium. In this chapter our attention is paid to the low ionosphere, whose monitoring is based on rocket and radar measurements (Strelnikova and Rapp, 2010; Chau et al., 2014), and on technology involving the propagation of very low- and low-frequency (VLF/LF) radio waves. Here we discuss big databases and their applications obtained by the technique that uses data related to amplitude and phase of VLF/LF signals recorded at the ground after the signal deflection from the ionosphere at heights below 90 km.

These databases have very important roles in both scientific and practical use. In the first case they are needed in detection of different astrophysical and geophysical events as well as for modeling plasma parameters under different conditions. In addition, there are indications of relationship between ionospheric perturbations and natural disasters like earthquakes. Some of these investigations show possible ionospheric variations before disasters which point out possible application of collected data in predictions of large catastrophes. Practical applications are primarily important in telecommunication. Namely, induced ionospheric disturbances may directly affect human activities on Earth that are related to radio communications, planned networks of mobile communications satellites, high-precision applications of global navigation satellite systems, etc.

The chapter is organized as follows. In Section 23.2 we describe observations of the low ionosphere and experimental setup used for this monitoring. Explanations of the collected big database applications in the detection of astrophysical and geophysical events, and in modeling plasma parameters are given in Sections 23.3 and 23.4, respectively, while practical applications are shown in Section 23.5. Finally, a short summary of this study is presented in Section 23.6.

23.2 EXPERIMENTAL SETUP AND OBSERVATIONS

The main way to analyze the low ionosphere is based on properties of propagating VLF/LF electromagnetic waves. Namely, during propagation from a transmitter to a receiver within the so-called Earth–ionosphere waveguide, these signals are reflected from the low ionosphere. The wave reflection height depends on the low ionospheric plasma properties (primarily the electron

Knowledge Discovery in Big Data from Astronomy and Earth Observation. https://doi.org/10.1016/B978-0-12-819154-5.00035-7

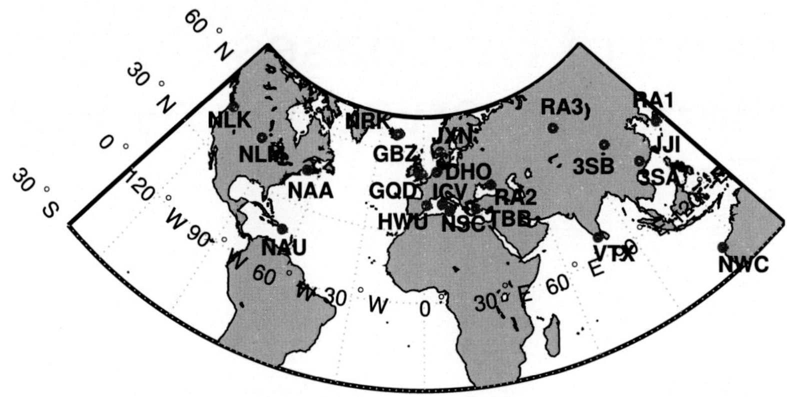

FIG. 23.1 Map of VLF and LF transmitters given in Table 23.1.

density space distribution) and its temporal variations induce changes in wave attenuation between a transmitter and receiver. Consequently, time variations of the recorded signal amplitude and phase visualize nonconstant ionospheric characteristics and can be used in their analyses.

In the following text we describe a global experimental setup and present an example of a VLF/LF receiver and collected data.

23.2.1 Global Experimental Setups

The very important characteristics of this technique for the low ionospheric monitoring relevant to applications of the collected database relate to the observational space, and temporal resolution of the collected data. They can be given in the following two items:

- The global setup of this method consists of numerous worldwide located transmitters and receivers of VLF/LF radio waves.

  In addition to power of an emitted electromagnetic wave, intensity and quality of received data depend on geographical locations of the considered transmitter and receiver. A map of geographical location and information on some VLF and LF transmitters are given in Fig. 23.1 and Table 23.1, respectively. On the other side, receivers are usually grouped in the global networks of radio receivers like Atmospheric Weather Electromagnetic System for Observation Modeling and Education (AWESOME) (Scherrer et al., 2008), South America VLF NETwork (SAVNET) (Raulin et al., 2009), and Antarctic-Arctic Radiation-belt (Dynamic) Deposition – VLF Atmospheric Research Konsortium (AARDDVARK) (Clilverd et al., 2009), which are designed to make continuous long-range observations of the lower ionosphere. Each transmitter emits energy into the entire surrounding space but attenuation of the waves near the Earth's surface is large and their intensities are weak at large distances. On the other side, sky waves which are reflected from the ionosphere propagate several thousand kilometers within the Earth–ionosphere waveguide with sufficient intensity, giving the most important contribution in the recorded signal at relevant frequencies and allowing us to use them in ionospheric observations. The fixed and unique frequency of a transmitter gives information about signal propagation path and space for which the collected data can be used in analyses, while the possibility of several detections by one receiver allows simultaneous monitoring of different regions of the low ionosphere from one position. In addition, numerous receivers provide the large number of databases with complementary information and their comparison practically cover whole latitude-longitude locations in the low ionosphere with high-quality data.
- Continuous emission and reception of radio signals with a very good time resolution (it can be 10 μs) allow for the detection of sudden, and consequently not precisely predicted, events as well as the detection of short-term ionospheric reactions. This property of the obtained databases is very important because many phenomena cannot be predicted and consequently the monitoring of their detection and induced effects cannot be planed. Also, durations of many events and effects that they cause, like lightnings (Salut et al., 2013) and gamma ray bursts (Nina et al., 2015), are very short (they can take only several ms) and they can be detected only by measurements

TABLE 23.1
Characteristics of some VLF and LF transmitters. The data for transmitters are found in the file AWESOME Transmitters.pdf on the web site http://nova.stanford.edu/~vlf/IHY_Test/TechDocs/.

Latitude (°)	Longitude (°)	Frequency (kHz)	Sign	Location
50.07	135.6	11.905	RA1	Komsomolsk-na-Amur, Russia
45.4	38.18	12.649	RA2	Krasnodar, Russia
55.76	84.45	14.881	RA3	Novosibirsk, Russia
59.91	10.52	16.4	JXN	Kolsas, Norway (NATO)
8.47	77.4	18.2	VTX	Katabomman, India
52.71	−3.07	19.6	GBZ	Anthorn, Great Britain (NATO)
−21.8	114.2	19.8	NWC	North West Cape, Australia (USA)
40.88	9.68	20.27	ICV	Isola di Tavolara, Italy (NATO)
25.03	111.67	20.6	3SA	Changde, China
39.6	103.33	20.6	3SB	Datong, China
40.7	1.25	20.9	HWU	Rosnay, France
20.4	−158.2	21.4	NPM	Lualualei, Hawaii, USA
40.7	1.25	21.75	HWV	Le Blanc, France (NATO)
52.4	−1.2	22.1	GQD	Anthorn, Great Britain (NATO)
32.04	130.81	22.2	JJI	Ebino, Japan
53.1	7.6	23.4	DHO	Rhauderfehn, Germany (DHO)
44.65	−67.3	24	NAA	Cutler, Maine, USA
48.2	−121.9	24.8	NLK	Jim Creek, Washington, USA
46.35	−98.33	25.2	NLM	LaMoure, North Dakota, USA
37.43	27.55	26.7	TBB	Bafa, Turkey
65	−18	37.5	NRK	Grindavik, Iceland (USA)
18	−67	40.75	NAU	Aguado, Puerto Rico (USA)
38	13.5	45.9	NSC	Sicily, Italy (USA)

with high time resolution like in the case of the considered setup.

These properties make the considered technique more suitable for monitoring local, short-term, and sudden events from techniques that are limited to localized observations and do not supply continuous measurements. However, unlike the rocket and radar measurements, propagation based on VLF/LF propagation techniques provides data that contain integral information along the entire waveguide between the transmitter and receiver. Due to the large distance between these devices and the atmospheric noise caused by various events, it is often not possible to isolate and analyze individually occurred phenomena. In some cases, this problem can be resolved by comparing multiple databases, as will be explained later.

23.2.2 Example of VLF/LF Receiver and Collected Data

Here, we present the AWESOME VLF/LF receiver and describe collected data at the Institute of Physics in Belgrade, Serbia, in the time period from 2008 to the present.

Receiver description. Collection of data by the AWESOME receiver is enabled by two wire-loop antennas, each sensitive to the component of the magnetic field in the direction orthogonal to the plane of the loop. They are set in North-South (NS) and East-West (EW) directions and connected to the computer through a preamplifier, long cable, GPS antenna (residing outdoors), and line receiver and ADC (residing indoors). The block diagram of the AWESOME receiver's main

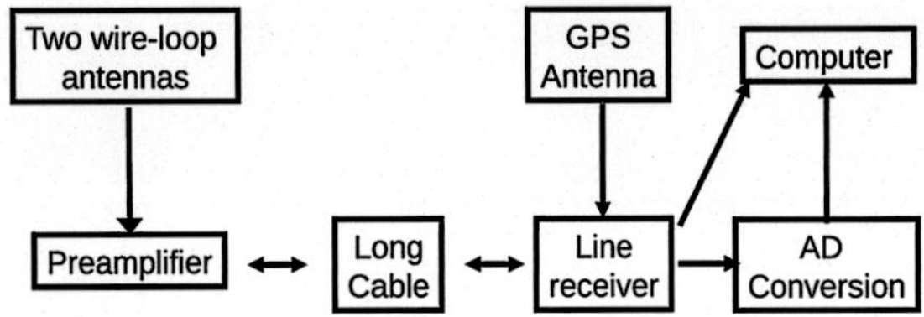

FIG. 23.2 Block diagram of the AWESOME receiver's components (a detailed description is given in Cohen et al., 2010).

components is shown in Fig. 23.2 while a detailed description is given in Cohen et al. (2010).

Collected data characteristics. This monitor provides two types of data: narrowband and broadband. The recorded data are separately systematized and their processing requires different techniques. The main characteristics of data collected by this receiver are:

- **Narrowband data.** This data type provides information about low ionospheric states. Namely, the relevant data relate to values of the amplitude and phase of the considered signals which are emitted by the VLF/LF transmitters a few hundred or thousand kilometers away from the receiver. As already said, the received signals are the so-called sky waves (waves closer to surfaces quickly attenuate) which are deflected from the low ionosphere and their variations in real-time primarily reflect the nonstationary physical and chemical conditions in the ionospheric plasma.
  The Belgrade AWESOME receiver can simultaneously record 15 signals emitted by different transmitters at fixed frequencies. This allows monitoring of the large area and detection of both local and global perturbations. The quality of the received data depends on the characteristics of the transmitter (the power of the emitted radiation) as well as the medium through which the signal propagates (the length of the route, the physical and chemical characteristics of the atmosphere, etc.). Thus, the signal noise increases with the length of the signal propagation path because of influences of more perturbations, while transmitted power affects the recorded amplitude level. In Table 23.2 properties of the most frequently considered signals in studies based on data collected by this VLF/LF receiver are presented. The optimal characteristics has the signal emitted from Germany (near Rhauderfehn) due to a relatively short propagation path and large power of transmitted waves.
  Time resolutions of the recorded data can range from 1 ms to 1 s, which is applicable for detection of very short-term perturbations lasting several ms.
  By default, observations are divided into two time periods during the day. For each time period four files are generated for one particular transmitter: one file with a high and one file with a low time resolution for each of the two antennas (NS and EW). So, if we monitor all 15 signals, 120 files should be generated per day. They are sorted in a folder indicating the receiver station, subfolder indicating the year and, finally, subsubfolder indicating the day of the year. The file name gives information about the transmitter, date, and start time, as well as the antenna and data resolution, while the obtained database can reach several tens of gigabyte/day, i.e., several terabyte/year.

TABLE 23.2
Transmitters characteristics and path length of the analyzed VLF/LF signals. The data for transmitters are found in the file AWESOME Transmitters.pdf on the web site http://nova.stanford.edu/~vlf/IHY_Test/TechDocs/.

Sign	Location	Frequency (kHz)	Power (kW)	Length (km)
DHO	Rhauderfehn Germany	23.4	800	1304
GQD	Anthorn UK	22.1	200	1935
ICV	Isola di Tavolara Italy	20.27	20	976
NRK	Grindavik Island	37.5	800	3230
NAA	Cutler Maine, USA	24.0	1000	6548
NWC	North West Cape Australia	19.8	1000	11974

- **Broadband data.** Much better data time resolution is provided by the broadband measurements which, unlike those of the narrowband data, have data only for the amplitude and are related to the detection of both the natural and transmitter-emitted electromagnetic waves. The time resolution of the broadband

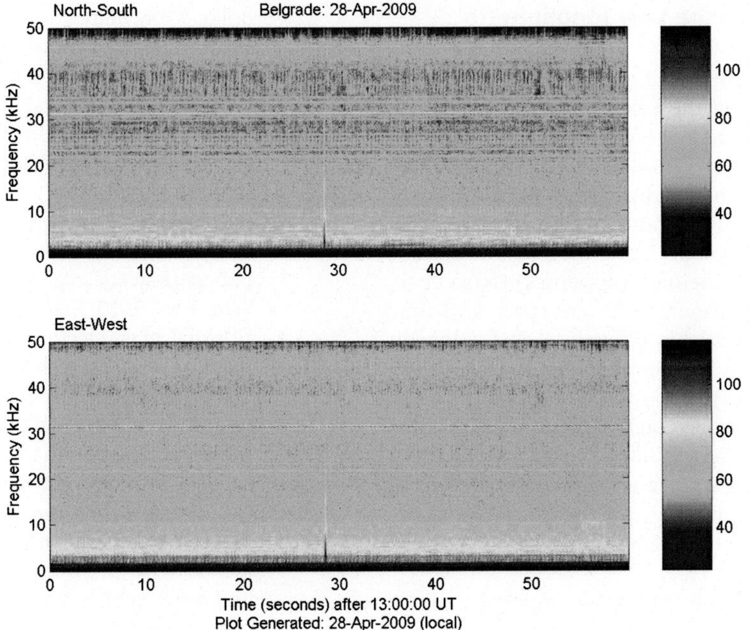

FIG. 23.3 FFT of broadband data collected by NS and EW antenna of the Belgrade AWESOME ELF/VLF/LF receiver.

data is 10 μs for waves in frequency domain from about 1 kHz to 47 kHz, i.e., in extremely low-, very low-, and low-frequency (ELF/VLF/LF) domains. The collected data reach about 32 gigabyte/day, or more than 10 terabyte/year.

The collected data show amplitude of the integral power of the electromagnetic waves in the indicated wave frequency domain. Extracting data for one particular frequency (e.g., very narrow frequency domain) can be done by using fast Fourier transform (FFT), which clearly shows significantly larger amplitudes at frequencies of the transmitter-emitted signals (see Fig. 23.3). For this reason we can connect the relevant data for the low ionospheric properties and use them in the analyses similarly as in the case of the narrowband data. However, there are two important differences in applications of these two types of data for ionospheric studies. First, the absence of information about phase in the broadband monitoring does not allow modeling of plasma parameters by methods based on both the amplitude and phase (Grubor et al., 2008) or only on phase values (Pacini et al., 2006). On the other side, the second difference indicates a better applicability of the broadband data in detections of very short-therm ionospheric disturbances induced by, for example, lightnings, whose durations can be significantly shorter than 1 ms.

23.3 APPLICATION OF DATABASES IN DETECTIONS OF ASTROPHYSICAL AND GEOPHYSICAL EVENTS

One of the most important applications of the considered databases is in the detection of different terrestrial and extraterrestrial events. In the following text we give a short review of phenomena that affect the low ionosphere and explain why a long-time monitoring of the low ionosphere with a good time resolution and different considered signals related to observations of various areas (and, consequently, big database) are necessary for detailed analyses of the mentioned detections. We will describe analyses of the signal perturbations in time and frequency domains, and their connection with a particular event or phenomenon. In addition, we will point out the importance of big databases in these investigations.

23.3.1 Sources of the Low Ionospheric Perturbations

As already said, the origins of events which disturb the low ionosphere are in the Earth's layers (atmosphere and lithosphere) as well as in outer space. Characteristics of their influence (intensity, durations, disturbed area) are dependent on the properties of radiation (radiation spectra, duration), medium of propagation (physical and chemical properties), and the low ionospheric plasma (physical and chemical properties) (Nina et al., 2018). All of these characteristics are space and time variables and can be used in different classifications in scientific studies (see, for example, Nina et al., 2017a and references therein).

There are several reviews on the phenomena that disturb the low ionosphere (Clilverd et al., 2009; Silber and Price, 2017). Here we give a classification of some of them according to induced variations important to detection and, consequently, data mining in the collected databases.

- **Extraterrestrial events.** The most important source of ionospheric perturbations is the Sun. However, some events that occurred in the deep universe can also significantly affect the upper atmosphere. Generally, these events can produce periodical and sudden ionospheric disturbances whose duration could range from 1 ms to several years.
  - Periodical perturbations. These perturbations are induced by variations in incoming solar radiation in the atmosphere due to:
    - Earth's rotation: daily variations which provide the most intensive signal changes reflected in larger amplitude during nighttime periods and increase and decrease going from sunrise to noon and from noon to sunset, respectively. Here it is important to point out excitation of acoustic and gravity waves induced by the solar terminator (Nina and Čadež, 2013).
    - Earth's revolution: seasonal variation which changes durations of daytime and nighttime signals as well as the amplitude intensity.
    - Solar cycle: the 11-year variations induced by changes in the number of sunspot during this period.

    In addition to the mentioned reviews, explanations of these perturbations are also given in Nina et al. (2017).
  - Sudden perturbations. These perturbations can last from 1 ms to several hours or days. They are induced by both terrestrial and extraterrestrial phenomena.
    - Solar X-ray flares. These events induce the most important variations in the ionospheric D-region, which last from a few tens of minutes to several hours. They can cause problems in communications by radio signals (so-called black-out) and additional deviations of satellite signals and variations of plasma parameters of several orders of magnitude. This influence is the subject of numerous scientific papers (Bajčetić et al., 2015; Nina et al., 2017; Schmitter, 2013; Srećković et al., 2017).
    - Coronal mass ejection (CME). Generally, impacts of relativistic charge particles (primarily electrons and protons) from outer space and upper atmosphere in the low ionosphere is significantly larger in the polar regions than in middle and especially equatorial areas due to properties of the geomagnetic field. These relativistic particles can induce disturbances in ionospheric plasma, lasting several minutes and more, which can be detected by the radio receiver considered in this chapter (Clilverd et al., 2007).
    - Gamma ray bursts and flares. This high-energy radiation with origin in the deep outer space arises during a supernova or hypernova as a rapidly rotating, high-mass star collapses to form a neutron star, quark star, or black hole, or appears to originate from a different process like the merger of binary neutron stars. The detected low ionospheric reactions range from a few tens of ms (Nina et al., 2015) to more than one hour (Inan et al., 2007b).
- **Terrestrial events.** In addition to the mentioned extraterrestrial radiation, the low ionosphere can also be affected by volcanic eruptions, earthquake phenomena, and different meteorological conditions. These events provide aperiodic perturbations whose duration is usually shorter than in the cases of the influences from outer space.
  - Lightning-induced perturbations. They are the most frequently analyzed perturbations induced by processes in the Earth. Their duration is significantly shorter than in the previous case and a very high time resolution is usually required for their detection. These processes can be divided in
    - lightning-induced electron precipitation (LEP),
    - early VLF events, and
    - long recovery events,

    and their analyses are given in Silber and Price (2017) and references therein, where they are connected by secondary events like sprites, elves, etc. In addition to narrowband monitoring, these perturbations are studied from the broadband observations (see Cheng and Cummer, 2005 and references therein).

Here it is also important to say that these phenomena can be associated with tropical cyclones and that lightning can be also generated during volcanic activities (see below). For these reasons their monitoring can be very important for investigations of natural disasters, especially because their intensification can be a precursor for the development of tropical depressions in hurricanes (Price et al., 2007).

- Earthquakes. There are many studies which point out to the ionospheric disturbance in periods around earthquake occurrences (Hayakawa, 2007; Pulinets and Boyarchuk, 2004). Their timescale is different and, in some cases, ionospheric variations are detected before the earthquake occurrences, which indicates a practical importance of relevant monitoring.
- Tropospheric not-short-term disturbances. These disturbances are not induced by lightnings and typical examples of them are periods around tropical depression beginnings (Nina et al., 2017b) and tropical cyclones (Kumar et al., 2017), lasting several hours. Like in the case of earthquakes and hurricanes there are some indications that these ionospheric disturbances can be precursors of intensive tropospheric motions.

23.3.2 Detections of the Low Ionospheric Perturbations

23.3.2.1 Time-Domain Analyses

To investigate signal variations and the ionospheric disturbances it is necessary to know answers to some of the following questions:

1. How do we know that the ionosphere is under the influence of an event?
2. Can we associate detected perturbations with the observed perturber?
3. How can we detect locations of local effects?
4. How can we connect perturbations with rare phenomena whose impact has not been sufficiently explored?

Descriptions of the importance of big databases in consideration of these items are explained in the text that follows where we present procedures related to (1) confirmation of an event occurrence, (2) association of detected perturbation with the considered event, (3) determinations of the local event location, and (4) detection of rare phenomena.

1. **Confirmation of an event occurrence.** As said in the introduction, the ionosphere is simultaneously exposed to influences of numerous natural and artificial events. Consequently, recorded signal characteristics which indirectly reflect ionospheric plasma properties are subject to noise and different tendencies which become of prime importance in the detection of particularly weak perturbations. Also, in addition to periodical and sudden variations in ionospheric plasma conditions, characteristics of signals like mutual locations of transmitter and receiver, power of transmitted signal, and geographical area of signal propagation affect the recorded signal properties. Namely, the intensity of the received signal amplitude depends on the emission power and on the distance between the transmitter and receiver. In the former case, a more intense emission induces a larger amplitude of the received signal than the emitted signal with lower power, but in the latter case the considered relationship is not so simple because the increase of the electron density in the low ionosphere does not necessarily induce amplitude rise. The numerical modeling of the signal propagation clearly indicates possibilities of the amplitude increase, decrease, and saturation, which is explained in detail in Nina et al. (2017a).

   For these reasons, using a database or databases which indicate the event occurrence is necessary for the analysis of possible sudden ionospheric disturbance (SID) detections. As examples of used databases in the low ionospheric investigations are those collected by satellites like GOES[1] and Swift,[2] which provide information about solar events and the gamma ray bursts, respectively (see, for example, Selvakumaran et al., 2015; Nina et al., 2015). Even in the case of intensive SIDs and evident detection of their influences on the low ionosphere and signal propagation, these database are important for research of different influences (like, for example, periodical variations in the ionosphere due to daytime, season, and solar cycle changes) or radiation properties (Cresswell-Moorcock et al., 2015; Nina et al., 2018) on the considered detections.

   Here it is important to say that inclusion of the other databases is not always easy and, in some cases, it is impossible. There are many reasons for

[1] https://satdat.ngdc.noaa.gov/sem/goes/data/.

[2] https://gcn.gsfc.nasa.gov/swift_grbs.html.

that, like data availability, complicated administration to access information, long time to access data, and cost of data.

2. **Association of detected perturbation and considered event.** Despite the numerous impacts on the ionosphere, in some cases individual events can have a dominant influence on local plasma properties. Although signal time evolution can have different shapes (see the explanation in Nina et al., 2017a), in this case the relationship between a particular event and the corresponding SID detection is evident. Typical examples of the clear visible connections are sufficiently intensive solar X-ray flares, which are the most frequently studied influences from outer space on the daytime ionospheric D-region.

   However, in some cases intensity of the SID is low and its effects on signal time evolution cannot be extracted from the noise in a particular case or the shape of the signal variation caused by the considered event is the same as or very similar to those induced by some other phenomena occurring in the same time period. Also, it is possible that the reaction is very short or that it does not induce clearly visible changes in recorded signal properties. In these cases, we cannot extract the low ionospheric reaction induced by some considered events but analyses of more events and applications of adequate statistical procedures can be used for confirmation of possible effects induced by the considered phenomena. There are different procedures for the detection of weak SIDs. Here we present three methods based on data processing from the VLF/LF database and a comparison with other relevant databases: the extraction of amplitude peaks, the comparison perturbed with relevant quiet period, and the superposed epoch technique.

   - **Extraction of amplitude peaks.** This technique is used and described in Nina et al. (2015), where the short-lasting low ionospheric reaction on gamma ray bursts is confirmed in statistical analysis of 54 of these high-energy events.

     The main goal of such a procedure is examination of changes in the peak numbers before and after registration of the considered event by some other relevant measurement. For example, in the mentioned study, the times of gamma ray burst occurrence are obtained from the database collected by the Burst Alert Telescope (BAT) on the Swift satellite. In this method, a significant increase in the peak numbers after detection of the considered events indicates the existence of the low ionospheric reaction on this phenomenon. In addition to a pure confirmation of medium responses, variations of the width of time bins (used to divide the analyzed period) can be applied for analyses of time delay in the low ionospheric responses as well as confirmations of the secondary effects of the considered phenomenon in the ionospheric plasma.

     This method is applicable to different datasets but it cannot be used for confirmation of the detectable variations induced by a particular event.

     The existence of big databases in this type of analysis is important for the following four reasons: (1) the considered phenomena are generally unpredictable and measurements should be continuous; (2) the low ionospheric disturbances and, consequently, signal peaks are very short, and hence require high time resolution of data; (3) the frequency of studied phenomena is not necessarily high, and hence a long observation period is required to obtain an adequate number of samples for a quality statistical analysis; and (4) the same or very similar variations of signal characteristics can be induced by different phenomena and the sample used in a study should be formed from the events which occurred in time periods when the effects of the other phenomena are very low, which additionally reduces the sample and extends the observation time period.

   - **Comparison of the perturbed with the related quiet period.** Information about the existence of sudden disturbances of the ionospheric plasma can be obtained using a comparison of signal characteristics from time periods with practically the same conditions, but in the absence of SIDs. This procedure is very useful in the cases without sudden strong variations in the signal-characteristic time evolutions. For this reason some large deviations of the signal time evolution from the expected values (estimated, for example, according to data relevant for the same time period but for the other, unperturbed, days) are not visible if someone analyzes only the considered time period.

There are several limitations and required confirmations that must be taken into account in applications of this method:

- For the consideration of signal deviations lasting up to a few hours it is very important to choose the relevant periods for several very close and quiet days. Namely, in some periods typical shapes of the signal amplitudes significantly vary from day to day (Cresswell-Moorcock et al., 2015) and it is necessary to find these tendencies and, according to them, to estimate the expected values for the considered time periods.
- Even if we can see unexpected time evolutions it is necessary that they stop for some time after the end of the perturbation source influence.
- In some cases, when it is necessary to analyze several days or months it is important to compare the obtained changes with possible variations during the same part of the previous or following years due to seasonal variations. Here we can have the influence of the solar cycle variations and this tendency should also be analyzed.

These limitations are a consequence of periodical variations in the low ionospheric characteristics.

For this model it is also important to perform a statistical analysis because similar effects on the signal propagation can be induced by some other phenomena too. Also, similarly to the previous method, this procedure is based on comparisons of datasets within the same database with additional use of the database of the other type for detection of the considered event.

As one can expect, it is very hard to make all these steps. The reasons are different: from technical (for example, a lack of data) to influences of many other phenomena. However, even if they can not be completely performed, it is very important to present parts that can be studied, first of all, because these analyses often refer to precursors of natural disasters.

As an example for this procedure we can quote the analyses of the periods around the solar terminator in search for a possible relationship between SIDs and tropical depressions (Nina et al., 2017b) and earthquakes (Molchanov et al., 1998). Similar changes are recorded for more cases in a statistical analysis presented in Nina et al. (2017b) which indicates possibilities for the low ionospheric variations in the period around tropical depression beginnings and opens a need for more detailed analyses.

- **Superposed epoch technique.** This technique is applicable for a weak perturbation which is not clearly visible in one particular case but which is repeated under the same influence, like, for example, the transmitter-induced precipitation of inner radiation belt electrons in the low ionosphere (Inan et al., 2007a). It is based on averaged sums of data (representing time series of the signal amplitude relevant for many events) within the same database and their comparison for time periods of occurrence and absence of the considered influences (known from some other database). Here it is important to say that the perturbations are not short-term with respect to the considered time period. Namely, in the case when amplitude changes are short-term with respect to the considered time interval, and not frequent, this method cannot confirm them although their occurrences are in the same period after the influence of the perturber. This conclusion is obtained for short-term responses of the low ionosphere on gamma ray bursts (Nina et al., 2015) whose detection is confirmed using the procedure of extraction of amplitude peaks.

3. **Determinations of local event location.** In addition to the SID differences with respect to their intensity, they can be divided into global and local perturbations. Global SIDs result from extraterrestrial events which affect a large part of the Earth's atmosphere. Contrarily, the local ionospheric perturbations are induced by terrestrial phenomena such as lightnings and they are usually located within a relatively small area. Because of this spatial property, intensity and times of detected disturbances by various receivers are different and can be used for estimation or determination of the considered event locations and/or location of the perturbed low ionospheric part. Precision of the location determination primarily depends on the number of used receivers, i.e., collected databases. Also, the procedure to do that is different for one and several receivers. However, in both cases we can keep in mind that the signal propagates within the Earth–ionosphere waveg-

uide and that some events which can significantly ionize the atmosphere, like lightnings, affect signal propagation not only in the ionosphere but also at lower altitudes. For this reason, when we speak about the perturbation location we think of longitude and latitude, and not altitude.

- **Detection by a single receiver.** These measurements can be applied only for estimation of perturbation locations. The procedure is based on analyses of several signals emitted by different transmitters located on various geographical locations which gives information about changes within different areas. Localization of event occurrences can be made by comparison of detected disturbances in the same (or very close) times by different signals which can be classified in two areas with respect to the receiver position and signal propagation paths:
  - **Disturbances near a receiver.** In this case the signal perturbation is visible for all considered paths.
  - **Disturbances far from a receiver.** In this case the absence of perturbations in the signal coming from one side indicates the SIDs above propagation paths of the perturbed signals.
- **Detection by at least three receivers.** The possibility to use several receivers allows us to find the perturbation location with high precision. There are several receiver networks whose databases are processed in the goal of detection of some events and their application is based on time differences in detected disturbances in different signals.

  An example of these networks is the World Wide Lightning Location Network (WWLLN), which is used for lightning monitoring by more than 70 VLF receivers and produces regular maps of lightning activity over the entire Earth. This network is also used as the global volcanic lightning monitor, which provides databases containing information about activities of close to 2000 volcanos worldwide.

4. **Detection of rare phenomena.** In the case of rare, especially local and unpredictable phenomena, like earthquakes, big databases are required because a long time period is needed for the collection of information of enough events to perform statistical analyses. In addition, in these studies comparative analyses of the different databases are important because more events can be included in the analyses.

23.3.2.2 Frequency-Domain Analyses

The frequency-domain analyses are primarily used for processing broadband data. However, in these analyses we can speak about integral information of the ionospheric variations and different excited electromagnetic waves because, as already said in Section 23.2.2, the broadband data provide information about both the natural (can be excited everywhere) and man-made (emitted by transmitters for the ionospheric monitoring) waves. There are numerous relevant studies which analyze the detection of different phenomena related to disturbances induced by lightnings, earthquakes (Cohen and Marshall, 2012), etc. They are based on application of FFT or wavelet transformations of data in time domain.

On the other side, processing of narrowband data and their analysis in frequency domain provides information on hydrodynamical waves excited in the low ionosphere by sufficient intensive events. As an example of a perturber we can give the solar terminator which generated acoustic and gravity waves in this ionospheric part (Nina and Čadež, 2013). In the presented procedure extraction of waves is based on the Fourier amplitude increase at some frequency in the time period after the event occurrence with respect to its value in the period before. In this type of investigations confirmation of induced waves by the considered phenomena is possible by using two criteria: the amplitude of the excited wave attenuates in time after the event influence, and the detected excitations and attenuations are repeating at many occurrences of the considered phenomenon.

23.4 APPLICATION OF DATABASES IN MODELING LOW IONOSPHERIC PLASMA PARAMETERS

As said in the introduction, the ionosphere is permanently exposed to influences of numerous events whose intensity is space- and time-dependent. Consequently, variations in the ionospheric plasma parameters can be very large. They can reach several orders of magnitude, which significantly changes conditions, and validity of approximations in different models becomes questionable. For this reason, the existence of different databases is of crucial importance in modeling the ionospheric parameters. In this sense, the VLF/LF databases are of great importance in modeling which can be used not only in scientific studies but also in practical applications (see Section 23.5). An advantages of the big

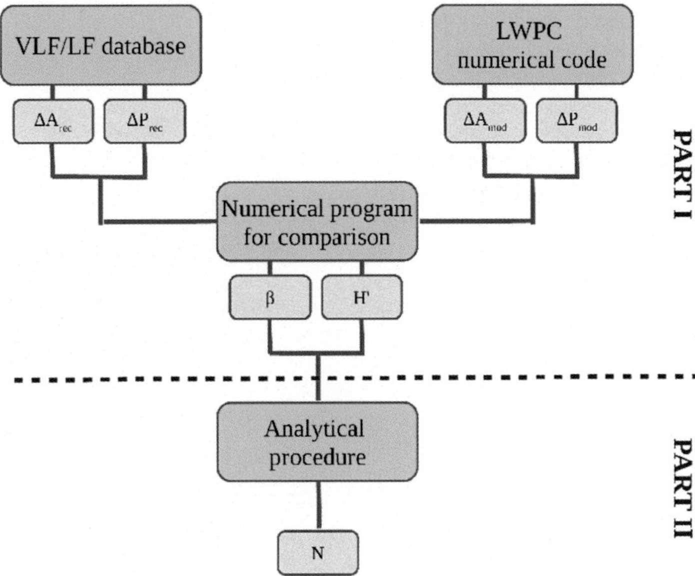

FIG. 23.4 Schematic representation of a database application in electron density modeling.

VLF/LF databases with respect to other similar databases is that they contain continual information about a very large part of the low ionosphere with a good time resolution in long time periods. These properties allow for the application of different numerical and analytical procedures in calculations of plasma parameters and their responses to numerous astro- and geophysical phenomena. The major disadvantage of these data is that they provide information about an entire area between a transmitter and receiver. However, using various procedures in modeling can reduce imprecisions caused by this characteristic.

23.4.1 Modeling of Low Ionospheric Plasma Parameters

Although one can find numerical and analytical procedures for a lot of parameters, the electron density determination is of primary importance because it is necessary for calculations of numerous other parameters. There are a few procedures for determination of the electron density from the data collected by the VLF/LF receiver. They are based on processing of values of amplitude values (Thomson, 1993), phase (Inan et al., 2007a), or both amplitude and phase (Grubor et al., 2008), and implementation of numerical models like LWPC (Ferguson, 1998) or Modefinder/Modesearch (Morfitt and Shellman, 1976). Finally, the parameters obtained from these numerical programs are involved in analytical expressions for calculations of time and space electron density values.

The knowledge of $N_e(h,t)$ allows us to use different analytical expressions for calculations of plasma parameters like the effective recombination coefficients (Žigman et al., 2007; Nina et al., 2012b; Nina and Čadež, 2014), temperature (Bajčetić et al., 2015) and, total electron content in the D-region (Todorović Drakul et al., 2016), as well as comparisons of their properties with characteristics of other values such as, for example, variation in the X-ray radiation spectrum during a solar X-ray flare (Nina et al., 2018).

23.4.2 Example of VLF/LF Database Application in Modeling

Here we present a procedure for the calculation of the electron density time evolution for the D-region altitudes h within the range between 60 km and 90 km during disturbed periods. This procedure, shown schematically in Fig. 23.4, is based on Wait's model of the ionosphere (Wait and Spies, 1964) and can be divided in two parts: numerical determination of Wait's parameters and analytical calculation of the electron density.

23.4.2.1 Numerical Determination of Wait's Parameters

As one can see in Fig. 23.4, implementation of the collected VLF/LF database is required in this part of the procedure for the electron density modeling. Namely, the input values in the numerical program for comparison of experimentally and numerically obtained values are data for the signal amplitude and phase deviations from values relevant to quiet ionospheric conditions ($\Delta A_{\rm rec}$ and $\Delta P_{\rm rec}$) obtained from the considered database. This program is based on criteria given in Grubor et al. (2008), which allow for the determination of time evolutions of White's parameters "sharpness" β and reflection height H'. These criteria require the best fit of the recorded and modeled (by the LWPC model) amplitude ($\Delta A_{\rm rec}$ and $\Delta A_{\rm mod}$, respectively) and phase ($\Delta P_{\rm rec}$ and $\Delta P_{\rm mod}$, respectively) variations relative to their initial conditions, i.e.,

$$\Delta A_{\rm mod}(\beta, H') \approx \Delta A_{\rm rec}(t), \tag{23.1}$$

$$\Delta P_{\rm mod}(\beta, H') \approx \Delta P_{\rm rec}(t). \tag{23.2}$$

Time evolution of White's parameters can be obtained when we apply the presented procedure for a discrete set of recorded data within the considered time period.

23.4.2.2 Analytical Calculation of Electron Density

The electron density time evolution $N(t, h)$ at a fixed altitude h can be derived from time evolutions of White's parameters "sharpness" β and reflection height H' and the following expression (Thomson, 1993):

$$N_e(h,t) = 1.43 \cdot 10^{13} e^{-\beta(t)H'(t)} e^{(\beta(t)-\beta_0)h}, \tag{23.3}$$

where N_e is in m^{-3}, H'(t) and h are in km, β is in km^{-1}, and $\beta_0 = 0.15$ km^{-1}. One example of the obtained electron density time and altitude distribution is given in Fig. 23.5, where we visualize the presented procedure during the influence of the solar X-ray flare that occurred on 5 May 2010. The collected data relate to the 23.4 kHz signal emitted by the DHO transmitter and received by the AWESOME receiver station located in the Institute of Physics in Belgrade, Serbia.

The explained method is used in numerous studies (see for example references in the review paper Radovanović, 2018). The obtained values are generally in good agreement with those obtained from other models or in measurements by rockets (Schmitter, 2013) especially within altitude domains around the signal reflection height (H').

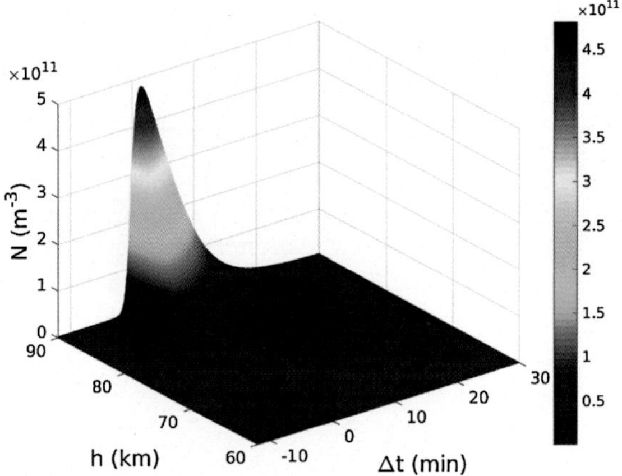

FIG. 23.5 Surface plot of the electron density time and space distribution during the analyzed solar X-ray flare. The Δt axis shows time with respect to its values for maximum X-ray intensity recorded by the GOES-14 satellite in the wavelength domain between 0.1 nm and 0.8 nm.

23.5 PRACTICAL APPLICATIONS

In addition to significance in scientific research, the collected databases can find their important role also in practical applications. There are two major types of these applications: in research and possible prediction of natural disasters and in telecommunications.

23.5.1 Natural Disasters

Although solar activity has a dominate influence on the ionosphere, violent processes in the Earth's layers can be sufficiently intensive to significantly perturb the ionospheric plasma. For some phenomena, like for example lightnings, detection of these SIDs is confirmed and many studies describe their properties. However, there are numerous analyses which indicate the possibility of connections of SIDs with some events. The most important of these indications are related to natural disasters: earthquakes and cyclones.

23.5.1.1 Earthquakes

Investigation of the SID connections with large earthquakes started in the second half of the 20th century (see Hayakawa, 2007 and references therein). These studies present variations in signal characteristics for particular earthquake events as well as statistical analyses of more events.

In the literature, we can find a few possible hypotheses on the mechanism of coupling between the lithospheric activity and ionosphere. Primarily, they relate to geochemical processes and acoustic waves.

- Chemical processes can result in variation of parameters like temperature and radon concentration near the Earth surface, which induce perturbations in the atmospheric conductivity and further, through the atmospheric electric field, the ionospheric disturbances (Pulinets and Boyarchuk, 2004; Sorokin et al., 2006).

- The perturbations of the parameters like temperature and pressure at the Earth's surface in a seismo-active region excites oscillations in the atmosphere which can move up to the ionospheric heights (Molchanov et al., 2001; Miyaki et al., 2002).

Here it is interesting to mention that heating and additional ionization can also be induced by radio waves generated in the lithosphere. However, the analysis given in Molchanov et al. (1995) indicates that this mechanism is not so important because the electromagnetic emission in the lithosphere is weak at all radio frequencies.

The most important conclusion of these studies is a possibility for earthquake prediction in the period of several days before the disasters. Significance of this possible practical application of the VLF/LF technology results in a big importance of relevant databases. Investigations are based on analyses of databases collected by a particular receiver as well as databases incorporated in networks of VLF and LF radio receivers. These networks are formed in the Pacific region and Europe, and their detailed descriptions are given in Hayakawa (2007) and Biagi et al. (2011), respectively. A global Pacific VLF/LF network consists of receivers in Japan, Russia, and Taiwan. Receivers in Italy, Greece, Turkey, Romania, and Portugal form the European VLF/LF radio network. All of them monitor several signals emitted by different transmitters and comparisons of the collected databases enable a better determination of perturbation location than in the case of a single receiver.

23.5.1.2 Cyclones

The most important atmospheric violent motions which induce natural disasters are tropical cyclones. Similarly to the case of earthquake events, electrical and electromagnetic effects (Isaev et al., 2002; Sorokin et al., 2005; Thomas et al., 2010) and the acoustics and gravity waves (Xiao et al., 2007) are considered as mechanisms of the troposphere–ionosphere links. The relevant low ionospheric perturbations are studied through analyses of the VLF/LF signals variations before (Price et al., 2007) and during (Nina et al., 2017b) a tropical depression and during tropical cyclones (Peter and Inan, 2005; Thomas et al., 2010; Kumar et al., 2017). So, for this phenomenon too, there are indications that the VLF/LF signals and, consequently, collected databases could be used for the prediction of natural disasters of a tropical cyclone.

In the research of both the earthquake and cyclone events, data processing is based on different procedures, including statistical analysis, data mining, and wavelet transformations. The timescales of disturbances are different. Consequently, in some cases analyses require high time resolutions while some of them study long time periods. In addition, a continuous monitoring by different transmitters and comparisons of databases obtained by different receivers are necessary for these studies because of low frequencies of these events, unperiodical and still unpredictable occurrences, different locations, and a relatively small radius of possible SIDs. For all of these reasons it is clear why big databases are needed for application of this Earth observation technique.

23.5.2 Telecommunication

Generally speaking, the ionosphere is a medium which affects propagation of electromagnetic waves. This property is used in many practical applications, such as information transfers over thousands of kilometers by telecommunication signals in a waveguide bordered by the Earth's surface and the ionosphere, and detection of different events using telecommunication signals (see Section 23.3).

On the other side, changes of the signal propagation paths and its attenuation can cause different types of errors and problems in communications, including radio black-out. These influences are altitude- and time-dependent due to variations in electron density height distribution. In addition, the intensity of these influences depends on signal frequency. The role of the low ionosphere is more important for signals emitted from the ground at the VLF and LF domains than for satellite signals of GHz frequency which primarily deviate in the F-region. The E-region has a smaller but still nonnegligible role in signal propagation properties while the lowest D-region is usually ignored in corresponding modeling. However, during periods of intensive perturbation, the D-region electron density can be sufficiently increased so that its influence on signal deviations cannot be ignored in modeling of the Global Navigation Satellite System (GNSS) and synthetic-aperture radar (SAR).

For these reasons one can see that big VLF/LF databases are also important in applications in telecommunication and, consequently, in other areas of human activity, like geodesy and land surveying, emergency responses, precision agriculture, and all forms of transportation (space stations, aviation, maritime, railroad, and mass transit).

23.6 SUMMARY

This chapter presents an example of application of big databases obtained in the Earth observations. Our attention was focused on the low ionosphere, which is the medium containing a lot of information relevant to geo- and astrophysics as well as for different practical applications. Here we described databases which are obtained in the low ionospheric monitoring by VLF and LF radio waves and point out their importance for:

- detection of different astro- and geo-phenomena,
- modeling of the ionospheric plasma parameters, and
- practical application for possible natural disaster prediction and in telecommunication.

In this text we showed the major reasons why big databases obtained in observations by VLF/LF radio waves are required. Primarily, they relate to continuous observations of different geographical locations with good time resolutions. Also, various procedures for processing the observed bases and the necessity of their connection are explained.

Finally, we want to emphasize that each of these databases is unique in the sense that it refers to observation of certain areas. Bearing in mind the importance of the information they contain, it is very important to point out the need to increase the number of the related receivers (primarily on the Southern Earth hemisphere, where they are not too numerous), as well as the development of their mutual connection, in order to provide a wider and clearer picture of phenomena and processes they describe.

ACKNOWLEDGMENT

This work was supported by the COST project TD1403. Also, this study was carried out under the grants III 44002 and 176002 of the Ministry of Education, Science and Technological Development of the Republic of Serbia. The author is grateful to Vladimir M. Čadež for very useful suggestions and comments. Requests for the VLF data used for analysis can be directed to the corresponding author.

REFERENCES

Bajčetić, J., et al., 2015. Ionospheric D-region temperature relaxation and its influences on radio signal propagation after solar X-flares occurrence. Thermal Science 19 (Suppl. 2), S299–S309. https://doi.org/10.2298/TSCI141223084B.

Balan, N., et al., 2008. Magnetosphere-ionosphere coupling during the CME events of 07–12 November 2004. Journal of Atmospheric and Solar-Terrestrial Physics 70, 2101–2111. https://doi.org/10.1016/j.jastp.2008.03.015.

Biagi, P.F., et al., 2011. The European VLF/LF radio network to search for earthquake precursors: setting up and natural/man-made disturbances. Natural Hazards and Earth System Sciences 11 (2), 333–341. https://doi.org/10.5194/nhess-11-333-2011.

Bochev, A.Z., Dimitrova, I.I.A., 2003. Magnetic cloud and magnetosphere - ionosphere response to the 6 November 1997 CME. Advances in Space Research 32, 1981–1987. https://doi.org/10.1016/S0273-1177(03)90636-3.

Chau, J.L., Röttger, J., Rapp, M., 2014. PMSE strength during enhanced D region electron densities: Faraday rotation and absorption effects at VHF frequencies. Journal of Atmospheric and Solar-Terrestrial Physics 118, 113–118. https://doi.org/10.1016/j.jastp.2013.06.015.

Cheng, Zhenggang, Cummer, Steven A., 2005. Broadband VLF measurements of lightning-induced ionospheric perturbations. Geophysical Research Letters 32 (8), L08804. https://doi.org/10.1029/2004GL022187.

Clilverd, Mark A., et al., 2007. Energetic particle precipitation into the middle atmosphere triggered by a coronal mass ejection. Journal of Geophysical Research: Space Physics (ISSN 2156-2202) 112 (A12). https://doi.org/10.1029/2007JA012395.

Clilverd, M.A., et al., 2009. Remote sensing space weather events: Antarctic-Arctic Radiation-belt (Dynamic) Deposition-VLF Atmospheric Research Konsortium network. Space Weather 7, S04001. https://doi.org/10.1029/2008SW000412.

Cohen, M.B., Inan, U.S., Paschal, E.W., 2010. Sensitive broadband ELF/VLF radio reception with the AWESOME instrument. IEEE Transactions on Geoscience and Remote Sensing 48, 3–17. https://doi.org/10.1109/TGRS.2009.2028334.

Cohen, M.B., Marshall, R.A., 2012. ELF/VLF recordings during the 11 March 2011 Japanese Tohoku earthquake. Geophysical Research Letters 39 (11). https://doi.org/10.1029/2012GL052123.

Collier, A.B., et al., 2011. Source region for whistlers detected at Rothera, Antarctica. Journal of Geophysical Research 116, A03219. https://doi.org/10.1029/2010JA016197.

Cresswell-Moorcock, Kathy, et al., 2015. Techniques to determine the quiet day curve for a long period of subionospheric VLF observations. Radio Science (ISSN 1944-799X) 50 (5), 453–468. https://doi.org/10.1002/2015RS005652.

Ferguson, J.A., 1998. Computer Programs for Assessment of Long-Wavelength Radio Communications, Version 2.0. Space and Naval Warfare Systems Center, San Diego, CA.

Fishman, G.J., Inan, U.S., 1988. Observation of an ionospheric disturbance caused by a gamma-ray burst. Nature 331, 418–420. https://doi.org/10.1038/331418a0.

Grubor, D.P., Šulić, D.M., Žigman, V., 2008. Classification of X-ray solar flares regarding their effects on the lower ionosphere electron density profile. Annales Geophysicae 26, 1731–1740. https://doi.org/10.5194/angeo-26-1731-2008.

Hayakawa, Masashi, 2007. VLF/LF radio sounding of ionospheric perturbations associated with earthquakes. Sensors (ISSN 1424-8220) 7 (7), 1141–1158. https://doi.org/10.3390/s7071141.

Heki, K., Ping, J., 2005. Directivity and apparent velocity of the coseismic ionospheric disturbances observed with a dense GPS array. Earth and Planetary Science Letters 236, 845–855. https://doi.org/10.1016/j.epsl.2005.06.010.

Inan, U.S., Golkowski, M., et al., 2007a. Subionospheric VLF observations of transmitter-induced precipitation of inner radiation belt electrons. Geophysical Research Letters 34 (2), L02106. https://doi.org/10.1029/2006GL028494.

Inan, U.S., Lehtinen, N.G., et al., 2007b. Massive disturbance of the daytime lower ionosphere by the giant γ-ray flare from magnetar SGR 1806-20. Geophysical Research Letters 34, L08103. https://doi.org/10.1029/2006GL029145.

Isaev, N.V., et al., 2002. Electric field enhancement in the ionosphere above tropical storm region. In: Seismo Electromagnetics: Lithosphere - Atmosphere - Ionosphere Coupling. TERRAPUB, Tokyo, pp. 313–315.

Kolarski, A., Grubor, D., Šulić, D., 2011. Diagnostics of the solar X-flare impact on lower ionosphere through seasons based on VLF-NAA signal recordings. Baltic Astronomy 20, 591–595. https://doi.org/10.1515/astro-2017-0342.

Kumar, Sushil, et al., 2017. Perturbations to the lower ionosphere by tropical cyclone Evan in the South Pacific Region. Journal of Geophysical Research: Space Physics (ISSN 2169-9402) 122 (8), 8720–8732. https://doi.org/10.1002/2017JA024023.

Lei, J., et al., 2005. Variations of electron density based on long-term incoherent scatter radar and ionosonde measurements over Millstone Hill. Radio Science 40, RS2008. https://doi.org/10.1029/2004RS003106.

McRae, W.M., Thomson, N.R., 2004. Solar flare induced ionospheric D-region enhancements from VLF phase and amplitude observations. Journal of Atmospheric and Solar-Terrestrial Physics 66, 77–87. https://doi.org/10.1016/j.jastp.2003.09.009.

Miyaki, K., Hayakawa, M., Molchanov, O.A., 2002. The role of gravity waves in the lithosphere - ionosphere coupling, as revealed from the subionospheric LF propagation data. In: Seismo Electromagnetics: Lithosphere - Atmosphere - Ionosphere Coupling. TERRAPUB, Tokyo, pp. 229–232.

Molchanov, O.A., et al., 1998. Precursory effects in the subionospheric VLF signals for the Kobe earthquake. Physics of the Earth and Planetary Interiors (ISSN 0031-9201) 105 (3–4), 239–248. https://doi.org/10.1016/S0031-9201(97)00095-2.

Molchanov, O.A., Hayakawa, M., Miyaki, K., 2001. VLF/LF sounding of the lower ionosphere to study the role of atmospheric oscillations in the lithosphere-ionosphere coupling. Advances in Polar Upper Atmosphere Research (ISSN 1345-1065) 15, 146–158. https://ci.nii.ac.jp/naid/110000037544/en/.

Molchanov, O.A., Hayakawa, M., Rafalsky, V.A., 1995. Penetration characteristics of electromagnetic emissions from an underground seismic source into the atmosphere, ionosphere, and magnetosphere. Journal of Geophysical Research: Space Physics 100 (A2), 1691–1712. https://doi.org/10.1029/94JA02524.

Morfitt, D.G., Shellman, C.H., 1976. MODESRCH, an improved computer program for obtaining ELF/VLF mode constants in an Earth-ionosphere waveguide. Naval Electronics Laboratory Center, USA.

Nina, A., Čadež, V.M., 2013. Detection of acoustic-gravity waves in lower ionosphere by VLF radio waves. Geophysical Research Letters (ISSN 1944-8007) 40 (18), 4803–4807. https://doi.org/10.1002/grl.50931.

Nina, A., Čadež, V.M., 2014. Electron production by solar Ly-α line radiation in the ionospheric D-region. Advances in Space Research 54 (7), 1276–1284. https://doi.org/10.1016/j.asr.2013.12.042.

Nina, Aleksandra, Čadež, Vladimir M., et al., 2017a. Diagnostics of plasma in the ionospheric D-region: detection and study of diferent ionospheric disturbance types. The European Physical Journal D (ISSN 1434-6079) 71 (7), 189. https://doi.org/10.1140/epjd/e2017-70747-0.

Nina, A., Čadež, V.M., et al., 2018. Analysis of the relationship between the solar X-ray radiation intensity and the D-region electron density using satellite and ground-based radio data. Solar Physics 293, 64. https://doi.org/10.1007/s11207-018-1279-4.

Nina, A., Čadež, V., Bajćetić, J., et al., 2017. Responses of the ionospheric D-region to periodic and transient variations of the ionizing solar Lyα radiation. Journal of the Geographical Institute "Jovan Cvijic" SASA 67 (3), 235–248. https://doi.org/10.2298/IJGI1703235N.

Nina, A., Čadež, V., Srećković, V.A., et al., 2011. The influence of solar spectral lines on electron concentration in terrestrial ionosphere. Baltic Astronomy 20, 609–612. https://doi.org/10.1515/astro-2017-0346.

Nina, A., Čadež, V., Srećković, V., et al., 2012a. Altitude distribution of electron concentration in ionospheric D-region in presence of time-varying solar radiation flux. Nuclear Instruments & Methods in Physics Research. Section B 279, 110–113. https://doi.org/10.1016/j.nimb.2011.10.019.

Nina, A., Čadež, V., Šulić, D., et al., 2012b. Effective electron recombination coefficient in ionospheric D-region during the relaxation regime after solar flare from February 18, 2011. Nuclear Instruments & Methods in Physics Research. Section B 279, 106–109. https://doi.org/10.1016/j.nimb.2011.10.026.

Nina, Aleksandra, Radovanović, Milan, et al., 2017b. Low ionospheric reactions on tropical depressions prior hurricanes. Advances in Space Research (ISSN 0273-1177) 60 (8), 1866–1877. https://doi.org/10.1016/j.asr.2017.05.024.

Nina, Aleksandra, Simić, Saša, et al., 2015. Detection of short-term response of the low ionosphere on gamma ray bursts. Geophysical Research Letters (ISSN 1944-8007) 42 (19), 8250–8261. https://doi.org/10.1002/2015GL065726.

Pacini, Alessandra Abe, Raulin, Jean-Pierre, 2006. Solar X-ray flares and ionospheric sudden phase anomalies relationship: a solar cycle phase dependence. Journal of Geophysical Research: Space Physics 111 (A9). https://doi.org/10.1029/2006JA011613.

Peter, W.B., Inan, U.S., 2005. Electron precipitation events driven by lightning in hurricanes. Journal of Geophysical Research: Space Physics 110, A05305. https://doi.org/10.1029/2004JA010899.

Pham Thi Thu, H., Amory-Mazaudier, C., Le Huy, M., 2011. Time variations of the ionosphere at the northern tropical crest of ionization at Phu Thuy, Vietnam. Annales Geophysicae 29, 197–207. https://doi.org/10.5194/angeo-29-197-2011.

Price, C., Yair, Y., Asfur, M., 2007. East African lightning as a precursor of Atlantic hurricane activity. Geophysical Research Letters 34, L09805. https://doi.org/10.1029/2006GL028884.

Pulinets, S., Boyarchuk, K., 2004. Ionospheric Precursors of Earthquakes. Springer, Berlin.

Radovanović, M., 2018. Investigation of solar influence on the terrestrial processes: activities in Serbia. Journal of the Geographical Institute "Jovan Cvijic" SASA 68 (1), 149–155. https://doi.org/10.2298/IJGI1801149R.

Raulin, J.-P., Bertoni, F.C.P., et al., 2010. Solar flare detection sensitivity using the South America VLF Network (SAVNET). Journal of Geophysical Research: Space Physics 115, A07301. https://doi.org/10.1029/2009JA015154.

Raulin, J.-P., Correia de Matos David, P., et al., 2009. The South America VLF NETwork (SAVNET). Earth, Moon, and Planets 104, 247–261. https://doi.org/10.1007/s11038-008-9269-4.

Salut, M.M., et al., 2013. On the relationship between lightning peak current and Early VLF perturbations. Journal of Geophysical Research: Space Physics 118 (11), 7272–7282. https://doi.org/10.1002/2013JA019087.

Scherrer, D., et al., 2008. Distributing space weather monitoring instruments and educational materials worldwide for IHY 2007: the AWESOME and SID project. Advances in Space Research 42, 1777–1785. https://doi.org/10.1016/j.asr.2007.12.013.

Schmitter, E.D., 2013. Modeling solar flare induced lower ionosphere changes using VLF/LF transmitter amplitude and phase observations at a midlatitude site. Annales Geophysicae 31 (4), 765–773. https://doi.org/10.5194/angeo-31-765-2013.

Selvakumaran, R., et al., 2015. Solar flares induced D-region ionospheric and geomagnetic perturbations. Journal of Atmospheric and Solar-Terrestrial Physics (ISSN 1364-6826) 123, 102–112. https://doi.org/10.1016/j.jastp.2014.12.009.

Silber, Israel, Price, Colin, 2017. On the use of VLF narrowband measurements to study the lower ionosphere and the mesosphere–lower thermosphere. Surveys in Geophysics 38 (2), 407–441. https://doi.org/10.1007/s10712-016-9396-9.

Sorokin, V.M., Isaev, N.V., et al., 2005. Strong DC electric field formation in the low latitude ionosphere over typhoons. Journal of Atmospheric and Solar-Terrestrial Physics (ISSN 1364-6826) 67 (14), 1269–1279. https://doi.org/10.1016/j.jastp.2005.06.014.

Sorokin, V.M., Yaschenko, A.K., et al., 2006. DC electric field formation in the mid-latitude ionosphere over typhoon and earthquake regions. In: Recent Progress in Seismo Electromagnetics and Related Phenomena. Physics and Chemistry of the Earth. Parts A/B/C (ISSN 1474-7065) 31 (4), 454–461. https://doi.org/10.1016/j.pce.2005.09.001.

Srećković, V., et al., 2017. The effects of solar activity: electrons in the terrestrial lower ionosphere. Journal of the Geographical Institute "Jovan Cvijic" SASA 67 (3), 221–233. https://doi.org/10.2298/IJGI1703221S.

Strelnikova, I., Rapp, M., 2010. Studies of polar mesosphere summer echoes with the EISCAT VHF and UHF radars: information contained in the spectral shape. Advances in Space Research 45, 247–259. https://doi.org/10.1016/j.asr.2009.09.007.

Thomas, J.N., et al., 2010. Polarity and energetics of inner core lightning in three intense North Atlantic hurricanes. Journal of Geophysical Research: Space Physics 115, A00E15. https://doi.org/10.1029/2009JA014777.

Thomson, N.R., 1993. Experimental daytime VLF ionospheric parameters. Journal of Atmospheric and Terrestrial Physics 55, 173–184. https://doi.org/10.1016/0021-9169(93)90122-F.

Todorović Drakul, M., et al., 2016. Behaviour of electron content in the ionospheric D-region during solar X-ray flares. Serbian Astronomical Journal 193, 11–18. https://doi.org/10.2298/SAJ160404006T.

Utada, H., et al., 2011. Geomagnetic field changes in response to the 2011 off the Pacific Coast of Tohoku Earthquake and Tsunami. Earth and Planetary Science Letters 311, 11–27. https://doi.org/10.1016/j.epsl.2011.09.036.

Voss, H.D., et al., 1998. Satellite observations of lightning-induced electron precipitation. Journal of Geophysical Research 103, 11725–11744. https://doi.org/10.1029/97JA02878.

Wait, J.R., Spies, K.P., 1964. Characteristics of the Earth-ionosphere waveguide for VLF radio waves. NBS Technical Note 300. National Bureau of Standards, Boulder, CO.

Xiao, Zuo, et al., 2007. Morphological features of ionospheric response to typhoon. Journal of Geophysical Research: Space Physics (ISSN 2156-2202) 112 (A4), A04304. https://doi.org/10.1029/2006JA011671.

Žigman, V., Grubor, D., Šulić, D., 2007. D-region electron density evaluated from VLF amplitude time delay during X-ray solar flares. Journal of Atmospheric and Solar-Terrestrial Physics 69, 775–792. https://doi.org/10.1016/j.jastp.2007.01.012.

CHAPTER 24

Influence on Life Applications of a Federated Astro-Geo Database

CHRISTIAN MULLER, DR

24.1 INTRODUCTION

24.1.1 History of the Influence of Geophysical Parameters on Health

The scientific study of the influence of environmental parameters on life and human health began with Hippocrates of Kos when it was discovered that hygiene, food, and sunlight when well controlled led to an improvement in the general condition of an individual. The Hippocratic method was based on an examination of the patient including his/her geographical location, climate, age, gender, habits, and diet (Kleisiaris et al., 2014). These point already to a geographical information system (GIS), which is an element of large data that are explicit in the Hippocrates (400 BCE) treaty: "Airs, Waters, Places." Aristotle introduces classifications in biology as he did for the other sciences, introducing also an element of Big Data. Before the Pasteurian development of microbiology, it was thought that diseases were caused by miasmas, which were like the subtle fluids of the pre-Maxwellian physics. The relation between these miasmas and dirty environments and bad weather was empirically discovered and was exemplified by the construction of the Hippocrates hospital in Kos, with permanently fair weather, view on the sea, and cleanliness of the Asclepios priests, who were always dressed in white. They practiced also other aspects of medicine, including surgery and prescription of medication. See Fig. 24.1.

The revival of Hippocratic theories during the renaissance led the European physicians to have an interest in meteorology. These medical doctors unfortunately resorted also to astrology to such a point that Galileo taught astronomy in Padua essentially to medical students who used horoscopes as part of their diagnostics (Fermi and Bernardini, 2003). It is in this context, in the 18th century, that medical doctors began to consider electricity without knowing what it was and attempted the use of "animal magnetism" to perform curative actions. In the second half of the 19th century, the progresses of the scientific method brought more rationality to medicine. Especially the Pasteurian view that the prevention of most diseases went through the identification of their agents and transmission vectors has led to the cure of hundreds of pathologies in plants, animals, and humans. The 19th century brought also a form of Big Data by the health statistics related to the selection of conscripts in the European armies; also, systematic mortality and morbidity reports began to be collected and allowed comparison between diseases and behavioral or environmental elements. Beginning at the end of the 20th century, the electronic monitoring of hospital patients and emergency services statistics added new elements.

Unfortunately, the preservation of hospital electronic records is still far from being a standard or even an accepted procedure and it is almost certain that correlations between health and geophysical, astronomical, behavioral, and sociological parameters are still to be discovered.

24.2 METEOROLOGY AND CLIMATE (TEMPERATURE, HUMIDITY, ETC.) AND APPLICATION TO DISEASE PROPAGATION

The entire ecosystem is dependent on the timing of events constrained by meteorology. Pollination depends for example on insects; if blooming occurs before the insects are present in a sufficient number, the entire cycle of plant reproduction is perturbed. In the same respect, if migratory birds arrive before the insects are present, they also miss their reproductive cycle due to a lack of food. Climatological parameters thus influence local situations, and this can only be assessed by the cross-comparison of very different data sources.

The priority was the study and forecast of the presence of disease vectors as the mosquitoes responsible for the development of malaria. Their presence depends on temperature, humidity, surface water, and altitude. Malaria itself is also dependent on the human pop-

Knowledge Discovery in Big Data from Astronomy and Earth Observation. https://doi.org/10.1016/B978-0-12-819154-5.00036-9

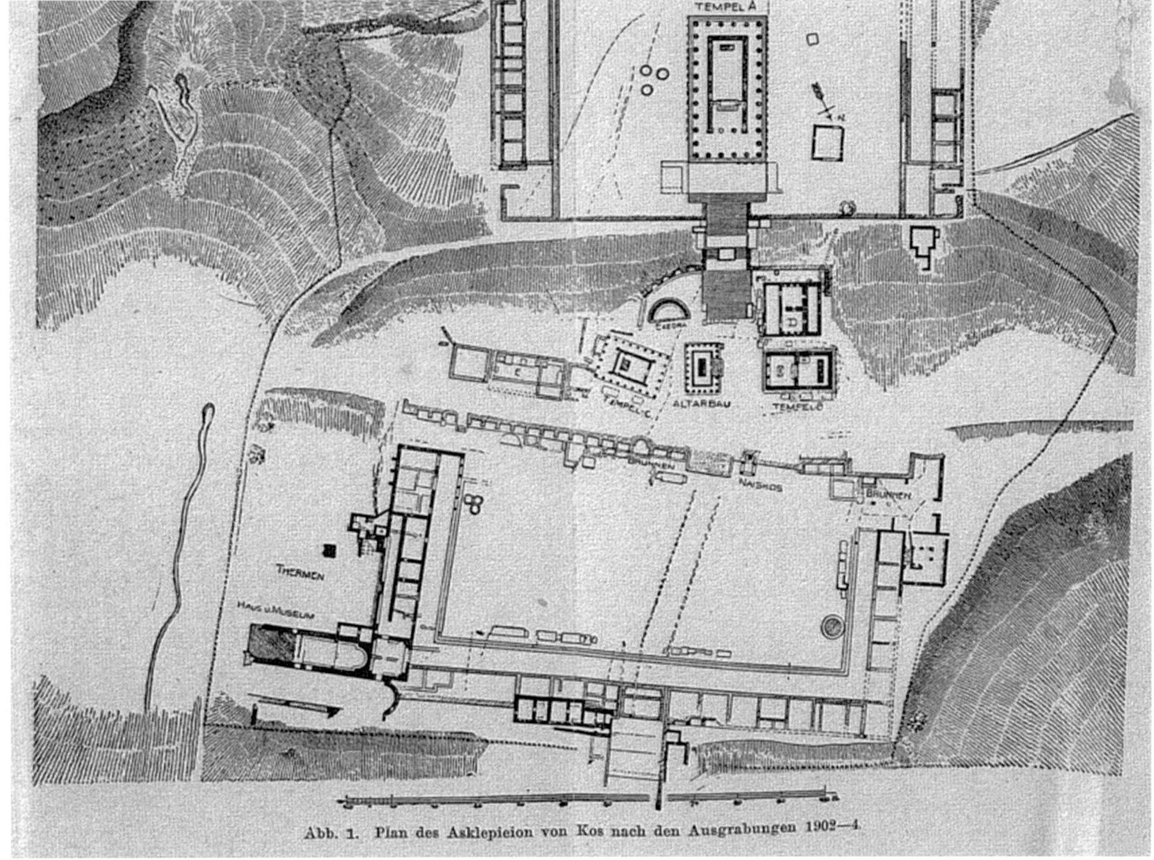

FIG. 24.1 Drawing of the Asklepion of Kos following the excavations of 1902–1904 (Herzog, 1907). Asklepions were the ancestors of modern hospitals. The cure involved purification baths and a special diet, empirically eliminating disease vectors. It involved also exercise and psychological aspects as the observation of dreams.

ulation; this requires the collection of a considerable amount of data, which became possible in the satellite era.

An early example is given by the treatment of Africa by Rogers et al. (2002), where low and medium spatial resolution sensors are used to make an elaborate modeling of the different parameters affecting malaria.

Higher spatial resolution leads to a local mapping of the propagation zones even in scarcely populated areas near the Southern border of French Guyana (Fig. 24.2, Adde et al., 2016). This kind of mapping will definitively require Big Data if extended to the entire malaria-infested part of the planet. Figs. 24.3 and 24.4 present the World Health Organization 2016 report on the status of malaria; more than 200 million new cases appear every year, around 90% of which occur in Africa. Europe and North Africa are clearly close to having eradicated the disease. The surface of the affected regions is however huge compared to the French Guyana example of Fig. 24.2, and a generalized high-resolution treatment of the whole area would mean an unprecedented dataset. The same rationale applies when a population of around 1000 persons is compared to the several billions in the countries where the disease is still endemic.

Malaria is the most evident example of an environmentally controlled infectious disease, but the same rationale applies to other water-borne and vector-borne diseases, as reviewed by Singh et al. (2015). Climatic changes might extend the zone where the vector can be

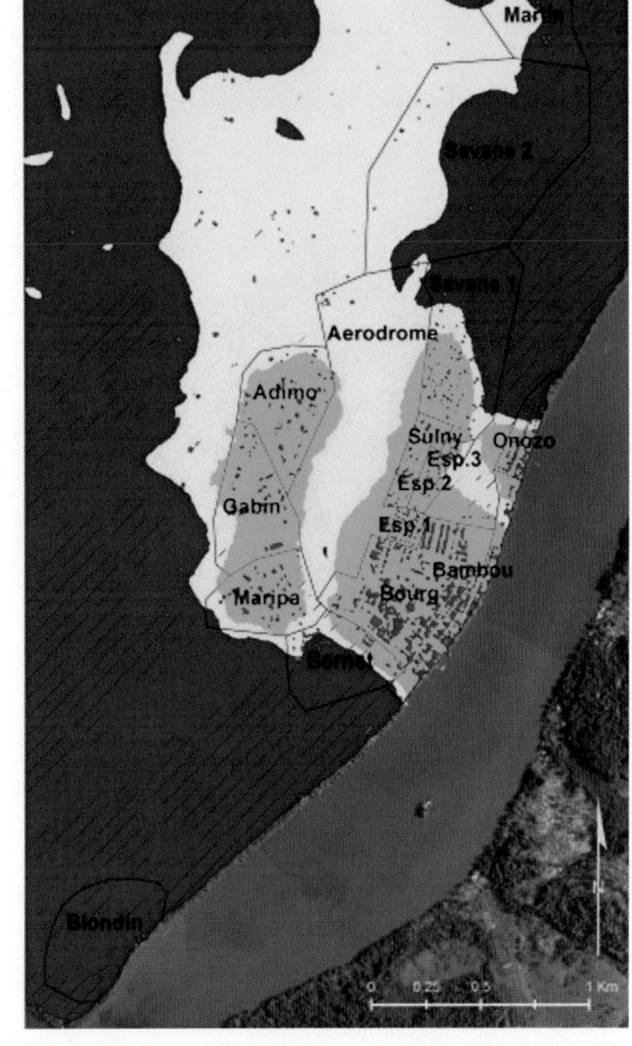

FIG. 24.2 Mosquito modeling of a French Guyana settlement along the Oyapock river using high-resolution satellite imagery. Individual housings are visible. High- and medium-density zones are defined with a resolution better than 10 m.

active as was already observed for the West Nile virus in Europe and North America. For each case, the specific conditions could be monitored from space using existing Earth observation satellites. The combination of geographical information systems (GISs), environment and climate, entomology, microbiology, veterinary medicine, and social sciences requires Big Data from various sources to produce epidemic forecasts and thus deploy protective and curative strategies.

24.3 INFLUENCE OF EXTRATERRESTRIAL SOURCES: SOLAR ACTIVITY, GALACTIC COSMIC RAYS, AND GEOMAGNETISM

The first inference of an influence of the sun on a biology-related indicator was the study of the correlation between sunspots and the price of wheat on the Windsor market by William Hershell (1801) just after the first statistics application to economics by Adam

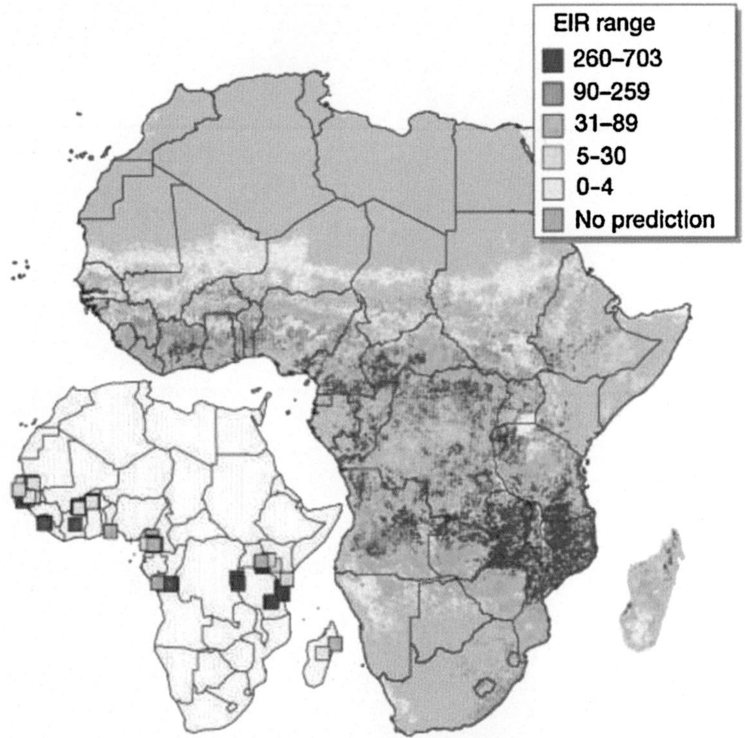

FIG. 24.3 Satellite-derived predictions of entomological malaria inoculation rate (EIR) in Africa. EIR data were grouped into five approximately equal-sized classes of mean levels of malaria challenge. Insufficient training data were available to define EIR in those parts of the continent marked gray (Rogers et al., 2002). Training is the ground truth or validation process that permits to control the quality of data deduced from satellite observations.

Smith: "The result of this review of the foregoing five periods is, that, from the price of wheat, it seems probable that some temporary scarcity or defect of vegetation has generally taken place, when the sun has been without those appearances which we surmise to be symptoms of copious emission of light and heat." Hershell did not discover the 11 year cycle in this analysis as his series still included the Maunder minimum of the 17th century which dominated the 11 year cycle. Schwabe (1844) was the first to put it in evidence after three textbook solar cycles. The Hershell anti-correlation was met with criticism since its first publication and revisited in an appendix of the Carrington (1863) analysis of sunspots. Besides being a superior solar observer, Richard Carrington was also a brewer and had his income tied to the cereals price. He concluded that in the 19th century, wheat prices, and especially the import duties, were dependent on tax laws passed by the British parliament. These taxes were amounting to import subsidies in order to have British farmworkers leave their job to participate in the new industrial expansion in the cities and be sufficiently fed with minimal wages. However, the work of Hershell after more than 200 years is still open for discussion both in the study of the influence of geophysical parameters on economy and in the use of statistics to determine causality (see, for example, Love, 2013). These points are at the basis of the reflection concerning the use of Big Data in current forecasts.

The relations between solar activity, magnetism, and galactic cosmic rays were discovered later in the 19th and 20th century. Large scale biological effects are still under discussion; the 1859 large solar flare event known as the Carrington effect led to significant perturbations of the telegraphic network, but any influence on biological parameters does not show up in the morbidity and mortality record with the exception of an increase in the birth rate in the Paris region nine months after the event, which was at the time attributed to the weather conditions (Muller, 2014). See Fig. 24.5.

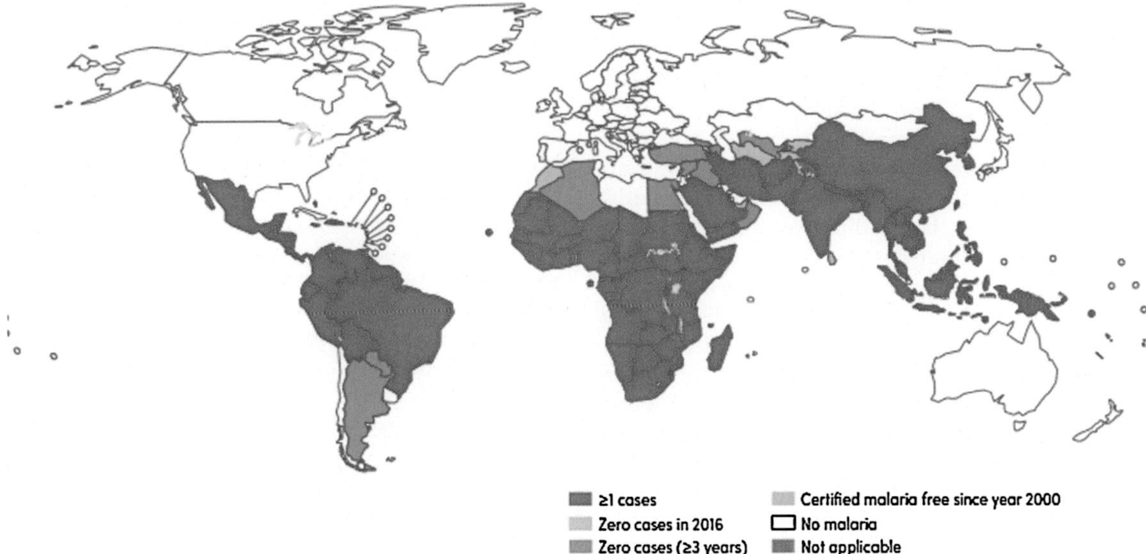

FIG. 24.4 Countries and territories with indigenous cases in 2000 and their status by 2016. Countries with zero indigenous cases over at least the past three consecutive years are eligible to request certification of malaria-free status from the WHO. All countries in the WHO European Region reported zero indigenous cases in 2016. Kyrgyzstan and Sri Lanka were certified malaria-free in 2016. Source: WHO database (latest full report: WHO, 2015).

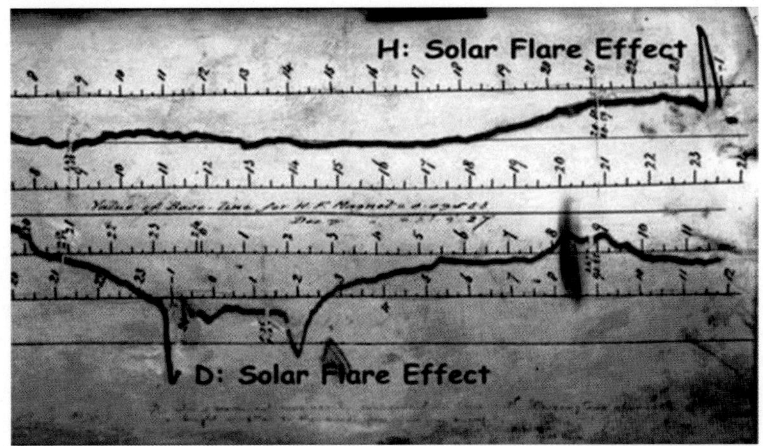

FIG. 24.5 Trace of the horizontal component of the Earth's magnetic field from Kew Observatory for 1–2 September 1859, showing the magnetic crochet at 11:15 UTC on 1 September and the great geomagnetic storm that followed 17.6 h later and drove the record off-scale (Cliver and Svalgaard, 2004). The geomagnetic storm generated auroral displays that were observed as close to the equator as the Caribbean islands.

In the 20th century, several less important solar storm events occurred and had consequences on electric transport lines as well as on communication networks, but again, the relation with public health records has not been universally studied. However, studies related to flight and space medicine have shown, especially in the Russian hearth morbidity statistics, a relation between space weather and hearth-related emergencies (Breus et al., 1995). The verification of these correlations would necessitate a worldwide study of hospital electronic patient records at the time of a solar event.

Galactic cosmic rays anticorrelate with solar activity; when the sun is low in activity, the heliosphere is weaker and thus the extrasolar flux is more important. The health risks are rarely considered, except for astronauts and flight crews, for whom several studies exist, and even a European Union directive imposes to inform crews of the radiation doses occurring during flights. Up to now, no direct causality relation between high-altitude flight and disease has been established (Lim, 2002). However, further studies might lead to regulations limiting the number of flight hours for specific categories such as pregnant women or imposing a lowering of flight altitude. In the case of astronauts, the small number of the cohort, a little more than 500, cannot lead to a definitive conclusion. A recent review (Hodkinson et al., 2017) lists the radiation risks encountered by the current astronauts. However, in the case of flight to Mars and beyond, the situation becomes more complex as for this long duration, statistical data are not a sufficient element to characterize the risk and both shielding procedures and pharmaceutical protection are under study.

24.4 SOLAR UV AND LIFE

Until the middle of the 20th century, medicine followed the Hippocrates hypothesis that exposition to solar radiation was beneficial. However, these effects could only be quantified when physicists developed both lamps and UV sensors and used them in a hospital environment (Latarjet, 1937). Subsequent studies showed damages comparable to the effects of ionizing radiation on plants and test animals and ended up establishing a relation between UV and the genesis of human skin cancer (Latarjet and Wahl, 1945; Latarjet, 1959). The progress of UV sensors led not only to this new development on the biological effects of solar radiation but also to studies of what was called at the time "the UV limit of the solar spectrum." These were precursors of the determination of the UV effects on atmospheric chemistry and its climate effects.

From a biological point of view, UV radiation is divided between UV-A (400–315 nm), UV-B (315–280 nm), and UV-C (280–100 nm). It is well known that UV-A is necessary in moderate doses to initiate metabolic processes as for example the production of vitamin D in the human skin.

Stratospheric ozone entirely filters UV-C and the largest part of UV-B. Ozone is not the only effective UV filter; carbon dioxide filters all UV below 220 nm, but a reduction in ozone would expose the biosphere to UV-B with consequences on both plants and animals.

The decrease of the ozone layer observed since the 1970s and especially since the Antarctic spring "ozone hole" has led to the 1987 Montréal protocol on ozone depleting species. Solar variations play a role in the equilibrium of the ozone layer, but this effect is buffered by the fact that atomic oxygen produced by the dissociation of molecular oxygen more than compensates for the destruction of ozone by UV radiation of longer wavelengths. Thus, the solar climate has very little effect on the UV radiation received at the Earth's surface; however, again, scientific research should never overlook the effects of perturbations and never consider a part of the Earth's system as definitively autostabilizing.

The Montréal protocol, while reversing the trend in ozone depletion, had very little effect on skin cancer incidence, which keeps increasing (for example, https://melanoma.canceraustralia.gov.au/statistics, 2018). Again, this shows the difficulties in treating causality when diverse effects intervene in the final result.

24.5 APPLICATIONS TO AGRICULTURE

The study of agriculture parameters from space began as early as the first military reconnaissance satellite as the worldwide status of crops gave important information on the food market in the near future. Early studies were made on both the US and the Soviet side but proved inconclusive until the concept of validation was introduced; validation consists in the comparison between ground truth and forecasts on a few carefully chosen test sites. A lot of these early attempts have been lost until the NASA civilian LANDSAT program began in 1972, evolving from an experimental digital multispectral mapping program to a fully operational Earth resource monitoring system. It is now managed by the United States Geological Survey, it has launched its eighth satellite, and other agencies have built similar programs with elements of compatibility with LANDSAT. The agriculture applications have been recently reviewed by Leslie et al. (2017) and range from academic research to operation products directly applied to actual farm operations. It is for example at the basis of the CropScape data portal of the US Department of Agriculture, as illustrated in Fig. 24.6.

CropScape can be accessed at https://nassgeodata.gmu.edu/CropScape/ (Han et al., 2012) and gives access to a complete mapping of several crops, which could be zoomed in on with a 200 m spatial resolution. These data can thus be exploited both locally and globally.

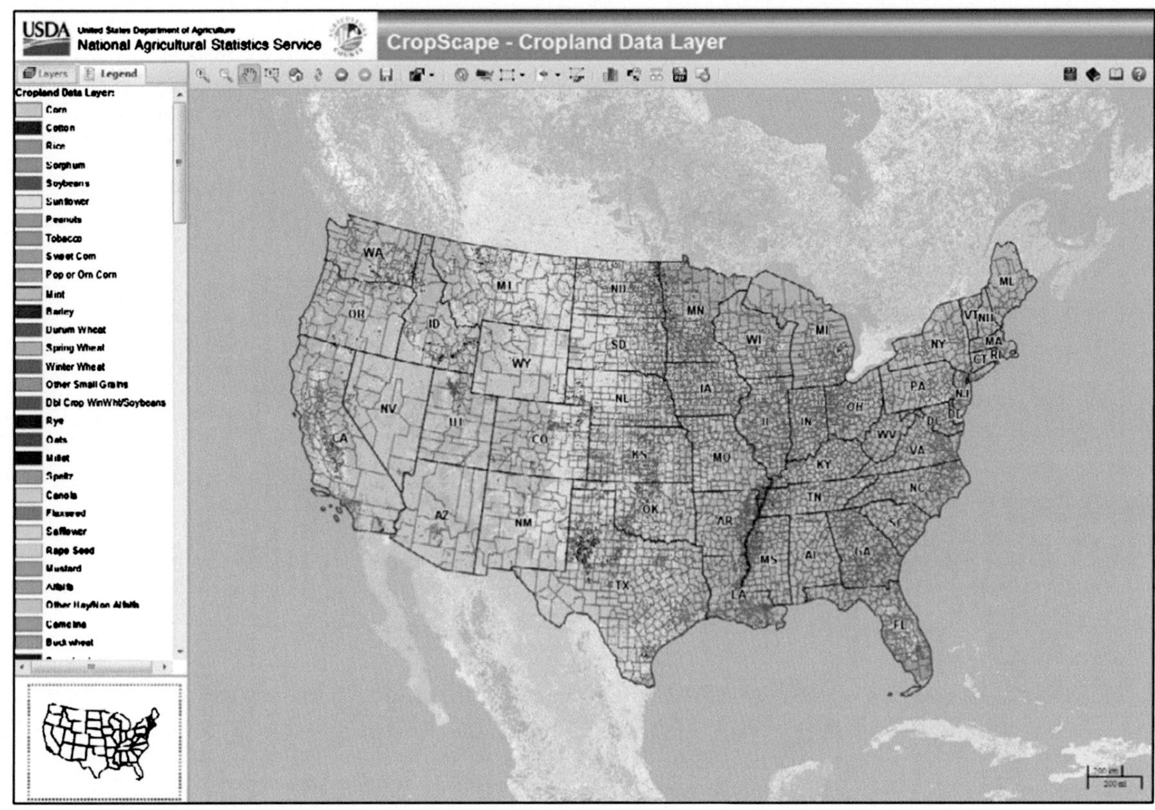

FIG. 24.6 The CropScape data portal, which allows for the display of the Cropland Data Layer by year from 1997 to the present; derived products such as the Cultivated and Crop Frequency Layers; and supporting layers like boundaries, water, and roads (USDA National Agricultural Statistics Service, 2017).

A color code identifies the number of years in which crops have been exploited and is thus a precious tool for forecasts. Zone maps can also be used to assess the efficiency of agricultural practices as fertilizer application and irrigation.

The US federal agricultural services have an international branch called the Foreign Agricultural Service (FAS). Originally, this service searched potential markets for US agriculture; now it has involved in a collaboration agency with other space faring nations in the frame of the Global Agricultural Monitoring System (GAMS). It also provides data to the United Nations Food and Agricultural Organization (FAO).

Parallel efforts are accomplished by other space agencies. ESA uses its Earth observation satellites for agricultural applications. As an example, the ESA Soil Moisture and Ocean Salinity (SMOS) mission monitoring satellite is used in combination with NASA green vegetation data from the MODIS instrument to make forecasts of the conditions leading to the reproduction of locusts. See Fig. 24.7.

Early forecasts and detection of locust swarms ensure the deployment of countermeasures to preserve crops essential for human alimentation. The current forecasts made by CIRAD use a lower spatial resolution than the maximum available from the sensors. Big Data techniques taking full advantage of high resolution might in the future enable to destroy locusts at the constitution of groups.

Soon, precision agriculture will use satellite data to the level of the individual plant (European Parliamentary Research Service, 2016). This will represent a new upscaling of the Big Data in agriculture, including real-time operations in terms of irrigation, phytopharmacy, and even harvest. This new paradigm will mean a very large extension of what has been outlined here.

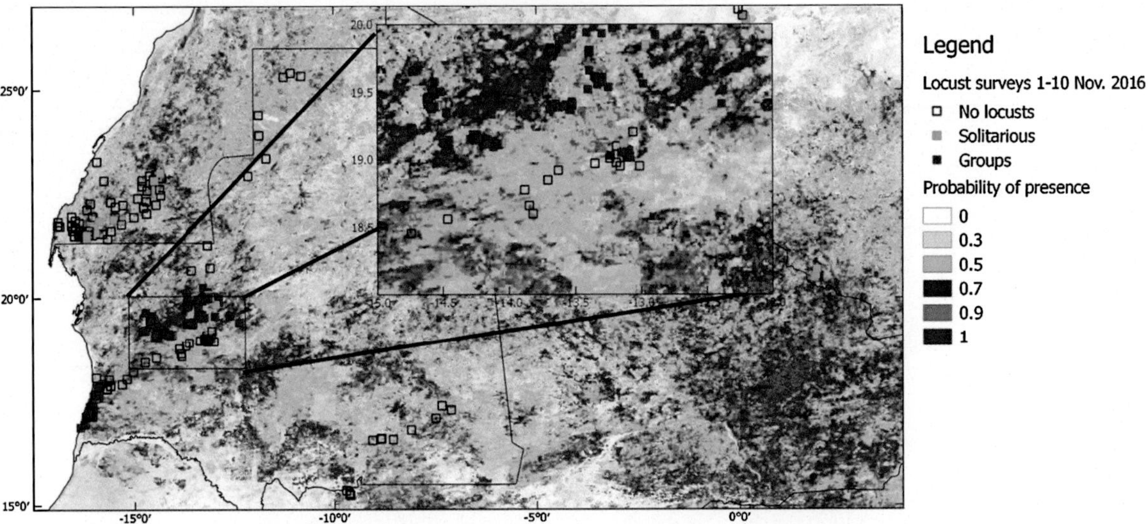

FIG. 24.7 Soil moisture data from the SMOS satellite and the MODIS instrument were used by the French CIRAD institute to create this map showing areas with favorable locust swarming conditions in red (FAO document, http://www.fao.org/news/story/en/item/895920/icode/).

24.6 CONCLUSIONS: FUTURE OF SPACE OBSERVATIONS, BIODIVERSITY, AND ASTROBIOLOGY

The different examples in agriculture and disease forecasts are going to expand rapidly in the coming years with the increase in sensor spatial resolution and repeat rates. Climatic change will modify the temperature and humidity conditions rapidly, as observed in the 20 last years. The adaptation of finely tuned ecological systems involving plants, insects, birds, and other animals is not evident and space data will be ideally positioned to observe the evolution and see if biodiversity is threatened. The threat comes from both a loss of synchronization between the different biological processes and new conditions favoring invasive species; only a large-scale inventory involving high-spatial resolution data will allow to quantify the changes in biodiversity defining new specific indices. Currently, several biodiversity indices have been proposed, and all depend on the statistics of the different populations in an ecological system. The issue is defining an inventory, the proportion of each of its elements, and applying a statistical formula to the results. The international definition of these indicators is linked to the United Nations Convention on Biodiversity, which is active since 1993 (https://www.cbd.int/convention/). A list of indicators is published on the convention's site (https://www.cbd.int/doc/strategic-plan/strategic-plan-indicators-en.pdf); each of them involves a different methodology and a few of them have not yet been identified. This field will necessarily need the techniques developed in the Big Data field for its development.

Finally, in a more tentative perspective, the discoveries of exoplanets keep increasing and their astronomical biomarkers analysis and inventories will come soon when spectroscopy of these distant worlds will become possible. Due to the very large number of planets, compared to the numbers of stars, this analysis will require techniques developed for Big Data, including artificial intelligence.

In these two cases, the techniques proposed in this book will prove fundamental to the study of life-related questions for both the Earth and the universe.

REFERENCES

Adde, A., Roux, E., Mangeas, M., Dessay, N., Nacher, M., Dusfour, I., Girod, R., Briolant, S., 2016. Dynamical mapping of Anopheles darlingi densities in a residual malaria transmission area of French Guiana by using remote sensing and meteorological data. PLoS ONE 11 (10), e0164685.

Breus, T., Cornelissen, G., Halberg, F., Levitin, A.E., 1995. Temporal associations of life with solar and geo-physical activity. Annales Geophysicae 13, 1211–1222. https://doi.org/10.1007/s00585-995-1211-8.

Carrington, R.C., 1863. Observations of the Spots on the Sun. Williams and Norgate, London, UK, pp. 1–248.

Cliver, E.W., Svalgaard, L., 2004. The 1859 solar–terrestrial disturbance and the current limits of extreme space weather activity. Solar Physics 224, 407–422. https://doi.org/10.1007/s11207-005-4980-z.

European Parliamentary Research Service, 2016. Precision agriculture and the future of farming in Europe. Scientific Foresight Study, PE 581.892. Brussels. Available at: http://www.europarl.europa.eu/RegData/etudes/STUD/2016/581892/EPRS_STU(2016)581892_EN.pdf.

Fermi, L., Bernardini, G., 2003. Galileo and the Scientific Revolution. Dover, Mineola, New York.

Han, W., Yang, Z., Di, L., Mueller, R., 2012. CropScape: a web service based application for exploring and disseminating US conterminous geospatial cropland data products for decision support. Computers and Electronics in Agriculture 84, 111–123. https://doi.org/10.1016/j.compag.2012.03.005.

Herzog, R., 1907. Aus dem Asklepieion von Kos. Archiv für Religionswissenschaft, vol. 10. Teubner, Leipzig, pp. 201–228.

Hippocrates, 400 BCE. Airs, Waters, Places. Available at MIT-classics: http://classics.mit.edu/Hippocrates/airwatpl.mb.txt.

Hodkinson, P.D., Anderton, R.A., Posselt, B.N., Fong, K.J., 2017. An overview of space medicine. British Journal of Anaesthesia 119 (suppl_1), i143–i153. https://doi.org/10.1093/bja/aex336.

Kleisiaris, C.F., Sfakianakis, C., Papathanasiou, I.V., 2014. Health care practices in ancient Greece: the Hippocratic ideal. Journal of Medical Ethics and History of Medicine 7, 6.

Latarjet, R., 1937. Le dosage des rayonnements ultraviolets utilisés en thérapeutique. Thèse de Doctorat: Sciences: Université de Lyon. Bosc et Riou, Lyon, p. 144.

Latarjet, R., 1959. Radiations in relation to carcinogenesis and mutation. In: Genetics and Cancer. University of Texas Press, pp. 119–132.

Latarjet, R., Wahl, R., 1945. Précisions sur l'inactivation des bacteriophages par les rayons ultraviolets. Annales de L'Institut Pasteur 71, 336–339.

Leslie, C.R., Serbina, L.O., Miller, H.M., 2017. Landsat and agriculture—Case studies on the uses and benefits of Landsat imagery in agricultural monitoring and production: U.S. Geological Survey. Open-File Report 2017–1034, p. 27.

Lim, M.K., 2002. Cosmic rays: are air crew at risk? Occupational and Environmental Medicine 59, 428–432.

Love, J.J., 2013. On the insignificance of Herschel's sunspot correlation. Geophysical Research Letters 40, 4171–4176. https://doi.org/10.1002/grl.50846.

Muller, C., 2014. The Carrington solar flares of 1859: consequences on life. Origins of Life and Evolution of the Biosphere 44 (3), 185–195. https://doi.org/10.1007/s11084-014-9368-3. Published online 2014 Oct 30.

Rogers, D.J., Randolph, S.E., Snow, R.W., Hay, S.I., 2002. Satellite imagery in the study and forecast of malaria. Nature 415 (6872), 710–715. https://doi.org/10.1038/415710a.

Schwabe, H., 1844. Sonnen-Beobachtungen im Jahre 1843. Astronomische Nachrichten 21, 233–236.

Singh, R., Ranjan, K., Verma, H., 2015. Satellite imaging and surveillance of infectious diseases. Journal of Tropical Diseases. https://doi.org/10.4172/2329-891X.1000S1-004.

WHO, 2015. World Malaria Report 2015. World Health Organization, Geneva. Available at: http://www.who.int/malaria/publications/world-malaria-report-2015/report/en/.

Index

A

Abell catalogue, 62
Absorption events, 372
Accuracy
  CCD, 198
  classification, 125, 241, 308, 309
  estimate, 198, 200, 202
  for classification, 310
  improvement, 392, 404
  measurements, 239
  sensors, 413
Active galactic nucleus (AGN), 66
  classification, 236
Active learning (AL), 228
Active region (AR), 356
Advance Land Observing Satellite (ALOS), 115
Advanced
  computing technologies, 159
  data management techniques, 159
  satellite technology, 405
  telescope equipment, 61
  visualization techniques, 109
Advanced Composition Explorer (ACE)
  data, 363
  satellite, 349
Advanced Cyberinfrastructure (ACI), 119
Advanced Earth Observing Satellite (ADEOS), 115
Aerological sounding, 406, 407, 412
Aerological sounding data, 406, 411
Aggregation data, 273
Airborne GNSS, 35
Airborne laser scanning (ALS), 33
Airborne Snow Observatory (ASO), 188
Amazon Web Service (AWS), 105
Ames Stereo Pipeline (ASP), 42
Ant colony optimization (ACO), 292
Antarctic stations network, 25
Antigen presenting cell (APC), 293
Apache Spark, 155, 160, 162, 242
Architecture and Data Committee (ADC), 109
Artificial immune system (AIS), 293
Artificial neural network (ANN), 127, 225, 251, 297, 308
ASCA satellite, 79
AstroGeoInformatics, 17
  ideas, 19
  projects, 18, 19
Astrographic Catalog (AC), 61
ASTROIDE, 165, 167, 168
Astronomical
  applications, 163, 216
  biomarkers, 442
  calculations, 41
  catalogues, 91, 96, 97, 232
  community, 43, 44, 49, 236
  data, 48, 50, 75, 88, 159, 167, 169, 232, 298, 299, 307, 320, 331
    center, 96
    cubes, 48
    mining tasks, 237
    processing, 331
    volumes, 75
  databases, 57, 88, 91, 165
  datasets, 236, 331
  experience, 50
  objects, 18, 57, 58, 63, 95, 173, 235
  objects database, 91
  objects redshifts, 236
  observations, 23, 65, 331
  phenomena, 43
  science, 173
  software, 48
  sources, 183, 235
  SSAP, 48
  surveys, 57, 77, 159, 232, 233
  systems, 168
  time series, 183
Astronomical Data Query Language (ADQL), 39
Astronomy
  gamma ray, 78
  interferometry, 42
  multiwavelength, 57, 307
  radio, 87
  surveys, 183
Astronomy Common Object Model (ASCOM), 44
Astroparticle satellites, 90
Astrophysics Data System (ADS), 96, 221
Astropy, 41, 45, 236
Atmospheric
  monitoring, 419
  sciences, 114
  state information, 405
Automated
  classification, 308, 310
  data reduction, 339
  detection, 332
  galaxy classification algorithms, 309
  galaxy spectral classification, 234
  morphological classification, 307, 310
  objects rejection, 332
Automatic
  classification, 268
  determination, 203
  galaxy classification, 234
Automatic Plate Measuring (APM), 73
  Galaxy Survey, 66

B

Backbone optimizer, 168
Background galaxies, 233
Balchik data, 359
Baryon Oscillation Spectroscopic Survey (BOSS), 67
Basic Formal Ontology (BFO), 13
Basic Linear Algebra Subprograms (BLAS), 155
BeppoSAX catalogues, 90
BeppoSAX satellite, 78
Berlin Big Data Center (BBDC), 114
Big Data Integrator (BDI), 10
Binary
  classification, 186, 231, 312
  morphological classification, 311
  stars, 186, 203
Braude Radioastronomical Observatory, 87
Bright
  galaxy, 64, 66
  galaxy catalogues, 95
  nonstellar objects, 61
  stars, 68
Brightest
  elliptical galaxies, 314
  galaxy, 62
  galaxy clusters, 81
Brightness
  equalization, 332, 334, 340
  in astronomy, 207
  measurements, 341
Broadband
  data, 422, 423, 428
  monitoring, 423
  observations, 425
Burst Alert Telescope (BAT), 81, 320, 426
Byurakan Observatory, 63

C
Canadian Advanced Network For Astronomical Research (CANFAR), 39
Canadian Astronomy Data Centre (CADC), 90
Canadian Earth Observation Network (CEONet), 107
Canadian Space Agency (CSA), 113
Cape Photographic Durchmusterung (CPD), 60
Catalina Sky Survey (CSS), 183
Catalogue
  clusters, 62
  data, 72
  galaxies, 315
  Hipparcos, 70
  quasars, 62
  Tycho, 70
Catalogue of Galaxies and Clusters of Galaxies (CGCG), 62
Catalogued
  celestial bodies, 91
  databases, 57, 307
  objects, 87
  values, 328
Catalyst optimizer, 168
CCD
  accuracy, 198
  astrograph catalogue, 329
  frames, 328, 329, 334, 337, 339, 340, 344
  frames series, 337
  surveys, 65
Celestial objects, 58, 59, 91, 167, 173, 340
Center for Astrophysics (CfA) catalogue, 64
Chandra Source Catalog (CSC), 81
Classification
  accuracy, 125, 241, 308, 309
  algorithms, 227, 307
  automated, 308, 310
  automatic, 268
  binary, 186, 231, 312
  errors, 307
  features, 186
  models, 9, 229, 237, 238
  morphological, 257, 308–310, 312
  network, 253
  objects, 18, 233
  problems, 227, 268, 312
  quasars, 233
  schemes, 16, 231
  score, 257
  signal, 268
  spectral, 127
  stars, 63, 233
  system, 234
  task, 8, 237–239
Cloud computing, 105, 112, 113, 117, 119, 137, 140, 145–147, 149, 154
Cloud infrastructure, 130
Cluster feature (CF), 241
Clusters, 62, 81, 88, 94, 140, 142, 145, 239, 241, 279, 280, 297, 333
  catalogue, 62
  galaxies, 62, 80, 81, 91–93, 235
  objects, 241
  stars, 59
CoLiTec software, 331–335, 337, 339–341, 344
  features, 332
  for automated detection, 332
CoLiTec virtual observatory, 332
Committee on Earth Observation Satellites (CEOS), 107, 108, 130
  data cube, 34
Compton Gamma Ray Observatory, 85
Computer-assisted design (CAD), 44
Continuous wavelet transform (CWT), 348, 367
Convolutional networks, 280
Convolutional neural network (CNN), 255–258, 279, 310
Coordinate reference systems (CRS), 41
Coronal mass ejection (CME), 349, 351, 424
Cosmic microwave background (CMB), 85, 313, 314
COST action BigSkyEarth, 367
CropScape data portal, 440
Cryospheric science domains, 119
Cyclone events, 431

D
DASCH project, 77
Data
  access, 106, 121, 139, 143, 148, 155, 162, 163, 165, 169, 374, 376
  access points, 112
  acquisition, 31, 34, 37, 268
  analysis, 160, 169, 332, 380, 387
  analysis tools, 267
  analytics, 12, 14, 103, 105, 119, 137, 153, 154, 169
  analytics methodologies, 36
  analytics software, 154
  astronomical, 48, 50, 75, 88, 159, 167, 169, 232, 298, 299, 307, 320, 331
  broadband, 422, 423, 428
  catalogue, 72
  centers, 96, 146–148, 178, 371, 386, 387
  channels, 273
  collection, 25, 31, 33, 251
  control, 332
  DPOSS, 184
EO, 103, 105, 107, 113–115, 117, 119, 122
from LEO satellites, 33
from NMDB, 381
geomagnetic, 347, 348, 355, 357–359, 361, 364
geosciences, 243
geospatial, 39, 40, 48, 103, 105, 107
GNSS, 385, 386, 405, 406
GPS, 406
ionospheric, 358, 369
lifecycle, 169
locality, 159
management, 113, 114, 137, 143, 147, 150, 165, 169
  process, 169
  system, 163
  technology, 163
meteorological, 24, 28, 406
mining, 1, 3–9, 14, 16, 104, 105, 119, 225, 236, 239, 242, 278, 331, 333, 347, 424, 431
  algorithms, 12, 15, 16, 276
  industry, 6
  methods, 3, 16
  process, 16
  project, 5, 9
  systems, 242
  tasks, 5, 9, 15, 16
  techniques, 1, 9, 16, 232
  tools, 5, 9, 234
  workflows, 14
model standards, 44
modeling, 31
NASA, 116
NMDB, 377, 381
observational, 254, 332, 388, 389, 406
organization, 162, 169
partition, 169
photometric, 341
processing, 10–12, 16, 18, 137, 150, 155, 159, 162, 339, 388, 426, 431
  operations, 12
  paradigm, 150
  task, 10, 11
project, 18
quality, 169, 372, 377, 382
quality control, 376
raw, 2, 3, 17, 69, 75, 90, 105, 119, 333, 357, 374
reduction, 331
redundancy, 374
samples, 259, 260
satellites, 111, 116, 441
schemas, 161
science, 104, 120, 254
science projects, 5
signal, 272, 273
sources, 161, 166, 169, 281, 435
stations, 374
storage, 33, 34

stream, 341
stream processing, 331
structure, 33, 161, 163, 164, 169, 272, 277
volume, 161, 169
Data Mining OPtimization Ontology (DMOP), 14
Database
characteristics, 163
for stars, 91
NMDB, 374
observational, 28
SDSS, 92
synoptic, 24
DCB data, 388
Decameter radio astronomy, 87
Decision Maker (DM), 284
Deep
learning, 154, 155, 186, 251, 253–256, 258, 268, 279, 281, 317, 348
algorithms, 253, 254, 279
approaches, 251, 258
methods, 186, 312
models, 253, 255, 262, 279
neural networks, 251, 254, 257
surveys, 80
Detection
automated, 332
objects, 335, 337
probability, 371
services, 385
time, 181
Determination
automatic, 203
tropospheric delay, 405, 407, 409, 411
Differential evolution (DE), 291
Differential GNSS (DGNSS), 406
Digital elevation model (DEM), 33, 114
Digital object identifier (DOI), 44
Digital Palomar Observatory Sky Survey (DPOSS), 72, 74, 183, 184
data, 184
observations, 184
Digital Sky Survey (DSS), 73
Digitized sky survey, 90
Dimensionality reduction, 125, 129, 185, 186, 232, 233, 235, 243, 261
Directed acyclic graph (DAG), 150, 159
Discrete anisotropy radiative transfer (DART), 124
Discrete Fourier transform (DFT), 207
Discrete wavelet transform (DWT), 124, 348, 368
Discriminator network, 261
Distant
active galaxies, 82
astronomical objects, 42
objects, 67
radio galaxies, 87
stars, 70
Distributed Active Archive Centers (DAAC), 105
Distributed stream computing platforms (DSCP), 176
DOGS project, 315
Dwarf galaxies, 61, 62
catalogue, 62
Dynamic time warping (DTW), 278

E

Earth, 31–33, 35, 42, 45, 104, 105, 112, 114, 117, 270, 329, 348–350, 354, 371, 377, 416, 419, 424, 428, 440, 442
halo, 356
lithosphere, 419
surface, 33, 431
Earth Observation (EO), 31, 33, 103, 104, 108, 116, 117, 130, 152, 154, 155, 159, 169, 347, 348, 431, 432
data, 103, 105, 107, 113–115, 117, 119, 122
data analytics, 115
data discovery, 112
datasets, 108
satellite missions series, 115
satellites, 103, 108, 110, 115, 121, 437, 441
systems, 120
Earth Observing System (EOS), 105, 118
EarthCARE, 112
EarthCube, 119
Earthquakes
activity, 35
analysis, 124
events, 430, 431
occurrences, 425
phenomena, 424
prediction, 35, 431
EarthServer
solution, 119
technology, 119
EBOSS project, 68
Eclipsing binary stars, 203, 216
EGRET catalogue, 78
Electromagnetic signal, 405
tropospheric delay, 413
Elliptical galaxies, 81, 300, 310, 312
brightest, 314
Enhanced Thematic Mapper (ETM), 122
ENVI standards, 44
ESO Schmidt telescope, 64, 74
Esper, 175–177, 179
documentation, 179
EPL, 177, 179
syntax, 177
EUMETSAT, 26, 28, 110, 112, 121
European Commission (EC), 33, 104, 109
European Space Agency (ESA), 33, 70, 71, 78–80, 85, 86, 90, 104, 130
European Space Agency Science Data Center (ESDC), 80
Event stream processing (ESP), 176
Events
earthquakes, 430, 431
solar, 425
Evolution strategies (ES), 290
Evolutionary programming (EP), 291
Evolutionary strategy (ES), 299
EXOSAT observatory, 79
EXOSAT satellite, 79
Extragalactic
objects, 59, 61, 62, 73, 74, 80, 91, 314, 317
surveys, 60
Extreme learning machine (ELM), 242
Extreme machine learning (EML), 301

F

Faint
moving objects, 344
moving objects automated detection, 332
objects, 236
stars, 77
False alarm probability (FAP), 209
Far Ultraviolet Spectroscopic Explorer (FUSE), 82
Fast Fourier transform (FFT), 191, 423
Features
classification, 186
CoLiTec software, 332
morphological, 309
signal, 274
Federal Geographic Data Committee (FGDC), 107
standardization activities, 107
standards, 107
Final Markov classes (FMC), 128
First Byurakan Survey (FBS), 63
Flexible Image Transport System (FITS), 45, 334
Forbush decrease (FD), 376
Foreign Agricultural Service (FAS), 441
Fourier transform (FT), 122, 191
Fully convolutional network (FCN), 280

G

Gaia catalogue, 71
Galactic cosmic ray (GCR), 371
Galaxies
automatic morphological classification, 308
bright, 64, 66
catalogues, 62, 315
clusters, 62, 80, 81, 91–93, 235
catalogues, 64, 73
databases, 81
radio observations, 92
distribution, 313, 317, 318
elliptical, 81, 300, 310, 312

environments, 66, 81
formation, 70, 307
groups, 62, 94
images, 310
massive, 94, 316
morphological, 310
morphological classifications, 234, 307, 308
morphology, 234, 307, 310, 312
observations, 61, 317
parameters, 312
properties, 66, 317
radio, 77, 235
redshift catalogues, 64
redshift survey, 66
sample, 66, 233, 308
SDSS, 307–312
spectra, 234
spectral features, 312
spiral, 310, 312, 314
superclusters catalogues, 64
surveys, 233, 315
Zwicky catalogue, 63
Galaxy and Mass Assembly (GAMA), 69
Galaxy Evolution Explorer (GALEX), 83
Galaxy Zoo
catalogue, 308
classification, 310
dataset, 310
project, 308, 309
Galileo, 23, 34–37, 111, 183, 385
satellite, 36
services, 35
signals, 35
Gamma ray, 57, 78, 92, 307, 319, 320, 419
astronomy, 78
flashes, 78
observatory, 78
telescope, 320
Gamma ray burst (GRB), 77, 419, 420, 424–426
detection, 181
observations, 77
Gated recurrent unit (GRU), 186, 258, 279
General Formal Ontology (GFO), 14
Generative adversarial network (GAN), 260, 317
Generative networks, 259, 260, 262
Genetic algorithm (GA), 290
Genetic programming (GP), 291
Geographic information system (GIS), 34, 118, 435, 437
Geography Markup Language (GML), 106
Geoinformatics, 31, 33, 34, 115, 116, 118, 121, 191, 283, 284, 290, 296
area, 31
community, 31, 33
Geomagnetic
data, 347, 348, 355, 357–359, 361, 364
field, 347–349, 351, 353–357, 371, 378, 381, 424
data from observatory SUA, 350
from observatory PAG, 350
registration data, 356
variations, 355
vector, 354
wavelet analysis, 358
network, 25
observatory, 357, 361
signals, 347
storms, 347–349, 351, 353, 354, 356–358, 361
Geophysical
events, 419
network, 25
Geosciences, 31, 39–44, 48, 89, 225, 227, 232, 237, 238
community, 48
data, 243
node, 40
workflows, 239
Geospatial
Big Data management, 163
data, 39, 40, 48, 103, 105, 107, 169
data analysis, 117
information, 103, 107, 110
information management, 107
standards, 44, 107
Geospatial Data Abstraction Library (GDAL), 40
Geostationary
satellites, 112
weather satellites, 116
Geosynchronous Earth orbit (GEO), 33
German Astronomical Virtual Observatory (GAVO), 39
Global Agricultural Monitoring System (GAMS), 441
Global Navigation Satellite System (GNSS), 35, 111, 385, 405, 406, 431
data, 385, 386, 405, 406
measurements, 407, 413, 414
networks, 405
observations, 35, 387, 388, 394, 397, 406, 407, 411–416
satellite, 385, 389, 393
satellite signals, 414
signals, 35, 36
stations, 392, 404, 409, 416
technologies, 405, 416
Global Positioning System (GPS), 33, 405
data, 406
meteorology, 405
observations, 405
satellites, 385, 403
satellites observations, 406
Google Earth, 39, 46, 105, 117, 119, 121
Google Earth Engine, 117
Google File System (GFS), 149
Graphical user interface (GUI), 109, 237
Gravitational search algorithm (GSA), 302
Gravity Recovery and Climate Experiment (GRACE), 188
Greatest common divisor (GCD), 215
Greenwich observatory, 23
Ground
geomagnetic data, 357, 363
observatory, 361
PAG data, 363
Ground Level Enhancement (GLE), 376, 377
events, 378
Group on Earth Observation (GEO), 107
Guo Shoujing Telescope project, 68

H

Hadoop Distributed File System (HDFS), 150, 159
Hamburg Quasar Survey (HQS), 64
Harvard College Observatory (HCO), 63
Herschel Space Observatory, 85
Hidden Markov model (HMM), 128
Hierarchical Data Format (HDF), 105
Hipparcos
catalogue, 70
observations, 21
satellite, 61
Hitachi Streaming Data Platform (HSDP), 176
HPC networks, 150
HSI classification, 237
Hubble Legacy Archive (HLA), 70, 90
Hubble Space Telescope, 70, 73, 85, 90
Hubble Telescope, 70
HYPERION satellite, 122
Hyperspectral
classification, 127
satellites, 105, 119
Hyperspectral image (HSI), 122, 124, 128, 129
classification, 128, 262
classification methods, 126

I

ICRF quasars, 71
IGS network, 388
IGS stations network, 388
IGS/EPN network, 387
Imaging mode (IM), 81
Indian Remote Sensing (IRS) satellite system, 115
Indian Space Research Organisation (ISRO), 115
Indian Space Science Data Centre (ISSDC), 115

Information
extraction, 129
geospatial, 103, 107, 110
loss, 129
remote sensing, 117
science infrastructure, 118
source, 114
spectral, 122, 126, 129
technologies, 406
Infrared Astronomical Satellite (IRAS), 83, 84
catalogues, 84
joint catalogue, 84
Infrared telescope, 85
Infrastructure
data, 108
network, 148, 382
space, 111
virtual observatory, 31
Instantaneous field of view (IFOV), 32
Integrated water vapor (IWV), 413
International Astronomical Union (IAU), 75
International Cartographic Association (ICA), 107
International Celestial Reference System (ICRS), 329
International Data Corporation (IDC), 107
International Geophysical Year (IGY), 26
International LOFAR Telescope, 88
International Planetary Data Alliance (IPDA), 45
International Standardization Organization (ISO), 44, 106
International Virtual Observatory Alliance (IVOA), 39, 43, 75, 96, 161, 167, 178, 340
executive committee, 96
recommendations, 340
standards, 45, 96, 178, 179
Interplanetary magnetic field (IMF), 347, 349, 356
data, 348, 358, 364
Ionization field, 392, 393, 395–398, 401, 402
restoration, 398, 401, 402, 404
Ionosound
data, 347, 363
stations, 349, 358
Ionosphere, 35, 349, 354–356, 385, 386, 388, 392, 393, 403, 419, 425, 428–431
characteristics, 386
monitoring system, 385
parameters, 385
state, 404
studies, 385
TEC, 404
Ionospheric
data, 358, 369
data registration, 355
disturbances, 385, 386, 419, 425, 426, 431
monitoring, 36, 428
observations, 420
parameters, 349, 354, 355, 357, 361, 367, 388, 428
plasma, 355, 363, 367, 419, 422, 424–426, 428–430
plasma parameters, 367, 428, 432

J

Joint Research Center (JRC), 109
projects, 112

K

Kepler catalogue, 89
Kernel orthogonal subspace projection (KOSP), 127
Knowledge discovery, 1–5, 16, 31, 114
community, 9
from satellite images, 114
in databases, 1
project, 4
Knowledge discovery process (KDP), 1–4, 6, 10, 19

L

Landsat, 33, 104, 111, 120, 122, 126, 440
archive, 113
archive initiative, 114
missions, 113, 114
satellites, 104, 116
series, 104
Large Area Telescope (LAT), 78
Large Sky Area Multi-Object Fibre Spectroscopic Telescope (LAMOST), 68
Large Synoptic Survey Telescope (LSST), 331
Las Campanas Redshift Survey (LCRS), 64
Leaf area index (LAI), 124
Leiden Observatory, 87
Level of detail (LOD), 273
Local area network (LAN), 302
Low Earth orbit (LEO), 32, 117
satellites, 33
LSTM networks, 279
Luminous red galaxy (LRG), 67
Lunar Reconnaissance Orbiter (LRO), 90

M

Machine learning (ML), 31, 34, 113, 117, 130, 139, 154, 155, 187, 225–227, 242, 251, 252, 254, 256, 267, 280, 281, 290, 301, 302, 307, 308, 310, 312, 314, 316
algorithms, 34, 229, 230, 239, 308
classifiers, 186
scalable, 155
techniques, 316
technologies, 281
Magnetic observatory, 354
Magnetometer data, 269, 271
Magnetospheric storm, 354
ionosphere, 354
Markov random field (MRF), 280
Math kernel library (MKL), 155
Maximum brightness, 213, 214
Maximum likelihood classifier (MLC), 237
Mercator Ocean International (MOI), 121
Message Passing Interface (MPI), 145
MESSENGER data, 268, 269, 280
MESSENGER dataset, 280
Messier catalogue, 59
Metadata
description, 161
standards, 57
Meteorological
data, 24, 28, 406
networks, 385
observations, 25, 26
satellites, 110
stations, 411
stations data, 415
Meteorology, 24, 25, 43, 112, 114, 405, 411, 416, 435
GPS, 405
synoptic, 24
Milky Way, 61, 67, 69, 77, 79, 82, 83, 173, 312–317
Millennium Galaxy Catalogue (MGC), 70
Minimal spatial bound (MSB), 161
Minimum bounding rectangle (MBR), 163
Minnesota Automated Plate Scanner (MAPS), 74
Monitoring
atmospheric, 419
broadband, 423
hurricane developments, 33
ionospheric, 36, 428
remote, 416
system ionosphere, 385
Morphological
classification, 257, 308–310, 312
classification techniques, 233
features, 309
galaxies, 310
galaxy classification, 310
Multifrequency Snapshot Sky Survey (MSSS), 88
Multiwavelength
astronomical databases, 318
astronomy, 57, 307
observations, 314
surveys, 316

N

Naive Bayes (NB), 229
Narrowband data, 422, 423
Narrowband monitoring, 425
NASA Herschel Science Center (NHSC), 85
National Aeronautics and Space Administration (NASA), 78, 85, 89, 90, 92, 104, 116, 119, 121, 126, 334, 441
  astrophysics satellite, 82
  data, 116
  exoplanet archive, 89
  Goddard Space Flight Center, 116, 121
National Radio Astronomy Observatory (NRAO), 86
National Remote Sensing Centre (NRSC), 115
National Science Foundation (NSF), 105, 119, 131
National Spatial Data Infrastructure (NSDI), 107
Natural language processing (NLP), 186
Naval Observatory, 25, 61
Navigational satellite ephemeris, 387
Network
  architectures, 251
  classification, 253
  components, 143
  connection, 148
  coverage, 409
  geomagnetic, 25
  infrastructure, 148, 382
  outputs, 279
  processing, 411
  sensors, 254
  software, 388
  stations, 25
  synoptic, 25
  technologies, 145
  topologies, 142
Neural network, 155, 253, 258, 261, 276, 279, 281
Neutral atmosphere, 405, 406
Neutron Monitor Database (NMDB), 372–374, 376, 378, 380–382
  consortium, 374
  data, 377, 381
  data providers, 374
  database, 374
  hardware, 374
  software, 374
  stations, 380
  status, 377
New General Catalogue (NGC), 59, 308
NMDB Event Search Tool (NEST), 376
NOAA, 26, 28
Nonstellar objects, 74
  bright, 61
NoSQL data model, 160
NRAO VLA Sky Survey (NVSS), 87
Nuclear Spectroscopic Telescope Array (NuSTAR), 81
  catalogue, 81
Numerical weather prediction (NWP), 405

O

Objects
  astronomical, 18, 57, 58, 63, 95, 173, 235
  catalogued, 87
  catalogued list, 71
  classification, 18, 233
  clusters, 241
  comprehensive classification, 63
  data, 11
  dataset, 239
  detection, 335, 337
  distant, 67
  faint, 236
  virtual observatory, 18
Observational
  data, 254, 332, 388, 389, 406
  data amount, 267
  database, 28
  surveys, 307, 317
Observations
  astronomical, 23, 65, 331
  astronomical data catalogues, 91
  broadband, 425
  DPOSS, 184
  galaxies, 61, 317
  GNSS, 35, 387, 388, 394, 397, 406, 407, 411, 412, 414, 416
  GPS, 405
  Hipparcos, 21
  ionospheric, 420
  meteorological, 25
  multiwavelength, 314
  photographic, 60, 65
  photometric, 67, 332, 340
  radio, 92, 319
  satellites, 406, 407
  spectral, 67
  spectroscopic, 317
Observatory
  gamma ray, 78
  geomagnetic, 357, 361
  ground, 361
  plate collection, 74
  Spaceship Earth, 381
Ontology of Biomedical Investigation (OBI), 14
Ontology of Data Mining (OntoDM), 14
Ontology Web Language (OWL), 13
Open Geospatial Consortium (OGC), 39, 44, 103
Operational Decision Manager (ODM), 176
Orbital Data Explorer (ODE), 40, 45
Orthogonal subspace projection (OSP), 127
Oschin Schmidt telescope, 61

P

PAG observatory, 361
Palomar Norris Sky Catalog (PNSC), 74
Palomar observatory, 74
Palomar Oschin Schmidt telescope, 61
Partitioned global address space (PGAS), 145
Payload ground segment (PGS), 114
Periodogram analysis, 203, 209, 211, 215, 217
Personal navigation system (PNS), 34
Photographic
  catalogues, 58
  observations, 60, 65
  projects, 61
  sky surveys, 71
  spectral surveys, 63
  surveys, 63, 72
Photometric
  data, 341
  data from CCD frames, 325
  data storing, 340
  observations, 67, 332, 340
  redshifts, 236, 318
  redshifts for galaxies, 236
Photon Counting (PC), 81
Piecewise aggregate approximation (PAA), 277
Planck satellite, 86
Planck space observatory, 85
Planetary Data Access Protocol (PDAP), 45
Planetary Data System (PDS), 40, 45, 89
Planetary Science Archive (PSA), 45, 90
POSS surveys, 71
Precipitable water vapor (PWV), 405
Precision Measuring Machine (PMM), 71
Predictive Model Markup Language (PMML), 12
Principal component analysis (PCA), 124, 186, 195, 218, 232
Probabilistic classification algorithm, 308
Project
  data, 18
  data mining, 5, 9
  knowledge discovery, 4
  management activities, 6, 10
  plan, 7, 8
  roadmap, 4
Proper orthogonal decomposition (POD), 124
Proprietary geospatial databases, 117
Public Regulated Service (PRS), 111
Pulsating stars, 183, 186

Q

QSO catalogue, 62
Quality
analysis, 397
characteristics, 106
data, 169, 372, 377, 382
indicators, 337
measures, 106
research, 104
Quality model (QM), 107
Quasars, 62, 66, 67, 69, 72, 74, 77, 82, 87, 236
catalogue, 62
classification, 233
photometric redshifts, 236
redshifts, 66
Query
astronomical data, 167
complexity, 167
cost, 162
engine, 169
evaluation, 162
evaluation plans, 162
execution, 168
engine, 161
times, 166, 167
optimizer, 160, 162, 165–167
paradigms, 159
pattern, 274
plan, 160–163, 168
processing, 160, 165, 169
processing engines, 168
processor, 161
RasDaMan, 48
time, 163
Query by example (QBE), 268
Query execution plan (QEP), 165, 166, 168

R

Radio
astronomy, 87
astronomy observatory, 86, 87, 319
astronomy research, 93
decameter astronomy, 307
galaxies, 77, 235
loud galaxies, 92
observations, 92, 319
signal propagation, 411
signals, 420, 424
sources, 57, 87, 88, 232, 235, 257, 385
surveys, 88, 184
telescope, 87, 88, 319
Random forest (RF), 231
RasDaMan, 39, 46, 48, 50
Raw
data, 2, 3, 17, 69, 75, 90, 105, 119, 333, 357, 374
data gathering, 2
datasets, 331
Recurrent neural network (RNN), 186, 258, 279
Redshifts, 61, 64, 66, 80, 88, 235, 310, 316
photometric, 236, 318
quasars, 66
spectroscopic, 236, 317
Remote
monitoring, 416
sensing, 16–19, 32–36, 39, 42, 43, 103, 104, 115, 116, 118, 225, 237, 254, 256, 258, 261, 300–302
catalogues, 115
imagery, 300
information, 117
instruments, 32
quality data, 32
satellites, 104, 115, 116
science, 116
supervised classification, 237
stations, 382
Remote Procedure Call (RPC), 45
Research Infrastructure (RI), 45
Resilient distributed dataset (RDD), 162, 242
Ridge regression (RR), 301
Roentgen satellite ROSAT, 79
Rover stations, 407, 414
in RTK observations, 409
Royal Observatory, 66, 75

S

SAS software tools, 9
Satellites
agencies, 105
antennas, 388
archives, 107
capabilities, 105, 119
data, 111, 116, 441
data archive, 111
data products, 105, 118
datasets spot, 118
downlink stations, 105, 119
EO, 103, 108, 110, 115, 121
GNSS, 385, 389, 393
GPS, 385, 403
hyperspectral, 105, 119
imagery, 32–34, 107, 114, 262
images, 33, 103, 114, 117
Landsat, 104, 116
lifetime, 35
meteorological, 110
missions, 82, 108, 130
movements, 398
navigation, 35
navigation systems, 385
observations, 406, 407
overflights, 105
positions, 404
receiver, 385
remote sensing, 36, 104, 115, 116
sensors, 104, 105, 108, 119
signals, 424, 431
surveying platforms, 104
system, 35, 36
technology, 32, 416
Scalable data processing algorithms, 4
Scalable distributed data structure (SDDS), 33
Schmidt telescope, 61, 63, 68, 73, 78, 314
Science
astronomical, 173
data, 104, 120, 254
remote sensing, 116
Scientific data, 267, 280, 334
Second Byurakan Survey (SBS), 63
SEGUE project, 67
Semantic Resolver for Coordinate Reference Systems (SECORE), 46, 50
Sensors
accuracy, 413
network, 254
satellites, 104, 105, 108, 119
Sentiment classification, 279
Sentinel Australasia Regional Access (SARA), 114
Sentinel satellites, 110, 113
Sentinel series, 110
Sequential pattern mining (SPM), 242
SETI project, 90
Seyfert galaxies, 64
Shanghai Astronomical Observatory, 75
Signal
amplitude, 335, 427, 430
annotation, 276
characteristics, 426, 430
classification, 268
data, 272, 273
decomposition, 348
deflection, 419
delay, 406
features, 274
frequency, 431
noise, 280, 422
patterns, 274
peaks, 426
processing, 125, 347, 348
propagation, 425, 427, 428, 431
propagation paths, 420, 422, 428, 431
space, 274
variations, 425
Simple Application Messaging Protocol (SAMP), 45
Simple Spectral Access Protocol (SSAP), 48
Singular Spectrum Analysis (SSA), 220
Singular value decomposition (SVD), 124, 154, 218
Sky survey, 31, 49, 65, 72, 74, 75, 87, 88, 90, 183, 184, 329
Sloan Digital Sky Survey (SDSS), 66, 67, 69, 70, 87, 90, 165, 233, 236, 307, 308
bright galaxies, 308
database, 92

galaxies, 307–312
Great Wall, 312
Legacy Survey, 67
spectrographs, 67
voids, 312
Small computer system interface (SCSI), 145
Smithsonian Astrophysical Observatory (SAO), 97
catalogue, 23
Snow water equivalent (SWE), 188
Software tools, 4–7, 9, 89, 95, 96, 105, 119
Solar
activity, 349, 371, 372, 377, 378, 380, 381, 419, 430, 437, 438, 440
events, 425
system, 42, 45, 70–72, 77, 82, 87, 89, 90, 173, 179, 331, 371
terminator, 349, 359, 364, 424, 427, 428
terminator influence, 363
Solar cosmic rays (SCR), 371
Solar energetic particle (SEP), 372
events, 372, 381
Solar system objects (SSO), 332
Southern Galaxy Catalogue (SGC), 64
Space
agencies, 109, 113, 118, 130, 267, 441
infrastructure, 111
signal, 274
telescope, 73, 78, 85, 89, 90
telescope for ultraviolet astronomy, 81
telescope science institute, 90
weather events, 377
Space Infrared Telescope Facility (SIRTF), 85
Space Science Data Center (SSDC), 90
Spark, 150, 154, 160, 162, 242
components, 243
core, 242
core engine, 242
ecosystem, 242
engine, 242
extension, 168
framework, 160, 165
GraphX, 243
MLLib, 243
programs, 242
SQL, 160–162, 242, 243
SQL parsing, 161
streaming, 242
Streaming, 242
streaming, 243
Spatial access method (SAM), 163
Spatial Data Infrastructure (SDI), 109
Spectral
classification, 127
information, 122, 126, 129
observations, 67
photographic observations, 63
Spectral angle mapper (SAM), 127
Spectral vegetation indices (SVI), 124
Spectroscopic
catalogues, 66
observations, 317
redshifts, 236, 317
surveys, 70
Spiral galaxies, 310, 312, 314
Spitzer Space Telescope (SST), 85
Square Kilometre Array (SKA), 147, 232
Standard Analysis Software System (SASS), 79
Standard deviation (SD), 402
Standards
activities, 107
association, 44
databases, 4
geospatial, 44, 107
international, 106
IVOA, 45, 96, 178, 179
WCPS, 50
Stars
binary, 186, 203
bright, 68
catalogues, 21, 22
classification, 63, 233
clusters, 59
distant, 70
faint, 77
galaxy, 233
galaxy classification, 233
photometry quality, 337
pulsation, 178
Stations
data, 374
GNSS, 392, 404, 409, 416
meteorological, 411
network, 25
NMDB, 380
remote, 382
SULP, 413
tropospheric delay, 409
STEC data, 389
STEC values, 393, 397, 402
Stochastic Neural Analog Reinforcement Computer (SNARC), 225
Storage area network (SAN), 144
Storage class memory (SCM), 147
Structured Query Language (SQL), 39
query, 161, 162
query processing, 160
SUA data, 359
SUA geomagnetic data, 359
Sudden ionospheric disturbance (SID), 424–426
Sudliche Bonner Durchmusterung (SBD), 59
Suggested Upper Merged Ontology (SUMO), 13
SuperCOSMOS Sky Survey (SSS), 74
Supervised classification, 237, 238
remote sensing, 237
Supervised machine learning, 268
Support vector machine (SVM), 127, 231, 252
Surveys
astronomical, 57, 77, 159, 232, 233
astronomy, 183
CCD, 65
galaxies, 233, 315
multiwavelength, 316
observational, 307, 317
photographic, 63, 72
radio, 88, 184
spectroscopic, 70
Sustainable development goals (SDG), 103
Swift observatory, 81
Sydney University Molonglo Sky Survey (SUMSS), 87
Synoptic
database, 24
meteorology, 24
network, 25
sky surveys, 31

T

Table Access Protocol (TAP), 39, 44
Technologies
GNSS, 405, 416
information, 406
machine learning, 281
network, 145
Telescope
aberration, 333
gamma ray, 320
infrared, 85
radio, 87, 88, 319
space, 73, 78, 85, 89, 90
system, 317
Temporal convolutional network (TCN), 279
TERRA satellite, 33
Terrestrial
atmosphere, 419
events, 424
Thematic exploitation platform (TEP), 122
Thematic Mapper (TM), 126
Time Domain Spectroscopic Survey (TDSS), 68
Total electron content (TEC), 385
data, 355
values, 355, 358, 387–389
Trained
classification models, 237
deep convolutional neural network, 257
neural networks, 251
Transcarpathian Region, 409, 415, 416
Transiting Exoplanet Survey Satellite (TESS), 90
Tropical Rainfall Measuring Mission (TRMM), 115
Tropospheric delay, 405–407, 409, 411
accuracy from GNSS observations, 407
determination, 405, 407, 409, 411

determination accuracy, 407
from GNSS data, 406
stations, 409
Tycho catalogue, 70

U

Uhuru satellite, 79
Ultraviolet astronomy, 81
Ultraviolet telescopes, 80
Unified Modeling Language (UML), 11
Unified Parallel C (UPC), 145
Unmanned aerial vehicle (UAV), 33
Unsupervised artificial immune classifier (UAIC), 302
Uppsala General Catalogue (UGC), 61
User definition function (UDF), 161
USNO catalogues, 72

V

Variable Infiltration Capacity (VIC), 188
Variable stars, 71, 77, 173, 174, 186, 192, 202, 216, 331, 333, 339, 341
VASCO project, 77
Very Large Array (VLA), 87
Virtual observatory (VO), 18, 39, 45, 46, 77, 95, 178, 332, 344
infrastructure, 31
objects, 18
standards, 45
Virtual reality (VR), 186
VLF data, 432
VLF/LF databases, 429
VLF/LF signals, 431
VTEC values, 389

W

Wavelet analysis, 124, 125, 192, 214, 348, 356, 358
geomagnetic field, 358
Wavelet transform (WT), 124
Web Coverage Processing Service (WCPS), 39, 46, 48, 106, 131
coverages, 47
servers, 48
standards, 50, 106
Web Coverage Service (WCS), 105
Web Feature Service (WFS), 105
Web Map Service (WMS), 39, 46, 105, 131
Web Time Series Service (WTSS), 31
Whole Earth Blazar Telescope (WEBT), 320
Wide Area Augmentation System (WAAS), 35
Wilkinson Microwave Anisotropy Probe (WMAP), 86
catalogues, 90
William Herschel Telescope, 88
Windowed Timing (WT), 81
Wisconsin Diagnosis Breast Cancer (WDBC), 8
World Data Center (WDC), 25, 354, 367, 371, 373
World Wide Lightning Location Network (WWLLN), 428

X

XPM catalogue, 329

Z

ZAKPOS network, 397, 406, 408, 409, 411, 413–416
Zenith tropospheric delay, 405, 407, 409
Zwicky galaxies, 64
Zwicky Transient Facility (ZTF), 174, 183

Printed in the United States
By Bookmasters